Woody Plants for the Central and Northern Prairies

By Walter T. Bagley and Richard K. Sutton

Illustrated by Nancy L. Scott

THE BLACKBURN PRESS

Woody Plants for the Central and Northern Prairies

ISBN-10: 1-930665-50-4
ISBN-13: 978-1-930665-50-7

Library of Congress Control Number: 2002102517

THE BLACKBURN PRESS
P. O. Box 287
Caldwell, New Jersey 07006 U.S.A.
973-228-7077
www.BlackburnPress.com

Preface

Woody Plants for the Central and Northern Prairies is designed for people of the prairies—a region of the United States that is more hospitable to grasses and associated forbes than it is to woody plants. We do not exclude inhabitants of other regions. Readers will find useful plant descriptions and perhaps find other tidbits of helpful information.

We do not include all woody plants. Species that were selected for inclusion—many native to other continents—are likely to be useful for many purposes including yard and garden beautification and creating windbreaks, woodlands, orchards and wildlife habitat.

We admire the perseverance and industry shown by prairie homesteaders. They staked their claims, constructed simple abodes, plowed a few acres for crop and garden, and, almost always, planted trees, shrubs and vines for food, shelter and home beautification. Their guide for selecting plants was experience they had gained in other climes. With luck and after years of weeding and watering they enjoyed shade, protection and bountiful fruits before drouth, hail, unseasonable temperatures or combinations thereof took their toll. And so they planted again, and again, attaining greater success each time as they identified which plants were most likely to survive and prosper.

The dearth of knowledge of the inherent characteristics of most woody plants is shameful in this enlightened age of high technology. Further research is needed to determine site adaptability and potential for providing useful products. We have much to learn about culturing native and exotic plants in unnatural environments.

We have elected to omit color photos because quality images of many woody plants can be found on compact disks (CD-ROM) that can be viewed on home computers. Web pages from university and government sources also often display plant images, though these pages come and go and are not always available. The following CD's contain images that may be of interest.

Michael A. Dirr's Photo-library of Woody Plants on CD-ROM. 1997. Over 7500 color images of woody plants and cultivars hardy to North America are on this searchable CD-ROM operating on MS-DOS or Windows.

***Horticopia: Trees, Shrubs & Groundcovers*-Edition II.** 1998. Eight thousand color images of over 3000 woody plants and cultivars. Plants may be searched for by characteristics and name. Requires a Windows operating system and a 2X CD-ROM.

ProSeries Plant Database for CD-ROM. 1995. A searchable base of 300 color images of trees, shrubs and perennials with an audio pronunciation guide. Windows operating system required.

Plantmaster 5.0. 1997. A 5000 plant image collection of woody and herbaceous material and data base. Windows NT or 95 or Power MacIntosh operating systems.

Landscaping Plants and Trees. 1998. More than 1000 color images covering 1000 trees, shrubs and perennials commonly used in landscaping. A searchable with an audio pronunciation guide. Operates on Windows 3.1 or higher and MacIntosh 7.0 or higher.

Common Plants of Canadian Landscapes. 1995. Canadian Nursery Trades Association. Color images of 350 common landscape plants found in Canadian landscapes and basic cultural requirements. Requires Windows 3.1 or higher or 68020 MacIntosh or later.

Acknowledgments

We are especially grateful to Tim Richley, former professor of horticulture, University of Nebraska School of Technical Agriculture, Curtis, for a thorough review of our proposed plant list. His comments about experience in growing many of these plants in southwestern Nebraska adds greatly to the regional credibility of this book. Max Peterson, farmer and lilac fancier in Perkins County, Nebraska, provided useful observations about his successes and failures in growing trees and shrubs on a western prairie site now devoted primarily to wheat production.

We are indebted to Craig Freeman of the Kansas Biological Survey for his pertinent suggestions following a thorough review of our manuscript in its early stages. The late Laird Wolfe, farmer and conservationist, provided plant material and valuable information from Webster County, Nebraska. Horticulturist Edward Rassmussen provided plant materials for our artist and helpful observations about woody plants growing in his extensive collection at a Washington County farm north of Omaha.

Nurserymen Jon and Gregory Morgenson related their experiences in growing uncommon plants and provided specimens. The late Roger Uhlinger, professor of horticulture, University of Nebraska–Lincoln (UNL), and Gene Howard, former superintendent of the Cheyenne Horticultural Field Station, provided insights from their experiences in growing woody plants in southeastern Wyoming. James Locklear, former curator of Dyck Arboretum, Hesston, Kansas, provided plant specimens and information relevant to woody plants of central Kansas.

Nancy Rosen and Richard Eckersley, University of Nebraska Press, provided helpful assistance to the authors and artist during the early stages of manuscript preparation. We appreciate the helpful comments of professors Edward Hasselkuss, James Klett and Steven Rodie in their manuscript reviews.

Others who have contributed to the success of this projcet in a variety of ways include Erling Jacobsen, Curt Twedt, Paul Read, Gary Hergenrader, Tom Wardle, Stephen Ernst, James Brandle, Donald Steinegger, William Gustafson, Jr., James Stubbendieck, Charles Butterfield, Margaret McVicker, Dale Lindgen, David Nuland, Gary Koller, John Pair, Charlene Kaus, Margaret Bollick, Larry Group, Jim Siffring, Steven Jungst and the late Sylvan Runkel. Many more not mentioned here provided encouragement and suggestions.

Furthermore, we owe thanks to our spouses, Virginia, Lisa and Randy, for their patience and moral support during these several years of effort.

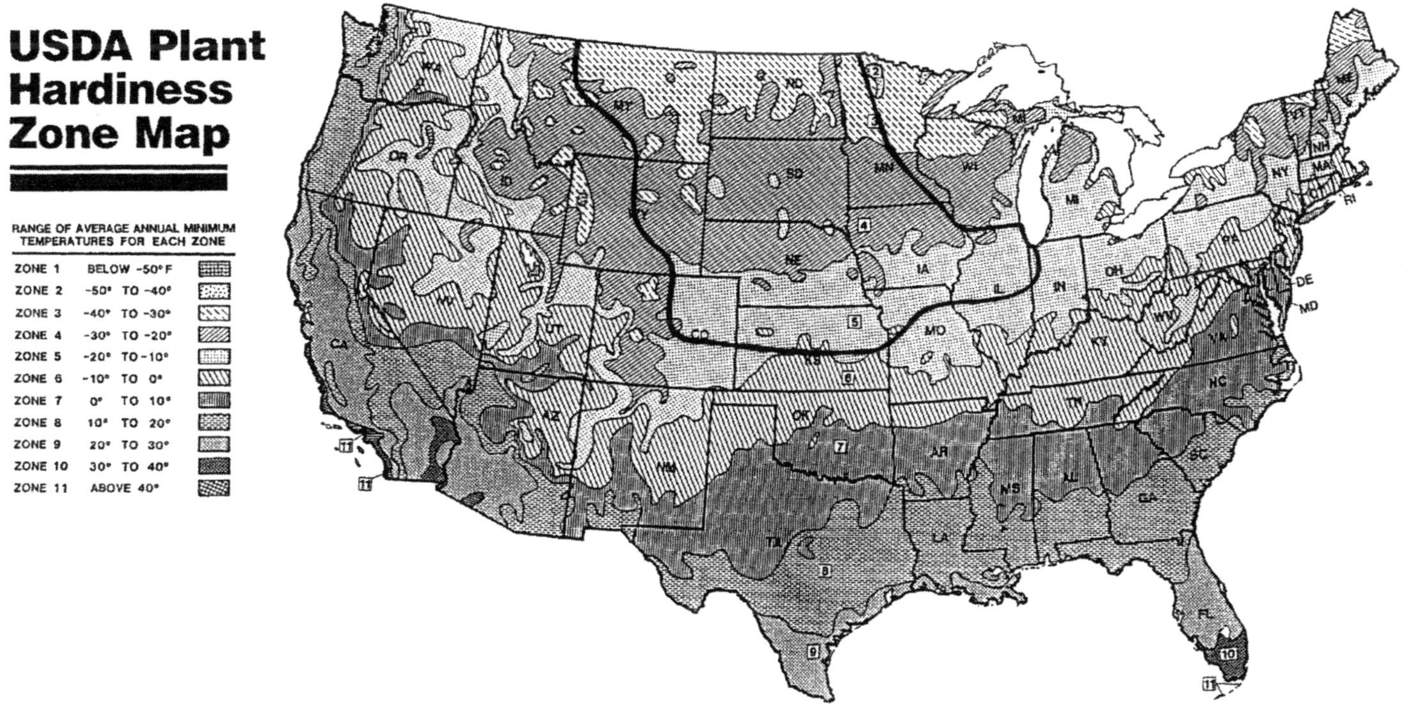

Boundary of the Central and Northern Prairie Region in the United States

Contents

Contents

Contents

Introduction

About Our Book

Woody Plants for the Central and Northern Prairies attempts to build upon the foundation of knowledge provided by prairie pioneers. We present gleanings from published research and the notes of professional and amateur horticulturists, landscape designers and foresters. We strive to provide factual information for professionals, students, farmers and urban dwellers in the heartland of the United States.

Our personal experience is centered in Nebraska and adjoining states. Nebraska places that will be mentioned many times are Maxwell Arboretum on East Campus, University of Nebraska–Lincoln (UNL); Horning State Farm, Plattsmouth; Prairie Pines, Lincoln; School of Technical Agriculture, Curtis; Bessey Nursery, Halsey; and University of Nebraska Research and Development Centers (UNARDC), at Ithaca, North Platte and Scottsbluff.

Also frequently mentioned is the W. D. Holley Plant Environmental Research Center (PERC), Fort Collins, Colorado, an important site for woody landscape plant research trials in the short-grass prairie (Fairchild and Klett 1993). The semiarid climate is sunny and often windy. Soils are clay loam with a pH of about 8.

Many people are frustrated by their inability to determine the identity of a woody plant growing in their backyards, neighborhood parks, along a roadside, or at a vacation site. **Woody Plants for the Central and Northern Prairies** is our attempt to assist them by providing descriptions and sketches of most of the native and exotic plants that they are likely to find or wish to grow.

Information such as pests and their control is not presented, nor are detailed descriptions of cultivars. Periodicals, bulletins and university educators are readily available for up-to-date assistance about culture and cultivars.

Diversity of soils and climate throughout the prairie region presents a task that we approach with great humility. We cite experiences of people growing woody plants throughout the prairies. But we are limited by a paucity of experience and research information about many species and their ability to adapt to different site conditions.

We remind you that taxonomy—the classification and naming of plants and animals according to their relationships—is an evolving science. We place organisms in labeled pigeonholes for purposes of communication. As more is learned about them, they are moved into different groupings, and, sometimes, their names are changed to clarify inter-relationships. Plant species are often impossible to describe in precise terms because individuals vary in appearance just as we of the human species do.

Species are arranged in sequence beginning with gymnosperms (Pinophyta), the most primitive, and progressing through the dicots (Magnoliopsida) and the monocots (Liliopsida) of the angiosperms (Magnoliophyta). The angiosperms are arranged in a sequence based upon one of the latest theories of plant evolution (Cronquist 1981), a theory that deviates considerably from the phylogenetic sequence presented by Rehder (1940), whose manual we rely upon as a valuable taxonomic reference.

Subclasses, orders and families are arranged in phylogenic sequence. Within each family and subfamily we have taken the liberty of alphabetically arranging genera. Species are listed alphabetically under each genus unless easily discernible characteristics—such as evergreen versus deciduous—permits grouping by common characteristics to aid in identification.

Taxonomic manuals generally include descriptive keys to assist in identification. Keys may be based on characteristics and arrangement of leaves, winter buds, flowers, fruits, or a combination thereof. Keys are often frustrating to use because they may require knowledge of plant parts not available at the time of identification. For example, you may have leaves and twigs but no information about flowers or fruits.

The many variants—subspecies, varieties and hybrids—are usually not included in a key and sometimes not described in a text. True identity remains uncertain. Be consoled that many times, even when you seek assistance from a professional taxonomist, you may not obtain a precise reply.

Complete keys are not included in this book. We use grouping keys as aids to narrow the search for a plant's identity. Illustrations will be helpful. Keys can be found in the following publications: Duncan and Duncan (1988); Great Plains Flora Association (1986); Bailey (1949); Elias (1980); Harlow (1946, 1954); Hoag (1965); Krussmann (1985); Preston (1968); Preston and Wright (1976); Rehder (1940); Sargent (1922); Stephens (1973); Van Der Linden and Farrar (1984); and Viertel (1970).

A concise paragraph titled **Planning** is included for each species to provide information about plant size and habit, date of bloom, foliage color, site requirements, zone of adaptation and landscape use. We reluctantly include the coldest zone to which a plant is adapted based on the **USDA Plant Hardiness Zone Map** (Cathey 1990). This map shows an estimate of hardiness based on the best information we have about a species, its varieties and its cultivars.

Hardiness zones are determined by average minimum temperature. Extremes of temperature which may occur occasionally in a zone add a variable to the plant survival equation that is not incorporated into the data. Couple that variable with differences in microclimate of a niche protected from the ravages of weather extremes and one can see that zone information is a weak crutch, at best, for determining plant adaptability (Refer to section **Soils, Climate and Heredity**).

The height of Nebraska and National champion trees is included to provide an unscientific guide for determining the ultimate size that a tree may attain (Kuhns 1991, American forestry Association 1992). Since trunk diameter and crown spread are included with height in the formula for determining a champion, a tall tree of small diameter may be outranked by a shorter tree with a larger stem and crown spread. For example, the Nebraska champion eastern redcedar is taller than the National champion, but its trunk is about one-half of the diameter of the National champion growing in Georgia.

Plant descriptions are included in a paragraph titled **Recognition**. Descriptions of leaves, flowers, fruit, twigs, buds, bark and wood are included insofar as they are helpful in determining identity. Characteristics that are consistent for a given order are not repeated for a family within that order. The same rule applies for family and genus. One should first determine the correct order, family and genus of a plant before attempting to identify its species.

Plants are described in non-technical language with the corresponding botanical term often added in parentheses to enhance the instructional value of this book. However, botanical names are used for plant parts such as petioles for leaf stalks and pistils for female reproductive organs. A complete glossary is included for readers who would like to increase their vocabulary.

Sketches by Nancy Scott are from plant specimens collected by the authors and artist, with the exception of a few borrowed from the herbariums of the Departments of Horticulture and Agronomy, UNL. The leaves and twigs of many species are included, but flowers and fruits are not always shown.

A Bit of History

The first settlers of the prairies cleared woodlands along the major streams soon after their arrival. Wood was used for construction and fuel. Land supporting trees was considered to be the most fertile for growing crops and, consequently, little thought was given to perpetuating the woods, at least in the early years. However, the dwindling trees soon became inadequate to meet the needs of the growing population. Lumber and fuel became scarce, and people longed for the protection and other amenities that trees provide. Thoughts turned to planting trees.

The need to control wandering livestock prompted one of the earliest shipments of nursery stock into the prairie region. The first osage-orange seedlings (*Maclura pomifera*) from an Illinois nursery arrived in Kansas in 1843 (Loucks 1984). Many hundreds of miles of osage-orange hedgerows soon marked the boundaries of fields and farms in southern Illinois, Missouri, Kansas and southern Nebraska, first serving as fences and then, with the invention of barbed wire, as a handy and renewable supply of indestructible fence posts. Osage-orange also provided wood for the hearth, habitat for wildlife and a barrier against the wind.

Only remnants of these hedgerows remain. They were removed when roads were widened and fields were enlarged. The need for osage-orange fence posts declined as creosoted pine posts became available. Pine was easier to cut, treat, and staple.

Societies were organized during the years of early settlement to disseminate information about the value of horticultural plantings, and about which species and varieties to plant and culture. These organizations were potent lobbyists at state legislatures for promoting the adoption of subsidies that encouraged tree planting.

Kansas provided an annual bounty in 1865 for planting and cultivating trees for wood, erosion control, and as roadside markers (Strickler 1971). The bounty was discontinued in 1873 when the U.S. Congress passed the Timber Culture Act that allowed individuals to acquire 160 acres of government land if 40 of these acres were planted to trees. The required number of tree acres was reduced to ten in 1878, and although the Act was much abused and misused before its repeal in 1891, remnants of "tree claims" still survive. The law repealing this act also halted the sale of public land and authorized the president of the U.S. to establish forest reserves, the forerunners of our national forests (Allen and Sharpe 1960).

Another example of early attempts to encourage tree planting was the creation of the office of Commissioner of Forestry for Kansas in 1887. The commissioner was empowered to establish two stations to grow tree seedlings to be distributed to the public free of charge. Several million trees were delivered through this office before it was replaced by the Division of Forestry at Kansas State College, now Kansas State University (Loucks 1984). The Fort Hays Branch Experiment Station near Hays, Kansas, established experimental plantings of forest tree species in 1902 (Strickler 1971).

Charles E. Bessey, a botanist who came from Iowa State University to the University of Nebraska in 1884, was a pioneer in the search for trees that could withstand the rigors of the Great Plains climate. He prevailed upon the U.S. Bureau of Forestry to establish trial plantations in the Nebraska Sandhills on the Lawrence Bruner property north of Swan Lake in southern Holt County. Trees were planted in 1892 and 10 years later the Bureau pronounced them so successful that President Theodore Roosevelt set aside two forest reserves in Nebraska. The Nebraska National Forest, the largest man-made forest in the United States, is west of the confluence of the Middle Loup and Dismal Rivers near Halsey. The Samuel R. McKelvie National Forest is west of the confluence of the Niobrara and Snake Rivers in Cherry County south of Nenzel. A Forest Service tree nursery bearing Bessey's name remains in operation near Halsey and has been a source of millions of trees for windbreak, wildlife and woodlot plantings in Nebraska and adjoining states.

N. E. Hansen, South Dakota State College professor of horticulture and noted plant explorer at the turn of the 20th century, recognized the importance of trees and shrubs to the early settlers of the Plains. Three South Dakota Experiment Station bulletins, listed in the bibliography, demonstrate Hansen's remarkable knowledge of woody plants. Prolific tree planter Robert W. Furnas, Nebraska's governor from 1873 to 1875, promoted fruit growing in Nebraska and sold osage-orange and other tree seedlings from his Brownville property. J. Sterling Morton, a contemporary of Furnas's and Bessey's, was responsible for establishing the first Arbor Day in 1873. Morton landscaped the spacious grounds of his mansion at Nebraska City with many species of trees and shrubs. His estate is now Arbor Lodge State Historical Park.

Despite, or because of, the dearth of trees in the plains and prairies of the mid-continent, an interest in forestry education developed among the early settlers. Elementary forestry courses appeared in the curricula of agricultural colleges of the plains states from 1874 to 1900 (McDonald 1954). In 1905 the University of Nebraska became one of the first universities in the United States to establish a degreed four-year forestry curriculum.

Carlos Bates, one of the first graduates of the University of Nebraska forestry program, published the results of this country's first research on the influence of windbreaks during his tenure at the Lakes States Forest Experiment Station in St. Paul, Minnesota. Another Nebraskan, G. B. McDonald, earned B.S. and M.S. degrees in Forestry in 1907 and 1914, respectively. "Prof Mac" taught forestry in the Department of Horticulture at Iowa State College in 1910, and eventually headed the forestry curriculum at that school until his retirement in 1948. W. J. Morrill, another Nebraska forestry graduate, went to Colorado Agricultural College (now Colorado State University) as a teacher and eventually became head of the Forestry Department there until his retirement in 1940.

Several other Nebraska forestry graduates left the Plains for "greener forests" and made their marks in forestry and related professions. Unfortunately, tight budgets and other considerations forced the demise of the University of Nebraska forestry program in 1915. The Forestry Department at Iowa State University continues and alumni number over 2,000 (Thomson 1979).

The Clarke-McN[illegible] Act of [illegible]4, one of the most significant, durable actions of the federal government to enc[illegible]age estab[illegible]ment of farm forests and windbreaks, provided seedling trees at cost to landowners. The U.S. Forest Service research program was launched by the McSweeney-McNary Act in 1928.

The Lakes States Forestry Experiment Station in Minnesota was responsible for forestry research pertinent to the northern Great Plains region. Raphael Zon, director of the Lake States Station, was an advisor to President Roosevelt during the inception of the Great Plains Shelterbelt Project. After the project was established in 1933, Zon served as the technical director and the Lakes States station initiated supporting shelterbelt research projects.

Carlos Bates directed crop influences research for the shelterbelt project with a field crew headed by Warren Barnes. Crew members Ronald Woolery, W. T. Bagley, John Molberg and Donald Usher collected samples from wheat, corn and cotton fields from Texas to North Dakota. The results of this research were reported by Joseph Stoeckeler (1962).

The Prairie States Forestry Project, as the shelterbelt project was later named, was terminated in 1942. But, by then nearly 223 million trees and shrubs in over 18,000 miles of field windbreaks were planted as a result of the project (Droze 1977).

During the first 100 years of European man's occupancy of the prairies, tree planting was an act of love and faith triggered by a desire to beautify homesteads, provide food for the larder, ameliorate the heat of the midsummer sun, and tame persistent winds. Thousands of trees, shrubs and vines were planted for ornamentation, protection, fences and useful products. Many species, ill-adapted to the prairie environment, were short lived. Others were so successful that they escaped cultivation, became naturalized and, eventually, weedy nuisances. Little scientific basis for tree and shrub plantings existed in the region except for some research, mostly after the fact, as part of the shelterbelt project of the 1930s.

The Lake States Forest Experiment Station initiated research to determine the value of windbreaks for preventing wind erosion, controlling snowdrifts, ameliorating temperature, and reducing evaporation, and to gauge their ultimate effect on crop yields (Bates 1944, Stoeckeler 1962).

The Kansas Agricultural Experiment Station cooperating with the USDA Agricultural Research Service conducted wind tunnel and field studies to determine the effects of wind barriers on wind velocity, evaporation, house heating, snowdrifting and air temperature (Woodruff 1954; Woodruff et al. 1959; Woodruff et al. 1963).

Establishment of the Rocky Mountain Forest and Range Experiment Station Research Center at UNL in 1952 was a significant milestone in the search for knowledge of tree culture in the prairies. Tree improvement was furthered by establishment of provenance (seed source) plantations of several coniferous species and the selection of genetically superior trees. Insect and disease research expanded knowledge of tree pests and their control. Noise abatement studies demonstrated how woody plants could be used to reduce traffic noise (Cook and Van Haverbeke 1971).

The McIntire-Stennis Cooperative Forestry Research Program, initiated in 1962, provided much-needed funds to state agricultural stations for tree research. Projects were initiated to determine the benefits of protection plantings, to breed and select woody plants adapted to prairie sites, and to learn how to avoid or control insect and disease injuries.

The next significant tree-related legislation of importance to the Prairie region was the Forestry Title of the 1990 Farm Bill which authorized the establishment of the Center for Semiarid Agroforestry at the Forest Sciences Laboratory of the U.S. Forest Service, Lincoln, Nebraska. In 1995, the Center's name was changed to the National Agroforestry Center (NAC) operating under the partnership of the Forest Service and the Natural Resources Conservation Service. National and international cooperation with other agencies, universities and interested organizations will encourage more agroforestry research.

Technical information will be distributed to enable establishment of appropriate conservation and/or production systems for farms, ranches and communities. If adequately implemented and sustained, this program could become the most significant development in the tree planting history of the prairies.

The bequest of Ticia Blanche Horning that gave a 240-acre farm near Plattsmouth, Nebraska, to "the School of Agriculture as a fruit farm or for forestry purposes" was opportune for the Nebraska forestry research program. Horning State Farm became available in 1959 at a time when agricultural experiment stations of the North Central Region of the United States were establishing a cooperative tree improvement project. Fourteen provenance plantations have been established on Horning State Farm in cooperation with the regional research project, the U.S. Forest Service and the Natural Resources Conservation Service. A like number of tree and shrub plantings are devoted to varietal evaluations and pest control for a total of over 100 acres of plantations.

The University of Nebraska Agricultural Research and Development Center (UNARDC) at Ithaca, and a 160-acre tract at the former Naval Ammunition Depot (granted to the Rocky Mountain Forest and Range Experiment Station) near Hastings, Nebraska, became sites of additional forestry research devoted to windbreak influences and tree improvement. Scotch and Austrian pine seed orchards of superior parentage have been established. Scotch pine and selected shrub varieties, grown from seed collected from the several seed orchards, can now be purchased through the distribution program of the Nebraska Forest Service.

Six 40-acre research plots protected by trees planted in "tic-tac-toe" windbreak designs were established in the early 1960s on a section of land at UNARDC to determine environmental changes in the lee of shelterbelts and consequent effects on plant growth and crop yield. Yields have been monitored on protected and unprotected plots annually since the first year of tree planting. This windbreak research laboratory is the first and only facility in the world established for long-term crop influences research (Miller et al. 1974, 1975; Brandle et al. 1992).

A project to select suitable plants for roadside beautification, initiated in the 1960s, has broadened knowledge of adaptability of selected tree and shrub varieties and cultivars to sites across Nebraska. Complementing this research, plantation trials established in cooperation with the North Central Plant Introduction Station in Ames, Iowa, have provided a wealth of information about adaptability of new varieties and cultivars to prairie sites.

Fifty clones of pecans, planted in the 1980s, are being evaluated for hardiness and productivity in the Northern Pecan Cultivar and Provenance Evaluation Research Project of the Department of Horticulture, UNL.

The Center for Development of Hardy Landscape Plants, a cooperative non-profit corporation, was established in 1990 among 70 institutions from across North America and Scandinavian and Baltic countries. The Center's goal is to develop stress-tolerant plants. One important project studies the impact of urban soil temperatures on tree growth (Graves 1994).

We do not have the space to enumerate the many other woody plant research projects currently under way at agricultural research and extension centers in other prairie states. Taken together, a strong foundation will be created for significant advances in the search for woody plants adapted to the prairies.

The continuing growth of organizations dedicated to woody plant research and promotional programs indicates keen public interest in woody plants. Whereas, prior to the 1950s, technical forestry assistance for landowners and communities in Nebraska was provided by one extension forester in the Department of Horticulture at UNL, today the State Forester and his professional staff serve landowners in five districts and many urban communities.

The Natural Resources Conservation Service and the Nebraska Natural Resource Districts provide assistance in windbreak design and planting. The Nebraska Game and Parks Commission encourages wildlife habitat plantings by providing plants and monetary assistance to landowners. Other prairie states are experiencing similar growth in their forestry programs. The Nebraska Nut Growers' Association supports the pecan research program at UNL in cooperation with the Northern Nut Growers' Association. A growing membership in Christmas Tree Growers' Associations in Nebraska and other prairie states is an indication of interest in Yule trees as an alternative crop.

Federated Garden Clubs promote beautification projects and support research efforts. Citizen and organization support has prompted the addition of three landscape architecture positions at UNL. In their role as teachers, designers of campus landscapes, and community beautification activists, they assist in creating an awareness in their students and the public that prairie landscapes need not be drab.

The list of enthusiastic advocates of trees for the Plains who have provided invaluable leadership in all kinds of tree-related activities is too long to include here. The list would include, however, cooperating farmers in the Prairie States Forestry Project, whose successful plantings serve to remind us that we can indeed modify our environment and diversify our prairie landscape if we so desire. We are grateful to those who have passed on and to those who are still among us for the enduring legacy of beauty and protection afforded by trees that we and future generations will enjoy.

In the 1920s, C. M. Redmond of Fremont selected a pyramidal hybrid linden (*Tilia ×euchlora*) which bears his name. Extension forester Earl Maxwell, promoter of protection plantings for farmsteads, planted a collection of trees and shrubs during the late 1930s that is now the nucleus of the present Maxwell Arboretum on the UNL East Campus. Glenn Viehmeyer, a self-educated Sandhills scientist noted for his development of new varieties of *Chrysanthemum* and *Penstemon*, contributed significantly to our knowledge of native woody plants in Nebraska and adjoining states. A Burt County farmer, J. J. Lydick, demonstrated that concolor fir, spruce, and sycamore will thrive in eastern Nebraska windbreak plantations.

A unique concept of arboreta which exemplifies the pioneer spirit of prairie pioneers was initiated with the founding of the Nebraska Statewide Arboretum (NSA) in 1978. Rather than choosing just one central arboretum, the founders chose to encourage establishment of arboreta in communities across the state of Nebraska, each one reflecting the character and heritage of that community and its founders. Each arboretum must meet standards defined by the NSA executive board to become accepted as an affiliate site. Testing plants at a wide variety of sites in cooperation with affiliate sites is one of NSA's many activities.

Hobbyists are carrying on the tradition of the pioneers by dedicating their spare time and energy to collecting, growing and evaluating species and cultivars of woody plants. With apologies to "plant people" of Nebraska and adjoining states who are omitted because of ignorance of their activities, we name a few, such as the late Ralph Steyer, Gene Solomon and Ed Rassmussen (all of Washington County, Nebraska) who have and are providing the leadership that places the Blair community on the forefront in arboreta development.

Leaders in the search, propagation and promotion of outstanding native woody plants include the late Allen Wilke and his son Evan of Columbus, Shirley and Harlan Hamernik of Clarkson, and Jon Morgenson of Lincoln (all in Nebraska), and Greg Morgenson of Bismark, North Dakota. In addition to his outstanding collection of lilacs, Max Peterson, Perkins county farmer, has successfully introduced to his land many trees and shrubs unusual to Nebraska's high plains. John D. Williams of Denver, former Iowan and later Nebraskan from Dundy and Frontier counties, continues his search for edible fruit plants adapted to the western prairies.

Soils, Climate and Heredity

The interaction of heredity and environment—soil, water, climate—determines a plant's ability to survive. Each plant that grows from seed carries genes different from those of its brothers and sisters. The environment determines which genotypes survive and thrive. Daylength and temperature are the most important environmental cues that trigger acclimation.

For example, progeny of a green ash population that have survived and thrived in Oklahoma for thousands of years may not survive South Dakota winters. Seedlings of southern origin planted in a more northerly latitude take advantage of longer days of the growing season and often grow more rapidly than their northern relatives. Leaves remain green and growth processes continue late into autumn. Yet these trees are vulnerable to injury by sharp, sometimes unseasonable freezes. Early snows accumulating on leafy branches can cause extensive damage. Meanwhile the native green ash population has ceased to grow, leaves have colored and are prepared to drop. They have "learned" through natural selection that a certain daylength signals the approach of freezing temperatures.

When progeny of native South Dakota green ash are moved south to Oklahoma, they drop leaves and become dormant long before danger from freezing weather occurs. Consequently, they grow less during the summer than they would in their native habitat despite a longer growing season. Fewer hours of light per day, the magic numbers on the biologicial clock that trigger dormancy, arrive much earlier on the calender than they do in the northerly native habitat.

Photoperiod-sensitive plants can be deacclimated in spring by critical daylength before danger of freezing temperatures is passed. Plants of southern origin growing in a latitude north of their natural habitat are vulnerable to injury from late spring freezes that may not be unusual for that site.

Prolonged swings in temperature during mid-winter (from freezing to unseasonably mild weather) may allow buds to break dormancy and make them vulnerable to injury from the inevitable freezing weather that follows.

Some plant species rely on temperature as a cue to commence acclimation. In addition to absolute temperatures, cumulative temperatures throughout a growing season can be very critical for a plant's survival. Many species native to hot summer climates can withstand very cold winters, but, if moved to climates with cool summers and mild winters, they may be injured during winter by low temperatures that they could normally withstand.

Raulston and Tripp (1994) cite the example of crape myrtle (*Lagerstroemia indica*) that is winter hardy in North Carolina where sunlight is intense, summers are long and hot, and night temperatures are high during the growing season. During winter, temperatures often drop to 0 degrees F. (-18 degrees C). The authors state that crape myrtle does not survive the winters of England where light is low and summers are cool even though temperatures rarely fall below 10 degrees F. (-12 degrees C). Scott (1998) reports crape myrtle surviving in Omaha on a protected site with precautionary measures such as mulching taken to acclimate the species as early as possible.

These examples illustrate the value of provenance or "origin" research. Throughout the text we mention results of and need for information that can be obtained from provenance research. Unfortunately, we can quote results for only a few of the hundreds of species with potential for improvement.

Provenance research is basic to plant selection and breeding. Seed collection from a plant species throughout its natural range insures genetic diversity among its progeny. Seedlings from such a collection are grown in a nursery and then planted in widely scattered test sites. Response to a common environment provides an estimate of genetic variability. A plant or a group of plants can then be selected for desirable characteristics and adaptability to a specific site. Cross-breeding and further selection provides desirable plants adapted to a variety of sites.

Climate and soil combine to create an imposing array of environments throughout the prairie region. Hundreds of soil types range in texture from sands to clays and in pH from acid to alkaline with shallow to deep profiles on terrain varying from steep to flat. They range from well-drained to swampy. Most of the soils in the prairie region were formed under grasses, short to tall.

The eastern boundary of the prairie was a mosaic of grassland and forest in pioneer days. Forests were found along the streams and grassland occupied the uplands (Illinois and adjoining states). In the Great Plains thin ribbons of the eastern deciduous forest stretch westward along rivers coursing generally from west to east. A unique situation occurs along the Niobrara River in north central Nebraska where species of deciduous forest mingle with those of the Rocky Mountain forests. Western forest species are found along the river bluffs from Rock to Sheridan counties, with outliers in Garfield, Loup and Custer counties in central Nebraska. A mixture of forest species are found on the several groups of wooded hills and low mountains in the northern prairie. The most prominent are the Black Hills of South Dakota, the Turtle Mountains of northern North Dakota, and the Cypress Hills of southern Saskatchewan. **The Natural Geography of Plants** (Gleason and Cronquist 1964) is an excellent reference.

Generally, average temperatures are not greatly different across the prairies from east to west, but the west is more prone to temperature extremes over a span of a few days or even a few hours. Annual precipitation ranges from near 40 inches (100 cm) in the east to 12 inches (30 cm) or less in the west. Humidity decreases and average wind speed increases from east to west.

Micro-environments within the transitional zone from tall-grass to short-grass prairies enable trees and shrubs to invade via the predominantly east/west-oriented stream valleys. Sharp escarpments and steep slopes and the direction they face are major determinants of composition of natural plant communities. The aspect and amount of exposure to sun and wind are also important factors in determining the makeup of plant populations.

On one hand, many plants commence growth in spring in large cities as much as a week earlier than their counterparts in adjacent rural settings and a hard freeze may injure new growth of these urban plants while still-dormant country plants escape. On the other hand, freezing temperatures may arrive in rural areas during the transitional seasons of spring and autumn causing plant injury while the thermometer stays above the freezing mark in the adjacent city.

A plant can be injured by temperatures that fall below the plant's maximum cold hardiness limit. Hardiness zone maps delineate the boundaries of areas where these temperatures are likely to occur. However, a plant's ability to tolerate freezing temperatures is complicated by other factors. A plant must be acclimated or "hardened off" to survive the minimum temperatures that it is genetically capable of surviving. In order for a plant to become acclimated, temperatures must drop during the proper season at an appropriate rate.

Periods of above-freezing air temperatures, bright sunlight and wind combine to increase transpiration from leaves of evergreens. Frozen soil prevents water replenishment; leaves desiccate and may die.

Bright winter sun shining on south-facing tree trunks causes expansion and contraction in response to rapidly changing temperatures. This can cause cambial cells to die, especially on thin-barked trees. Shading and protection from wind reduces winter scorch.

Nomenclature — Naming Plants

Classification of the plant kingdom used here is based upon a theory of the evolution of plants by Cronquist (1981). This theory is illustrated by the classification of American elm as follows:

Division	Magnoliophyta
Class	Magnoliopsida
Subclass	Hamamelidae
Order	Urticales
Family	Ulmaceae
Genus	*Ulmus*
Species	*americana*
Variety	*floridana*

The common name, angiosperm, is used in reference to a flowering plant belonging to division Magnoliophyta; dicot refers to class Magnoliopsida; monocot refers to class Liliopsida.

Seed-bearing plants which do not bear true flowers are placed in Division Pinophyta (gymnosperms). Within this division are classes Ginkgoopsida, Coniferopsida and Taxopsida (Welch 1991). *Ginkgo* is the single genus in the first-mentioned class; those in the other two classes are commonly known as conifers.

Plants were first named to suit the fancy of a person or persons in a community. A tree discovered on the Tigris River was perhaps named the "tiger tree." However, that same tree found in Egypt could be the "sphinx tree." And so it was, until the number of names for known plants led to considerable confusion. But to this day we cling to that naming method by attaching common or English names to many of the plants in our everyday life. For example, American, soft, water and white are one and the same elm tree. Ponderosa, yellow, bull, blackjack, and rock are all common names for the same pine species.

Systematic studies to bring order to human understanding of the plant world, begun by the Greeks and continued by the Romans, were allowed to lapse into disarray during the Middle Ages. The Renaissance brought new attention to the problem and resulted in the publication of "herbals" during the 16th century. Latinized adjectives were used to indicate differences between plants, a method which became cumbersome as the number of named plants increased.

In 1753 Linnaeus published **Species Plantarum** in which he described a uniform and stable naming system for plants of the world—a binomial system that gives each plant two latinized words. These names have been standardized by the International Code of Botanical Nomenclature and can be understood and used by persons of all nationalities. The first name of the binomial represents genus; the second is the specific epithet. Two names in combination designate a species. Thus, the species name for American elm is *Ulmus americana*. A given specific epithet may be used with any number of different generic names, as in *Tilia americana* or *Fraxinus americana*.

A group of related genera is termed a family. With only a few exceptions, a family came can be recognized by the "-aceae" ending. The initial letter of a family and generic name is always capitalized. Letters of the specific ephithet are lowercase in today's accepted terminology—as in *Prunus besseyi* or *Ulmus americana*—even if the name may be a proper noun commemorating a person or place that would normally be capitalized.

Variety is used to subdivide a species and is written as the third name in a species designation. It may be preceded by var., as in *Ulmus americana* var. *floridana*. (Subspecies [ssp.] is sometimes used instead of variety.) However, the distinction between these terms is not always clear (Porter 1967).

Plants selected from a species population because of distinctive characteristics are given a non-Latin "fancy" name to distinguish them as cultivars. These cultivars are named according to rules of the International Code of Nomenclature of Cultivated Plants.

Before adoption of the Cultivated Code, Latin-form names were often given to selections and written after the specific name which confuses cultivars with varieties. Only those Latinized names in use before January 1, 1959, may still be used; they must be distinguished by abiding by the typographical rules specified in the Code (Welch 1991). Thus, a cultivar must be written as follows: *Ulmus americana* cv. Moline or *U. americana* 'Moline'. A plant name, especially in nursery trade listings, will often have the specific epithet omitted, such as *Malus* 'Dolgo'.

A hybrid in the binomial system is identified by placing a times sign (×) before the specific epithet, as in *Populus ×canadensis*, to designate a hybrid poplar. It can also be written *Populus deltoides × nigra* to show parentage. By convention the female parent is listed first.

The name of the person who initially described, named and published information about a plant follows the species name. It is usually abbreviated, as in *Ulmus americana* L., for which the authority is Linnaeus. If two authors are listed, as in *Toxicodendron radicans* (L.) Kuntze, the name in parentheses identifies the original author; the second name denotes the person making a change in nomenclature (such as a change caused by moving a species from one genus to another).

If nearly identical plants have been described and named by different taxonomists, the first valid published name takes precedence. Any other names of the same plant become invalid synonyms. If an author publishes the name of a plant proposed by another author, the plant name is shown thus: *Penstemon digitalis* Nutt. ex Sims. In this case Sims proposed the name *P. digitalis*, but Nutall published it. Consult Rehder (1940) or Hortus Third (1976) to identify authors.

Publications of widely different years may use different scientific names for the same plant. For example, Rehder (1940) lists a tree commonly called red oak as *Quercus borealis* Michx. f., while Stephens (1973) lists it as *Q. borealis* Michx. f. var. *maxima* (Marsh) Ashe and Little (1971) and Elias (1980) use the name *Q. rubra* L.

It is easiest to learn scientific or biological names for each plant by using those names as much as possible (they are prominently shown in the subtitles of the text). Although it is not our purpose to eliminate the use of many colorful and descriptive names, one common name is assigned to each plant species (**Nebraska Statewide Arboretum** 1982).

Morphology — The Nuts and Bolts of Plants

Knowledge of plant structure (morphology) and taxonomic terminology is necessary for efficient use of most plant reference publications. Leaves, twigs and buds are most often used in woody plant identification. Flowers are distinctive, but their usefulness is limited by their short life span (or the plant may not be of flowering age). The fruits of many woody plants, helpful aids to recognition, persist for several weeks or even months in recognizable form, either on the plant or on the ground below. Trunk bark, though often difficult to describe, can provide clues for identification, especially in the forest where twigs and leaves may be out of reach. Wood anatomy is useful in determining the family (and often the genus) to which a plant belongs, but identifiable differences among species are often difficult to detect. **Tree Anatomy** (Shigo 1994) is an illustrated reference that explains the marvelous structure of the tree system.

Habit

Differences in appearance of uncrowded woody plants range from subtle to striking. Silhouettes vary from very narrow or columnar to rounded and spreading. Habit is greatly affected by spacing and environment and it may change throughout the year or with age.

A woody plant is considered to be a tree if one stem is dominant and mature height is at least 12-16 feet (3.5-5 m). A tree's basic framework may consist of a single stem or trunk—as in spruce (*Picea*), pine (*Pinus*), pin oak (*Quercus palustris*) and aspen (*Populus tremuloides*)—or a many-branched stem—like that of cottonwood (*Populus deltoides*), green ash (*Fraxinus pennsylvanica*) or bur oak (*Quercus macrocarpa*).

A shrub is defined as a woody plant which habitually branches near the ground into a number of stems. If only the plant base is woody, and the top is herbaceous (i.e., it dies back at the end of the growing season) the plant is a suffrutescent shrub.

A vine climbs by clasping with tendrils or twining around trees, trellises or similar objects. Some vines attach themselves to walls by aerial rootlets or adhesive discs; aerial rootlets will establish in soil if a wall or appropriate substitute is not at hand.

Tree

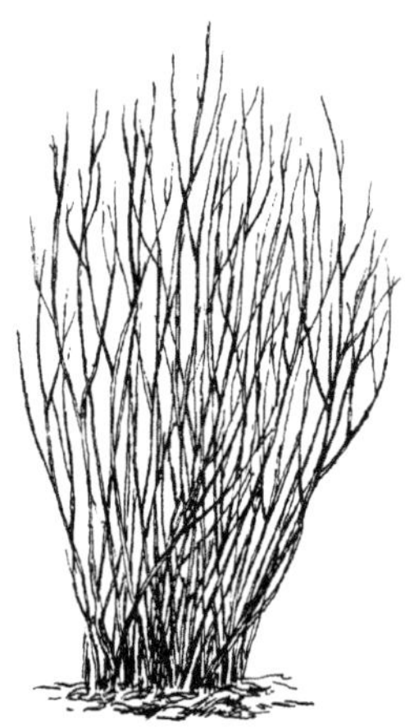

Shrub

Vine

CONIFER LEAVES

Platycladus orientalis

Scale-like

Juniperus communis

Awl-shaped (subulate)

Picea pungens

Needle-like

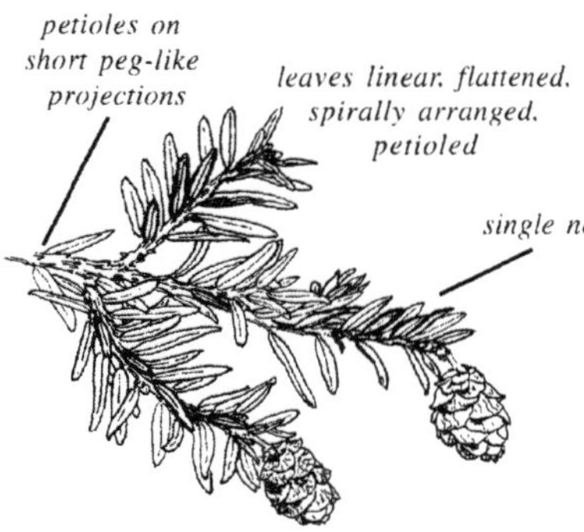

Tsuga canadensis

Pseudotsuga menziesii

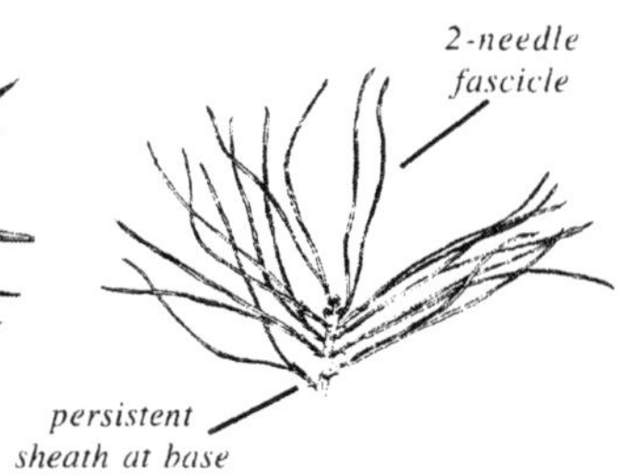

Pinus virginiana

Needle-like

Leaves

Leaves can be very helpful in identification but sometimes can cause frustration because they tend to be quite variable in shape and size among plants of a species, and even on the same plant, as in the oaks. Sprout and shade leaves often are abnormal.

Most flowering plants (angiosperms) and a few conifers growing in temperate climates lose their leaves in autumn after becoming dormant. They are deciduous plants. Plants whose leaves remain functional and withstand freezing temperatures are classified as evergreen. These plants may retain leaves for several years.

Many of the leaves of conifers are evergreen. Their leaf shape varies from scale-like to awl-shaped or needle-like. Needle-like leaves may be borne singly on the twig or in groups of 2 or more (fascicles). Scale-like leaves may completely cover small twigs.

Introduction Morphology

Arrangement of angiosperm leaves on a twig is often a valuable first clue in identification. In their absence, leaf scars and winter buds can suffice. Avoid spur shoots, water sprouts, and twig tips in making a determination.

Stipules are leaf-like basal appendages of petioles present on some species. They may completely encircle the twig (*Platanus, Magnolia, Liriodendron*). Others are often very small and they may fall early in the growing season leaving a barely discernible scar as the only clue to their presence.

SIMPLE ANGIOSPERM LEAVES

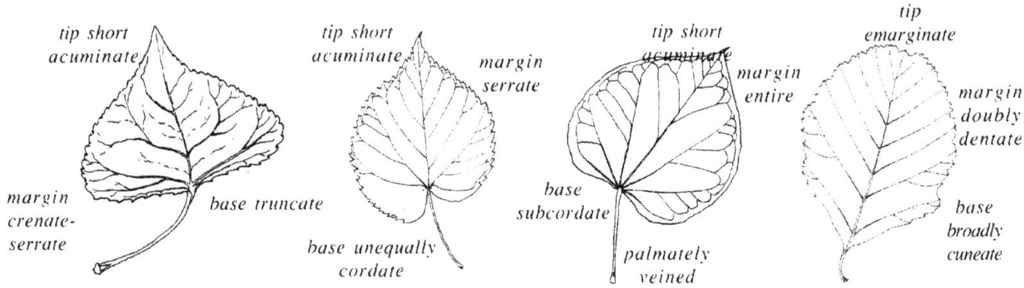

Populus deltoides **Deltoid** — ***Tilia americana*** ***Cercis canadensis*** **Broad ovate** — ***Alnus glutinosa*** **Oval**

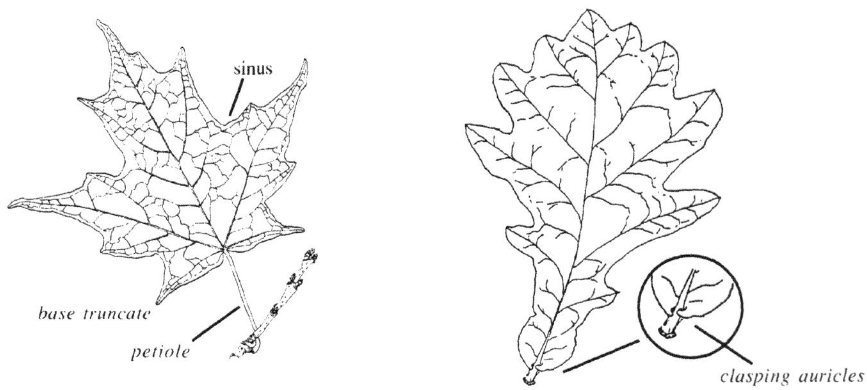

Acer saccharum
Suborbiculor-palmately lobed

Quercus robur
Obovate-pinnately lobed

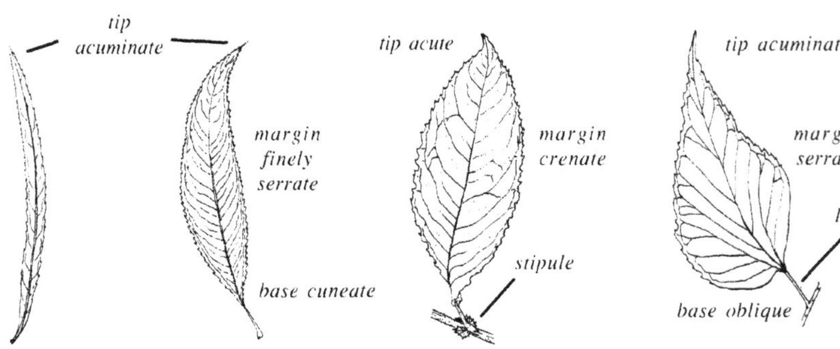

Salix interior **Linear** — ***Salix amygdaloides*** **Lanceolate** — ***Salix humilis*** **Ellipical** — ***Celtis occidentalis*** **Ovate-lanceolate**

Leaves of the angiosperms consist of the blade (a broad, green, flattened portion) and the petiole (a narrow stalk). These are simple leaves. Leaves composed of separate blades (leaflets) are compound.

Leaflets can be confused with simple leaves. Winter buds at the base of a petiole identify a leaf. Buds are never found on the stalk of a compound leaf at the base of a leaflet.

COMPOUND ANGIOSPERM LEAVES

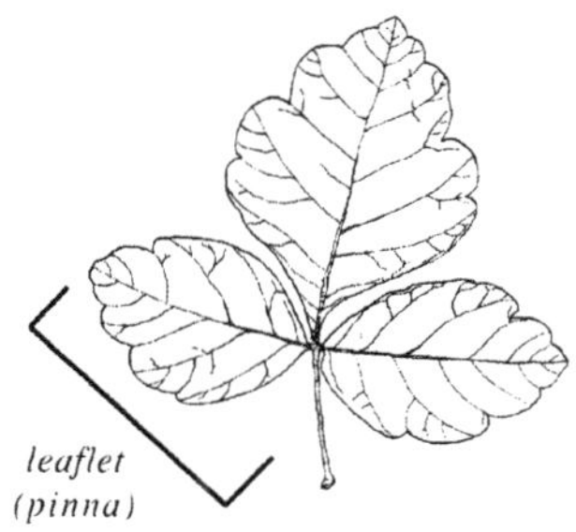

Rhus aromatica

Trifoliate
(Ternately compound)

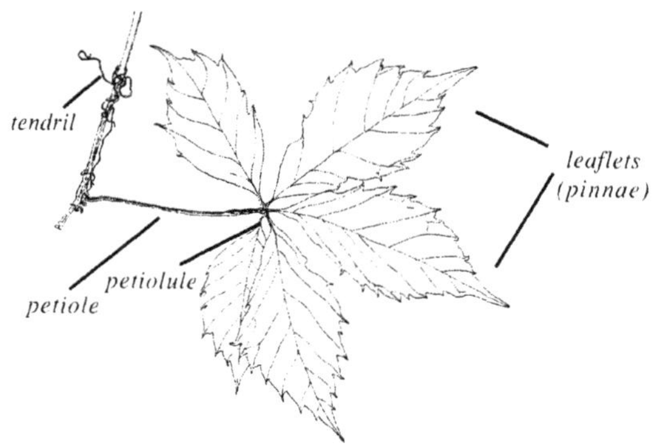

Parthenocissus inserta

Palmately compound

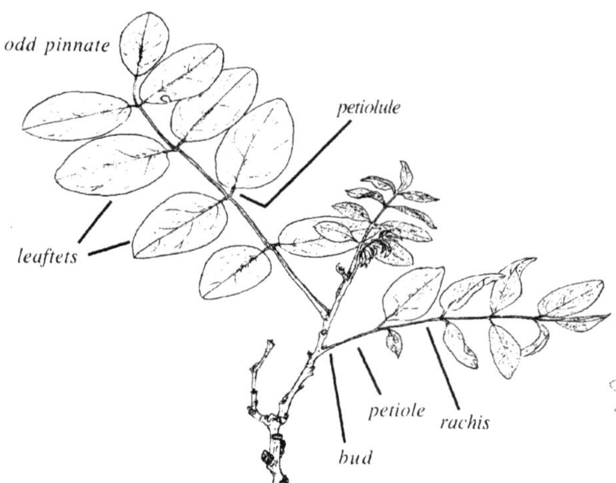

Maackia amurensis

Pinnately compound

Gleditsia triacanthos

Bi-pinnately compound
(Twice pinnate)

CONIFER FLOWERS

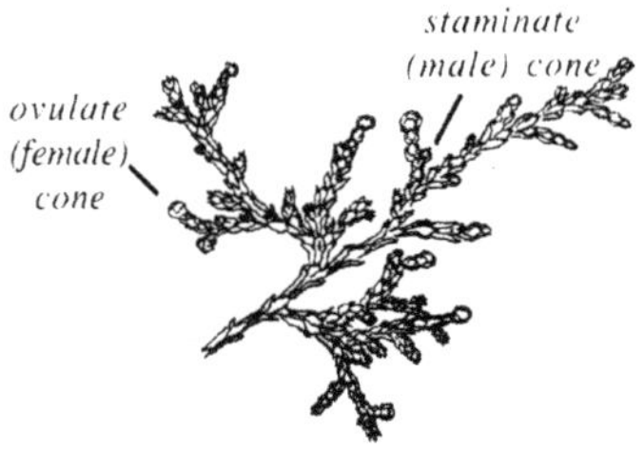

Platycladus orientalis

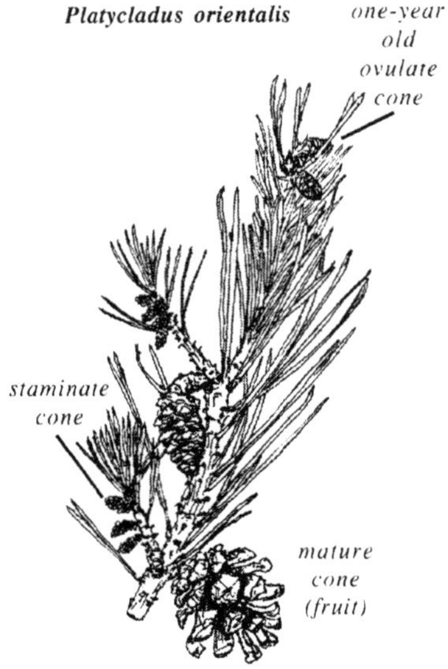

Pinus sylvestris

Flowers

The reproductive organs of *Ginkgo* and the conifers are not true flowers. Short catkins produce the pollen of *Ginkgo* and ovules are borne in pairs on tips of fruit stalks.

Separate male and female parts of conifers are strobili, commonly called cones. They are usually borne on the same plant (monoecious). Exceptions include species of *Juniperus* and *Taxus* in which the male and female strobili are borne on separate plants (dioecious). The non-woody male cones are short-lived, disappearing soon after disseminating their pollen. The ovulate strobili, which vary greatly in size and appearance, are cones except in the yews (Taxaceae). Complete angiosperm flowers bear a perianth (sepals and petals) and male and female reproductive organs. Incomplete flowers lack sepals and/ or petals known collectively as calyx and corolla, respectively. Bracts are modified leaves associated with a flower or cluster of flowers (inflorescence).

Bisexual (perfect) flowers bear male organs (stamens) and female organs (pistils). A stamen consists of a slender stalk (filament) and a pollen-bearing sac (anther). The basal portion of a pistil is the ovary (becomes the fruit) which is generally topped by a distinct style and stigma. Unisexual (imperfect) flowers bear the organs of one sex—male (staminate) or female (pistillate).

Monoecious plants bear unisex flowers (male and female) on the same plant, and dioecious plants bear flowers of one sex only. If bisexual and unisexual flowers are present on the same plant, it is termed polygamous.

The above-mentioned flower characteristics are usually adequate to identify a plant. Descriptions in greater detail can be found in **Taxonomy of Flowering Plants** (Porter 1967).

ANGIOPSERM FLOWERS

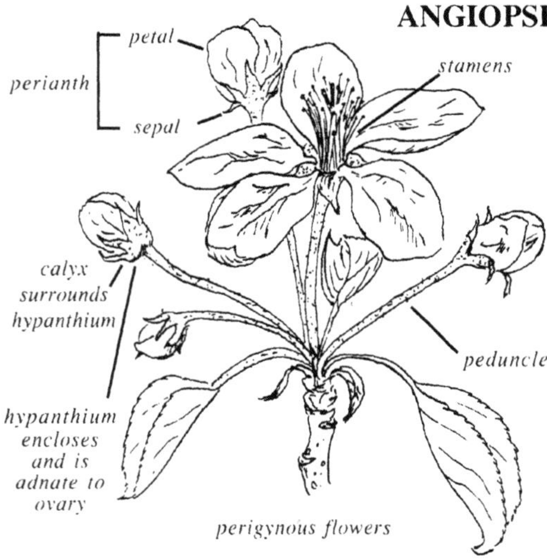

Malus pumila

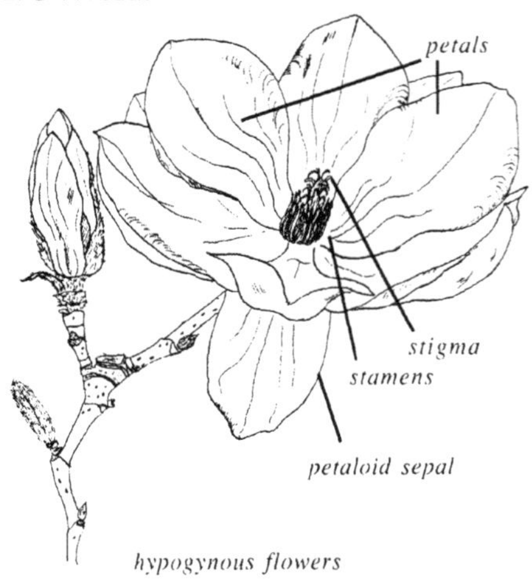

Magnolia ×soulangeana

Flower Clusters (angiosperms)

The arrangement of flowers on a plant may be very simple or highly complex and difficult to define. They may be borne singly or in clusters (inflorescences). Sequence of blooming may be significant. In a determinant inflorescence, the oldest flower terminates the main axis and the general progression of blooming is downward or outward. If the sequence of blooming is upward or inward, blooming is indeterminate. However, the sequence is sometimes irregular or nearly simultaneous and is thus of little value in identification.

ANGIOSPERM INFLORESCENCES

Amelanchier arborea

Raceme

Ligustrum vulgare

Panicle

Pyrus calleryana

Corymb

Viburnum ×rhytidophylloides

Cyme

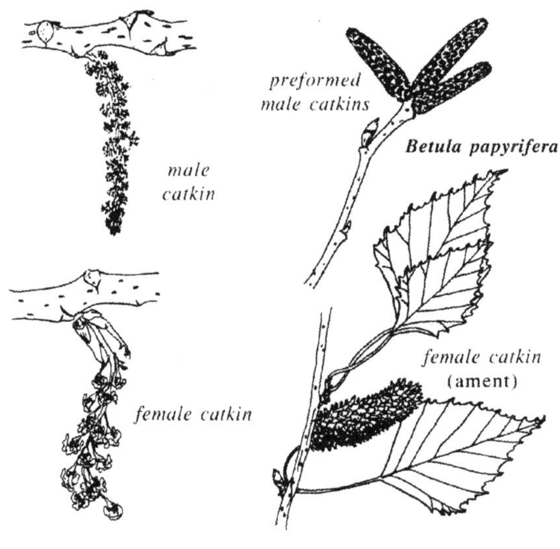

Populus deltoides *Betula pendula*

Catkin (Ament)

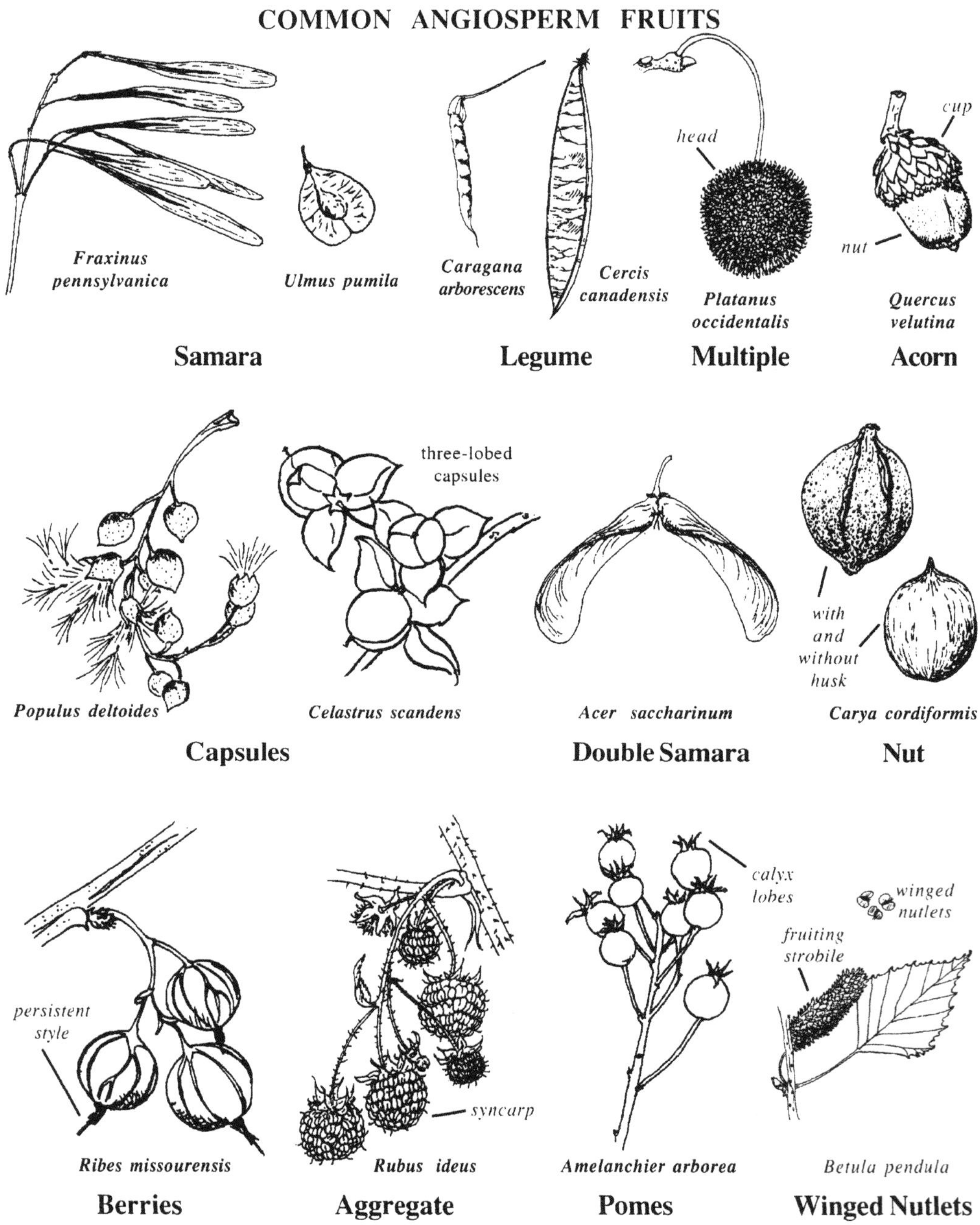

Fruits

A fruit is a matured ovary that has one or more seeds (mature ovules) and closely associated flower parts. Thus a grain of corn, a pod of beans, a walnut, an acorn, a tomato and an apple are all fruits.

Fruit forms of angiosperms are amazingly diverse. Structure and arrangement of female reproductive parts determine fruit type. Fruits of most species are simple. They develop from flowers with single pistils, as in *Prunus*. Aggregrate fruits of raspberry (*Rubus*) and osage-orange (*Maclura*) develop from flowers with more than one pistil, and multiple fruits of mulberry (*Morus*) and sycamore (*Platanus*) develop from a mass of several flowers.

COMMON CONIFER FRUITS

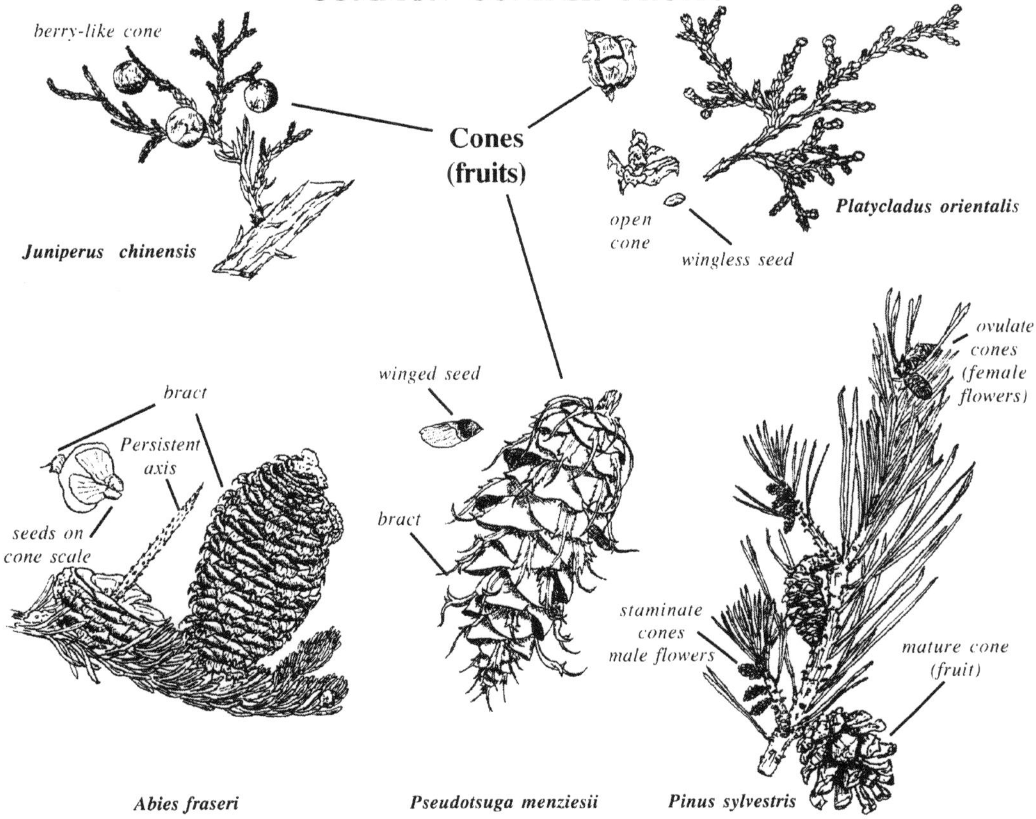

Cone scales deciduous **Cone scales persistent**

Fleshy fruits include drupes, berries and pomes. The single seed of a drupe is surrounded by a thick bony covering that is, in turn, surrounded by juicy or mealy flesh, as in *Prunus*. Berries are entirely fleshy with two or more seeds distributed throughout the pulp, as in *Vitis*. Pomes differ from berries in that their seeds are confined to the center of the fruit with a thin, cartilaginous wall separating them from the outer pulp (hypanthium), as in *Pyrus*.

Dry fruits that split open along seams (dehiscent fruits) are follicles, legumes and capsules. Follicles are derived from a superior ovary of one carpel and split open along one seam, as in *Paeonia*. A one-chambered fruit of two seams, usually with several seeds, is a legume (Fabaceae). It may split along only one of the seams. Capsules are derived from an ovary of two or more carpels. They have two or more chambers and two or more seams, as in *Catalpa*, *Populus* and *Syringa*.

Nonsplitting (indehiscent) fruits are achenes, samaras or nuts. Achenes are tiny and one-seeded, as in Asteraceae. Some are aided in dispersal by long cottony hairs or prickly hooks. The dry membranous wings of samaras allow them to catch the wind, as in *Acer*, *Fraxinus* and *Ulmus*. Nuts are one-seeded with thick, woody, hard shells separate from the seed. Some are partially or wholly enclosed by husks, cups or a similar structure, as in *Quercus*, *Carya* and *Castanea*. Tiny and nut-like describe the fruits (nutlets) of the birches (*Betula*) and American hophornbeam (*Ostrya*).

The ovulate strobili of conifers mature after fertilization to become the reproductive structures which we commonly refer to as cones. The cones of junipers (*Juniperus*) are berry-like. However, the fleshy fruits of yew (*Taxus*) are not cones. The seed of *Ginkgo* is embedded in a pulpy fruit-like structure resembling a cherry. For convenience all of these seed-bearing structures are classified as "fruits" in our descriptions.

Introduction Morphology

Twigs (angiosperms)

We define a twig as the current year's growth of a stem. Twig characteristics are useful in determining the identity of a species during all seasons of the year. Buds are present on twigs except during the period of stem elongation. They "house" twigs, leaves and/or flowers. The shape, size and scale characteristics are often distinctive for a species.

A true terminal bud may be absent. A false terminal bud is a lateral bud located at the twig end caused by stem dieback, as in *Ulmus*, *Platanus* and *Salix*. The absence of vascular traces identifies a scar that is the result of twig breakage.

Mixed buds produce both leaves and flowers. Flower buds are often different than leaf buds in shape and size. Some buds are embedded in the twig (hidden). Thin, chaffy structures called scales protect delicate tissues during dormancy. Naked buds lack scales and are often covered by protective hairs.

Arrangement, size and shape of leaf scars and the number and arrangement of the vascular bundles or leaf traces are often clues to a species identity. These traces are evident as scars after leaves fall, but may not be easily visible on some species.

A cross-section of a twig starting from the outside reveals the bark (epidermis, periderm and phloem), cambium, wood (xylem) and pith. Pith is especially useful for identification. It may be round to star-shaped in cross-section, or exceptionally large or small. Color can be distinctive. Longitudinal section may reveal pith to be either continuous, chambered as in *Juglans*, or diaphragmed as in *Magnolia*. The stem also may be hollow, as in some species of *Lonicera*.

ANGIOSPERM WINTER TWIGS

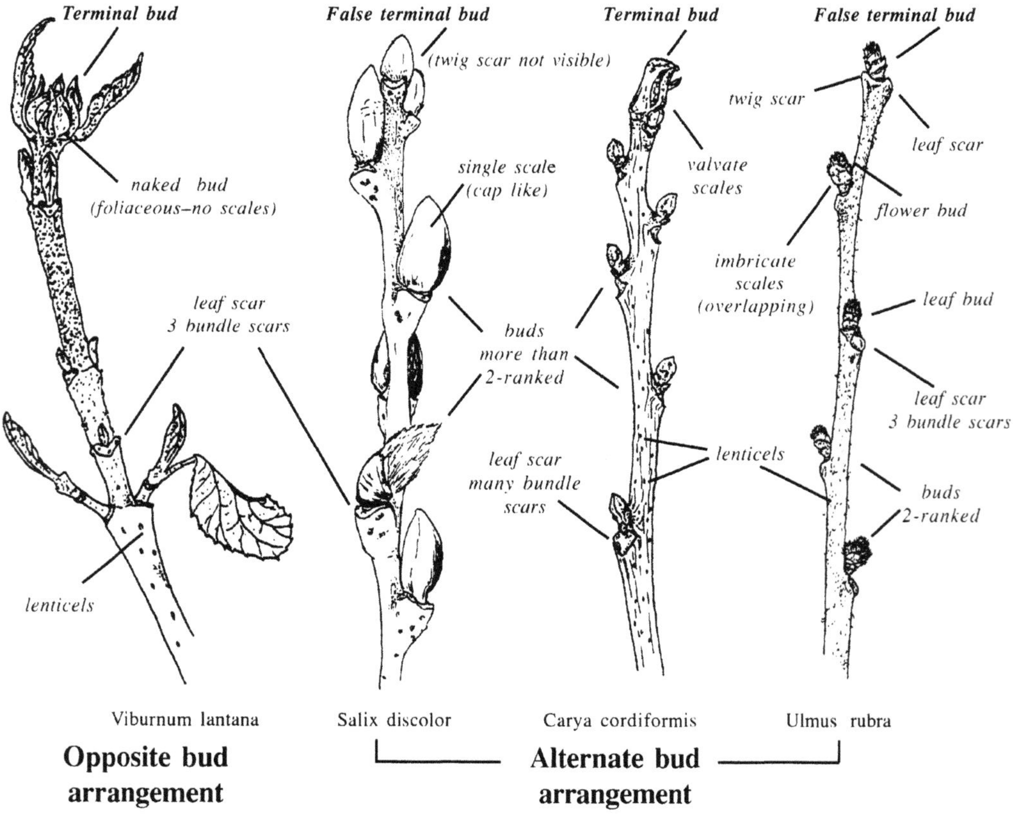

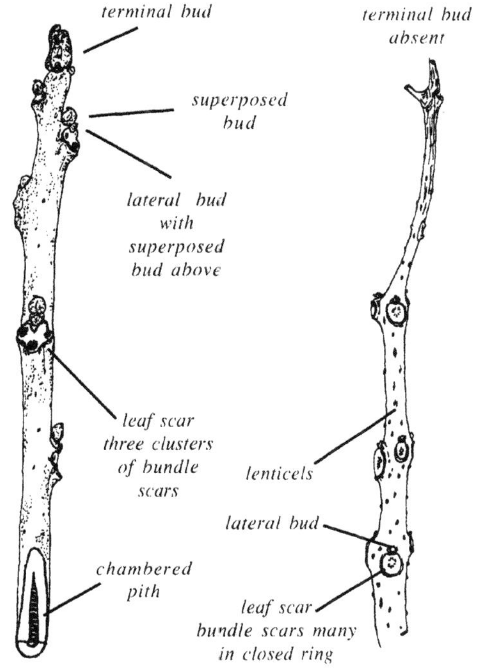

Juglans nigra

Alternate bud arrangement

Catalpa speciosa

Whorled bud arrangement

Lenticels are small groups of loosely arranged cells which form in the periderm, usually during the first season's growth. In some plants lenticels disappear as the stem enlarges. They do, however, persist in some woody plants and become greatly elongated, as in *Betula.*

Specialized structures include true thorns, which are modified twigs. They do not bear buds or leaf scars, as in *Gleditsia triacanthos* and *Crataegus.* Sharply pointed twigs with buds and leaf scars are found on a few species, such as *Prunus americana* and *Malus.* Spines are sharp-pointed woody stipules which occur in pairs on either side of a leaf scar, as in *Robinia pseudoacacia.*

Tendrils of climbing plants originate from leaves or stems, as in *Parthenocissus* and *Vitis.* Spur shoots are short, stubby twigs covered with leaf scars that bear a terminal bud, as in *Malus, Betula* and *Ginkgo.* Catkins of some species become visible in summer and remain through winter, as in *Betula, Ostrya* and *Corylus.*

Bark on Main Stems

The outer covering of a twig usually changes greatly as the stem ages. Sometimes the bark is a distinctive feature of a woody plant. It varies from smooth to deeply furrowed and ridged. In some species it exfoliates, as in *Platanus* and *Pinus bungeana.* Patterns and colors are often difficult to describe, but some species can be recognized from afar by bark alone.

Wood

Minute and gross structural characteristics, hardness, weight, color and many other features are useful in determining the identity of woody plants or products therefrom. The science of wood technology is beyond the scope of this book. Turn to Panshin and De Zeeuw (1964) or a similar publication for wood identification.

Roots

A plant's root system is seldom seen except during transplanting. Consequently roots are not usually described in taxonomic publications. Large taproots are typical of the oaks (*Quercus*), hickories (*Carya*) and walnuts (*Juglans*). Elms (*Ulmus*), maples (*Acer*) and cottonwoods (*Populus*) are examples of species with roots that grow laterally, often reaching out far beyond crown spread. Roots of junipers (*Juniperus*) tend to be much-divided and fibrous. Bright yellow distinguishes the roots of mulberry (*Morus*), and orange-red is the distinctive color of bittersweet (*Celastrus*) roots. Soil texture and water modify root systems; rocks and impervious subsoils limit growth and affect soil water.

Other Books about Woody Plants

Dirr's **Manual of Woody Landscape Plants** (1990) is one of the most complete books on trees, shrubs and vines that is applicable to plants adapted to the eastern prairies. It includes information on culture, propagation, troublesome pests and cultivar descriptions. Since Dirr does not include keys and plants are listed alphabetically by genus, it can be time-consuming to use the manual for identification purposes without some prior notion of the identity of the plant.

The Complete Trees of North America by Elias (1980) is a readable reference for identification and general information about trees. Its illustrated keys use non-technical language to aid identification.

Rocky Mountain Trees (Preston 1968) is a nicely illustrated, concise handbook of trees native to the west of our region. Shrubs are not included in these two references and those that follow, unless specifically mentioned.

Woody Landscape Plants for the High Plains (Fairchild and Klett 1993) evaluates the performance of plants in the arboretum of the W. D. Holley Plant Environmental Research Center (PERC) at Colorado State University, Fort Collins. It is a valuable guide for selecting trees and shrubs for planting on alkaline sites in a semiarid climate. This technical bulletin includes 255 trees and shrubs, the best of the more than 600 taxa in the arboretum. Plants are grown on a clay loam soil that has a pH of about 8. An overhead sprinkler system provides water as needed (the site receives 14 to 16 inches or about 36 to 38 cm of precipitation annually).

Forest and Shade Trees of Iowa (Van Der Linden and Farrar 1984) is an inexpensive paperback edition which includes most of the trees found in the eastern prairie. The book presents keys for both winter and summer conditions as well as excellent line drawings and black and white photographs to aid in identification.

A non-technical guide to tree identification, **The Audubon Society Field Guide to North American Trees** (Little 1980) in two volumes, is illustrated by color photographs of leaves, flowers and fruits. The eastern volume will provide sufficient information for most of the prairie region.

Pool's **Handbook of Nebraska Trees** (1951), which has a summer condition key, is an inexpensive and useful reference that includes native Nebraska trees and a few widely planted exotics. Some tree names have been changed since it was published. Technical terms are used sparingly, but the handbook has no glossary.

Trees and Shrubs for Northern Gardens (Snyder 1980), **Trees of the Southeastern United States** (Duncan and Duncan 1988) and **Trees and Shrubs for the Northern Plains** (Hoag 1965) provide cultural information and assistance in identifying plants adapted to those regions.

Trees, Shrubs and Vines: a Pictorial Guide to the Ornamental Woody Plants of the Northern United States Exclusive of Conifers (Viertel 1970) is an illustrated leaf key that can be a useful non-technical guide to identification of many broad-leaved woody plants that grow in the prairie also.

Trees, shrubs and vines are included in **Woody Plants of the North Central Plains** (Stephens 1973), a well-illustrated technical book. It is an excellent aid to identification of native and naturalized woody plants, but does not include exotic plants that can grow in the region.

The **Manual of Cultivated Trees and Shrubs** (Rehder 1940) is the most comprehensive aid to identification of all woody plants, trees, shrubs and vines that are hardy in North America. It is not illustrated, and highly technical language discourages the beginning taxonomist. To use the keys requires a good knowledge of all plant parts.

For up-to-date nomenclature, **Hortus Third**, originally compiled by L. H. and Zoe Bailey and revised in 1976 by the Liberty Hyde Bailey Hortarium staff, is a ponderous publication including most cultivated plant species from around the world.

Maples of the World (Gelderen et al. 1994) is an illustrated comprehensive treatment of genus *Acer* that includes history of maple names, native habitats and distribution, pests and diseases, propagation, evolution, reproduction and taxonomy. This book includes a proposed revision of the genus and describes hybrids and cultivars.

The **Manual of Cultivated Conifers** and the **Manual of Cultivated Broad-leaved Trees and Shrubs** (Krussmann 1985) are comprehensive illustrated books that describe many species and cultivars used in landscapes. The language is technical.

The Conifer Manual, Volume I, (Welch 1991) is well illustrated with guides to identification in lieu of keys. Cultural information and the uses and value of conifers is included along with descriptions of many varieties and cultivars. Volume II has not been published.

Landscape Plants for Eastern North America (Flint 1982) is a useful illustrated guide to the selection of landscape plants with emphasis on size, habit, functions and seasonal interest. Many cultivars are described, but the publication is not intended as a manual for identification.

Trees, Shrubs and Vines for Urban and Rural America (Hightshoe 1987) includes exhaustive tables, sketches and photos depicting visual characteristics that are useful in landscape design. Ecological relationships, cultural requirements and a phenological calendar are included, but it lacks exotics, a key and many important native plants of the western prairies.

Identification of Southeastern Trees in Winter (Preston and Wright 1976) is a helpful key for identification of trees without leaves—including many which grow in the prairies. It includes many of those that grow in our region.

The **Fruit Key to Northeastern Trees** (Harlow 1946) and the **Twig Key to the Deciduous Woody Plants of Eastern North America** (Harlow 1954) are useful references for tree identification.

The **Trees of North America** (Mitchell 1987) describes over 500 species of native and introduced trees illustrated by color drawings of silhouettes, leaves, flowers, fruits, seeds, twigs and bark.

Division Pinophyta - gymnosperms

Gymnosperms are among the world's most useful and interesting plants. Their lineage is more ancient than the flowering plants or angiosperms (Magnoliophyta), the other division of vascular plants.

The gymnosperms exhibit naked seeds commonly subtended by scales, as opposed to the angiosperms which have seeds enclosed by an ovary. The reproductive structures of gymnosperms will be referred to as flowers although, technically, the reproductive organs are enclosed in a cone.

Pinophyta consists of two classes, Coniferopsida and Taxopsida. By tradition, ginkgo, the sole species in class Ginkgoopsida, though not a gymnosperm, is included under division Pinophyta.

Recognition

Habit: trees, shrubs; mostly evergreen
Leaves: needle-shaped, scale-like or rarely fan shaped
Flowers: unisexual, on the same plant (monoecious) or different plants (dioecious); ovules not enclosed in an ovary

Grouping Key for Gymnosperms

Gymnosperm genera described in this book are grouped according to habit, twig and leaf characteristics.

Tree; deciduous; leaves wedge-shaped, fan-veined; mainly 2-lobed on long stalks: ***Ginkgo***

Trees; evergreen; leaves needle-like in clusters of 2, 3 or 5 (rarely more): ***Pinus***

Trees; evergreen or deciduous; second-year twigs woody; leaves thin or needle-like; single, spiral and widely spaced on twigs and in clusters of 10-30 on short spurs: ***Cedrus, Larix, Pseudolarix***

Trees; evergreen or deciduous; second-year twigs woody; leaves emerging singly, often 2-ranked by a twist of the leaf; linear, flattened, laterally compressed or 4-angled; leaving scars on twigs as they fall: ***Abies, Metasequoia, Picea, Pseudotsuga, Taxodium, Tsuga***

Trees or shrubs; evergreen; second-year shoots remain green; leaves emerging singly, often 2-ranked; linear, flattened, usually twisted to appear 2-ranked: ***Taxus***

Trees; evergreen; twigs usually lying in flat sprays; leaves scale-like in opposite, alternate pairs at right angles to each other (decussate), lying smoothly against the twig, only the tips free; buds hidden: ***Chamaecyparis, Platycladus, Thuja***

Trees or shrubs (spreading to prostrate); evergreen; twigs around the main shoot, not in flat sprays; leaves of mature foliage scale-like in opposite, alternate pairs at right angles to each other, lying smoothly against the twig, only tips free; buds hidden; juvenile foliage awl-shaped in pairs or whorls of 3: ***Juniperus chinensis, J. horizontalis, J. sabina, J. scopulorum, J. virginiana***

Trees or shrubs (spreading to prostrate); evergreen; twigs round the main shoot; leaves in whorls of 3 or 4: ***Juniperus communis, J. procumbens, J. squamata***

Class Ginkgoopsida - gingko

Order Ginkgoales

The leaves of the gingko are broad and deciduous. The tree does not bear true flowers but shares its reproductive system with the fern.

Ginkgoaceae Engl. - ginkgo family

The ginkgo family is the sole occupant of class Ginkgoopsida and order Ginkgoales. The family is represented by one surviving species, a native of southeastern China.

Ginkgo L.

Ginkgo biloba L. - ginkgo

Ginkgos have been cultivated for centuries in the temple gardens of China and Japan. Although seldom found in the wild today, fossil imprints show that the ginkgo ranged throughout the world some 400 million years ago (the Devonian period of the Paleozoic era). Survival during the drift of continents, the rise of mountain ranges and the comings and goings of ice ages must be attributed to the ginkgo's tenacity and sound design.

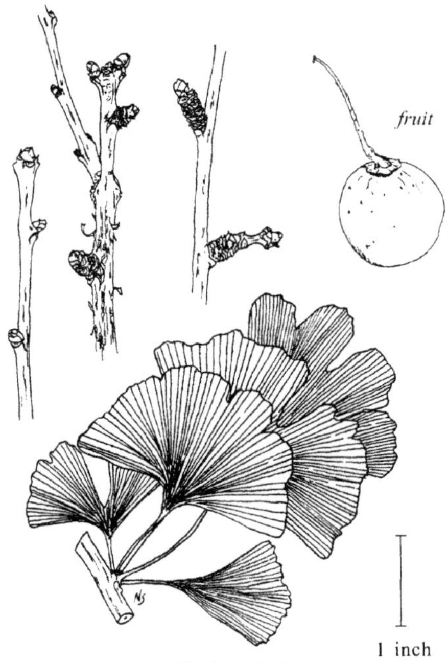

Ginkgo biloba

The ginkgo has outlived enemies that surely must have existed sometime in its history. Today, it is not seriously bothered by insects, diseases or environmental pollution. The ginkgo is widely planted in the U.S. and survives a range of site conditions in North America from Minnesota to Georgia and west to the Rockies.

The greatest hazard to ginkgo is sudden severe cold in autumn while the tree is still actively growing. Though injury usually occurs only during the seedling years, ginkgo overcomes this problem when sprouts growing from the roots finally attain the stature and accompanying physiological conditions necessary to survive the first sharp freezes of autumn.

One year in May the crown of a young 10-foot (3 m) ginkgo at Prairie Pines Arboretum, Lincoln, was killed. A sharp freeze injured the leaves as they were emerging. Though most angiosperms can produce a new crop of leaves from adventitious buds after such a calamity, apparently the ginkgo cannot. However, in this case a sprout from the root crown is now a 20-foot (6 m) tree.

The fleshy covering of the edible nut of the female ginkgo tree becomes malodorous and messy when ripe, a reason for discrimination against the female of the species in landscape plantings. Yet, it is possible that we are overlooking a potential food crop by favoring male trees in our ornamental plantings because this large silvery edible nut is considered a delicacy in the Orient. In addition, one of the world's largest ginkgo plantations, recently established in South Carolina by a Dutch pharmaceutical firm, will soon supply leaves for the manufacture of medicinal chemicals.

We know of no research designed to determine genetic variability, growth rate, or adaptability of ginkgo to various soils and climates.

Autumn foliage, though short-lived, is a beautiful clear yellow, but this condition is not always attained because of the frivolities of weather. Male cultivars include 'Princeton Sentry', an upright, somewhat columnar type that Flemmer (1985) describes as long-lived, and 'Autumn Gold', described by Dirr (1990) as a handsome, broad conical tree.

Planning

Height: up to 100 ft (30 m); Nebraska champion, 1981, 50 ft (15.2 m), Nebraska City
Habit: tree; deciduous; pyramidal, branches ascending to spreading; bright yellow autumn foliage
Site: sun, part shade; pH and soil adaptable; tolerates air pollution; zone 4
Use: specimen, shade, fruit
Pests: pest resistant

Recognition

Leaves: spirals on young twigs, clusters on short spurs on older growth; fan shaped, about 2 in (5 cm) across, variously divided or incised at summit with no midrib; veins parallel, forking in pairs
Flowers: unisexual, borne on separate trees (dioecious); male in clusters (catkin-like), female solitary or in pairs on long stalks; fertilization by free-swimming sperm cells
Fruits: plum-like, 1 in (25 mm) across, ripening yellow to orange, fleshy outer layer malodorous when ripe; nuts silvery, kernels sweet, edible when roasted
Twigs: light brown, becoming gray, with stringy, peeling bark; with short compressed dark colored branchlets (spurs)
Buds: broad, conical, short pointed; scales overlapping, brown

Class Coniferopsida

The only order of Coniferopsida is Coniferales.

Order Coniferales - conifers

Coniferous forests provide turpentine and resin and much of our paper and construction lumber. They serve as protectors of our watersheds, tame the winds as valuable components of our windbreaks, and beautify our landscapes.

In addition to the common name "conifer," they are also known as "evergreens" though a few species drop their leaves with the advent of winter. Lumbermen identify them as "softwoods" as distinguished from "hardwoods" (the broad-leaved trees of the dicots of the angiosperms). However, the wood of some softwoods may actually be harder than the wood of some hardwoods.

Five families are included in Coniferales. Pinaceae, Taxodiaceae and Cupressaceae include species native to the prairies and valuable exotics. Species of Podocarpaceae and Araucariaceae are native to tropical and subtropical regions.

Recognition

Habit: trees, shrubs; often evergreen
Leaves: needlelike, scale-like, awl-shaped, or linear; borne singly or in clusters (fascicles); spirally arranged, opposite, whorled, or on short spur shoots
Flowers: unisexual, male and female on the same or different plants
Fruits: cones woody (rarely fleshy as in Juniperus)

Pinaceae Lindl. - pine family

The pines are the largest and most important family of the conifers. Most of the ten genera and over 200 species are found in the northern hemisphere (Welch 1991). Some taxonomists include species of the redwood family (Taxodiaceae) and the cypress family (Cupressaceae) in the Pinaceae.

Recognition

Habit: resinous trees, rarely shrubs; deciduous or evergreen
Leaves: solitary or in clusters (fascicles), needle-like or linear
Flowers: cone-like, female with bracts and distinct seed scales, ovules 2 at base of cone scale
Fruits: cones woody, pendent or upright; maturing in 1 to 3 seasons; seeds winged or wingless; remain intact, some disintegrate at maturity

Abies Mill. - fir

The firs, also known as true firs, are majestic trees residing in hilly to mountainous habitats of the Northern Hemisphere where they often grow in stands to the exclusion of other species. The firs resemble spruces in form but are strikingly different in detail. Their leaves are not prickly and they cling to drying twigs, whereas the prickly leaves of spruce drop quickly from a drying bough. Fir twigs are smooth whereas minute pegs remain on spruce branches after leaf fall. The fir's attractive upright cones frustrate the cone collector by disintegrating at maturity leaving only the persistent axis pointing skyward. Each species possesses its own distinctive scent emanating from resin gum exuded from bark blisters and crushed leaves.

Nine of about 50 species are found in North America. Seven of these occur in our western states and Canadian provinces. In addition to species described in detail below, several may be worth trying in your landscape if you have a suitable niche.

A fir identified as *A. nordmanniana* (Stev.) Spach, Nordmann fir, is growing at Arbor Lodge State Park, Nebraska City, on a protected, fertile site. Gene Solomon, Blair, Nebraska, has fine young specimens of *A. koreana* Wils., Korean fir, and *A. lasiocarpa* var. *arizonica* (Merriam) Lemm., corkbark fir, in his landscape. *A. cilicica* (Ant. & Kotschy) Carr., Cicilian fir, and *A. homolepis* Sieb. & Zucc., Nikko fir, are species, among others, which adventurous persons may wish to try on their grounds. A healthy young specimen of Nikko fir is growing north of Omaha (Rassmussen 1989).

Most firs, somewhat shade tolerant, thrive only in humid, cool climates on moist, well-drained, acid soils.

Recognition

Habit: trees; evergreen
Leaves: spiral on branches, often spreading in 2 ranks (pectinate); persisting 7 to 10 years; linear, usually flattened and grooved above, usually 2 white bands of stomata beneath; rarely 4-angled with stomata on all 4 sides
Flowers: solitary from buds of previous season; male strobili in leaf axils on underside of branches near the crown, female cones erect on crown branches above male flowers
Fruits: cones woody, maturing at end of first season; upright, egg-shaped to oblong-cylindric with closely overlapping thin, leathery scales subtended by narrow, exserted or enclosed bracts; scales falling at maturity from persistent axis; seeds ovoid to oblong, winged
Twigs: smooth becoming woody the second year, leaf scars circular
Buds: usually blunt and resinous
Bark: usually smooth and thin, becoming thick and furrowed on old trees

Abies balsamea (L.) Mill. - balsam fir

Balsam fir, a native of the northeastern U.S. and much of Canada east of the Rockies, is considered to be a cold-climate tree requiring abundant moisture.

Products from the native balsam fir forests include wood for boxes and light construction, pulp, and oleresin, sometimes used in cementing lenses and mounting microscope specimens. This species is grown in great numbers for the Christmas tree market in the northeastern U.S.

Balsam fir is not widely planted in prairie regions because climatic extremes require that protection be provided for successful establishment. However, healthy 30-year-old trees grow at Horning State Farm near Plattsmouth and at Prairie Pines Arboretum near Lincoln on fertile, non-irrigated sites protected from excessive winds. These two stands have not fallen prey to the pests that bedevil balsam fir in its native habitat.

Several cultivars have been named, but most are not widely in use. 'Nana', a globose bush, is the most commonly marketed of the several prostrate and bushy forms of balsam fir.

A. fraseri (Pursh) Poir., Fraser fir, a native tree of the high mountains of Virginia, North Carolina and Tennessee, closely resembles balsam fir except for bracts of the cone scales, which are longer than the scales and reflexed.

Twenty-five-year-old Fraser firs at Horning State Farm and Prairie Pines are beautiful trees, but are now showing signs of decline. They have produced cones since age 15 years, whereas healthy balsam firs at these locations remain barren at twice that age.

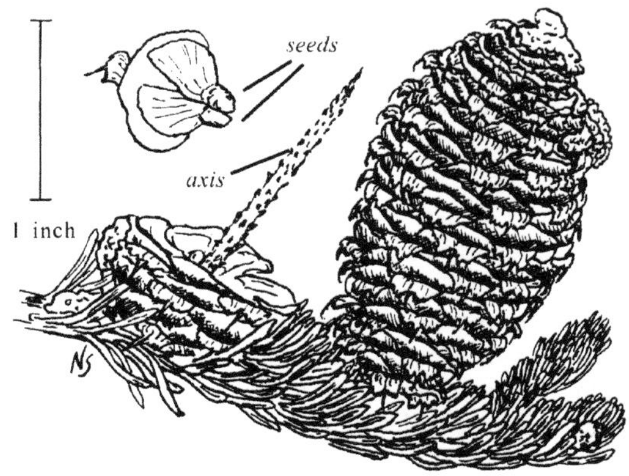

Abies fraseri

Successful Fraser fir plantings in the prairies are limited to moist, fertile sites protected from excessive winds.

'Canaan' fir, native to the Canaan valley in West Virginia, is touted as more tolerant of wet sites than balsam fir of other origins and, perhaps, even more drouth resistant. It is sold as *A. balsamea* var. *phanerolepis* or *A. intermedii.*

Considerable genetic variation is possible in the balsam–Fraser populations. Provenance research could be rewarding.

Planning

Height: up to 50 ft (15 m); National champion, 1962, 116 ft (35.1 m), Michigan
Habit: tree; pyramidal, narrow conical
Site: sun, part shade; prefers well-drained, moist, acidic soils; tolerates drouth, requires protection from excessive wind; zone 4
Use: specimen, Christmas tree

Recognition

Leaves: linear, flattened, 2 ranked, about 1 in (25 mm) long; tip blunt or notched; leaves on upper branches shorter, tip pointed; dark shiny green above, two silver bands of stomata beneath
Fruits: cones woody, ovoid-cylindrical, 2-4 in (5-10 cm) long; dark violet at first, becoming various shades of blue, gray and green; bracts hidden or slightly showing, blade rounded, apex awl-shaped; seeds triangular, wedge-shaped, 0.2 in (5 mm) long, wings 0.4 in (1 cm) long, shiny purplish
Twigs: slender, grayish, hairy; leaf scars flat and rounded
Buds: globular, reddish, very resinous in winter
Bark: thin, dull green to grayish except for raised resin blisters, becoming reddish brown with scaly plates
Wood: light-colored, soft, weak, nearly without odor or taste

Abies concolor (Gord. and Glend.) Lindl. - **white fir**

Two varieties of white fir are recognized. Variety *concolor*, which is best adapted to our region, is a large tree of the mountains of Utah, southwestern Colorado, Arizona and New Mexico. Variety *iowiana* grows in the Pacific coastal region of California and Oregon and is distinguished by dark green leaves about 1.5 in (4 cm) long arranged pectinately in two ranks to form a "v" (Welch 1991). Intermediate forms can be found in Arizona and California.

Though trees grown from seed collected in Arizona and New Mexico may grow quickly at the latitudes of the northern prairies, they also are subject to unsightly deformation because of bud and leaf injury caused by sharp temperature changes that are often experienced in our region. Although not substantiated by provenance research, observations of Nebraska plantings indicate that white fir grown from seed collected from native Colorado white fir, though slower growing, is more likely to be successful in Nebraska (Bagley 1998).

Extensive windbreaks planted by J. J. Lydick on his farm in Burt County, Nebraska, during the 1930s are composed in part of white fir grown from seed that was probably collected from native Colorado trees. White fir trees from Lydick's farm were transplanted to the north and and south approaches to the Nebraska State Capitol Building in the late 1930s. These trees subsequently died and are now being replaced by trees of the same species. The firs on the Lydick farm continue to thrive.

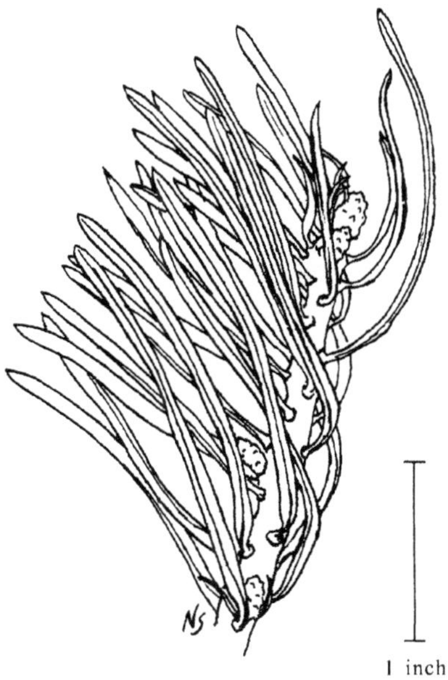

Abies concolor

Thirty-year-old white fir of uncertain origin are beautiful 25-foot (7.6 m) trees at both the Horning State Farm, Plattsmouth, and Prairie Pines Arboretum, Lincoln. They are essentially the same height or taller than adjacent ponderosa, Austrian and Scotch pine trees of the same age. Outstanding specimens of white fir are growing in southwestern Nebraska (Richley 1988).

A common complaint of nurserymen and Christmas tree growers has been the poor survival rate of white fir seedlings after transplanting. The sparse root systems of two-year old seedlings, undoubtedly a major cause of this difficulty, are largely corrected by growing the seedlings in nursery transplant beds for one or two years before moving them to field plantings (Bagley 1990).

A provenance study designed to provide information about genetic variability within the species and adaptability to various sites would contribute significantly to the success of white fir in our region. Welch (1991) describes 18 cultivars, most of which are dwarf or weeping forms.

Planning

Height: up to 60 ft (18 m); Nebraska champion, 1985, 54 ft (16.4 m), Falls City
Habit: tree; conical, branches ascending to horizontal, rigid
Site: sun, part shade; prefers neutral to slightly acidic, moist, well-drained soils; tolerates drouth and can exist on thin rocky soils; does not tolerate water-logged soil; zone 4
Use: specimen, windbreak, wildlife habitat, Christmas tree
Caution: avoid water-logged soils

Recognition

Leaves: extending at nearly right angles from twig, but varying from obscurely 2-ranked to tightly curving out and up; 1-3 in (25-75 mm) long; flattened; covered with a waxy bloom; light green to silvery blue; pores (stomata) on both sides faint, divided by a faint keel beneath
Fruits: cones woody, short-stalked or sessile, usually not appearing before age 30 years; cylindrical, 3-6 in (7.5-15 cm) long, pale green to purplish before maturity; scales broadly fan-shaped, toothed; bracts hidden, one third as long as cone scales
Twigs: hairless to slightly downy, yellow-green to gray; leaf scars circular
Buds: globe-shaped, blunt, resin-covered
Bark: smooth, resin-bearing pockets becoming scaly and finally deeply furrowed on old trees

Abies veitchii Lindl. - Veitch fir

Veitch fir from the slopes of Mt. Fuji is one of several exotic fir species that deserves more trial plantings in our region where soils are not highly alkaline and excessive wind can be avoided.

Widespread plantings have been very successful and its use is especially popular in Britain (Welch 1991). A few young trees purported to be Veitch fir can be found in Lincoln. An excellent specimen is growing in Blair, Nebraska (Solomon 1990).

Planning

Height: up to 50 ft (15 m)
Habit: tree; broadly pyramidal
Site: sun, part shade; prefers moist, well-drained, slightly acidic soils but tolerates slight alkalinity; zone 4
Use: specimen, screen, grouping

Recognition

Leaves: crowded; in ranks, those in top rank short, directed forward and steeply upward; lower ranks more or less spreading (pectinate); flat, 0.5-1.0 in (12-25 mm) long, tip notched; dark shiny green and grooved above; pores (stomata), 2 highly visible white bands beneath
Fruits: cones woody, 2-3 in (5-8 cm) long, sessile; bluish purple, becoming brown; bracts 3-lobed, slightly exserted, tip pointed, slightly reflexed
Twigs: densely brown-hairy
Buds: globular, resinous, purplish to grayish brown
Bark: smooth, grayish, with prominent resin blisters

Cedrus Trew. - cedar

Cedar is the name attached to several species of different genera because of their spice-scented wood, but *Cedrus* is the true cedar. Three of the four species come from the shores of the Mediterranean—Atlas cedar, *C. atlantica* Manetti, is from the Atlas mountains of Algeria and Morocco; *C. libani* Loud. and the closely related but little-known *C. brevifolia* (F. Hook) Henry are natives of Asia Minor and Cyprus, respectively. Deodar cedar, *C. deodara* (Roxb.) Loud., is a native of the western Himalayas.

The four cedar species are so similar that some authorities have regarded them as a single species. The crown shape of mature trees is a good diagonistic feature (Welch 1991). Unfortunately for those of us in the mid-continent, winters are too harsh for all but possibly one species of these magnificant trees—*C. libani*—to survive. Many cultivars have been selected from these species (Welch 1991).

Cedrus libani Loud. - cedar of Lebanon

A grove of 400 dar of Lebanon trees estimated to be 2500 years old remains on the slopes of Mount Lebanon, little changed from the first descriptions of it by writers of Biblical times. Dirr (1990) describes this species as "a specimen tree of unrivaled distinction, uniting the grand with the picturesque." A specimen of variety *stenocoma* (Schwarz) Davis thrives on the Purdue University campus, West Lafayette, Indiana (Dirr 1990). Seedlings grown from seed of the Purdue tree by Gregory Morgenson survived two winters at Prairie Pines but gradually declined in vigor and succumbed (Bagley 1990). However, more trials are warranted for the southeastern part of the central Great Plains.

Planning

Height: up to 50 ft ((15 m)
Habit: tree; evergreen; pyramidal, becoming flat-topped and wide-spreading
Site: sun; prefers soils neutral to slightly acidic; dry atmosphere; zone 5
Use: specimen

Recognition

Leaves: spirally arranged, scattered on long shoots; densely bunched on spurs; needle-like, stiff, triangular in cross-section, pointed; bright to dark green; 30 to 40 per twig spur; 0.7-1.2 in (2.5-3 cm) long

Fruits: cones woody, upright, stalked, solitary; barrel-shaped, purple-brown; 3-4 in (8-10 cm) long, about 2 in (5 cm) across; maturing in second or third year; scales numerous, densely appressed, overlapping, dense woolly-hairy (tomentose); seeds oblong-ovoid, narrowed to base; wings thinly membranous, 1 in (25 mm) long

Twigs: long growth shoots numerous, densely arranged (more so than those of *C. atlantica*), spreading in horizontal plane; lateral spurs short

Bark: gray, smooth, becoming fissured and scaly

Larix Mill. - larch

About 12 species of larch, valuable as timber trees, thrive as natives of the colder, often mountainous regions of the Northern Hemisphere. These deciduous trees, more important to forest industry than to horticulture, have rarely been planted as ornamentals and few clones have been described (Welch 1991). Though the three larches indigenous to North America are winter hardy, they are ill-suited to hot dry summers. Occasional infestations of larch sawfly, *Pristiphora erichsoni*, cause defoliation but usually not serious injury.

Golden larch, *Pseudolarix kaempferi* Gord., a native of China, deserves more use in landscapes (Dirr 1990). It is similar to *Larix* species in many characteristics. Light green leaves become golden in autumn before falling, cones disintegrate at maturity, and spur shoots are longer than those of *Larix* with a distinct constriction between annual rings.

A golden larch seedling at Prairie Pines Arboretum survived several years but died after making little growth. Additional plantings will be required to determine its adaptability to the prairies of the Great Plains.

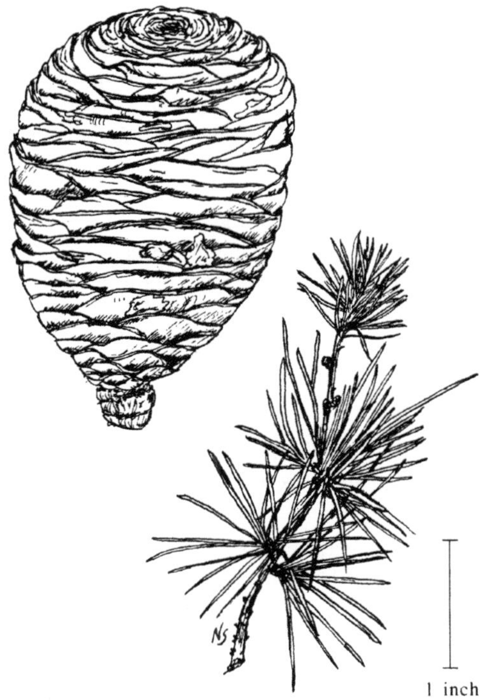

Cedrus libani

Recognition

Habit: trees; deciduous; crowns often irregular; flowers appearing with or ahead of leaves

Leaves: linear; cross-section triangular or sometimes 4-angled; spirally arranged on new shoots and clusters of smaller slender leaves on short spur branches of 2-year-and-older wood; lines of pores (stomates) numerous on all surfaces, most abundant beneath

Flowers: solitary, terminal; male (staminate) strobili yellow to yellowish green, several to many short stamens; female (ovulate) cones erect, few to many, usually scarlet; scales with 2 ovules borne on axils of larger bracts

Fruits: cones woody, maturing in one season, erect, short-stalked; scales thin, persistent, subtended by sharp pointed bracts; seeds triangular, wings terminal

Twigs: slender, drooping

Buds: nearly round, scales overlapping, nonresinous

Larix decidua Mill. - European larch

This exotic from northern and central Europe provides a delightful contrast to other trees with its feathery green leaves of April and glowing golden color of autumn. Variability in form is expressed by cultivars of European larch, such as columnar 'Fastigiata' with short ascending branches and 'Pendula' with a drooping demeanor. Insects and diseases are not generally serious problems for this widely planted tree.

Peterson (1987) reports that a European larch growing at his Meadowlark Hill Farm south of Ogallala, Nebraska, thrives in the windy High Plains climate of highly variable temperatures that is tempered by well-designed farmstead windbreaks. It loses leaves earlier than normal in the fall after a dry summer. Irrigation comes to the rescue during long drouths.

Dunkeld larch— *L. ×eurolepis* Henry (*L. kaempferi × decidua*), a natural hybrid—came from seed of a side-by-side planting of European and Japanese larches on the banks of the River Tay near Dunkeld castle, Scotland, in the late 1800s. This fast-growing, pest-resistant tree has been widely planted.

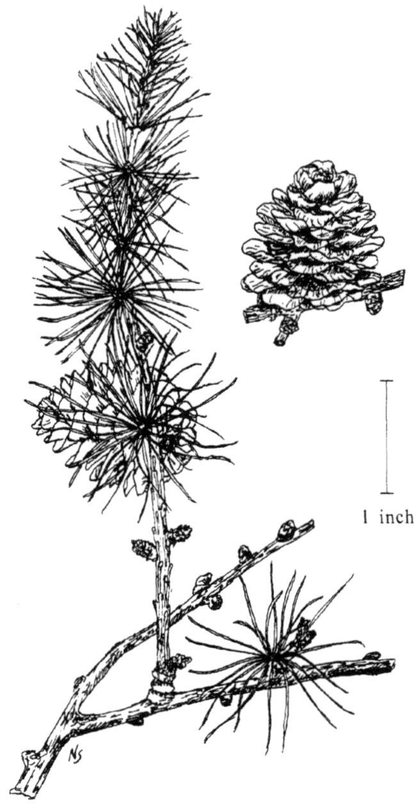

Larix decidua

Planning

Height: up to 60 ft (18 m); Nebraska champion, 1981, 56 ft (17 m), Omaha; National champion, 1991, 83 ft (25 m), Connecticut
Habit: tree; pyramidal, becoming irregular; branches horizontal to drooping; flowers ahead of leaves
Site: sun; prefers moist, slightly acidic soil; zone 2
Use: specimen, screen
Caution: shelter from excessive wind

Recognition

Leaves: single on growth shoots, up to 2.5 in (6 cm) long; keeled below, blunt; clusters of 30 to 40 on spurs, 0.6 to 1.4 in (15-35 mm) long
Flowers: male near twig tips, yellow; female on older twigs, usually reddish
Fruits: cones woody, globe-shaped, 1-1.5 in. (25-38 mm) long; scales round, loosely appressed, entire above, striated, margin sometimes wavy, outer surface soft-hairy; bracts equal to or shorter than scales
Twigs: terminal often yellowish, furrowed, without hairs; short spurs dark, marked with annual rings
Buds: terminal globose, short-pointed; lateral small, blunt
Bark: thin, scaly, becoming gray-brown and deeply fissured exposing reddish brown inner bark

Larix kaempferi (Lamb.) Carr. - Japanese larch

Japanese larch, a species very similar in appearance to European larch, is quite variable in growth characteristics despite its limited natural range in the high mountains of central Honshu Island, Japan. Though beautiful in its early years, this species becomes rather rag-tag with age. Some references list it as *L. leptolepis* (Sieb. & Zucc.) Gord.

Research plots of seven seed origins were planted in 1960 in eight states of the north central region of the U.S., including Nebraska. Read (1970) reported that trees in the Nebraska plantation were larger than those in the other seven states at age 10 years. However, survival and growth in these other plantations, including those in northern Minnesota and Michigan, were satisfactory (Farnsworth, et al. 1972). All sources of Japanese larch broke dormancy earlier in the spring in these trials than is normal for conifers. The provenances differed in dates of leaf flush, some much earlier than others. However, Read recommended that all origins tested could be satisfactory components of windbreaks in eastern Nebraska.

A few years after these reports were published, death of the earlier flowering provenances of Japanese larch at Horning State Farm raised a caution flag in selection of seed sources for planting in our region. Injury was caused by a warmer than normal February followed by a sharp freeze in early March.

Larix kaempferi

Those seed sources which initiate growth later in the spring also drop leaves earliest in the fall and grow more slowly. However, they are more likely to survive the vagaries of the Great Plains climate.

To insure against repetition of this calamity we must either establish seed orchards of the hardy provenances or obtain seed from specified stands in Japan. Information concerning the location of these stands can be obtained from the Rocky Mountain Forest and Range Experiment Station, U.S. Forest Service, Fort Collins, Colorado.

Japanese larch hybridizes readily with European larch, *L. decidua*, and Dahurian larch, *L. gmelinii* (Rupr.) Kuzeneva. The latter is native to the northeastern area of Asia that extends north of the Arctic Circle. Dahurian larch is quite variable. Consequently, several geographical varieties have been named (Welch 1991). Trial plantations of this species and its hybrids are warranted for our region.

Planning

Height: up to 60 ft (18 m)
Habit: tree; pyramidal, pendulous branchlets
Site: sun; pH and soil adaptable; prefers moist sites; zone 5
Use: specimen, screen
Caution: protect from excessive wind

Recognition

Leaves: 1-1.3 in. (25-33 mm) long in clusters of 40 on spurs; up to 2 in (5 cm) long on leading shoots; pores (stomates) conspicuous on under side, 2 bands, 5 rows each
Flowers: inconspicuous; male dark red-brown on new shoots, female on older wood
Fruits: cones woody, egg-shaped (ovoid), 0.7-1.2 in (15-35 cm) long; scales rolled back (reflexed), tip curled under, forming a rosette; bracts concealed
Twigs: furrowed, waxy bloom; with or without soft brownish hairs
Bark: thin, scaly; becoming thick, deeply fissured

Larix laricina (Du Roi) K. Koch - **tamarack**

This tree of the northlands naturally ranges from Alaska to New Foundland and dips down into the United States from Minnesota to Maine. Tamarack inhabits cool swamps in the south of its range and moist uplands in the north, making it an unlikely candidate for planting in the central and southern prairies. However, the presence of a cone-bearing, healthy, 25-foot (7.5 m) tamarack tree on the parking strip along North 61st Street in the 3600 block, Lincoln, and two smaller transplants from Minnesota growing in a yard near Roca, Nebraska, provide us with enough evidence to encourage additional tamarack trials.

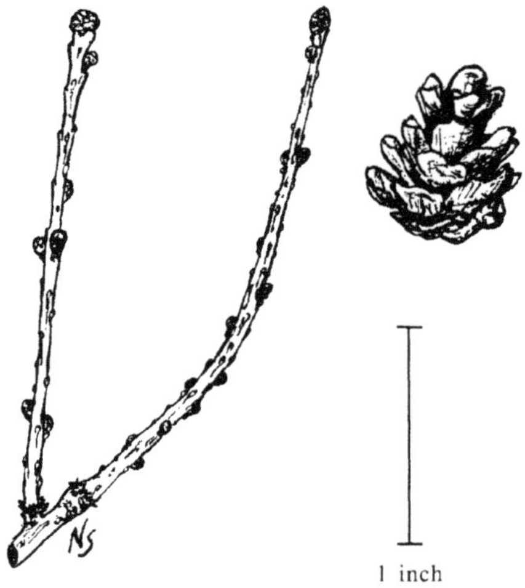

Larix laricina

Planning

Height: up to 40 ft (12 m)
Habit: tree; pyramidal, becoming rounded, branches horizontal and drooping; autumn foliage yellow
Site: sun; prefers moist, acidic soils; zone 2
Use: grouping, specimen

Recognition

Leaves: 3-angled, bright blue-green 1.0-1.2 in (25-30 mm) long on leading shoots; strongly keeled; bundles of 12-30 on spur shoots, slender; stomata beneath only
Flowers: female red
Fruits: cones woody, oblong, less than 1 in (25 mm) long; violet-red, becoming yellow-brown; scales rounded, erect margin curving inward, sparingly toothed, barely longer than broad; bracts visible only near base
Twigs: slender, gray-bloomy, becoming glossy, pale brown
Buds: globose, glossy dark red
Bark: thin, smooth, reddish brown, becoming silver-gray; scaly

Picea A. Dietr. - **spruce**

Spruce derives from the old English name, "pruce," for the kingdom of Prussia, a home of verdant spruce forests. Spruce inhabits cool, temperate regions of the Northern Hemisphere, often growing on wet sites with shallow soil. Despite this evidently preferred habitat, some species are drouth tolerant and succssfully grown throughout the prairies. Spider mites, aphids and bagworms are potential nuisances.

About 40 species have been recognized, seven of which are native to North America while 18 are found only in China. Minor differences among some species coupled with the many selected forms often lead to confusion about their true identities. Undoubtedly other species than those in this text may be adapted to parts of the prairie region. For example, Gene Solomon of Blair, Nebraska, has healthy young *P. asperata* Mast. growing in his yard.

Spruces are generally large evergreen trees of conical outline which retain their leaves and branches for a long time. They are attractive trees, well suited for windbreaks, sound barriers, and as Christmas trees. The tendency to drop leaves suddenly after reaching a certain point of dryness limits the usefulness of spruce for cut decorations. Immediate and continued immersion of the cut stem in water will prolong the usefulness of spruce Christmas trees and boughs.

The soft and odorless wood is strong for its weight and is used for special purposes such as sounding boards for musical instruments, and spruce is one of our most important sources of pulp for newsprint and other paper. Native Americans brewed a beverage from newly formed spruce buds. Wildlife use twigs, leaves and seed for food (though deer and rabbits prefer the twigs and leaves of other woody species).

Recognition

Habit: trees; evergreen
Leaves: needle-like, often sharp; spirally arranged, comb-like (pectinate); borne on peg-like stalks (sterigmata); usually 4-angled, pores (stomates) on all sides, or flattened with stomates only on upper (inner) side
Fruits: cones cylindric, hanging; maturing at end of first growing season; becoming dry, papery or woody; not shattering when ripe, sometimes persisting through winter
Twigs: roughened by peg-like leaf bases
Buds: egg-shaped or conical, scales overlapping, resinous or not
Bark: thin, scaly

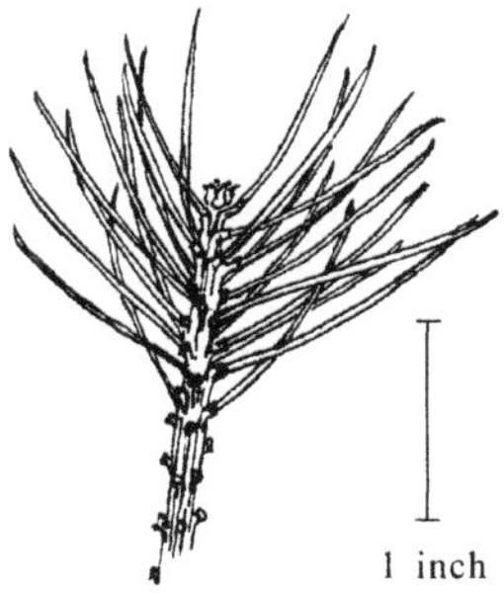

Picea

leaves are borne on peg-like stalks (sterigmata)

Picea abies (L.) Karst. - Norway spruce

Norway spruce, Europe's tallest native tree, is a valuable source of wood for building material and probably one of the original Christmas trees. This quite variable species has made possible the selection and naming of many off-beat forms, varieties or cultivars with varying degrees of droopiness, height and spread. For instance, the new shoots of 'Argenteo-spica' are creamy white before turning green. 'Clanbrassiliana' and 'Procumbens' form a low, dense, flat-topped mound often wider than high. 'Nidiformis', birds nest spruce, gets its name from a depression in the center of a low, broad and dense plant.

Norway spruce is widely planted in North America for ornament and protection. Its natural habitat is on well-drained, fertile rather sandy soils in a cool climate. However, many older 30- to 40-foot trees are in good condition on the UNL campus and other sites in the eastern prairies on fine textured soils with subsoils of high clay content. Undoubtedly, future research will provide improved matchups of provenances and sites.

Siberian spruce, *P. obovata* Ledeb., is similar in character to Norway spruce but is not likely to become popular (Hoag 1965).

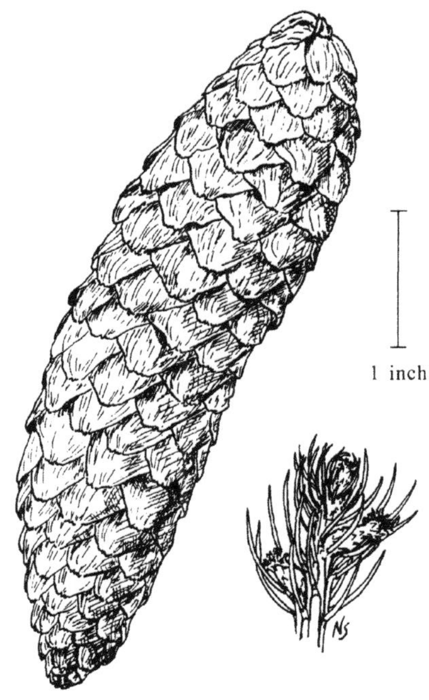

Picea abies

Planning

Height: up to 60 ft (18 m); Nebraska champion, 1987, 75 ft (22.7 m), Plattsmouth; National champion, 1991, 116 ft (35.4 m), Pennsylvania
Habit: tree, pyramidal, branchlets pendulous; shrub, procumbunt
Site: sun, part shade; prefers moist, sandy, acidic soils; cool climate, but pH and soil adaptable; zone 3
Use: specimen, windbreak, wildlife habitat

Pinaceae *Picea*

Recognition

Leaves: 4-sided, 0.5-1.0 in (12-25 mm) long; stiff with blunt horny point; pores (stomata), 2 to 3 lines on each side
Flowers: female terminal, reddish pink on branch tips
Fruits: cones cylindrical, pendulous, 4-6 in (10-15 cm) long; scales thin, wedge-shaped (truncate), sometimes notched, without undulations
Twigs: becoming pendulous on old trees, hairless or nearly so
Buds: pointed, light brown, not resinous; scales often with spreading tips forming rosettes
Bark: thin, gray, scaly

Picea glauca (Moench) Voss - **white spruce**

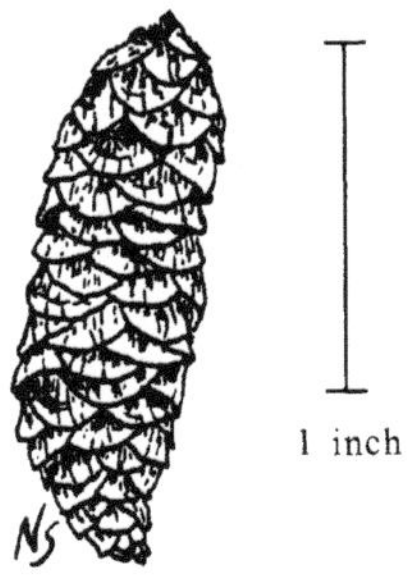

Picea glauca

White spruce, a tree of the far north, ranges across Canada from Labrador into Alaska and barely dips into the U.S. in Montana, the Great Lake states and the New England states. A colony in the Black Hills is a remnant of the recent glacial age. Tough and pliable roots provided laces to Native Americans for their birchbark canoes. Today, white spruce supplies soft white wood for crates and general construction and pulp for paper. It is planted extensively throughout much of the northern prairies in ornamental and protection plantings.

Hoag (1965) considers *P. glauca* var. *densata* Bailey, Black Hills spruce, a highly ornamental tree with a dense crown, to be better adapted to the northern Great Plains than other varieties of the species. However, this native of the Black Hills is not recognized as a variety by some authorities (Stephens 1973, Great Plains Flora Association 1986). It is doubtful that differences between white spruce and Black Hills spruce are sufficient to warrant a varietal name for the latter. Provenance research could determine adaptability of various seed sources to different sites.

Several distinctive varieties of white spruce have been described. *P. glauca* var. *conica* Rehd., dwarf white spruce (commonly known as dwarf Alberta spruce), is a natural dwarf discovered near Lake Laggan, Alberta, Canada. Dwarf white spruce foliage is very dense and the needles radiate around the stem giving it an appearance much different from other spruces. It is easily propagated from cuttings and is widely available from nurseries. Dwarf white spruce grows only 2-4 in. (50-102 mm) per year, reaching 10 feet in 30 years if you are lucky enough to keep it living that long in our central and southerly prairies. It requires a moist soil with protection from drying winds, summer and winter.

Variety *albertiana* (S. Br.) Sarg. is one of many varieties that probably resulted from natural hybridizing with *P. engelmannii* (Parry) Engelm., Engelmann spruce, where the ranges of these species overlap (Wright 1955, Garman 1957).

Planning

Height: up to 50 ft (15 m); National champion, 1975, 128 ft (39 m), Minnesota
Habit: tree; pyramidal, dense, branches ascending; flowers inconspicuous
Site: sun, part shade; pH adaptable, prefers moist loams; tolerant of heat, cold and drouth; zone 2
Use: specimen, hedge, windbreak, wildlife habitat

Recognition

Leaves: 4-sided, 0.4-0.8 in (10-20 mm) long; tip horny, not sharp; crushed foliage often pungent; light green or covered with waxy bloom; pores (stomates), equal bands on each side
Flowers: male pale red, female purple
Fruits: cones light brown, 1-2 in (25-50 mm) long; scales thin, flexible, usually rounded, smooth or slightly notched at apex; seeds 0.1 in (3 mm) long, wings 0.3 in (8 mm) long
Twigs: smooth, sometimes white-bloomy
Buds: egg-shaped, obtuse; scales sometimes reflexed, ragged
Bark: thin, scaly, ash-brown to silvery

Picea omorika (Pancic) Purkyne - **Serbian spruce**

This native of the mountains of eastern Yugoslavia (Serbia) adapts to a wide range of site conditions (Dirr 1990). Kaittinen et al. (1991) discovered much genetic variation in Serbian spruce despite its small native range. One accession of Serbian spruce fared poorly in 10-year trial plantings in Kansas, Nebraska and South Dakota (Widrlechner et al. 1992). Trials which include a more complete sample of the population might discover provenances that are better adapted to the prairies.

This slim, graceful tree adapts to city life and needs less space than widely planted blue spruce.

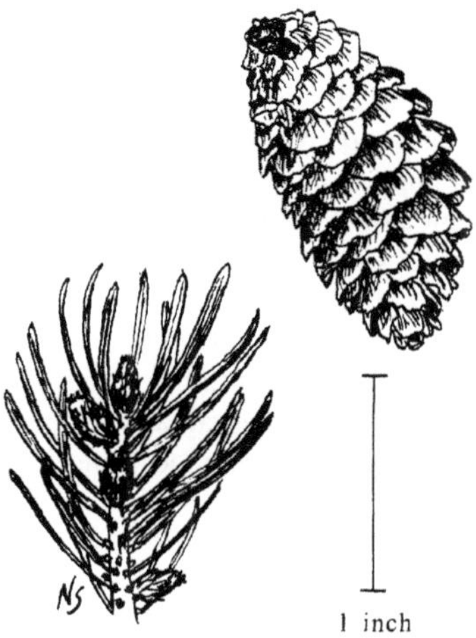

Picea omorika

Planning

Height: up to 50 ft (15 m)
Habit: tree; crown narrow, pyramidal, pendulous branchlets
Site: sun, part shade; pH and soil adaptable; zone 5
Use: specimen, grouping, street

Recognition

Leaves: compressed, 0.5-0.8 in (13-20 mm) long; abruptly tipped with tiny point on young plants, rounded on older; keeled on both sides; dark green on one side, bloomy white on other side with 4-6 distinct bands of pores (stomates) on each side of midrib
Fruits: cones cylindric to egg-shaped, 1.3-2.0 in (30-50 mm) long; blue-black turning shiny brown when ripe; scales circular, fine toothed
Twigs: brown, hairy
Buds: not resinous; scales at base long, awl-shaped
Bark: thin, brown, scaly

Picea orientalis (L.) Link. - **Oriental spruce**

Oriental spruce, a graceful tree from the Caucasus mountains in Asia Minor, tolerates infertile gravelly soils where winters are not excessively cold or dry (Dirr 1990). This species merits trials on prairie sites that are protected from drouth and desiccating winds.

Cultivars include 'Gowdy', narrow columnar; 'Weeping Dwarf', slow-growing with pendulous branches; and 'Gracilis', small, dense and conical.

Planning

Height: up to 50 ft (15 m)
Habit: tree; compact, narrow, pyramidal, slow growing
Site: sun, part shade; pH adaptable, tolerates gravelly soils; zone 5
Use: specimen
Caution: protect from excessive wind

Recognition

Leaves: 4-sided, stout, obtuse, 0.2-0.4 in (5-10 mm) long; crowded and appressed, hiding the stem on top, lustrous dark green; pores (stomates), 1 to 4 lines on each side
Fruits: cones egg-shaped to cylindrical, 2-4 in (5-10 cm) long; reddish purple, turning brown at maturity; scale margins smooth (entire)
Twigs: pale brown, short-hairy
Bark: brown, scaly

Picea pungens Engelm. - **blue spruce**

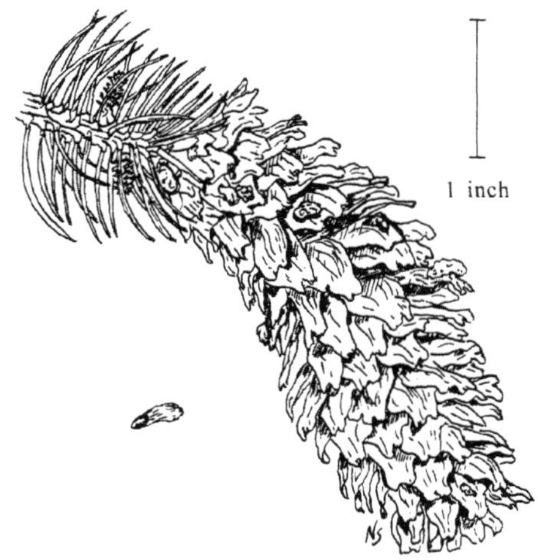

Picea pungens

This ubiquitous tree, native to the central Rocky Mountain region of the U.S., often usurps the frontyards of city dwellings. Although blue spruce grows naturally in mountains along stream-banks in cool humid valleys, it is one of the most adaptable and drouth-tolerant spruces. Foliage color ranges from green through all shades of blue-green to extremely silvery blue, the "shiners" of the nursery trade. Many cultivars have been selected, each vying with the others as queen of the show. The dense pyramidal 'Hoopsii' and pendulous 'Koster' are good choices if silvery-blue foliage is desired.

Great genetic variability is demonstrated among the 43 provenances of blue spruce growing at Horning State Farm. Van Haverbeke (1984) concluded from analysis of this planting that populations from southern Colorado, New Mexico and Arizona were faster growing and foliage was bluer than those of more northerly origins. Because of variability within stands, greater uniformity of planting stock can be obtained by collecting seed from selected trees. Cultivars for landscape trees are vegetatively propagated.

Engelmann spruce, *P. engelmannii* (Parry) Engelm., a native of the Rocky Mountains at higher altitudes and more northerly latitudes, so closely resembles some forms of blue spruce that it is difficult to distinguish between the two. However, even in areas where they intermingle, there is little evidence of natural hybridization between the two species because of various prefertilization incompatibilities (Kossuth and Fechner 1973). They can be hybridized, however, by controlled pollination (Garman 1957).

Engelmann spruce is considered to be a softer, more graceful tree of less layered appearance than blue spruce. Its leaves lack the sharp, acid taste of those of the latter species and its tips are not as sharp. Cones of Engelmann spruce are shorter with relatively broader scales than those of the blue spruce. The Engelmann is slow growing and long-lived in its natural, high altitude habitat. The few known in Lincoln grew only two feet (60 cm) in seven years, even with the best of care, including screening from wind (Sutton 1998).

Planning

Height: up to 60 ft (18 m); Nebraska champion, 1981, 58 ft (17.6 m), Arbor Lodge Historical Park, Nebraska City; National champion, 1991, 122 ft (37.1 m), Utah

Habit: tree; pyramidal; branches horizontal, stiff; foliage green to silveryblue

Site: sun; pH and soil adaptable; drouth tolerant; zone 3

Use: specimen, windbreak, wildlife habitat

Recognition

Leaves: 4-sided, sharp, extending at right angles to the twig; 0.7-1.2 in (18-30 mm) long; pores (stomates) equal on each side; sharp pungent acid taste

Fruits: cones leathery, hanging; cylinder-shaped, tapering to tip; 2.2-4 in (6-10 cm) long; scales loosely appressed, flat, irregularly toothed on outer margin

Twigs: stout, hairless, orange-brown

Buds: slightly resinous or not; broadly conical, blunt; scales loosely appressed, turned back (revolute) at apex

Bark: thin; pale gray, becoming reddish brown; scaly, becoming furrowed

Pinus L. - pine

Pines are generally tall, pyramidal trees. About 90 species are native to the Northern Hemisphere, from the equator to the northern limits of tree growth. Pines are most easily distinguished by their relatively long, evergreen, needle-like leaves, usually grouped in clusters (fascicles) of two to five.

The terminal winter bud flushes to produce a central leader, while the surrounding lateral buds produce stems which radiate outward. One flush each season is normal, except under unusual circumstances such as plentiful water during the latter part of the growing season. Exceptions to this pattern of growth are found in jack pine (*P. banksiana* Lamb.) which exhibits multinodal growth. Jack pine has lateral buds that develop between the nodes and immediately break dormancy, forming lateral branches.

Christmas tree growers routinely shear pines to reduce growth and thereby increase foliage density while shaping the tree to attain similarity to the classical shape of firs and spruces (the trees first brought into homes at Yuletide). New growth is shortened by pruning the stem at any point where leaves remain below the cut. Adventitious buds will form at the base of needles near the cut unless the stem has become too woody. The best time of year for shearing varies with species but is generally June and early July at the mid-latitude of the prairie region. Pines that have been sheared for this purpose are not desirable for landscape planting because they become rather shrubby in appearance for several years after shearing is discontinued.

Pinus virginiana

Trees of most pine species are significantly variable in form, foliage, pest resistance and other characteristics. It is desirable to plant seedlings of an origin known to be adapted to your site and design requirements. Vegetative propagation, a costly method for obtaining the desired plant characteristics, is warranted for landscape plantings.

In addition to pine species described herein, Virginia pine, *P. virginiana* Mill., which grows naturally on dry soils of low fertility in the lower altitudes of the Appalachians, may be satisfactory for planting south of the Nebraska border. It is grown for the Christmas tree market in southern states. Table-mountain pine, *P. pungens* Lamb., another tree from the Appalachians, has been growing successfully, though slowly, for over 15 years at Horning State Farm. This pine is easily recognized by cones armed with stout, hooked spines that persist on the trees for many years.

Recognition

Habit: trees; evergreen

Leaves: juvenile or first seedling leaves solitary, spirally arranged, usually early deciduous; adult or secondary leaves, needle-like, usually borne in clusters of two, three or five (rarely one to eight), enclosed at base with a persistent or deciduous sheath; triangular, semicircular or rarely circular in cross-section; pores (stomates) in lines on each surface

Flowers: unisexual (monoecious); male cones (strobili) oblong, crowded clusters at base of current year's growth; female cones globe-like, scales spirally arranged, overlapping, each containing 2 seed-producing structures (ovules) and a bract

Fruits: cones woody, maturing in 2 (rarely 3) seasons; opening at maturity or persisting and closed on branches for years; scales variable, exposed part (apophysis) of each scale thickened, showing a protuberance (umbo) often armed with a prickly point or stout hook; seeds 2 to each scale, wings terminal (rarely wingless)

Twigs: annual whorls or multinodinal

Buds: scaly, variable in size, shape and color

Grouping Key for the Pines

Leaves in clusters of 5 (occasionally 1 to 4); sheath deciduous on mature clusters; cross-section shows 1 fibrovascular bundle; cone scales usually unarmed; wood usually soft, transition from spring to summer wood gradual: **subgenus *Haploxylon* (soft pines)**

Leaves in clusters of 2 to 3 (occasionally 5); sheath persisting on mature clusters; cross-section shows 2 fibrovascular bundles; cone scales usually armed with a spine; wood usually hard, close grained, annual rings distinct: **subgenus *Diploxylon* (hard pines)**

Subgenus *Haploxylon* - soft pines

A handsome species of the soft pine group, *P. wallichiana* A. B. Jacks. (*P. griffithii* McClelland), native to the Himalaya mountains, has been planted in eastern Nebraska with little success. However, a search for hardier provenances may be fruitful. An excellent specimen grows in Morton Arboretum (Dirr 1990). A closely related species, *P. armandii* Franch., native to China and Korea, is also worthy of trial in the southern prairies.

Pinus aristata Engelm. - Rocky Mountain bristlecone pine

Rocky Mountain bristlecone pines grow on rocky soils of the high country of Colorado, New Mexico and Arizona. At timberline bristlecone pines become gnarled and twisted by high winds and deep snow. It is hardy and long-lived on sunny, dry, infertile sites but often does not thrive in the humid eastern prairies. An attractive healthy specimen grows at the Holly Plant Environmental Research Center (PERC) in Fort Collins (Fairchild and Klett 1993).

Native stands of the Rocky Mountain bristlecone include some of the oldest living trees, exceeded only by individuals of a close relative, *P. longaeva* D. K. Bailey, Great Basin bristlecone pine, that are estimated to be 5,000 years old. The Great Basin bristlecone, a native of Utah, Nevada and a slice of southeastern California is considered to be a variety of the Rocky Mountain species by some authorities.

Pinus aristata

Planning

Height: up to 20 ft (6 m); National champion, 1985, 76 ft (23 m), New Mexico
Habit: tree, shrubby in cultivation; crown irregular
Site: sun; pH and soil adaptable; zone 4
Use: specimen, rock garden, foundation
Caution: lives longer in dry climates

Recognition

Leaves: fascicles of 5, 1-1.5 in (25-38 mm) long, crowded toward branch tips; shiny dark green, dotted with conspicuous white resin flecks
Flowers: male oval, clustered near branchlet tips; female, single or in pairs
Fruits: cones woody, 3 in (8 cm) long; scales armed with bristle-like prickles; seeds 0.2 in (5 mm) long, oval, terminal wing 3 times longer than seed

Pinus bungeana Zucc. - **lacebark pine**

This native of China takes its name from its beautifully textured exfoliating bark. The lacebark's beauty is further accentuated by a tendency to be multi-stemmed. A 53-foot (16 m) specimen grows on the north side of Love Cooperative Residence Hall, UNL East Campus, where it borders a parking lot and is shaded by ponderosa and Austrian pines. It has survived under these inhospitable conditions for more than 40 years and has provided scions for successful grafts to ponderosa pine seedlings through the late Hans Burchardt's skill as a propagator.

Despite apparent propagation problems and rather slow growth, this small, long-lived pine could effectively grace the average residential lot in much of the southern prairie region. We have noted no insect or disease problems.

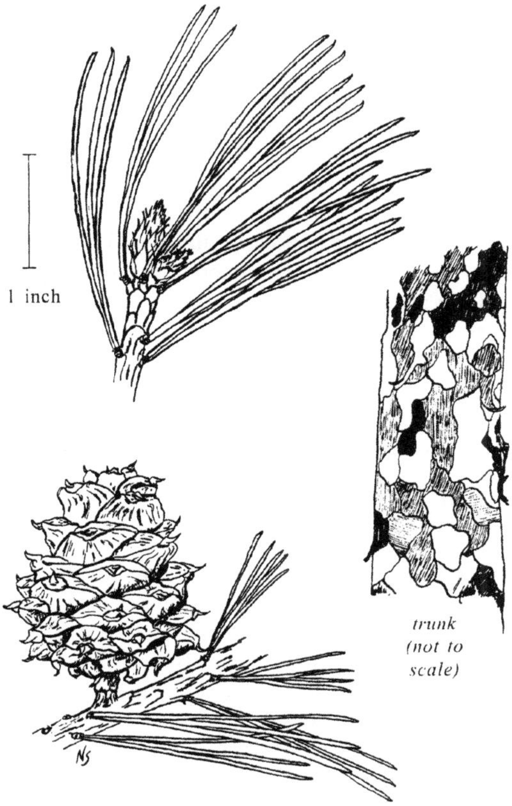

Pinus bungeana

Planning

Height: up to 40 ft (12 m); Nebraska champion, 1980, 53 ft (16 m), East Campus, UNL
Habit: tree; rounded, sometimes multiple-stemmed
Site: sun; pH and soil adaptable; zone 5
Use: specimen

Recognition

Leaves: clusters of 3, 2-3 in (5-8 cm) long; dark green, lustrous, sharp
Fruits: cones woody, egg-shaped, short-stalked, 2 in (5 cm) long; scales with strong, reflexed spines; seeds winged
Twigs: green, shiny
Buds: reddish brown, scale tips spreading
Bark: brownish, exfoliating in patches revealing light yellow-green or cream-colored inner bark; mottling increases with age

Pinus edulis Englm. - **pinyon pine**

This native tree of the semi-arid Southwest is one of a group of pines known as piñons or nut pines. Pinyon pine grows in groves or as scattered trees asssociated with juniper species on dry, rocky or gravelly canyon soils.

The seeds of the pinyons were once an important source of food for Native Americans and wildlife of the Southwest, but now most of the pinyon harvest is sold as an edible delicacy. Pitchy pinyon logs provide high energy fuel and a distinctive aura to the atmosphere of the Southwest during cool weather, and its fragrant foliage makes it a tree of choice for Christmas decorations by the local populace.

In addition to its obvious adaptation to the southwestern prairies, pinyon pine has been grown with mixed success on sunny sites in eastern Nebraska. A series of cool growing seasons with above normal rainfall has contributed to needle diseases and general decline of some older trees.

Pinyon pine is winter hardy and adapted to the soils at PERC (Fairchild and Klett 1993). Pest problems are minor but fruit production is not reliable or abundant.

Planning

Height: up to 30 ft (11 m); National champion, 1982, 69 ft (21 m), New Mexico
Habit: tree; rounded, shrubby; fragrant foliage
Site: sun; soil and pH adaptable, drouth tolerant; zone 5
Use: specimen, food for man and animals
Caution: short-lived in humid climates

Recognition

Leaves: clusters of 2, occasionally 1 or 3, 0.5-1.5 in (12-50 mm) long, persisting 3 or more years; rigid, sharp, fragrant, bluish green, becoming yellow-green; margins smooth, sheath partially deciduous
Flowers: clusters near branch tips; male dark red, female purple
Fruits: cones woody, nearly round, 1.5-2 in (38-50 mm) long; scales few, thick, blunt; seeds wingless, narrowly egg-shaped, 0.5 in (13 mm) long, sweet, oily, thin-shelled; 1,900 seeds per lb.

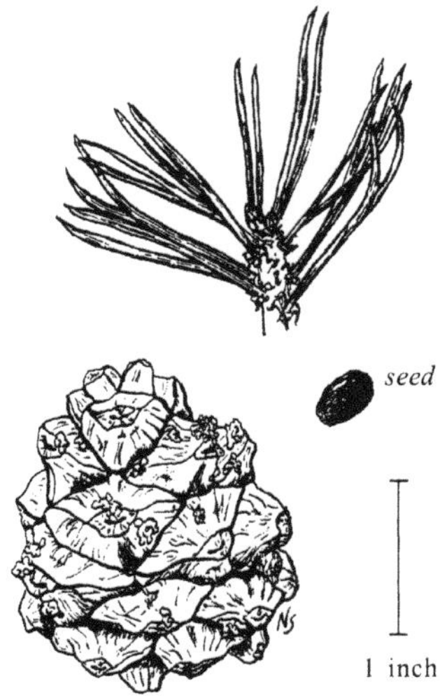

Pinus edulis

Pinus flexilis James - limber pine; *P. strobiformis* Engelm. - southwestern white pine

Older publications consider the second named species to be a variety of *P. flexilis*, or consider the two as one. The major difference between the two species is geographic distribution. The first-named is native to the area north of 35 degrees north latitude in New Mexico and Arizona, north into Canada. The latter is found naturally only to the south of that parallel in the two aforementioned states and Mexico (Little 1971). Morphological differences between the two are less well defined. Cone scales of *P. strobiformis* are more strongly reflexed than those of *P. flexilis*, and *strobiformis* matures into a much taller tree. Trees growing in northern Arizona and New Mexico exhibit intermediate characteristics.

A study of the growth and morphology of limber and southwestern white pine trees grown from seeds collected from throughout the native range of both species was conducted in provenance research plantations at various agricultural experiment stations including Horning State Farm. This research has provided valuable information for selecting the best source of seed for plantations.

Pinus flexilis

Van Haverbeke (1983), reporting on the Nebraska research, found that survival and growth rate of *P. strobiformis* was much better than for *P. flexilis*. At age 17 years the average heights were 27 feet (7.1 m) and 8 feet (2.5 m), respectively. Sources of *P. flexilis* from southeastern Wyoming near the Nebraska border and from central Colorado on the eastern slope of the Rockies performed better than those of the same species from other parts of the native range. However, the specimens were much shorter than *P. strobiformis* trees from central Arizona and New Mexico of the same age. Many of the *P. flexilis* trees have since become unhealthy and died.

Unfortunately, we do not have knowledge of their potential for growth in the western prairies. The general good health of the native stand of limber pine growing in extreme western Nebraska (in Kimball County) indicates that some provenances may be well adapted to similar sites in the western prairies.

Most trees originating from central Arizona and New Mexico (*P. strobiformis*) in the provenance plantation and at Prairie Pines Arboretum remain healthy at age 30 years. Of some concern is the recent "blowdown" of apparently healthy 20-foot (6 m) trees, one at each location, by strong winds that did not damage trees of other species (Bagley 1998).

An unfortunate sequel to these findings is that many trees grown from seed commercially collected from southwestern white pine in Arizona and New Mexico suffer cold injury during winter. Injury usually occurs when a second flush of growth in late summer fails to set winter buds, and the tender shoots succumb to freezing temperatures. Late flush does not usually occur on the surviving trees of the provenance plantation. We can only surmise that the commercial seed was collected from trees growing at lower altitudes than the seed parents of the provenance plantation trees.

In order to obtain southwestern white pine trees well adapted to Nebraska, Van Haverbeke (1983) recommends seed collected from elevations above 7600 feet (2300 m) in central Arizona and New Mexico. An alternative is to establish orchards for seed and scion wood production from parent trees of known hardiness.

Limber pine–southwestern white pine trees have been free from pests, except for a needle blight which is apparently caused by *Dothistroma pini*, a fungus which also causes leaf disease on Austrian and ponderosa pines. The fungus is not likely to be a problem to limber–southwestern pines unless the trees are infected in the nursery or are planted in close proximity to infected trees. Timely spraying with a copper fungicide should keep the disease in check.

Planning

Height: up to 50 ft (15 m): limber pine—Nebraska champion, 1989, 78 ft (23.8 m), Lincoln; National champion, 1968, 58 ft (17.7 m): Utah; southwestern white pine—National champion, 1974, 111 ft (33.6 m), Lincoln National Forest, New Mexico
Habit: tree; crown rounded, spreading; branches flexible
Site: sun; pH and soil adaptable; limber-zone 3, southwestern white-zone 5
Use: specimen, Christmas tree, windbreak
Pests: dothistroma needle blight

Recognition

Leaves: clusters of 5, 2-3 in (5-8 cm) long; slender, dark, blue-green
Flowers: male reddish, female reddish purple
Fruits: cones woody, cylindrical, 3-7 in (8-18 cm) long; scales thick, tips slightly bent back, or more so for *P. strobiformis*; seeds 0.4 in (1 cm) long, winged
Twigs: stout, tough, flexible, drooping with age
Bark: smooth, light gray, becoming dark, deeply furrowed

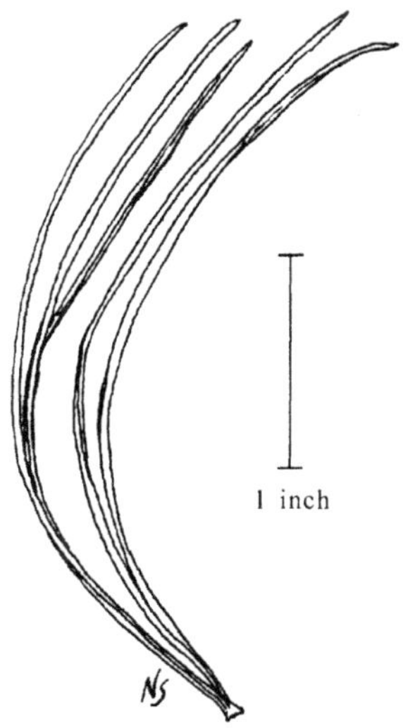

Pinus koraiensis

Pinus koraiensis Sieb. & Zucc. - Korean pine

Korean pine, little known in our part of the world, is rated as one of the hardiest pines by Dirr (1990). Dense blue-green foliage maintained close to the ground makes this five-needled pine a desirable specimen tree as well as a candidate for many other uses, including Christmas trees. Its large seeds are edible.

Swiss stone pine, *P. cembra* L., the European look-alike, is sufficiently hardy for the northern Great Plains (Hoag 1965). The only easily distinguishable characteristic between the Swiss stone pine and the Korean pine is cone size. The cones of the former are one-half the length of the latter's cones. We recommend provenance research for both species.

Planning

Height: up to 40 ft (12 m)
Habit: tree; crown pyramidal
Site: sun; pH and soil adaptable; zone 3
Use: specimen, screen, windbreak, food for man and animals

Recognition

Leaves: clusters of 5, 2.5-4.5 in (65-115 mm) long; lustrous dark green, white stomatal lines on 2 surfaces, margins toothed entire length
Fruits: cones woody, short-stalked, cylindric-conical, 3.5-6.0 in (9-15 cm) long, yellow-brown; scales with recurved obtuse apex
Twigs: reddish brown, hairy
Buds: oblong egg-shaped
Bark: scaly, gray-brown

Pinus peuce Griseb. - Macedonian pine

Macedonian pine, often extolled as one of the most handsome pines, has been widely used in landscape plantings elsewhere, especially in Europe, but it is apparently not widely planted in our prairies. A 25-year-old trial planting at Horning State Farm contains most of the Macedonian pine known to be growing in Nebraska.

The naturally dense, narrow crowns in this stand are a Christmas tree grower's delight (attested to by the high priority placed upon them by Christmas tree poachers). Foraging deer consider the Macedonian's foliage to be a delicacy. This native of the Balkan mountains in Eastern Europe deserves additional trial plantings, at least in the eastern prairies.

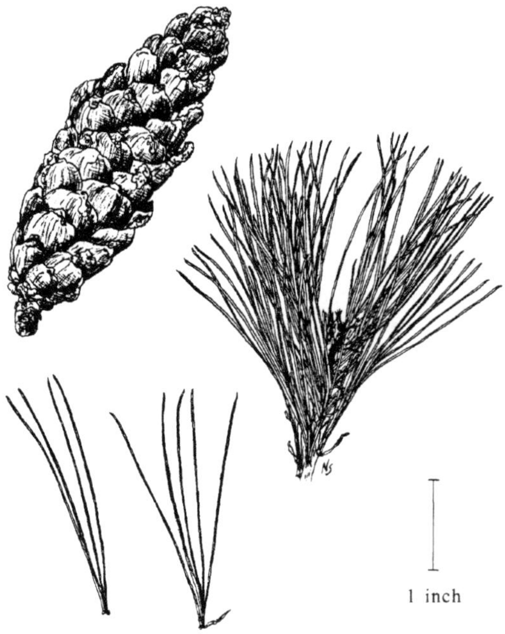

Pinus peuce

Planning

Height: up to 40 ft (12 m)
Habit: tree; crowns columnar to narrow pyramidal
Site: sun; pH and soil adaptable; zone 5
Use: specimen, screen, Christmas tree

Recognition

Leaves: clusters of 5 persisting 3 years, 3-4 in (8-10 cm) long; dark green above, grayish below, margins finely toothed
Fruits: cones woody, nearly cylindrical, about 6 in (15 cm) long; light brown, resinous; seeds small, wings long
Twigs: green, turning gray-brown; glossy
Buds: egg-shaped (ovoid)
Bark: gray-brown, becoming scaly

Pinus strobus L. - eastern white pine

This prize lumber tree of the 19th Century was a significant resource in the development of our nation east of the Mississippi River. The original stands of this largest of the northeastern conifers were destroyed by logging, but white pine's ability to reproduce naturally has reforested thousands of acres of abandoned farm land in the northeastern U.S., where it is known as old field pine. Also, many landscapes, Christmas tree plantations, and forests have been planted to this versatile species.

Eastern white pine is adapted to most sites in the eastern prairies but can grow amazingly well in the west if protected from excessive wind where soils are not highly alkaline. Max Peterson (1987) notes that his 35-foot white pine is exceptionally healthy in the midst of wheat country at Meadowlark Hill, Perkins County, Nebraska. He attributes his success in growing unusual woody plants to the protection afforded by windbreaks. Eastern white pine's vulnerability to persistent winds is demonstrated by the northerly pointing "flags" that you see on older trees growing in the open.

Provenance studies show genetic differences expressed in growth rate, drouth and disease resistance. Plants of southern origin tend to grow faster on northern sites than those originating from more northerly latitudes. For example, Sprackling and Read (1976) reported that eastern white pine trees of Appalachian origin in a provenance plantation at Horning State Farm survived and made excellent growth despite the fact that all origins were from latitudes south of the plantation site. Trees with northern Georgia origins are growing quite well more than 400 miles north of their native habitat, where winter temperatures are quite often lower than -9 degrees F (-23 degrees C). The authors reported that the eastern white pines that showed the greatest growth came from Polk County, Tennessee, and those with the best form came from West Virginia.

Numerous cultivars have been introduced including 'Compacta', dense, rounded and slow growing; 'Fastigiata', narrowly upright and columnar; and 'Pendula', weeping.

Pests that are a problem within the native range of white pine—such as white pine blister rust, weevils, and bark beetles—are not common in most of the prairies. Eastern white pine tolerates shading much better than most other pines and will grow quite well on the north side of a building. Lower leaves are retained under partial shade. Mulch formed by falling needles provides a weed-free groundcover early in the life of a plantation.

Planning

Height: up to 60 ft (18 m); Nebraska champion, 1981, 94 ft (28.5 m), Arbor Lodge Historical Park, Nebraska City; National champion, 1984, 201 ft (61 m), Michigan

Habit: tree; crowns pyramidal becoming irregular, spreading

Site: sun, part shade; soil adaptable; zone 3

Use: specimen, screen, Christmas tree

Caution: intolerant of air pollution; protect from strong winds

Recognition

Leaves: clusters of 5, 2-4 in (5-10 cm) long; slender, bluish green turning bronze in cold weather; deciduous in autumn of the second or third year

Fruits: cones woody, pendent, 6-8 in (15-20 cm) long, cylindrical, often curved, resinous; bright green, becoming light brown at maturity in second year; scales flexible, not spine-tipped

Twigs: slender, tough, flexible

Buds: egg-shaped, sharp-pointed; some scales free at tips

Bark: thin, greenish, becoming dark and deeply furrowed

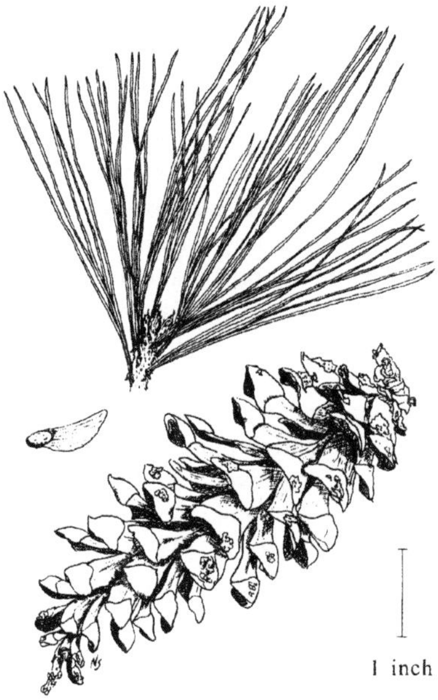

Pinus strobus

Subgenus *Diploxylon* - hard pines

Pinus banksiana Lamb. - jack pine

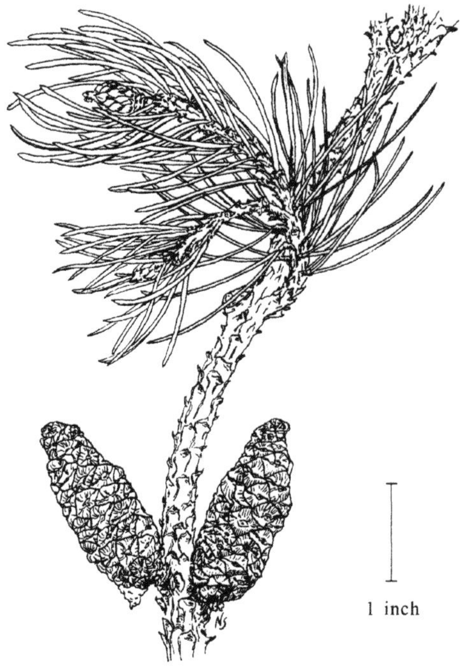

Pinus banksiana

Natural stands of jack pine are found where extremes in temperature and precipitation are great. This species survives on a wide range of soils from drouthy sands to clays, but is most at home on loamy sands. Jack pine heights of 80 feet (24 m) are attained in the northern Great Lakes region.

Jack pine can be found across Canada, from the foothills of the Rockies near the Arctic Circle to New England and down into the Great Lakes region. Trees growing naturally over such a large area exhibit great genetic variability that is most easily noticed in form and growth rate. Jack pine may be tall and straight stemmed, or "character" trees with unpredictable stem and branching habits.

Jack pine was successfully planted in the Sandhills of Holt County in 1891 and in the Nebraska National Forest near Halsey in the early 1900s. Considerable mortality among these trees during the dry 1930s prompted reduced planting of the species in the western prairies for a few years. However, jack pine has since proved to be quite well adapted to a wide range of sites throughout the prairies and is now a part of many protection plantations.

This pioneer tree is particularly well adapted for reforestation after fire because cones cling to the trees for many years. Cones often remain closed until heat from fire pops them open to carpet the newly bared soil with viable seed. We observed this phenomenom on the Nebraska National Forest near Halsey after the disastrous fire of May 1965. A study in 1987 determined that the average stand of the reproduction on this burn was 30 jack pine per 100 square meters, with a basal area of 0.25 sq. meters (Sutton 1987).

Experimental plantings established under the cooperative Regional Tree Improvement Project (NC-99), and involving several state experiment stations, documented genetic differences in jack pine. A study by Sprackling and Read (1975a) on a jack pine plantation located at Horning State Farm confirmed excellent survivability on silty clay loam soil. They also found that trees grown from seed collected from seven of the provenances studied combined rapid growth with aesthetic appeal. As a result of these findings, seedlings presently grown at Bessey Nursery, Halsey, for the Clark-McNary tree distribution program, originate from these stands.

Lodgepole pine (*P. contorta* Dougl. ex Loud. var. *latifolia* Engelm.) is another pioneer species similar in appearance to jack pine. It is native to the Pacific Coast region and the Rocky Mountains from Colorado north to Alaska at elevations of 1500 to 11,500 feet (460 to 3515 m). Lodgepole pine also has the ability to establish itself quickly on burned-over sites. Trial plantings of this pine grown from seed collected in Colorado have been slow-growing and bedeviled by insects and diseases. The lodgepole's susceptability to pine tip moth (*Rhyacionia* sp.) and needle diseases found in humid conditions limits its potential for successful planting, especially in the eastern prairies.

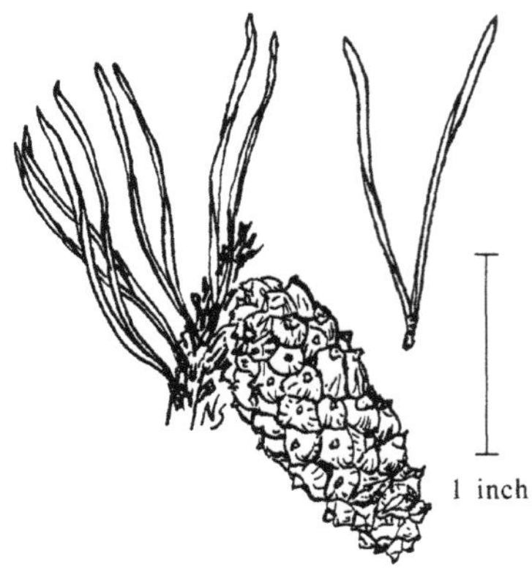

Pinus contorta

Planning

Height: up to 60 ft (18 m); National champion, 1981, 84 ft (25.5 m), Michigan
Habit: tree; crown rounded, irregular
Site: sun; pH and soil adaptable including very sandy soils; drouth tolerant; zone 2
Use: windbreak, wildlife habitat, specimen

Recognition

Leaves: clusters of 2, persistent 2 or 3 years, 0.8-1.5 in (19-38 mm) long; flat, stout, often twisted, yellow-green
Flowers: male oblong, yellow; female globe-shaped, dark purple near branch tips
Fruits: cones woody, variable in shape, often strongly incurved, light brown, shiny; armed with a minute prickle; persistent several years, often remaining closed for years; seeds 0.1-0.2 in (3-5 mm) long, triangular, rough, black, wings 0.3 in (8 mm) long
Twigs: multinodinal (not whorled as in most other pines)
Buds: oblong-egg-shaped, resinous
Bark: thin, reddish brown

Pinus densiflora Sieb.& Zucc. - Japanese red pine

Japanese red pine, a native of Japan, Korea and China, with reddish bark and asymmetrical form, reminds us of Scotch pine. Although Japanese red pine is grown in experimental plantings in eastern Nebraska, they are not proven substitutes for Scotch pine. Leaves are often injured by winter cold in Fort Collins (Sutton 1998). Yet Japanese red pine at the U. S. Forest Service Experiment Station, Cloquet, Minnesota, are hardy (Dirr 1990).

Many cultivars have been selected which exhibit unusual form and foliage because of hereditary differences. Adventuresome gardeners may wish to add Japanese red pine to their landscapes with the caution that its many genetic types have not been thoroughly tested in all parts of the prairie region.

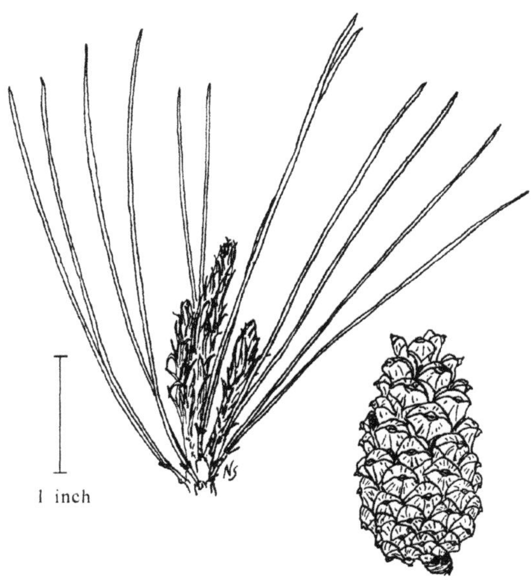

Pinus densiflora

Planning

Height: up to 50 ft (15 m)
Habit: tree; crown irregular, stem often crooked, widespreading
Site: sun; prefers moist, slightly acidic, well-drained soil; zone 4
Use: specimen, bonsai

Recognition

Leaves: cluster of 2, 3-6 in (7-15 cm) long; lustrous, slender, twisted; margins finely toothed
Fruits: cones woody, short-stalked, 1.3-2 in (3-5 cm) long; scales with or without short prickles; persistent 2 to 3 years
Twigs: initially green, becoming orange-yellow
Buds: brown, egg-shaped; may be resinous
Bark: orange-red, thin scales peeling; gray and fissured in old age

Pinus mugo Turra - **mugo pine**

Mugo pine is a highly variable species from the mountains of central and southern Europe valued chiefly for its dwarf cultivars that are used for foundation and mass plantings. Other trees of this species may reach heights of 50 feet (15 m.). Some authorities use the name *P. montana* Mill.

Mugo pine serves quite well as a foundation plant on the sunny side of a wall. For example, one of the larger of the shrub varieties graces both the east and west facades of the Nebraska State Capitol. Mugo pine tolerates a wide range of soils (including those highly calcareous) and is adapted to most sites in the northern prairies. It is susceptible to its fair share of pests, but normally not serious ones.

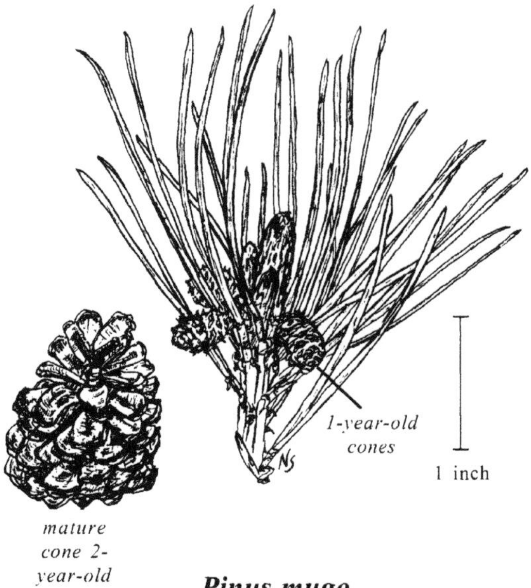

Pinus mugo

Planning

Height: up to 15 ft (4.5 m)
Habit: shrub, tree; usually broad, spreading, bushy, but varies from prostrate to pyramidal
Site: sun, part shade; pH and soil adaptable, drouth tolerant; zone 3
Use: foundation, grouping

Recognition

Leaves: clusters of 2, persisting 5 or more years; 1-3 in (3-7 cm) long; rigid, dark green; margins finely toothed
Fruits: cones woody, nearly round, very short-stalked or sessile; 1-1.4 in (25-35 mm) long; dark gray at maturity; scales with umbo bordered by dark ring, bearing remnant of short spine
Twigs: smooth, green, becoming brown
Buds: scales brown, resin encrusted, closely appressed
Bark: gray, scaly

Pinus nigra Arnold - **Austrian pine**

Austrian pine, sometimes known as European black pine, is a southern Europe native ranging from Spain and Morocco through Austria and Italy to eastern Turkey. Because it grows on a wide range of soils, including sandy and calcareous, Austrian pine has been planted widely in most of the U.S. as an ornamental and as a component of windbreaks (including the Prairie States Forestry Project of the 1930s). Because Austrian pine is more resistant than ponderosa pine to attacks of western pine tip moth (*Rhyacionia bushnelli*), it has often replaced the latter in plantings where the pine tip moth is rampant.

A cooperative regional tree improvement project (NC-99) compared Austrian pines grown from seed collected from locations throughout its natural range. This study confirmed and refined earlier information concerning variability in growth rate, disease resistance, and needle characteristics, enabling us to select origins best adapted to specific sites. Read (1976) reports that provenances from southern France and Greece did not survive in a planting at Horning State Farm. Other origins grew slowly and some were very susceptible to *Dothistroma pini*, a needle blight disease. A Yugoslavian provenance was highly resistant to needle blight.

Tip blight has caused serious injury, usually to Austrian pines over 30 years of age. The causal organism, *Sphaeropis sapinea*, usually inflicts the greatest injury by first infecting 2-year-old seed cones and then moving to new shoots before progressing to older stems, eventually causing great damage or death (Peterson 1981). Timely applications of copper fungicides effectively reduce infections of these blights.

Diseases are less likely to be a problem for Austrian pine in the dry climate of the western prairies. Fairchild and Klett (1993) report that a tree in the PERC site at Fort Collins is remarkably free of insects and diseases. An Austrian pine seed orchard composed of trees resistant to dothistroma needle blight is established at UNARDC, Ithaca, Nebraska.

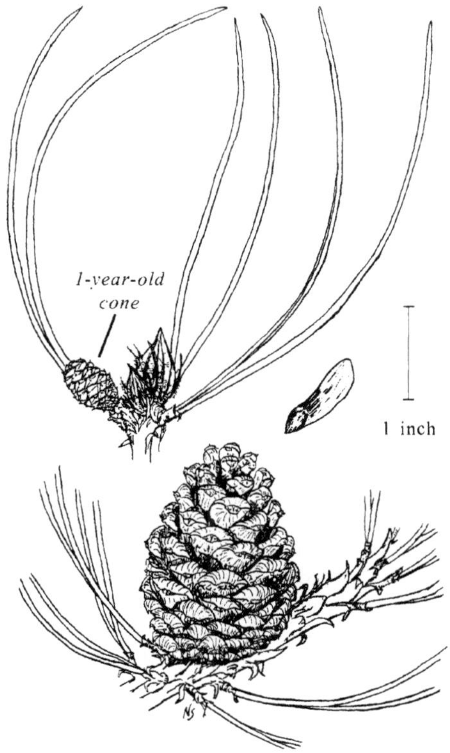

Pinus nigra

Planning

Height: up to 50 ft (15 m); Nebraska champion, 1986, 44 ft (13.3 m), Panama

Habit: tree; crown pyramidal, becoming flat-topped with stout spreading branches

Site: sun; pH and soil adaptable, drouth tolerant; zone 4

Use: specimen, screen, windbreak, wildlife habitat

Recognition

Leaves: clusters of 2, very rarely 3, persisting 2 to 3 years; 4-6 in (10-15 cm) long; stiff, tip sharp; margins minutely toothed, dark green

Flowers: male yellow, female yellow-green

Fruits: cones woody, symmetrical, egg-shaped; 2-3 in (5-8 cm) long; shiny yellow-brown; scales sometimes with short prickles, not sharp to touch

Twigs: smooth, yellow brown

Buds: silvery, shiny; abruptly contracted to a sharp point

Bark: dark gray-brown, plated, deeply furrowed

Pinus ponderosa Laws. - ponderosa pine

This important lumber-producing pine grows in western North America including the western sections of the Dakotas and Nebraska. Isolated eastern outliers of ponderosa pine are found in Garfield and Holt counties, Nebraska. Ponderosa is planted extensively throughout the western prairies for windbreaks and ornamentals.

Superior seed sources of ponderosa pine have been identified from information gleaned from 17 research plantations throughout the Great Plains region including Plattsmouth, Hastings and Alliance in Nebraska (Read 1983). Based on 10-year results, Read recommends native stands of pine along the Niobrara River from south of Nenzel eastward and those in the drainage of the South Fork of the White River near Rosebud, South Dakota, as sources of seed for Great Plains plantations. Trees of these provenances resist two potentially serious pests: western pine tip moth, *Rhyacionia bushnelli*, and *Dothistroma* needle blight. Trees grown from seed collected west of the Continental Divide will not survive, or at best will grow very poorly in the Great Plains.

Closely related species are *P. jeffreyi* Grev. & Balf., jeffrey pine, and *P. washoensis* M. & S., washoe pine. The latter is possibly a hybrid between ponderosa and jeffrey pine. These species may be adapted to sites in the western prairies.

Planning

Height: up to 50 ft (15 m); Nebraska champion, 1987, 93 ft (28 m), Pine Ridge near Harrison; National champion, 1974, 223 ft (67.5 m), California
Habit: tree; crown pyramidal, becoming irregular, branches stout
Site: sun; pH and soil adaptable, drouth tolerant; zone 3
Use: specimen, windbreak, wildlife habitat

Recognition

Leaves: clusters of 2 and 3, persisting 2 to 3 years; 5-10 in (12-24 cm) long; rather twisty, margins minutely toothed; yellowish green, tip sharp
Flowers: male yellow, female red
Fruits: cones woody, egg-shaped, 3-6 in (7-15 cm) long; scales thin, flexible, armed with recurved spine; seed winged
Twigs: smooth, orange brown becoming dark
Buds: oblong, cylindrical; scales reddish brown, closely appressed
Bark: dark brown to black, becoming yellowish red with age

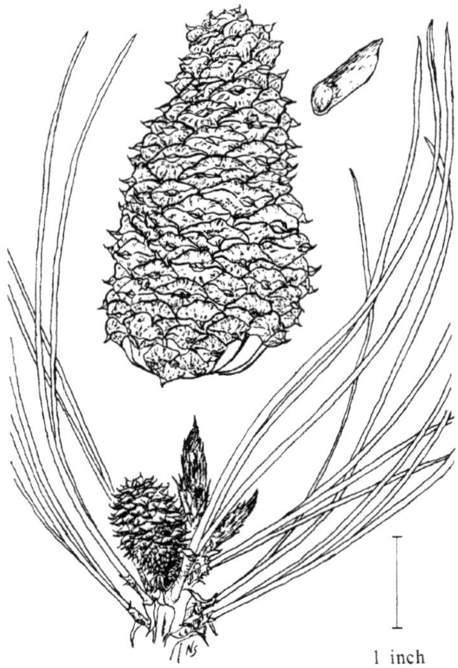

Pinus ponderosa

Pinus resinosa Ait. - red pine

Red pine, a native sometimes called Norway pine, is a beautiful tree with a dark green oval crown and straight trunk. Its limited natural latitudinal range, primarily between the 40th and 50th parallels in North America, is probably the reason it has relatively uniform growth and form characteristics. Within this 500-mile-wide band it ranges from New Brunswick west to Minnesota.

A 1963 provenance plantation of 77 origins at Horning State Farm demonstrates that red pine can be grown successfully in eastern Nebraska. Sprackling and Read (1975b) found that form, foliage color and needle length were very similar among all trees in the plantation at age 11 years. Height among origins ranged from 12 to 19 feet. Though these trees were reportedly free of insect and disease damage, red pine's susceptibility to pine tip moth (*Rhyacionia* sp.) limits its usefulness where this insect is prevalent.

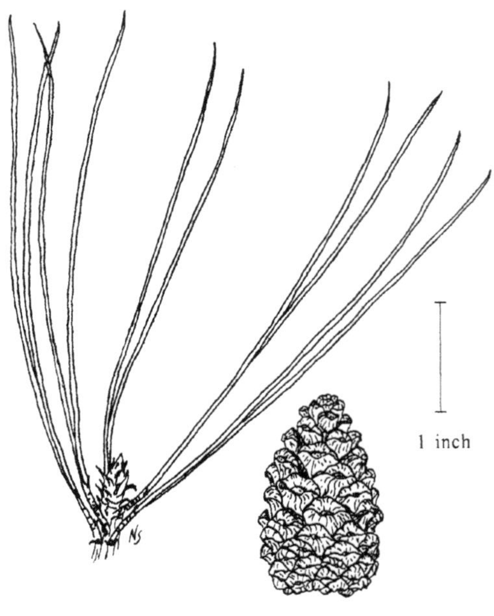

Pinus resinosa

Chinese pine, *P. tabulaeformis* Carr., is very similar in appearance to red pine. Both have a unique characteristic that causes a fresh green needle to snap or break when bent. Chinese pines over 20 years old are healthy and vigorous at Horning State Farm. However, there is no known reason, other than enlarging a pine collection, for using this species in lieu of the red pine.

Planning

Height: up to 50 ft (15 m); National champion, 1985, 154 ft (46.6 m), Michigan
Habit: tree; crown pyramidal, becoming oval, spreading
Site: sun; prefers acid soils; tolerates low fertility and drouth; zone 2
Use: specimen, windbreak, wildlife habitat

Recognition

Leaves: clusters of 2, 5-6 in (12-15 cm) long; slender, straight, dark yellow-green; snapping when bent
Fruits: cones woody, 1.5-2 in (4-6 cm) long, not prickly; seeds winged
Twigs: smooth; orange-brown, turning reddish
Buds: narrow, cone-shaped; some scales free at tips
Bark: thin, light reddish brown with scaly, flat ridges

Pinus sylvestris L. - Scotch pine

Scotch pine grows in natural stands throughout Europe and Asia, from Spain to Turkey, from Scotland to Finland, and across Russia into Manchuria. Consequently, this most widely distributed of all pines is highly variable in many heritable characteristics.

The orange-tinted upper stems are often gnarled to form a picturesque pattern against the sky. Many veterans can be found in farmsteads, cemeteries, villages and cities throughout the prairies. A plantation on a north-facing hillside overlooking North Platte, Nebraska, included Scotch, Austrian and ponderosa pines; at 40 years of age, Scotch pine was the most healthy of all three species with the fewest insect and disease problems (Bagley 1998). Scotch pines over 100-years survive from Bessy's first sandhills plantation near Swan Lake, Holt County, Nebraska (Sutton 1998)

Pinus sylvestris

Scotch pine was the subject of genetic research dating back to the 19th Century. Scientists were searching for trees with straight trunks and desirable wood characteristics to enhance value for lumber. After learning in the 1940s that Scotch and other pines could be sheared to form beautiful spruce-like Christmas trees, it became apparent that heritable characteristics are very important to success. Foliage color, stem straightness and resistance to cold injury varied greatly throughout the Scotch pine population.

Seed collected from throughout the natural range of Scotch pine in the late 1950s by Jonathan Wright of Michigan State University provided seedlings to establish provenance studies by several state experiment stations, including Nebraska. Read (1971), reporting results from the Horning State Farm plantation, stated that Scotch pines with blue-green winter foliage originated from southern France, Spain and Turkey. A Scotland provenance with green winter foliage and desirable growth characteristics was rated highly as an all-around tree for eastern Nebraska. Provenances from Belgium and the nearby Vosges Mountains of France and Germany were excellent trees for windbreaks because of their fast growth. Northern varieties, especially those from Siberia, change to a golden yellow in midwinter to offer designers oportunities to create striking contrasts in landscape color.

Pinaceae *Psuedotsuga*

David Van Haverbeke's selections from the Horning State Farm Scotch pine plantation are growing in a seed orchard at UNARDC, Ithaca, Nebraska. Seed is collected that produces Scotch pine trees desirable for Christmas tree, landscape and protection plantations. Testing and selection will continue to increase quality and uniformity as well as provide trees adapted to prairie sites.

Nurserymen are marketing Scotch pine cultivars of unique character for ornamental purposes; such as 'Argentea', with silvery leaves; 'Fastigiata', of columnar form; and 'Watereri', which is dense and slow-growing.

Planning

Height: up to 50 ft (15 m); Nebraska champion, 1988, 49 ft (17.6 m), Beatrice; National champion, 1983, 64 ft (19.4 m), Michigan

Habit: tree; crown pyramidal, irregular, becoming flat-topped, spreading; winter foliage blue-green to orange-yellow

Site: sun, part shade; prefers acid soils, but pH and soil adaptable; zone 3

Use: specimen, windbreak, wildlife habitat, Christmas tree

Recognition

Leaves: clusters of 2, persisting 2-3 years; 1.5-3 in (4-8 cm) long; often twisted

Fruits: cones woody, 1.5-3.0 in (4-8 cm) long; varying from symmetrical to oblique; falling at maturity; scales thickened at tip, obtuse

Twigs: smooth, green becoming brown

Buds: pointed; scales fringed, upper ones free at tips

Bark: upper stems orange-red, thin, flaky; becoming reddish brown and fissured, in long, scaly plates

Psuedotsuga Carr. - Douglas-fir

These long-lived, valuable timber trees are represented by only five species worldwide: one native to Japan, two to continental Asia, and two to North America, one of which is adapted to our region.

Pseudotsuga menziesii (Mirb.) Franco var. *glauca* (Beiss) Franco - Douglas-fir

Douglas-fir is the largest tree in the northwestern U.S., and, in this country, second only to the giant sequoias of California. It has been variously classified as a spruce, true fir, and hemlock after its discovery by Archibald Menzies in 1791. David Douglas, intrepid plant explorer from the London Horticultural Society, first questioned its relationship to these species and sent seeds home to Great Britain in 1827. Trees grown from these seeds are living today.

Unfortunately, Elie A. Carriere coined the name *Pseudotsuga*, "false hemlock," for a tree that in no way resembles a hemlock. Its common name commemorates Douglas and serves to distinguish this species from the true firs. Two geographic varieties are recognized: *menziesii* of the North Coast and Sierra regions and *glauca* of the Rocky Mountains. The former is not adapted to the harsh climate of the interior U.S.; the latter is widely grown east of the Rockies for Christmas trees and ornament.

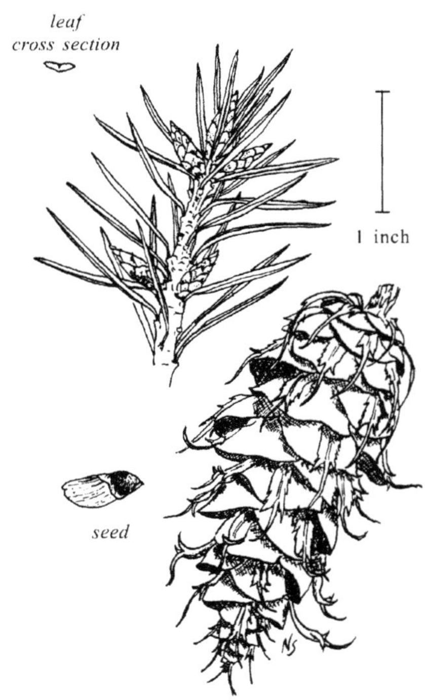

Pseudotsuga menziesii

Significant genetic variability among trees of the Rocky Mountain variety is demonstrated by trees growing in provenance plantations at several locations throughout the eastern U.S., including Horning State Farm in eastern Nebraska. Trees originating from southern Arizona and New Mexico have continued to grow faster than those with more northerly origins in the 1965 Plattsmouth plantation despite considerable winter injury to buds with consequent development of multiple leaders (Read and Sprackling 1976). Southern provenances lack the classic pyramidal form desired in ornamentals and Christmas trees. The authors conclude that slow-growing, winter hardy Colorado provenances are best for planting in the eastern portion of the Great Plains.

Planning

Height: up to 60 ft (18 m); Nebraska champion, 1987, 69 ft (21 m), Lydick farm, Tekamah; National champion, 1984, 158 ft (47.9 m), Oregon

Habit: tree; evergreen; crown pyramidal

Site: sun, part shade; prefers moist neutral to slightly acidic, soils; tolerates drouth; zone 4 (varies with seed source)

Use: specimen, grouping, wildlife habitat, Christmas tree

Caution: protect from strong winds; a favorite deer browse

Recognition

Leaves: linear, flattened; spirally arranged on short stalks but appearing 2-ranked; persistant for several years; 0.7-1.3 in (19-32 mm) long; apex usually rounded; shiny, dark green above; pores (stomates) aranged in 2 white bands beneath

Flowers: unisexual (monoecious)

Fruits: cones woody, pendent from branch tips; maturing in 1 season, remaining intact; 3-4 in (8-10 cm) long; bracts exserted, 3-lobed, fork-like

Twigs: smooth, marked with oval scars after leaves fall

Buds: spindle shaped, sharp-pointed; scales overlapping, lustrous brown

Bark: gray, smooth on young stems except for resin blisters; becoming very thick, divided into reddish brown ridges

Tsuga Carr. - hemlock

Four of the 10 species of hemlock are found in North America. The remaining six are native to Asia. Hemlocks are evergreens that thrive on well-drained, cool, moist, acid soils. Forest seedlings normally start in the shade of other trees where they grow slowly until they eventually dominate the site. Hemlocks can live for hundreds of years.

Hemlock bark is high in tannin content and has been an imporant source of that product. Hemlocks have also been a valuable source of lumber, and their many forms can contribute beauty to ornamental plantings.

Western hemlock, *T. heterophylla* (Raf.) Sarg., is one of the major lumber-producing trees in the Pacific Northwest (Harlow and Harrar 1968). *T. mertensiana* (Bong.) Carr., mountain hemlock, another tree of the Northwest (including southeastern Alaska), is seldom used for lumber. Ornamental plantings of mountain hemlock are limited to very cool climates such as that found in the Netherlands, where hemlocks prosper (Dirr 1990). Carolina hemlock, *T. caroliniana* Engelm., of the Blue Ridge Mountains of North and South Carolina, is better adapted than eastern hemlock for city planting but is little used (Dirr 1990). Exotic species of hemlock are potential landscape plants but do not compete because of similarities to our native species.

Tsuga canadensis (L.) Carr. - eastern hemlock

This attractive tree, native to the Great Lakes States, the Appalachians and New England, grows in cool moist valleys, primarily on northern and eastern slopes (which gives a clue about the requirements for successful prairie plantings).

A host of cultivars have been selected from its wide range of forms—fastigiate, weeping, prostrate, globose and variegated.

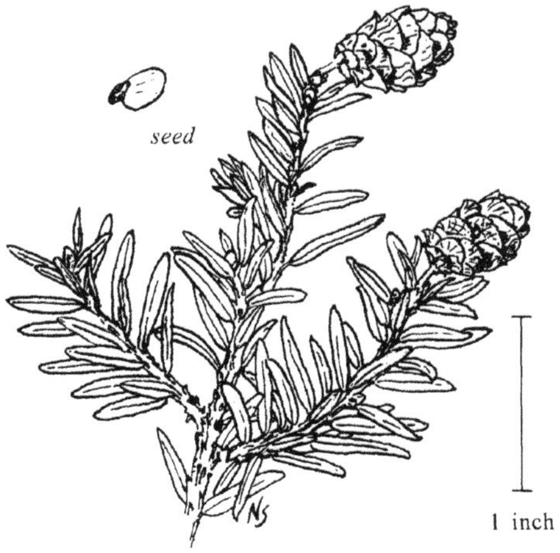

Tsuga canadensis

Planning

Height: up to 50 ft (15 m); Nebraska champion, 1985, 55 ft (16.8 m), Ft. Calhoun; National champion, 1979, 123 ft (37.3 m), West Virginia

Habit: tree, shrub; evergreen; crown pyramidal with nodding leader, branches drooping; cones persist into second season

Site: shade, full to partial; prefers high humidity, moist, well-drained, acidic soils; zone 4

Use: specimen, screen, hedge

Caution: protect from excessive wind

Recognition

Leaves: 2-ranked, 0.3-0.7 in (7-16 mm) long; flattened, rounded to notched at tip, margin finely toothed; pores (stomata) beneath in 2 whitish bands; conspicuously petioled

Flowers: unisexual (monoecious); male strobili axillary, globose, yellowish; female strobili erect, terminal on lateral branches; scales with 2 basal ovules subtended by bracts of nearly equal length, green

Fruits: cones woody, oblong-ovoid; 0.5-0.8 in (12-20 mm) long; short-stalked, pendent; scales suborbicular, margin smooth; seeds 0.6 in (1 mm) long, wings terminal, 0.2 in (6 mm) long

Twigs: slender

Bark: scaly, reddish; becoming deeply furrowed between broad flat scales

Taxodiaceae Warm. - redwood and swamp cypress family

Members of this family formed many forests throughout the world during the mild, humid climate of the Miocene epoch of nine million years ago. Many species are now extinct. Although 16 of the species in ten genera are found throughout the world, only three genera are native to North America: *Sequoia*, *Sequoiadendron*, and *Taxodium*. These three genera, noted for their great size and long life, produce valuable decay-resistant lumber. Otherwise, they are morphologically diverse with little in common except man's desire to find a convenient "pigeonhole" for them in the classification system. Some authorities do not recognize this family and place the three genera in the pine family.

Umbrellapine, *Sciadopitys verticillata* Sieb. & Zucc., a native of China, is found in a wild state only in Japan's Valley of the Kiso-gawa and in the vicinity of Koya-san in Hondo (Dirr 1990). This unusual, distinctive conifer was once considered a member of Pinaceae, but more recently has been placed in Taxodiaceae (Welch 1991). Umbrellapine is easily recognized by whorls of linear, flat leaves 2 to 5 inches (5-13 cm) long, crowded at the branch tips.

Trial plantings of umbrellapine in the prairies may be worthwhile. Dirr (1990) recommends a planting site with moist, acidic soils with plenty of sun and protection from excessive wind.

Recognition

Habit: trees; evergreen or deciduous
Leaves: linear to awl-shaped, spirally arranged or opposite
Flowers: unisexual (monoecious); bracts and egg-bearing (ovuliferous) scales partially fused; ovules erect, 2-12 per scale
Fruits: cones woody; scales several, shield-shaped (peltate)

Metasequoia Miki. - metasequoia

One species is native to China.

Metasequoia glyptostroboides H.H. Hu & Cheng - dawn-redwood

Fossil remains of this tree were first discovered in Tokyo by a Japanese scholar in 1941. That same year, half-way up the Yangtze River near Chungking, a botanist discovered three living dawn-redwoods; specimens were collected three years later. A thorough search disclosed more trees, and seeds collected in 1948 were forerunners of dawn-redwoods planted worldwide. Like the Gingko, this survivor of many ice ages appears to have outlived its enemies, though untimely freezes may cause damage because it tends to grow late into autumn at the mid-latitude of the prairies.

'Sheridan Spire' is narrow, columnar and densely branched, and 'National' is a seedling selected at the U.S. National Arboretum with characteristics similar 'Sheridan Spire' (Welch 1991).

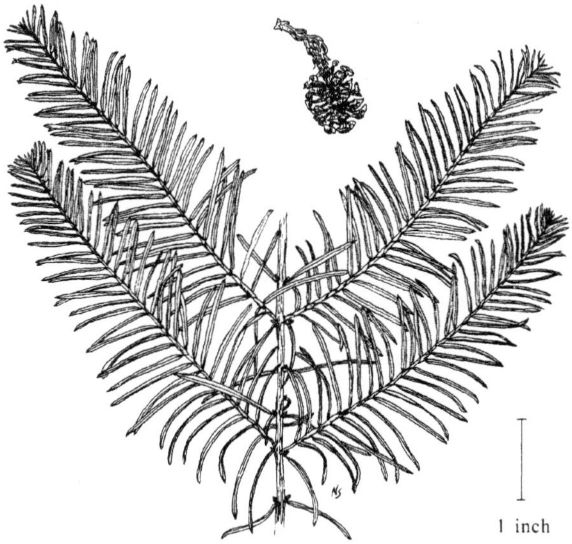

Metasequoia glyptostroboides

Planning

Height: up to 70 ft (21 m)
Habit: tree; deciduous; crown pyramidal, usually becoming broad
Site: sun; prefers moist, acidic, well-drained soil on sites with good air drainage to avoid "low pocket" frosts in autumn; zone 5
Use: specimen, screen, grouping

Recognition

Leaves: opposite, 0.5 in (13 mm) long; linear, flattened; mid-vein grooved above, slightly raised below; pores (stomates) beneath obscure
Flowers: unisexual (monoecious); male clustered, female solitary
Fruits: cones woody, long stalked; mature first year
Twigs: 2-ranked, opposite; persistent twigs bear deciduous twigs but few leaves
Buds: pairs at base of deciduous twigs, sometimes solitary, egg-shaped, about 0.25 in (6 mm) long; scales reddish brown, keeled
Bark: reddish brown, becoming dark gray; fissured and peeling in long strips; a fluted, flaring trunk develops with old age

Taxodium Rich. - baldcypress

Taxodium, widely distributed in Europe and North America during prehistoric times, is now represented by *T. mucronatum* Ten., Montezuma baldcypress, a native to southern Texas and south to Guatemala, and by *T. distichum* described below. A variety of the latter, *nutans* (Ait.) Sweet., pondcypress, is also recognized as *T. ascendens* Brongn. by some taxonomists (Dirr 1990). The pondcypress is distinguished from the species by its erect branches and appressed or incurved leaves.

Taxodium distichum (L.) Rich. - baldcypress

Curiously, this tree, at home on wet, moist soils of the southeastern U.S. and the lower Mississippi Valley, also thrives in much colder and drier climes on soils ranging from acidic to alkaline (Fowells 1965). The unique "knees" are formed only in shallow water from the roots of trees growing nearby.

Baldcypress in its natural habitat lives to be several hundred years old and attains dimensions that produce valuable logs. The durable wood is good for use in conditions that lead to decay or where exposed to the weather.

Baldcypress is an interesting and functional tree for use in landscapes with large spaces. Cultivars include 'Monarch of Illinois', widespreading in form, 'Pendens', pyramidal with drooping branches, and 'Shawnee Brave', very narrow pyramidal (Dirr 1990).

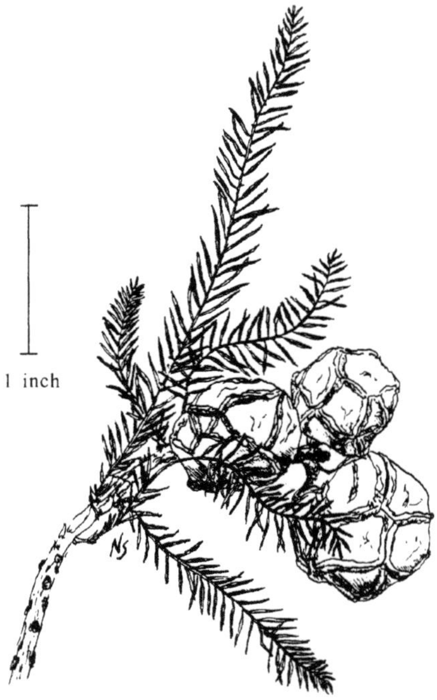

Taxodium distichum

Planning

Height: up to 60 ft (18 m); Nebraska champion, 1981, 64 ft (19.4 m), UNL East Campus, Lincoln; National champion, 1981, 83 ft (25.2 m), Louisiana

Habit: tree; deciduous; crown conical to spreading, becoming flat-topped; leaves of spring bright yellow-green, summer sage green, autumn russet to orange-brown

Site: sun; soil adaptable, preferably acid, tolerates slight alkalinity; drouth tolerant; zone 4

Use: specimen, grouping, lumber

Recognition

Leaves: linear, 0.5-0.8 in (12-19 mm) long; spiral on the twig, appearing 2-ranked

Flowers: unisexual (monoecious), male in drooping 4-5 in (10-13 cm) long panicles; female subglobose cones, each scale bearing 2 ovules

Fruits: cones woody, 0.7-1 in (19-25 mm) across; wrinkled, green, becoming brown; usually disintegrating at maturity (1 year); scales peltate, 4-angled in cross-section, terminating in a short tip (mucro); seeds irregularly 3-angled, 3-winged

Twigs: slender, green branchlets alternate on the twigs and fall with leaves in autumn

Buds: subglobose; scales overlapping, sharp-pointed

Bark: reddish brown, fibrous or scaly

Wood: greasy, odor often rancid; color and weight variable; heartwood resists decay

Cupressaceae Bartl. - **cypress family**

The cypress family, comprised of over 140 species, has representatives widely scattered throughout forests of the world. Five of 15 genera are native to North America.

Worthy of mention is ×*Cupressocyparis leylandii*, Leyland cypress, the result of a natural cross between *Cupressus macrocarpa* Hartw., Monterey cypress, and *Chamaecyparis nootkatensis* (D. Don) Spach, Alaska-cedar. The intergeneric cross occurred in 1888 on the grounds of a Welsh country house when a Monterey cypress pollinated an Alaska-cedar. This beautiful and fast-growing growing conifer in Great Britain, easily propagated by cuttings, has become one of the country's most useful landscape plants.

It promises to be a valuable addition to landscape plantings in coastal areas of the U.S. but, unfortunately, is not adapted to the climatic extremes of the midsection prairies.

Most *Cupressus* species bear two types of leaves: juvenile (which are usually awl-shaped and found on seedlings) and scale-like (which often dominate on old plants). Occasionally both types of leaves are found on the same plant, even on trees at an advanced age. Even more disconcerting is the retention by some plants—usually dwarf forms—of juvenile foliage for the life of the plant. In fact, this characteristic gave rise to a separate but now unused genus *Retinospora*.

Recognition

Habit: trees, shrubs; evergreen

Leaves: scale-like or awl-shaped, pressed against and hiding the branches; former mostly opposite (decussate), latter often in 3s (ternate) and dominant on young plants (both may be present on the same plant)

Flowers: unisexual, male and female may be found on the same or different trees; bract and egg-bearing (ovuliferous) scale completely united; ovules erect, 2 to many on each scale

Fruits: cones woody, leathery, or semifleshy; scales opposite, shield-shaped (peltate) or thin and flattened; seeds laterally winged or wingless

Chamaecyparis Spach - **falsecypress**

These evergreens inhabit cool moist regions in North America and eastern Asia. They provide valuable lumber and have been widely planted in landscapes. The highly decorative *C. lawsoniana* (A. Murr.) Parl., Port-Orford cedar or Lawson cypress, though native in a small area of coastal Oregon and northern California, is planted worldwide where soil moisture is plentiful and humidity is high.

Over 250 cultivars, from dwarfs to full size trees, have been selected from this species, with many variations in leaf shape and arrangement, color, and growth habit. Port-Orford cedar and two other North American indigenous species, *C. thyoides* (L.) B.S.P. and *C. nootkatensis* (D. Don) Spach, are unable to withstand dry climates, winds and temperature extremes. Although they cannot survive most prairie sites, dwarf forms of two Asian species can be planted in our more humid areas if special protective measures are taken.

Recognition

Habit: trees; evergreen, pyramidal

Leaves: needle-like, spreading on young plants and terminal shoots; scalelike, pointed, lateral pairs boat-shaped, facial pairs flattened on older twigs

Flowers: unisexual (monoecious), terminal; male and female on separate branches; male cylindrical, female sub-globose

Fruits: cones erect, globose, leathery or semifleshy; maturing in 1 or 2 years; scales shield-shaped (peltate); seeds 2-5, laterally winged

Twigs: slender, green

Cupressaceae *Chamaecyparis*

Chamaecyparis obtusa (Sieb. & Zucc.) Endl. - **Hinoki falsecypress**

Hinoki falsecypress, native to Japan and Formosa, has been used both for its lumber and ornamental value. Interestingly, it provides bark used for the unique roofs of Zen temples. Tree-like forms grace many gardens in the cool, humid regions of North America and Europe.

The many Hinoki falsecyrpess cultivars include dwarf forms that can be used as evergreen shrubs on protected sites in the more humid areas of the prairies. Dirr (1990) reports that variation in the seedling population is endless and that "production of new and different cultivars of falsecypress is about as simple as sowing seed." Perhaps we could discover forms that are better adapted to our climate if a well-designed breeding project were conceived.

Many cultivars are so similar that it is impossible to distinguish among them. 'Nana gracilis' is a popular dark green, bush form. Other dwarf forms of note are the green-leaved 'Kosteri Nana' and 'Ericoides' that has bluish foliage.

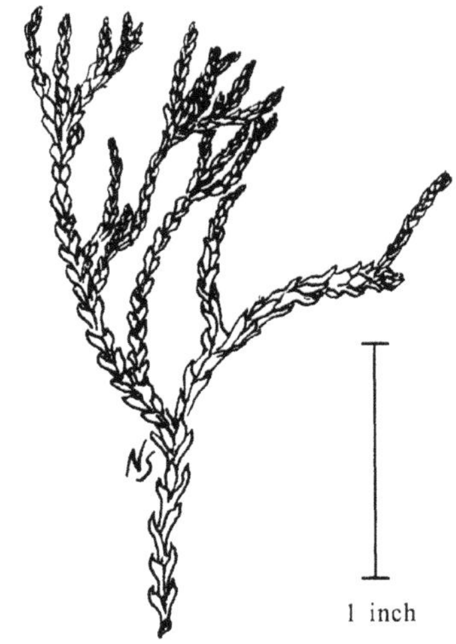

Chamaecyparis obtusa

Planning

Height: up to 50 ft (15 m)
Habit: tree (dwarf and shrubby forms); crown slender, pyramidal, branchlets drooping
Site: sun; moist, well-drained, acidic soil; zone 5
Use: specimen
Caution: protect from strong winds

Recognition

Leaves: adult foliage closely pressed, green above, white marks below, not glandular; lateral pair larger, boat-shaped, smaller pair triangular with thickened apex; juvenile leaves awl-shaped, spreading
Flowers: male yellow, female solitary
Fruits: cones nearly round, about 0.5 in (12 mm) across
Branches: arranged in horizontal planes, slightly drooping at tips
Bark: red-brown, peeling in thin strips

Chamaecyparis pisifera (Sieb. & Zucc.) Endl. - **Sawara falsecypress**

This Japanese species with many cultivars has identity problems because the foliage of most of its cultivars is so different from that of the species. Some are string-leafed and others are decked with quite variable juvenile leaves.

Cultivars include 'Filifera', with drooping stringy branches, 'Squarrosa', with soft, gray-green almost needle-like leaves, and 'Boulevard', with needle-like leaves, that are silvery green in summer (Dirr 1990).

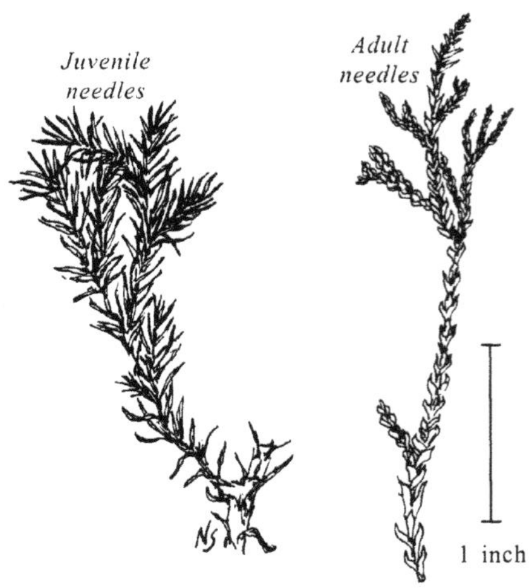

Chamaecyparis pisifera

Planning

Height: up to 50 ft (15 m)
Habit: tree (dwarf and shrubby forms); crown pyramidal, open; branchlets feathery
Site: sun; prefers moist, loamy, acidic soils; zone 4
Use: specimen, rock gardens
Caution: protect from excessive winds

Recognition

Leaves: adult appressed, long pointed with slightly spreading tips, obscurely glandular, dark green above, white lines beneath; juvenile needle-like, spreading
Fruits: cones globose, 0.25 in (6 mm) across
Twigs: flattened, 2-ranked, arranged in horizontal planes
Bark: smooth, reddish brown; peeling in thin strips

Juniperus L. - juniper

Between 40 and 70 juniper species inhabit the Northern Hemisphere, depending upon the taxonomic authority used. Many species, varieties and cultivars are so similar that professional taxonomists are hard pressed to discern differences among them (which explains the discrepancy in the number of recognized species).

Whereas so many coniferous species prefer moist, acid soils, we are beholden to these sunlight-loving, drouth-hardy plants (many of which are tolerant of lime) for filling niches in our landscapes where few other woody plants can exist. These workhorses of prairie windbreak and landscape plantings range in habit from prostrate shrubs to small trees. They provide a wide range of color and texture. All species are readily grown from seed and most cultivars can be propagated from cuttings.

Phomopsis and *Cercospera* blights can be serious diseases for some juniper species found in humid sites. Some species act as an alternate host to cedar-apple rust, which causes unsightly galls on the juniper and is a menace to apple growers. Most junipers are hosts to bagworms, spider mites and other insects, though the former is the only insect likely to cause serious injury if it occurs in large numbers.

Recognition

Habit: shrubs, trees; evergreen, aromatic
Leaves: scale-like to linear, opposite or in whorls of 3; sometimes in 2 forms (dimorphic) on some species; juvenile awl-shaped or linear, adult scale-like; persist for several years
Flowers: unisexual, axillary or terminal, solitary or clustered, usually dioecious; male spherical to egg-shaped; female of 2-6 scales, lower sterile, terminal sometimes fertile
Fruits: cones berry-like, fleshy to leathery; ripening at end of 1, 2 or 3 growing seasons; seeds 1-6, egg-shaped, wingless
Twigs: slender
Buds: tiny, hidden
Bark: thin, scaly or stringy
Wood: soft, close-grained, white or reddish, aromatic; heartwood decay resistant

Grouping Key for Junipers

Group one - leaves all ternate and acicular

Group two - leaves wholly or partly scale-like, usually opposite or opposite and ternate on the same plant

Group one

Juniperus communis L. - common juniper

Common juniper, one of the most widely distributed woody plants, occurs naturally in much of northern North America, Europe and Asia. Often found on dry, sterile sites, it withstands high wind and temperature extremes. Common juniper is usually considered to be a scrubby plant. Its berries are used as a source of medicinal oil or flavor for gin.

The common juniper's great variability in growth and habit has prompted taxonomists to assign a dozen specific epithets and many varietal names to this complex population. The most recent solution is to treat these junipers as varieties of *J. communis* (Welch 1991). Welch suggests that varieties be limited to *depressa* Pursh, *hemispherica* (Presl) Parl, *hornibrookii* Hornibr., and *montana* Ait.

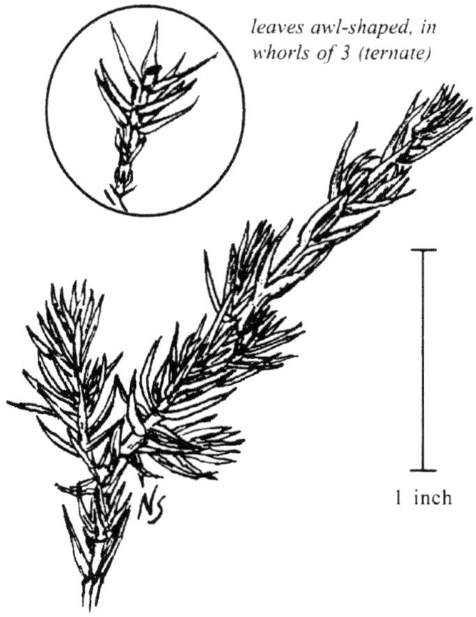

Juniperus communis

Variety *depressa* is a dwarf, broad, prostrate shrub found in the eastern North American segment of the species. Stephens (1973) states this is the plant native to the North American prairies, and it is the only variety recommended by Hoag (1965) for use in the northern Great Plains (he believes that it is best adapted to shade or sheltered sites).

Innumerable cultivars have been selected, varying from prostrate to upright. Among them are 'Repanda', a cultivar of low mound growth habit and soft to the touch, recommended for the Great Plains (Klett 1977); 'Hibernica', Irish juniper, that is dense and upright; 'Compressa', a very slow-growing, cone-shaped form; and 'Green Carpet', a groundcover form about 0.5 ft (15 cm) tall (Dirr 1990).

Planning

Height: up to 15 ft (4.5 m)
Habit: tree-like to prostrate
Site: sun, part shade; pH and soil adaptable; drouth tolerant, zone 3
Use: groundcover, naturalizing, wildlife habitat
Caution: protect from excessive wind

Recognition

Leaves: awl-shaped, in whorls of 3 (ternate), persisting 3 years; curved and ascending, never clasping; tapering from base to spiny point; 0.6 in (15 mm) long; concave above, with broad white band, sometimes divided by a green midrib at base; lower surface convex and bluntly keeled
Flowers: unisexual, dioecious, axillary; male egg-shaped, about 0.15 in (4 mm) long, scales brown; female cone-like, 0.1 in (2 mm) long, scales green
Fruits: cones nearly round, 0.2-0.3 in (5-7 mm) across; green, becoming black upon ripening in second or third year; seeds 2-3, rough, ridged, with 3-9 resin glands, apex usually sharp-pointed
Twigs: green, soft, becoming brown and hardening the second year; ridge-like resin gland extending from leaf base to node below
Buds: naked
Bark: gray-brown, peeling in papery shreds

Juniperus procumbens (Endl.) Miq. - **Japgarden juniper**

Japgarden juniper, a handsome groundcover native to Japan, is considered a variety of *J. chinensis*, *J. horizontalis* or *J. squamata* Lamb. by various authorities (Welch 1991). It resembles *J. squamata*, singleseed juniper, except that the leaves of Japgarden juniper are about twice as long and its fruits are larger with two to three seeds.

Variety *nana* Grootend., a dwarf clone, was originally introduced as *J. japonica* D. Hill in 1904 (Welch 1991). Its slightly differing forms have been named 'Nana Glauca' and 'Bonin Isles'.

Japgarden juniper and var. *nana* tolerate the soil and climate of PERC (Fairchild and Klett 1993).

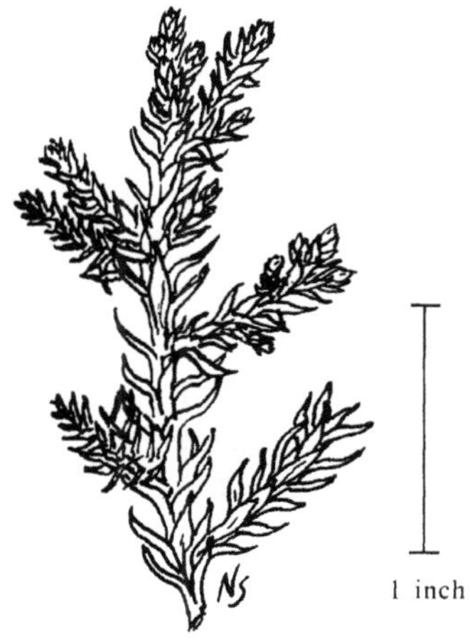

Juniperus procumbens

Planning

Height: up to 2 ft (60 cm) and spread to 15 ft (4.5 m)
Habit: shrub; branches stout, stiff, tips ascendant
Site: sun; pH and soil adaptable; drouth tolerant; zone 5
Use: groundcover, erosion control

Recognition

Leaves: in threes, spiny pointed, 0.3 in (7 mm) long; concave above and whitened with a green midrib, bluish below with 2 white spots near the base
Fruits: cones subglobose, about 0.3 in (7 mm) across; seeds 2-3

Juniperus squamata Lamb. - **singleseed juniper**

The singleseed juniper can be found at a great diversity of site conditions within its native range, which extends from the Himalayas and Afghanistan into western China. This diversity has been a contributing factor to the great variability of the species, whose form varies from prostrate to columnar (Dirr 1990). Dirr mentions that cultivars of singleseed juniper prosper in a cool European climate but languish in the heat and humidity of the southeastern U.S. Provenance research is needed to determine the potential for planting this juniper in the prairies because named cultivars have been selected primarily with regard to foliage color and form.

Most cultivars have been selected for their rich, blue-green foliage. They include low-growing 'Blue Carpet' and 'Blue Star', the taller-growing 'Blue Swede' and 'Chinese Silver', and the columnar 'Loderi' (Dirr 1990). It is quite difficult to distinguish these plants from many of the cultivars of *J. procumbens* unless cones are present, even though the leaves of the latter are somewhat longer.

***Juniperus squamata* 'Blue Star'**

A 'Meyeri' tree growing at PERC exhibits an unusual vase shape with blue to bluish white needles that become purplish in winter (Fairchild and Klett 1993).

Planning

Height: up to 5 ft (1.5 m)
Habit: shrub; decumbent to erect; foliage bluish green to gray-green
Site: sun; pH and soil adaptable if well-drained; drouth tolerant; zone 5
Use: groundcover, foundation, specimen

Cupressaceae *Juniperus*

Recognition

Leaves: awl-shaped, whorls of 3, about 0.15 in (4 mm) long; pressed together to slightly spreading, sharp-pointed; grayish green, concave above and whitened; convex below, grooved; persisting after becoming dry and brown

Fruits: cones ripening second year; ellipsoid, about 0.3 in (7 mm) long; reddish brown, becoming purplish black; 1 seed

Bark: reddish brown, exfoliating

Group two

Juniperus chinensis L. - Chinese juniper

Chinese juniper is almost never sold as or even recognized by its specific name. It is usually marketed as a cultivar because this very complex species is quite variable. Hardiness and adaptability vary with cultivar or variety. Form ranges from narrow and pyramidal to low and wide-spreading. Descriptions of the many cultivars and varieties may be found in Dirr (1990), Welch (1991), and other references.

'Pfitzeriana', a wide-spreading male cultivar and one of the most widely planted junipers, is derived from a union of Chinese Juniper and Savin juniper (*J. sabina* L.), but might more properly be dubbed J. ×*media* Van Melle according to Johnson (1973) and Welch (1991). In contrast, *J. chinensis* 'Hetzii', a fruiting, vase-shaped spreader, provides an added benefit: winter food for robins, cedar wax-wings and other fruit-eating birds. Both cultivars are capable of growing on calcareous, occasionally dry soils. These sun-loving plants will grow in partial shade, but they may become somewhat scraggly in appearance and gradually die if shading increases, as it sometimes does, when nearby adjoining trees in the landscape grow.

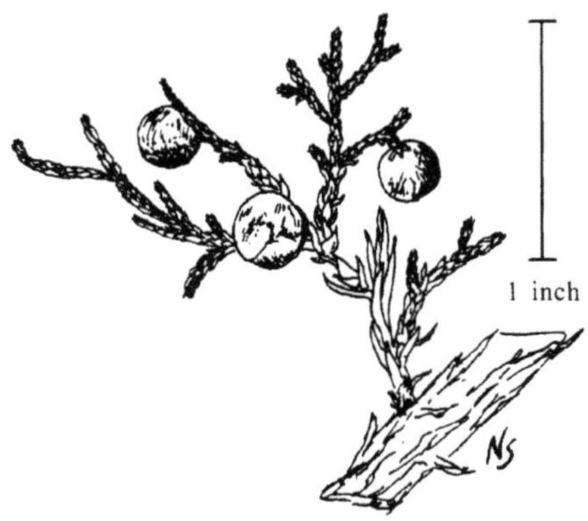

Juniperus chinensis

Cultivars thriving at PERC include 'Admiral', a narrow upright shrub; 'Ames', a spreading shrub that produces fruit; 'Armstrong', a spreading male clone; 'Blue Point', a broad, pyramidal blue-green shrub; 'Monlep', a compact bright green shrub; 'Old Gold', which is compact with gold new growth; 'Pfitzeriana'; and 'Robusta Green', a tree form with blue fruit (Fairchild and Klett 1993).

Planning

Height: up to 50 ft (15 m)

Habit: trees, shrubs; crowns conical to low, wide-spreading; fruits persist into winter

Site: sun; pH and soil adaptable, prefer well-drained; drouth tolerant; zone 3

Use: specimen, screen, hedge, groundcover, foundation, wildlife habitat, windbreak

Recognition

Leaves: adult in 4 ranks of opposite pairs; closely appressed, marked with a glandular depression on back; juvenile spiny-pointed, awl-shaped, in whorls of 3, or opposite pairs; green midrib with 2 whitish bands above

Flowers: unisexual, usually dioecious

Fruits: cones globose, 0.2-0.3 in (6-8 mm) across, whitish blue; ripening second year to dark brown; seeds usually 2-3, sometimes 1 or 4

Juniperus horizontalis Moench. - **creeping juniper**

Creeping juniper, a low-growing creeping spreader native to southern Canada and the northern U.S.—including northwest Nebraska and Colorado—can be used as a groundcover for a wide range of sites including hot, dry situations and slightly alkaline soils. It tolerates soil textures ranging from heavy to gravelly and rocky, and it is even found growing in swamps. Creeping juniper can be readily identified by its habit, unique among juniper species, of rooting at the nodes of the trailing evergreen stems.

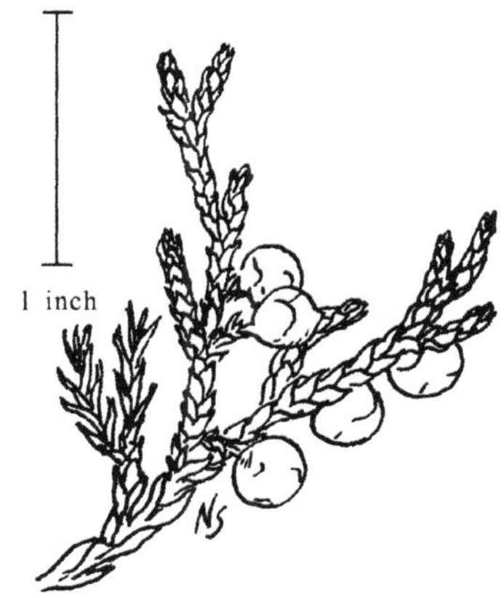

Juniperus horizontalis

Numerous clones have been selected, many of which can hardly be distinguished from one another without a label. Based on trials at McCrory Gardens, South Dakota State University, Klett (1977) recommended 'Glauca', 'Blue Rug' or 'Wiltoni' and 'Plumosa' (Andorra juniper) as superior groundcovers for the Great Plains.

Creeping junipers performing well at PERC include 'Bar Harbor', 'Blue Chip', 'Blue Horizon', 'Emerald Isle', 'Emerald Spreader', 'Hughes', 'Livingston' (female), 'Plumosa', 'Wisconsin' (female), and 'Yukon Belle' (Fairchild and Klett 1993).

'Sutton', discovered growing near Sutton, Nebraska, by J. O. Young and Hans Burchardt in 1965 had been planted in 1940 on a south-facing road cut of Nebraska Highway 6. It was selected for trial plantings because of its beauty, vigor and size under competition with smooth brome grass. Roadside beautification plots have proved that 'Sutton' is adapted to diverse Nebraska sites.

Another creeping juniper clone with Nebraska connections, dubbed 'Emerson', was found in 1915 in the Black Hills by G. A. Marshall of Marshall Nurseries, Arlington, Nebraska, and Professor R. A. Emerson, University of Nebraska. The potential for bearing fruit by this female cultivar enhances its value for wildlife habitat plantings. Foliage of this dense, very prostrate plant is bluish.

Two cultivars selected from native plants found in the Nebraska Sandhills by the late Allen Wilke, 'Oglala' and 'Green Spreader', have recently been added to the list of juniper cultivars. 'Oglala' and 'Sutton', growing side by side at Prairie Pines, appear visually identical (Bagley 1990).

Planning

Height: up to 2 ft (60 cm) and 20 ft (6 m) spread
Habit: shrub; procumbent, rooting at nodes; cultivated plants seldom fruit
Site: sun; pH and soil adaptable; tolerant of drouth and temperature extremes; zone 3
Use: groundcover, foundation, erosion control, wildlife habitat

Recognition

Leaves: 4-ranked; mostly scale-like, closely appressed; some awl-shaped, tip sharp, base sessile, clasping, margin entire; color variable, usually shades of blue-green; gland on dorsal side shorter than distance from the gland to leaf tip
Flowers: terminal, dioecious; male cylindric, light brown, scales in 4 rows, 4 scales each row; female on short reflexed branches, 2-6 ovaries; scales modified, distinct, soft
Fruits: cones on recurved leafy pedicel, scales fused, mature second year; 0.3 in (7 mm) across, blue to purple, with white bloom; seeds 2-3, irregularly egg-shaped
Twigs: tips greenish, becoming brown

Cupressaceae *Juniperus*

Juniperus sabina L. - Savin juniper

The Savin juniper's form varies from low and spreading to a large shrub. This native of the mountains of central and southern Europe and much of Asia tends to appear ragged. However, low-growing types, many of them selections of variety *tamariscifolia* Ait., make handsome foundation plants and groundcovers. Unfortunately, 'Tamariscifolia' is very prone to juniper blight caused by *Phomopsis juniperovora* (Welch 1991).

Trials at McCrory Gardens, South Dakota State University, indicate that 'Arcadia', 'Broadmoor', and 'Skandia' are well adapted to the Great Plains (Klett 1977). 'Buffalo', 'Monna', and 'Von Ehron' have performed well at PERC (Fairchild and Klett 1993).

Planning

Height: up to 7 ft (2 m)
Habit: shrub; branches ascending or procumbent, tips curving upward
Site: sun; pH and soil adaptable; drouth tolerant; zone 4
Use: groundcover, foundation, mass planting

Recognition

Leaves: scale-like in 4 ranks; usually bearing a resin gland; awl-shaped leaves present on young plants and some older twigs, about 0.2 in (4-5 mm) long; concave blue-white upper surface split with conspicuous midrib
Flowers: unisexual, monoecious or dioecious
Fruits: cones ripening the first year or the next spring; nearly round, 0.2 in. (5 mm) across, brown or blue-black, bloomy; seeds 1-3
Twigs: slender; strong odor when bruised

Juniperus scopulorum Sarg. - Rocky Mountain juniper

The native range of this Rocky Mountain juniper extends into the western part of the Great Plains. Van Haverbeke (1968) characterized the Great Plains juniper as a hybrid population, *J. scopulorum* × *J. virginiana*, with individuals in the north and west resembling the Rocky Mountain juniper and those in the south and east more like eastern redcedar. Rocky Mountain juniper is the most valuable of the species for shelter and ornamental plantings in the northern plains (Hoag 1965).

Several research plantations (Regional Research Project, GP-13) were established in 1980-82 from Texas to North Dakota to determine site condition adaptability of various origins of the Rocky Mountain–eastern redcedar complex. Genetic variability of this juniper population will be identified and a basis established for making improvements in the species by controlled breeding. A fifth-year analysis of these research plantations indicates that seed should be collected from northwest Nebraska or central Montana for planting in the Great Plains west of the 100th Meridian and north of west-central Kansas (Van Haverbeke and King 1990).

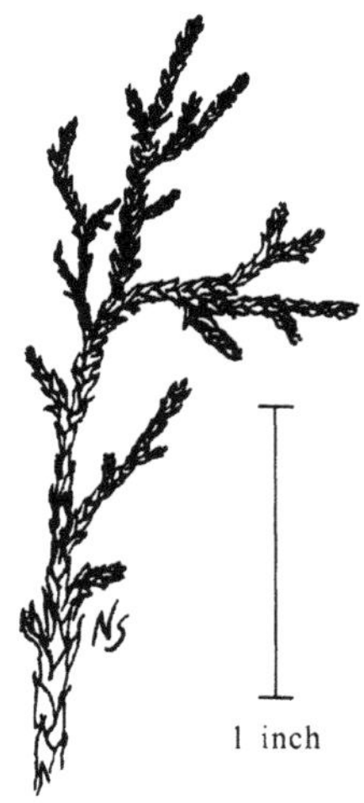

Juniperus scopulorum

Little is known about site adaptability of the many cultivars that are selected primarily for desirable form and/or foliage color. The entire range of these characteristics can be obtained in a few cultivars (Hoag 1965). Hoag recommends 'Medora'and 'Moffet', bluish and semi-columnar plants; 'Welchii', blue-green and pyramidal; 'Pathfinder', a silvery blue type; 'Cologreen', a green form selected by Marshall's Nursery, Arlington, Nebraska; and 'Hillborn Globe', very broad, roughly globe-shaped. Another cultivar worth trying in our region is a female plant, 'Tabletop Blue', selected by nurseryman Clayton Berg from a Montana mountain population (Sutton 1998). 'Skyrocket' is one of the most narowly columnar trees in existence but tends to lose its shape with age (Welch 1991).

Cultivars growing at PERC include 'Cupressifolia Erecta', a pyramidal form with green foliage; 'Gray Gleam', dense with gray foliage; 'Macfarlane', columnar with bluish green foliage; 'Sutherland', pyramidal with green foliage; 'Tolleson's Weeping', fruitful and weeping with blue to gray-green foliage; 'Wichita Blue', pyramidal with blue foliage; 'Winter Blue', a spreading shrub with blue foliage; and 'Woodward Columnar', narrow columnar with green foliage (Fairchild and Klett 1993).

The Rocky Mountain juniper can be effectively used as a shrub row in a windbreak and as an attractive landscape plant. However, its use is limited in the eastern prairies because of its susceptability to *Cercospera* blight, a foliage disease that spreads rapidly under humid conditions. This juniper is reasonably long-lived on sunny, breezy sites. Rocky Mountain juniper has survived and remained in good condition growing as the west, south or east rows of windbreaks at Prairie Pines Arboretum for nearly 30 years without spray application.

Cercospera blight can be effectively controlled by two applications per year of a copper fungicide such as bordeaux. On the plus side, Rocky Mountain juniper is less likely than eastern redcedar to be an alternate host to cedar-apple rust.

Planning

Height: up to 40 ft (12 m)
Habit: tree, sometimes shrubby; pyramidal, often multiple stemmed; year-around foliage green to blue-green
Site: sun; pH and soil adaptable; drouth tolerant; zone 4
Use: screen, hedge, foundation, windbreak, wildlife habitat
Pests: *Cercospera* blight at humid sites

Recognition

Leaves: scale-like, tightly appressed, not or slightly overlapping; tips blunt, margins entire; recessed gland on dorsal side longer than distance from gland to leaf tip; on juvenile branches, spreading to appressed
Flowers: unisexual; dioecious, rarely monoecious, solitary at branchlet tips; male yellowish brown, sessile, ellipsoid; female subglobose, cone-like
Fruits: cones ripening the second year; globose, somewhat fleshy; pulp resinous, sweetish; 0.2 in (4 mm) across, bluish with whitish cast; seeds 1-2, egg-shaped, smooth, weakly ridged toward base

Juniperus virginiana L. - eastern redcedar

Eastern redcedar, a native tree on sites ranging from the swamps of the Atlantic coast of the U.S. to the dry outcrops of western Nebraska, Kansas and South Dakota, has been planted widely and successfully in windbreak and wildlife habitat plantations in much of the prairie region. Its drouth resistance was demonstrated by its excellent survival through the Dust Bowl of the 1930s (Pool 1939). To the dismay of stockmen, eastern redcedar invades grasslands whenever a seed source is nearby and when moisture conditions are favorable.

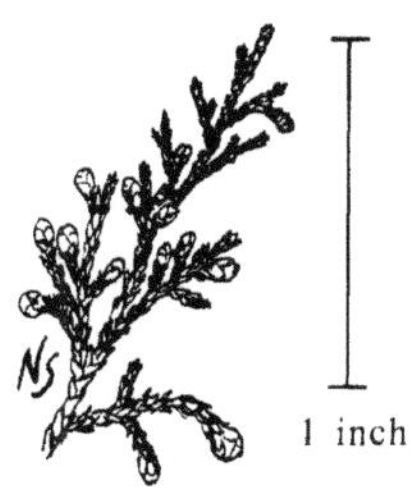

Juniperus virginiana

Many cultivars of varying form and foliage color have been selected, mostly from northern variety *crebra* Fern., for use in landscape plantings. Cultivars vary in form, from the shrubby pfitzer-like 'Grey Owl' to the narrow-crowned 'Taylor' that bears gray-green summer foliage which turns reddish green in winter. 'Taylor' was discovered and propagated by the late Allen Wilke, nurseryman of Columbus, Nebraska. 'Manhattan Blue', a pyramidal tree with bluish green foliage that becomes green in winter, grows at PERC (Fairchild and Klett 1993).

We have learned by failed plantations and through research that the eastern redcedar of a southerly origin cannot be successfully planted very far north of its native habitat. A fifth-year analysis of a *Juniperus* provenance plantation indicates that *J. virginiana* seed should be collected in east-central Nebraska for planting throughout the Great Plains region (Van Haverbeke and King 1990).

Eastern redcedar heartwood resists decay. Products include fence posts and lumber for crafts and trunks, closet linings and interior room trim. Unfortunately, depletion of large redcedar trees has reduced its availability for lumber. Leaves, twigs and wood provide oils used in the manufacture of perfumes and medicines. It is also used as a Christmas tree, especially in the southeastern U.S.

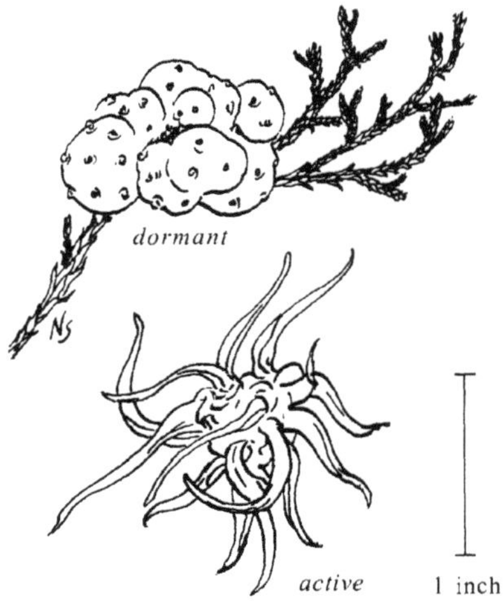

Cedar-apple rust galls

Planning

Height: up to 50 ft (15 m); Nebraska champion, 1989, 72 ft (22 m), 32 in (81 cm) dbh, Verdel on the Niobrara River; National champion, 1989, 55 ft (16.7 m), 67 in (170 cm) dbh, Georgia

Habit: tree; crown pyramidal, broad to narrow, sometimes shrubby; summer foliage green, winter foliage reddish to purplish brown

Site: sun, part shade; pH and soil adaptable; drouth tolerant; zone 3

Use: screen, windbreak, topiary, hedge, wildlife habitat, erosion control, lumber

Recognition

Leaves: scale-like, 4-ranked; tightly appressed, free at the tips, overlapping, tapering to a point, margin entire; dorsal glands oval, shorter than distance from gland to leaf tip; needle-like juvenile often present on adult trees

Flowers: unisexual, usually dioecious, terminal; male globose or cylindric with 4 vertical rows of 3 peltate scales, each scale yellow-brown, bearing 4-6 yellow anther locules; female sub-globose, scales 4, green, fused at base, center fleshy with 1-3 ovules

Fruits: cones globose, 0.2-0.3 in (5-8 mm) across, dark blue with heavy bloom; scales fused, resinous; seeds 1-3, egg-shaped, ridged, pitted; maturing in 1 season

Platycladus Spach - **Oriental arborvitae**

Platycladus is a monotypic genus.

Platycladus orientalis (L.) Franco - **Oriental arborvitae**

Thuja orientalis L. is a commonly used synonym for Oriental arborvitae, and less common is the synonym *Biota orientalis* L. Oriental arborvitae is very similar in appearance to northern whitecedar (*T. occidentalis*) except for its distinctive cone scales with horn-like hooks. A prostrate Siberian form marketed for use as groundcover is *P. microbiota*.

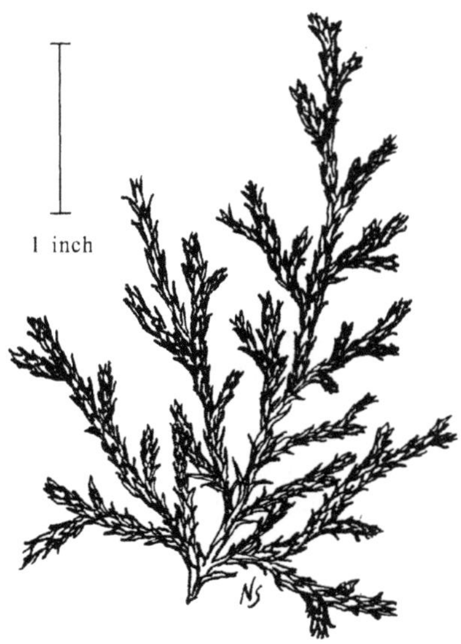

Platycladus microbiota

'Excelsa' and 'Aurea Nana' do not suffer winter injury and adapt to alkaline soils at Fort Collins, but extra water and protection from winter winds is recommended (Fairchild and Klett 1993). The former is an upright, green shrub; 'Aurea Nana' is a dwarf plant on which the new growth is golden yellow.

Seedlings of unknown origin have performed well in southeastern Nebraska. This species may be a reasonable substitute for eastern redcedar as a windbreak where cedar-apple rust is a concern of apple growers. Oriental arborvitae has been successfully used as an understock for juniper grafts.

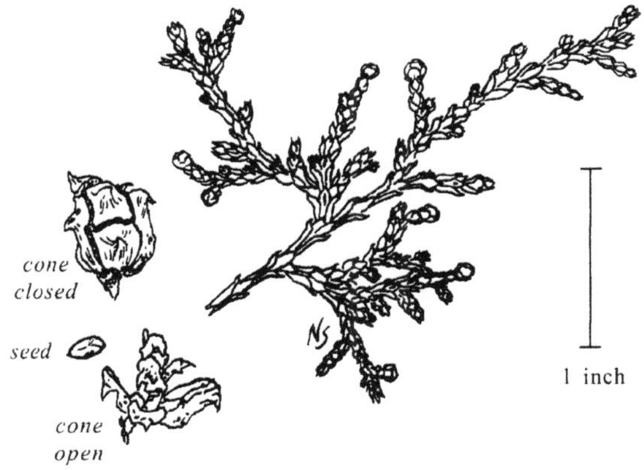

Platycladus orientalis

Planning

Height: up to 20 ft (6 m); National champion, 1991, 41 ft (12.5 m), Maryland
Habit: tree, shrub; evergreen; dense, conical, becoming loose, open
Site: sun, part shade; pH and soil adaptable; prefers moist, well-drained soils; zone 5
Use: screen, windbreak, wildlife habitat
Caution: winter watering may be necessary in dry climates

Recognition

Leaves: appressed, scale-like. 0.06 in (2 mm) long; pores (stomates) minute
Flowers: unisexual, monoecious, solitary, near twig tips
Fruits: cones egg-shaped, fleshy, green, becoming woody, about 0.8 in (2 cm) long; scales 6-8, thick with protuding curved beak; seeds wingless

Thuja L. - arborvitae

Two of the five species of *Thuja* are native to North America where they are commonly known as cedars. Western redcedar, *T. plicata* Donn, a large tree valued for its decay-resistant wood, is used for shingles, siding and poles.

A western redcedar, age 13, protected from winter winds grows in Fort Collins with winter watering (Fairchild and Klett 1993). A 12-inch western redcedar seedling at Prairie Pines Arboretum that survived its first winter, courtesy of one foot of snow cover throughout the winter, made excellent growth the following summer, but succumbed during the next winter of little snow. Dry soil may have been responsible for its demise.

Recognition

Habit: trees, shrubs; evergreen
Leaves: scale-like, closely pressed against flattened, fan-shaped sprays of twigs; alternating pairs at right angles to each other (decussate)
Flowers: unisexual, monoecious, terminal
Fruits: cones leathery or woody, usually solitary, erect; maturing in one season; scales thin; seeds laterally 2-winged

Thuja occidentalis L. - northern whitecedar

Northern whitecedar, a tree of the cool regions of eastern Canada and the northeastern U.S., has been planted far from its native climes for a wide variety of uses. Although it is a tree of the swamps and low wet forests, it is also often found on soils of limestone origin. Thus, it tolerates our alkaline soils, but foliage can be singed by winter winds and dry soil. Insect pests and diseases are not serious problems for this tree.

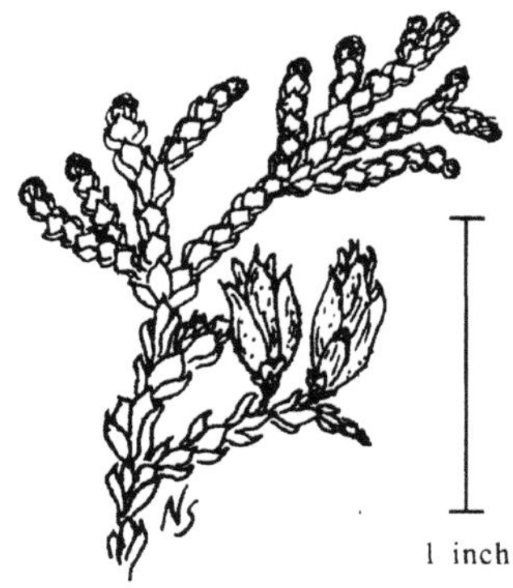

Thuja occidentalis

Over 100 cultivars of northern whitecedar have been selected, varying in foliage color from golden to dark green, and varying in form from dwarf and globose to tall and pyramidal.

Dark green cultivars 'Techny' and 'Nigra' grow well on most sites in our region if partially shaded and protected from extreme winds. 'Pyramidalis' was once commonly planted in eastern Nebraska, but many succumbed to the Armistice Day freeze of 1940. A few of these tall sentinels remain alive in cemeteries and private grounds.

Six cultivars growing at PERC are adapted to alkaline soils and are winter hardy with no insect or disease problems (Fairchild and Klett 1993). 'Hetz Midget' maintains a formal globe shape with dense, medium foliage. 'Nigra' is a broad oval with dense, dark green foliage. 'Pyramidalis' is broad and columnar with multiple leaders and dense green foliage. 'Smaragd', from Denmark, has a narrow pyramidal shape with medium foliage. 'Techny' is broadly pyramidal with dense dark green foliage. 'Woodwardii' is globe shaped with dense, dark green foliage.

Planning

Height: up to 40 ft (12 m); Nebraska champion, 1985, 36 ft (10.9 m), Falls City; National champion, 1978, 113 ft (34.2 m), Michigan

Habit: tree, shrub-like; crown broad pyramidal, branches ascending, dense foliage

Site: sun, part shade; pH adaptable, moist, well-drained soil; zone 4

Use: specimen, hedge, foundation, screen

Caution: protect from excessive wind; keep soils moist in winter

Recognition

Leaves: yellow-green; on leading shoots 0.2-0.3 in (5-7 mm) long, conspicuously glandular, long-pointed; on lateral twigs scale-like, flattened, 0.1 in (2 mm) long, glandular pits obscure

Flowers: solitary; male globose, stamens 3-6 pairs alternating at right angles, filaments scale-like; female at twig tips; scales 8-12, each with 2-3 ovules

Fruits: oblong cones, 0.3-0.5 in (7-12 mm) long; scales opposing pairs, tips rounded or short-spined, usually 4 fertile; seeds 2-winged, wings as wide as body

Twigs: flattened into broad fan-shaped sprays

Bark: thin, light reddish brown, shreddy

Wood: fine-textured, light, soft

Class Taxopsida

The class Taxopsida consists of one order.

Order Taxales

Two families are members of Taxales. Species of Cephalotaxaceae—which includes the genus *Torreya* Arn., formerly included in Taxaceae (Welch 1991)—are not hardy in our prairies.

Taxaceae Lindl. - **yew family**

The yew family comprises three genera. Only one, *Taxus* L., has species adapted to our prairies.

Taxus L. - **yew**

Yews are "classy" landscape shrubs and small trees. The dark evergreen foliage is highly visible in winter and the plants adapt to shearing for hedges and topiary. Slow growth reduces the necessity for frequent pruning. Yews are relatively free of insect and disease problems and are long-lived where adapted. Life is short where they are surrounded by concrete and exposed to sun and wind or to splashing salt water from city streets in winter.

Many yews, especially the hybrids, are sold by cultivar name only, such as *Taxus* 'Hicksii'. The specific epithet is omitted. Except for form and, perhaps, foliage color and needle density, the only significant difference among most yew hybrids is the label. If sheared or pruned they lose their identity and cannot be readily distinguished from one another. Indiscriminate naming has caused much confusion and prompted the Ohio Agricultural Research and Development Center in Wooster to plant a collection of cultivars for evaluation (Keen and Chadwick 1976).

Three of the eight species of yew are native to North America. Two species—trees found in Florida and in the Pacific coastal region—are not hardy in the prairies.

The foliage, bark and seeds of the yew are toxic to humans and animals, but the fleshy part of the fruit is not. Unless the seed is chewed, it is passed undigested. *T. brevifolia* Nutt., Pacific yew, is a source of taxol, a bark extract in great demand for the treatment of breast cancer.

Recognition

Habit: trees, shrubs; evergreen

Leaves: spirally arranged, often appearing 2-ranked; linear, flattened, 2 light-colored bands beneath; short-stalked, extending downward from point of attachment (decurrent)

Flowers: axillary, at junction of leaf and twig; unisexual, male and female usually on separate trees; male globe-shaped, scales overlapping, 4-8 pollen sacs; female solitary, appearing as a small, stalked, conical bud; ovule 1, base surrounded by circular, fleshy disc on the uppermost of several overlapping scales

Fruits: seeds with bony shells surrounded by fleshy, bright red cups (arils)

Buds: minute, rounded; scales thin, overlapping

Bark: reddish brown, scaly

Wood: hard, heavy, straight-grained, orange to brown

Taxaceae *Taxus*

Taxus baccata L. - English yew

This picturesque native of Europe, northern Africa and western Asia has been cultivated for hundreds of years. Over 100 English yew cultivars have been recognized. Unfortunately this patriarch among landscape plants is not capable of withstanding winters in most of the prairies.

When crossed with its Asian relative, *T. cuspidata,* the offspring, *T. ×media,* which most often resemble the latter in appearance, are adapted to most prairie sites if given reasonable care.

Planning

Height: up to 40 ft (12 m)
Habit: tree, sometimes shrubby; broad rounded or shrubby, dense
Site: sun, shade; pH adaptable; prefers fertile, moist, well-drained soil: marginal zone 5
Use: hedge, screen, foundation
Caution: protect from excessive wind

Recognition

Leaves: spirally arranged; spreading all around on erect shoots, appearing 2-ranked on horizontal shoots; 0.5-1.3 in. (13-32 mm) long; abruptly narrowed into short greenish petiole, tapering at apex to a narrow horny point; dark green and lustrous above
Flowers: unisexual, usually dioecious; male on twigs of previous year, globose, 6-14 stamens; female green
Fruits: solitary; seeds broad-ellipsoid, slightly angled, slightly compressed; 0.2 in (6 mm) long
Twigs: surrounded at base by persistent, brownish bud scales
Buds: scales blunt, not keeled

Taxus canadensis Marsh. - Canadian yew

This straggly native to the northeastern U.S. and adjoining parts of Canada is the hardiest of the yews but the least desirable for landscape purposes. The foliage displays a brownish tinge in winter. Few cultivars have been selected, but it is potentially a hardy parent of hybrids.

Planning

Height: up to 7 ft (2 m)
Habit: shrub; ascending, often prostrate, straggly
Site: shade; pH adaptable; prefers moist, well-drained soil; zone 3
Use: groundcover
Caution: protect from excessive wind

Recognition

Leaves: often indistinctly 2-ranked; less 0.1 in (2 mm) wide, 0.5-0.8 in (13-20 mm) long; apex abruptly pointed; dark yellow green above, 2 pale green bands below about as broad as green margin; very short-stalked
Fruits: solitary; seed broader than high, fleshy coating (aril) light red
Twigs: green, becoming reddish brown
Buds: scales sharp-pointed, keeled

Taxus cuspidata Sieb. & Zucc. - Japanese yew

Japanese yew, native to Japan, Korea and Manchuria, is a valuable evergreen landscape plant with dense dark green foliage. This favorite in formal landscapes can withstand heavy pruning and can be used as a hedge and foundation plant. Cultivars range from prostrate creepers such as 'Nana' to small conical trees such as 'Capitata'. 'Densa', one of the best dwarf forms according to Dirr (1990), is a low shrub with extremely dark green leaves.

Planning

Height: up to 30 ft (9 m)
Habit: tree, shrub; single to multiple stems, upright or spreading, sometimes irregular
Site: shade, part shade; pH and soil adaptable; prefers moist, well-drained soil; endures dust and smoke; zone 5
Use: hedge, screen, foundation, grouping, erosion control, bonsai
Caution: protect from excessive wind

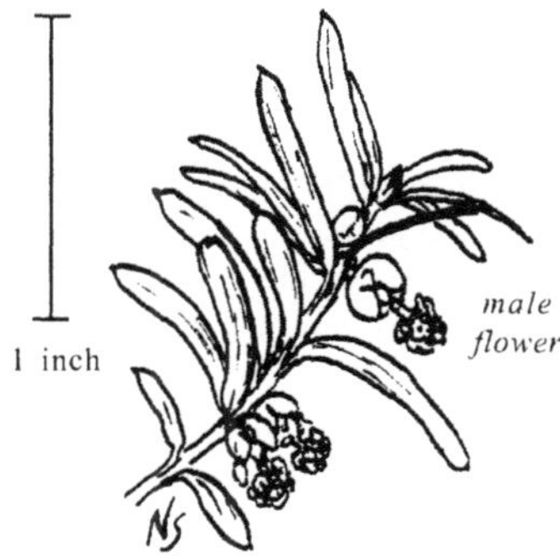

***Taxus cuspidata* 'Capitata'**

Recognition

Leaves: not distinctly 2-ranked; over 0.1 in (2-3 mm) wide, 0.5-1.0 in (13-24 mm) long; abruptly sharp pointed; dark green with prominent midrib above; 2 yellowish bands below, twice as broad as green margin; distinct yellowish stalk
Flowers: dioecious
Fruits: solitary; seeds egg-shaped, about 0.3 in (8 mm) long with slight angles
Twigs: reddish brown
Buds: egg-shaped, chestnut brown; scales overlapping, egg-shaped, keeled on back

Taxus ×media Rehd. - **Anglo-Japanese yew**

This hybrid, *T. baccata* × *T. cuspidata*, bred in Massachusetts sometime around 1900, and its many cultivars have become a valuable group of landscape evergreens. Anglo-Japanese yews range in habit from spreading shrubs to small trees and vary in adaptability to various sites. They exhibit characteristics ranging from similar to each parent to forms intermediate between the two, hence the hybrid name *media*.

Many cultivars have been selected from this hybrid population because of differing form or habit, foliage color and needle density. Those seen quite frequently in our region are dense shrub-like 'Densiformis', rounded 'Brownii', wide-spreading 'Wardii', columnar 'Hicksii', pyramidal 'Hatfieldii', and low-growing 'Everlow'.

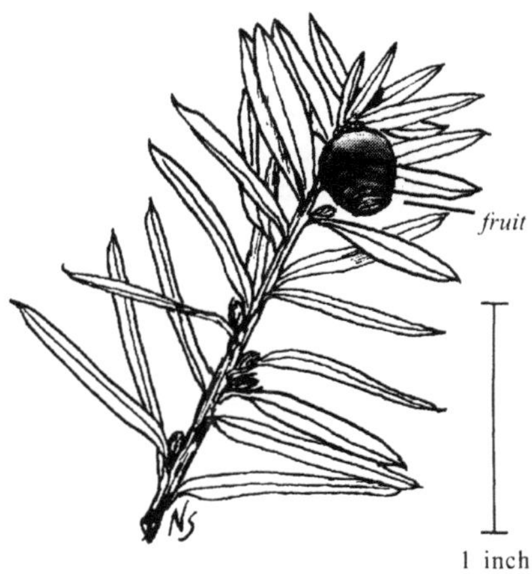

Taxus ×media

Planning

Height: up to 30 ft (9 m)
Habit: tree, shrub; broad, pyramidal to spreading multistemmed
Site: sun, shade; pH and soil adaptable; prefers moist, well-drained soil; zone 5
Use: hedge, screen, foundation, mass planting
Caution: protect from excessive wind

Recognition

Leaves: 2-ranked, dark green, lighter beneath
Twigs: olive, not becoming brown second year
Buds: scales blunt

Division Magnoliophyta - angiosperms

The angiosperms are a widely distributed, complex group of plants numbering about 250,000 species. They are of more recent origin and are structurally more advanced than the conifers. The reproductive structures of angiosperms are true flowers.

Some angiosperm plants complete their life cycle in a matter of weeks while others live one year (annuals), two years (biennials), or longer than two years (perennials). Trees, shrubs and vines of the angiosperms are perennial, with well-developed usually broad leaves that typically die in autumn in temperate regions but may remain alive and green throughout the year (or longer) in tropical or subtropical climates.

The dicots (Magnoliopsida) show two seed leaves at germination and netted leaf veination, as contrasted to the single seed leaf and parallel veination of the monocots (Liliopsida). Most monocots are herbaceous as in the well-known grass family, Poaceae. Very few monocots, such as the palms (Arecaceae), are tree size.

Class Magnoliopsida

The dicots, most of which are deciduous in temperate climates, include the majority of species of our North American woody plants. The trees are commonly referred to as hardwoods.

Recognition

Habit: herbaceous or woody plants
Leaves: usually flattened, few are linear or scale-like; veins in netted patterns
Flowers: bisexual or unisexual with both sexes on the same plant (monoecious) or on different plants (dioecious): petals and sepals, if present, typically arranged in 4's or 5's

During much of the year, stems are the only part of a deciduous woody plant that is readily seen in the prairies—especially the central and northern regions. Leaf and fruit remnants are sometimes present on the twigs or on the ground beneath. The search for identity of a woody plant is simplified by grouping dicot families according to bud arrangement and leaf characteristics.

Grouping Key for Magnoliopsida (Dicot Families) by Leaf

A family may be found in more than one group. The groups include only the species described in this book. Leaves with wavy and undulate margins are included in the group having entire margins. Toothed margins range from serrulate and denticulate to serrate and dentate. Dissected leaves are included with lobed leaves. Plants that usually retain green leaves year-around in the central and northern prairies are designated evergreen. Habit of woody plants—tree, shrub, vine— within families is included as an additional aid to identification.

Leaves evergreen; alternate, simple; margin entire

Araliaceae	vine	p466	Rosaceae	shrub	p258
Brassicaceae	shrub	p 220	Thymelaeaceae	shrub	p 374
Ericaceae	shrub	p 223			

Leaves deciduous; alternate, simple; margin entire

Anacardiaceae	shrub	p454	Lauraceae	tree, shrub	p 82
Annonaceae	tree	p 79	Magnoliaceae	tree	p 73
Aristolochiaceae	vine	p 84	Malvaceae	shrub	p 196
Asteraceae	shrub	p 545	Menispermaceae	vine	p 100
Buddlejaceae	shrub	p 478	Nyssaceae	tree	p 377
Cactaceae	shrub	p 179	Onagraceae	shrub	p 376
Caesalpiniaceae	tree	p 349	Polygonaceae	vine	p 183
Chenopodiaceae	shrub	p 180	Rhamnaceae	tree, shrub	p 409
Clethraceae	shrub	p 221	Rosaceae	tree, shrub	p 258
Cornaceae	tree, shrub	p 378	Sapotaceae	tree, shrub	p 234
Ebenaceae	tree	p 235	Schisandraceae	vine	p 86
Elaeagnaceae	tree, shrub	p 368	Solanaceae	shrub	p 472
Euphorbiaceae	shrub	p 407	Styracaceae	tree, shrub	p 237
Fabaceae (*Genista*)	tree, shrub	p361	Tamaricaceae	shrub	p 197
Fagaceae	tree	p 139	Thymelaeaceae	shrub	p 374

Leaves evergreen; alternate, simple; margin toothed

Aquifoliaceae	tree, shrub	p399	Ericaceae	shrub	p223
Berberaceae	shrub	p 94	Rosaceae	shrub	p 258
Buxaceae	shrub	p 405			

Leaves deciduous; alternate, simple; margin toothed

Actinidiaceae	vine	p187	Onagraceae	shrub	p376
Asteraceae	shrub	p 545	Rhamnaceae	shrub	p 409
Betulaceae	tree, shrub	p163	Rosaceae	tree, shrub	p 258
Buxaceae(*Pachysandra*)	shrub	p406	Salicaceae	tree, shrub	p 199
Celastraceae	shrub, vine	p 389	Schisandraceae	vine	p 86
Clethraceae	shrub	p 221	Styracaceae	tree, shrub	p 237
Eucommiaceae	tree	p 109	Symplocaceae	shrub	p 239
Grossulariaceae	shrub	p 249	Theaceae	shrub	p 185
Hamamelidaceae	tree, shrub	p 104	Tiliaceae	tree	p 191
Moraceae	tree	p 121	Ulmaceae	tree	p 110
Nyssaceae	tree	p 377	Vitaceae	vine	p 417

Group five–leaves evergreen; alternate, simple; lobed

Araliaceae	tree, shrub	p466

Group six–leaves deciduous; alternate, simple; lobed

Aristolochiaceae	vine	p 84	Moraceae	tree	p121
Asteraceae	shrub	p 545	Myricaceae	shrub	p 137
Fagaceae	tree	p 139	Platanaceae	tree	p 102
Grossulariaceae	shrub	p 249	Rosaceae	tree, shrub	p 258
Hamamelidaceae	tree, shrub	p 104	Tiliaceae	tree	p 191
Lauraceae	tree, shrub	p 82	Vitaceae	vine	p 417
Magnoliaceae	tree	p 73			

Magnoliopsida Key

Leaves deciduous; alternate, compound; margin entire

Anacardiaceae	tree, shrub	p454	Mimosaceae	shrub	p 348
Caesalpiniaceae	tree, shrub	p 349	Rosaceae	tree, shrub	p 258
Fabaceae	tree, shrub	p 354	Rutaceae	tree, shrub	p 462
Lardizabalaceae	vine	p 99	Sapindaceae	tree, shrub	p 427

Leaves deciduous; alternate, compound; margin toothed

Anacardiaceae	tree, shrub	p454	Ranunculaceae (*Xanthoriza*)	shrub	p 94
Araliaceae	tree, shrub	p 466	Rosaceae	tree, shrub	p 258
Caesalpiniaceae	tree, shrub	p 349	Sapindaceae	tree, shrub	p 427
Juglandaceae	tree	p 125	Simaroubaceae	tree	p 461

Leaves deciduous; alternate, compound; lobed

Anacardiaceae	tree, shrub	p454	Sapindaceae	tree, shrub	p427
Paeoniaceae	shrub	p 184	Vitaceae	vine	p 417
Rosaceae	tree, shrub	p 258			

Leaves evergreen; opposite or whorled, simple; margin entire

Apocynaceeae	shrub	p471	Lamiaceae (*Lavandula*)	shrub	p477
Buxaceae (*Buxus*)	shrub	p405	Rubiaceae	shrub	p 507

Leaves deciduous; opposite or whorled, simple; margin entire

Bignoniaceae (whorled)	tree	p501	Elaeagnaceae (*Shepherdia*)	tree, shrub	p367
Calycanthaceae	shrub	p 80	Oleaceae	tree, shrub	p 480
Caprifoliaceae	shrub	p 509	Rhamnaceae (*R. cathartica*)	shrub	p413
Clusiaceae	shrub	p 189	Rutaceae	tree, shrub	p 462
Cornaceae	tree, shrub	p 378			

Leaves evergreen; opposite or whorled, simple; margin toothed

Celastraceae	shrub, vine	p389

Leaves deciduous; opposite or whorled, simple; margin toothed

Buddlejaceae	shrub	p478	Lamiaceae (*Elsholtzia*)	shrub	p477
Caprifoliaceae	shrub	p 509	Oleaceae	tree, shrub	p 480
Cercidiphyllaceae	tree	p 101	Rutaceae	tree, shrub	p 462
Hydrangeaceae	shrub	p 241	Verbenaceae (*Callicarpa*)	shrub	p474

Leaves deciduous; opposite or whorled, simple; lobed

Aceraceae	tree, shrub	p434	Hydrangeaceae	shrub	p241

Leaves deciduous; opposite, compound; margin entire

Oleaceae	tree, shrub	p480	Rutaceae	tree, shrub	p462
Ranunculaceae (*Clematis*)	vine	p 87			

Leaves deciduous; opposite, compound; margin toothed

Aceraceae	tree, shrub	p434	Oleaceae	tree, shrub	p480
Bignoniaceae	shrub	p 501	Rutaceae	tree, shrub	p 462
Hippocastanaceae	tree	p 430	Staphyleaceae	shrub	p 426

Grouping Key for Magnoliopsida Subclasses
(Cronquist 1981)

Leaves opposite or alternate, simple, sometimes compound; flowers usually regular, ovary usually superior (hypogynous); perianth parts (tepals) distinct, seldom fused, sepals and petals often similar or petals missing, stamens usually numerous: **subclass Magnolidae**

Leaves alternate, sometimes opposite, simple, sometimes compound; flowers mostly small and inconspicuous, mostly wind pollinated; commonly without petals, often without sepals, if present small; stamens 2 to many: **subclass Hamamelidae**

Leaves alternate, opposite, simple, commonly entire, sometimes reduced to spines, stipules usually absent; flowers usually bisexual and regular, sepals distinct or connate, sometimes petal-like, or sepals and petals not clearly differentiated: **subclass Caryophyllidae**

Leaves alternate or opposite, simple, seldom compound, margins variable; flowers polypetalous, rarely without petals, sometimes sympetalous: **subclass Dilleniidae**

Leaves alternate or opposite, simple or compound; flowers usually bisexual, polypetalous, sympetalous or sometimes reduced or wanting: **subclass Rosidae**

Leaves alternate, opposite, sometimes whorled, simple or less often compound; flowers bisexual or unisexual, regular or irregular, petals sympetalous, tubular, strap-shaped or 2-lipped or wanting; usually well developed, often showy; fruits usually dry: **subclass Asteridae**

Subclass Magnolidae

Magnolidae conists of eight orders, 39 families and about 12,000 species (Cronquist 1981).

Order Magnoliales

Ten families and about 3,000 species have been recognized in this order.

Recognition

Habit: trees, shrubs, vines
Leaves: opposite or alternate, simple, entire, or lobed
Flowers: solitary or open, few-flowered clusters; perianth with mostly distinct parts

Magnoliaceae Juss. - magnolia family

Twelve genera and over 180 species are recognized in the magnolia family of trees and shrubs of temperate and tropical climates. Two genera are native to North America: *Magnolia* with six species, and *Liriodendron* (yellow-poplar) with one species. The magnolias are widely planted for their fragrant, showy flowers and yellow-poplar, one of the tallest native deciduous species, is a valuable lumber tree.

Magnoliaceae *Liriodendron*

Recognition

Habit: trees, shrubs; deciduous or evergreen

Leaves: alternate, simple, stipulate (stipules enclose buds and leave conspicuous encircling scars on twigs)

Flowers: terminal or axillary, regular; usually bisexual; sepals 3 (rarely 4) often similar to petals; petals 6 or more, rarely wanting; stamens and pistils arranged spirally

Fruits: cone-like aggregates, follicles or samaras

Liriodendron L. - tuliptree

Only two species remain of a genus that fossil records show was once widespread over the Northern Hemisphere. Both *L. chinense* (Hemsl.) Sarg., a native of China and Viet Nam, and *L. tulipifera* L., a native of the eastern U.S., are important timber trees.

Liriodendron tulipifera L. - tuliptree, yellow-poplar

Tuliptree (a name popular with horticulturists) or yellow-poplar (coined by foresters) is a native of the U.S. in an area bounded by a line joining Michigan and Louisiana eastward to the Atlantic. We have reports of this species being cultivated as far west as Kearney and Curtis in Nebraska.

In its native habitat tuliptree is distinctive because of its tall, straight trunk. The leaves quiver in the slightest breeze and large, showy flowers that attract bees are borne high in the tree, partially hidden by leaves unfolding ahead of them.

Several cultivars which vary in leaf and form include 'Aureo-marginatum', with yellow leaf margins; 'Compactum', a dwarf form; and 'Fastigiatum', with a narrow crown and upright branches (Dirr 1990). We could probably learn much about tuliptrees from provenance research.

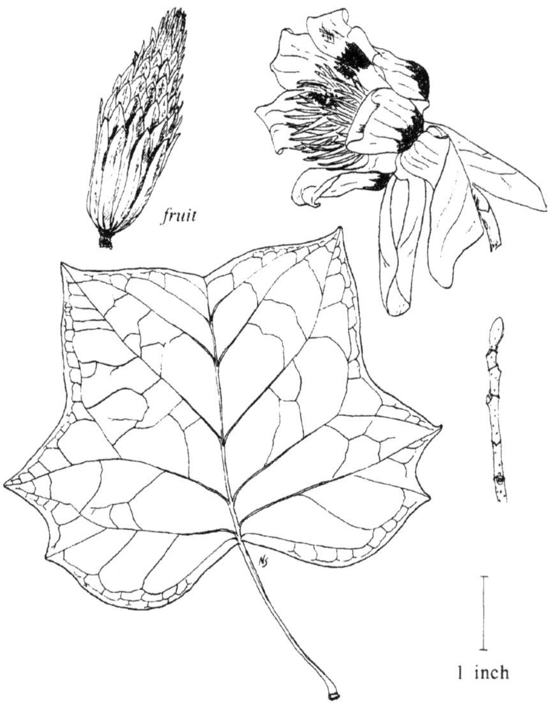

Liriodendron tulipifera

Planning

Height: up to 80 ft (24 m);

Nebraska champion, 1987, 80 ft (24.4 m), Oakland; National champion, 1986, 146 ft (44.2 m), Virginia

Habit: tree; oval crown, deciduous; late May bloom after leaves unfold; autumn foliage yellow; fruits persist through winter

Site: sun; pH and soil adaptable; prefers fertile, moist, well-drained, slightly acidic soil; zone 4

Use: shade, specimen on large properties, lumber

Caution: protect from strong winds

Recognition

Leaves: squarish, broadest near base, mostly 4-lobed, more or less squared off at base and apex (truncate), margin smooth (entire); surfaces glabrous; petioles slender, about 3 in (8 cm) long; stipules conspicuous, encircling twig

Flowers: solitary; about 1.6 in (4 cm) wide, cup-shaped; bisexual; sepals 3; petals 6 in 2 rows, light greenish yellow; stamens and pistils numerous, spirally arranged

Fruits: cone-like aggregates of samaras about 3 in (8 cm) long, persistent

Twigs: red-brown, glabrous, lenticels pale; pith diaphramed, white; leaf scars nearly circular, bundle scars many, small; stipular scars conspicuous, encircling twig

Buds: terminal 0.5 in (12 mm) long; lateral smaller, flattened, rounded at tip (duck's bill), glabrous; scales 2, not overlapping (valvate)

Bark: smooth and gray with black v-markings below branches becoming deep-furrowed; inner bark bitter and aromatic

Wood: straight-grained, moderately light, easily worked

Magnolia L. - magnolia

About 75 species of magnolia are found mostly in warm temperate and tropical regions of the world. Of the six species native to the U.S., only one is adapted to a significant portion of the southern prairie. The soft, rather brittle wood has limited use. Many cultivars of native and exotic species are widely planted for ornamental purposes.

Magnolias are divided into three types. An example of the evergreen type is *M. grandiflora* L., bull bay, a handsome tree of the southeast U.S. that will not survive the rigorous climate of much of our prairies. The other two types are deciduous: one blooms in spring and the other in summer. The former are mostly small trees or shrubs from the Far East that are either tulip-flowered or star-flowered.

To enjoy magnolia blooms to their fullest extent, plant the trees in sheltered locations. They enjoy a light, woodland shade. Ed Rassmussen has followed this advice by planting many cultivars under the canopy of an old farmstead windbreak north of Omaha, a young plantation that is surviving remarkably well.

The superabundance of magnolia cultivars is beyond the scope of this publication. We recommend membership in the Magnolia Society for those who desire more information. **World of Magnolias** (Callaway 1994) is an illustrated reference of all known species, including many cultivars.

Recognition

Habit: trees, shrubs; deciduous or evergreen

Leaves: entire, unlobed; stipules united with petioles, scar encircling twig

Flowers: solitary, terminal; sepals 3, often petal-like, petals 6-15; stamens numerous

Fruits: cone-like aggregates of follicles; seeds 1-2

Buds: terminal large, conspicuous; scale 1

Magnolia acuminata L. - cucumbertree

This hardiest of our native magnolias ranges from Louisiana and eastern Oklahoma north to western New York. Cucumbertrees prefer deep loamy fertile soils on the acidic side, but will tolerate calcareous soils and may grow reasonably well in the east central and southern prairie if protected from extreme winds and drouth. Cucumbertrees are best suited to large properties because of the potential great height and crown spread.

M. cordata Michx. is a closely related shrub or small tree distinguished by densely hairy twigs and bearing canary yellow flowers.

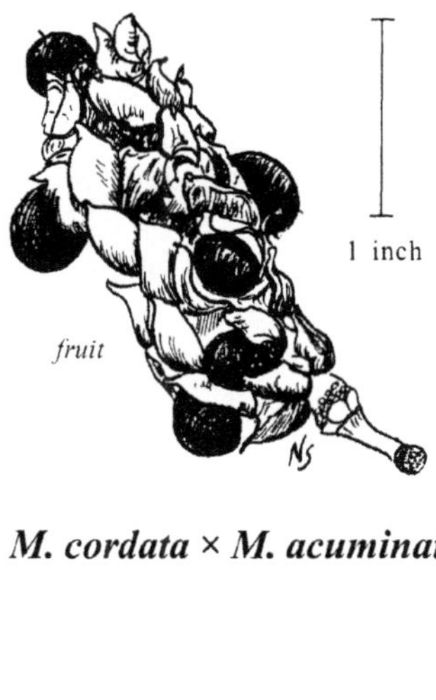

M. cordata* × *M. acuminata

Crosses with other magnolia species have produced several cultivars including 'Elizabeth', with a pyramidal habit; 'Evamaria', with a distinctive flower color; 'Woodsman', with larger than normal purple-tinged flowers; and 'Yellow Bird', with yellow flowers (Dirr 1990).

Planning

Height: up to 50 ft (15 m); Nebraska champion, 1980, 48 ft (14.5 m), Lincoln; National champion, 1985, 75 ft (22.9 m), Iowa
Habit: tree; deciduous; crown rounded, becoming broad; late May bloom hidden by leaves
Site: sun, part shade; pH and soil adaptable, prefers fertile, moist, well-drained soils, slightly acidic; zone 4
Use: shade for large areas

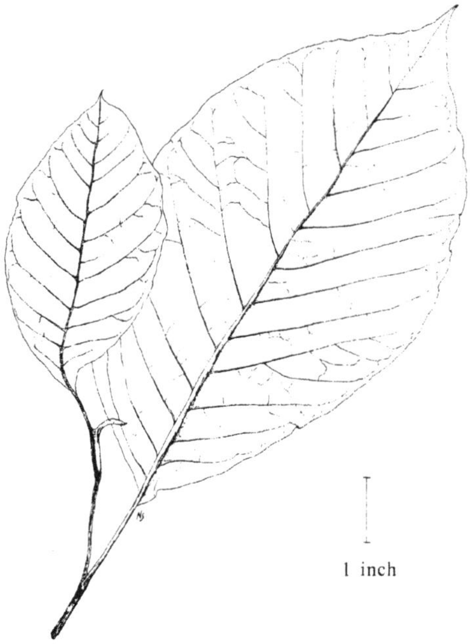

Magnolia acuminata

Recognition

Leaves: elliptic to egg-shaped or longer, usually broadest below the middle, tapering to uneven base, tip pointed; upper surface glabrous, dark green, shiny, pale green covered with soft hairs beneath; petioles about 1.2 in (3 cm) long
Flowers: bisexual; cup to bell-shaped, upright; 9 petal-like structures, greenish yellow; odorless
Fruits: cucumber shaped, reddish, about 2 in (5 cm) long; seeds about 0.4 in (1 cm) long, reddish
Twigs: brittle, reddish brown, glabrous or nearly so, bitter aromatic; leaf scars U-shaped, bundle scars many
Buds: silky-hairy; terminal about 0.6 in (1.5 cm) long
Bark: fissured; narrow, flaky ridges
Wood: straight-grained, moderately heavy, fairly hard, greenish yellow or darker

Magnolia denudata Desrouss. - Yulan magnolia

Yulan magnolia, native to central China, appears to be prospering in a few southeastern Nebraska arboreta. Though the Yulan takes longer to reach flowering age than some other magnolias, it may well be worth the wait. Large, pure white, fragrant blossoms appear very early in the spring, a trait that will occasionally subject them to freeze injury in the prairie climate.

Planning

Height: up to 30 ft (9 m)
Habit: tree; deciduous; round-topped; early spring fragrant flowers
Site: sun, part shade; prefers moist soil, slightly acidic; zone 5
Use: specimen

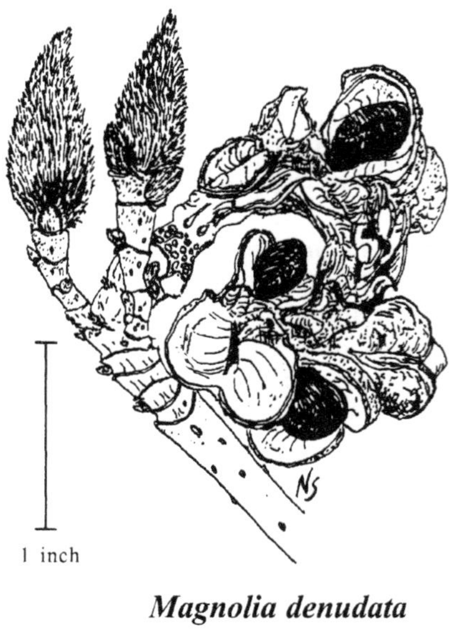

Magnolia denudata

Recognition

Leaves: oblong, widest above the middle (obovate), abruptly pointed, tapering at base; about 5 in (13 cm) long; dark green with sparse hairs above; lighter below, hairy
Flowers: solitary, white, fleshy, about 5 in (13 cm) across; sepals and petals alike, totaling 9, wider above middle, more than 0.8 in (2 cm) wide
Fruits: cylindric, asymmetrical; about 4 in (10 cm) long
Twigs: hairy
Buds: hairy; scales 2

Magnolia kobus DC. - Kobus magnolia

Kobus magnolia, one of the hardiest and most adaptable of the magnolias, seldom blooms before age 15 years. This small-to-medium-sized tree, native to Japan, tends to develop multiple stems, a trait that may reduce winter sunscald injury on open grown trees. A young Kobus magnolia at Prairie Pines Arboretum near Lincoln has suffered injury to the sunny side of the trunk but seems to be recovering by producing sprouts from the lower part of the stem. Richley (1988) reports specimens low in vigor at Curtis in southwestern Nebraska.

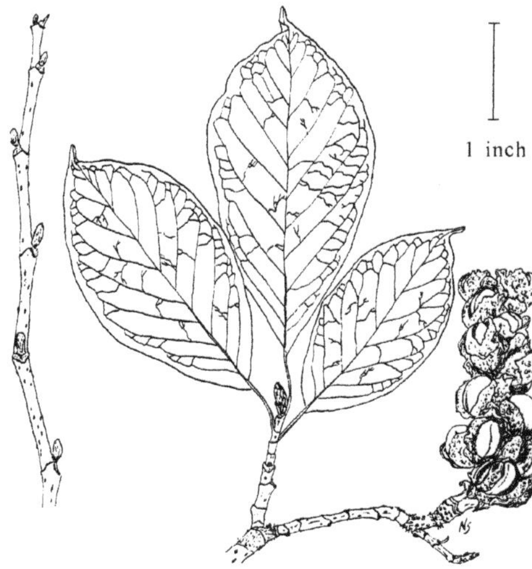

Magnolia kobus

Planning

Height: up to 30 ft (9 m)
Habit: tree; deciduous; crown rounded; early spring bloom
Site: sun, part shade; pH and soil adaptable, prefers fertile, moist, well-drained soils, slightly acidic; zone 5
Use: specimen

Recognition

Leaves: See *M. denudata*
Flowers: slightly smaller than *M.denudata*; sepals 3, distinct, greenish; petals 6
Fruits: cylindric, usually curved
Twigs: hairless
Buds: leaf, slightly hairy; flower, densely hairy
Bark: gray, smooth becoming rough with age

Magnolia ×*soulangeana* Soul. (*M. denudata* × *M. quinquepeta*) - saucer magnolia

This harbinger of spring is one of the most widely planted magnolias in the central prairies. Blooming saucer magnolias give us a lift from the winter doldrums except in the occasional years when untimely freezes injure flower buds. Much confusion exists about the correct identity of the numerous cultivars that have been selected (Dirr 1990). Date of blooming and flower color, size, and form vary considerably among them. Colors range from white to rosy purple.

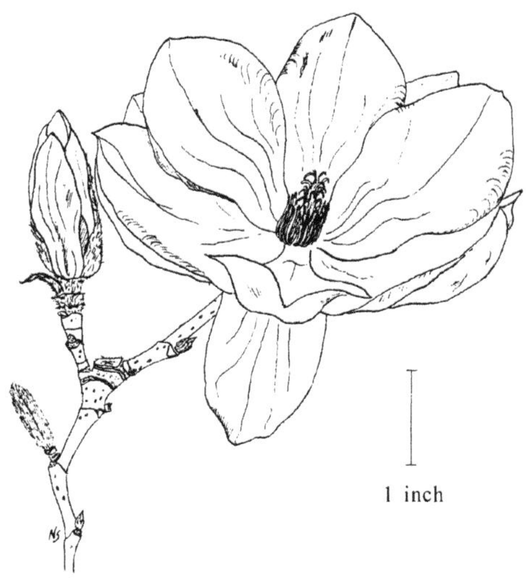

Magnolia ×soulangeana

Magnoliaceae *Magnolia*

Although Richley (1988) gives all magnolias a poor mark in the Curtis, Nebraska, area, Peterson (1987) states that *M. ×soulangeana* 'Lennei' is hardy at Meadowlark Hill south of Ogallala and "blooms every year." His secret to success is a protected site with soil "adjusted to slightly acid and maintained with acid fertilizers."

Planning

Height: up to 20 ft (6 m)
Habit: tree, shrub; deciduous; spreading branches; early spring bloom
Site: sun, part shade; prefers moist, fertile, well-drained soil, slightly acidic; zone 5
Use: specimen

Recognition

Leaves: see *denudata*
Flowers: bell-shaped (campanulate), 5-10 in (13-25 cm) across; white to pink to purplish; sepals resemble petals but about half as long
Fruits: seldom present
Twigs: glabrous; lenticels grayish; stipular scars prominent; fragrant
Buds: silky-hairy
Bark: gray, usually smooth

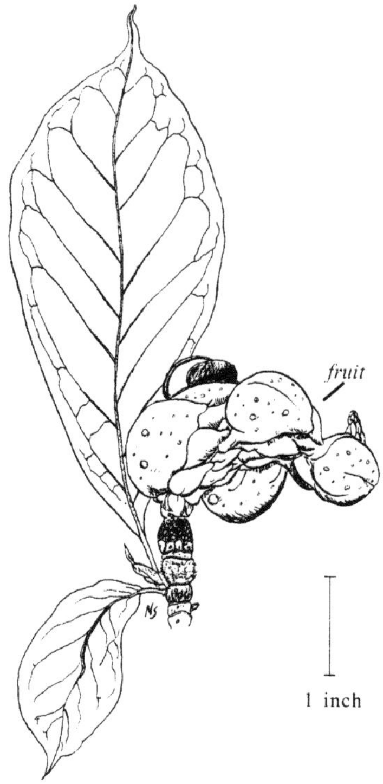

Magnolia ×loebneri

Magnolia stellata (Sieb. & Zucc.) Maxim.- **star magnolia**

Star magnolia appears to be better adapted to the prairie climate than saucer magnolia, but it is not as popular. Although it may attain tree form, star magnolia is usually a rather small shrub because of its slow growth, though it does bloom at a tender age. Richley (1988) gives star magnolia fair marks for performance at Curtis, Nebraska.

Several cultivars are available that differ in flower color and shape and plant habit. Dirr (1940) states that 'Royal Star' is "possibly the best and quite common in commerce."

Loebner magnolia, *M. ×loebneri* Kache (*M. kobus × M. stellata*), is a vigorous hybrid with flowers similar to those of star magnolia, but the Loebner is much faster growing. Several cultivars have been selected including 'Ballerina'and 'Merrill'. Both bloom late, thereby escaping many spring frosts.

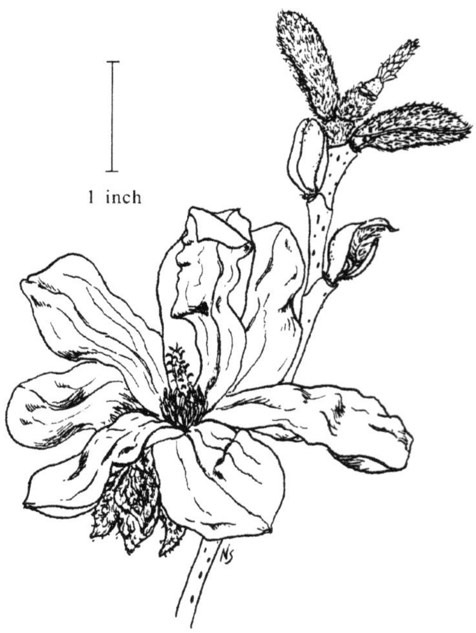

Magnolia stellata

Planning

Height: up to 15 ft (4.5 m)
Habit: shrub, tree; deciduous; crown rounded, multistemmed; young plants provide April bloom before leaves appear
Site: sun, part shade; prefers fertile, moist, well-drained soil, slightly acidic; zone 4
Use: specimen, foundation
Caution: protect from strong winds

Recognition

Leaves: elliptic to oblong, usually wider above the middle; rounded at apex, gradually tapering at base; hairless above, veins net-like beneath, smooth except for hairs along veins
Flowers: solitary; usually white, about 3 in (8 cm) across; fragrant; petals and sepals similar, 12-18, up to 0.6 in (15 mm) broad, becoming reflexed
Fruits: cylindrical, about 2 in (5 cm) long, twisted; seeds few
Twigs: slender, hairy becoming smooth, brown
Buds: densely hairy

Annonaceae Juss. **- custard apple family**

Many members of this mostly tropical family of 80 genera and some 800 species produce fleshy, edible fruit. Two genera are found in North America. One species of genus *Annona*, the pond apple, is native to Florida.

Recognition

Habit: trees, shrubs, vines; deciduous or evergreen
Leaves: simple, alternate, aromatic
Flowers: bisexual
Fruits: mostly berry-like

Asimina Adans. **- pawpaw**

Eight species (all shrubs but one) are native to the eastern U.S., mostly the coastal region.

Recognition

Habit: shrubs, rarely trees; deciduous or evergreen
Flowers: solitary or small clusters; axillary; short-stalked, nodding; petals 6; sepals 3, smaller than petals, early deciduous
Fruits: berries, oval to oblong; seeds large, compressed

Asimina triloba (L.) **Dun. - pawpaw**

Pawpaw is found in the river bottom forests of southeastern Nebraska where the deep fertile soils provide a proper niche. Its natural range extends east to the Atlantic Coast and south to eastern Texas. Paleo-botanists theorize that pawpaw migrated into temperate regions during warm, interglacial periods via some now extinct, fruit-eating animals.

Pawpaw produces few flowers and even fewer fruits, but are nonetheless relished by raccoons, opossums and bears. Deer browse twigs and leaves. Potential cancer fighting agents have been discovered in the pawpaw's leaves and twigs. Cultivars selected for their large fruits include 'Sunflower', 'Overleese' and 'Davis'.

Planning

Height: up to 25 ft (7.5 m); National champion, 1986, 60 ft (18.3 m), Mississippi

Habit: tree, large shrub; deciduous; crown pyramidal in the open, spreading in shade, suckering; autumn foliage yellow; flowers few but interesting on growth of previous season

Site: sun, full shade; prefers moist, fertile, slightly acidic soil; zone 5

Use: naturalizing, wildlife habitat, fruit

Recognition

Leaves: 2-ranked; broadest near tip (obovate), up to 12 in (30 cm) long; apex pointed, base gradually tapered; upper surface hairy on midrib, rusty hairy beneath, becoming nearly glabrous; petioles strongly grooved above, hairy; stipules none

Flowers: broadly bell-shaped, stems stout; velvety with red-brown hairs; calyx lobes 3, hairy; petals 6, arranged in 2 series, deep purple, wrinkled (rugose), black hairy on outer surface, glabrous inside, outer 3 reflexed, inner 3 erect or spreading

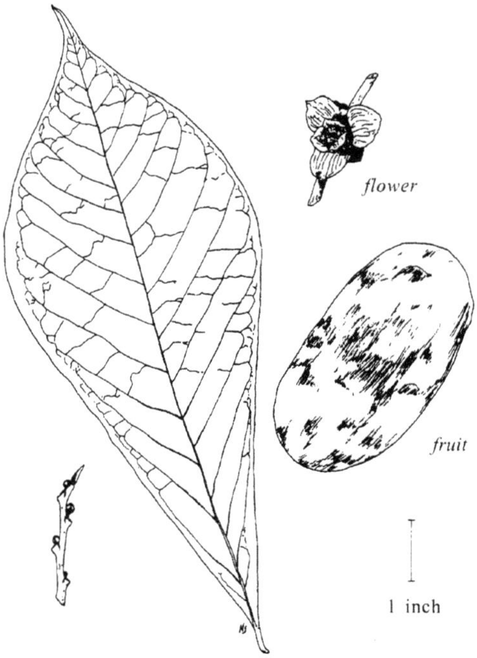

Asimina triloba

Fruits: cylindric, irregular; about 4 in (10 cm) long, 1.5 in (4 cm) thick; greenish yellow becoming brown, soft at maturity, sweet, edible; seeds embedded in flesh, oval, flattened, up to 0.8 in (2 cm) long

Twigs: red-hairy, becoming smooth; pith diaphramed, becoming chambered; leaf scars horseshoe-shaped

Buds: naked; terminal flat, elongated, rusty hairy, about 0.5 in (12 mm) long; lateral about 0.1 in (3 mm) long, accessory bud above true lateral (superposed); flower buds stalked, globose, rusty-hairy

Bark: thin; irregular fissures

Order Laurales

Recognition

Habit: trees, shrubs, woody vines; deciduous or evergreen

Leaves: opposite or alternate, simple, seldom compound; margins entire, rarely toothed or lobed

Calycanthaceae Lindl. - sweetshrub family

The sweetshrub family is comprised of two genera.

Recognition

Habit: shrubs; deciduous or evergreen
Leaves: opposite; entire; without stipules
Flowers: solitary; bisexual; perianth parts (sepals and petals) not differentiated, numerous; stamens 5-30, anthers facing outward
Fruits: akenes, capsule-like at maturity; seeds 1
Bark: aromatic

Calycanthus L. - sweetshrub

Four species, all native to North America, have been recognized. Dirr (1990) states that the three eastern species are so similar that characteristics of all three species can be found in any one population. The West Coast species, *C. occidentalis* Hook. & Arn., differs primarily in its exposed leaf buds, whereas the buds of the other species are concealed by the petiole base.

Calycanthus floridus L. - Carolina allspice

This native of the East Coast of the U.S. is a trouble-free plant (Dirr 1990). Richley (1988) rates it as a good shrub for southwestern Nebraska if soil can be kept moist. A young plant growing in a shady location in Maxwell Arboretum at UNL is thriving. 'Katherine' is a yellow-flowered cultivar with exceptional fragrance (Dirr 1990).

Planning

Height: up to 8 ft (2.5 m)
Habit: shrub; deciduous; dense, upright, rounded top; June bloom, fall foliage yellowish
Site: partial to full shade; soil adaptable; zone 5
Use: hedge, border

Recognition

Leaves: egg-shaped to elliptic, up to 5 in (12 cm) long; apex often long-pointed, base wedge-shaped to rounded; glabrous but slightly rough above, densely hairy below, fragrant
Flowers: solitary in leaf axil, 2 in (5 cm) across; perianth dark reddish brown, fragrant
Fruits: capsule-like, urn-shaped structures containing many akenes
Twigs: glabrous, aromatic, compressed at nodes; leaf scars U-shaped, raised
Buds: terminal lacking; brown, hairy, without scales (naked), concealed by petiole base

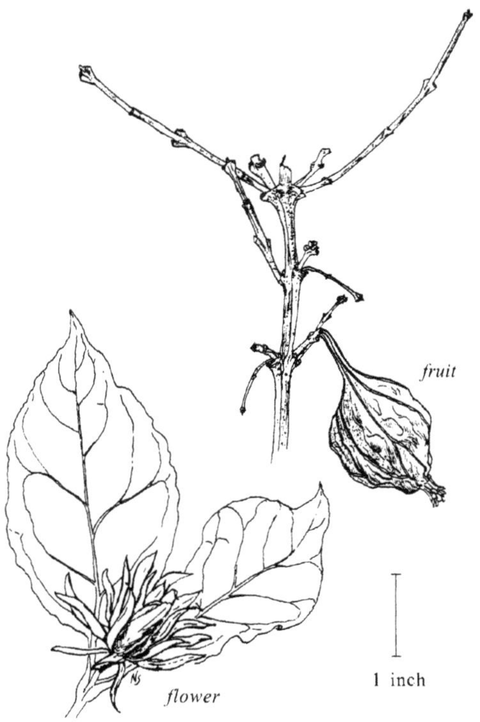

Calycanthus floridus

Lauraceae Juss. - laurel family

This family of 35 genera and 2,500 species (mostly tropical) is represented in North America by seven genera in the southeastern U.S. and one found in Oregon and California. Tropical members furnish fruits, such as the avocado; spices and fragrances such as cinnamon, camphor and perfumes; and valuable lumber suitable for cabinets and furniture. Most plant parts contain oils that provide a pleasant fragrance when broken or crushed.

Recognition

Habit: trees, shrubs; evergreen, rarely deciduous
Leaves: mostly alternate; simple; margins mostly entire, aromatic
Flowers: clusters; usually bisexual, sometimes unisexual; sepals 3, petals 3, look alike; stamens in 3-4 whorls of 3, some often sterile (staminodial), pistil 1, usually superior (hypogynous)
Fruits: drupes or berry-like

Sassafras Nees - sassafras

One of the three species of sassafras is native to North America; the other two are found in China and Taiwan.

Sassafras albidum (Nutt.) Nees - sassafras

Many senior citizens remember grandmother's prescription for almost every childhood ailment that ailed them as children from spring malaise to bronchitis—tea steeped from bark of sassafras tree roots. Sassafras tea is now considered a health hazard. Sassafras oil is used as a flavoring agent in beverages and for fragrance in perfumes. The bark is reputed to have insect repellant qualities. Chewing the bark supposedly aids in breaking the tobacco habit. Sassafras wood is light, durable and aromatic, and the grain is beautiful. It is easily shaped and takes a good finish—qualities that are desirable in furniture and carving. Paddles made of sassafras wood are prized possessions of many avid canoers.

Sassafras is a tree of the open woodlands and a pioneer in abandoned fields. It is native to the eastern U.S. from Maine to Florida, west to Michigan and southwest to eastern Oklahoma and Texas. In the southeastern prairie sassafras is found as a small tree or large shrub. Richley (1988) reports that chlorosis can cause problems for sassafras in southwestern Nebraska. Success of plantings and fall foliage color are undoubtedly largely determined by seed origin. Provenance research could identify varieties adapted to specific sites.

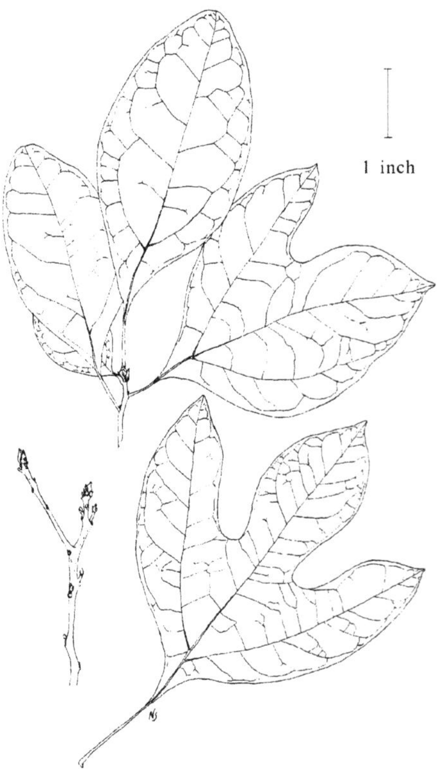

Sassafras albidum

Planning

Height: up to 65 ft (20 m); Nebraska champion, 1981, 58 ft (17.5 m), Arbor Lodge Historical Park, Nebraska City; National champion, 1982, 76 ft (23 m), Kentucky
Habit: tree; deciduous; crown irregular; April bloom before leaves; colorful autumn foliage
Site: sun; pH 6.0-7.0; soil adaptable; zone 5
Use: naturalizing, wildlife habitat

Recognition

Leaves: alternate; egg-shaped, up to 6 in (15 cm) long; entire or with 1-3 mitten-like lobes; 3-nerved at base; tips variable, base wedge-shaped; hairy young becoming glabrous, waxy bloom beneath; petioles hairy
Flowers: clusters (corymbs) terminal; unisexual, usually dioecious; fragrant; hairy bract at base; perianth yellow; male with 9 stamens, anthers 4-valved; female with 6 sterile stamens (staminodes)
Fruits: olive-like drupe, dark blue; about 0.5 in (12 mm) long; stalks scarlet, expanded, forming cup-like base; ripening in September and falling quickly
Twigs: hairy or glabrous; yellow-green becoming red-brown
Buds: egg-shaped, scales 3-5; terminal about 0.5 in (12 mm) long, lateral about 0.1 in (2 mm) long
Bark: greenish brown, smooth, becoming dark red-brown and fissured
Wood: moderately heavy and hard, aromatic; durable when in contact with soil

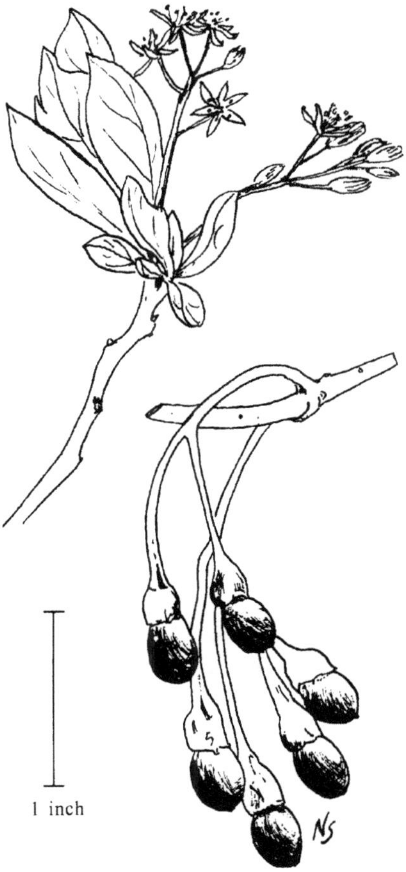

Sassafras albidum

Lindera Thunb. - spicebush

Two of the 60 species of *Lindera* are native to North America. The others are found in eastern and southern Asia. Dirr (1990) notes that Japanese spicebush, *L. obtusiloba* Bl., is a beautiful, large shrub or small tree that withstands temperatures to -17 degrees F (-25 degrees C). Japanese spicebush is distinguished by leaves that are three-nerved at the base and its black fruit.

Recognition

Habit: trees, shrubs; deciduous or evergreen
Leaves: alternate; entire or 3-lobed
Flowers: axillary clusters, with involucre of 4 deciduous scales; unisexual, or both sexes on same plant (polygamous); sepals usually 6, petals 0; male with 9 stamens; female with globe-shaped ovary, 9-15 sterile stamens (staminodes)
Fruits: drupes, nearly round

Lindera benzoin (L.) Bl. - spicebush

Spicebush is native to Maine and Ontario west to Michigan and south to southeastern Kansas and Texas, and east to the coast. It inhabits low, moist woodlands, especially along streams. This spicy, fragrant shrub with brilliant red fruits should thrive on moist protected sites in the southeastern prairies.

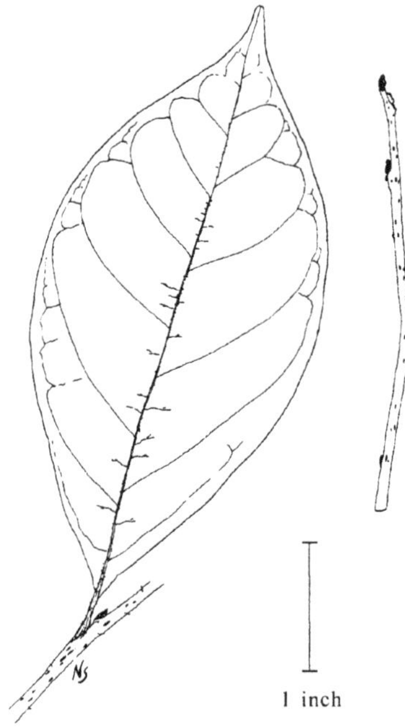

Lindera benzoin

Planning

Height: up to 10 ft (3 m)
Habit: shrub; deciduous; rounded to open; April bloom yellow, before leaves; September fruit red, persistent
Site: full sun, partial shade; pH 5.5-6.5; moist soil preferred; zone 5
Use: naturalizing, border

Recognition

Leaves: oblong, up to 5 in (13 cm) long, often wider above middle; veins branching along central midrib (penniveined); tip pointed, base wedge-shaped; surfaces sparingly hairy, sometimes hairs along margins (ciliate); petioles about 0.5 in (12 mm) long; stipules none; aromatic
Flowers: clusters dense, nearly sessile; unisexual (dioecious); perianth greenish yellow
Fruits: elliptic-oblong, about 0.4 in (1 cm) long; scarlet, smooth and glossy; strongly aromatic
Twigs: green or olive-brown, glabrous; lenticels prominent, pale
Buds: terminal lacking; leaf bud narrow egg-shaped, single or clustered; flower bud globose, stalked, slightly above leaf buds
Bark: gray-brown, smooth; lenticels raised

Order Aristolochiales

This order is comprised of one family.

Aristolochiaceae Juss. - birthwort family

Most members of this family of eight to ten genera and about 600 species inhabit tropical and subtropical regions. The genus *Aristolochia* accounts for most of the species; many of the remainder are found in the genus *Asarum* (Cronquist 1981).

Recognition

Habit: herbs, shrubs, often twining; deciduous or evergreen
Leaves: alternate, simple, mostly entire; often palmately veined; without stipules
Flowers: solitary or clusters; bisexual; calyx 3-lobed to lobeless, corolla wanting or much reduced; stamens 4 to many, usually free
Fruits: usually dry, splitting (capsules)

Aristolochia L. - birthwort

Birthwort is a genus of about 580 species found in the temperate and tropical regions of both hemispheres.

Recognition

Habit: herbs, shrubs, often twining; deciduous or evergreen

Flowers: axillary, symmetrical; perianth usually a long curved tube inflated at base, constricted at mouth, then enlarged and spreading; stamens usually 6 fused to the style

Fruits: capsules, 6-valved

Aristolochia durior Hill. - Dutchman's pipe

The heyday of Dutchman's pipe passed with the demise of the front porch and the advent of air conditioning. Dirr (1990) reports seeing many sunscreens fashioned from trellises covered with this vigorous vine. Dutchman's pipe is native to the eastern U.S., from Pennsylvania to Georgia, and west to Kansas and Minnesota.

Stephens (1973) states that *A. tomentosa* Sims is the species native to eastern Kansas. Leaves are smaller than those of *A. durior* and are hairy. All flower parts are hairy and the lobes (limbs) of the perianth are yellowish. Twigs are covered with dense woolly hairs (tomentose). We have no evidence that this species is better adapted to the prairies than *A. durior* is.

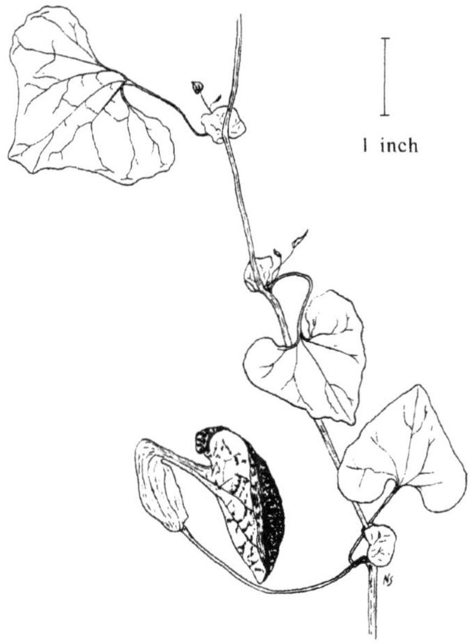

Aristolochia durior

Planning

Height: up to 30 ft (9 m)

Habit: vine, twining; deciduous; vigorous; blooms May-June

Site: sun, partial shade; soil adaptable; zone 5

Use: trellis, screen

Recognition

Leaves: heart- or kidney-shaped, 4-10 in (10-25 cm) across, tip rounded to pointed; glabrous above, hairy below, becoming glabrous; petioles about 1.5 in (4 cm) long

Flowers: 1-2 in leaf axils, stalk about 1.5 in (4 cm) long with a bract below the middle; perianth u-shaped about 1.2 in (3 cm) long, yellow-green and glabrous outside, constricted at base spreading into 3 brownish purple lobes (limbs) about 0.8 in (2 cm) across

Fruits: 2-3 in (50-75 mm) long

Twigs: green, glabrous, nodes swollen; leaf scars horseshoe-shaped, 3 bundle scars; pith white, greater than half the diameter of the stem

Buds: 3, one above the other (superposed), partially enclosed in stem tissue and dense white hairs (tomentum)

Order Illiciales

This order is represented by two families, Illiciaceae and Schisandsraceae, and about 90 species (Cronquist 1981). Illiciales is not recognized by Rehder (1940).

Recognition

Habit: trees, shrubs, vines; deciduous or evergreen
Leaves: alternate, simple; stipules none
Flowers: solitary or in small clusters, usually axillary; bisexual or unisexual; regular, perianth parts 5 to many, not clearly differentiated into sepals and petals; ovary superior (hypogynous)

Schisandraceae Bl. - schisandra family

This family of two genera is represented in the southeastern U.S. by a single species, *Schisandra glabra* (Brickell) Rehd. (Cronquist 1981). Schisandraceae is not recognized by Rehder (1940).

Recognition

Habit: vines; evergreen or deciduous; margins usually toothed
Fruits: berries on elongated receptacles forming a drooping spike

Schisandra Mich. - **schisandra**

About 12 species are placed in this genus (Rehder 1940). Rehder places this genus in Magnoliaceae.

Recognition

Habit: vines, twining; evergreen or deciduous
Leaves: entire or denticulate; without stipules
Flowers: axillary clusters, few flowered; dioecious; sepals and petals not differentiated
Fruits: berries on elongated receptacle forming drooping spike
Buds: scales several, overlapping (imbricate)

Schisandra chinensis (Turcz.) Baill. - **Japanese schisandra**

This native to Japan and northeastern Asia has not been widely planted in the Great Plains region. In 1991 the Nebraska Statewide Arboretum established trial plantings of Japanese schisandra in Nebraska. Seedling plants in Lincoln and Holdrege, Nebraska, survived the winter of 1991-92 with slight dieback and made excellent growth the following summer (Gibson 1992). These plants continue to survive in a shaded site at Prairie Pines Arboretum near Lincoln, but growth is slow (Bagley 1998).

Planning

Height: potential unknown
Habit: vine; deciduous
Site: experimental
Use: trellis

Schisandra chinensis

Recognition

Leaves: broadest at middle or above; tip pointed, base wedge-shaped; margin fine-toothed (denticulate); shiny above, green beneath
Flowers: clusters axillary; perianth creamy white or pinkish, fragrant; stamens 5
Fruits: scarlet
Twigs: glabrous with vertical ridges, lenticels prominent, pith spongy, green; leaf scars raised, bundle scars 3, closely grouped
Buds: solitary, stalkless, 0.1 in (2 mm) long, egg-shaped; scales overlapping, sparsely short-hairy with hairy margins (ciliate)

Order Ranunculales

Eight familes and about 3,200 species occupy the order Ranunculales (Cronquist 1981).

Recognition

Habit: herbs, vines, shrubs, seldom small trees; deciduous
Leaves: alternate, sometimes opposite; simple or compound; stipules not present or very small and early deciduous
Flowers: variously clustered; bisexual or unisexual; perianth variable, ovary superior
Fruits: various

Ranunculaceae Juss. - buttercup family

This worldwide family of 50 genera and 2,000 species is comprised mostly of herbaceous plants that are widely used as ornamentals. A few are woody climbers or shrubs (Cronquist 1981).

Recognition

Habit: herbs, vines, shrubs
Leaves: alternate or opposite; simple, often dissected or variously compound; stipules usually wanting
Flowers: variously clustered, sometimes solitary; usually bisexual; petals variable in number or wanting; sepals often petal-like (petaloid), often early deciduous; stamens free and numerous
Fruits: dry or fleshy, usually follicles, achenes or berries

Clematis L. - clematis

Herbaceous and woody forms are included in this genus. About 230 species native to the temperate regions of the Northern Hemisphere are found in *Clematis*. Only those woody shrubs and vines that are most likely to be grown successfully in our prairies are described in detail here.

C. texensis Buckl., scarlet clematis, and *C. pitcheri* Torr. & Gr., purple leatherflower, are native to the southern prairie. Both are difficult to propagate and are not widely commercially available.

Clematis texensis

Exotics that will likely adapt to parts of the prairies include *C. montana* DC., anemone clematis, a vigorous white-flowering species, and *C. viticella* L., vinebower, with purple to violet flowers. Many cultivars have been selected from these species.

Recognition

Habit: shrubs, vines; deciduous
Leaves: opposite; usually pinnately compound, often ending in tendrils (twining petioles); rarely simple
Flowers: solitary or clusters (panicles or cymes); bisexual or rarely unisexual and dioecious; without petals; sepals usually 4, rarely 5-8, petal-like (petaloid); stamens numerous, petal-like staminodes (sterile stamens) sometimes present; carpels many
Fruits: dry, 1-seeded, non-splitting (akenes); styles persistent, usually long and plumose
Twigs: woody, to woody only at the base (suffrutescent)

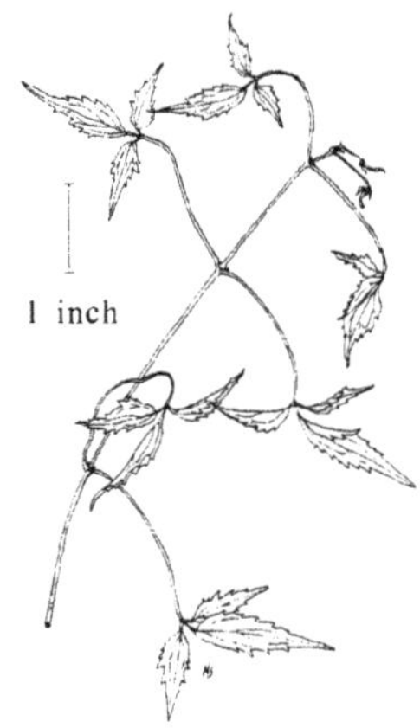

Clematis montana

Grouping Key for *Clematis*
(Rehder 1940)

Subshrubs; leaves simple or ternate, entire or toothed; flowers solitary and terminal or axillary clusters; sepals upright, margin narrow, or broad and spreading above middle; stamens upright, usually hairy: **section Viorna**

Climbers; leaves ternate to twice-ternate; flowers axillary, solitary, nodding, bell-shaped; petaloid staminodes; stamens hairy: **section Atragene**

Climbers; leaves simple to twice-pinnate; leaflets entire or somewhat lobed; flowers 1-3, axillary or terminal; sepals spreading, 4-8; stamens glabrous or a few hairs below the anthers: **section Vitacella**

Upright herbs to climbing shrubs; leaves simple to twice-pinnate; flowers white, pink or yellow; sepals not overlapping (valvate), margin usually narrow; stamens spreading, glabrous or hairy at base: **section Flammula**

Section Viorna

Clematis heracleifolia var. *davidiana* (Decne.) Hemsl. - David's fragrant tube-flower clematis

This Asian species differs from typical members of *Clematis* in its form and flower characteristics. Flowers of variety *davidiana* are larger and more showy than those of the species, and the female and male flowers are borne on separate plants (dioecious). Sepals are light blue and spreading rather than tubular. Persistent plumose styles on the fruits of female plants add interest in autumn and winter though they become somewhat frazzled with age.

This clematis is more appropriate as a part of an herbaceous perennial border than for use with shrubs because of its tendency to die back in winter. Pruning back to live wood is a spring chore that will be amply rewarded with vigorous growth and a profusion of light blue flowers in August. Plants at Prairie Pines Arboretum near Lincoln are well adapted to sunny and partly shaded sites (Bagley 1998).

Planning

Height: up to 3 ft (1 m)
Habit: subshrub; dies back, woody at base, vigorous regrowth; wide spreading; August flowers fragrant
Site: sun, part shade; moist soil; pH 6.5-8.0; zone 5
Use: specimen, border, grouping

Recognition

Leaves: compound; leaflets 3, broad egg-shaped, up to 6 in (15 cm) long; slightly hairy; base wedge-shaped; coarsely toothed, often slightly lobed
Flowers: clusters axillary; unisexual or bisexual on same plant (polygamous); sepals light blue, tubular at base, spreading above, up to 1 in (25 mm) long, hairy outside; stamens glabrous or sparingly hairy, filaments about as long as anthers
Fruits: clusters of akenes; styles plumose, persistent
Twigs: green, hairy, becoming glabrous

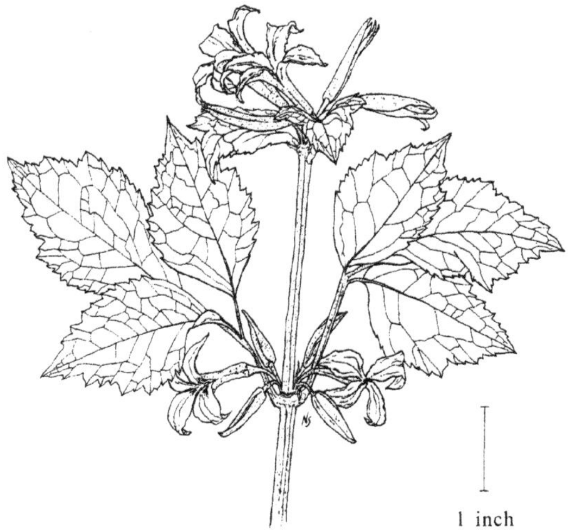
1 inch

Clematis heracleifolia

Clematis integrifolia L. - simple-leaved clematis

This low, bushy plant is considered to be herbaceous in the northern prairies. Although used in breeding, it is rarely cultivated (Hoag 1965).

Planning

Height: up to 3 ft (1 m)
Habit: subshrub; upright
Site: experimental
Use: naturalizing

Recognition

Leaves: simple; egg-shaped, margins entire; sometimes hairy beneath, thin; stalkless (sessile)
Flowers: solitary, terminal; bisexual; bell-shaped, about 1.5 in (4 cm) long, violet to bluish
Fruits: akenes; styles plumose, persistent

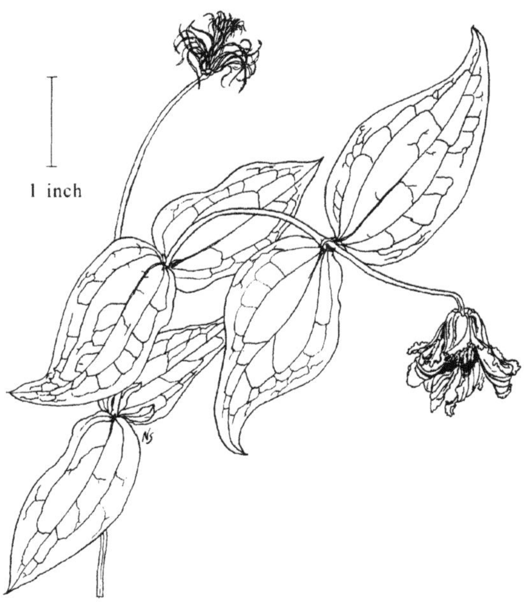
1 inch

Clematis integrifolia

Section Atragene

Clematis macropetala Ledeb. - **large-flowered clematis**

Large-flowered clematis occupies a space on a trellis in Maxwell Arboretum at UNL. It is closely related to *C. alpina* (L.) Mill.

Clematis macropetala

Planning

Height: undetermined
Habit: shrubby climber; styles plumose
Site: experimental
Use: trellis

Recognition

Leaves: ternately compound; leaflets deeply 3-lobed, margins coarsely toothed
Flowers: solitary, axillary; violet, staminodes widest at base (lanceolate) tapering to a long point (acuminate), nearly as long as sepals, longer than stamens
Fruits: akenes; styles plumose

Section Vitacella

Clematis ×jackmanii T. Moore - **Jackman clematis**

Jackman clematis, a gardener's favorite, originated in 19th-century England as a cross between *C. lanuginosa* Lindl. and *C. viticella* L. (Rehder 1940). Dirr (1940) has divided Jackman clematis cultivars into five groups, each with different parentage: Jackman, Lanuginosa, Patens, Viticella and Florida. Since many cultivars have been derived from the Jackman hybrids—more than can be adequately described in this text—we recommend **Clematis, The complete Guide** by Ruth Gooch (1996) and/or **Clematis, Queen of the Climbers** by Jim Fisk (1994).

'Madame Edouard Andre', with its rich crimson, pointed sepals and yellow stamens, is a long time favorite cultivar. 'Henryi' displays large white flowers on old wood in late spring and on new wood in late summer.

Large white, pink or purple flowers bloom from June until the first frost of autumn if soil is cool and moist and leaves receive lots of sun. Deep organic mulches are desirable to protect roots from temperature extremes and moisture fluctuations. Cultural methods to promote best flowering and growth differ among the cultivars. It is important to know the characteristics of each cultivar and cultural techniques best suited for each—points that are covered in the two reference sources mentioned above.

Clematis viticella

Planning

Height: up to 12 ft (4 m)
Habit: twining vine; dies back in winter, summer blooms on new wood
Site: sun, partial shade; pH 6.0-7.5; moist soil; zone 5
Use: trellis, rock wall, fence

Recognition

Leaves: pinnately compound, upper often simple; leaflets egg-shaped, often slightly hairy beneath, margins entire
Flowers: solitary or clusters of 3, 4-7 in (10-18 cm) across; bisexual; sepals 4, sometimes 6, egg-shaped, spreading, usually violet-purple, showy; stamens divergent, filaments often shorter than anthers, sometimes hairy
Fruits: akenes; plumose styles persistent, up to 2 in (5 cm) long
Buds: egg-shaped; scales, 1-3 pairs, hairy
Twigs: stems slender, ridged; pith angular or star-shaped, white

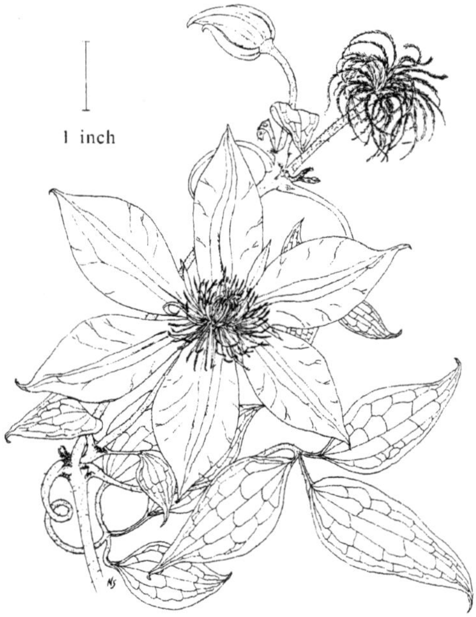

***Clematis* ×'Henryi'**

Section Flammula

Clematis terniflora DC. - sweetautumn clematis

This import from Japan is a "rampant, rampaging vine which engulfs every structure in sight" (Dirr 1990). Sweetautumn clematis, a sleeping beauty during most of the spring and summer, lives up to its name from August through September when it explodes with a fragrant blizzard of small, white blossoms.

Richley (1988) says that sweetautumn clematis is a very nice vine in southwestern Nebraska. Synonyms are *C. paniculata* (Rehder 1940) and *C. maximowicziana* (Dirr 1990) and *dioscoreifolia* var. *robusta* (Carr.) Rehd.

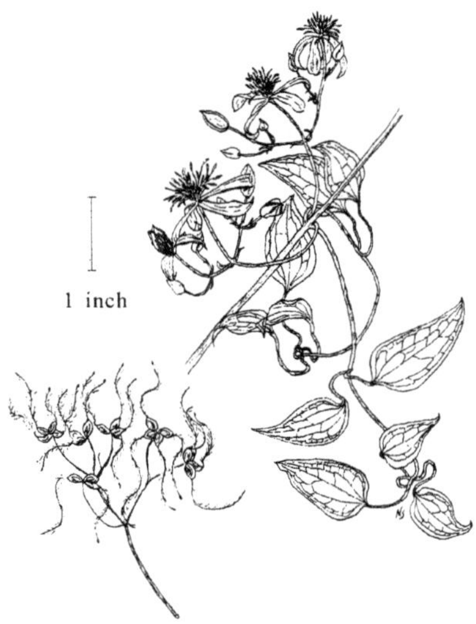

Clematis terniflora

Planning

Height: up to 10 ft (3 m) long
Habit: vine; climbing by twining petioles, August flowers fragrant, white
Site: sun, partial shade; moist soil; pH 6.5-8.0; zone 5
Use: fence, trellis

Recognition

Leaves: pinnately compound; leaflets 3-5, egg-shaped, up to 4 in (10 cm) long; apex pointed, base rounded to heart-shaped; margin entire or sometimes lobed
Flowers: clusters (panicles) terminal and axillary; bisexual; white, about 1.2 in (3 cm) across; sepals 6-8, spreading, oblong, nearly glabrous outside; stamens divergent, filaments longer than anthers, glabrous or nearly so
Fruits: akenes margined, narrowed at base, thinly appressed

Ranunculaceae *Clematis*

Clematis ligusticifolia Nutt. - western virginsbower

Western virginsbower is native to the western U.S., from New Mexico north through western Nebraska to Alberta and British Columbia, Canada, and west to California. It grows on wooded or brushy hillsides and creek banks under a variety of soil conditions, and climbs up to 30 ft (10 m) long over rocks and bushes by twining and using its tendril-like leaf petioles (Stephens 1973). White flowers and misty plumed fruit add interest to the landscape from mid- through late summer. Nearby male plants are necessary for fruit development.

Planning

Height: up to 30 ft (10 m)
Habit: vine, much branched, twining; woody at base (suffrutescent); summer flowers, July-September fruit
Site: sun, partial shade; pH 6.5-7.5; soil adaptable; zone 4
Use: naturalizing, wildlife habitat, trellis, screen

Recognition

Leaves: pinnately compound, leaflets 5-7, egg-shaped, up to 2.5 in (65 mm) long; 3-5 veins from base connecting with lateral veins from midrib; tips long-pointed, margins coarsely toothed or lobed; surfaces glabrous or with few straight stiff hairs; petioles ribbed, short hairy, rib on underside extending down onto stem
Flowers: clusters axillary, hairy bract at base of each branch; unisexual (dioecious); sepals 4-5, elongated, white, reflexed or spreading, minutely hairy on both sides; male, stamens, divergent, glabrous, anthers elliptic, several times longer than filaments; female with numerous sterile stamens, ovary silky-hairy
Fruits: akenes flat, short-hairy; style elongated, white, plumose
Twigs: somewhat 6-ribbed, glabrous
Buds: egg-shaped, dark brown; scales, outer slightly keeled
Bark: shreddy, gray-brown; inner yellow

Clematis orientalis L. - Oriental clematis

Oriental clematis, a native of the Himalayas, has escaped cultivation and become naturalized in the central Rockies. This slender climber with attractive yellow flowers has survived temperatures to -30 degrees F (-34 degrees C) according to records of the North Central Plant Introduction Station (1970).

Planning

Height: up to 15 ft (5 m)
Habit: vine, twining; decorative flowers and fruit
Site: sun, partial shade; pH 6.5-7.5; moist soil; zone 4
Use: naturalizing, wildlife habitat, screen

Recognition

Leaves: pinnate to trifoliate; leaflets egg-shaped to much longer than wide, up to 2 in (5 cm) long; lobed to coarsely toothed; surfaces glabrous or minutely hairy
Flowers: solitary or few, about 1.5 in (4 cm) across; bisexual; sepals elliptic to oblong, pointed, spreading becoming somewhat reflexed, yellow, hairy inside and usually outside; stamens divergent, filaments hairy, longer than anthers
Fruits: akenes hairy, styles plumose

Clematis tangutica (Maxim.) Korsh. - golden clematis

This native to Mongolia and northwestern China is noted for its showy yellow flowers and feathery seed heads.

Planning

Height: up to 10 ft (3 m)
Habit: vine; yellow summer bloom, feathery autumn fruits
Site: part to full shade; soil adaptable; zone 4
Use: trellis, screen

Recognition

Leaves: pinnately compound; leaflets long, narrow, wider at bottom (lanceolate); coarsely toothed (serrate), sometimes deeply lobed; surfaces glabrous
Flowers: usually solitary, stalks hairy, 3-6 in (8-15 cm) long; flowers bright yellow; sepals narrow egg-shaped, about 1.5 in (4 cm) long; open bell-shaped, becoming spreading
Fruits: akenes in a head about 2 in (5 cm) across; styles plumose

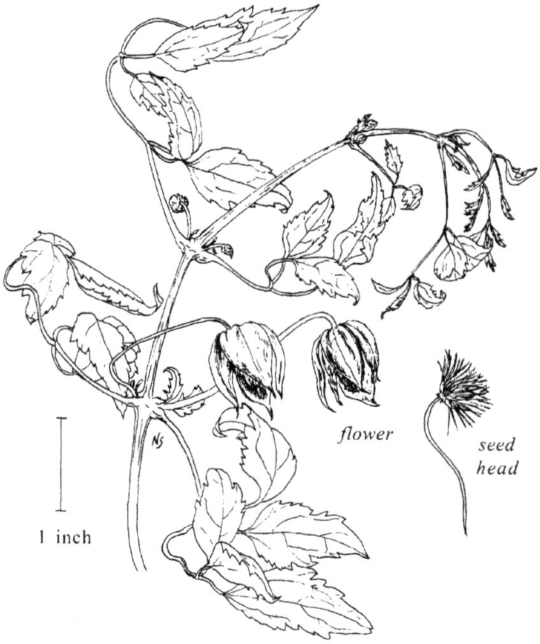

Clematis tangutica

Clematis virginiana L. - virginsbower

This native of the northeastern U.S. and southeastern Canada can also be found in the eastern prairies among woodland thickets.

Planning

Height: up to 15 ft (4.5 m)
Habit: vine, twining; woody at base (suffrutescent); decorative autumn fruits
Site: sun, partial shade; moist soil; zone 4
Use: naturalizing, wildlife habitat, screen

Recognition

Leaves: compound (ternate), rarely in 5's; leaflets egg-shaped, up to 3.5 in (9 cm) long; tips long-pointed, base rounded or somewhat heart-shaped, coarsely toothed, nearly glabrous
Flowers: clusters axillary, leafy; about 1 in (25 mm) across; unisexual (dioecious); dull white; sepals spreading, oblong, wider above the middle to spoon-shaped (spatulate), hairy outside; stamens divergent, anthers several times longer than filaments, elliptic
Fruits: akenes in a head, about 2 in (5 cm) across, hairy; styles plumose
Twigs: straw-colored, finely hairy

Xanthorhiza Marsh. - yellowroot

The only species is native to North America from New York to Kentucky and Florida.

Xanthorhiza simplicissima Marsh. - yellowroot

Yellowroot, a source of yellow dye for early Native Americans, has been slighted by modern American gardeners (Dirr 1990). This little-known plant is a potential groundcover for shady areas where the soil is cool and moist. The interesting, oddly colored flowers and mass of bright green foliage is attractive.

Planning

Height: up to 2 ft (0.6 m)
Habit: shrub; deciduous; flat-topped, stems erect; thicket forming
Site: part sun, shade; avoid high pH; moist, cool soil; zone 5
Use: naturalizing, groundcover, streambank erosion control

Recognition

Leaves: alternate, clustered, pinnately or bi-pinnately compound; leaflets egg-shaped to approaching oblong, up to 2.7 in (7 cm) long; toothed, lustrous bright green
Flowers: clusters (compound racemes) terminal, often before leaves; unisexual and bisexual (polygamous); star-shaped, about 0.2 in (4 mm) across; petals wanting, sepals brownish purple; stamens 5-10
Fruits: dry, splitting along one suture (follicles); seeds one
Twigs: greenish gray; inner bark yellow
Roots: yellow, bitter

Berberidaceae Juss. - barberry family

This family of 13 genera is widespread in the temperate regions of the Northern Hemisphere. *Berberis*, the largest genus, also extends south through the Andes to the Straits of Magellan (Cronquist 1981).

Intergeneric crosses between *Berberis* and *Mahonia*, two closely related genera, have produced a genus dubbed ×*Mahoberberis* Schneid. Several species have been produced from this hybrid with quite variable foliage and sparse to nonexistent flowers and fruits (Dirr 1990). These plants, though botanically interesting, are not widely sold by nurseries because of their undesirable characteristics.

Recognition

Habit: herbs, shrubs; evergreen or deciduous
Leaves: alternate; simple or compound, often spiny-toothed; petioles dilated at base or with small vestigial stipules
Flowers: solitary or in clusters (cymes, racemes or panicles); bisexual; sepals and petals overlapping, often similar, usually in whorls of 3, rarely 2 or 4, petals sometimes lacking; stamens 4-9, opposite petals; ovary superior (hypogynous)
Fruits: usually berries, seldom dry
Twigs: spiny or unarmed

Berberis L. - barberry

Barberries, with their sharp spines and dense growth, are excellent barriers. Their handsome summer foliage, colorful flowers, bright autumn foliage and fruits enjoyed by birds are excellent characteristics for border plantings and wildlife habitat groups. Because barberries become unkempt with age, removal of old canes is desirable.

Many of the 175 barberry species, including American barberry (*B. canadensis* Mill.) and common barberry (*B. vulgaris* L.), are alternate hosts for stem rusts that affect cereals. Propagation and distribution of these species is banned in regions where wheat and other grains are important agricultural crops. Barberries described here, mostly of Asian origin, are immune or highly resistant to these diseases.

Evergreen species (such as *B. candidula* Schneid., paleleaf barberry) are omitted from this text even though some are adapted to the southern prairies (zone 6).

Recognition

Habit: shrubs, rarely small trees; evergreen or deciduous
Leaves: simple, often clustered at nodes with spines on short axillary spurs
Flowers: solitary or in clusters; yellow; sepals 6; petals 6, often smaller than sepals; stamens 6; stigma sessile or on short style
Fruits: berries, red to black; seeds 1 to several
Twigs: bearing spines, often 3-parted; inner bark and wood yellow

Berberis koreana Palib. - Korean barberry

Korean barberry, exotic to the U.S., came from plants found north of Seoul, Korea. It is well adapted to most soil and climatic conditions of the prairies, according to results of extensive field trials of the North Central Plant Introduction Station (1970). Korean barberry thrives at Fort Collins (Fairchild and Klett 1993).

A related species, *B. gilgiana* Fedde, from North Central China is an attractive barberry worthy of trial in the prairies.

Planning

Height: up to 6 ft (2 m)
Habit: shrub; deciduous; multistemmed oval, dense, suckering; thorny; yellow flowers mid-May after leaves; autumn foliage reddish purple to crimson-orange; bright red fruits persistent through winter
Site: full sun, part shade; drouth tolerant; soil well-drained; pH 6.0-8.0; zone 4
Use: hedge, barrier, mass, specimen

Recognition

Leaves: elliptic or wider above middle (obovate), up to 3 in (75 mm) long; apex rounded, base wedge-shaped; minutely toothed, somewhat spiny; surfaces glabrous, net-veined below (reticulate)
Flowers: clusters (racemes), up to 4 in (10 cm) long; yellow
Fruits: red, egg-shaped, about 0.3 in (7 mm) long
Twigs: grooved, glabrous, reddish to purple-brown, grayish waxy coating (bloomy); spines usually simple

Berberis ×mentorensis L. M. Ames - Mentor barberry

This hybrid (*B. julianae* Schneid. × *B. thunbergii*) combines the evergreen leaves of the former with the hardiness of the latter. Mentor barberry is half-evergreen in the central prairies, where foliage remains green until late fall when it becomes brown, usually without bright color. Mentor barberry is subject to frequent and sometimes severe freeze-back in the northern prairie (Hoag 1965).

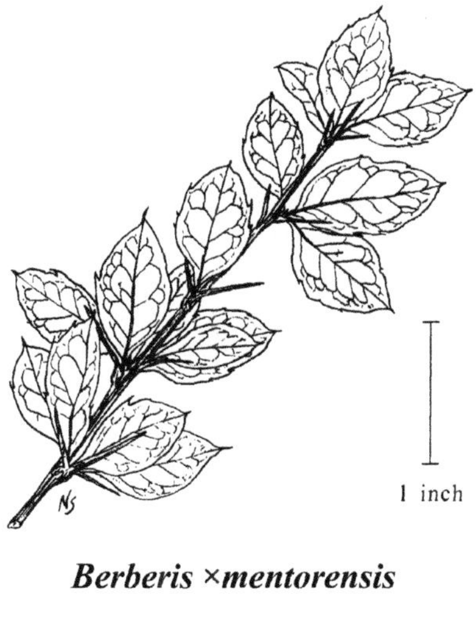

Berberis ×mentorensis

Berberidaceae *Berberis*

A foundation planting on the west side of the Plant Industry Building, East Campus, UNL, maintained excellent health and vigor for over 30 years with only an occasional removal of all top growth. Flowers and fruits were never evident.

Planning

Height: up to 6 ft (2 m)
Habit: shrub; semi-evergreen; spiny; upright, stiff; inconspicuous flowers late April, early May
Site: sun, partial shade; pH 6.0-7.5; heat tolerant; zone 5
Use: hedge, barrier, grouping, foundation

Recognition

Leaves: somewhat leathery, elliptic to egg-shaped; apex minutely pointed, sparingly spinulose toothed; surfaces glabrous; persistent into late fall
Flowers: solitary or paired, yellow
Fruits: berries dull red, ellipsoidal
Twigs: grooved, glabrous, greenish red, becoming brown; 3-spined

Berberis thunbergii DC. - Japanese barberry

Japanese barberry, a versatile work horse of the barberry family, is adapted to a wide variety of site conditions. The extensive natural range (from southern Europe to Japan) contributes to its great genetic diversity. It withstands drouth and tolerates urban conditions better than most shrubs (Dirr 1990). However, high alkalinity will cause chlorosis (North Central Plant Introduction Station 1968). Hoag (1965) reports some winter dieback of Japanese barberry in the northern Great Plains, especially if moisture is limited.

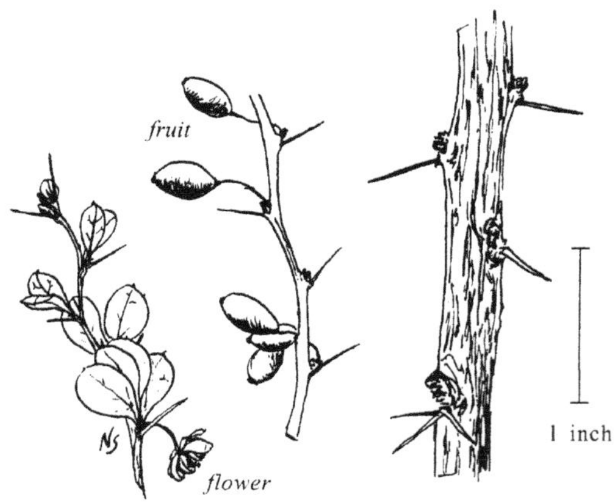

Berberis thunbergii

Many cultivars of differing growth rates and branching habits have been selected from this normally multistemmed compact shrub. Summer foliage color ranges from bright green to the red-purple of variety *atropurpurea*. 'Erecta', an upright green-leafed cultivar, suffers minor twig dieback and patches of minor chlorosis, but growth is dense and fruit set is consistent at PERC, Fort Collins (Fairchild and Klett 1993).

'Kobold'and 'Globe', dwarf plants with rounded, compact growth and bright green leaves, are substitutes for Japanese holly and boxwood on sunny sites in the southeastern prairies. 'Jewell' is a nearly thornless plant with the excellent foliage of the species. 'Aurea', with its leaves of bright yellow, and 'Variegata', with leaves of various colors, can add variety to our landscapes.

Selections of variety *atropurpurea* include the dwarf 'Crimson Pygmy' that is adapted to the southern Great Plains. They suffer from winter dieback at Fort Collins. 'Rose Glow', a taller cultivar selected for its variegated foliage, has been grown successfully on sites from Fort Collins to Urbana, Illinois, to Athens, Georgia (Fairchild and Klett 1993, Dirr 1990). 'Gold Ring', with its dark red, green-margined leaves and 'Marshall', with its dark red foliage, are abundantly fruitful at the PERC site (Fairchild and Klett 1993).

Planning

Height: up to 5 ft (1.5 m)
Habit: shrub; deciduous; rounded, often broader than tall; foliage green changing to shades of red and purple in autumn (var. *atropurpurea*, summer foliage red-purple); persistent bright red fruit
Site: sun, partial shade; soil and pH adaptable; drouth tolerant; zone 4
Use: hedge, barrier, grouping, wildlife habitat

Recognition

Leaves: oblong to wider above middle, size quite variable; margin entire; tip obtuse, base narrowed into petiole; bright green, reddish purple or variegated, slightly covered with waxy bloom beneath (glaucescent)
Flowers: solitary or clusters 2-4; yellow, up to 0.4 in (1 cm) across
Fruits: berries bright red, ellipsoidal; up to 0.4 in (1 cm) long
Twigs: grooved, purplish red becoming brownish; spines usually simple
Buds: egg-shaped, solitary, sessile; scales 6

Mahonia Nutt. - grapeholly

This genus of 50 species is native to Asia, North America and Central America. Its fruits are grape-like and the holly-like leaves are evergreen. The similarity of grapeholly to evergreen true hollies (*Ilex*) is often a cause for mistaken identity even though the leaves of the former are compound and the latter simple. Some authorities do not recognize genus *Mahonia* and place these plants in genus *Berberis* (Stephens 1973).

Recognition

Habit: shrubs, rarely small trees; evergreen
Leaves: alternate, odd-pinnately compound or rarely with 3 leaflets; lateral leaflets sessile, spiny toothed
Flowers: usually many in a cluster; bisexual; yellow; sepals 9, petals 6
Fruits: berries; dark blue, rarely red or whitish; bloomy
Twigs: unarmed; wood yellowish

Mahonia aquifolium (Pursh) Nutt. - Oregongrape

The Oregongrape's glossy, evergreen foliage and bright yellow flowers made it an instant favorite as an ornamental shrub in the East after the Lewis and Clark Expedition brought it from "Oregon Country." The fleshy berries have been used in jellies and jams.

Oregongrape is reasonably prosperous in the southeastern prairies if planted in shady, protected areas. Dirr (1990) states that it is severely injured at -20 degrees F (-28 degrees C). Several cultivars have been selected including 'Compactum', which grows to about 3 feet (1 m). 'Golden Abundance' is very fruitful, and the young foliage of 'Orange Flame' becomes bright bronze-orange in full sun (Dirr 1990).

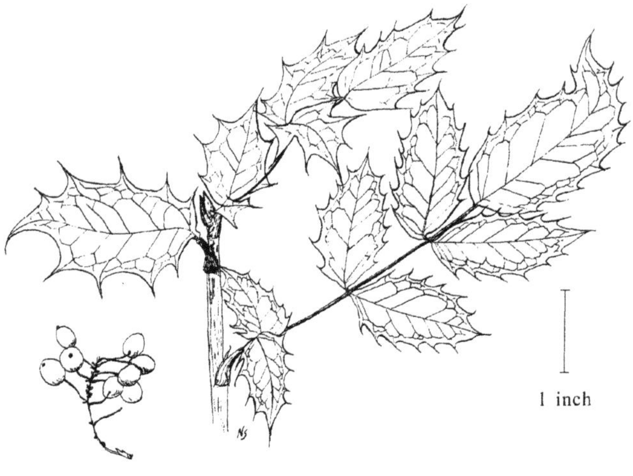

Mahonia aquifolium

Berberidaceae *Mahonia*

Planning

Height: up to 7 ft (2 m)
Habit: shrub; upright, stoloniferous; April bloom; fruit persists into late autumn
Site: shade, partial sun; moist soil; zone 6
Use: foundation, border, specimen

Recognition

Leaves: leaflets 5-9, egg-shaped, up to 3 in (75 mm) long; margins wavy, spiny-toothed; glossy green above, often purplish in winter; stiff and leathery
Flowers: clusters (racemes) fascicled, erect, terminal, up to 3 in (8 cm) long; yellow
Fruits: berries blue-black, bloomy, about 0.3 in (8 mm) across
Twigs: stout; leaf scars half encircling stem
Buds: lateral small, rounded, green; terminal up to 0.4 in (1 cm) long

Mahonia repens (Lindl.) G. Don - **creeping mahonia**

This stoloniferous groundcover occupies rocky niches on thinly wooded slopes of Nebraska's Pine Ridge, at the Black Hills, and in the Rocky Mountains west to Northern California and Oregon. The wide latitudinal natural range of creeping mahonia provides an excellent opportunity for selection of genetically diverse provenances.

Protection from the drying effect of winter wind and sun is essential for creeping mahonia to thrive. It is used in rock gardens and can be interplanted with tulips, daffodils and other spring blooming plants for an interesting springtime landscape effect (Sutton 1998).

Planning

Height: up to 1 ft (30 cm)
Habit: shrub; erect single stems, stoloniferous; fall foliage purple, fruits persistent
Site: partial shade; soil adaptable, pH 6-7.5; zone 4
Use: groundcover, rock garden
Caution: protect from wind

Recognition

Leaves: leaflets 3-7, egg-shaped, up to 2.4 in (6 cm) long; margins wavy and thickened, spiny-toothed (spinose-serrate); dull green above, purplish in winter; stiff and leathery
Flowers: clusters (racemes) fascicled, erect, terminal, up to 2.3 in (6 cm) long; flowers yellow, 3 egg-shaped bracts at base
Fruits: berries top-shaped; dark blue with heavy bloom, about 0.3 in (8 mm) across; seeds 5-7
Twigs: stout, leaf scars half encircling stem
Buds: lateral small, rounded, green; terminal up to 0.4 in (1 cm) long; scales green, often brown-edged, linear tooth on each side
Bark: yellow-brown, inner bright yellow

Lardizabalaceae Decne. lardizabala family

Eastern Asia and central Chile are home to this family of eight genera.

Recognition

Habit: vines, shrubs; deciduous or evergreen
Leaves: alternate, compound
Flowers: clusters axillary; unisexual (monoecious or dioecious), seldom bisexual; regular; sepals 3, 6 to many; petals 6, smaller than sepals or wanting; male with 6 stamens opposite petals, filaments often fused (connate); female, ovary superior (hypogynous), with or without 6 sterile stamens (staminodes)
Fruits: fleshy, often edible; splitting or not

Akebia Decne. - akebia

Two species of *Akebia* grow in eastern Asia.

Akebia quinata (Houtt.) Decne. - fiveleaf akebia

This vine, native to central China, Japan and Korea, is deciduous or evergreen, depending upon the site's winter temperatures. Richley (1988) rated fiveleaf akebia "a lovely, delicate vine" well adapted to southwestern Nebraska. Dirr (1990) says it is extremely adaptable and requires minimal care, "a fine choice for covering trellises, fences and other structures." A healthy specimen grew for many years on the south wall of Joslyn Castle, Omaha, until removed by the new owners.

Akebia trifoliata (Thunb.) Koidz., threeleaf akebia, bears three leaflets that may be toothed or entire. It is very similar to fiveleaf akebia in appearance, but is less vigorous and perhaps not as cold tolerant.

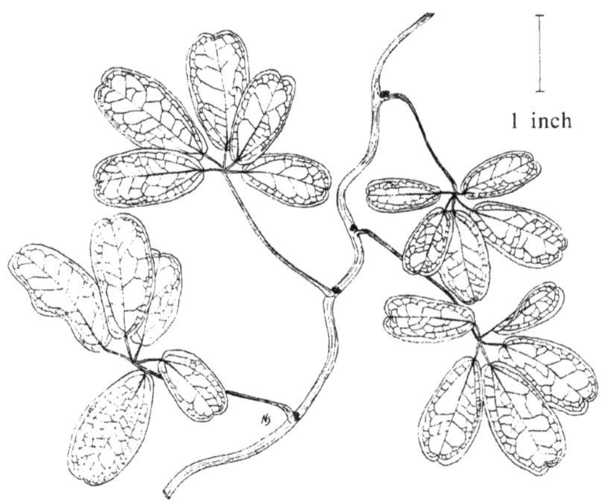

Akebia quinata

Planning

Height: up to 30 ft (9 m)
Habit: vine, twining; deciduous or evergreen; May flowers fragrant, seldom fruits
Site: sun, part shade; pH and soil adaptable; zone 5
Use: trellis, fence

Recognition

Leaves: palmately compound; leaflets 5, oblong or sometimes broadest above the middle (obovate); tip slightly notched (emarginate), base rounded or wedge-shaped (cuneate); surfaces glabrous, waxy bloom beneath
Flowers: clustered; unisexual (monoecious), sometimes polygamous; fragrant; sepals 3, petals none; female at base of raceme, about 1 in (25 mm) across, purplish brown; male smaller, at tip of raceme, rosy purple, stamens 6
Fruits: fleshy pods opening along ventral suture, about 3 in (8 cm) long; purple, bloomy, showy; seeds black, numerous, several rows
Twigs: rounded (terete); glabrous, green becoming brown, lenticels many; leaf scars raised, traces 6 or more
Buds: without stalks (sessile), egg-shaped; scales many, overlapping, glabrous

Menispermaceae Juss. - moonseed family

Most of the 70 genera (400 species) of the moonseed family are native to the tropics. In addition to *Menispermum canadense* (described below), two other species grow naturally on the fringe of the southeastern prairie—*Calycocarpum lyoni* (Pursh) Gray, cupseed, and *Cocculus carolinus* (L.) DC., snailseed (Stephens 1973).

Menispermum L. - moonseed

M. dauricum DC., a native of East Asia, is very similar to the North American species described below.

Menispermum canadense L. - moonseed

This twining vine grows along streams and ravines throughout much of eastern North America including much of the eastern prairie. Moonseed is a fast-growing, attractive vine with grape-like fruits that are potentially toxic to some people (Stephens 1973).

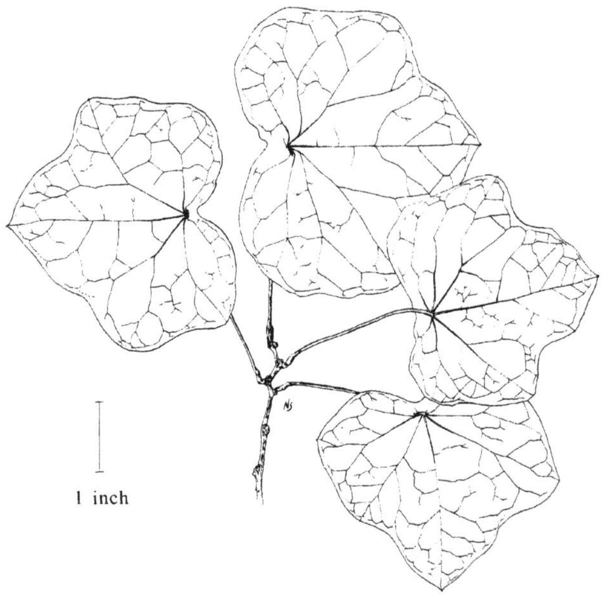

Menispermum canadense

Planning

Height: up to 50 ft (15 m)
Habit: vine, twining; deciduous
Site: sun; pH and soil adaptable; zone 5
Use: trellis

Recognition

Leaves: alternate, simple; nearly circular, often 3-7 lobed, some with an abrupt sharp tip (mucronate); margin not toothed; petioles 2-5 in (5-13 cm) long, attached to blade near the base (peltate)
Flowers: loose, drooping clusters (panicles) axillary, attached above the leaf; unisexual (dioecious); regular; sepals distinct, 5-6, greenish white; petals 5-6, white, shorter than sepals
Fruits: drupes; clusters grape-like; nearly round, about 0.3 in (8 mm) across; flesh thick, juicy, skin tough; seeds yellowish, circular with a triangular wedge missing, flat with ridge near margin
Twigs: round (terete), shiny with longitudinal ridges; leaf scars circular, bundle scars 3 groups
Buds: superposed, hairy; lower depressed and usually hidden under leaf scar, upper a flower bud, barely visible

Subclass Hamamelidae

Hamamelidae consists of 11 orders, 24 families and about 3,400 species. Most species belong to Urticales and Fagales (Cronquist 1981).

Order Hamamelidales

Five families are recognized in the order Hamamelidales.

Recognition

Habit: trees, shrubs; deciduous, evergreen
Leaves: alternate, sometimes opposite, simple; entire or toothed or palmately lobed
Flowers: mostly in small clusters; bisexual or unisexual; perianth present or absent; petals mostly small or narrow if present
Fruits: dry; splitting or not splitting (indehiscent)

Cercidiphyllaceae Engl. - katsuratree family

The katsuratree family, which consists of a single genus of one species, is native to eastern Asia (monotypic).

Cercidiphyllum Sieb. & Zucc. - cercidiphyllum

Cercidiphyllum japonicum Sieb. & Zucc. - katsuratree

This tree of China and Japan is lauded by Johnson (1973) and Dirr (1990) as one the most beautiful large trees. Richley (1988) states that it grows fairly well at Curtis, Nebraska. A few young trees are growing on UNL grounds. Further trials are warranted.

Dirr (1990) writes it is "one of my favorite trees, overwhelming in overall attractiveness; if I could use only one tree this would be my first tree." His observations are based on specimens at the universities of Illinois, Georgia, and Maine, and at several aboreta including Morton and Arnold. Apparently insects and diseases are not serious problems to this tree. One type, a tree with drooping branches, is named 'Pendula'.

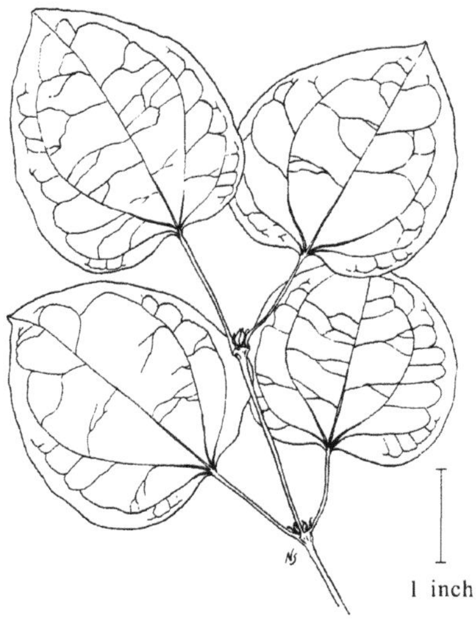

Cercidiphyllum japonicum

Katsuratree leaves emerge a reddish purple, become blue-green through summer, then change to a beautiful yellow-orange mixture of colors and develop a spicy cinnamon odor in autumn. We would not be surprised if fall coloration is disappointing in some parts of our region. We recommend extensive exploration in katsuratree's native habitat to obtain seed from as wide a latitude as possible for provenance research to identify varieties better adapted to a number of sites.

Planning

Height: up to 40 ft (12 m)
Habit: tree; deciduous; branches upright, sometimes spreading; autumn foliage yellows and reds
Site: sun; pH and soil adaptable, prefers moist soils; zone 5
Use: shade, street, specimen

Recognition

Leaves: simple, opposite or nearly so (subopposite), solitary on spurs; nearly round to egg-shaped, palmately veined; apex blunt, base heart-shaped (cordate), margins obscurely toothed (crenate-serrate); purplish when unfolding, becoming dark bluish green above; somewhat bloomy beneath (glaucescent)

Flowers: on short spurs before the leaves, unisexual (dioecious); female, 4 green sepals, no petals, styles long, purple; male, calyx minute, stamens several, anthers red

Fruits: dry, splitting (pods), about 0.7 in (15 mm) long; seeds many, winged

Twigs: brownish, glabrous, swollen at nodes

Buds: terminal lacking; lateral appressed; scales 2, not overlapping, reddish, glabrous

Platanaceae Dumort. - sycamore or planetree family

Sycamore, a family of a single genus, consists of six or seven species, three of which are native to North America (Cronquist 1981). Sycamores are easily recognized by their mottled, light-colored, flaking bark, "button ball" flowers and fruits, and large broad leaves with pointed lobes.

Platanus L. - sycamore or planetree

These large, fast-growing trees are valued as ornamentals the world over. Most sycamore species tolerate and thrive in urban environments, though they are not suitable for lining narrow streets or shading small properties. The hard, tough sycamore wood resists breakage and the tree's ability to sprout new buds and a mass of branches just below a cut facilitates pollarding or coppice—practices that are used in producing grotesque landscape forms and for biomass production in energy plantations.

Recognition

Habit: trees; deciduous

Leaves: alternate, simple; 3-7 lobed, main veins originating at base (palmate); petioles cover axillary buds; stipules usually encircle twig above the leaf

Flowers: ball-shaped heads, dense; unisexual (monoecious), male and female on separate stems; sepals 3-8, petals 3-8

Fruits: globe-shaped, tightly packed 1-seeded nutlets (multiple syncarps) surrounded by long hairs; often persisting into winter

Twigs: encircled by stipular scars; leaf scars surround buds

Buds: terminal lacking, lateral solitary, conical; scale 1, cap-like

Bark: whitish, smooth, peeling in thin plates to produce mottling; becoming gray-brown with shallow, narrow furrows and broad scaly plates

Wood: moderately heavy, hard, with interlocking grain; difficult to split; ray cells distinctive, conspicuous, producing beautiful patterns in quarter-sawn boards

Platanus ×*acerifolia* (Ait.) Willd. - London planetree

The first record of London planetree's existence dates to 1663 in England. The tree is believed to be a cross between *P. orientalis* L. and *P. occidentalis* L. (Johnson 1973). London planetree, very common in Europe, has been widely planted in temperate climates throughout the world excluding the U.S. It is reputed to be resistant to anthracnose, a fungus disease that appears in spring before the leaves and kills the tips of one-year-old twigs.

London planetree seedlings grown from seed collected at Manhattan, Kansas, did not survive beyond three years of age in a 1966 plantation at UNDARC, Ithaca, Nebraska (Bagley 1998). However, Dirr (1990) reports that 'Bloodgood' cultivar is resistant to anthracnose, is rapid growing and is tolerant of soil compaction, heat and drouth.

Planning

Planning: up to 70 ft (21 m)
Habit: tree; crown rounded, becoming wide-spreading; fruits persist into winter
Site: sun; pH and soil adaptable, tolerates pollution; zone 5
Use: shade for large areas

Recognition

Leaves: lobes about as wide as long
Fruits: usually in pairs on same stem, sometimes single or in 3s

Platanus occidentalis L. - American sycamore

American sycamore is native to most of the eastern U.S. including extreme southeastern Nebraska, the eastern third of Kansas, and Oklahoma south into Texas. It is one of the largest trees in the hardwood forest, found on fertile streambanks where it tolerates fluctuating groundwater levels. American sycamore adapts to upland soils if water is adequate. Its rapid growth and ability to produce stump sprouts makes it a good candidate for short-rotation biomass production.

Anthracnose-resistant American sycamore trees have been selected from a research plantation established in 1966 at UNDARC, Ithaca, Nebraska. Progeny of 17 American sycamore trees growing in plantations and natural stands throughout southeastern Nebraska were included in the planting. Trees originating from seed collected in a field shelterbelt on the Burt County farm of the late J. J. Lydick are especially vigorous and are less severely injured by anthracnose than the progeny of most trees of other parentage. This study suggests potential for developing disease-resistant sycamore through selection and breeding (Bagley 1998).

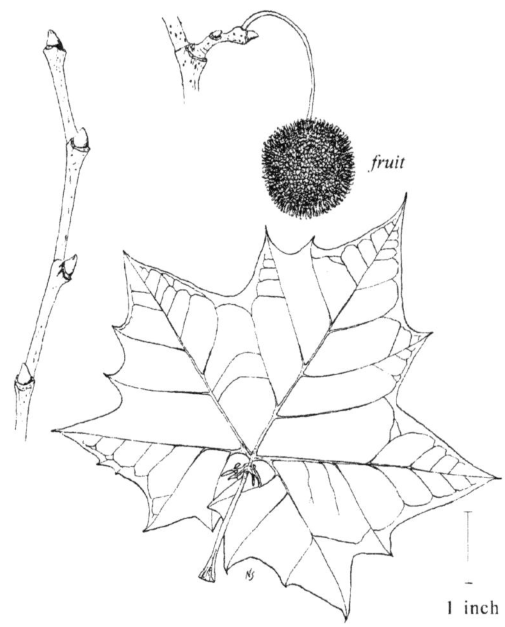

Platanus occidentalis

Planning

Height: up to 80 ft (24 m); Nebraska champion, 1980, 91 ft (27.6 m), Brownville; National champion, 1974, 129 ft (39.1 m), Ohio
Habit: tree; crown rounded, becoming broad
Site: sun; pH and soil adaptable, prefers fertile, moist, well-drained soils; tolerates drouth; zone 4
Use: shade, windbreak, lumber, biomass production

Recognition

Leaves: shallow lobes, wider than long
Fruits: usually solitary

Hamamelidaceae R. Brown - **witch-hazel family**

The 26 genera of the witch-hazel family—including over 100 species—are natives of Asia and, to a lesser extent, Africa, Australia and North America (Cronquist 1981). These distinctive plants are not easily confused with plants of other families. Several species produce high quality lumber. Extracts are used in liniments and several are cultivated as ornamentals.

Recognition

Habit: trees, shrubs; deciduous (temperate climates) or evergreen
Leaves: alternate, rarely opposite, simple; usually with stipules
Flowers: clusters (heads or racemes); unisexual or bisexual; usually regular, 4-,5-parted; petals and/or sepals sometimes missing; stamens 4-5, sometimes more; ovary inferior (epigynous) or surrounded by a hypanthium (perigynous)
Fruits: dry, splitting open along one side (capsules); seeds sometimes winged
Buds: scaly or naked

Fothergilla L. - **fothergilla**

Three species of this deciduous shrub are native to the southeastern U.S.

Fothergilla major (Sims) Lodd. - **large fothergilla**

This native of the Allegheny Mountains—from Virginia to South Carolina—requires an acidic soil which limits its use in the prairies. Autumn foliage is not likely to be consistently colorful in northern latitudes because of its southerly origin.

F. gardenii Murr., a coastal plains species, is a low-growing shrub recommended as a companion for azaleas and rhododendrons (Dirr 1990).

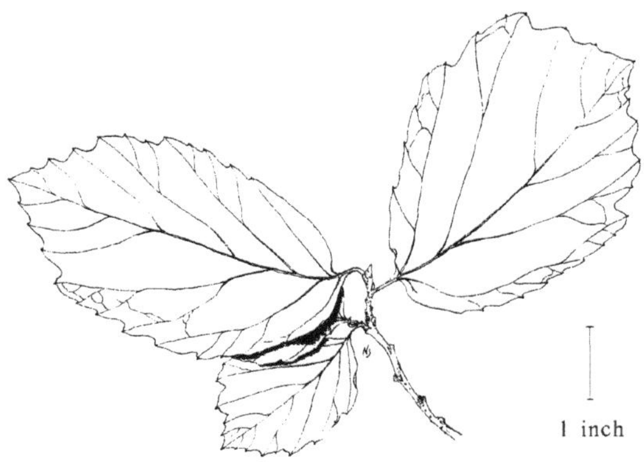

Fothergilla major

Planning

Height: up to 10 ft (3 m)
Habit: shrub; deciduous; erect stems, rounded, dense; spring bloom fragrant, autumn foliage colorful
Site: sun, part shade; pH 5.0-6.5; zone 5
Use: border, massing, foundation
Caution: protect from winter sun and wind
Pests: resistant

Recognition

Leaves: nearly round to oval, up to 4 in (10 cm) long; base heart-shaped or squared (truncate), variably toothed; dark green and glabrous above, whitish bloom (glaucous) and star-shaped hairs (stellate) beneath; leathery; petioles hairy, up to 0.4 in (1 cm) long; stipules present
Flowers: clusters (bottlebrush-like spikes) terminal, about 1.5 in (4 cm) long; bisexual; petals none, calyx 5-7 lobed; stamens, about 24, showy, filaments white, anthers yellow
Fruits: capsules, stout upright beaks, 0.4 in (1 cm) long; light brown inside with red markings on inner suture
Twigs: star-shaped hairy (stellate-tomentulose)
Buds: scales 2, falling prematurely (caducous), dense star-shaped hairs (stellate-pubescent)

Hamamelis L. - witch-hazel

Two of the five or six species of witch-hazel are native to North America. These pest-resistant plants with attractive foliage and flowers are of special interest because of their unusual blooming dates.

Recognition

Habit: shrubs, trees; deciduous

Leaves: teeth shallow, often obtuse; base oblique; short-petioled; stipules sometimes early deciduous (caducous)

Flowers: clusters axillary, few-flowered; bisexual; petals ribbon-like, 4-parted; calyx lobes spreading, obtuse, egg-shaped, hairy outside; stamens 4, alternating with scale-like, sterile stamens (staminodes), filaments short; styles distinct, short

Fruits: short, stout, 2-valved capsules, horned on top; seeds 2, shiny, black, oblong; at maturity, forcibly ejected from capsule

Twigs: star-shaped hairs (stellate), sometimes becoming glabrous, lenticels pale, inconspicuous; leaf scars half round, margin raised; bundle scars 3

Buds: stalked, naked (no scales)

Hamamelis mollis Oliv. - Chinese witch-hazel

This central China native ushers in spring with its fragrant, bright yellow, long-lasting flowers and welcomes autumn with golden yellow foliage. Because flower buds of Chinese witch-hazel can be injured by temperatures that fall below 0 degrees F. (-18 degrees C), blooming may not be reliable in much of the central and northern prairies.

An ever-growing list of hybrids between Chinese witch-hazel and Japanese witch-hazel (*H. japonica* Sieb. & Zucc.) provides plants with hybrid vigor and a wide range of flower and fall foliage color. Dirr (1990) sees a great potential for many of these lovely, maintenance-free plants.

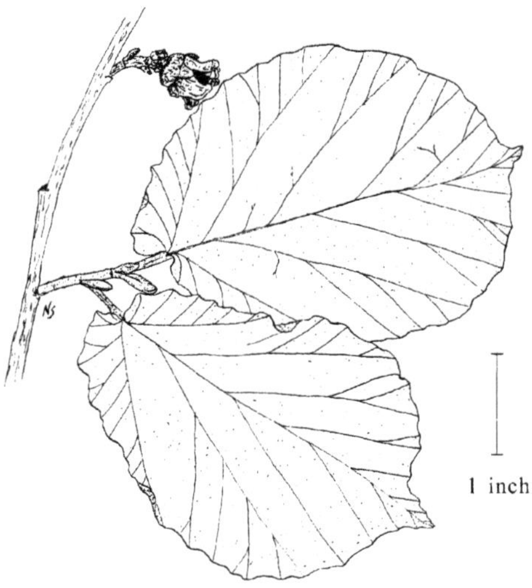

Hamamelis mollis

Planning

Height: up to 20 ft (6 m)

Habit: shrub, tree; crown oval to rounded, upright; March bloom fragrant, long lasting; autumn foliage golden yellow

Site: sun, part shade; prefers acidic soil; zone 5

Use: border, specimen

Recognition

Leaves: roundish, sometimes wider above middle, up to 6 in (15 cm) long; base obliquely heart-shaped (cordate); hairy above, densely woolly (lanate) beneath; petioles hairy, about 0.2 in (5 mm) long

Flowers: calyx rusty-hairy outside, purplish red inside; petals yellow, reddish at base, about 0.7 in (12 mm) long, not wavy

Fruits: one-third enclosed by calyx

Twigs: densely woolly-hairy (lanate)

Buds: leaf densely hairy; flower on a crooked stalk, in clusters 2-4, egg-shaped

Hamamelidaceae *Hamamelis*

Hamamelis vernalis Sarg. - vernal witch-hazel

The native range of this very early flowering shrub extends west from the Missouri Ozarks to eastern Oklahoma and south to Louisiana. Date of flowering varies with latitude. Spangler and Ripperda (1977) state that flowers appear in late March in Michigan. Dirr (1990) notes that vernal witch-hazel is used effectively in groupings on the University of Illinois campus.

Hamamelis vernalis

Planning

Height: up to 7 ft (2 m)
Habit: shrub; upright, stoloniferous; late winter bloom, autumn foliage golden yellow
Site: sun, part shade; pH 5.5-7.5; tolerates poorly drained soils; zone 5
Use: border, unpruned hedge, grouping

Recognition

Leaves: egg-shaped, wider above middle (obovate), up to 5 in (13 cm) long; tip obtuse, base broad tapered to squared-off (truncate), rarely heart-shaped; toothed above middle; surfaces glabrous, slightly whitish-bloomy beneath; 4-6 vein pairs; petioles hairy, about 0.3 (8 mm) long
Flowers: petals yellow to reddish; calyx lobes inner surface usually red
Fruits: egg-shaped, broader above middle (obovoid); calyx slightly below middle
Twigs: densely hairy, becoming glabrous
Buds: leaf hairy; flower stalked, rounded, 3-4 per cluster

Hamamelis virginiana L. - witch-hazel

Witch-hazel is native to much of the eastern U.S. extending westward to northeastern Iowa, the Ozarks of Missouri, and into southeastern Oklahoma. It grows best on moist, deep, fertile soils, but tolerates a wide variety of conditions along forest edges and in the understory. Blooms appear during October through November; seeds are forceably ejected from the capsules the next autumn when the fruits ripen.

Oil from the leaves, twigs and bark of witch-hazel is used in liniments and patent medicines. A uniformly branched twig of witch-hazel is a favorite divining rod for "water-witching."

A 15-foot (4.5 m) specimen growing in Maxwell Arboretum, UNL, is healthy and blooms each October. Provenance research coupled with trial plantings of witch-hazel throughout the prairies has the potential to increase the area of usefulness of this species.

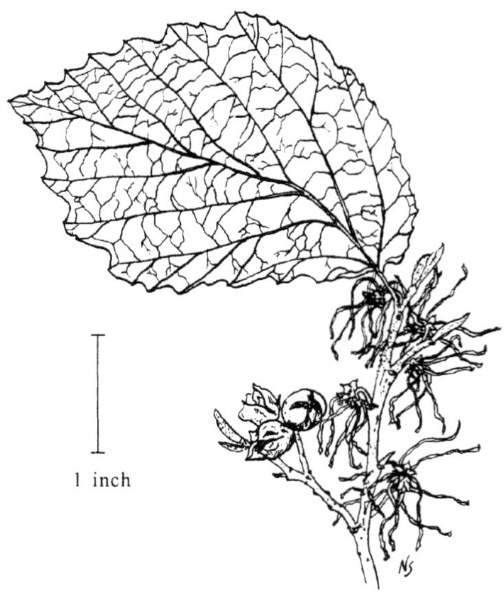

Hamamelis virginiana

Planning

Height: up to 20 ft (6 m); National champion, 1984, 35 ft (10.7 m), Virginia
Habit: shrub, tree; autumn bloom fragrant, autumn foliage yellow
Site: sun, part shade; pH 6.0-7.5, moist; zone 4
Use: border, naturalizing, specimen

Recognition

Leaves: egg-shaped to elliptic, sometimes broader above middle; base uneven, round to wedge-shaped; margin entire to toothed or somewhat scalloped; glabrous above, hairy on veins beneath; 5-7 vein pairs; petioles about 0.5 in (12 mm) long, hairy becoming glabrous

Flowers: groups of 2-3; petals bright yellow, about 0.7 in (18 mm) long; calyx lobes bent back, hairy, dull brownish yellow; stamens inconspicuous

Fruits: persistent, broadest about middle, calyx slightly below middle

Twigs: hairy, becoming glabrous; lenticels pale, inconspicuous; pith pale green

Buds: leaf, terminal stalked, flattish, hairy, about 0.4 in (1 cm) long; lateral smaller; flower globose, 2-4 on bent stalk

Liquidambar L. - sweetgum

One of the four or five species of sweetgum native to North America is described below. An exotic species, Formosan sweetgum, *L. formosana* Hance, is not likely to withstand the climate of much of the prairie region. Although it is a handsome tree, Dirr (1990) places it second to our native species.

Liquidambar styraciflua L. - sweetgum

Sweetgum ranges from southern New York south to Florida and southwest through the Ohio River Valley to eastern Texas. It grows best on moist, fertile, alluvial soils but also tolerates poorly drained soils and swampland. Sweetgum invades and can become a dominant species on abandoned cropland in the southern U.S. Its large trees are commercially valuable for lumber and veneer. Gum obtained from the bark can be used in the manufacture of soaps, drugs and adhesives. Sweetgum is a popular landscape tree noted for colorful autumn foliage.

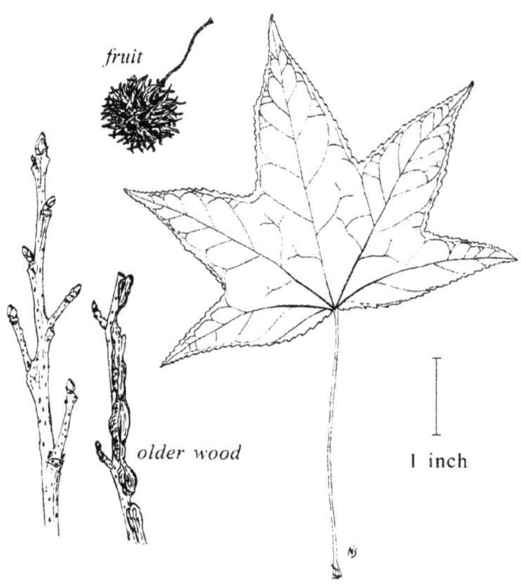

Liquidambar styraciflua

Richley (1988) reports severe chlorosis and winter injury to sweetgum growing in southwestern Nebraska. Several sweetgums growing in Lincoln and vicinity were undoubtedly planted for their reputed autumn beauty. However, the fall color of this healthy populaton is disappointing in most years.

'Moraine' is the most likely cultivar to display autumn color in the central prairies. This tree displays excellent form and foliage at the shade tree evaluation plots, Ohio Agricultural Research and Development Station, Wooster (Dirr 1990). Provenance research would be helpful in determining the varieties best adapted to the prairies.

Planning

Height: up to 65 ft (20 m); Nebraska champion, 1986, 36 ft (11 m), Lincoln; National champion, 1986, 136 ft (41 m), North Carolina

Habit: tree; deciduous; crown pyramidal becoming oblong; colorful autumn foliage; fruits persist into winter

Site: sun, part shade; moist, well-drained soils, somewhat acidic; tolerates poorly drained soils; zone 5

Use: shade, street tree, specimen, lumber

Pests: resistant

Hamamelidaceae *Parrotia*

Recognition

Leaves: star-shaped, deeply 5- to 7-lobed; up to 7 in (18 cm) across, about as long; tips long-pointed, base squared to somewhat heart-shaped; margin fine-toothed; surfaces glabrous except for hairy tufts at vein junctions; petioles slender, glabrous, up to 6 in (15 cm) long

Flowers: globose heads; unisexual (monoecious); male small heads in racemes, without perianth, intermixed with small scales; female terminal and solitary, calyx minute, no corolla, 4 non-functioning stamens

Fruits: heads of capsules, globose, woody, about 1.4 in (35 mm) in diameter; spiny from the persistent styles; seeds terminally winged, 1-2 each capsule

Twigs: green, finely hairy, becoming glabrous and brown, often with irregular, corky wings; pith angled in cross-section; leaf scars raised, triangular to half-round, 3 bundle scars

Buds: ovoid to conical, pointed; scales 6-9, shiny, orange to reddish brown, margins hairy (ciliate); adjoining (collateral) buds often present

Bark: thick, deeply furrowed

Wood: moderatley hard, growth rings and cellular structure inconspicuous without magnification

Parrotia C. A. Mey. - parrotia

This genus with a single species is native to Iran.

Parrotia persica C. A. Mey. - Persian parrotia

Very few of these interesting, pest-free trees have been planted in the prairies. Although there may be limitations, it would be worthwhile to know where and how they can be successfully added to prairie landscapes.

Planning

Height: up to 35 ft (11 m)

Habit: tree; deciduous; broader than tall; March-April bloom before leaves, colorful autumn foliage; bark variegated, exfoliating, attractive

Site: sun, part shade; prefers soils somewhat acidic; zone 5

Use: specimen, street tree

Pests: resistant

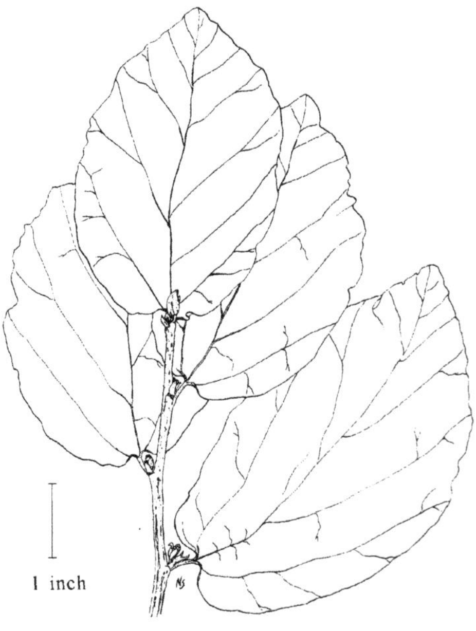

Parrotia persica

Recognition

Leaves: oval to egg-shaped and wider above the middle (obovate), up to 4 in (10 cm) long; tip obtuse, base rounded to slightly heart-shaped; margins toothed above the middle (crenate-dentate); glabrous above, sparse hairs beneath; petioles up to 0.2 in (5 mm) long

Flowers: heads dense, surrounded by large, brownish, hairy bracts; appearing before leaves; bisexual; sepals 5-7 enclosing about 50 percent of ovary, petals none; stamens 5-7, opposite sepals, pendulous, crimson, showy

Fruits: egg-shaped (ovoid), about 0.4 in (1 cm) long; 2-lobed at apex with recurved beaks; calyx margin near the base; seeds, 1 in each cell, oblong, shiny

Twigs: hairy, becoming glabrous

Buds: leaf stalked; scales 2, brownish, delicately woolly; flower globose, about 0.3 in (7 mm) across; hairy, brownish black

Bark: exfoliating on old branches; gray, green, white and brown revealed

Order Eucommiales

A single family occupies this order.

Eucommiaceae Engl. - eucommia family

This family of one genus with a single species is native to western China.

Eucommia Oliv. - eucommia

Eucommia ulmoides Oliv. - eucommia

This small elm-like tree with rubbery sap, often known as the hardy rubber tree, does not contain enough latex for commercial rubber production. However, it is worthy of trial plantings in the prairies for ornamental purposes. Dirr (1990) states that eucommia is an excellent shade tree with outstanding summer foliage and complete freedom from pests. It grows well in Ohio and Illinois and survives at temperatures as low as -20 degrees F. (-29 degrees C.).

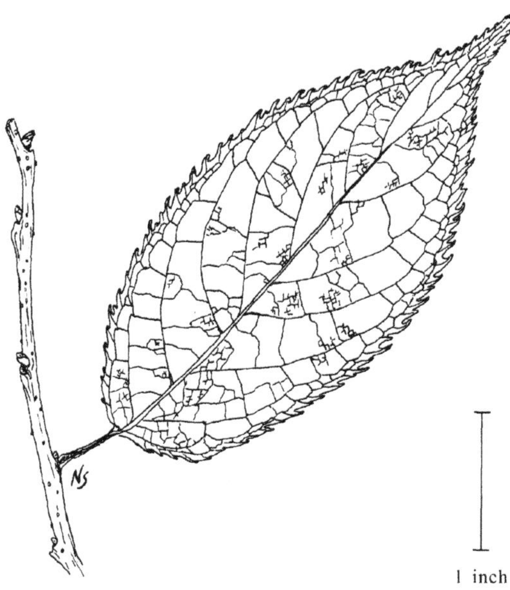

Eucommia ulmoides

Planning

Height: up to 40 ft (12 m)
Habit: tree; deciduous; crown rounded
Site: sun; pH and soil adaptable, tolerates drouth; zone 5
Use: shade

Recognition

Leaves: alternate, simple; variably egg-shaped in outline (ovate), tip long-pointed, base rounded, margin toothed; shiny, glabrous above; sap rubbery; stipules none
Flowers: solitary, axillary in bracts at base of young twigs; unisexual (dioecious); perianth wanting
Fruits: nutlets (samaras), 1-seeded, oblong, compressed, winged, notched at apex; about 1.5 in (4 cm) long
Twigs: glabrous, bloomy; pith laminated; sap rubbery, stringy
Buds: terminal lacking; lateral egg-shaped; scales overlapping, brown
Bark: furrowed

Order Urticales Lindl.

Six families and 2,200 species are placed in the order Urticales (Cronquist 1981).

Recognition

Habit: trees, shrubs, vines, herbs
Leaves: alternate, sometimes opposite, mostly simple; base very often oblique; sometimes deeply lobed, seldom palmately compound (as in *Canabis*)
Flowers: clusters, seldom solitary, individually small and inconspicuous, usually unisexual; petals none, stamens numerous
Fruits: various, often multiple

Ulmaceae Mirb. **- elm family**

Eighteen genera and more than 150 species are included in this family that inhabits temperate and tropical regions throughout the world. Five genera are native to the U.S., including two—*Celtis*, the hackberries, and *Ulmus*, the elms—that are native to much of the prairie region.

Recognition

Habit: trees, rarely shrubs; deciduous (rarely evergreen)
Leaves: simple, 2-ranked, alternate, usually oblique at base, margins often toothed (serrate); stipules present
Flowers: bisexual or unisexual (monoecious); calyx lobes 4-8; a stamen opposite each lobe; ovary usually 1-celled
Fruits: winged (samaras), dry; or fleshy and berry-like (drupe); seeds 1

Grouping Key for Ulmaceae

(Cronquist 1981)

Leaves pinnipalmately veined, with 3 main veins from the base; fruits drupaceous, seeds globose: **subfamily Celtidoideae**

Leaves strictly pinnately veined; fruits dry, most commonly a samara, seeds flattened: **subfamily Ulmoideae**

Subfamily Celtidoideae

Celtis L. **- hackberry**

Roughly 70 species of hackberry are widely distributed throughout the temperate and tropical areas of the world. Six or seven species—trees and shrubs—are native to the U.S. Elias (1980) states, "Individual trees often display considerable variation in their leaves, flowers, and fruits. Because of this and the presence of individuals with intermediate features, species of hackberries are sometimes difficult to distinguish." According to Stephens (1973), more research is needed to better describe *Celtis* species.

Native hackberry species of minor importance include: *C. tenuifolia* Nutt., dwarf hackberry—native to the eastern U.S. including southern Kansas; *C. lindheimeri* Engl., Lindheimer hackberry—native only to the Edwards Plateau in central and southern Texas; *C. douglasii* Planch.—native to the Southwest and arid parts of the northwestern U.S.; and *C. pallida* Torr., spiny hackberry—a semi-evergreen native of the extreme Southwest and adjoining Mexico.

Hackberries are not widely planted in landscapes despite their adaptability to a wide range of site conditions. They lack aesthetic qualities that are desirable in ornamental plants and are susceptible to pests that can cause visual deformities. Most species are too small to produce lumber.

Dirr (1990) reports that *C. jessoensis* Koidz., Jesso hackberry, is a beautiful, pest-free tree from Korea and Japan. Several of these upright but spreading trees grow in the Arnold Arboretum. Their bark is smooth gray and the pith of their twigs is not chambered. Trial plantings of Jesso hackberry in the Great Plains might be rewarding.

Recognition

Habit: trees, shrubs; deciduous or semi-evergreen

Leaves: egg-shaped (ovate); margins toothed or smooth (entire); 3-nerved at base; apex long-pointed, base unequal or oblique; stipules early deciduous (caducous)

Flowers: in leaf axils of new growth, small, greenish; bisexual and unisexual (polygamo-monoecious); male on lower branches in clusters (fascicles), stamens 4-6; bisexual and female solitary on high branches; calyx 4- to 6-lobed

Fruits: drupes, thin-fleshed, sweetish, edible, rounded; mature in autumn, persistent into winter; seed, net-like pattern on surface

Twigs: slender, zigzag, seldom spiny; pith round, finely chambered at nodes only or continuous; leaf scars oval to crescent-shaped, bundle scars 3

Buds: terminal absent; lateral egg-shaped, pointed, appressed to stem; visible scales 4 in 2 ranks with longitudinal striations

Bark: smooth, gray, often roughened with corky or warty growths

Celtis laevigata Willd. - sugarberry

Sugarberry, native to the southeastern U.S. including the tip of southeastern Kansas, is typically not hardy north of its natural range (Stephens 1973). However, five large trees grow at the Lincoln Regional Center Arboretum, an affiliate site of the Nebraska Statewide Arboretum.

Planning

Height: up to 50 ft (15 m); Nebraska champion, 1983, 51 ft (15.5 m), Lincoln Regional Center Arboretum; National champion, 1984, 81 ft (23.6 m), South Carolina

Habit: tree; deciduous; crown rounded, branches spreading

Site: sun, part shade; pH and soil adaptable; zone 5

Use: shade

Pests: resistant to the mite that causes "witches broom"

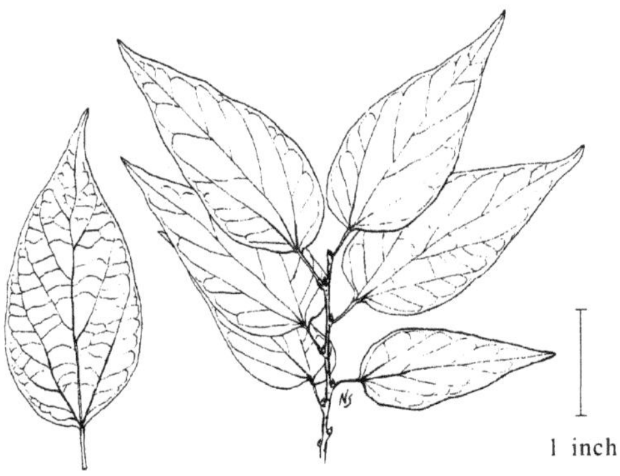

Celtis laevigata

Recognition

Leaves: egg-shaped to ovate-lanceolate, more than twice as long as wide, margin entire or with few teeth; tip long-pointed, often curved (falcate); surfaces glabrous except for hair tufts in vein axils beneath

Flowers: appearing with the leaves; unisexual (monoecious); male solitary or in clusters 2-3 at base of new twigs; female solitary or in 2s toward apex of new twigs

Fruits: orange to brownish red, smooth, sweet

Twigs: red-brown, shiny, glabrous

Buds: egg-shaped, pointed, appressed to twig; glabrous, chestnut brown

Bark: light gray with broad, flat furrows and short, narrow warty ridges

Celtis occidentalis L. - hackberry

This highly adaptable tree grows naturally on a wide variety of sites, from ridgetops to valleys throughout most of the prairies. Hackberry is found from Oklahoma to North Dakota, east through Tennessee to Virginia and north to the southern New England states. Outliers occur in adjoining states to the south and in southern Manitoba, Quebec and Ontario provinces of Canada.

Hackberry has been extensively planted as a part of windbreaks throughout the Great Plains with mixed results. Read (1958) reported little hackberry mortality in the Great Plains Shelterbelt Project. Growth and vigor in these shelterbelts were only moderately satisfactory, probably due in part to their central location in 10-row belts where they were usually competing with adjacent rows of green ash and honey locust. Seed origin may have been a contributing factor. Growth was poor on sandy soils; best on deep loam soils.

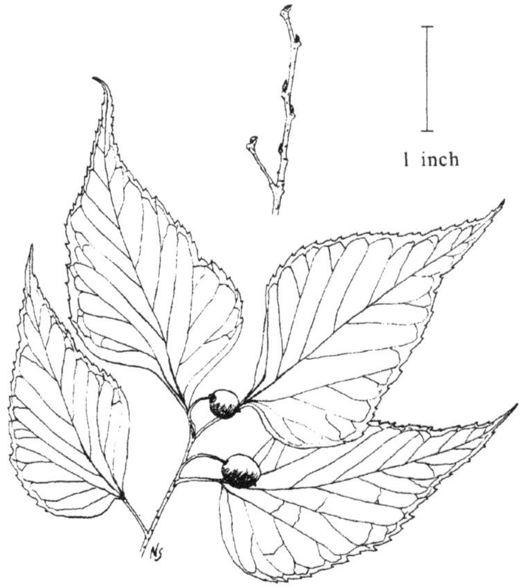

Celtis occidentalis

Hackberry continues to be recognized as an important component of windbreaks in the western prairies. Provenance plantations incorporating progeny of native trees from throughout the Great Plains region have been established as a regional research project of the various agricultural experiment stations of the region, GP-13.

Bagley (1979) reported the results from a source study of nine Great Plains hackberry provenances collected by the Natural Resources Conservation Service and planted in 1963 at three locations: Brookings, South Dakota, Plattsmouth, Nebraska, and Manhattan, Kansas. At Horning State Farm, Plattsmouth, trees originating from seed collected from northeastern Kansas grew faster than those grown from seeds collected from native trees of South Dakota and Oklahoma. Trees of Oklahoma origin were slowed by winter injury in the Brookings plantation.

The slow growth of hackberry seedlings planted in Nebraska in recent years may be due to genetic characteristics of the parent trees. Much of the seed on the commercial market during the past few years has been collected in the northern Great Plains (Morgenson 1992). Daylength triggers earlier-than-normal dormancy in fall which in turn causes reduced growth for the season. However, trees of northern Great Plains origin will probably never be injured by an early fall freeze in Nebraska, and they may be exceptionally drouth tolerant.

A few hackberry cultivars have been selected and there is potential for many more. 'Chicagoland' has a single upright leader; 'Prairie Pride', a rapid-growing tree, is purported to be resistant to the mite that causes "witches broom"; and 'Windy City' is an upright, spreading tree with attractive foliage (Dirr 1990).

Hackberry fruits, though small and thin-fleshed, are a source of food for birds and small mammals. The hard seed is distributed far and wide, contributing to the great natural range of the species. The heavy, but weak, soft wood is a common component of furniture. It is not durable in contact with moisture.

Minor pests include nipple galls—unsightly outgrowths from the leaf surface caused by a small insect, and "witches broom"—caused by a mite. Lace bugs may cause yellowing and premature leaf drop.

Planning

Height: up to 70 ft (21 m); Nebraska champion, 1985, 60 ft (18.3 m), South Sioux City; National champion, 1989, 111 ft (33.6 m), Wisconsin
Habit: tree; deciduous; crown vase-shaped, developing broad top
Site: sun, part shade; pH and soil adaptable; drouth, wind and pollution tolerant; zone 3
Use: shade, windbreak, wildlife habitat, lumber, firewood

Recognition

Leaves: egg-shaped to ovate lanceolate; margins sharply toothed, sometimes not below the middle or at the base; young densely hairy (tomentose), becoming glabrous on upper surface or sometimes with stiff hairs (scabrous) and slightly hairy below; petioles slender, about 0.4 in (1 cm) long
Flowers: appearing with leaves; unisexual (monoecious); male solitary or in clusters of 2-3 at base of new branch, female solitary or in 2s toward apex
Fruits: brownish to dark purple, diameter 0.2-0.4 in (6-10 mm); stalks about 0.5 in (12 mm) long
Twigs: brownish; hairy sometimes becoming smooth
Buds: egg-shaped, pointed, flattened, light brown; scales 2-ranked
Bark: gray; corky growths, warty or ridged becoming scaly
Wood: light colored, moderately heavy, soft, without a characteristic odor or taste

Celtis reticulata Torr. - netleaf hackberry

The native range of netleaf hackberry—Oklahoma and Texas—overlaps that of *C. laevigata,* which causes the identities of the two species to sometimes be confused. Netleaf hackberry is also native to western Kansas, eastern Colorado and as far north as Frontier County, Nebraska, where it grows in ravines, sometimes on sandy soil (Sutton 1998; Stephens 1973). The mountainous areas of the western states of Arizona, New Mexico, Colorado, Wyoming, Utah and Idaho are also home to netleaf hackberry and a closely related form, named *C. douglasii* Planch. by some authorities (Preston 1968).

Netleaf hackberry, a virtual unknown to the nursery trade, could be a desirable tree for planting in the southwestern prairie.

Planning

Height: up to 30 ft (9 m)
Habit: shrub, small tree; deciduous; crown open and irregular, branches usually crooked
Site: sun; pH and soil adaptable; drouth tolerant; zone 5
Use: wildlife habitat, naturalizing

Recognition

Leaves: broadly egg-shaped, 1 to 3 in (25mm to 75 mm) long; tip sharp-pointed to obtuse, base obliquely rounded or heart-shaped; margin entire or with a few broad teeth; dark green and rough or smooth above, somewhat hairy with conspicuous reticulate veinlets below; petioles grooved above, hairy
Flowers: axillary on a short new twig with the leaves; stalks slender, glabrous or hairy; unisexual; male flowers toward twig base, female toward apex; calyx lobes linear, pointed, hairy or glabrous, margin fringed or scarious
Fruits: orange-red to red-brown
Twigs: green and hairy at first, becoming red-brown, hairy or glabrous; lenticels small, pale, abundant
Buds: egg-shaped, hairy

Subfamily Ulmoideae

Hemiptelea Planch. - hemiptelea

One species is present in this genus.

Hemiptelea davidii (Hance) Planch. - hemiptelea

This native of northeast Asia is a small spiny tree.

Planning

Height: small
Habit: tree; deciduous; shrubby, spiny
Site: experimental throughout most of the prairies
Use: hedge

Recognition

Leaves: oblong to elliptic, margins coarsely toothed; scattered hairs above on new leaves, leaving dark circular depressions; sparse hairs on veins below; petioles up to 0.1 in (3 mm)
Flowers: appear with the leaves; unisexual and bisexual on same plant (polygamous), clusters of 1-4; calyx 4-5 parted, stamens 4
Fruits: nutlets, obliquely cone-shaped, 2-edged, crest-like wing on upper edge
Twigs: spiny, chiefly on young plants; shaggy hairy (pilose), becoming grayish brown
Buds: minute, round to egg-shaped

Ulmus L. - elm

The elms—trees of the temperate regions of the Northern Hemisphere—have been separated into 20 to 30 species, six of which are native to North America. The elms provide us with valuable wood for furniture, paneling and construction; rope and coarse cloth have been made from the bark fibers; and its seeds, buds and twigs provide food for wildlife.

Elms are among the first plants to herald the advent of spring with millions of flowers bursting with pollen, an event quickly noted by people who are sensitive to the fine powder and by bees venturing from their hives with an appetite for proteinaceous food.

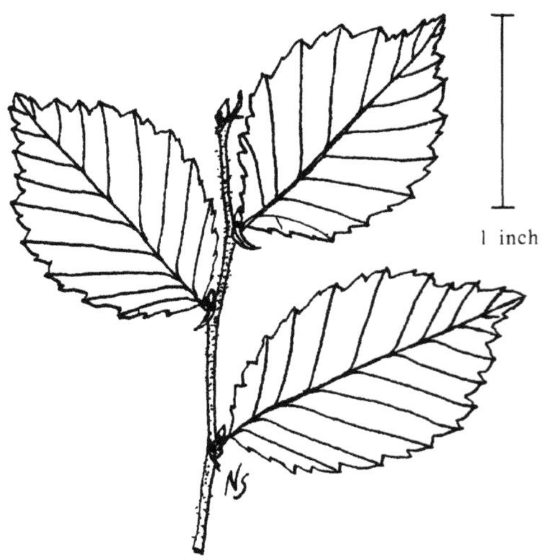

Ulmus japonica

U. procera Salisb., English elms, were planted in many northeastern U.S. cities during colonial days. A tree on East Campus, UNL, was identified as English elm, primarily because it produced a profusion of root suckers and didn't quite meet the description of any other elm species. Though healthy, it was in the path of progress and was removed in the early 1960s.

Dirr (1990) relates problems encountered when describing English elm and remarks that it may not be a valid species, but a hybrid. It reproduces only by root suckers, according to Johnson (1973). Rehder (1940) notes the suckering tendency but describes flowers and fruit with no indication of sterility. Rehder also lists many varieties of English elm, an indication of its great genetic variability.

Dutch elm disease (DED), an import from Europe and Asia caused by the fungus *Ceratocystis ulmi*, is carried by bark beetles. It was first detected in Ohio in 1930. Thirty years later it entered the central Great Plains and wreaked havoc among our native elms. Fortunately, isolated large trees and many seedlings escaped the first wave of infestation. Elms are now repopulating their native habitat where protected from grazing animals. The new populations may be ravaged as they age, but some will survive and, eventually, may develop resistance to DED.

Asian elms are resistent to DED. Two Asian species in addition to those herein described deserve trials on prairie sites. *U. davidiana* Planch., David elm, with its broad vase-shape and tolerance of highly alkaline soil is a promising tree for plantings in urban environments (Ware 1991). *U. japonica* (Rehd.) Sarg., Japanese elm, is a related species that Hoag (1965) recommends for additional trial in the northern Great Plains. He cites the success of established specimens at the Great Plains Field Station in Mandan, North Dakota.

Recognition

Habit: trees; deciduous, rarely half-evergreen

Leaves: tips pointed, margins usually doubly saw-toothed (serrate); pinnate veins run directly to teeth; surfaces smooth to hairy

Flowers: clusters small, in leaf axils; appear in spring before the leaves or in autumn; calyx bell-shaped, 4-9-lobed with a stamen for each lobe

Fruits: nutlets (samaras), flattened; encircled by thin papery wings often notched opposite the stem; maturing late spring or autumn a few weeks after flowering

Twigs: slightly zigzag, often hairy, sometimes corky winged; leaf scars semicircular, slightly to one side of bud, bundle scars 3

Buds: terminal lacking; lateral usually solitary; scales 4-8 visible, in 2 ranks

Bark: furrowed

Ulmus americana L. - American elm

American elm, native to the eastern U.S. and southern Canada, is so well known and recognized that it could be proclaimed the national tree except for the mortality caused by Dutch elm disease during the past sixty-odd years. Environmentally hazardous control measures to limit DED's effects strained many a city budget, but to little avail. However, as the elm population thinned DED waned, and we learned a lesson. We now diversify our tree plantings to guard against such calamities. American elm might be safely planted again in moderate numbers.

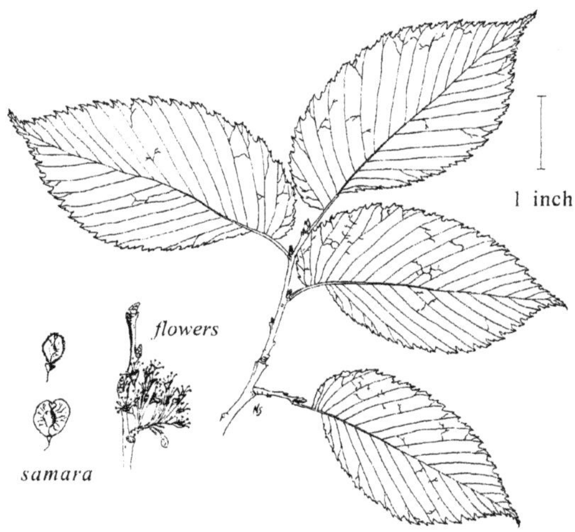

Ulmus americana

'Moline', once common in the nursery trade, was purported to be resistent to DED and was one of the last elms to succumb to its ravages on the University of Illinois campus (Dirr 1990). Resistant hybrids have been developed, but to date none has qualities that endear. A number of narrow-crowned cultivars have been marketed including 'Ascendens', 'Augustine' and 'Littleford'.

American elm was once an important source for lumber, slack cooperage, boxes and crates. The wood is strong and tough with superior bending qualities, making it desirable for bent furniture parts like rockers and arm pieces.

Ulmaceae *Ulmus*

Planning

Height: up to 70 ft (21 m); Nebraska champion, 1981, 90 ft (27.4 m), Nemaha; National champion, 1991, 100 ft (30 m), Kansas
Habit: tree; deciduous; crown vase-shaped; autumn foliage yellow
Site: sun, part shade; pH and soil adaptable; drouth tolerant; zone 3
Use: shade, street, windbreak, lumber
Caution: susceptible to DED, phloem necrosis, and several nuisance insects

Recognition

Leaves: elongated egg-shaped, often broadest near the middle tapering to a sharp point; upper surface glabrous to stiff hairy, usually hairy beneath; petioles about 0.3 in (7 mm) long, glabrous to slightly hairy
Flowers: clusters of 3-5 on slender drooping stalks, appearing in spring before leaves
Fruits: encircled by thin wings, notched at tip; egg-shaped, about 0.5 in (12 mm) long; glabrous on both sides, margin often hairy (ciliate)
Twigs: slender, drooping, reddish brown, smooth or sparsely hairy
Buds: chestnut brown; leaf buds narrow egg-shaped, flower buds broad, sharp pointed, glabrous or slightly hairy
Bark: furrowed with alternating dark and light layers in cross-section
Wood: hard, heavy, light brown, difficult to split

Ulmus carpinifolia Gleditsch - **smoothleaf elm**

This elm, a native of Europe and North Africa, somewhat resembles American elm with its slender ascending branches. Many crosses (*U. ×hollandica* Mill.) have been made between smoothleaf elm and Scotch elm, *U. glabra* Huds. From these, a number of cultivars have been selected including 'Christine Buisman', 'Bea Schwarz', 'Dampien', 'Groeneveld', 'Regal' and 'Pioneer', all of which are somewhat resistant to DED (Dirr 1990). Some of these cultivars have been extensively planted in the U.S. including in Lincoln where several can be seen on East Campus, UNL. These hybrids are not well adapted to western Nebraska (Richley 1988).

'Urban'—a hybrid resistant to DED—was developed at the Agricultural Research Service Shade Tree Laboratory in Delaware, Ohio. Though lacking the aesthetic qualities of American elm, 'Urban' is said to be tolerant of drought, pollution and soil compaction.

Planning

Height: up to 70 ft (21 m)
Habit: tree; deciduous; ascending branches, suckering roots
Site: sun, half shade; pH and soil adaptable; zone 5
Use: shade

Recognition

Leaves: elliptic to egg-shaped, sometimes wider above middle; surfaces glabrous, shiny, dark green above, axillary tufts below; petioles about 0.3 in (8 mm) long, usually hairy
Flowers: clusters dense, not pendulous; appear in spring
Fruits: elliptic or egg-shaped, not hairy; seed near to closed notch at tip
Twigs: glabrous or nearly so
Buds: dark brown to black, egg-shaped, soft hairy, scale edges often hairy (ciliate)
Bark: deep-fissured

Ulmus glabra Huds. - Scotch elm

This large, wide-spreading, vase-shaped tree—native to central and northern Europe and western Asia—is one of the parents of *U. ×hollandica* (see *U. carpinifolia*). Notable selections include 'Pendula' and 'Camperdownii'. The latter is listed in **Hortus III** as a selection of *U. ×vegeta* (Loud.) Rehd., the Huntingdon elm, another hybrid of Scotch and smoothleaf elms.

Planning

Height: up to 75 ft (23 m)
Habit: tree; deciduous; wide crown
Site: sun, part shade; zone 5
Use: shade (intermingle with species of other genera)

Recognition

Leaves: elliptic to oblong and wider above the middle; tip abruptly pointed, base unequal; sharply and doubly toothed (serrate); stiff hairy (scabrous) above, usually hairy beneath; petioles about 0.2 in (5 mm) long
Fruits: broad winged, about 1 in (25 mm) long; smooth
Twigs: bristly hairy
Buds: dark brown; scales stiff hairy

Ulmus parvifolia Jacq. - Chinese elm

This native of northern and central China and Korea and Japan is an unusual elm that differs greatly in general appearance from most other elm species. A striking feature is its beautifully mottled bark with combinations of gray, green, orange and brown that develop as the stems enlarge. Flowers appear in September and fruits ripen in late October, if the tree manages to reach that stage before a killing freeze. Dirr (1990) reports that leaves of the Chinese elm in Georgia exhibit a reddish purple color in late November. In the central prairies leaves are usually damaged by freezing before reaching that stage.

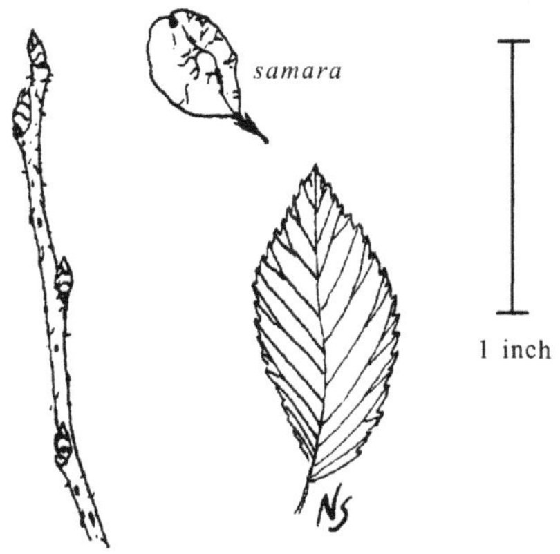

Ulmus parvifolia

Chinese elm is a "tough, durable tree for about any situation" (Dirr 1990). It shows tolerance to infertile soils and extremes of pH, and is resistant to most elm pests including DED. Dirr also reports that Chinese elm was only slightly damaged from a February ice storm in the Athens, Georgia, area that devastated many other trees.

Two Chinese elms of unknown origin in Maxwell Arboretum, one over 30 years of age and the other less than half that age, exhibit a significant difference in dates of fall dormancy and fruit ripening. Both trees suffered severe injury during the winter of 1991-92, injury so severe that the younger tree was removed. Two ten-year-old seedling progeny of the younger tree in Prairie Pines Arboretum near Lincoln have suffered significant dieback.

Selected cultivars that may be reasonably hardy in the central prairies include 'Prairie Shade', 'King's Choice' and 'Dynasty'.

Ulmaceae *Ulmus*

Planning

Height: up to 50 ft (15 m)
Habit: tree; deciduous; crown rounded, branches somewhat pendulous; mottled bark
Site: sun, part shade; pH and soil adaptable; zone 5
Use: shade, windbreak
Pests: DED resistant

Recognition

Leaves: more or less egg-shaped to elliptic; unequally rounded at base, tips variable; margins usually simply toothed (serrate); shiny dark green and glabrous above, hairy beneath when young, becoming rather leathery; petioles about 0.3 in (8 mm) long
Flowers: clusters in leaf axils, appearing in late summer
Fruits: egg-shaped (ovate), 0.3-0.4 in (8-10 mm) long; glabrous, orbicular wing notched at apex, seed in middle
Twigs: slender, glabrous or slightly hairy
Buds: brown, slightly hairy
Bark: smooth, mottled

Ulmus pumila L. - Siberian elm

This tall, vigorous native of eastern Siberia, northern China and adjoining areas is both praised and damned. In the semi-arid plains its tenacity under heat and drouth on soils ranging from sand to clay has been a boon to "tree-starved" people looking for protection from wind and sun.

Hoag (1965) details the history of the introduction of Siberian elm into North America and the development of such hardy strains as 'Dropmore' and 'Chinkota', and the part that this species plays in hybrid development. Read (1958) reported excellent survival and growth of Siberian elm in shelterbelts of the Prairie States Forestry Project. In more humid regions its agressive, pestiferous seedlings coupled with brittle branches and dead twigs that strew the lawn after every windstorm combine to sully its reputation as a shade tree.

Ulmus pumila

Despite its hardiness, Siberian elm has suffered serious mortality in Nebraska on at least two occasions: Armistice Day, 1940, and Halloween, 1991. Mild autumn weather abruptly terminated by sudden temperature drops below freezing were responsible. At this latitude Siberian elm is not acclimated to withstand freezing temperatures in early autumn. Other species were also injured. Most of the apple orchards in southeastern Nebraska also were wiped out by the 1940 freeze.

Planning

Height: up to 80 ft (24.5 m); Nebraska champion, 1977, 106 ft (32 m), North Platte; national champion, 1991, 146 ft (44 m), Michigan
Habit: tree; deciduous; crown rounded to spreading
Site: sun, part shade; pH and soil adaptable, drouth tolerant; zone 3
Use: shade, windbreak
Pests: DED resistant

Recognition

Leaves: elliptic, base oblique to evenly rounded, tip pointed; surfaces glabrous with hairy tufts on vein axils beneath; petioles hairy, about 0.1 in (2 mm) long
Flowers: clusters before leaves
Fruits: circular, glabrous, about 0.5 in (12 mm) across; wing notched at apex with rounded overlapping tips
Twigs: slender, flexible, drooping, slightly hairy to glabrous, gray or gray-green
Buds: leaf egg-shaped, flower nearly round; scales dark brown to red-brown, margins hairy (ciliate)
Bark: gray, shallow furrowed
Wood: hard, red-brown, difficult to split

Ulmus rubra Muhl. (*U. fulva* Michx.) - slippery elm

Slippery elm gets it name from the sticky slime found in its inner bark. This liquid served as a thirst quencher for pioneer woodsmen and a remedy for almost any ailment, from sore throat to constipation. The dried bark of slippery elm was used as a poultice applied to sores and inflammations. Slippery elm wood, like that of most other elms, is strong and tough which makes it useful for products that must withstand rough duty such as bridge planking. Unlike most other elms, it splits with relative ease.

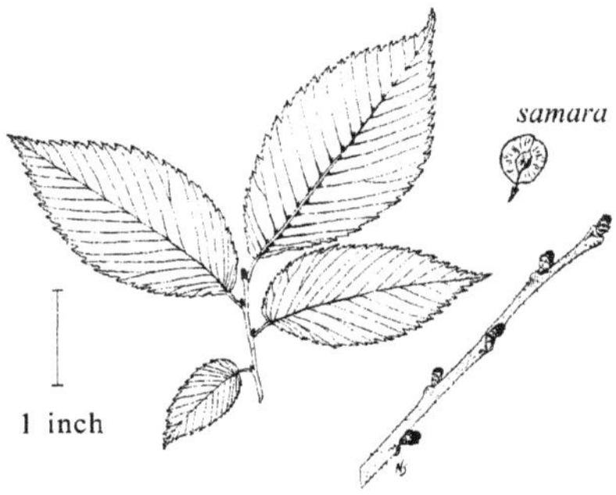

Ulmus rubra

Slippery elm resembles American elm in habit. Close examination reveals that slippery elm leaves are sandpapery rough and bud scales are reddish-hairy. A cross-section of a bark ridge reveals a uniform red-brown color, differing from the alternating light and dark layers of American elm bark.

DED has thinned the ranks of slippery elm. Though native throughout much of the eastern Great Plains eastward to the Atlantic, it was never as common in our native woodlands as American elm was, especially in the prairies, and it has seldom been planted. All of the slippery elms on the UNL grounds have succumbed to disease and we know of no others in the city of Lincoln. A 20-year-old specimen grows in the Prairie Pines Arboretum near Lincoln.

Planning

Height: up to 65 ft (20 m); National champion, 1989, 100 ft (30.5 m), Ohio
Habit: tree; deciduous; vase-shaped, crown rounded
Site: sun, part shade; pH and soil adaptable; zone 3
Use: shade, windbreak, lumber
Pests: DED susceptible

Recognition

Leaves: egg-shaped to oval, broadest below or near middle; abruptly tapering to long point, base oblique, rounded; upper surface dark green, stiff hairy (scabrous), hairy below; petioles about 0.3 in (7 mm) long, hairy
Flowers: clusters short, crowded, few-flowered, appearing before leaves
Fruits: nearly circular, about 0.5 in (12 mm) across; seed area hairy, encircled by a thin glabrous wing with smooth margin, notched at tip
Twigs: slender, reddish brown, hairy
Buds: leaf, narrowly egg-shaped; flower, broadly egg-shaped; both types of buds are dark brown, obtuse, rusty hairy; scales 2-ranked
Bark: long-furrowed, uniformly reddish brown in cross-section, inner bark mucilaginous
Wood: hard, heavy, dark reddish brown; easy to split

Ulmus thomasi Sarg. - rock elm

The native range of rock elm reaches the eastern fringes of Nebraska and Kansas, and includes Tennessee north to the Great Lakes and east to Vermont. Scattered trees were growing in the Salt Creek valley south of Lincoln until killed by DED. These tall, narrow-crowned trees produce valuable lumber. Growth is best on fertile, loamy soils, but rock elm is also found on rocky slopes of limestone origin. Rock elm has seldom been cultivated.

Planning

Height: up to 65 ft (20 m); National champion, 1989, 117 ft (35.5 m), Michigan
Habit: tree; deciduous; crown narrow
Site: sun; pH and soil adaptable; zone 4
Use: lumber, shade
Caution: DED susceptible, intermingle with species of other genera

Recognition

Leaves: broadly elliptic or oval, tip pointed, base rounded; upper surface glabrous, lower often slightly hairy; petioles about 0.3 in (7 mm) long, hairy
Flowers: clusters 3-9, on slender stalks; appearing before leaves
Fruits: densely hairy, about 0.5 in (12 mm) long; seed encircled by thin wing, slightly longer than wide, shallow notch at tip
Twigs: hairy becoming glabrous, developing corky wings in 2-3 years
Buds: narrowly egg-shaped, pointed, dark brown, more or less hairy
Bark: furrowed similar to *U. americana*; uniform color in cross-section
Wood: hard, heavy, pale brown, difficult to split

Zelkova Spach - zelkova

Four species are of Asian origin. They resemble their "cousins," the elms, in form and leaf and they resemble the hackberries in fruit characteristics. There are good reasons for placing *Zelkova* in either one of the elm subfamilies (Cronquist 1981).

Z. sinica Schneid., Chinese zelkova, is growing at Morton Arboretum (Dirr 1990). This species may be adapted to part of the prairies. Trial plantings are recommended. *Z. carpinifolia* (Pall.) K. Koch and *Z. schneideriana* Hand.-Mazz. are not likely to be hardy in most of the prairie region.

Recognition

Habit: trees, shrubs; deciduous
Leaves: margins toothed (serrate), veins pinnate
Flowers: unisexual and bisexual on the same plant (polygamous); appear with leaves in spring; male clusters in axils of lower leaves, female solitary in axils of upper leaves; calyx 4-5-lobed; stamens 4-5
Fruits: drupes ripening in autumn

Zelkova serrata (Thunb.) Mak. - Japanese zelkova

Specimens of this native of Korea and Japan are growing on the campus of UNL. Richley (1988) reports that plants in southwest Nebraska are slow to establish and suffer unpredictable dieback. Japanese zelkova shares both the good and the bad characteristics of American elm. These graceful, well-proportioned trees are bedeviled by many of the same insects as the elms but are resistant to Dutch elm disease. The thin, flaking bark, which is somewhat similar to that of *U. parvifolia*, is distinctive and attractive.

'Spring Grove' is a handsome vase-shaped tree propagated from an 80-foot tree growing in Cincinnati, Ohio (Dirr 1990). Some of the exceptionally fast growing cultivars such as 'Green Vase' and 'Village Green' are likely to be susceptible to injury by the untimely early fall freezes that plague much of the prairies. The species and two cultivars, 'Spring Grove' and 'Halka', failed to establish at PERC in Fort Collins (Fairchild and Klett 1993). Provenance trials could be very productive in identifying hardy cultivars.

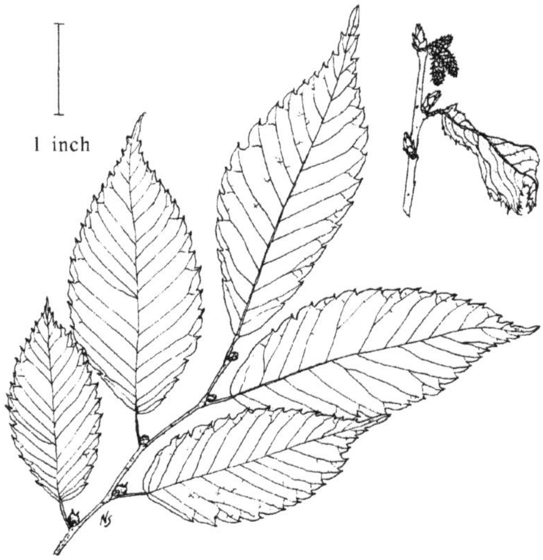

Zelkova serrata

Planning

Height: up to 60 ft (18 m)
Habit: tree; vase-shaped
Site: sun, part shade; pH and soil adaptable, tolerates drouth; zone 5
Use: shade, street

Recognition

Leaves: egg-shaped to oval, widest near the middle; margins sharp-toothed, tip pointed, base rounded; surface somewhat rough above, nearly glabrous below; petioles about 0.1 in (3 mm) long
Fruits: about 0.2 in (5 mm) in diameter
Twigs: hairy, becoming glabrous, brown
Buds: diverging from the stem; egg-shaped, pointed; scales overlapping, dark brown, glabrous
Bark: smooth, gray initially; becoming flaky

Moraceae Link - **mulberry family**

This family of about 40 genera and 1,000 species are largely inhabitants of tropical and subtropical regions. They are mostly woody plants, and many are economically important. Figs, mulberries and breadfruit are important in the food and beverage industries. Species of *Ficus* and *Castilla* are tapped for their latex for rubber making.

Paper mulberry, *Broussonetia papyrifera* (L.) Vent., has been introduced into the United States and has become naturalized in the East and Southeast.

Recognition

Habit: trees, shrubs, vines, rarely herbs; deciduous or persistent
Leaves: alternate or opposite, mostly simple; margins smooth (entire), toothed or lobed; stipules present, sometimes persistent
Flowers: clusters axillary, usually spikes or heads or affixed to variously shaped receptacles; unisexual (monoecious or dioecious), petals wanting
Fruits: fleshy, variable in size and shape; a flower cluster may form an agregate or compound fruit

Maclura Nutt. - **osage-orange**

The only *Maclura* species is a North American native. Its rather limited native range includes southeast Oklahoma and southern Arkansas, extending southwest across Louisiana and Texas to near the Rio Grande River.

Maclura pomifera (Raf.) Schneid. - **osage-orange**

Osage-orange is best known for its large orange-like fruit, thorny twigs, and heavy, indestructible wood. It served as a impenetrable, permanent living fence on much of the prairies during the mid-1800s. The advent of barbed wire spelled its decline as a living fence, but it became a valuable source of long-lived, unbreakable fence posts and hot-burning firewood. Osage-orange lost favor for use as posts with the discovery that pine posts could be chemically treated to increase durability in contact with the soil.

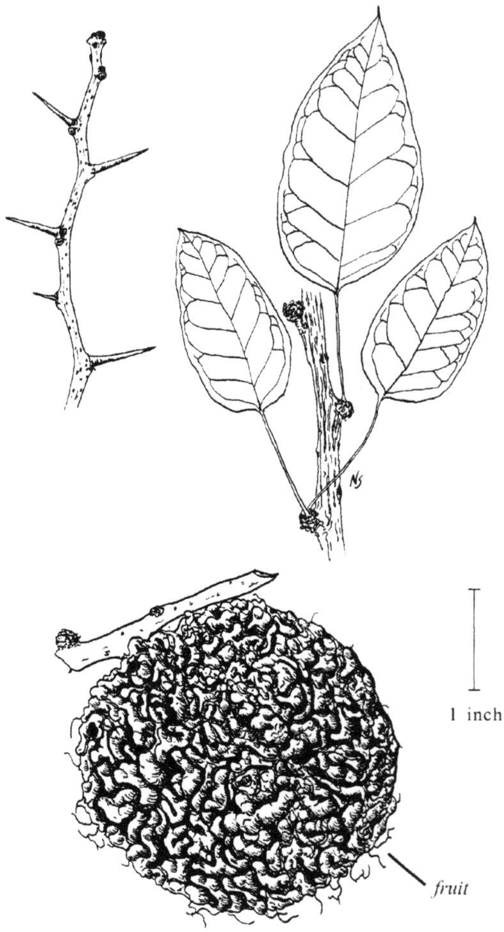

Maclura pomifera

Osage-orange wood was used for bows by Indians and named "bois d'arc" by French trappers. Its bark and wood were sources of bright yellow dye for early settlers. The sticky, unpalatable fruits are reputed to repel insects but are generally considered a nuisance (they await discovery of an important use).

Osage-orange hedgerows were quite common in Missouri, eastern Kansas, southern Iowa and southeastern Nebraska until powerful bulldozers were developed. When petroleum shortages develop, osage-orange may come back in favor for use as posts and firewood.

Though seldom used as a landscape plant, a few large osage-orange trees can be found in Lincoln yards. Most are tall and single-stemmed, much different from the multistemmed thorny trees of the hedgerows.

The late John Pair, Kansas State University horticulturist, selected and propagated thornless osage-orange trees with some success. One of these selections, now about 10 years old, grows at Horning State Farm near Plattsmouth. Thorns were present during the tree's first few years but twigs of this 10-foot (3 m) tree are now thorn free (Bagley 1998). The availability of thornless male cultivars will undoubtedly increase its use as a landscape plant.

Planning

Height: up to 40 ft (12 m); Nebraska champion, 1977, 60 ft (18.3 m), Brownville; National champion, 1987, 68 ft (20.6 m), Virginia
Habit: tree; deciduous; crown rounded, irregular; thorny, sometimes thornless
Site: sun; pH and soil adaptable; tolerates drouth, wind, pollution, infertile soils; zone 5
Use: shade, windbreak, living fence, fence posts, fuel, specialty products
Pests: none

Recognition

Leaves: alternate or clustered at end of short spurs; egg-shaped to much longer than wide, broadest below middle tapering to a long point, base rounded to somewhat heart-shaped; margin smooth (entire), somewhat wavy; surfaces dark green, glossy, glabrous; petioles hairy, about 1 in (25 mm) long, juice somewhat milky
Flowers: unisexual (dioecous), appearing late June after leaves; male clusters many-flowered at spur ends; female clusters dense, globe-shaped at leaf axils
Fruits: globose, orange-like with yellow-green rind, about 4 in (10 cm) in diameter; milky juice; many-seeded (syncarp of drupes)
Twigs: rigid, glabrous with stout spines about 0.5 in (12 mm) long, sometimes spineless; sap milky
Buds: terminal absent; lateral partly concealed, rounded, often paired, red-brown; scale margins hairy (ciliate)
Bark: irregular furrows separating broad ridges, orange-brown
Wood: heavy, hard, strong, flexible, durable; bright orange, becoming brownish upon exposure; splits easily

Morus L. - mulberry

Only two of approximately a dozen species of mulberry are native to North America. Several species are grown in Europe and Asia for their sweet fruits. This practice is less common in the U.S. where they are considered valuable for providing food and habitat for wildlife. Roots and bark of some species yield a yellow dye.

A western Asian species, black mulberry, *M. nigra* L., has become naturalized in the U.S., especially in the southeast. Its leaves resemble those of red mulberry.

Recognition

Habit: trees, shrubs; deciduous
Leaves: alternate, lobed or undivided; margins toothed
Flowers: clusters (catkins) axillary, dense, on new growth; bisexual, monoecious or dioecious; greenish
Fruits: egg-shaped to cylindric, fleshy aggregrates (syncarps); red becoming white, purple or black
Twigs: unarmed, sap milky
Buds: terminal absent; lateral egg-shaped
Wood: soft, lightweight, moderately durable; splits easily

Morus alba L. - white mulberry

White mulberry—long cultivated in China as silkworm food—was introduced into the U.S. to promote the silk industry. That venture failed, but the trees thrived. White mulberry was also planted for shade and hedges. It escaped to become naturalized throughout the eastern U.S.

Natural crosses with our native red mulberry, *M. rubra*, produces hybrids that make precise identification of many of our mulberries nearly impossible. *M. alba* var. *tatarica* (L.) Ser., Russian mulberry, was planted extensively in the shelterbelts of the Great Plains Forestry Project, often in the shrub row, a function for which it was ill-adapted (Read 1958). Survival and growth was good except on the shallow soils of the uplands. Read noted freeze damage in northern Nebraska.

Moraceae *Morus*

Russian mulberry is undoubtedly the source of many of the naturalized white mulberry trees found in the central Great Plains. Credit is due to our fruit-eating birds for its wide distribution. These small, fast-growing, competitive trees are adapted to a wide range of soil and climatic conditions. They are seldom planted as ornamentals because of their unkempt appearance, but skillful pruning and patience can produce a presentable tree.

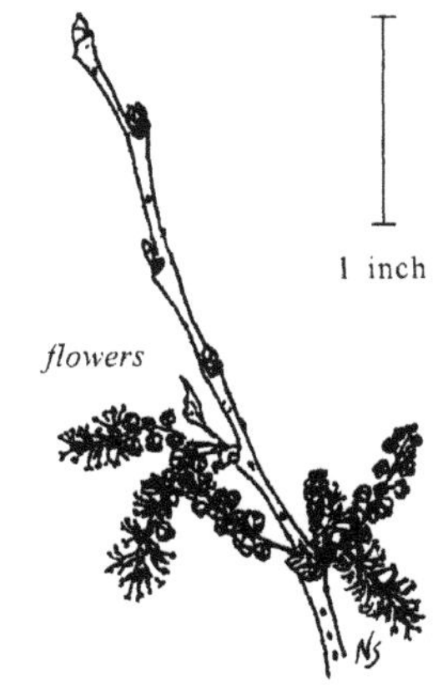

***Morus* species**

Planning

Height: up to 40 ft (12 m); National champion, 1983, 75 ft (22.9 m), Michigan

Habit: tree; crown rounded, often irregular; early summer fruit

Site: sun, part shade; pH and soil adaptable, tolerates drouth and pollution; zone 4

Use: wildlife habitat, windbreak

Recognition

Leaves: egg-shaped, variously lobed; tip pointed, base heart-shaped to rounded; upper surface glabrous, hairy beneath, mostly on veins; petioles grooved above, up to 1 in (25 mm) long

Fruits: ripen to white, purple or black; sweet but usually insipid

Twigs: glabrous, slightly zigzag; lenticels narrow, elliptic, light-colored; leaf scars half-round, bundle scars numerous; juice milky

Buds: appressed against twig, about 0.15 in (4 mm) long, typically as wide; pointed to rounded; scales 2-ranked, yellow-brown, glabrous with few marginal hairs

Bark: thin, orange-tinted becoming deeply furrowed, tight, yellow-brown ridges

Morus alba

Morus rubra L. - red mulberry

Red mulberry, one of two mulberry species native to North America, is found south of the Great Lakes, east and south to the Atlantic coast, west to the eastern half of Texas, and north to southeastern Nebraska. It prefers moist, fertile, lowland soils.

The other native, *M. microphylla* Buckl., is similar to red mulberry except that its leaves are smooth or finely hairy beneath and its fruits are smaller. It is found in the southwest, from Texas to Arizona, and in northern Mexico along streams and mountain canyons.

Although mulberries abound throughout the prairies, we have found few trees in Nebraska that are bona fide specimens of red mulberry. However, Freeman (1992) states that they are fairly common in the upland woodlands of the southern Great Plains.

Red mulberry has seldom been planted. Its fibrous bark has been used in making rope and coarse cloth, and its wood is used for fence posts and furniture. Fruits of the red mulberry are reputed to be more tasty than those of white mulberry. Using fruit taste as a criteria, we have discovered a few trees that may be closely related to red mulberry.

Planning

Height: up to 50 ft (15 m); Nebraska champion, 1978, 53 ft (16.1 m), Auburn; National champion, 1981, 72 ft (21.8 m), Michigan
Habit: tree; crown rounded
Site: sun, part shade; pH and soil adaptable, prefers moist, fertile soil; zone 4
Use: wildlife habitat, shade, fruit, lumber, fuel

Recognition

Leaves: broadly oval to nearly round, seldom lobed except on sprouts; tip pointed, base rounded to somewhat heart-shaped; upper surface smooth to somewhat wrinkled, lower usually densely short-hairy; petioles round, glabrous or short-hairy, about 1 in (25 mm) long
Fruits: ripen purple to black; sweet
Twigs: glabrous, slightly zigzag; lenticels elliptic, light-colored; leaf scars nearly round, concave, many bundle scars
Buds: about 0.25 in (6 mm) long, longer than broad, usually divergent from twig; scales dark-margined
Bark: flaky, plates loose at tips on large trunks

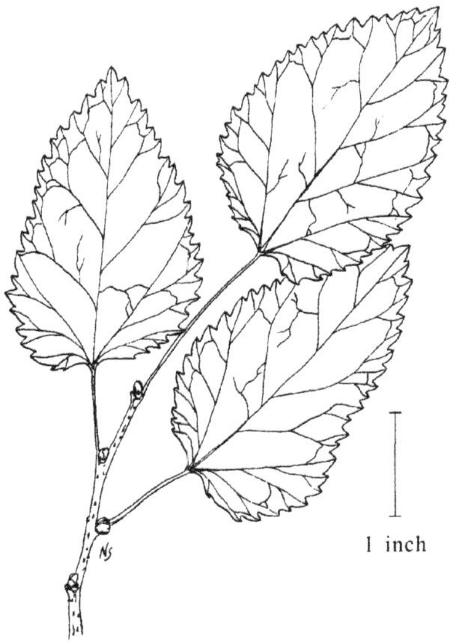

Morus rubra

Order Juglandales

Two families that include about 60 species form the order Juglandales (Cronquist 1981).

Recognition

Habit: trees, rarely shrubs; deciduous
Leaves: alternate, rarely opposite; compound; aromatic
Flowers: clusters, individually inconspicuous; unisexual or bisexual; sepals small, scale-like or none, petals wanting
Fruits: nuts or drupe-like

Juglandaceae A. Richard ex Kunth - walnut family

Only two of the walnut families' eight genera occur naturally in North America. Both of these genera—*Carya* and *Juglans*—are found in the prairie region. These valuable timber trees provide food for man and animals. Squirrels aid in their natural reproduction—a seedling often marks the burial location of nuts not needed or forgotten.

Direct seeding is an effective method of establishing these species unless squirrels abound. These animals have an uncanny way of locating buried nuts. Setting up feeding stations could reduce depredation.

The wind-pollinated flowers usually appear as leaves unfold, or soon thereafter. The sometimes sweet and tasty fruits mature at the end of the first growing season and drop with the onset of frosty weather. Production tends to be cyclic with good yields every two or three years, but crops are occasionally spoiled by late spring freezes that injure flowers, or cool, damp weather that hinders pollination.

Caucasian wingnut, *Pterocarya fraxinifolia* (Lam.) Spach, a broad, spreading tree is extolled by Dirr (1990) as a very handsome shade tree. The unusual winged fruits hang in long, pendulous strings. It is not likely to be hardy in the northern prairies, but a hybrid, *P.* ×*rehderiana* Schneid., might be worth a try.

Recognition

Habit: trees, rarely shrubs

Leaves: leaflets on each side of a common axis (rachis) with a single on the end (odd pinnately compound) or with 3 leaflets (trifoliate); alternate (except for two Central American genera); stipules none

Flowers: unisexual (monoecious), appearing with or after the leaves; male clusters (catkins) drooping, on previous year's growth; female solitary or few-flowered clusters (spikes or racemes) terminating new growth

Fruits: fleshy or pulpy surrounding stony shells enclosing kernels (nuts); husks sometimes splitting; cotyledons remain enclosed in shell at germination

Carya Nutt. - hickory

Although fossils of hickories have been found throughout the Northern Hemisphere, most of the present day species are found in the eastern U.S., adjacent to Canada, and in Mexico. Four species are native to southeastern Asia.

Great variability within the genus has prompted the naming of a number of varieties. Hybridization, both natural and man-induced, has compounded confusion in nomenclature and identification. Pecan-hickory hybrids are named hicans. The most important are *C. illinoensis* × *C. cordiformis* and *C. illinoensis* × *C. ovata*.

The very strong, heavy and hard wood of the true hickories is less in demand now than it was in the days of our grandfathers, when tough shock-resistant wood was needed for tool handles, ladders and various agricultural implements. Several species—important sources of energy in the form of wood and charcoal—provide us with tasty fruit. Most species of these durable, long-lived, handsome trees are planted for nut production. They are adapted to most of the eastern prairies but will develop chlorosis if planted in alkaline soils.

Hickories are easily distinguished from walnuts by the homogenous, pale pith of their twigs, and by their splitting (dehiscent) fruit husks and smooth nut surfaces. Seedlings develop long taproots with very little top growth during the first few years. They become difficult to transplant after one year of age.

Recognition

Habit: trees

Leaves: 3-17 leaflets, each with unequal base; margins finely toothed

Flowers: male on 3-branched (ternate) catkins, stamens 3-10; female on 2- to 10-flowered spikes; ovary enclosed in a whorl of 4 bracts (involucre), stigmas 2

Fruits: nuts nearly round to oblong, smooth or slightly wrinkled (rugose), often angled; encased in woody husks splitting along 4 sutures when ripe

Twigs: covered with star-shaped hairs (stellate pubescence); lenticels pale, conspicuous; pith solid (homogenous); leaf scars heart-shaped to 3-lobed, bundle scars many

Buds: terminal larger than lateral; scales overlapping (imbricate) or non-overlapping (valvate)

Bark: hard, compact, smooth, gray, becoming scaly, ridged or plated

Grouping Key for *Carya* (hickories)

(Harlow and Harrar 1968)

Leaflets 5-17; fruit husks winged at sutures; bud scales 4-6, meeting at edges (valvate), scurfy, brown to yellow: **subgenus *Apocarya*** (pecan hickories)

Leaflets 3-9 (mostly 5-7); fruit husks not winged, occasionally ribbed at sutures; bud scales overlapping (imbricate): **subgenus *Eucarya*** (true hickories)

Subgenus *Apocarya* - pecan hickories

Carya cordiformis (Wangh.) K. Koch - bitternut hickory

The natural range of bitternut hickory includes at least portions of nearly every state east of a line extending from Minnesota through eastern Nebraska south through eastern Texas. As the common name implies, its kernels are bitter and probably seldom eaten by animals. Bitternut hickory is commonly found in woodlands in association with the oaks. This hickory is a relatively fast-growing, beautiful tree with bright yellow autumn foliage, but it is seldom used for landscape purposes.

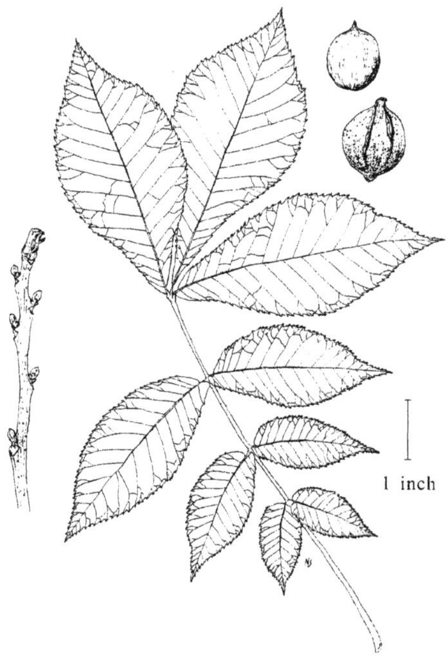

Carya cordiformis

Planning

Height: up to 80 ft (24 m); Nebraska champion, 1977, 101 ft (30.6 m), Barada; National champion, 1982, 137 ft (41.5 m), Michigan
Habit: tree; crown cylindrical, slender, branches ascending; brilliant yellow autumn foliage
Site: sun, part shade; pH neutral to slightly alkaline, soil adaptable, drouth tolerant; zone 5
Use: shade, wildlife habitat, lumber, fuel

Recognition

Leaves: leaflets 5-9; lateral sessile, lance-shaped (lanceolate); tip long-pointed, base uneven, rounded to tapering; terminal stalked, egg-shaped (obovate); toothed (serrate); glabrous, yellow-green above, hairy beneath; petioles hairy
Flowers: male clustered in slightly hairy catkins, female single or in pairs on short terminal spike
Fruits: rounded (globose), up to 1.2 in (3 cm) long, tip sharp-pointed; husk yellow-green, slightly scurfy, 0.1 in (1-2 mm) thick, 4 sections slightly winged, partial splitting at maturity; nut smooth, pale brown, slightly egg-shaped, laterally compressed, tip short, sharp-pointed; shell thin, kernel bitter
Twigs: stout, somewhat downy, becoming glabrous, greenish brown
Buds: somewhat angled, sulfur yellow, scurfy; scales not overlapping (valvate)
Bark: close and firm, remaining smooth for many years, becoming narrow ridged

Carya illinoensis (Wangh.) K. Koch - **pecan**

Pecan, the tallest of the hickories, is native to the Mississippi and Ohio River valleys from Iowa, Illinois and Indiana, south to the Gulf and west into central Texas, Oklahoma and eastern Kansas, with outliers in Alabama and central Mexico. It is planted worldwide for its fruit where warm summer days and long frost-free periods combine to allow fruits to mature.

Carya illinoensis

Over 500 pecan cultivars have been selected from a highly variable population resulting from mutations and natural and controlled pollinations. In addition to the fruit's edible value to man and wildlife, the wood is valued for interior finishing and veneers. The wide-crowned pecans serve as shade and ornamental trees.

Fruit production is erratic throughout much of the central prairie region due to vagaries of climate. Fortunately, animals can utilize immature nuts. Like all members of Juglandaceae, the pecan is best established by direct seeding or transplanting at one or two years of age.

The Northern Pecan Research Project at UNL is designed to evaluate fruit productivity of pecans in southeastern Nebraska. Eight cultivars have been selected that mature their fruits within a normal frost-free period of 150 to 185 days (Gustafson 1990). These include 'Lucas', 'Starking Hardy Giant', 'Peruque', 'Colby', and 'James', with frost-free requirements ranging from 158 days for 'Lucas' to 176 days for 'James'.

Planning

Height: up to 85 ft (26 m); Nebraska champion, 1977, 90 ft (27.3 m), Brownville; National champion, 1980, 143 ft (43.3 m), Tennessee
Habit: tree; crown wide-oval, symmetrical
Site: sun; pH neutral to slightly alkaline, prefers deep, fertile, moist soil; zone 5
Use: fruit, shade, wildlife habitat, lumber

Recognition

Leaves: 12-20 in (30-50 cm) long; leaflets 9-17, lance-shaped (oblong-lanceolate), often curved (falcate), long-pointed, base unequal; upper surface nearly glabrous, dark yellow-green; paler, glabrous to slightly hairy beneath
Flowers: male catkins usually in 3's, stamens 3-6, anthers yellow; female spikes terminal, flowers 2-4
Fruits: oblong, rounded at base, tip pointed, 1-2 in (25-50 mm) long; husk 4-winged entire length, thin, splitting at maturity; nut thin-shelled; kernel oily, sweet
Twigs: stout, hairy, becoming glabrous; lenticels orange-brown
Buds: hairy, yellow-brown, becoming scurfy
Bark: thick, light brown, becoming darker, ridged and scaly
Wood: hard, heavy, somewhat brittle

Subgenus *Eucarya* - true hickories

Carya glabra (Mill.) Sweet - pignut hickory

Pignut hickory, though common in the eastern hardwood forests from Michigan and Massachusetts to the Gulf states, barely reaches into the southeastern prairie region of Illinois and Missouri. Seldom planted, it deserves trial plantings to determine its use for shade and wildlife.

C. ovalis (Wangh.) Sarg., red hickory, is recognized by some authorities as a separate species. The principal difference between it and *C. glabra* is that the husk of the former promptly splits to the base. Many trees are intermediate in character (Little 1953).

Planning

Height: up to 50 ft (15 m); National champion, 1985, 190 ft (57.6 m), North Carolina
Habit: tree; crown oval, open
Site: sun; pH neutral to slightly alkaline, prefers fertile, deep, moist soil but grows on dry ridges; zone 5
Use: wildlife habitat, lumber, fuel

Recognition

Leaves: leaflets 5-7, the largest at tip; lance-shaped, narrow, widest at base or near middle; margin sharp-toothed; dark yellow-green and glabrous above, paler beneath, midrib and veins hairy
Flowers: male catkins hairy, stamens 4; female clusters few-flowered
Fruits: oval to widest near tip, 1-2 in (25-50 mm) long; husk smooth, about 0.1 in (2-3 mm) thick, not splitting uniformly at maturity; nut thick-shelled, 4-ribbed to smooth; kernel sweet, often bitter
Twigs: slender, glabrous with numerous light-colored pores (lenticels)
Buds: terminal less than 0.5 in (13 mm) long, egg-shaped to subglobose; lateral smaller, somewhat divergent; outer scales reddish brown, nearly glabrous; shedding in winter, exposing inner scales
Bark: smooth becoming ridged and sometimes scaly with age, similar to shagbark hickory

Carya laciniosa (Michx. f.) Loud. - shellbark hickory

This slow-growing, long-lived tree is native to the upper Mississippi and Ohio River valleys including southeastern Kansas to southeastern Iowa, east to Pennsylvania and south to Tennessee, with outliers much farther south and east. It has also been found in southeastern Nebraska (Pool 1951; Stephens 1973).

Nut fanciers have planted shellbark hickories throughout the eastern prairies for their large, edible nuts. Patience and long life are requisites for this task. 'Ross', 'Stephens', and 'Totten' are some of the most promising cultivars.

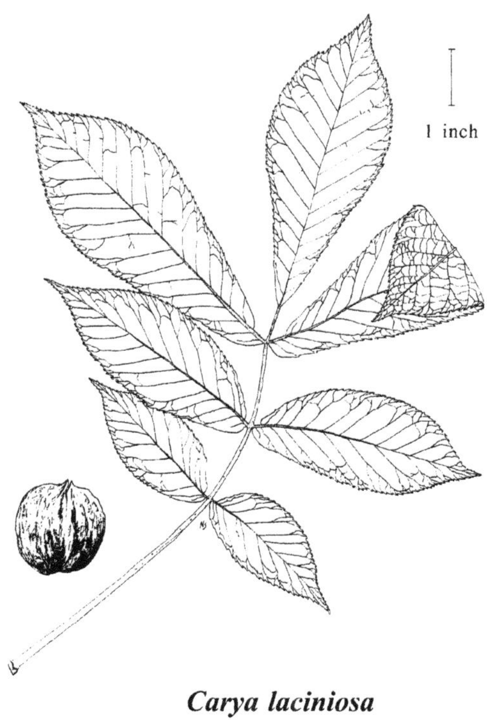

Carya laciniosa

Juglandaceae *Carya*

Planning

Height: up to 65 ft (20 m); National champion, 1986, 105 ft (31.8 m), Virginia
Habit: tree; crown narrow, oblong
Site: sun, shade as a seedling; pH neutral to slightly alkaline, best on deep, fertile, moist soils, but adaptable; zone 5
Use: fruit, shade, wildlife habitat, lumber, firewood

Recognition

Leaves: leaflets 5-9, usually 7; egg-shaped (obovate to oval); toothed (finely serrate) with single hairs along edge (ciliate); dark yellow-green, shiny above, paler and velvety hairy below; petioles often persistent on trees after leaflets fall
Fruits: solitary or in pairs; subglobose to obovoid, about 2 in (5 cm) long; husk woody, brown with minute scales or orange spots, 0.3-0.5 in (7-12 mm) thick, 4-valved, splitting to base at maturity; nut ridged, angled, tip pointed, base rounded, shell thick and hard; kernel sweet
Twigs: orange-brown, hairy, becoming glabrous; lenticels large, orange
Buds: terminal about 1 in (25 mm) long, lateral buds one-fourth of terminal; egg-shaped, blunt; outer scales keeled, tawny hairy, middle scales white-hairy; inner scales silky, orange-hairy, enlarging when buds open
Bark: gray, smooth, breaking into plates that curve away from trunk

Carya ovata (Mill.) K. Koch - **shagbark hickory**

Shagbark hickory grows in upland woodlands in association with oaks throughout most of the eastern U.S., including extreme eastern Nebraska, Kansas, Oklahoma and Texas. This picturesque, long-lived, slow-growing tree bears large crops of delicious nuts. The few that have been planted for ornament and shade in eastern Nebraska have proved to be hardy and free of serious pests.

Seedling transplants are difficult to establish. Success is greater by planting nuts directly into their permanent home. In fact, squirrels seem to establish these trees more successfully than man does.

Shagbark hickory does not normally bear fruit until after age 30 years, about the time the bark becomes shaggy. A tongue-in-cheek theory has it that the rough trunk evolved to foil climbing squirrels, thus allowing fruit to mature and fall before being harvested.

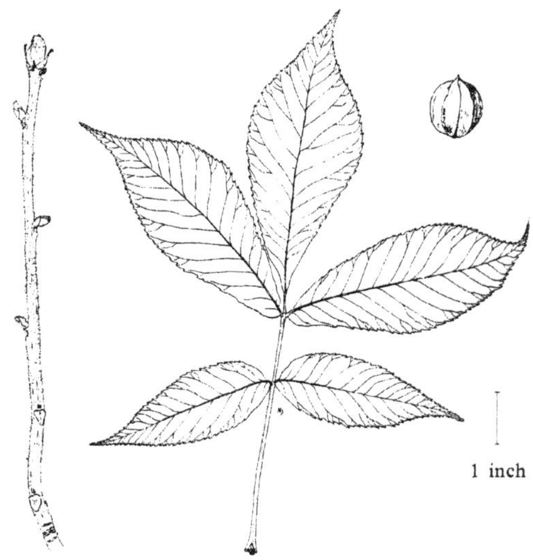

Carya ovata

Shagbark hickory nuts were an important food for early Native Americans. Though kernel extraction is tedious, the meat is a delicacy enjoyed by many present day Americans. This problem has been alleviated by identifying shagbark hickories which bear easily extractable nut meats. **Nut Tree Culture in North America** (Jaynes 1979) and **Nut Tree Cultivars for Nebraska** (Gustafson 1990) describe cultivars and provide cultural information.

Planning

Height: up to 80 ft (24 m); Nebraska champion, 1980, 86 ft (26 m), near Indian Cave State Park; National champion, 1981, 153 ft (46.4 m), South Carolina
Habit: tree; crown oblong to oval; bark shaggy on old trees
Site: sun, part shade; pH neutral to slightly alkaline, drouth tolerant; zone 4
Use: shade, wildlife habitat, lumber, firewood

Recognition

Leaves: leaflets 5-7 (usually 5); elliptic to longer (lanceolate) sometimes wider above middle (obovate), up to 6 in (15 cm) long; toothed with tufted hairs along the edges; upper surface yellow-green, glabrous; paler, occasionally hairy beneath, especially on veins
Flowers: appear after leaves; female solitary or 2-3
Fruits: globe-shaped to sometimes longer than broad, 1-1.5 in (25-38 mm) long; husk dark brown, granular, yellowish scales; 4-ribbed, readily splitting to base at maturity, 0.3-0.5 in (7-12 mm) thick; nut 4-angled, slightly flattened; kernel sweet
Twigs: stout, hairy, becoming glabrous; lenticels numerous, small
Buds: terminal broadly egg-shaped, about 0.7 in (18 mm) long; outer scales glabrous or hairy, keeled, tip often long-pointed; middle with short point (apiculate), white hairy; inner orange-hairy, persistent and enlarging as bud opens, becoming yellow-green tinged with red, dropping when leaves are half-grown; lateral much smaller
Bark: smooth and gray becoming shaggy as it breaks into thin plates which curve away from trunk

Carya tomentosa (Poir.) Nutt. - mockernut hickory

Mockernut hickory—so named because of the difficulty in extracting the small, sweet kernels from the nut—is native to most eastern U.S. states with the western range limit extending from southeastern Iowa to extreme eastern Kansas and south to eastern Texas. Pool (1951) states that it was reported to be, but not confirmed, growing naturally in a narrow belt along the eastern boundaries of Richardson and Cass counties in Nebraska.

Mockernut hickory is found on a wide variety of soil types including dry, sandy sites (which indicates that it might be adapted to some of the drier prairies). The wood is among the finest of the hickories with excellent bending qualities. Pioneers extracted a black dye from the mockernut by boiling the bark in vinegar solution (Elias 1980).

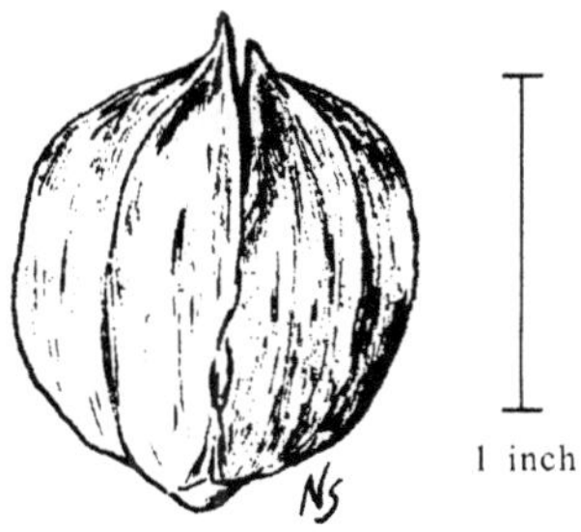

Carya tomentosa

Planning

Height: up to 65 ft (20 m); National champion, 1989, 156 ft (47.3 m), Massachusetts
Habit: tree; crown oval
Site: sun; pH neutral to slightly alkaline, prefers fertile, moist soils, tolerates dry, rocky soils; zone 4
Use: shade, wildlife habitat, lumber, firewood

Recognition

Leaves: leaflets 7-9 (rarely 5); up to 8 in (20 cm) long, widest above middle (obovate); toothed, tufted hairs along the edges; upper surface dark yellow-green, glabrous except on veins above, paler beneath with clustered hairs and red glands; aromatic when crushed

Flowers: appear after leaves are partly grown; female clusters usually 2-4 flowers

Fruits: nearly globe-shaped, 1-2 in (25-50 mm) long; husk 0.1-0.2 in (3-6 mm) thick, splitting to near the base, granular to hairy, minute clear glands; nut nearly round to egg-shaped, tip pointed, base rounded, 4-angled, slightly flattened; shell thick, hard; kernel sweet

Twigs: stout, hairy, becoming glabrous, lenticels conspicuous

Buds: terminal nearly round to egg-shaped, much larger diameter than twig, hairy, red-brown; outer scales keeled, deciduous in autumn, inner scales silvery brown, hairy, enlarging in spring; lateral much smaller

Bark: dark gray, shallow furrows, net-like pattern

Juglans L. - walnut

Six walnut species are native to the U.S. Fifteen to 20 other species inhabit temperate and tropical regions elsewhere in the world. Fossils indicate that walnuts once were numerous and more widely distributed.

Walnuts are noted for their valuable wood and edible nuts. Excellent sources of information about nut production and the many cultivars are **Nut Tree Culture in North America** (Jaynes 1979) and **Nut Tree Cultivars for Nebraska** (Gustafson 1990).

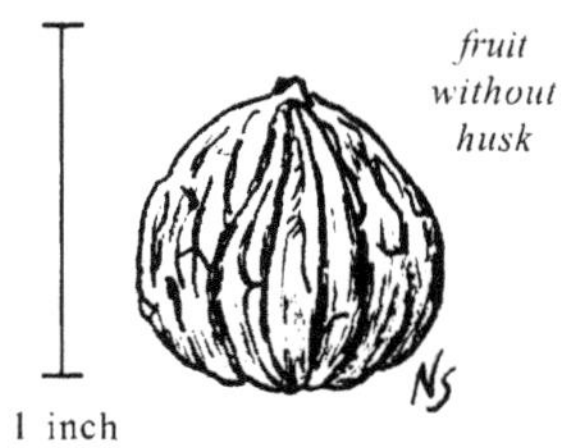

Juglans microcarpa

Juglans microcarpa Berland., Texas black walnut—a small, deep-rooted native of extreme southern Kansas and south into Mexico—grows in dry rocky ravines and floodplains in a region of periodic drouth. A young tree growing on East Campus of UNL has produced fruit. A specimen grows at the PERC in Fort Collins without any apparent winter injury or other serious problems (Fairchild and Klett 1993). This walnut is worthy of test plantings in the western prairies.

Manchurian walnut (*J. mandshurica* Maxim.) also deserves attention. The fruits are similar to those of the black walnut in shape and size, but their nuts are distinguished by eight prominent, sharply edged ridges. Fairchild and Klett (1993) state that this walnut is an attractive, adaptable tree with no problems in the Fort Collins area.

Recognition

Habit: trees; deciduous

Leaves: leaflets 9 to many, short stalked, long-pointed, unequal bases; margins finely toothed; aromatic

Flowers: appearing with or shortly after leaves; male catkins preformed, appearing in fall as scaly, cone-like buds; unbranched, pendulous; female clusters (racemes) 2- to 8-flowered, terminal

Fruits: globe-shaped, outer husk not splitting, or finally separating into 2 valves; nut variably furrowed; kernel sweet

Twigs: stout, pith becoming chambered at end of first season; leaf scars raised, shield-shaped to 3-lobed; bundle scars in 3 U-shaped clusters

Buds: terminal hairy, scales 2-4 visible; lateral smaller, often two, one above the other (superposed)

Bark: scaly or furrowed

Juglans ailantifolia Carr. **var.** *cordiformis* Mak. - **heartnut**

Heartnut, a variety of Japanese walnut, is named for its heart-shaped nut. Many cultivars have been selected for their hardiness, fruitfulness and good cracking characteristics (Jaynes 1979; Gustafson 1990).

Heartnut has been planted by a few nut fanciers in the southeastern prairies. It may suffer winter injury and tends to flush growth early in the spring which invites frost injury to flowers and foliage. Fruits may not always mature before freezing weather. Several small trees in Prairie Pines Arboretum survived with no injury during the difficult winter of 1991-92. A sharp freeze at Halloween, 1991, injured many woody species.

Heartnut crossed with butternut (*J. cinerea*) have been named buartnuts. The resulting fruits are said to be an improvement over the fruit of either parent. Heartnut has also been crossed with *J. regia* (Jaynes 1979).

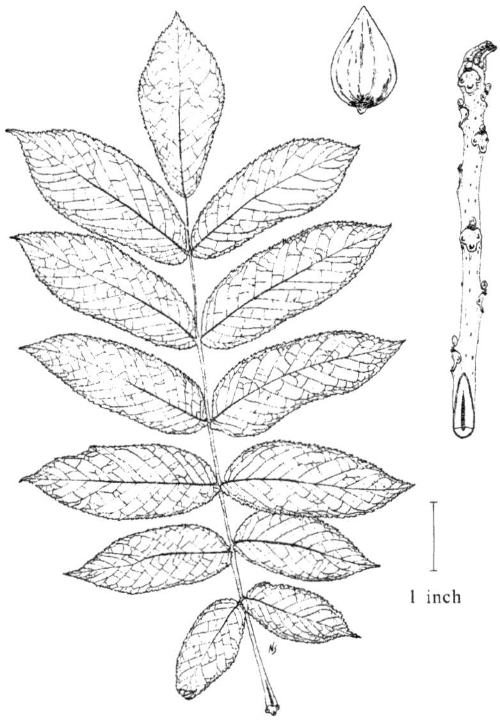

Juglans ailantifolia

Planning

Height: up to 50 ft (15 m)
Habit: tree; crown broad, round
Site: sun; pH and soil adaptable; zone 5
Use: shade, fruit

Recognition

Leaves: leaflets 17-23, toothed, glabrous above, sparsely hairy below, especially on midrib; leaf stem (rachis) sparsely hairy
Flowers: female clusters, 12-20 flowers
Fruits: globose to egg-shaped, hairy, sticky; nuts heart-shaped to egg-shaped, flattened, smooth with a shallow longitudinal groove in middle of flat side
Twigs: rusty, short hairy, becoming smooth; leaf scars shield-shaped, bundle scars 3 clusters; pith brown, chambered
Buds: hairy, terminal much larger than lateral

Juglans cinerea L. - **butternut**

This fast-growing, short-lived tree occurs naturally in the eastern U.S. hardwood forests on well-drained, bottomland soils. Most authorities list its western range limit as central Minnesota south through Missouri to northern Arkansas. However, Pool (1951) indicates that scattered butternut trees have been found growing naturally in southeastern Nebraska.

Although a number of cultivars have been selected, butternut has not been widely planted for ornamental purposes or for its nuts and wood. Black walnut is longer lived and produces more valuable wood than butternut, and nuts of the former are generally considered to be tastier (Jaynes 1979; Gustafson 1990). The kernel becomes rancid in a short time unless stored in a freezer. Butternut's inner bark has mild cathartic properties, and fruit husks have been used to make orange and yellow dyes.

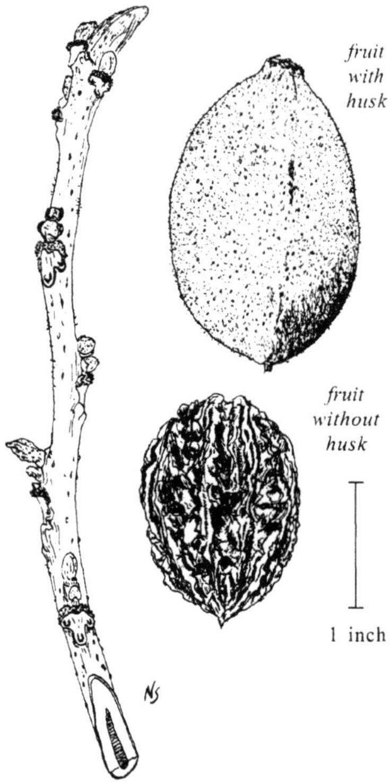

Juglans cinerea

Planning

Height: up to 50 ft (15 m); Nebraska champion, 1980, 46 ft (13.9 m), Brownville; National champion, 1989, 88 ft (26.7 m), Oregon
Habit: tree; crown rounded, spreading
Site: sun, part shade; pH adaptable, prefers deep, fertile, well-drained soil; drouth tolerant; zone 4
Use: fruit, wildlife habitat, naturalizing
Pests: virulent canker

Recognition

Leaves: leaflets 11-17, up to 4 in (10 cm) long; oblong to lance-shaped, sometimes wider above middle, toothed; yellow-green and slightly wrinkled above, paler below, sparsely hairy; leaf stem (rachis) rusty hairy
Flowers: male catkins hairy, yellow-green; female spike 4-7 flowers; at end of current growth
Fruits: egg-shaped to oblong; husk semi-fleshy, greenish bronze, sticky hairy; nuts egg-shaped, tips pointed, deeply corrugated with sharp ridges, 4-ribbed; kernels oily, sweet
Twigs: stout, pores (lenticels) conspicuous; leaf scars large, 3-lobed to shield shaped, 3 U-shaped clusters of bundle scars, downy pad above; pith dark, becoming chambered late in first year
Buds: terminal cone-shaped, 0.5-0.7 in (12-18 mm) long, whitish, hairy; lateral smaller
Bark: ashy gray, fissured, becoming a deeply furrowed, diamond-shaped pattern
Wood: light, soft

Juglans nigra L. - black walnut

Almost every tree fancier would love to own a grove of black walnuts 70 feet (21 m) tall with 24-inch (60 cm) diameter stems. Walnut veneer is valuable! Fruits would be a welcome bonus, especially to the squirrel population. Few homeowners desire black walnut as a yard tree, however, because of the litter produced by dingy leaves and messy fruits. But the tree reciprocates by not performing well in many urban yards because it needs deep, fertile soils.

This native of the eastern hardwood forests is found in almost every state east of the Great Plains, and it extends far into the Plains along many river bottoms. Black walnut has been planted throughout the prairie region for various purposes with mixed success. Even though it takes a human's lifetime to produce a prime veneer log on fertile soil, many people are planting black walnut as an investment for their children and grandchildren.

An agricultural crop planted between wide-spaced walnut rows (intercropping) allows good tree growth and wood quality and provides an early financial return from annual crops and nuts. Black walnut culture and utilization as an alternative farm crop is currently being studied at the University of Missouri (Kincaid, et al. 1982).

The black walnut's wide geographic range on a great variety of site conditions has led to a species of great genetic diversity. A University of Nebraska professor conducted one of the first provenance studies of the species (Emerson 1906). He found that differences in date of autumn leaf drop, twig maturation and amount of winter dieback were related to latitude of seed origin. Bey (1973) reports that black walnut trees moved 200 miles north of their origin grow as fast and often faster than trees from local or more northerly sources.

Improvements in wood and fruit quality and quantity and disease resistance are being made by selection and breeding. Nut growers are harvesting nuts from many of the 400 black walnut cultivars that they have selected for ease of cracking and high yield of tasty kernels (Jaynes 1979; Gustafson 1990).

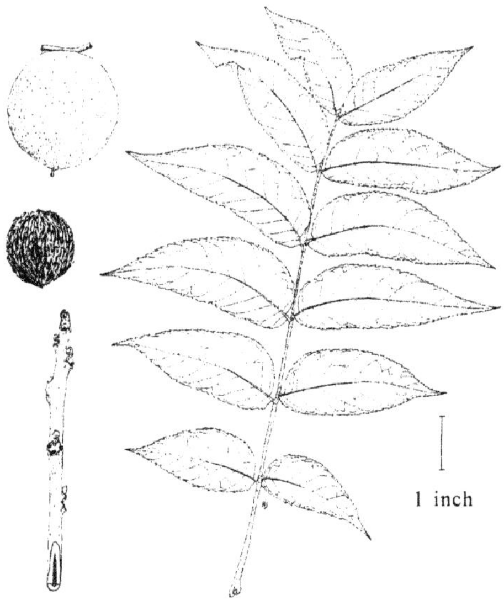

Juglans nigra

Black walnut trees can be effectively established by field planting nuts or transplanting one-year-old seedlings. Although insects and diseases are not normally serious problems for this species, their impact can be minimized by mixing walnut with other hardwoods that are compatible—oak, hickory, ash, maple and other native forest species. Problems caused by variable soils can also be reduced by a mixed-species plantation.

Planning

Height: up to 70 ft (21 m); Nebraska champion, 1977, 77 feet (23.3 m), Pauline; National champion, 1991, 130 ft (40 m), California

Habit: tree; crown oval to rounded

Site: sun, part shade; pH in the near neutral range; prefers fertile, well-drained soils; tolerates sands, low fertility, and drouth; zone 4

Use: lumber, fruit, wildlife habitat, windbreak, shade

Recognition

Leaves: leaflets 11-23, ovate-oblong to lance-shaped, up to 5 in (13 cm) long; tips pointed, margins toothed; terminal leaflet often poorly developed or missing; upper surface yellow-green and glabrous, pale and hairy below, mostly on midrib and veins; leaf stem (rachis) hairy, rusty

Fruits: rounded, single or in pairs; husk thick, yellow-green, granular, somewhat hairy, turning dark brown at maturity, not splitting; nuts deeply furrowed with sharp ridges; kernels 4-lobed, oily, edible

Twigs: hairy, becoming glabrous; leaf scars shield-shaped, bundle scars in 3 U-shaped clusters, no hairy fringe on upper margin; pith light brown, chambered

Buds: terminal egg-shaped, about 0.3 in (7 mm) long, silky hairy; lateral buds smaller, sometimes superposed with male flower buds below

Bark: scaly, becoming thick and deeply furrowed, dark brown to blackish

Wood: heavy, strong, hard, various shades of brown

Myricaceae *Juglans*

Juglans regia L. - Persian walnut

Most of us know fruits of this species as "English" walnuts that we often find in our Christmas stockings. This cherished, staple food of the ancient Greeks and Romans was introduced to much of the world through the travels and conquests of these two nations. The English walnut was named for the settlers who brought it to America. Persian walnuts were often injured by cold in the northeastern U.S., which meant that we had to wait for California to be settled to enjoy the fruit in quantity.

Persian walnut grows naturally from the Carpathian Mountains of eastern Europe into southern Russia, Turkey and east to Afghanistan. In the 1930s a concerted effort was made to introduce hardy strains from the rugged Carpathian Mountains of Poland to the eastern U.S. Hardy trees are growing in backyards and small orchards. A few minimally productive Persian walnut trees survive and occasionally produce fruit in southern Nebraska as far west as McCook and North Platte.

These small, round-topped trees are excellent street and shade trees. They require little maintenance and may yield fruit as a bonus at about eight years of age. Two or more trees should be planted to insure good pollination. Some are not self-fertile because female flowers are not receptive at time of pollen shedding. For the best chance of success, plant Persian walnuts on high sites with good air drainage to minimize injury from untimely freezes in spring and fall.

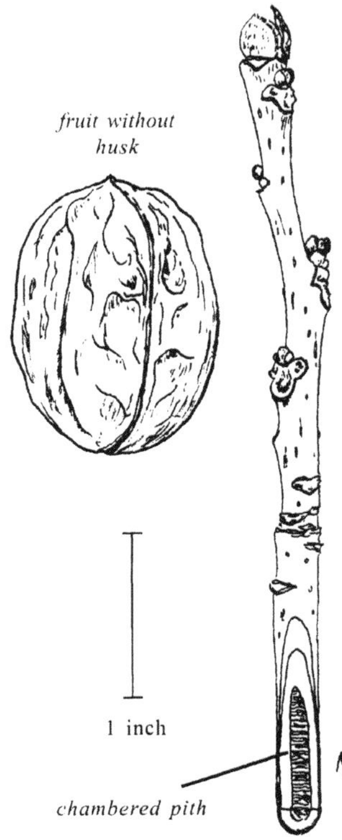

Juglans regia

Information about hardy cultivars of Persian walnut and their culture can be found in **Nut Tree Culture in North America** (Jaynes 1979). There is good reason to believe that a coordinated program of testing and breeding seed introductions from northern climates could result in cultivars hardy and fruitful throughout much of the U.S.

Planning

Height: up to 50 ft (15 m); Nebraska champion, 1979, 51 ft (15.5 m), Lincoln
Habit: tree; crown rounded to spreading
Site: sun, part shade; pH neutral to somewhat acidic; prefers loamy, well-drained soil, tolerates drouth; zone 5
Use: shade, fruit

Recognition

Leaves: pleasantly fragrant when crushed; leaflets usually 5-9, egg-shaped to elliptic-oblong, up to 5 in (13 cm) long; tips pointed, margins smooth (entire); surfaces glabrous except for a few tufts of hairs on veins beneath; leaf stem (rachis) glabrous
Fruits: nearly round, husk smooth; nuts more or less wrinkled, shell thickness variable; kernels edible
Twigs: glabrous, greenish brown, becoming gray; pores (lenticels) light-colored; pith chambered, brown; leaf scars 3-lobed, bundle traces in 3 clusters
Buds: terminal much larger than lateral; scales short-hairy, scurfy, brownish, barely overlapping; male flower buds small, often below leaf buds (superposed)
Bark: silvery gray, smooth, becoming rough with age

Order Myricales

One family occupies this order.

Myricaceae Bl. - bayberry family

Three genera in this family of about fifty species are widely distributed throughout temperate and tropical regions of the world. These small trees and shrubs are important because they can grow on soils of low fertility. Bayberries effectively control erosion on acidic sandy or peaty soils where they function as soil builders by adding nitrogen through root nodules.

Recognition

Habit: trees, shrubs; deciduous or evergreen

Leaves: alternate, simple; margins entire, toothed or pinnately cleft (pinnatifid); fragrant due to glandular hairs

Flowers: clusters (aments) in axils of bracts; unisexual (monoecious or dioecious); without petals

Fruits: dry (drupaceous), nut-like; coat leathery with waxy grains

Comptonia L'Her. - sweetfern

This genus is monotypic.

Comptonia peregrina (L.) Coult. - sweetfern

This native to eastern Canada and the northeastern U. S. colonizes by suckering profusely, thereby serving to effectively control soil erosion. For this plant with fern-like, aromatic foliage to be successful in the prairies, we must modify soils to its liking. Dirr (1990) grew a fantastic plant in a peat-soil mix in his Illinois garden.

Planning

Height: up to 4 ft (1.2 m)

Habit: shrub; round to flat top, suckering

Site: sun, partial shade; acidic sandy or peat soils; zone 3

Use: erosion control, specimen

Benefit: fixes nitrogen, a soil builder

Recognition

Leaves: deciduous, linear oblong, up to 4 in (10 cm) long and about 0.4 in (1 cm) wide; deeply cleft into oblique lobes, hairy, fragrant

Flowers: usually monoecious, yellow-green; male in cylindric catkins; female in globose to egg-shaped catkins, ovary surrounded by 8 persistent bracts

Fruits: nutlets, about 0.2 in (5 mm) long, olive-brown, lustrous

Twigs: covered with resin dots, green to reddish, hairy, becoming shiny, yellowish brown to reddish purple

Myrica L. - bayberry

We describe the species most likely to be successful in the prairie region.

Myrica pensylvanica Loisel. - northern bayberry

Northern bayberry, native to the northeastern North America seaboard and inland to Lake Erie, rarely reaches tree size. It should thrive throughout much of the prairie region, especially on sandy soils of low alkalinity where water is adequate (Richley 1988). Good fruit development requires both sexes to be planted within close proximity. Bayberry wax—used for scenting candles and soaps—is obtained from the fruits, which are also valuable food for wildlife.

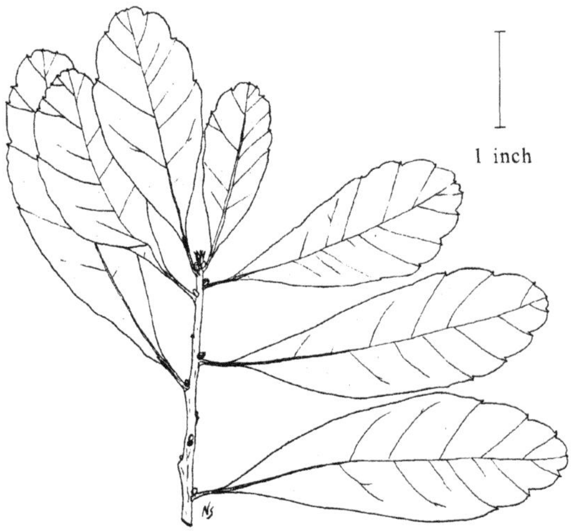

Myrica pensylvanica

Planning

Height: up to 10 ft (3 m)

Habit: shrub; deciduous or half-evergreen; crown rounded; suckering; fruits persist through winter

Site: sun, partial shade; pH adaptable up to moderately saline; thrives in sterile, sandy soils and heavy clays; tolerates salt spray; zone 3

Use: massing, border, foundation, wildlife, erosion control

Recognition

Leaves: aromatic; inverted egg-shape (obovate) to oblong, up to 4 in (10 cm) long; apex usually rounded, margin entire or shallow-toothed near apex; lustrous dark green above, resin-dotted, hairy beneath

Flowers: clusters before the leaves; mostly dioecious; calyx and corolla absent

Fruits: gray, waxy coated

Twigs: resin-dotted when young; pith continuous, green

Buds: terminal absent; lateral solitary, nearly round; scales 2-4

Order Fagales

Fagales consists of three families (Cronquist 1981). Most of the 900 species are oaks, beeches, chestnuts, birches and alders (Fagaceae and Betulaceae).

Recognition

Habit: trees, shrubs; deciduous or evergreen

Leaves: alternate, seldom otherwise, simple; veins branch from a central vein (pinnate)

Flowers: unisexual (mostly monoecious), seldom bisexual; male often in long clusters (catkins); female often in a cup-like structure (cupule), individually or in small clusters; perianth wanting or sometimes sepals scale-like

Fruits: nuts, samaras or drupes, often subtended by a hull

Fagaceae Dumort. - beech family

Members of this large, important family of broadleaved trees and shrubs are found naturally throughout temperate regions and mountainous tropical areas, except in Africa. Of the six or eight genera in the beech family (depending upon the authority cited), five are native to North America. *Lithocarpus* Bl. and *Castanopsis* Spach. are each represented by one species native to the west coast of the U.S.

Members of the beech family are valuable to the lumber industry and essential to the survival of many species of wild birds and animals. They provide food and habitat for wildlife and high quality lumber and food for man.

Recognition

Habit: trees, shrubs

Leaves: alternate; margins variable; petioles short, stipules present

Flowers: male clusters (heads or catkins); female solitary or clustered (heads or catkins), usually 1-4 flowers in bracts that persist and become woody

Fruits: nuts, 1-seeded, borne singly or in 2-3s; enclosed or partially enclosed in scaly, bristly or spiny bracts that have grown together; seed leaves (cotyledons) large and fleshy, remaining within seed coat upon germination (except *Fagus*)

Castanea Mill. - chestnut

The chestnuts, natives of the Northern Hemisphere, are economically important because of their durable wood and edible nuts. Chestnut blight disease is caused by *Cryphonectria parsitica*, a fungus indigenous to eastern Asia. The disease is a nuisance to some species and deadly to others, especially those that are native to North America.

Trial plantings of *C. ozarkensis* Ashe, Ozark chinkapin, conducted in the mid-prairies are inconclusive as to hartiness. Ozark chinkapin is native to southwestern Missouri, northwestern Arkansas and the eastern fringe of Oklahoma. *C. crenata* Sieb. & Zucc., Japanese chestnut, is resistant to chestnut blight, but would probably be hardy only in the southern prairies. *C. sativa* Mill., Spanish chestnut, a native of the Mediterranean region, is very susceptible to chestnut blight, and its importation into the U.S. is restricted (Dirr 1990). It is nonetheless a large, beautiful tree in Europe.

Fagaceae *Castanea*

Recognition

Habit: trees, shrubs; deciduous

Leaves: usually broadest near the base, tip pointed, coarsely toothed (serrate); veins numerous, prominent, parallel

Flowers: appear after leaves nearly full grown; creamy white; strong, unpleasant odor; borne on ascending male and bisexual clusters (catkins); female solitary or clusters of 2-3 on lower part of short, bisexual catkins on upper branches

Fruits: spiny; husk 4-parted, enclosing 1-3 edible nuts

Twigs: usually brown, smooth or hairy; lenticels raised, white; leaf scars half-round, bundle scars 3; stipular scars narrow, elongated; pith 5-angled, white

Buds: terminal lacking; lateral egg-shaped; scales 2-3 visible

Bark: fissured, brown

Castanea dentata (Marsh.) Borkh. - American chestnut

This once-dominant tree of the eastern deciduous forest would be extinct were it not for the persistent stump sprouting and isolated planted trees that have escaped the ravages of chestnut blight. The blight was introduced into the U.S. in 1904. Research continues in the hope that a disease-resistant American chestnut will be discovered.

In the meantime, people in much of the prairie and elsewhere outside of the blight areas can enjoy chestnut trees that are grown from seed collected from isolated trees such as those at Arbor Lodge State Historical Park in Nebraska City. Three other native American chestnut species, apparently more resistant to the blight, are economically unimportant small trees and shrubs native to the southeastern U.S.

The durable chestnut wood was once used for posts, poles, railroad ties, furniture and countless other products. Tannin was once obtained from its bark and wood. Blight-killed trees are salvaged for paneling. The nutritious nuts were once an important food for man and animals.

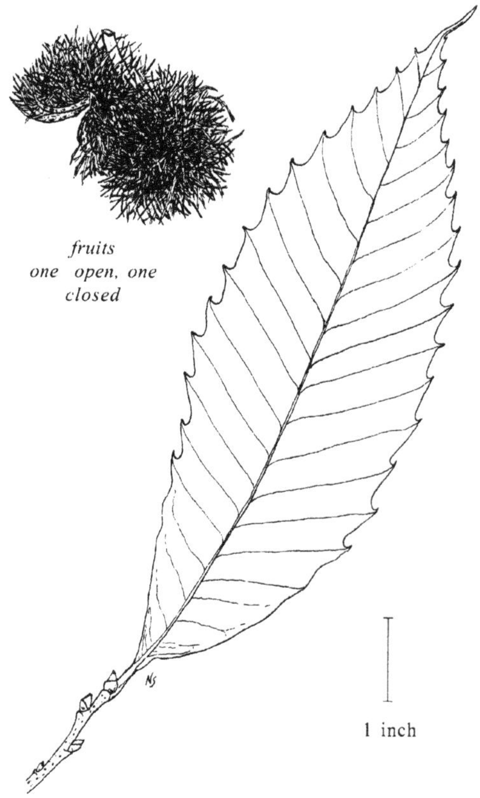

Castanea dentata

Planning

Height: up to 50 ft (15 m); Nebraska champion, 1982, 52 ft (15.8 m), Nemaha; National champion, 1993, 106 ft (32.3 m), Washington

Habit: tree; crown broad, rounded

Site: sun; prefers moist neutral to acidic soils; zone 4

Use: shade, fruit, wood

Pests: chestnut blight; not likely to be a problem on isolated prairie sites

Recognition

Leaves: oblong lance-shaped, 5-10 in (13-26 cm) long, 1.5-3.0 in (4-8 cm) wide; base wedge-shaped, tip pointed; dark green, shiny and glabrous above, hairy and glandular beneath when young, becoming glabrous and shiny; petioles becoming glabrous, stout, about 0.5 in (13 mm) long

Twigs: slightly hairy when young, becoming brown and glabrous; leaf scars half-round; bundle scars 3, frequently sub-divided

Buds: brown, glabrous or only slightly hairy

Bark: thick, dark brown, broad flat ridges separated by shallow irregular furrows

Wood: moderately light, durable, high in tannin

Castanea mollissima Bl. - Chinese chestnut

Blight-resistant Chinese chestnut trees are widely planted in the U.S. These trees are smaller than American chestnut trees and their nuts are considered less flavorful. Scientists are endeavoring to develop a hybrid that combines the disease resistance of the Chinese species with the larger tree size of the American species.

Planning

Height: up to 45 ft (14 m)

Habit: tree; crown rounded to broadly spreading

Site: sun; prefers moist, neutral to acidic soils, tolerates drouth; zone 5

Use: shade, fruit, wildlife habitat

Pests: resistant to chestnut blight

Recognition

Leaves: narrow elliptic to lance-shaped, up to 8 in (20 cm) long; dark green, shiny and glabrous above, silvery hairy beneath, without glands; petioles usually hairy

Twigs: first year hairy, yellow-green, becoming glabrous and darker second year

Buds: egg-shaped, yellow-brown, hairy

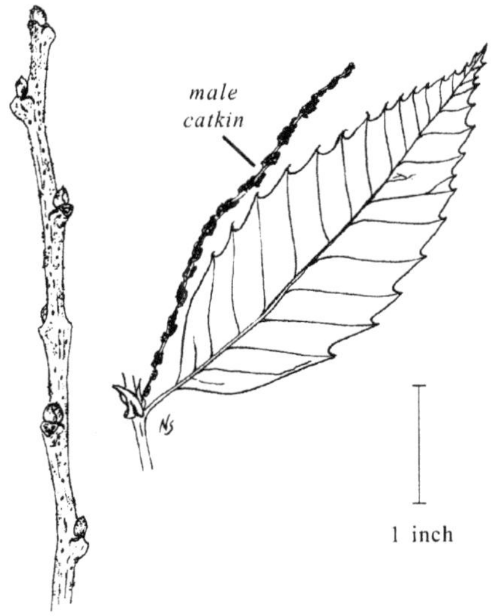

Castanea mollissima

Fagus L. - beech

The ten or so beech species are nearly identical in flower and fruit characteristics but differ markedly in form and leaf. The wide geographic range of these species throughout much of the north temperate zone indicates the possibility of great genetic variation within and between species. Planting in Europe and the U.S. has been largely limited to European beech, *F. sylvatica.*

The beech's dense canopy and fine network of shallow roots conspire to eliminate most of its competition. Hence, in areas where beeches are adapted they dominate the hardwood forest and become the climax species. Inadequate water and temperature extremes are factors that limit the natural distribution of beech trees and place constraints on planting them outside their natural range. Open grown beeches retain their lower branches, thereby protecting the thin-skinned bark and shallow roots from drying sunshine. Branch retention allows the beeches to become excellent sheared hedges.

Fagaceae *Fagus*

Recognition

Habit: trees; deciduous

Leaves: variable in shape, tips long-pointed, margins with sharp incurved teeth to nearly smooth (entire)

Flowers: appear with or after the leaves; male, slender-stalked heads; female, 2-4 flowered spikes surrounded by numerous bracts united into a stalked involucre

Fruits: nuts nearly round to triangular; 1-3 within a woody, 4-parted involucre covered with rough or prickly appendages; ripen in fall

Twigs: pith round, homogenous

Buds: elongated, narrow, sharp-pointed; scales overlapping, dark brown; terminal and lateral similar

Bark: smooth

Fagus grandifolia Ehrh. - **American beech**

American beech—one of the most common and distinctive trees of the lowland hardwood forest of the eastern U.S.—is not widely planted as an ornamental. The largest American beech trees are found in the Ohio and Mississippi River Valleys on cool north-facing slopes. Before and during European settlement, great flocks of now-extinct passenger pigeons descended upon beech forests in autumn to devour newly fallen seed and to rest on the branches overnight. Clearing the forests for agriculture was undoubtedly a factor in the demise of the pigeon.

The oily beech seeds are an important food for many species of wild animals and birds. They have been used for flour and as a source of oil. Seed flavor varies from tree to tree, but little attention has been given to selection of American beech trees for nut production.

Three races of American beech—gray, white, and red—have been described (Rushmore 1961). The gray race, found in the Great Lakes region and the mountains of North Carolina and Tennessee, grows mostly on neutral and slightly alkaline soils. The other two races grow on acidic soils. American beech trees of the gray race are the most likely to be adapted to prairie sites, but survival will depend upon adequate irrigation and effective protection from wind and sun.

The fine-grained, strong, hard wood of the American beech is used in furniture and specialty items such as butcher blocks because its wood imparts no taste to foods.

Planning

Height: up to 60 ft (18 m); National champion, 1984, 130 ft (39.4 m), Ohio

Habit: tree; crown wide; autumn foliage golden brown; shoots from roots

Site: sun, part shade; prefers moist, well-drained soils, pH 5.0-6.5; zone 4

Use: specimen

Caution: avoid water-logged compact soil; protect from strong winds

Recognition

Leaves: egg-shaped (ovate); silky when unfolding; up to 5 in (13 cm) long; tip long-pointed, base broadly wedge-shaped (cuneate); margin with sharp incurved teeth; pale green turning blue-green and glabrous at maturity; 9-14 vein pairs, often with hair tufts at junctions; petioles up to 0.6 in (15 mm) long, usually hairy; stipules strap-like, membranous, about 1 in (25 mm) long

Flowers: appearing after leaves unfold

Fruits: nuts, unevenly triangular

Twigs: slender, slightly zigzag; light green and hairy, becoming brown and glabrous; leaf scars half-round, not directly below bud; bundle scars 3 or more; stipular scars very small, nearly surrounding twig

Buds: over 0.7 in (2 cm) long; slender, lance-shaped, sharp-pointed, lustrous; scales overlapping, in 4 ranks, numerous

Bark: thin, smooth, light blue-gray, sometimes mottled; changing little with age

Wood: fine-grained, strong, hard

Fagus sylvatica L. - European beech

This important timber tree of central and southern Europe has long been planted in Europe and, since colonial days, in the U.S. as well. European beech is used as a hedge plant because it can withstand heavy pruning. This beech is best adapted to the northeastern U.S. from Boston to Washington, D.C., though it has been widely planted throughout much of the U.S., including the eastern prairie, with some success.

Dirr (1990) lists 42 cultivars of European beech ranging in form from columnar to spreading and weeping, with leaves that vary from green to golden and dark purple and are nearly entire in outline to undulate and cutleaf. A deep purple-leafed form, 'Riversii', and a variegated form are among those known to be growing in Lincoln. Some interesting and ornamental cultivars include 'Asplenifolia' with dark green cut leaves, 'Pendula' with widespreading drooping branches, 'Tortuosa' with twisted trunk and branches, and 'Fastigiata' with a narrow crown.

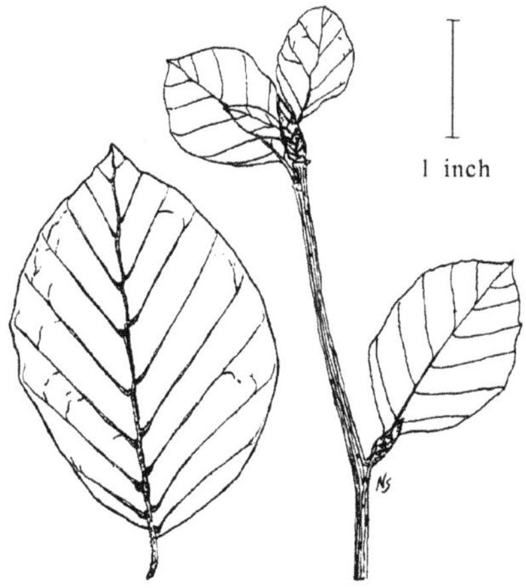

Fagus sylvatica

Planning

Height: up to 50 ft (15 m)
Habit: tree; crown dense, oval to spreading
Site: sun, part shade; moist fertile soil, neutral to acidic; zone 4
Use: specimen, hedge
Caution: shade trunks, protect from excessive wind

Recognition

Leaves: as in *F. grandifolia* except margins entire, undulate or large-toothed; colors variable including yellows and purples; 5-9 vein pairs
Flowers, Fruits, Twigs and Buds: similar to *F. grandifolia*
Bark: becoming dark and rough with age

Quercus L. - oak

About 500 oak species are native throughout much of the Northern Hemisphere. Oaks have been an essential commodity of industrial and domestic life in the evolution of man and in the struggles of nations for dominance (Thirgood 1971). Prior to the Industrial Revolution, oak was a major component of naval and merchant vessels. *Q. prinus* L., chestnut oak, and other species with barks of high tannic acid content were once important to the leather industry.

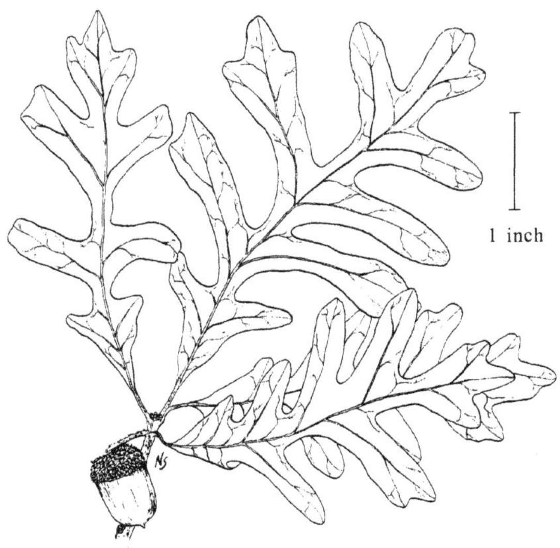

***Quercus* leaves and fruit**

Fagaceae *Quercus*

Oaks provide valuable wood products, from leak-proof barrel staves to fine furniture, flooring and trim. The thick bark of *Q. suber* L., an evergreen of the Mediterranean region, is stripped every 10-15 years to produce bottle corks and many other items. Acorns are an important element in the diet of animals, wild and domestic, and the sweeter ones are occasionally used as food for man.

***Quercus* male catkins**

Oaks grow on a wide variety of sites, from wet or swampy to dry upland. Many of these generally long-lived trees with their variable forms are substantial components of our landscape, protection and woodland plantings. We have included oak species that appear best adapted to the prairies. Undoubtedly there are others that may be useful in some parts of the region.

Northern pin oak, *Q. ellipsoidalis* E. J. Hill, a native on our eastern fringe, is a candidate for prairie planting. One specimen of *Q. turbinella* Greene, shrub live oak, has survived in Lincoln for several years (J. Morgenson 1992). This shrubby, slow-growing oak with persistent leaves is a native of an area from southwestern Colorado west to California. An exotic that probably deserves some attention, especially in the north, is *Q. mongolica* Turcz., Mongolian oak. Hoag (1965) indicates an interest in this native of Asia for the northern Great Plains.

Great genetic diversity is expressed in our oak populations. The number of species and hybrids is a point of disagreement among taxonomists. Many individual trees are difficult to identify because the characteristics listed for a named species are often not always displayed. An example of such an individual is a healthy oak with deep-cut leaves growing in Prairie Pines Arboretum near Lincoln. It is probably *Q.* ×*bebbiana* (*Q. alba* × *Q. macrocarpa*). The tree's acorns resemble those of bur oak (*Q. macroparpa*) without the fringe and its leaves drop in early autumn without changing to bright colors.

The availability of selected oak cultivars with desired attributes is limited because of the difficulty of propagation by grafting or other vegetative means. Rates of growth among individuals are quite variable due to hereditary factors interacting with site—latitude, soil type and water supply. Species noted for brilliant autumn foliage may not always provide color for the reasons explained in our introduction. Leaves of some species, or individuals within species, tend to persist on the tree late into winter or early spring while others may drop leaves soon after dormancy occurs or at first autumn freeze. These characteristics are important to the selection of the proper tree for a landscape.

Visits to a nearby nursery in fall and again in winter to select trees for purchase from their field is the most reliable method for obtaining the desired foliage characteristics. A less exact but often satisfactory method of tree selection is planting seed collected from trees with desirable characteristics. A significant number of offspring will be similar to the parents.

One- and two-year-old oak seedlings can be readily transplanted bare root with excellent success. Direct seeding—planting acorns where the tree is desired—is usually successful with an amazing rate of growth, especially after the first year. Fall is an excellent time for planting acorns. If necessary, planting may be delayed until spring by storing acorns in plastic bags at about 35 degrees F (2 degrees C) through winter. Some sprouting may occur, but with careful handling the acorns can be successfully planted if the emerging sprouts have not been injured. .

Acorn viability can be determined by carefully cutting away a side portion of the seed coat to avoid damaging the growing point at the tip. Acorns are viable if the cotyledon is firm, light-colored and free of insect damage. The ensuing scar will not damage the acorn if drying is prevented. In fact, the germination period may be shortened.

Squirrels, rabbits and deer are potential hazards to oak seeds and young seedlings. The most damaging insect pest to the oak is gypsy moth which has not yet become established in our region. Oak wilt disease has caused tree death in some woodlands but has not, and probably will not, cause significant damage in most of the prairie region.

Recognition

Habit: trees, shrubs; deciduous or evergreen

Leaves: shape variable; margin variably lobed, toothed or not toothed (entire); papery to tough and leathery; stipules usually deciduous

Flowers: appearing on old or new growth, with or after new leaves; male densely clustered in hanging catkins, calyx 4-7 lobed usually enclosing 6 stamens; female inconspicuous, solitary or on short few-flowered spikes from axils of new leaves, calyx 6-lobed partly enclosed in a whorl of bracts (involucre)

Fruits: acorns; basal cups (modified involucres) partially to almost completely enclosing nuts, variable shape and size, maturing at end of first growing season (white oaks) or at end of second growing season (red oaks)

Twigs: round or angular; lenticels small, pale; pith homogenous, 5-pointed in cross-section; leaf scars raised, half-round, bundle scars many, scattered; stipular scars small or absent

Buds: terminal present, several clustered near twig end; scales numerous

Quercus has been variously subdivided into subgenera, but the actual assignment of a certain species to one group or the other has not been easy. Rehder (1940) divides the oaks into three subgenera, but a more recent grouping based on wood anatomy is more logical (Harlow and Harrar 1968).

Grouping Key for *Quercus*

(Harlow and Harrar 1968)

Leaves mostly toothed or lobed with bristles extending beyond margins; or, if unlobed, the margins, tips or both are often with spines; fruits (acorns) usually maturing in two seasons, barely larger than flower buds at end of first season, inner surface of acorn cups often hairy (tomentose), seeds usually bitter; late wood pores large, distinct, round, thick-walled, open: **subgenus *Erythrobalanus*** (red oaks)

Leaves entire, toothed or lobed, not bristle-tipped; fruits maturing in one season on current season's twigs, inner surface of acorn cups glabrous, seed usually "sweet"; late wood pores many, indistinct, angled, thin walled, heartwood pores usually plugged by tyloses, heartwood typically pale, reddish brown: **subgenus *Leucobalanus*** (white oaks)

Subgenus *Erythrobalanus* - red oaks

Oaks of the red oak subgenus differ markedly from white oaks in leaf, fruit and wood characteristics. Acorns usually mature in two years and normally do not sprout until the following spring. Hence, red oak acorns can usually be stored successfully under refrigeration until spring planting season. If earlier germination is desired for greenhouse planting, removal of seed coat hastens germination.

Quercus acutissima Carruth. - sawtooth oak

Sawtooth oaks are sometimes mistaken for chestnut trees because of leaf similarity. This native of Asia—from the Himalayas through China, to Korea and Japan—flourishes in southeastern Nebraska. A 20-year-old tree on East Campus, UNL, produced prolific acorn crops from age 10 years. It was the mother tree of several sawtooth oaks in the Lincoln vicinity before its removal from the path of progress. Sawtooth oak in southwestern Nebraska has suffered twig dieback in winter (Richley 1988).

'Gobbler', is a small-seeded sawtooth oak released by Natural Resources Conservation Service at the Quicksand Plant Materials Center in Kentucky. Acorns are about one-half the size of normal. Wild turkeys evidently appreciate the small nuts, hence the name 'Gobbler'. The rather extensive range of sawtooth oak presents an opportunity for successful provenance research.

Fagaceae *Quercus*

Planning

Height: up to 40 ft (12 m)
Habit: tree; deciduous; crown oval, becoming broad
Site: sun; pH and soil adaptable, avoid highly alkaline soil, drouth tolerant; zone 5
Use: shade, wildlife habitat, windbreak

Recognition

Leaves: oblong, up to 7 in (18 cm) long; tip pointed, base broad tapering to rounded; bristly teeth along margin terminating 12-16 parallel veins; hairy when unfolding, becoming lustrous and glabrous above, lighter green and glabrous below except for tufts of hairs in axils of veins; petioles up to 1 in (25 mm) long
Flowers: male catkins golden, pendant
Fruits: about 1 in (25 mm) long, without stalk (sessile); cup with long, spreading and recurving scales enclosing about two-thirds of nut
Twigs: glabrous
Buds: hairy-woolly, gray-brown; scales overlapping
Bark: deeply ridged and furrowed

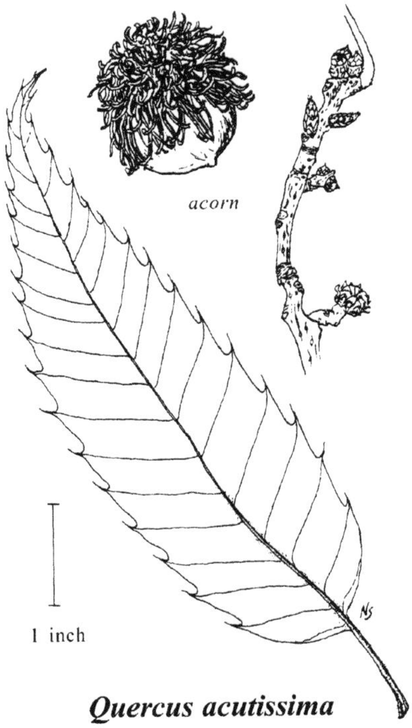

Quercus acutissima

Quercus coccinea Muenchh. - scarlet oak

Scarlet oak, a red oak/black oak look-alike, comes by its common name because of its reputed autumn leaf coloration. Brilliance of this display depends upon latitude of seed origin in relation to latitude of planting site. Scarlet oak grows naturally from southeastern Missouri eastward to Maine and Georgia.

Scarlet oak, occupier of drouthy claypan and sandy soils in its native habitat, should be adapted to much of the prairies except alkaline sites. Though capable of rapid growth on fertile soils, the scarlet oak is apparently not widely planted in the Great Plains. Several trees in Lincoln reputed to be scarlet oak are actually black oak, *Q. velutina,* based upon close examination of their acorns and bark. In fact, incorrect identification probably led Pool (1951), when commenting about scarlet oak, to state that "it has entered the southeastern corner of this state from Missouri forests and has spread along the river from Richardson County as far northward as Cass County."

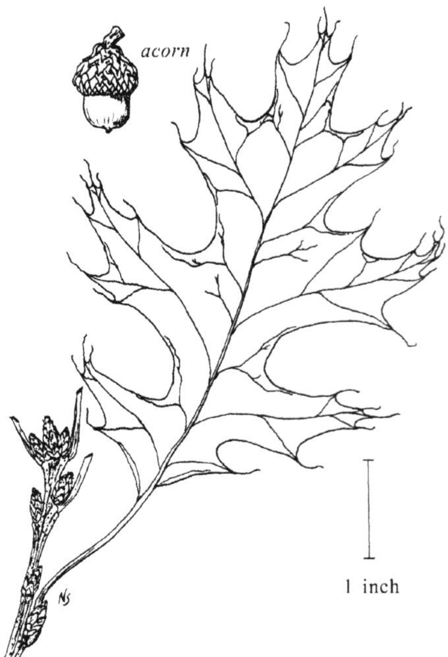

Quercus coccinea

Planning

Height: up to 60 ft (18 m); Nebraska champion, 1991, 50 ft (15.2 m), Lincoln; National champion, 1978, 150 ft (45.5 m), Alabama
Habit: tree; deciduous; crown rounded; autumn foliage scarlet
Site: sun; prefers fertile, moist soil, neutral to acidic, drouth tolerant; zone 4
Use: shade, windbreak

Recognition

Leaves: egg-shaped in outline; 3 to 4 large lobes each side, bristle-tipped, openings between lobes (sinuses) deep and round; hairy on both sides at first, becoming glabrous; dark green above, pale green below; hair tufts at vein junctions
Fruits: acorns about 1 in (25 mm) long, nearly egg-shaped, usually with 1 or more concentric rings around tip; cups thin, enclosing half of nut, scales tightly appressed, somewhat shiny
Twigs: hairy at first, becoming glabrous, reddish brown
Buds: egg-shaped, about 0.3 in (8 mm) long, tips covered with fine gray hairs; scales overlapping
Bark: rough, nearly black; inner bark reddish

Quercus imbricaria Michx. - shingle oak

Shingle oak—at first glance often not recognized as an oak—is most abundant in the central states, ranging from eastern Kansas to Pennsylvania. Pool (1951) states: "This interesting oak has been repeatedly reported from southeastern Nebraska, but we do not have any authentic specimens in the herbarium which were collected in this state." Shingle oak is commonly found on dry, upland soils and moist lowlands. Its ability to grow on heavy clay should make it a good candidate for urban soils. The few known shingle oaks in Lincoln are healthy and appear to have no serious pest problems. Provenance research should be rewarding.

Shingle oak lumber, sold as red oak, is not often of high quality. Early settlers gave the shingle its common name because the wood splits easily and was used as shingles. The shingle oak's small acorns are an important food source for squirrels, deer and wild turkey.

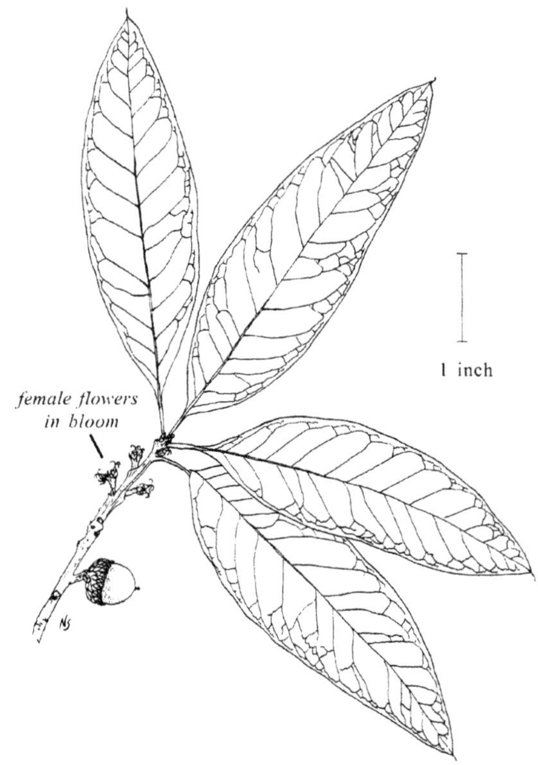

Quercus imbricaria

Planning

Height: up to 50 ft (15 m); National champion, 1989, 104 ft (31.5 m), Ohio
Habit: tree; deciduous; crown pyramidal to rounded, lower branches horizontal; autumn foliage usually reddish, often persisting through winter
Site: sun; prefers neutral to acidic, moist, fertile soils, tolerates drouth; zone 4
Use: street, shade, windbreak, wildlife habitat

Recognition

Leaves: oblong to elliptical, up to 6 in (15 cm) long, widest at the middle; margin entire, slightly wavy, with slender bristle tip; shiny, dark green above, fine red and white star-shaped hairs beneath; leathery; petioles about 0.5 in (12 mm) long

Fruits: acorns egg-shaped, about 0.6 in (15 mm) long; solitary or in pairs, stalks about 0.5 in (12 mm) long; cups bowl-shaped enclosing one-third to one-half of nuts, scales appressed, hairy

Twigs: shiny, dark green to light brown

Buds: egg-shaped, pointed, shiny smooth to hairy

Bark: gray-brown, broad ridges, shallow fissures; not scaly, resembling pin oak

Quercus marilandica Muenchh. - blackjack oak

The southeastern corner of Nebraska is the northwestern range limit of blackjack oak. It grows naturally—usually at the forest's edge—from eastern Kansas, most of Oklahoma, south in Texas almost to the Gulf of Mexico, and east to the Atlantic Ocean. The pioneer, scrubby blackjack oaks are, to some extent, products of the dry, barren soils upon which they grow. Many limbs and poor form spoil them for lumber, but the acorns are valuable wildlife food.

Though seldom planted, trials incorporating blackjack oak into windbreak plantings would seem worthwhile because of its drouth tolerance and tendency to retain leaves through the winter.

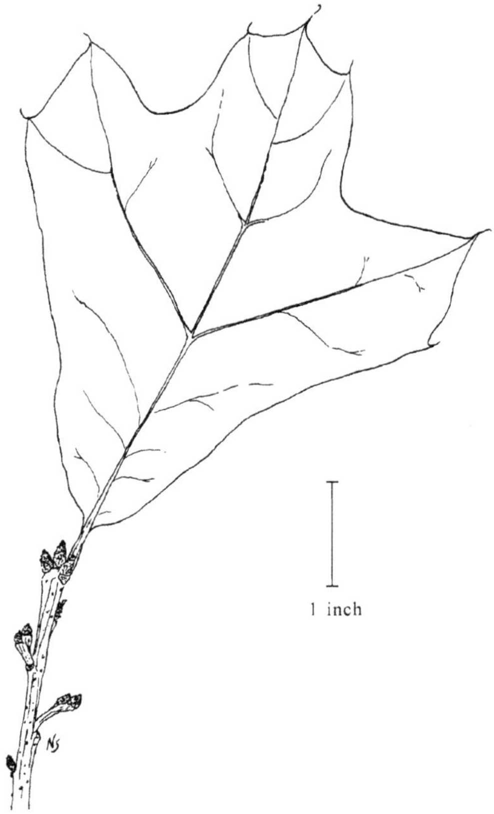

Quercus marilandica

Planning

Height: up to 30 ft (9 m); Nebraska champion, 1991, 40 ft (12.2 m), Lincoln; National champion, 1991, 90 ft (27.4 m), South Carolina

Habit: tree; deciduous; crown rounded, irregular; leaves persist through winter

Site: sun; pH and soil adaptable, tolerates drouth and infertile soil; zone 5

Use: windbreak, wildlife habitat

Recognition

Leaves: egg-shaped, wider near the tip (obovate), up to 6 in (15 cm) long, usually shallow 3-lobed; major veins and some large vein branches extending into minute, sharp points; dark green and glossy above, rusty-hairy beneath, becoming less so with maturity

Fruits: acorns about 0.8 in (2 cm) long, oblong to nearly round, about half enclosed in cups; cups thickened at upper rim, hairy inside, scales hairy; nut nearly round, about 0.5 in (13 mm) across, kernels yellow, not palatable

Twigs: red-brown, scurfy-hairy, becoming smooth

Buds: oval, somewhat angled; scales rusty-brown hairy

Bark: black, rough, blocky

Quercus palustris Muenchh. - **pin oak**

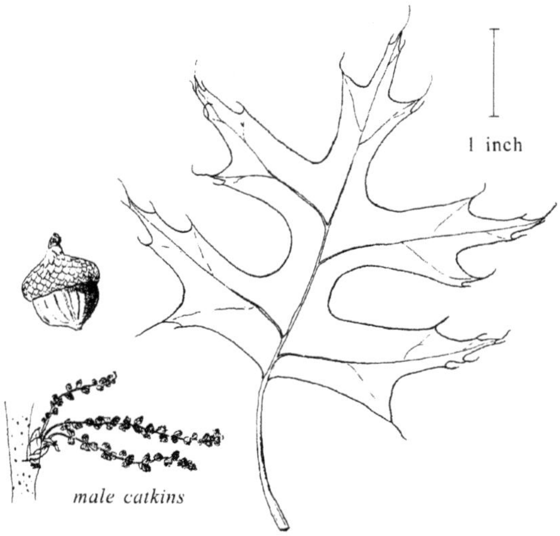

Quercus palustris

The handsome pin oak—widely planted as a lawn and street tree—is a native of the U.S. from eastern Kansas to the Atlantic coast and a north-south range from Tennessee to southern Michigan. Pool (1951) reports, "We may expect it to wander into our state [Nebraska] some day via the southeastern corner, and to become much more common." In fact, pin oak is commonly planted in our towns and cities and could easily escape to the countryside with the aid of industrious squirrels.

Pin oak grows naturally on poorly drained, claypan soils and withstands short periods of flooding. It can adapt to a wide range of site conditions, but foliage yellows and growth is often stunted if planted on calcareous soils. Chlorosis and stunting often occurs on city trees where topsoil has been removed or deeply covered, leaving a high lime subsoil as the planting medium. Calcareous soils do not provide an adequate iron supply for chlorophyll synthesis. Treatments—such as implantation of rusty nails and ferric ammonium citrate capsules in the trunk, spraying the foliage, or treating soil with ferrous sulfate—can temporarily alleviate the chlorotic condition.

Pin oak is a desirable ornamental because its moderately shallow, fibrous root system facilitates transplanting, and its tough branches resist storm damage. The conifer-like pyramidal crown with a central leader lends a formal aspect to the landscape. However, pin oak becomes open and oval to round-topped with age.

The pin oak's deeply dissected leaves provide a fine textured appearance. Autumn foliage is usually a brilliant red. However, a late October stroll along some of our cities' avenues reveals pin oaks dressed in magnificant reds alongside sister trees wearing the greens of summer or the browns of winter. By October some pin oaks will have dropped their leaves; later in winter we find many pin oaks fully leafed—now brown—a condition that can continue until leaf flush in spring.

Autumn leaf color and winter leaf retention are important characteristics to consider in selecting pin oaks for planting. Advice: observe pin oaks growing in your local nursery during fall and winter before selecting a tree.

Pin oak does not produce high quality lumber due to its many small persistent branches. Bounteous acorn crops produced every two or three years after age 20 furnish valuable food for wildlife. Pin oaks growing on well-watered, highly fertilized lawns may be much more fruitful, often to the homeowner's consternation and the squirrel's delight.

Planning

Height: up to 60 ft (18 m); Nebraska champion, 1989, 83 ft (25.2 m), Lincoln; National champion, 1991, 110 ft (33.6 m), Tennessee

Habit: tree; deciduous; crown pyramidal, lower branches drooping, middle horizontal, upper ascending; autumn foliage variable, green to shades of yellow and red

Site: sun, partial shade; prefers fertile, moist, neutral to acidic soils; zone 4

Use: street, windbreak, wildlife habitat, shade

Recognition

Leaves: ovate to elliptic in outline, up to 6 in (15 cm) long; 5-7 deep, bristle-tipped lobes; bright green and glabrous above, pale beneath, hair tufts at large vein axils

Fruits: acorns nearly round with short beaks, about 0.5 in (13 mm) across; cups saucer-shaped, covering the base only; scales thin, appressed, red-brown; kernels bitter

Twigs: slender, shiny, red-brown

Buds: conical, usually twice as long as broad, pointed; scales red-brown, usually glabrous, sometimes hairy at tip

Bark: gray-brown, smooth for many years, becoming rough with low scaly ridges

Quercus phellos L. - willow oak

Willow oak, a large, handsome tree of the southeastern U.S., is widely planted as a shade and street tree. It grows naturally on poorly drained bottomlands. Though rarely planted in Nebraska, the few willow oaks that we have observed planted here are healthy. Willow oaks of Missouri and Pennsylvania/New Jersey origin may be best adapted to the prairie climate, provided soils are acidic and water is adequate.

Richard Lodes discovered a large, healthy oak in a Lincoln yard with long, narrow shallow-lobed leaves and acorns very similar to those of willow oak. Acorns from this tree have produced very healthy, rapid-growing oaks at Prairie Pines Arboretum near Lincoln. Their leaves vary, from looking much like the parent tree to wider and more deeply lobed, indicating probable pollination by nearby pin or red oaks. Autumn foliage develops brilliant shades of red.

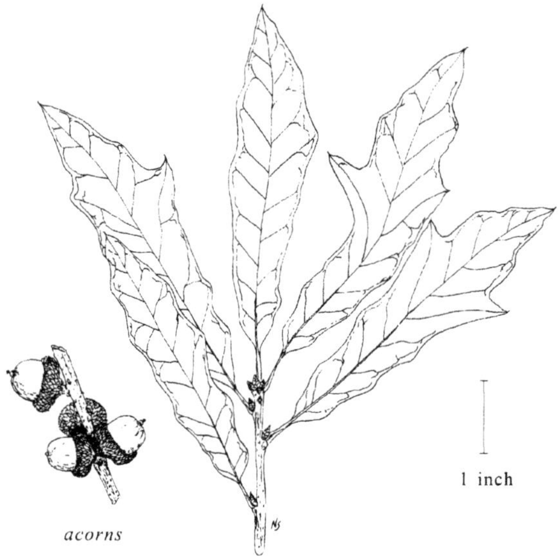

Quercus phellos

Planning

Height: up to 50 ft (15 m); National champion, 1991, 112 ft (34.3 m), Maryland

Habit: tree; deciduous; crown oval

Site: sun; prefers soil neutral to acidic, moist to wet; zone 5

Use: shade, specimen

Recognition

Leaves: linear, narrow, up to 5 in (13 cm) long and 1 in (25 mm) wide, gradually narrowing to a bristle tip and to the petiole; margin not toothed (entire), sometimes wavy, rarely few shallow lobes; glabrous above and below, sometimes hairy below; veins conspicuous; petioles short, stout

Fruits: solitary or in pairs; nuts nearly globular, about 0.6 in (15 mm) across; sometimes vertically striped, minutely hairy or glabrous; enclosed by saucer-shaped cups at base; cup scales thin, appressed, glabrous or hairy

Twigs: reddish, glabrous

Buds: egg-shaped, pointed; scales chestnut brown, margins pale

Bark: smooth, eventually rough and dark colored

Quercus rubra L. (***Q. borealis*** Michx. f.) - **northern red oak**

Northern red oak—native to states east of the Great Plains—grows naturally on the eastern fringes of the Plains. This round-topped oak can substutute for pin oak in many situations with less chance of becoming chlorotic on calcareous soils. Red oak's hard, strong wood is used for flooring, furniture, veneer and interior finishing. Acorns, produced after trees are about 20 years of age, are valuable wildlife food.

Genetic variation among red oak provenances and, to a lesser extent within provenances, affects growth rate, form, fall color and leaf retention. Provenance plantations established in the early 1960s in several north central states, including Nebraska, have provided and will continue to provide a wealth of information about red oak (Schlarbaum and Bagley 1981).

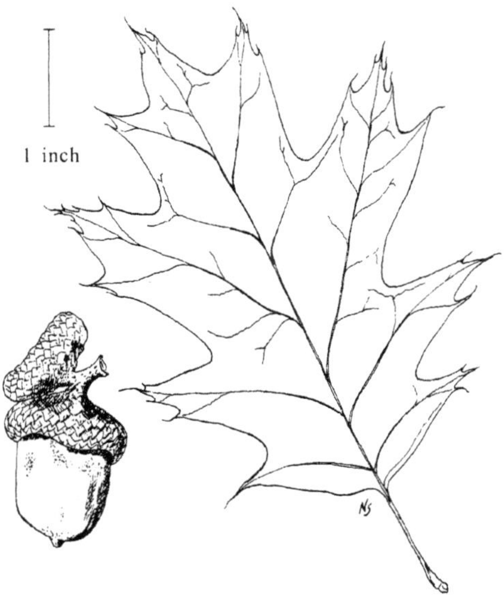

Quercus rubra

Northern red oak of Tennessee origin planted at Horning State Farm near Plattsmouth is fast growing and straight stemmed, but it seldom displays fall color. Leaves usually remain green until November, dropping soon after the first sharp freeze. Trees in this plantation that originated from northerly latitudes grow slower and usually display colorful autumn foliage in early to mid-October. Red oaks of some provenances (generally western) tend to retain leaves throughout the winter, a characteristic often desirable for windbreaks.

Planning

Height: up to 70 ft (21 m); Nebraska champion, 1982, 76 ft (23 m) tall and 13.7 ft (4.2 m) in circumference, Fremont; National champion, 1987, 66 ft (20 m) tall and 30.8 ft (9.3 m) in circumference, New York

Habit: tree; deciduous; crown rounded, spreading; autumn foliage reddish

Site: sun; prefers fertile, moist soils, neutral to acidic, drouth tolerant; zone 4

Use: shade, street, windbreak, wildlife habitat, lumber, firewood

Recognition

Leaves: oval in outline, sometimes wider above middle, up to 8 in (20 cm) long; lobes usually 7 to 9, tapering to narrower toward the outer end, bristle-tipped; base usually tapering to sometimes squared-off (truncate); openings between lobes (sinuses) moderate to shallow; glabrous and shiny above, paler beneath with occasional axillary tufts of hairs; petioles glabrous, up to 1.5 in (4 cm) long; stipules linear, early deciduous (caducous)

Fruits: acorns sessile or short-stalked, solitary or paired; nuts about 0.5-1.0 in (12-25 mm) long, broadly egg-shaped to nearly round, base broad, apex rounded, tip conical, covered with fine hairs (puberulent), kernel bitter; cup saucer-shaped, covering up to one-third of nut; scales tight, finely hairy and appressed, inside of cup nearly glabrous

Twigs: glabrous, greenish to reddish brown, becoming dark brown

Buds: egg-shaped, pointed; scales glabrous to slightly hairy near the tip, reddish brown, margins sometimes hairy (ciliate)

Bark: smooth, slate gray to grayish brown becoming nearly black; flat-topped ridges separated by shallow fissures; inner bark reddish

Quercus shumardii Buckl. - shumard oak

Shumard oak is often misidentified. It can be confused with northern red, black, scarlet and pin oaks. Leaf and acorn characteristics are quite variable. Hybridizing with many other red oak species is common (Elias 1980). Dirr (1990) opines that the glabrous, gray or pale, straw-colored buds are the only reliable aid to the shumard's identity.

Shumard oak grows mainly in bottomland forests, from eastern Kansas across Missouri, north into Indiana, then southeast to the Carolinas and west along the Gulf Coast into central Texas.

Though shumard oak is not widely used as an ornamental, it is considered a valuable, drouth-tolerant species suited for planting in Kansas, Oklahoma and Texas (Dirr 1990). On fertile sites shumard trees attain large size and produce superior lumber.

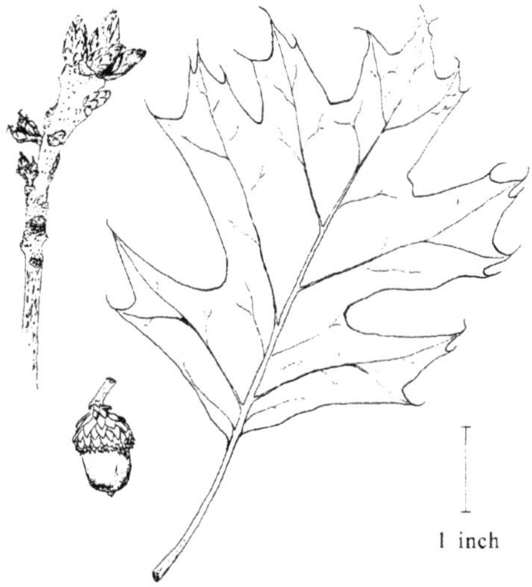

Quercus velutina

This look-alike of northern red oak may be growing more widely in the prairies than we realize. Provenance research is needed to determine its hardiness and potential for autumn coloration throughout the prairie region.

Planning

Height: up to 70 ft (21 m); National champion, 1991, 97 ft (29.4 m), Louisiana
Habit: tree; deciduous; crown various, pyramidal to spreading; autumn foliage scarlet
Site: sun; prefers fertile, moist soils, neutral to acidic, tolerates drouth; zone 5
Use: shade, street, wildlife habitat, lumber, firewood

Recognition

Leaves: as in *Q. rubra* except that lobes are generally wider at the tip than at the base with more bristles; hair tufts in the axils of the veins are prominent
Fruits: acorns, kernels edible; similar to *Q. rubra*
Twigs: red-brown, glabrous
Buds: egg-shaped, pointed; scales gray-brown to pale straw-colored, glabrous, tips sometimes hairy
Bark: inner reddish to gray

Quercus velutina Lam. - black oak

The natural site for black oak is a typically rocky, sandy to clay soil on a dry upland where trees are usually poorly formed. This oak approaches northern red oak in stature on fertile, bottomland soil. Black oak is native to Iowa, southeastern Nebraska, south into eastern Texas, and it grows in all the eastern states of the U.S. Provenance research is recommended to aid in selecting the best seed sources for prairies sites.

Planning

Height: up to 50 ft (15 m); Nebraska champion, 1980, 68 ft (20.6 m), Barada; National champion, 1989, 131 ft (39.7 m), Michigan
Habit: tree; deciduous; crown rounded; autumn foliage red, leaves often persist through winter
Site: sun; prefers fertile, moist soil, neutral to acidic; tolerates dry, sandy to clay soils; zone 4
Use: shade, windbreak, wildlife habitat, firewood

Recognition

Leaves: egg-shaped in outline; lobes 5-7, bristle-tipped, upper commonly widening at outer end, openings between lobes (sinuses) deep and round; base tapered, sometimes squarish; surface lustrous, dark green above; sometimes copper-colored with scurfy hairs beneath, tufts in vein axils often conspicuous; petioles stout, rusty hairy or becoming glabrous, about 2 in (5 cm) long; stipules long, narrow, hairy, early deciduous

Fruits: acorns sessile or on very short stalks; nuts egg-shaped, 0.5-0.7 in (12-18 cm) long, finely hairy; cup enclosing about one-half of nut; cup scales overlapping, velvety-hairy, spreading tips on upper half provide a fringe

Twigs: scurfy hairy becoming glabrous, brown; lenticels conspicuous

Buds: terminal 0.3-0.5 in (9-12 cm) long, egg-shaped, strongly angular, obtuse, gray to brown, woolly; lateral smaller

Bark: nearly black on old stems, deeply furrowed, blocky; inner bark bright orange to yellow, bitter, high in tannic acid

Subgenus *Leucobalanus* - white oaks

White oaks differ from the red oaks in their leaf, fruit and wood characteristics. Large trees, sound of trunk and free of knots, provide staves for leak-proof whiskey barrels. The wood is moderately durable in contact with the soil. White oak acorns mature in one year and will germinate soon after ripening. They can be fall planted, or stored in a refrigerator until spring.

Quercus alba L. - white oak

This valuable, long-lived tree is native to southern Minnesota, Iowa, south into eastern Texas and all states to the east. Pool (1951) states that white oak, though not abundant, has been found along the Missouri River in southeastern Nebraska as far north as Cass County. However, Stephens (1973) did not find evidence of this oak growing naturally in Nebraska.

Open-grown white oak is a sturdy tree, developing broad, rounded crowns with strong branches. Although it attains its greatest size on deep fertile soils, white oak is not restricted to these sites. It can be found on upland soils within its native range. Acorn crops can be large after age 20 and usually occur at 4- to 6-year intervals.

White oak is a desirable windbreak tree because the leaves of most of these trees persist throughout the winter. Unfortunately, we have little information about its adaptability to sites in the western prairies.

Splashes of rich, wine red color among wooded hills reward the late October traveler crossing Iowa. If you are a Nebraska resident, an acorn collection stop in that state, with permission of the landowner and resident squirrels, will provide an opportunity to brighten your fall landscape a few years after planting.

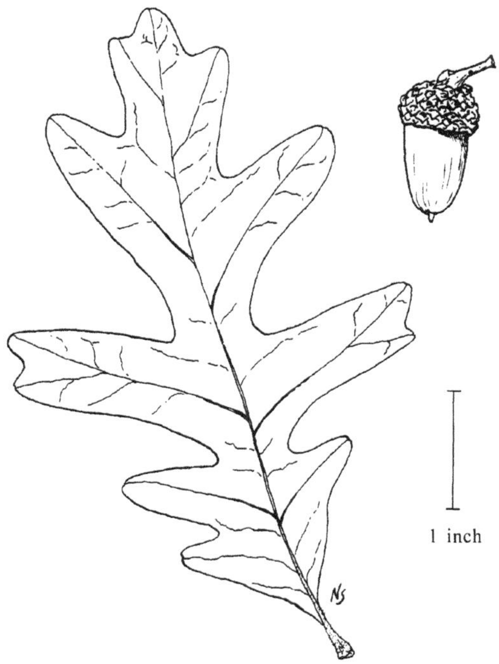

Quercus alba

Well-known white oak cultivars offered by nurseries may flourish, but the dormant date may not coincide with autumn temperatures for best foliage color. Two white oak trees in Maxwell Arboretum, UNL, are usually in the non-coloring category. However, the exceptionally mild, dry fall of 1988 allowed these two trees, a 'Wye' oak and another unnamed white oak to reach peak color by mid-November.

Planting white oak acorns in place is the easiest method of establishment because of the strong taproot. However, bare-root two- to three-year-old seedlings are moved without great difficulty. Rate of growth is not slow, as indicated by Dirr (1990), but equal to that of most long-lived tree species (Bagley 1998).

Planning

Height: up to 60 ft (18 m); Nebraska champion, 1979, 58 ft (17.6 m), Lincoln; National champion, 1945—and still champion in 1991—79 ft (24.2 m) tall, 31.2 ft (9.5 m) trunk circumference, 102 ft (31.2 m) crown spread, Wye Mills State Park, Maryland

Habit: tree; deciduous; crown rounded, becoming wide-spreading; autumn foliage a durable wine red

Site: sun; prefers fertile, moist soil, neutral to acidic, drouth tolerant; zone 4

Use: shade, windbreak, wildlife habitat, lumber, firewood

Recognition

Leaves: widest above the middle (obovate) up to 8 in (20 cm) long; lobes 7 to 10, rounded; surfaces glabrous, bright green above, lighter beneath, sometimes star-shaped hairy along midrib; petioles up to 0.6 in (15 mm) long, glabrous or hairy; stipules linear, hairy, early deciduous

Fruits: acorns with or without short stalks; nuts ellipsoid to egg-shaped, smooth, apex rounded, often indented around tip, 0.5-0.8 in (12-20 mm) long; cups bowl-like enclosing about one-fourth of nut; cup scales thickened, warty, upper small and appressed to nut, brown hairy inside; kernels sweet

Twigs: bright green and hairy becoming gray and glabrous

Buds: broadly egg-shaped, scales dark red-brown, glabrous or hairy, often hairy fringed (ciliate)

Bark: light, ashy gray, variably fissured, becoming blocky on old trees

Quercus bicolor Willd. - swamp white oak

Swamp white oak, a tree of lowlands and swamp edges, grows naturally from Iowa and Missouri eastward to the Atlantic coast. Elias (1980) and others credit northeast Kansas as a small outlier area. Stephens (1973) could find no evidence of swamp white oak growing naturally in Kansas and/or Nebraska as claimed by Pool (1951).

Swamp white oak can easily be mistaken for bur oak, *Q. macrocarpa*, or for white oak, *Q. alba*, due to leaf similarities. However, long-stalked acorns are its credentials. If acorns are not present, a close look at leaf margins, winter buds and bark is required. Another caution: *Q. robur*, English oak, also bears long-stalked acorns and leaves that are sometimes quite similar to those of swamp white oak.

Swamp white oak is a handsome shade tree with two-toned leaves, shiny green above and whitish beneath, suitable for homes and parks. The subdued yellow foliage of autumn quickly turns brown and persists late into winter on many trees. Like pin oak, swamp white adapts to sites that are much different than its native habitat.

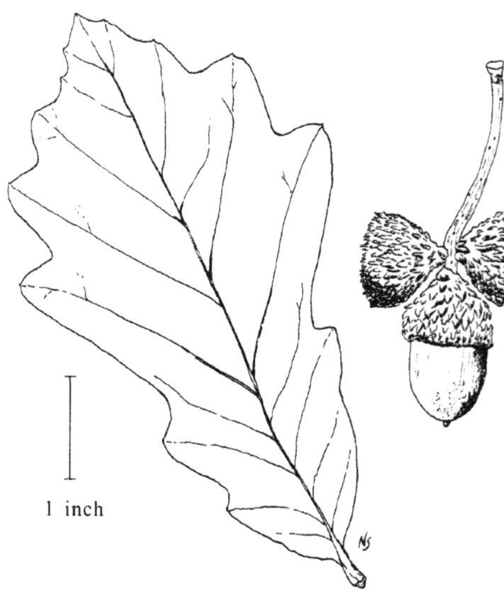

Quercus bicolor

This fast-growing tree thrives naturally on moist lowlands but adapts to upland sites and compacted urban soils. A shallow, spreading root system, minus a deep taproot, facilitates transplanting.

A stately row of swamp white oaks adorns the east side of Woods Park at 33rd and O streets, Lincoln. Bounteous, frequent acorn crops from this stand of trees provide seed for tree planters and contribute to the well-being of many squirrels.

Planning

Height: up to 60 ft (18 m); Nebraska champion, 1981, 69 ft (20.9 m), Lincoln; National champion, 1989, 144 ft (43.6 m), Michigan

Habit: tree; deciduous; crown rounded, spreading; autumn foliage subdued brownish yellows and reds, often persistent through winter

Site: sun; prefers moist, fertile, neutral to acidic soils, tolerates compact soils high in clay; zone 4

Use: shade, street, windbreak, wildlife habitat, firewood

Recognition

Leaves: similar to *Q. alba*, except lobes more shallow and contrast in color of upper and lower surfaces is greater, usually finely hairy beneath

Fruits: acorns on slender stalks, 1-4 in (2.5-10 cm) long; nuts broad egg-shaped, about 1 in (2.5 cm) long; cup covers about one-third of nut, upper scales often forming a fringe around the rim; kernels edible

Twigs: hairy, becoming smooth, yellow-brown

Buds: subglobose, tips blunt, yellow-brown

Bark: on young branches ashy gray, peeling; trunk grayish, scaly becoming furrowed with blocky ridges

Wood: heavy, strong, hard, moderately durable

Quercus gambelii Nutt. - Gambel oak

Gambel oak is abundant on the dry foothills, canyons and lower slopes of the mountains of Nevada, Utah, Colorado, Arizona, New Mexico, and the fringes of adjoining states and Mexico. This oak is known locally as scrub oak because of its slow growth and tendency to form thickets. The variable Gambel oak has frustrated many attempts to divide it into numerous species and varieties (Preston 1968).

Gambel oak is a tree suitable for small spaces and dry sites. Deer, wild turkey and squirrels relish its sweet nuts. The Gambel should be adapted to most high plains sites. Test plantings of Gambel oak have been established in western Nebraska wildlife habitat areas by the Nebraska Game and Parks Commission (J. Morgenson 1992).

Gambel oak acorns taken from the foothills south of Denver have produced 10-foot trees at 18 years of age at Prairie Pines Arboretum. An abundant 1988 acorn crop was quickly consumed by squirrels.

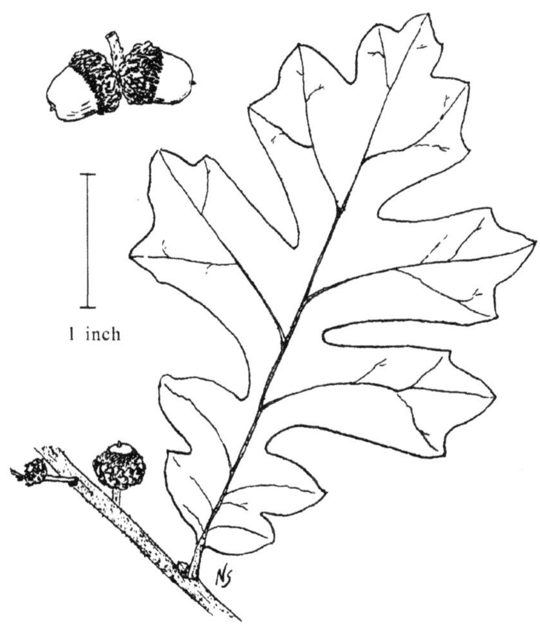

Quercus gambelii

Planning

Height: up to 20 ft (6 m); National champion, 1981, 47 ft (14.2 m), New Mexico
Habit: tree, shrub; deciduous; crown narrow, sparse
Site: sun; pH and soil adaptable, drouth tolerant; zone 4
Use: naturalizing, wildlife habitat

Recognition

Leaves: thick and leathery; oblong, wider above middle (obovate), up to 6 in (15 cm) long; lobes 5 to 9, moderate to deep; base tapering to rounded, apex rounded; glabrous above, pale, soft hairy beneath; petioles hairy becoming glabrous, up to 1 in (25 mm) long
Fruits: acorns sessile or short-stalked, usually solitary; nuts egg-shaped, about 0.6 in (15 mm) long; cup covering one-third to one-half of nut, scales thickened, hairy, glabrous inside; kernels edible
Twigs: densely hairy, becoming glabrous, orange-brown
Buds: broadest near base, tapering to a point, hairy
Bark: thin, scaly, light brown to gray
Wood: hard, heavy, close-grained

Quercus lyrata Walt. - overcup oak

Acorns and leaves of overcup oak are very similar to those of some variants of bur oak. This oak, native to the southeastern U.S., grows naturally on poorly drained, heavy clay soils; hence it may be adapted to similar urban sites.

A 50-foot (15 m) overcup oak growing at 35th and J streets, Lincoln, is a healthy oldster. Seedling progeny of this tree, now several years old, are growing at Prairie Pines Arboretum and several other Lincoln locations.

Seed for central prairie plantings should be collected from natural stands of overcup oak in the northernmost portions of its natural range—southern Illinois, Missouri or New Jersey—to reduce chances of cold injury. The overcup drops its leaves earlier than other southern oaks do, a characteristic desirable for avoiding the ill effects of early fall freezes (Dirr 1990).

Planning

Height: up to 50 ft (15 m); National champion, 1987, 156 ft (47.3 m), North Carolina
Habit: tree; deciduous; crown oval
Site: sun, part shade; neutral to acidic, adapts to a variety of soil conditions including poorly drained; withstands prolonged inundation; zone 5
Use: street, shade, wildlife

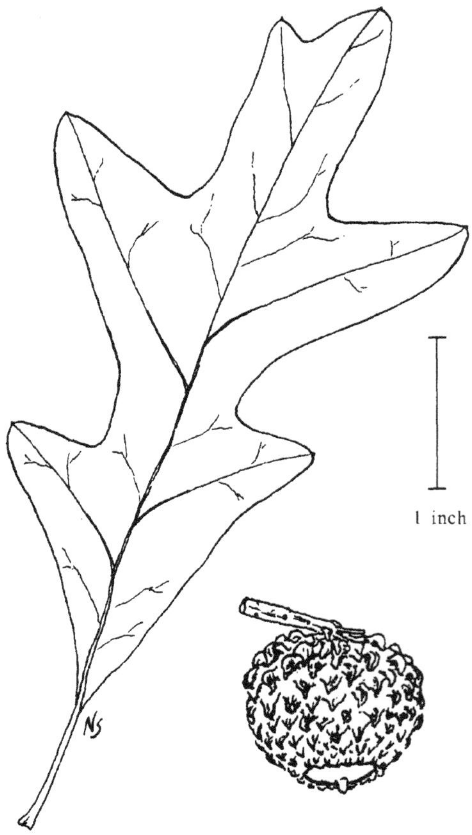

Quercus lyrata

Recognition

Leaves: oblong, broadest above middle; lobes several, shallow to deep, rounded, irregular; tip rounded, tapering to a narrow base; dark green, glabrous above, paler, downy beneath; petioles hairy to glabrous, up to 1 in (25 mm) long

Fruits: acorns solitary or paired, with or without stalks; nuts 0.5-1.0 in (12-25 mm) long, nearly globe-shaped; cup enclosing two-thirds or more of nut, sometimes fringed or ragged on rim; kernels edible

Twigs: green, hairy, becoming brownish, glabrous

Buds: rounded, gray to brown, usually hairy

Bark: brownish gray; large, irregular ridges

Quercus macrocarpa Michx. - **bur oak**

A bur-like fringe located around the rim of the acorn cup is responsible for the name of this majestic oak. Bur oak could qualify as the "state tree" for many prairie states. It grows naturally from southern Manitoba, the Dakotas, the eastern half of Nebraska and Kansas south through Oklahoma and east central Texas. The bur's natural range also extends across southern Canada to Maine, and southwestward through northern Arkansas, with scattered outliers on the periphery of this area.

Bur oak invaded and successfully competed with plants of the eastern prairie to create a savannah that is now converted largely to row crops. This transformation was possible because the thick bark of a mature bur oak resists injury by prairie fires.

Bur oak provenance research documents the genetic variability of the portion of the species native to the Great Plains region. Acorns collected from throughout the region varied in size from a small pullet egg nearly covered by a conspicuously fringed cup found in Oklahoma to a large pea with very little or no fringe found in North Dakota. A plantation at UNARDC, Ithaca, established from this seed collection showed that leaf and twig characteristics and growth rate are quite variable within and between Great Plains provenances (Long 1965; Dicke and Bagley 1980).

Oaks of Kansas origin grew faster than those of more northerly origin in this Nebraska planting. Crown dieback of unknown cause reduced the height of trees of the Oklahoma provenance at UNARDC, but oaks of Oklahoma origin at Prairie Pines Arboretum near Lincoln have made excellent growth without injury (Bagley 1998). Leaves of the Oklahoma oaks remain green until early November or the first severe freeze. Acorns on these trees—somewhat smaller than those of the mother trees growing in Oklahoma—do not ripen until mid-November. Angle (1985) found that a Nebraska provenance specimen was the most drouth resistant but also the slowest growing of trees originating from four sources: Nebraska, Oklahoma, North Dakota and Pennsylvania.

Bur oak can be a valuable component of windbreaks and forest plantations in much of the prairie region. This picturesque, durable tree is suitable for landscapes that can accommodate wide-spreading crowns. Massive, rugged branches reaching outward and upward into the winter sky compensate for the lack of bright color in the leathery leaves that flutter groundward after the first sharp freeze.

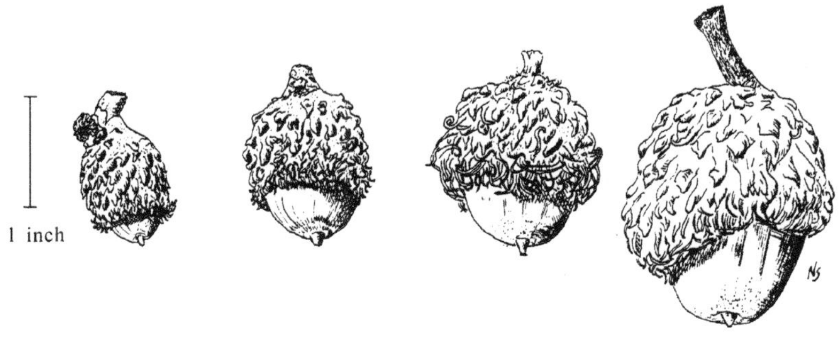

Quercus macrocarpa **acorns showing variation in size**

Fagaceae *Quercus*

The only insect likely to injure bur oak significantly is the gypsy moth. Oak wilt disease killed an insignificant number of bur oak d is not likely to become a serious threat. B oak is tolerant of a great variety of soil and moisture conditions and withstands the rigors of city life, except for soil disturmance around the roots of established trees (Sutton 1998).

Bur oak lumber is sold as white oak. Most of our majestic old trees provided barrel staves for whiskey distilleriess during the decade after World War II. Historic bur oaks include the Council Oak of Sioux City, Iowa, which marks the site of a meeting of Lewis and Clark and the local Native Americans, and the Post Office Oak of Morris County, Kansas, which served as a mail drop for west-bound immigrants in the early 1800s.

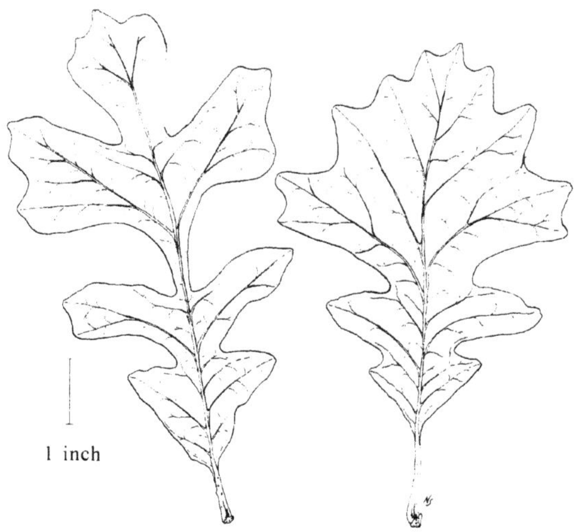

Q. ×bebbiana ***Q. macrocarpa***

Planning

Height: up to 75 ft (73 m); Nebraska champion, 1977, 71 ft (21.5 m), Johnson; National champion, 1985, 95 ft (29 m), Kentucky

Habit: tree; deciduous; crown rounded, wide-spreading

Site: sun; pH and soil adaptable, tolerates occasional poor drainage and inundation; zone 2 for nc hern provenances

Use: shade, windbreak, wildlife habitat, lumber, firewood

Recognition

Leaves: oblong, broadest above middle, apex rounded, base wedge-shaped to rounded, variably lobed, usually more deeply indented near the middle; dark green, glabrous above, variably white-downy beneath; petioles up to 1 in (25 mm) long; stipules linear, hairy, about 0.3 in (8 mm) long, early deciduous

Fruits: acorns solitary, sometimes in pairs, stalks 0.0-0.5 in (0-12 mm) long; nuts globe-shaped to somewhat oblong, 0.7-1.3 in (19-32 mm) long; cup enclosing from one-third of nut on trees of northern origins to nearly all of those on southern origins; northern origins with slight or no fringe, Nebraska and more southern origins usually conspicuously fringed; kernels variably edible

Twigs: stout, hairy, greenish, becoming smooth, variably corky ridged

Buds: conical to egg-shaped, variably hairy

Bark: becoming deeply ridged, usually dark gray

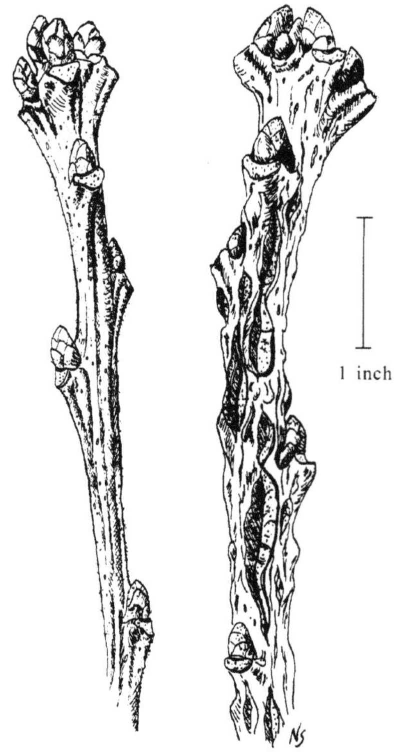

Quercus macrocarpa

Quercus muehlenbergii Engelm. (***Q. prinoides*** Willd. **var.** ***acuminata*** (Michx.) Gl. - **chinkapin oak**

Chinkapin oak generally grows on well-drained sites, from streambanks to dry ridges but with a preference for soils of limestone origin. It is seldom found on acidic soils. Chinkapin occurs naturally in a broad belt that stretches from central Texas, Oklahoma, and eastern Kansas to the northeastern U.S., excluding the Gulf Coastal Plain. Provenance research would be desirable to determine the chinkapin's range of adaptability and genetic variability.

Though chinkapin oak does not display bright fall colors, its neat, rounded crown provides summer shade and its shaggy bark adds a touch of character to the winter landscape. The sweet acorns were consumed by Native Americans and are excellent wildlife food. Chinkapin oak is an excellent candidate for windbreak and wildlife habitat plantings.

Clark (1989) concurs with Gleason and Cronquist (1963) that chinkapin oak is so closely related to *Q. prinoides* that it should be considered a variety of the latter species. These authors also concur that, if the alternative name is used, it should be spelled "muhlenbergii". We leave this matter for future generations to resolve. An alternative spelling for the common name is "chinquapin."

Quercus muehlenbergii

Planning

Height: up to 60 ft (18 m); Nebraska champion, 1978, 68 ft (20.6 m), Indian Cave State Park; National champion, 1991, 120 ft (36.7 m), Illinois

Habit: tree; deciduous; crown rounded

Site: sun; pH and soil adaptable, drouth tolerant; zone 4

Use: shade, windbreak, wildlife habitat, wood, firewood

Recognition

Leaves: narrow, oblong, broader near base, or above middle; numerous coarse, rounded to sharp incurved teeth along margin; tip pointed, base tapered to rounded; yellow-green, glabrous above, pale and finely hairy below; petioles about 1 in (25 mm) long; stipules narrow, hairy, early deciduous

Fruits: acorns solitary or in pairs, stalks 0.0-0.2 in (0-5 mm) long; nuts egg-shaped, 0.5-1.0 in (12-25 mm) long; cup variously hairy inside, enclosing one-fourth to one-half of nut; cup scales hairy, fringe short along margin; kernels edible

Twigs: glabrous to hairy, becoming glabrous; gray-brown

Buds: egg-shaped, scales glabrous except for fringed margin, chestnut brown

Bark: thin, gray, scaly with shallow fissures

Quercus prinoides Willd. var. *prinoides* - dwarf chinkapin oak

Dwarf chinkapin oak, a clump-forming shrub or bushy tree, grows most of the same places as *Q. muehlenbergii* grows. In a mixed population, it can be difficult to distinguish between the two species Gleason and Cronquist (1963) prefer to consider the two chinkapins a different varieties of the same species. See *Q. muehlenbergii* above.

A 12-foot (3.5 m) multistemmed dwarf chinkapin oak growing in a shady spot in Maxwell Arboretum, UNL, has shown little growth ove the past 30 years. Dwarf chinkapin oak can be trained into a single-stemmed tree or kept as a shrub.

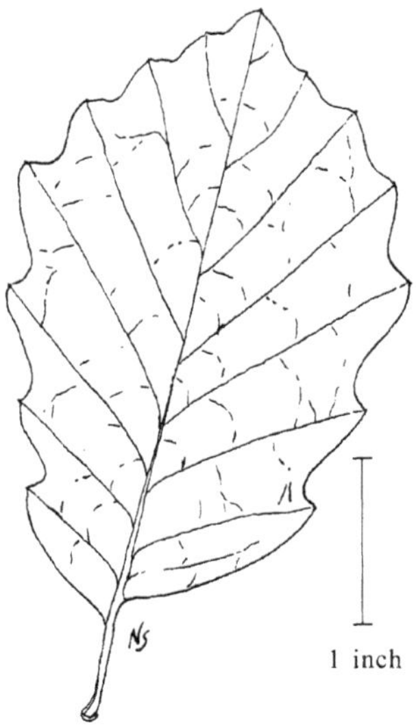

Quercus prinoides

Planning

Height: up to 15 ft (4.5 m)
Habit: shrub, small tree; deciduous; often multiple-stemmed
Site: sun, partial shade; pH and soil adaptable, drouth tolerant; zone 4
Use: naturalizing, wildlife habitat

Recognition

Leaves: elliptic, broader at middle or above; margin wavy to variousl toothed; yellow-green, scattered hairs above, pale and velvety or sparsely hairy below; petioles hairy, up to 0.5 in (12 mm) long; stipules linear, sparsely hairy, early deciduous
Fruits: acorns stalkless or nearly so; nuts egg-shaped, 0.5-1.0 in (12-25 mm) long, densely short hairy towards tip; cup densely hairy inside, enclosing one-third of nut, scales densely short-haired; kernels edible
Twigs: hairy, becoming glabrous
Buds: egg-shaped to globose; scales hairy, red-brown, tips rounded
Bark: smooth; lenticels horizontal; becoming gray with shallow fissures and thin plates

Quercus prinus L. (*Q. montana* Willd.) - chestnut oak

The westernmost edge of chestnut oak's natural range is near the Mississippi River in southern Illinois and Tennessee. From there it extends in a broad band northeastward to the New England states. Although chestnut oak reaches its largest size on deep, fertile, well-drained soils, it will grow better than most other oaks on dry sandy soils. It sprouts prolifically from cut stumps.

The chestnut oak's acorns, which are low in tannin, are relished by wild animals. Its tannin-rich bark, however, was once important in the leather tanning industry, which nearly caused the demise of the species by logging before the advent of manufactured substitutes for natural tannin.

Twenty-five-year-old chestnut oak trees at Prairie Pines Arboretum are 30 feet (9 m) tall and thriving (Bagley 1998). Provenance research could determine this oak's genetic diversity and adaptability to varied prairie sites.

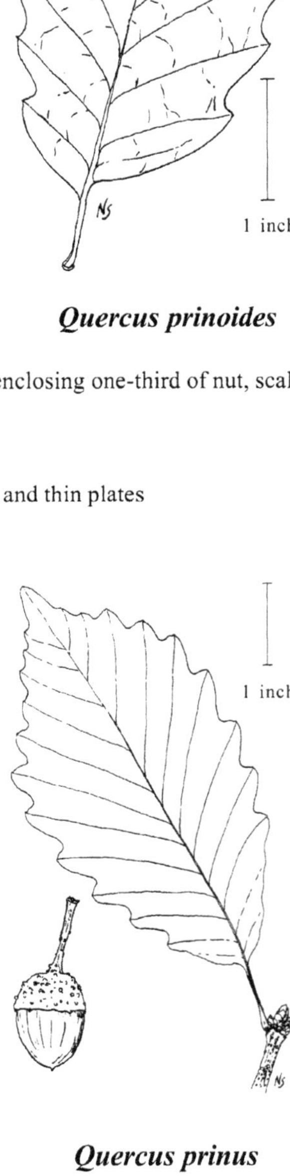

Quercus prinus

Planning

Height: up to 60 ft (18 m); National champion, 1983, 95 ft (29 m), New York
Habit: tree; deciduous; crown oval to rounded; autumn foliage yellow-orange
Site: sun; neutral to acidic soil, drouth tolerant; zone 4
Use: shade, wildlife habitat, windbreak, firewood

Recognition

Leaves: oblong, broadest above the middle; margin coarsely toothed, tips obtuse, base tapered to rounded; yellow-green and glabrous above, paler and often finely hairy beneath; petioles up to 0.6 in (12 mm) long

Fruits: acorns solitary or in pairs, stalks shorter than petioles; nuts egg-shaped to nearly round, 1.0-1.5 in (25-38 mm) long, very shiny; cup encloses about one-half of nut, scales thickened, rough, warty, hairy; kernels edible

Twigs: glabrous, reddish brown

Buds: broadest near base tapering to a point; scales shiny, dark brown

Bark: brown to nearly black, becoming deeply and coarsely furrowed

Quercus robur L. - English oak

English oak, native to Europe, East Asia and North Africa, is undoubtedly quite genetically variable because of its vast natural range. Crowns that vary in form from weeping to fastigiate and leaves and acorns that vary in shape and size lend credence to this genetic variability. English oak can be recognized by its long-stalked acorns, short-petioled leaves, and the occasional arrowhead-shaped (sagittate) base.

Q. petraea (Mattuschka) Lieblein, durmast oak, another valuable oak of northern Europe, is similar to and hybridizes with English oak. The leaves of durmast oak are long-stalked with no ear-like appendages (auricles) at the base and acorns that are nearly stalkless. Perhaps hybridization explains some of the variation in the species we call *Q. robur*.

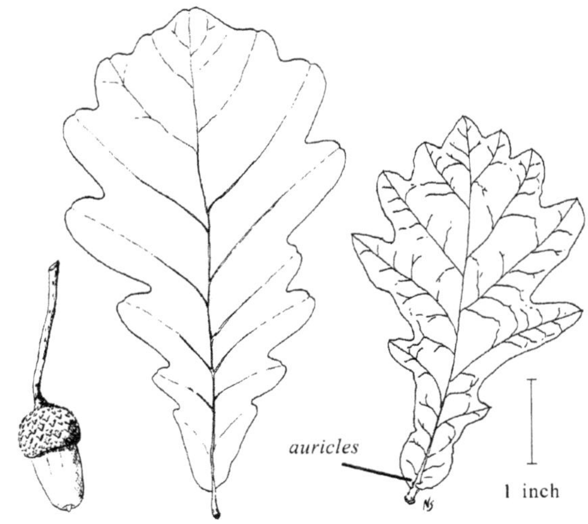

Quercus robur

A veteran oak located southeast of Filley Hall, UNL East Campus, known affectionately as the "Russian Oak," differs from the classic English oak. It bears shorter acorns than the classic species and leaves that are similar to bur oak leaves, but smaller. The large acorn crop that it produces almost every year is one characteristic that the "Russian oak" shares with two classic English oaks located in the nearby Maxwell Arboretum.

Trees growing at Prairie Pines Arboretum from acorns collected from several English oak trees at Michigan State University, Lansing, exhibit great differences in their leaves, fruits, form, growth rate and leaf retention (Bagley 1998). Some are exceptionally vigorous, others are slow growing with a hint of leaf chlorosis. Most of these oaks bore fruit at eight years of age.

English oak, especially those of a more northerly origin, are probably adapted to much of the prairie region. Cultivars include 'Asplenifolia', with leaves deeply cut; 'Atropurpurea', with purple leaves; and 'Concordia', with bright yellow leaves in spring. A fastigiate cultivar recently planted in Maxwell Arboretum exhibits many dead branches due to apparent cold injury. This tree retains green leaves into November. Since many of the fastigiate forms are grown from seed, others may be more winter hardy. Two 10-year-old seedling fastigiate oaks at Prairie Pines Arboretum have been spared significant winter injury (Bagley 1998).

Planning

Height: up to 65 ft (20 m); Nebraska champion, 1985, 65 ft (19.8 m), Omaha; National champion, 1984, 85 ft (25.9 m), Ohio

Habit: tree; deciduous; crown broad, rounded to narrow

Site: sun; pH and soil adaptable, drouth tolerant; zone 5

Use: shade, windbreak, wildlife habitat, firewood

Recognition

Leaves: oblong, widest above middle (obovate-oblong), up to 5 in (13 cm) long, variably lobed; dark green, glabrous above, paler and glabrous below; occasionally ear-like appendages (auricles) at base; petioles short

Fruits: acorns often in pairs, stalks 0.5-2.0 in (13-50 mm) long, sometimes longer; nuts globe-shaped to oblong, up to twice as long as the diameter; cup encloses one-fourth to one-third of nut; kernels edible

Twigs: glabrous, red-brown

Buds: egg-shaped; scales overlapping, fringed with hairs, chestnut brown

Bark: gray-black, deeply furrowed

Quercus stellata Wangenh. - post oak

Post oak, native to much of Texas, Oklahoma and eastern Kansas, and areas eastward to Rhode Island and south to central Florida, grows on a wide variety of soils from very acidic to slightly alkaline. Post oak is typically a small, shrubby tree growing on dry, sandy, gravelly or rocky ridges. But this oak is also found growing on sites that are alternately waterlogged and hard and dry. It has invaded grasslands in the semi-arid Southwest.

Even though post oaks are too small to be valuable timber trees, their wood is used for furniture, flooring and interior finish work. The post oak earned its common name because post-size trees sprout readily after logging, thereby lending themselves to perpetual post harvesting operations. The post's winter leaf retention and prolific fruiting are assets for wildlife and windbreak plantings.

Planning

Height: up to 50 ft (15 m); National champion, 1987, 85 ft (25.9 m), Virginia

Habit: tree, sometimes shrubby; deciduous; crown round-topped, spreading; autumn foliage red

Site: sun; pH and soil adaptable, drouth tolerant; zone 5

Use: shade, windbreak, wildlife habitat, posts, firewood, lumber

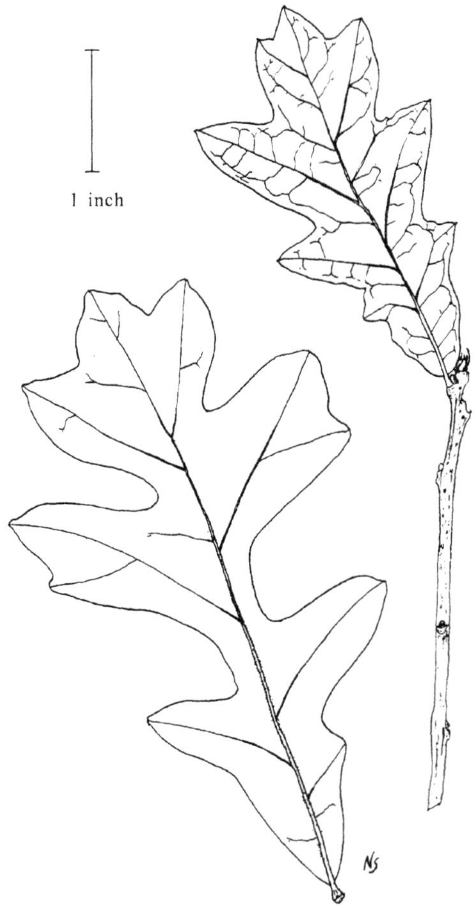

Quercus stellata

Recognition

Leaves: egg-shaped, usually widest above middle (obovate), lobes 5 to 7, wide, deep; tip rounded, base rounded; dark green, leathery (coriaceous); upper surface roughened by stiff, clustered, white or rusty hairs, also hairy beneath; petioles up to 1 in (25 mm) long

Fruits: acorns often in clusters of 2-4, with or without short stalks; nuts egg-shaped, 0.5-1.0 in (13-25 mm) long; cup enclosing one-third to one-half of nut, scales thin, hairy; kernels bitter

Twigs: hairy, reddish, becoming glabrous, gray or dark brown

Buds: broadly egg-shaped, tips rounded; scales often hairy

Bark: ashy gray to dark brown to black, deeply furrowed

Wood: hard, moderately durable

Betulaceae S. F. Gray - **birch family**

Five of the six genera of this nearly cosmopolitan birch family occur naturally in North America. Two shrub species of one genus, *Ostryopsis*, are native only to China. Dwarf shrubs are found in tundra and high mountain regions and large trees grow in lowlands. Some genera are important in the lumber industry, others are valued for their fruit, and some are used primarily for landscape beautification.

Recognition

Habit: trees, shrubs; deciduous

Leaves: simple, alternate, veins straight, margins toothed; stipules deciduous

Flowers: unisexual (monoecious), clustered; sepals 0-6, scale-like, petals none; male pendulous, elongate, formed the previous growing season (preformed) except on *Carpinus*; stamens commonly as many as sepals, sometimes many more; female pendulous or erect, short and firm, often woody; ovary 2-celled, styles 2

Fruits: winged (samaras) or nuts inclosed in bracts, tiny to large; seeds 1

Twigs: lenticels sometimes becoming horizontally elongated, pith homogenous; leaf scars with 3 bundle scars

Buds: terminal often missing; lateral with scales overlapping (imbricate) or sometimes meeting at edges (valvate)

Bark: usually thin and smooth, white to brown, sometimes peeling

Alnus B. Ehrh. - **alder**

Thirty alder species—trees and shrubs—are found primarily in north temperate regions, but they also grow at high elevations in Central and South America and northern Africa. Only one North American species—red alder (*A. rubra* Bong.), a native of the coastal Pacific Northwest and southern Alaska—grows large enough to be of value for lumber. The other seven species native to North America seldom reach tree size.

Alders are trees of moist, cool areas that are capable of symbiotically fixing soil nitrogen. For this reason, tree-form alders have been recognized as having potential for biomass production. They have not been widely used in landscape plantings despite their ability to thrive on infertile soils.

Two similar shrubby alders—speckled alder, *A. rugosa* (Du Roi) Spreng., a native to eastern Canada and the northeastern U.S. south to Virginia, and hazel alder, *A. serrulata* (Ait.) Willd., a native of the eastern U.S. from Maine to Florida and west to Kansas—are possible candidates for wet prairie sites. Speckled alder is considered by Stephens (1973) to be a subspecies of white alder, *A. incana* (L.) Moench, a tree native to Europe's cold, wet sites.

White alder is a subject of intensive research for biomass production (Hall et al. 1983). Cltivars of white alder include 'Aurea', with yellow leaves and orange male catkins; 'Pendula', with pendulous branches; and the cut-leaved 'Laciniata'.

Recognition

Habit: trees, shrubs **Leaves:** usually broadest near base or middle

Flowers: male clusters (catkins) elongated, each bract with 3 flowers, calyx 4-parted, stamens 4, rarely less; female catkins short, sometimes preformed, each bract with 2 flowers; calyx none

Fruits: cone-like, leathery to woody, composed of many persistent scales, forming an egg-shaped strobile; nutlets flattened, winged pairs in scale axils

Twigs: reddish, pith triangular in cross-section

Buds: stalked; scales 2-3, meeting at edges (valvate) or sessile with 3-6 overlapping scales

Alnus glutinosa (L.) Gaertn. - European alder

This vigorous native of Europe now reproduces naturally in low, moist areas of the eastern U.S. European alder usually attains tree form, but many planted in the northern prairies are multistemmed and shrubby due to cold injury. European alder provenance research at the Iowa Agricultural Experiment Station (Hall et al. 1983) shows great variability in winter hardiness and rate of growth. Northern provenances are more winter hardy and southern provenances are better adapted to drouthy sites.

European alder, an attractive tree with dark green foliage, is especially adapted to streambanks subject to flooding. It does survive, however, on dry upland sites, and, because of its nitrogen-fixing ability, thrives on infertile soils. Insects and diseases are not normally serious problems for this alder unless it is stressed by drouth.

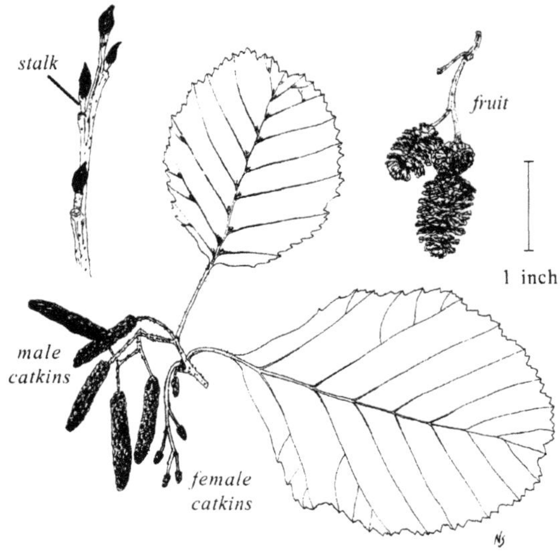

Alnus glutinosa

'Pyramidalis' might be considered a substitute for Lombardy poplar (Dirr 1990). 'Laciniata' and 'Imperialis' display incised leaves, and 'Aurea' leaves are golden yellow.

Planning

Height: up to 50 ft (15 m)
Habit: tree, sometimes shrubby; crown pyramidal to oblong
Site: sun, partial shade; slightly alkaline to acidic soil, tolerates dry and extremely wet infertile soils; nitrogen-fixing; zone 3
Use: biomass production, streambank erosion control

Recognition

Leaves: nearly round, about 3 in (8 cm) long; coarsely toothed, veins running to teeth tips; dark green and glabrous above, pale below with tufts of hair in vein axils; stipules prominent, eventually deciduous
Flowers: female catkins appear in autumn on twig ends (preformed)
Fruits: cone-like, egg-shaped, about 1 in (2.5 cm) long, often occurring in 3s, persistent through winter; nutlets compressed, laterally winged
Twigs: glabrous, green; lenticels tan, prominent
Buds: stalked; scales 2, purplish, meet at edges without overlap (valvate)
Bark: smooth, gray, becoming dark and scaly; lenticels light-colored, prominent

Betula L. - birch

Forty to 60 *Betula* species are widely scattered throughout the Northern Hemisphere. The uncertainty in species numbers is due to a highly variable population that doesn't lend itself to pigeonholing. About half of the 15 species native to North America attain tree size, three of which are valuable timber trees.

The birches are graceful trees (or shrubs) with small branches, thin twigs and often golden autumn leaves. Leaf size and shape and bark color and texture vary greatly among the birches. Birch bark is water repellent and exceptionally durable, remaining intact long after the wood has decayed.

Birches prefer moist, cool, well-drained soils, a requirement not easy to attain in much of the prairie region. A fertile site protected from wind with midday shade on the ground and adequate water is necessary to maintain the vigor that will resist bronze birch borer attacks.

Recognition

Habit: trees, shrubs

Leaves: triangular or egg-shaped in outline; margins with a double row of fine teeth (bi-serrate); tips pointed

Flowers: male catkins solitary or in clusters of 2 to 3; 3 flowers to each bract, calyx 4-parted; stamens 2, divided at apex (bifid); female catkins usually solitary, 3 flowers at each 3-lobed bract, individual flowers without calyx

Fruits: nutlets laterally winged, borne on erect or pendent strobiles resembling small cones; scales usually deciduous from persistent cone axis; nutlets released at maturity

Twigs: slender, some aromatic; lenticels conspicuous, becoming horizontally elongated; leaf scars half oval to triangular

Buds: terminal absent, sometimes present on short spurs; lateral egg-shaped, pointed; scales 3-4 visible, overlapping (imbricate)

Bark: white to brown, peeling or breaking into scaly plates

Grouping Key for *Betula*

(Rehder 1940)

Leaves with 7 or more pairs of veins; fruits (strobiles) in clusters (racemes), wings much broader than nutlets: **section Acuminatae**

Leaves with 3-7 pairs of veins; fruits (strobiles) solitary, wings less than or not much broader than nutlets: **section Costatae**

Leaves with 3-7 pairs of veins; fruits (strobiles) solitary, wings broader than nutlets: **section Albae**

Leaves with 3-7 pairs of veins; fruits (strobiles) solitary, wings much narrower than nutlets, rarely as broad: **section Humiles**

Section Acuminatae

Betula maximowicziana Reg. - **monarch birch**

Monarch birch is reputed to be resistant to bronze birch borer. Unfortunately, trees improperly bearing the monarch birch label have been inadvertantly sold by reputable nurseries. An investigation by Santamour and Meyer (1977) of the U.S. National Arboretum disclosed that most monarch birch found in arboreta are labeled incorrectly.

Trees purchased as monarch birch for Prairie Pines Aboretum in 1979 did not resemble Rehder's description and, more important, proved to not be resistant to the bronze birch borer. Most of these trees died within four years after planting (Bagley 1998). A recently planted birch tree in Maxwell Arboretum is apparently *B. maximowicziana* based on a comparison of its characteristics with Rehder's description (1940). If it thrives and fends off borer attacks, we will have a birch to recommend for planting on similar sites.

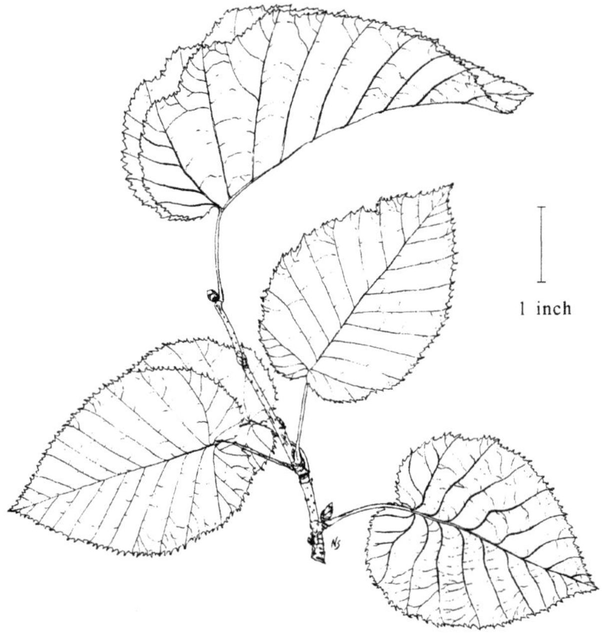

Betula maximowicziana

Planning

Height: up to 40 ft (12 m)
Habit: tree; crown oval to irregular
Site: sun, part shade; prefers soil neutral to slightly acidic, moist, fertile; zone 5
Use: naturalizing, specimen
Pests: resistant to bronze birch borer

Recognition

Leaves: over 3 in (8 cm) long, heart-shaped with two rounded lobes at base, deeply cleft between; surfaces hairy becoming glabrous; 10-12 pairs of veins
Fruit: strobiles up to 3 in (8 cm) long, cylindric; bracts with lateral lobes ascending and shorter than middle lobe; nutlet wings much broader than nutlet
Twigs: dark reddish brown, glabrous
Bark: dark gray or orange-gray, peeling in papery flakes

Section Costatae

Yellow birch, *B. alleghaniensis* Brit. (*B. lutea* Mich.), a native to the northeastern U.S., is a source of valuable lumber. Pioneers distilled oil of wintergreen from the twigs and inner bark, a distinctive flavor that can aid in this birch's identification.

Erman birch, *B. ermanii* Cham., an Asian species with creamy white, peeling bark, deserves trials to determine its resistance to borers (Solomon 1990). Dahurian birch, (*B. davurica* Pall.) which resembles river birch, Japanese cherry birch (*B. grossa* Sieb. & Zucc.), and dwarf birch (*B. nana* L.), are some of the many species not described herein that could be worthwhile ventures for gardeners looking for unusual landscape plants.

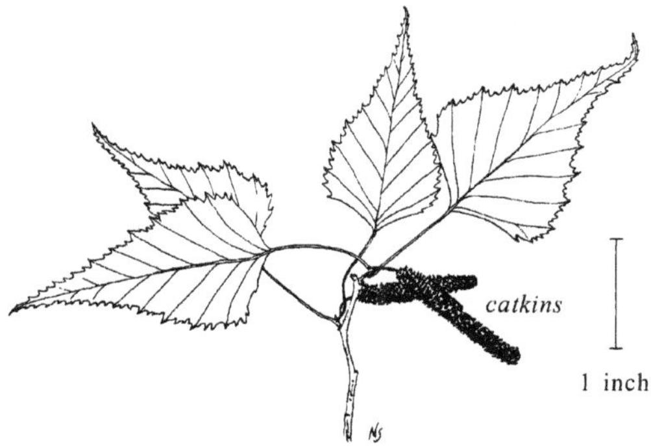

Betula platyphylla

Varieties *japonica* and *szechuanica* of Asian white birch, *B. platyphylla* Sukachev (*B. mandshurica* Nakai), have been cultivated for the nursery trade. 'Whitespire', a selection of *japonica* that has demonstrated resistance to bronze birch borer at the University of Wisconsin, is preferred over other progeny of the species (Dirr 1990).

Betula lenta L. - sweet birch

Sweet birch, sometimes called cherry birch because of the young trees' mahogany red cherry tree-like bark, is native to the Appalachians and southern Quebec and Ontario in Canada. Pioneer families harvested many young trees for their oil of wintergreen. Beer has been made by fermenting corn with the sap obtained from sweet birch trees tapped in late winter (Elias 1980). The strong, hard wood, which resembles mahogany after exposure to air, is used in the manufacture of fine furniture.

In its native habitat sweet birch grows best on moist soils with a northerly aspect, but a twenty-foot tree growing on city campus, UNL, is remarkably healthy and the parent of many seedlings planted in the Lincoln community.

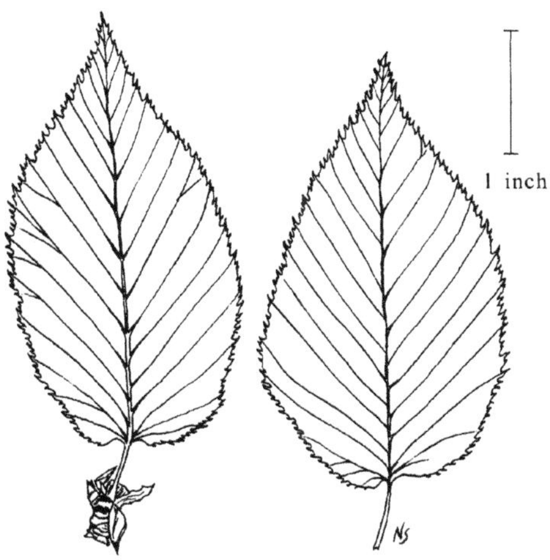

Betula lenta

Planning

Height: up to 40 ft (12 m); National champion, 1989, 78 ft (23.6 m), New Hampshire

Habit: tree; crown pyramidal, becoming rounded and spreading; autumn foliage yellow

Site: light shade when young, requiring more sun with age; prefers moist, fertile, neutral-to-slightly-acidic soils; somewhat drouth tolerant; zone 4

Use: naturalizing, specimen, lumber

Pests: resistant to bronze birch borer

Recognition

Leaves: egg-shaped (ovate), up to 6 in (15 cm) long; base heart-shaped, toothed (doubly serrate); glabrous above, hairy on veins beneath; petioles up to 1 in (25 mm) long, deeply grooved above, hairy
Flowers: appear before the leaves
Fruits: strobiles oblong, egg-shaped; bracts smooth, sometimes hairy at base, with lobes of nearly equal length; nutlet wings as broad as nutlet
Twigs: reddish brown, glabrous, shiny, strong wintergreen flavor; spurs numerous on old growth
Buds: conical, sharp-pointed, mostly hairless, shiny, divergent
Bark: smooth, reddish brown to nearly black, becoming brownish black, breaking into thin, irregular, scaly plates; lenticels horizontal, prominent
Wood: hard, heavy, strong

Betula nigra L. - river birch

River birch, native to the lowlands of the eastern U.S. from New England to Florida, and west to Kansas and Minnesota, is the birch most likely to endure the hot summers of the prairie region in areas of moderate precipitation. A river birch at Prairie Pines Arboretum near Lincoln is healthy after 30 years with no supplemental water except roof runoff from a low building which shades a portion of the root zone. A river birch tree in a somewhat less favorable site died at approximately 15 years of age (Bagley 1998).

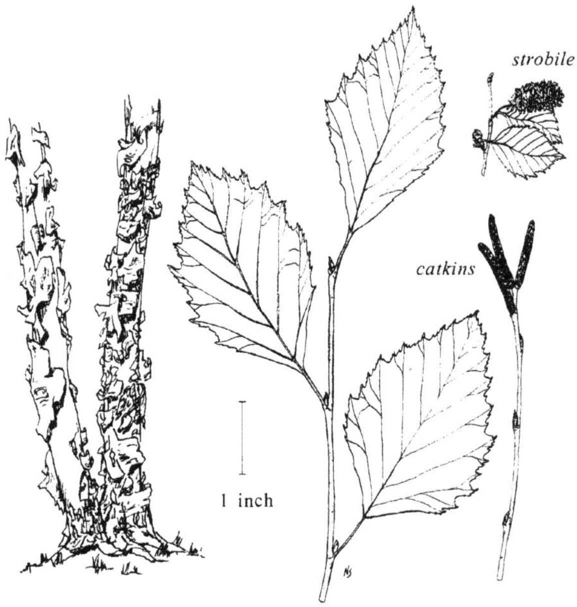

Betula nigra

River birch seed matures in late spring, an adaptation enabling it to establish seedlings following spring flooding. Considerable genetic variability within the species is indicated by differences amor ıdividual trees in bark color and text. ıd date of autumn leaf colora Heritage', a selection superior t ıf-the-mill seedlings, is adapte ıe wide range of climatic conditic ırring within zones 4 to 9 (Dirr 1' 'rovenance research could provide information useful in matching provenances to sites for best growth a.. ıongevity.

Planning

Height: up to 50 ft (15 m); Nebraska champion, 1988, 42 ft (12.8 m), Lincoln; National champion, 1988, 90 ft (27.4 m), Tennessee
Habit: tree; crown pyramidal, becoming rounded, often multistemmed; autumn yellow short-lived
Site: sun, part shade; prefers neutral to acidic, moist soil, heat tolerant; zone 4
Use: specimen, naturalizing
Pests: resistant to bronze birch borer

Recognition

Leaves: triangular to egg-shaped, up to 3 in (8 cm) long; tip sharp, base broad wedge-shaped; coarsely toothed (doubly serrate); glabrous or sparsely hairy above, hairy beneath, especially on midrib; petioles about 0.5 in (12 mm) long, hairy

Fruits: strobiles oblong-cylindric; bracts hairy with 3 upright, nearly linear lobes; thin, membranaceous wings nearly as broad as nutlets

Twigs: reddish brown, usually hairy, lenticels oval becoming horizontal

Buds: egg-shaped, pointed, red-gray, hairy

Bark: salmon pink to various shades of brown, papery; becoming brown, ridged and furrowed

Section Albae

Betula occidentalis Hook. (*B. fontinalis* Sarg.) - water birch

Water birch, a small, usually multistemmed and shrubby tree, is native to western North America from New Mexico and Arizona north to British Columbia and Saskatchewan, including the northern and western fringes of the prairies as far east as the Pine Ridge of Sioux County, Nebraska.

This birch can be distinguished by winter catkins that are shorter than those of young paper birches and by a leaf shape different from another shrubby birch, *B. glandulosa* Michx., a native of the Black Hills and west. Leaves of the latter are nearly round with rounded teeth along the margins (crenate). Though the dark brownish, smooth bark of water birch is bland in comparison to many of the other birches, it should be more widely tested as a component of prairie landscapes, especially in the northwest.

A fine planting of water birch graces the Montana State capitol grounds in Helena (Sutton 1998).

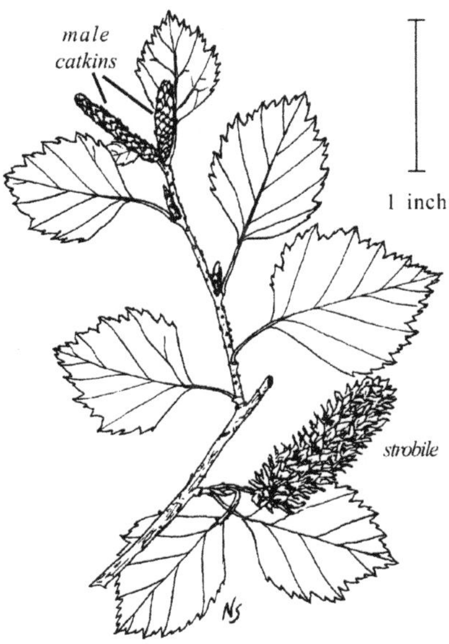

Betula occidentalis

Planning

Height: up to 15 ft (4.5 m)

Habit: tree, shrub; often multistemmed

Site: sun, part shade; prefers moist, fertile soils; zone 3

Use: naturalizing

Recognition

Leaves: egg-shaped, tip usually sharp, base rounded to wedge-shaped; margin toothed (serrate) except at base; shiny, dark green, glabrous above, paler beneath, sometimes dotted with resinous glands; petioles slender, glabrous or slightly hairy, grooved above, up to 0.6 in (15 mm) long

Fruit: strobiles cylindric, about 1 in (25 mm) long; bracts pointed, hairy-fringed (ciliate), lateral lobes broad, angular, ascending, shorter than narrow middle lobe; wings thin, as broad as nutlet

Twigs: reddish brown, covered with gray waxy bloom and resin glands; lenticels elliptic

Buds: egg-shaped, pointed; scales red-brown, often scurfy, margin hairy (ciliate)

Bark: thin, very dark on young, trees becoming reddish brown with conspicious horizontal lenticels, not peeling

Betula papyrifera Marsh. - paper birch

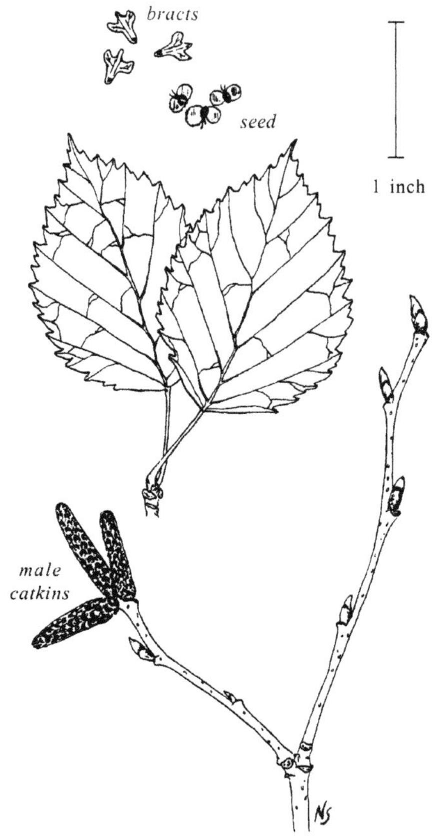

Betula papyrifera

This North American birch, which grows naturally throughout much of Canada and Alaska, extends into the northern tier of the U.S. with outliers in the Dakotas, Iowa and along the Niobrara River in Nebraska. The most southerly paper birch is a small stand in Colorado on Green Mountain (Weber 1990). These few trees are now protected in the city of Boulder's mountain parks.

The Nebraska Statewide Arboretum, through the efforts of the late Joe Young, its one time director, collected seed from native paper birch on the Niobrara River in 1980. The Niobrara seedlings were distributed to various arboretum sites for evaluation. We learned from this collection that the leaves and twigs of Niobrara birches are more pubescent than the leaves of most other baper birches. At this time it is too soon to determine whether the paper birch of Nebraska is better adapted to hot summers than those now in the nursery trade.

These graceful white-barked trees are widely planted despite their short life. The stress of hot, windy summers increases susceptability of paper birch to bronze birch borers and other insects. Paper birch of unspecified origin growing at PERC in Fort Collins require winter watering to reduce susceptability to borer attack (Fairchild and Klett 1993). The trees on this site display minor to moderate chlorosis on the alkaline soils of pH 8.

Hoag (1965) notes that the wide natural distribution of paper birch contributes to a quite variable population. A well-designed provenance study could identify provenances and/or individual specimens better adapted to the prairies.

Many nurseries do not distinguish between paper birch and European white birch, *B. pendula*, merely listing them as white-barked birch. We believe that many white-barked birches in the trade are of the latter species.

Planning

Height: up to 50 ft (15 m); Nebraska champion, 1981, 45 ft (13.7 m), Arbor Lodge Historical Park; National champion, 1979, 93 ft (28.2 m), Maine

Habit: tree; crown pyramidal, becoming rounded

Site: sun, part shade; neutral to acidic, prefers acidic, moist, cool, fertile soils; zone 3

Use: specimen, naturalizing

Pests: susceptible to bronze birch borer

Recognition

Leaves: egg-shaped, irregularly toothed (doubly serrate); long pointed, base wedge-shaped, rarely heart-shaped (cordate); surfaces variably hairy; petioles flattened or slightly grooved, glabrous or hairy and sometimes with black glands

Fruits: cone-like heads (strobiles) cylindric, brown, drooping, persisting into early winter; bracts hairy with lateral lobes ascending, rarely spreading, shorter than the middle lobe; wings as broad as nutlets

Twigs: reddish brown, sometimes hairy when young with prominent lenticels, variably glandular, no wintergreen odor

Buds: sharp-pointed, tapering from base to apex; scales brown, glabrous to hairy

Bark: thin, dark brown turning white, separating into papery strips becoming nearly black and fissured with old age

Betula pendula Roth - European white birch

The much planted European white birch, native to Europe in areas from the Balkan Peninsula to beyond the Arctic Circle, is often sold as "white-barked" birch. The often short life of this birch in prairie landscapes is usually attributed to injury by the bronze birch borer. The broad natural range of the species indicates probable great genetic diversity. Provenance testing could discover strains that are better adapted to prairie sites.

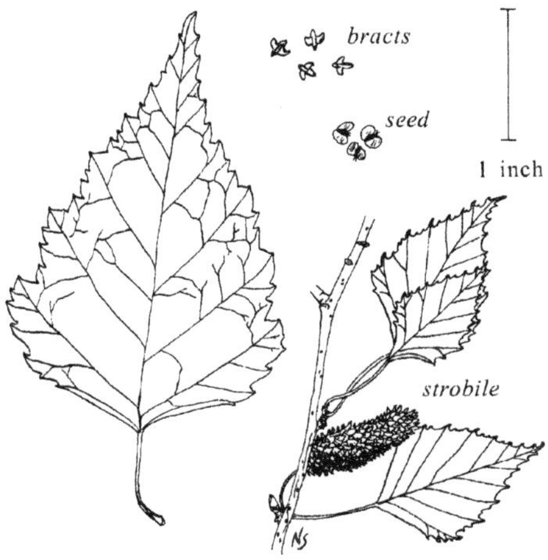

Betula pendula

The European white birch and paper birch are so similar in habit that most people will never know the true identity of their white-barked birch. Rehder (1941) uses the shape of the bracts enclosing nutlets as a key difference between the two species. The bark of the European species supposedly becomes white at an earlier age than paper birch's bark does.

European white birch usually develops a pendulous branch habit with age. However, cultivars have been selected with erect branches ('Fastigiata') to very drooping branches ('Elegans'). 'Dalecarlica' is an elegant tree with deeply lobed leaves, and 'Youngi' is pendulous without a leading stem.

Planning

Height: up to 45 ft (14 m); Nebraska champion, 1986, 62 ft (18.8 m), Cedar Rapids

Habit: tree; crown pyramidal, becoming rounded

Site: sun, part shade; slightly alkaline to acidic, moist, fertile soil; tolerates dry soil; zone 3

Use: specimen

Pests: susceptible to bronze birch borer

Recognition

Leaves: egg-shaped to diamond-shaped, tapering to long-pointed tips; base wedge-shaped, sometimes squared off (truncate), sometimes deeply lobed; surfaces glabrous, dark green; petioles minutely hairy or glabrous

Fruits: cone-like heads (strobiles) cylindric, breaking up in fall; bracts minutely hairy or glabrous, lateral lobes spreading or recurved, usually longer than the middle lobe (Rehder 1941); wings broader than nutlet

Twigs: brown, resinous-glandular, glabrous

Bark: creamy white, flaky or peeling

Betula populifolia Marsh. - gray birch

Gray birch, a small tree that invades abandoned fields in its native northeastern U.S. habitat, grows on sandy and gravelly soils (wet or dry) as well as on heavier soils. This birch grows fast and produces abundant root suckers from its shallow root system, enabling it to populate a rather inhospitable site very quickly.

This white-barked tree offers a means of obtaining contrast in a landscape if suckering can be tolerated. Though normally short-lived, gray birch may survive longer than many other birch species because of its resistance to bronze birch borers.

Gray birch at Hans Burchard Arboretum in Lancaster County, Nebraska, are thriving and reproducing on a wet site (Sutton 1998).

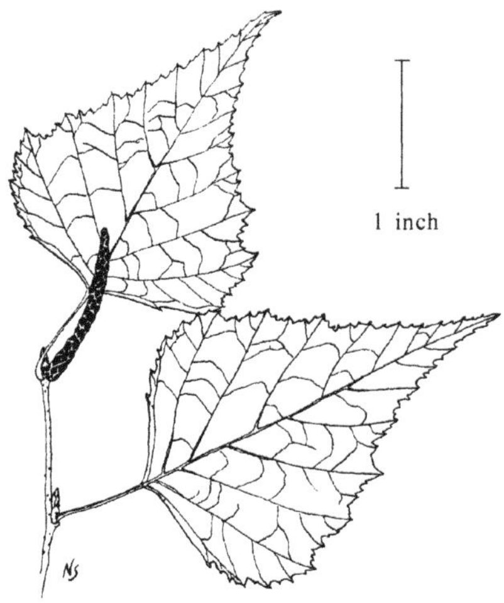

Betula populifolia

Planning

Height: up to 30 ft (9 m); National champion, 1989, 77 ft (23.3 m), Connecticut
Habit: tree; crown narrow conical; thicket forming
Site: sun; pH and soil adaptable, tolerates infertile, dry soil; zone 4
Use: naturalizing
Pests: resistant to bronze birch borer

Recognition

Leaves: triangular with an almost straight, broad base; tip long-pointed (acuminate); glaucous, lustrous dark green above, paler and glaucous below; petioles slender, dotted with glands
Fruits: drooping cylindric strobiles; scales hairy, lateral lobes spreading, longer than middle lobe; wings broader than nutlets
Twigs: slender, drooping, roughened with pale resinous glands
Buds: tapering both ways from middle or from the base; scales sometimes resinous, often with hairy margins
Bark: smooth, not peeling, reddish brown, becoming chalky white; black triangular patches below branch bases

Section Humiles

The shrubby birch species in section Humiles are not commercially important and are probably adapted only to the extreme northern prairie. For example, *B. glandulosa* Michx. is a native of the Black Hills, the northern U.S. and Canada.

Carpinus L. - hornbeam

Yokes, spokes, cogs, axles and other items were made from hornbeam wood before iron became plentiful and inexpensive. The hard, close-grained, tough wood is used for bowls and dishes because it doesn't crack or split. The most distinctive feature of these small trees is a fluted stem that appears to be flexing its "muscles."

Only one of these 25 shade-tolerant species grows naturally in North America. The others are native to Europe, Asia and Central America. Research to determine the worthiness of hornbeam species for prairie landscapes could be rewarding. The few hornbeams that we know of in Nebraska are relatively young and pest free.

In addition to the two species described herein, adventuresome gardeners can consider *C. cordata*, heartleaf hornbeam, *C. japonica*, Japanese hornbeam, and *C. orientalis*, Oriental hornbeam, among others, for their landscapes.

Recognition

Habit: trees, shrubs
Leaves: usually egg-shaped (ovate), margins fine-toothed
Flowers: appear in spring with leaves; male catkins pendulous, not preformed; calyx and corolla (perianth) wanting; stamens 3-10; female catkins terminal on twig, 2 flowers each bract subtended by 2 bractlets; perianth fused to ovary (adnate) with 6-10 teeth
Fruits: nutlets ribbed, covered by large leaf-like bracts; ripening in autumn
Buds: terminal absent; laterals pointed; scales overlapping, tips pointed
Bark: gray, smooth to scaly

Carpinus betulus L. - European hornbeam

European hornbeam, native to Europe and Asia Minor, is a useful hedge component because it tolerates light shade and withstands heavy shearing. Cultivars of this hornbeam include: 'Columnaris', densely branched with a central leader; 'Fastigiata' and 'Globosa', both with upright branches and no central leader; and 'Pendula' with pendant branches. Those with unusual foliage characteristics are 'Asplenifolia' with its deeply toothed leaves (some large enough to be called lobes), 'Purpurea' with its purplish new foliage, and 'Variegata' with its irregular yellow patches on leaves (Dirr 1990).

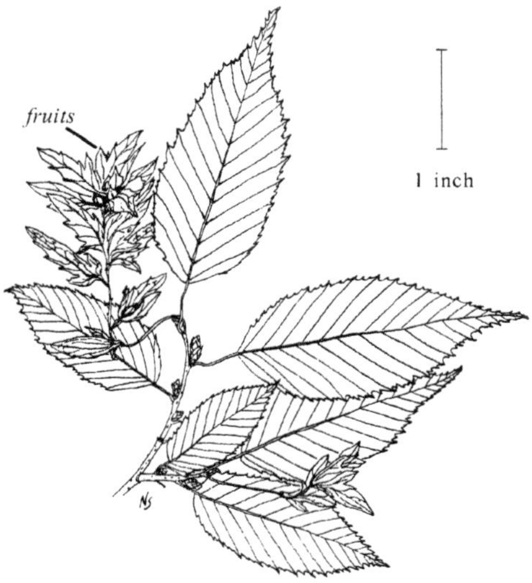

Carpinus betulus

Planning

Height: up to 50 ft (15 m)
Habit: tree; crown pyramidal, becoming rounded
Site: sun, part shade; pH and soil adaptable, zone 5
Use: screen, hedging, grouping
Pests: insect and disease resistant

Recognition

Leaves: usually egg-shaped (ovate); variably pointed, base heart-shaped or rounded; margin sharp toothed (doubly serrate); veins seldom forking at ends; surface glabrous except on veins beneath; petioles hairy, up to 0.5 in (12 mm) long
Flowers: female catkins; bracts conspicuous, 3-lobed, 3-5 nerved, margin entire or barely toothed, middle lobe about 1.4 in (4 cm) long
Fruits: 0.25 in (6 mm) long enclosed by leaf-like bract
Twigs: slender, glabrous or slightly pubescent when young
Buds: pointed, about 0.2 in (5 mm) long, usually appressed and curled around stem; scales overlapping, glabrous
Bark: gray, fluted

Carpinus caroliniana Walt. - **American hornbeam**

American hornbeam is classified as a "weed tree" by foresters because of its small size, but it has a place in our landscapes because of its unique trunk and fruits. American hornbeam is relatively free of insect and disease pests. The wide distribution of this native of the entire eastern U.S., extending west to Minnesota, Iowa, Missouri and eastern Texas, indicates that we might profit from provenance research because of its possible significant genetic variability.

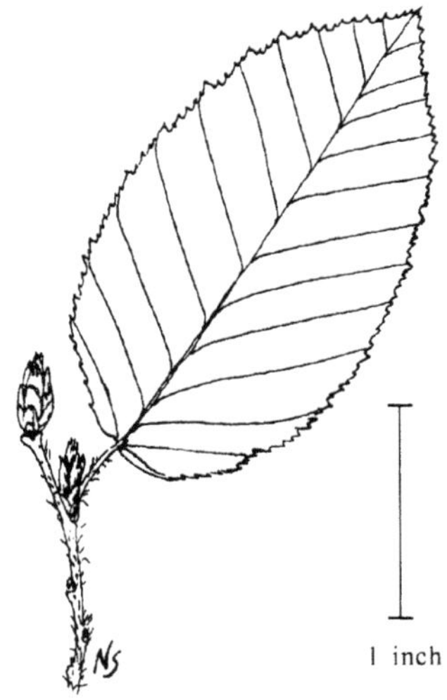

Carpinus caroliniana

Planning

Height: up to 30 ft (9 m); National champion, 1975, 69 ft (21.1 m), New York

Habit: tree; crown spreading to rounded, often multi-stemmed; autumn leaf color variable, from yellows to orange-red

Site: shade, sun; prefers fertile, moist acidic soils; tolerates flooding; zone 3

Use: naturalizing

Recognition

Leaves: usually egg-shaped (ovate), variably pointed; sharply toothed (doubly serrate); base wedge-shaped to rounded; veins running directly to teeth; surfaces glabrous except variably so on veins beneath

Flowers: female catkins; bracts conspicuous, 3-lobed, glabrous, 5-7 nerved, about 1 in (2.5 cm) long, margin toothed

Fruits: nuts ribbed, enclosed by lobed, leaf-like bracts

Twigs: slender, zigzag, glabrous or slightly hairy when young

Buds: 2 sizes, about 0.1 in (3 mm) long and 0.6 in (15 mm) long, respectively; not tightly appressed to twig; pointed, often a woolly patch at tip; scales overlapping, longitudinally striated, margins often hairy (ciliate)

Bark: gray, fluted

Wood: hard, heavy, tough

Corylus L. - **filbert (hazel)**

Filberts, natives of the North Temperate Zone, are of special interest because of their potential for producing edible fruit in addition to their ornamental value. Nine species—two of which are native to North America—are presently recognized. Many more have been described, but they have been found to be variants of the basic nine. Most *Corylus* species are self-sterile. Crosses between species have been made primarily for furthering nut production (Jaynes 1979).

Although filbert has been selected by plant scientists as the common name for *Corylus*, hazel remains the common name of choice in the eastern U.S. Folklore tales attribute mystic powers to filbert nuts. They were used as a symbol of fertility in marriage ceremonies. Filbert twigs were the first divining rods or dowsers used for finding underground water. Many herbalists used filberts in home remedies (Jaynes 1979).

Recognition

Habit: shrubs, rarely trees
Leaves: generally egg-shaped (ovate), more or less hairy
Flowers: male clusters cylindric, pendulous; preformed, overwintering as a closed catkin; without perianth, 4-8 stamens each bract; female from axillary buds formed the previous summer, head-like clusters enclosed in a small scaly bud with only red styles protruding
Fruits: clusters at twig ends; nuts nearly round, included or surrounded by leafy involucres; cotyledons thick, fleshy, remaining enclosed in shell at germination
Buds: scales many, overlapping (imbricate)

Corylus americana Walt. - American hazelnut

American hazelnut is native to southeastern Canada and the eastern U.S. including the eastern Great Plains, with fingers extending west along the Niobrara River and others to a lesser extent. Tasty fruits, sometimes harvested for local human consumption, are important food for wildlife, especially squirrels and other small mammals.

American hazel plants are self-sterile, requiring two or more plants in close proximity for good fruit production. Insects and diseases do not present serious problems. Fairchild and Klett (1993) report minor chlorosis on otherwise healthy, but inconsistently fruitful, hazel plants growing on alkaline soils in Fort Collins.

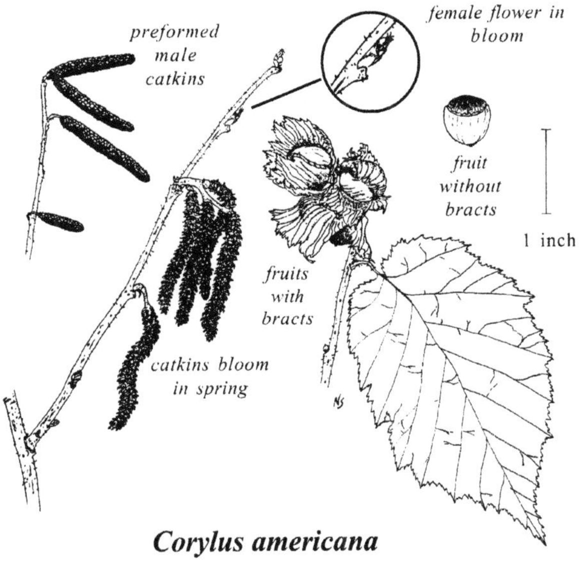

Corylus americana

Planning

Height: up to 10 ft (3 m)
Habit: shrub; multistemmed, leggy, rounded, colony-forming
Site: sun, part shade; prefers moist, well-drained soil, slightly acidic to moderately alkaline; zone 4
Use: naturalizing, wildlife habitat, food

Recognition

Leaves: generally egg-shaped (ovate), up to 4.5 in (11 cm) long; base rounded to heart-shaped (cordate); margins doubly serrate, sometimes with shallow lobes, fringed with hairs (ciliate); surface thinly hairy above, more densely hairy beneath; petioles glandular hairy
Flowers: female, 6-12 flowers per bud, in pairs, one in each pair sterile
Fruits: clusters of 2-4; bracts broad-toothed, at base of each cluster; 2 downy, lobed bracts attached at base of each nut enclose and extend beyond the nut for about twice its length, outer margin wavy, coarsely toothed; nuts nearly round, about 0.6 in (1.5 cm) in diameter, surface velvety smooth except lower one-third rough; tips tiny, shell hard; kernels edible
Twigs: red-brown, hairy, glandular; lenticels obvious
Buds: egg-shaped; scales overlapping (imbricate), hairy, red-brown, margins hairy-fringed (ciliate)

Corylus avellena L. - European filbert

European filbert—once the dominant vegetation of northern Europe after the last ice age—is now commercially grown for fruit in four areas of the world; Turkey, Italy, Spain and the coastal valleys of Oregon and Washington, near large bodies of water where summers are cool and winters mild. Healthy, vigorous European filberts growing at PERC in Fort Collins require supplemental watering but are not consistently fruitful (Fairchild and Klett 1993).

Ornamental cultivars of European filbert and hybrids involving other *Corylus* species include: 'Contorta' with twisted branches, sometimes called Harry Lauder's walkingstick; 'Fusco-rubra', a purple-leaf form; 'Heterophylla' with small, lobed leaves; and 'Pendula' with weeping branches. 'Contorta' has been grown with reasonable success in Lincoln.

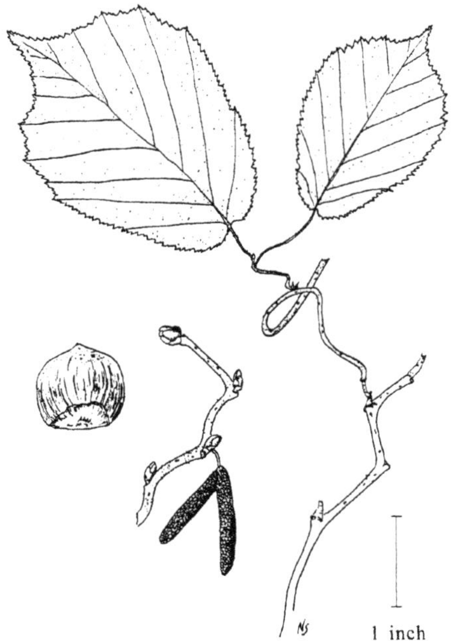

***Corylus avellena* 'Contorta'**

Planning

Height: up to 15 ft (4.5 m)
Habit: shrub; crown rounded
Site: sun; prefers moist, well-drained, fertile soil, slightly acidic to alkaline (pH 8); zone 5
Use: naturalizing, specimen, border

Recognition

Leaves: rounded to wider above middle (obovate), up to 4 in (10 cm) long; tip pointed, base heart-shaped; margin toothed (doubly serrate), often slightly lobed; slightly hairy above, hairy beneath, especially on veins; petioles glandular hairy
Fruits: clusters of 1 to 4; bracts (involucre) usually shorter than nuts, deeply and irregularly divided into lobes; nuts nearly round, kernels edible
Twigs: hairy, glandular ('Contorta' twigs contorted and twisted)
Buds: greenish

Corylus colurna L. - Turkish filbert

Turkish filbert, native to southeast Europe and western Asia, tolerates adverse sites—hot summers, cold winters, drouth and a wide range of soil pH and texture. A healthy tree growing on the grounds of the West Central Research and Extension Center, North Platte, Nebraska, provides evidence of adaptability to the western prairie.

Turkish filbert is deserving of wide use in prairie landscapes where small trees of formal character are desired. Trazels, a cross between Turkish and European filberts, have received mixed reviews in Nebraska.

Planning

Height: up to 40 ft (12 m)
Habit: tree; pyramidal, becoming broad and round-topped; early spring catkins
Site: sun; pH and soil adaptable, drouth tolerant; zone 5
Use: shade, street, windbreak

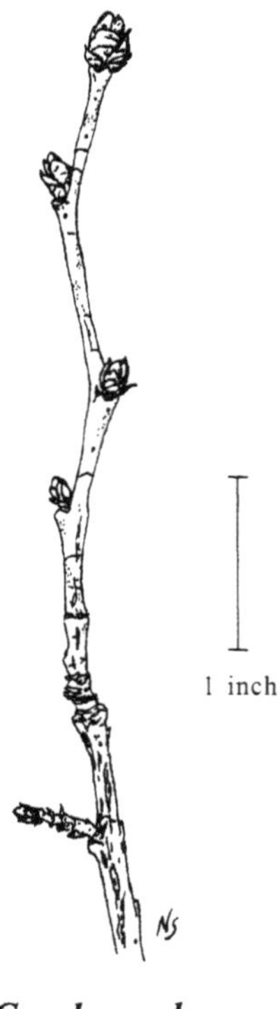

Corylus colurna

Recognition

Leaves: broad egg-shaped, up to 5 in (13 cm) long; tip pointed, base heart-shaped; margin toothed, often lobed; nearly glabrous above, hairy on veins below; petioles about 0.8 in (2 cm) long, glandular hairy when young
Fruits: clustered; bracts (involucre) bell-shaped (campanulate), glandular lobes recurved and deeply divided, fringed, about twice nut length; nuts nearly round, about 0.7 in (18 mm) across
Twigs: hairy, glandular, becoming vertically furrowed, corky
Buds: softly hairy
Bark: yellow-gray to brownish, flaky

Corylus cornuta Marsh. - beaked hazel

The unique beaked fruit of this shrub gives *C. cornuta* its common name. Beaked hazel is native to southern Canada, the east and west coast and northern U.S. including North and South Dakota and Minnesota. Beaked hazel has been crossed with European filbert to produce a hybrid dubbed "filhazel" (Jaynes 1979).

Planning

Height: up to 10 ft (3 m)
Habit: shrub; loose thickets
Site: sun; pH and soil adaptable; zone 3
Use: naturalizing, wildlife habitat, erosion control

Recognition

Leaves: egg-shaped, up to 4 in (10 cm) long; tip pointed, base rounded to heart-shaped; margin sharply toothed, often shallow lobed; densely hairy, becoming nearly glabrous above, hairs only on veins beneath; petioles slightly grooved to flattened, hairy, about 0.8 in (2 cm) long
Fruits: often clustered; bracts enclose nut and extend about 1.5 in (38 mm) beyond, tubular, much constricted above nut, beak straight or curved, bristly; nuts nearly round, about 0.5 in (12 mm) across, covered with fine short hairs, longer hairs at apex
Twigs: glabrous or a few hairs at nodes, becoming flaky with age
Buds: egg-shaped, brownish; scales hairy with marginal hairs (ciliate)
Bark: gray-brown, fissured, scales tight

Corylus maxima Mill. - giant filbert

Giant filbert, native to southeastern Europe and eastern Asia, and hybrids with European filbert are cultivated for nut production in Europe, but they are unimportant in the U.S. Purple giant filbert, variety *purpurea*, is worthy of trial as an ornamental for large areas in natural settings. The dark purple leaves fade to green in late summer.

Planning

Height: up to 15 ft (4.5 m)
Habit: shrub, sometimes small tree; leggy, round-topped
Site: sun; zone 5
Use: naturalizing, wildlife habitat

Recognition

Leaves: egg-shaped to wider above the middle (obovate); tip abruptly sharp, base heart-shaped; margin toothed, often shallow-lobed, (var. *purpurea*, leaves dark purple in early summer); petioles up to 0.6 in (15 mm), glandular-hairy
Fruits: often clustered; bracts (involucre) enclose nuts and extend about a length of nut beyond, short hairy, tubular, apex lobed and toothed; nuts slightly longer than wide
Twigs: glandular hairy

Ostrya Scop. - hophornbeam

Only three of the seven or eight recognized hophornbeam species are native to North America. One is described below. The other two native species are found only in restricted areas of the southwestern U.S.

Ostrya virginiana (Mill.) K. Koch - American hophornbeam

American hophornbeam, a small, unassuming, understory tree, grows in association with tall-growing trees throughout hardwood forests in the eastern U.S, and in parts of most of the Great Plains states, mostly along major rivers and in the Black Hills. The hophornbeam—also known as ironwood because its wood is extremely hard and tough—lacks showy flowers and its ordinary leaves are often drab in autumn. It is seldom noticed until it is old enough to display distinctive bark, catkins and fruits.

American hophornbeam grows in sun and shade on a variety of sites from moist, fertile soils to sands. Richley (1988) reports that it is adapted to southwestern Nebraska sites and Fairchild and Klett (1993) state that it grows in Fort Collins without problems. It is seldom found in the nursery trade, partially due to a seed dormancy that is difficult to break. This dormancy is controlled by a biochemical system that is only partially overcome by complicated seed treatments (Newman 1981).

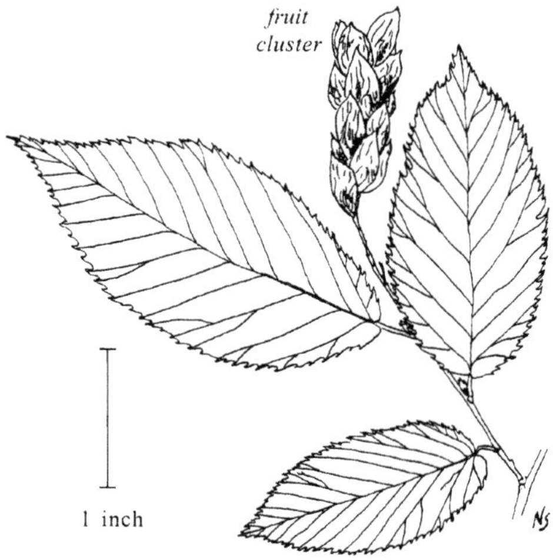

Ostrya virginiana

The extensive natural range of American hophornbeam indicates a possibility of great genetic variability that could be exploited. Dirr (1990) reports that *O. carpinifolia*, European hornbeam, and *O. japonica*, Japanese hornbeam, are very similar to our native species.

Planning

Height: up to 40 ft (12 m); Nebraska champion, 1977, 50 ft (15.2 m), Barada; National champion, 1976, 74 ft (22.6 m), Michigan

Habit: tree; crown pyramidal to rounded, preformed catkins visible in winter

Site: sun, part shade; pH and soil adaptable, drouth tolerant; zone 4

Use: naturalizing, street tree, screen, shade, specimen

Recognition

Leaves: elliptic to egg-shaped; tip long-pointed, base rounded to slightly heart-shaped; margin sharply toothed (biserrate); glabrous or sparsely hairy above, hairy below, especially at vein junctures; veins often forked to serve major teeth; petioles up to 0.2 in (5 mm) long, slender, hairy

Flowers: female catkins on new growth, appearing with leaves near twig tips; male catkins usually in 3's, formed the previous season

Fruits: elongated, hop-like clusters about 1.7 in (45 mm) long; nutlets flattened, enclosed in overlapping bladder-like papery sacs

Twigs: slender, slightly zigzag; at first green and hairy, becoming brown and glabrous; lenticels small, scattered; leaf scars half-round to oval with 3 bundle scars

Buds: reddish brown, divergent from twig, often hairy, sharp pointed; scales, about 8 visible, overlapping, often longitudinally striated (sisible under magnification)

Bark: thin, gray-brown; fissured into narrow, plate-like scales, sometimes loose at ends

Subclass Caryophyllidae

Subclass Caryphyllidae consists of three orders: Caryophyllales, Polygonales and Plumbaginales (Cronquist 1981). Plumbaginales, an order consisting of the Leadwort Family (Plumbaginaceae), is not included because its species are not likely to be adapted to the prairies.

Order Caryophyllales Benth. & Hook.

Twelve families and about 10,000 species are in this order.

Recognition

Habit: herbs, shrubs, trees or tree-like

Leaves: alternate, opposite or seldom whorled, simple; margins usually smooth (entire), sometimes reduced to spines; stipules usually absent, if present very small

Flowers: variously clustered, usually bisexual, usually regular; sepals distinct or joined below to form a cone (connate), sometimes petal-like; petals often missing; stamens 1 to many

Fruits: various, dry or fleshy

Cactaceae Juss. - **cactus family**

Hundreds of cacti species are native to the Americas. Though characteristically found in dry regions, they grow in other climates as well, including tropical rainforests. Towering saguaro (*Cereus giganteus* Engelm.), the symbol of the American Southwest desert, is the most famous cactus.

Spines replace cactus leaves which enhances water storage by reducing transpiration. They also protect the soft parts of the plant. Cactus spines rise from complex structures called areoles, which are formed from buds where leaves would otherwise arise.

Cacti provide water, food and shelter for many wildlife species. Humans also have used them for food and wound dressings. Fire is used to burn off spines during prolonged drouths to allow livestock to eat the cactus stems. Unfortunately, many cacti species are uprooted from their native habitats despite protective laws, and then sold for landscaping, which endangers the survival of many animals that rely on them for food and shelter.

Recognition

Habit: herbs, shrubs, trees

Leaves: alternate; scale-like or wanting, rarely flat and broad

Flowers: solitary, bisexual; sepals and petals many, similar; stamens many

Fruits: berries, somewhat dry or fleshy

Stems: fleshy, branched or unbranched, cylindrical, globose or flattened, often constricted or jointed, usually spiny; juice watery or milky

Opuntia Mill. - **pricklypear and cholla**

Forty-seven *Opuntia* species have been identified in the U.S. Those of us who grew up in the semi-arid prairies are very familiar with one or more of the low-growing pricklypears that have distinctive flattened stems and showy solitary flowers in shades of yellow to red. Less noticeable are the small globose pincushions, species of *Coryphantha* Lem., with their purplish or reddish flowers.

Round-stemmed *Opuntia* species are known as chollas or cane cacti. Their well-armed stems allow them to flourish on overgrazed ranges. Some chollas attain the stature of small bushes or trees. Most chollas are native to South America.

Recognition

Habit: shrubs, tree-like
Leaves: scale-like to short-linear, falling early (caducous); areoles in axils covered with barbed bristles, sometimes spiny
Flowers: on bristle-bearing part of areoles; sepals and petals spreading, wheel-shaped
Fruits: fleshy, rarely dry, some edible
Stems: jointed, flattened or round

Opuntia imbricaria (Haw.) DC. - tree cactus

The tree cactus, native to southwestern Kansas, Oklahoma, Texas, New Mexico and the plains of southeastern Colorado, blooms in early summer displaying rose-pink to purple flowers. The green branches become brownish with age, and, once the plant is dead, soft tissue decomposes leaving a hollow skeleton perforated with oval holes.

Planning

Height: up to 7 ft (2 m)
Habit: shrub; green stems become brown, soon after death become perforated skeletons
Site: sun; dry, sandy soil preferred; zone 5
Use: naturalizing, wildlife habitat

Recognition

Leaves: nearly circular in cross-section (terete), about 0.8 in (2 cm) long, smooth, green, early deciduous
Flowers: clustered at stem ends; sepals and petals several, similar (tepals), egg-shaped in outline, widest near tip, forming a cup, rose-pink to purple, outer tepals greenish
Fruits: short, cylindric, about 1 in (25 mm) long, about as wide, many seeded
Stems: rigid, green, terete, spongy and succulent, erect in summer, drooping in winter; aeroles depressed; spines 10-20 from base of each areole, slender, barbed, sheathed in white, papery covering with yellowish tip, base surrounded by white hairs; old stems dry and brownish covered with thin papery scales and spines
Buds: not visible
Wood: a hollow framework with oval holes spirally arranged; a skeleton after soft tissue decomposes

Chenopodiaceae Vent. - goosefoot family

The worldwide goosefoot family, best known for herbaceous plants—beets, spinach, lambsquarter, Russian thistle and others—also includes shrubs and subshrubs (suffrutescent plants). Many of the woody plants, found naturally associated with saline soils, are halophytes. Though seldom planted as ornamentals, these interesting plants are often useful because they thrive on dry sites of high salt content.

Recognition

Habit: herbs, shrubs, rarely small trees
Leaves: alternate, rarely opposite, simple; often fleshy, sometimes scaly; margins smooth (entire) or irregularly toothed; stipules wanting
Flowers: inconspicuous, bisexual or unisexual; petals none, sepals none to 5
Fruits: dry, rarely fleshy; enclosed by calyx or bracts

Atriplex L. - saltbush

About 150 species of this denizen of the salt flats grow naturally throughout the world (Cronquist 1981).

Recognition

Habit: herbs, shrubs
Leaves: scurfy, mealy or hairless and smooth
Flowers: clusters (spikes or panicles), usually terminal; unisexual, sometimes mixed with bisexual; calyx of male 3-5 parted; female flowers without calyx, enclosed by 2 bracts
Fruits: dry, thin-walled, usually splitting (utricles), enclosed by bracts; seeds 1

Atriplex canescens James - fourwing saltbush

Fourwing saltbush, native to the western prairies and the western U. S., is often found on dry, sandy, moderately saline or alkaline sites of the plains, foothills and intermountain valleys. It is one of the most valuable forage shrubs (Forest Service 1937). Its abundant pollen limits the saltbush's usefulness as a landscape plant for those with pollen allergies. *A. nutallii* S. Wats. and *A. confertifolia* (Torr.) S. Wats. are related but less common natives of the Great Plains (Sutton and Johnson 1974).

Atriplex canescens

Planning

Height: up to 3 ft (1 m)
Habit: shrub; rigid, dense; leaves usually cling through winter; flowers June to August
Site: sun; pH 7.0-8.5; prefers deep, sandy soil but adaptable; zone 4
Use: naturalizing, wildlife habitat, border

Recognition

Leaves: alternate or somewhat clustered; linear, or spatula-shaped, up to 2 in (5 cm) long and 0.3 in (8 mm) wide; margin entire; surfacees finely scurfy-hairy, gray-green; petioles up to 0.1 in (2 mm) or wanting; stipules wanting
Flowers: unisexual, usually dioecious; male in dense clusters (spike-like) near twig ends, calyx 5-parted, stamens 5; female in leaf axils, clustered, stalked; pistil enclosed by 2 united bracts
Fruits: bracts united nearly to apex forming a 4-angled structure; wings 4, rounded in outline, up to 0.6 in (15 mm) wide, thin, net-veined, edges usually wavy, toothed; tan, scaly
Twigs: scurfy-hairy, becoming glabrous, gray to yellow
Buds: egg-shaped, hairy or scaly

Eurotia Adans. - winterfat

Four Eurotia species are found in North America, Europe and Asia.

Eurotia lanata (Pursh) Moq. - winterfat

Winterfat, the dominant and most conspicuous plant in large areas of the western U.S., grows on dry and moderately alkaline soils in the plains, foothills and mountain valleys. This deep-rooted plant endures long periods of drouth and recovers rapidly after significant precipitation. Winterfat provides palatable and nutritious forage for livestock and wild animals. The overuse of winterfat has severely reduced or eliminated it on many ranges, but it responds well to regulated grazing and grows luxuriantly under cultivation (Forest Service 1937).

The long white hairs of winterfat shimmering in autumn sunlight prompts cutting for winter bouquets. Its gray foliage also provides a year-round contrast in landscapes. Some authors favor *Ceratoides lanata* (Pursh) T. J. Howell as the plant's proper name.

Eurotia lanata

Planning

Height: up to 2 ft (60 cm)
Habit: subshrub; erect, dry leaves cling through winter
Site: sun; pH 7.0-8.5; drouth tolerant; zone 4
Use: naturalizing, wildlife habitat, erosion control, mass planting

Recognition

Leaves: linear, about 1 in (25 mm) long and 0.1 in (2 mm) wide; margin entire, rolled under (revolute); densely hairy (simple and stellate hairs), gray-green; petioles about 0.1 in (2 mm) long or wanting
Flowers: clusters dense, in leaf axils, crowded toward twig ends; male above female on same plant (monoecious) or female only; male with 4-parted, hairy calyx, stamens 4; female without calyx, ovary enclosed by 2 united bracts
Fruits: dry, oval, flat, about 0.1 in (2 mm) long, white hairy
Twigs: densely covered with woolly hairs, white becoming rust-colored
Crown: woody, herbaceous above

Sarcobatus Nees. - greasewood

One of the two greasewood species native to western North America also grows in the northwestern prairies.

Sarcobatus vermiculatus (Hook.) Torr. - greasewood

Though more at home in the Great Basin region, greasewood also grows as far east as the Dakotas and northwestern Nebraska. It grows in alkaline soils on flats at the base of eroded hills and on flood plains along prairie creek banks. Greasewood's low, rigid, spiny branches provide excellent protection for small mammals and birds.

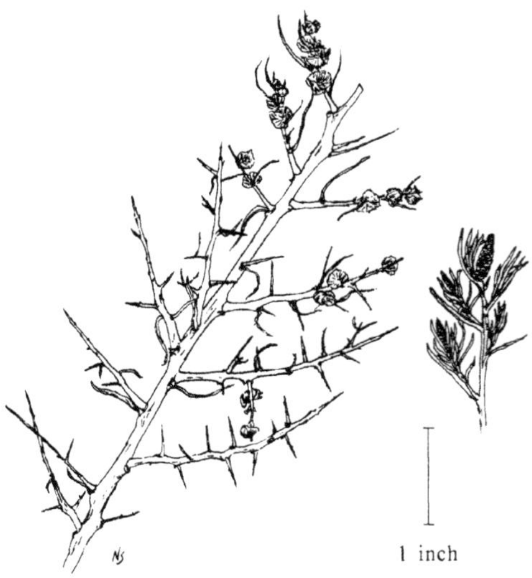

Sarcobatus vermiculatus

Planning

Height: up to 5 ft (1.5 m)
Habit: shrub; tardily deciduous; dense, spiny
Site: full sun; pH 7.0-8.5; drouth tolerant; zone 4
Use: wildlife habitat, naturalizing
Caution: foliage can be poisonous

Recognition

Leaves: linear, sessile, succulent, up to 1 in (25 mm) long; almost as thick as wide; lower leaves opposite, shorter and broader
Flowers: unisexual (monoecious); male in spike-like clusters (catkins); perianth wanting; stamens 1 to 3 under each scale; female stalkless in leaf axils; perianth inversely conical, lower part united to ovary
Fruits: enclosed by persistent perianth which enlarges to form lobed disk at right angles to ovary at the midway point; minutely hairy (stellate)
Twigs: becoming spiny and rigid; glabrous or hairy, grayish to yellow-white; ridge extends down from center of leaf scar
Buds: egg-shaped; set deeply into stem; 2 outer scales thick, valvate

Order Polygonales

One family is present in the order Polygonales.

Polygonaceae Juss. - buckwheat family

Thirty genera consisting of nearly 1,000 species of the buckwheat family grow mostly in the temperate portions of the Northern Hemisphere. Two varieties of *Eriogonum effusum* Nutt.—umbrella plant—are native subshrubs (suffrutescent) in the western Nebraska panhandle, west to Utah and southwest to Arizona (Stephens 1973). These drab plants grow on dry, rocky or gravelly soils and provide forage for game animals and livestock.

Recognition

Habit: herbs, shrubs, vines, trees
Leaves: alternate, few otherwise; simple, mostly entire, sometimes cleft
Flowers: variously clustered; bisexual or unisexual (usually dioecious); regular; perianth 2-6 parted, sepals and petals not clearly differentiated, green and herbaceous to colored, commonly persistent on fruit; stamens 2-9, most commonly 6, distinct or united at base
Fruits: dry (achenes) or small nuts

Polygonum L. - knotweed

The species of knotweed most likely to be adapted to the prairies is described here.

Polygonum aubertii L. Henry - silvervine fleeceflower

Silvervine fleeceflower, a plant with attractive foliage, is native to western China. It is easily propagated and grows almost anywhere; in fact, it grows "like a weed." However, it may suffer cold injury in the northern prairies.

Polygonum aubertii

The identity of silvervine fleeceflower and related species is somewhat confusing (Dirr 1990). Slight differences in flower color separate this species from *P. baldschuanicum* Reg., bokaravine fleeceflower.

Planning

Height: up to 15 ft (4.5 m)
Habit: vine; dense, twining; underground rhizomes; autumn flowers
Site: sun, shade; soil adaptable; drouth tolerant; zone 5
Use: trellis, groundcover

Recognition

Leaves: egg-shaped to oblong egg-shaped, up to 3.5 in (9 cm) long; apex variously pointed, base arrowhead shaped (hastate); margin wavy, reddish when young, becoming green; petioles up to 2 in (5 cm) long
Flowers: slender clusters (panicles); bisexual; white or greenish, sometimes pinkish, about 0.2 in (5 mm) across, 5-parted
Fruits: dry, 1-seeded (achenes); 3-angled
Twigs: green, glabrous

Subclass Dilleniidae

This subclass consists of 13 orders and 78 families containing about 25,000 species of woody and herbaceous plants (Cronquist 1981).

Recognition

Habit: trees, shrubs, herbs
Leaves: simple, seldom compound; margins entire or toothed, sometimes dissected
Flowers: petals separate (polypetalous), rarely wanting; petals sometimes united (sympetalous)

Order Dilleniales

About 400 species are present in the two families of Dilleniales.

Recognition

Habit: trees, shrubs, herbs
Leaves: alternate, seldom opposite; simple or ternately compound, sometimes deeply lobed; stipules wing-like and fused to petiole or wanting
Flowers: mostly bisexual and regular; sepals persistent, often leathery; petals distinct; stamens usually many; ovary superior, compound, styles distinct
Fruits: mostly dry, splitting; or fleshy

Paeoniaceae Rudolphi - peony family

Cronquist (1981) and the Bailey Hortorium (1976) consider peonies to be in a separate monogeneric family.

Paeonia L. - peony

Peonies have been cultivated in China for about 2,500 years. Most are herbaceous perennials, such as the long-lived *P. lactiflora* Pall., from which many cultivars have been selected. These fragrant plants are present in many prairie yards, cemetaries, and gardens, even surviving at abandoned homesteads.

Recognition

Habit: herbs, shrubs
Leaves: deciduous, alternate; simple and deeply parted or ternately compound
Flowers: terminal, usually solitary, sometimes several; variations of red, white or yellow; sepals 5, persistent, petals 5-10; stamens many
Fruits: dry, splitting along one suture (follicles); each with several large seeds

Paeonia suffruticosa Andr. - tree peony

Woody stems distinguish tree peony from its herbaceous kin. An 80-year-old plant grows in Lincoln, but suffers dieback during extremely cold winters (Sutton 1998).

Planning

Height: up to 7 ft (2 m)
Habit: shrub; deciduous; spreading, coarse; spring bloom
Site: sun, partial shade; fertile, well-drained soil, pH 6.0-7.5; zone 5
Use: specimen
Caution: protect from strong winds

Recognition

Leaves: compound; leaflets variably egg-shaped, broad to oblong, stalked or sessile; 3-5 lobed, rarely entire; slight waxy bloom underneath and glabrous or sparingly hairy
Flowers: solitary, 4-12 in (10-30 cm) across, white, rose or red, often double-whorled; stamens prominent, yellow
Twigs: stout; young are greenish, smooth; leaf scars v-shaped
Buds: pinkish green, scales few
Bark: rough, brown

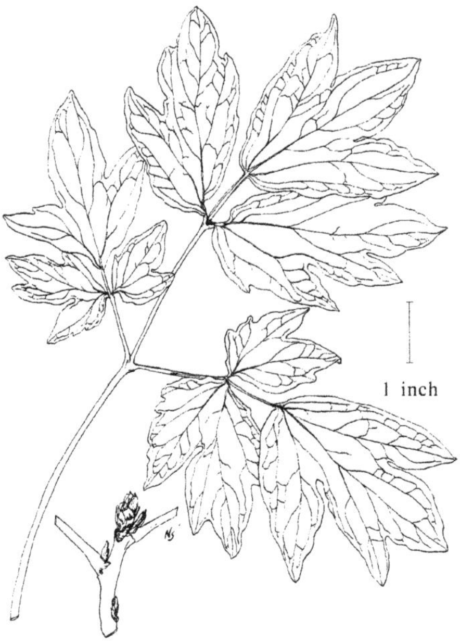

Paeonia suffruticosa

Order Theales

Eighteen familes and nearly 3,500 species are present in the order Theales (Cronquist 1981).

Recognition

Habit: trees, shrubs, vines, herbs
Leaves: alternate, opposite or whorled; simple, seldom compound; with or without stipules
Flowers: bisexual or unisexual, usually regular; sepals usually overlapping (imbricate), distinct or connate at the base or to the middle; petals distinct or sometimes connate at the base, seldom sympetalous; stamens 4 to many; ovary usually superior, styles distinct
Fruits: dry or fleshy, splitting or not

Theaceae D. Don - **tea family**

Theaceae, a family of 40 genera and 600 species, is found in tropical and subtropical regions with only a few species in warm-temperate regions. *Camellia sinensis* is the source of our most common teas. Horticulturally important plants in the family include camellias and gardenias.

Three genera grow naturally in the southeastern U.S.—the two described here and *Gordonia* Ellis.

Recognition

Habit: shrubs, trees; evergreen or deciduous

Leaves: alternate, simple; margins entire or toothed; often leathery

Flowers: solitary or in few-flowered clusters; axillary or subterminal, often showy; usually bisexual; sepals usually 5-7, free and overlapping (imbricate); petals 5, rarely otherwise, sometimes united near base; stamens many

Fruits: dry or drupaceous

Franklinia Marsh.

The genus *Franklinia* includes one species.

Franklinia alatamaha Marsh. - **Franklin tree**

Franklin tree has not been found in the wild since 1790. John Bartram discovered this species on the banks of the Altamaha River in Georgia and transplanted a few to his Philadelphia botanic garden in the 1760s. He named the tree for Benjamin Franklin. All Franklin trees now growing originated from this one garden.

Showy white flowers and colorful fall foliage combine to make Franklin tree a desirable specimen. However, in most of the prairie region its flowers may be injured by early frost and its leaves may not have an opportunity to attain their normal autumn glory. Selection for hardy plants will be hampered by the limited gene pool.

Planning

Height: up to 20 ft (6 m); National champion, 1968, 36 ft (10.9 m), Pennsylvania

Habit: shrub, tree; deciduous; upright, spreading, open and airy; summer and early autumn blooms; autumn foliage orange-red

Site: sun, part shade; pH 5.0-6.5; soil moist, well-drained; zone 6

Use: specimen

Recognition

Leaves: egg-shaped, widest above the middle (obovate), gradually narrowing into a short petiole; up to 6 in (15 cm) long; tip rounded to pointed, margin finely sharp-toothed; surfaces shiny above, hairy beneath

Flowers: short-stalked, solitary in junctions of upper leaves; about 3 in (8 cm) across; fragrant; petals white, widest above middle, tips rounded; sepals 5, unequal; stamens many, providing a yellow center

Fruits: dry, woody, 5-valved capsules, about 0.7 in (2 cm) in diameter, splitting into 10 segments; seeds many, angled, wingless

Twigs: silky-hairy, becoming glabrous; leaf scars half-round, bundle scar 1

Buds: hairy, pointed, reddish brown

Bark: smooth, gray to dark brown; broken by vertical fissures

Stewartia L.

Two of the 15 *Stewartia* species are native to the southeastern U.S.; the others grow in southeastern Asia. These attractive plants with their white summer blooms are typically found in the understory of deciduous forests in warm-temperate climates. The bark of these species often falls off in irregular patches, creating a handsome mottled effect.

Recognition

Habit: shrubs, trees; deciduous

Leaves: margins toothed (serrate); petioles short

Flowers: solitary, bisexual; cup-shaped, 1-2 bracts below the calyx; sepals 5, persistent; petals white, silky outside, fused at base (connate), 5, sometimes up to 8, nearly round, sometimes wider above middle, margins toothed (crenulate); stamens numerous, attached to petals; ovary superior, 5-celled, styles 5, free or fused (connate)

Fruits: capsules woody, splitting into 5 valves; seeds compressed, usually narrowly winged

Stewartia ovata

Stewartia ovata (Cav.) Weatherby - mountain stewartia

Mountain stewartia, native to the southeastern U.S., grows along wooded streambanks. Mountain stewartia is rare in its native habitat and may require protection as an endangered species (Elias 1980).

Japanese stewartia (*S. pseudocamellia* Maxim.) is hardier than our native U.S. species, but site requirements are similar (Dirr 1990, Spangler and Riperda 1977). It can be distinguished from mountain stewartia by its smaller leaves, flowers with united styles, seeds that are not winged, and mottled trunk bark.

Planning

Height: up to 20 ft (6 m); National champion, 1982, 25 ft (7.6 m), Tennessee

Habit: shrub, tree; spreading branches; summer bloom, autumn foliage orange-scarlet

Site: part shade; acidic, moist soil; zone 6

Use: specimen, border

Caution: part shade desirable at mid-day

Recognition

Leaves: elliptic or long egg-shaped, up to 4 in (10 cm) long; tip pointed, base rounded; dark green above, sparsely hairy beneath; petioles winged

Flowers: axillary or subterminal, 3 in (8 cm) across; petals 5, often 6, wider above middle; stamens with white filaments, anthers orange; styles free

Fruits: hairy, 5-angled, beaked, about 0.7 in (2 cm) long; calyx persistent

Twigs: red-brown, pith spongy; lenticels whitish, inconspicuous; leaf scars half-round, bundle scars 1

Buds: solitary or one above another (superposed), about 0.2 in (5 mm) long, spindle-shaped; scales 2, outer often hairy

Bark: smooth, eventually shreds; does not exfoliate in plates

Actinidiaceae Hutchens. - **Chinese gooseberry family**

Three genera, chiefly found in the tropics and mountains of Asia, are divided into nearly 300 species (Cronquist 1981).

Recognition

Habit: trees, climbing shrubs
Leaves: simple, alternate, toothed, rarely entire
Flowers: axillary or lateral clusters (basically cymose), sometimes reduced to 1 or few; bisexual, unisexual or both on same plant; 5-parted, sometimes 4; sepals generally overlapping, distinct or fused at base; petals with same features; stamens 10 to numerous
Fruits: berries or dry, splitting capsules

Actinidia Lindl. - **actinidia**

The best known of the 25 Asian species of *Actinidia* is *A. chinensis* Planch., Kiwi fruit. It is not adapted to the central and northern prairies.

Recognition

Habit: twining shrubs; deciduous
Leaves: toothed, rarely entire; stipules wanting
Flowers: solitary or in axillary clusters (cymes), unisexual and bisexual on same plant, petals and sepals usually in fives; stamens and styles many
Fruits: berries, seeds numerous

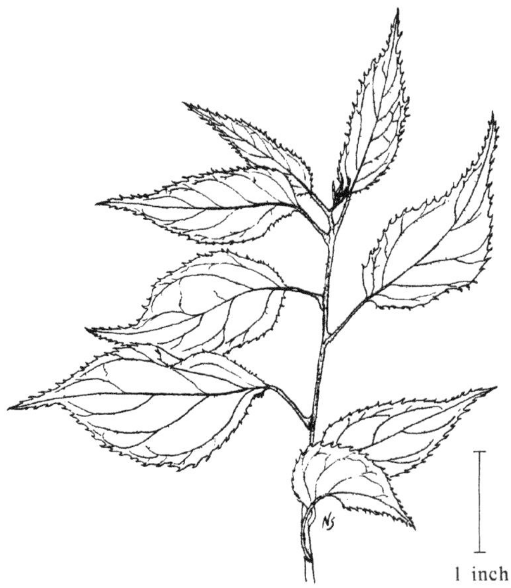

Actinidia arguta

Actinidia arguta (Sieb. & Zucc.) Miq. - **bower actinidia**

Bower actinidia, a vigorous, high climbing, twining vine, can grow rampant on fertile sites, necessitating considerable pruning to keep it within bounds. The edible fruit, a greenish yellow berry, varies greatly in taste ranging from insipid to pleasant (Dirr 1990). Nancy Scott (1998) reports that a bower actinidia in her Bellevue, Nebraska, garden bore sweet, grape-sized fruits with a sweet flavor for one year before it died during a late April freeze.

Silver-vine, *A. polygama* (Sieb & Zucc.) Miq., produces edible fruit, but it is not as vigorous as bower actinidia (Dirr 1990).

Planning

Height: up to 20 ft (6 m)
Habit: twining vine; requires support
Site: sun, part shade; pH and soil adaptable, tolerates infertile soil; zone 5
Use: arbor

Recognition

Leaves: broad egg-shaped to elliptic; abruptly pointed, base rounded or somewhat heart-shaped; margin toothed; midrib and petiole may be bristly beneath
Flowers: axillary in clusters of 3 or more, fragrant; sepals green, petals white tinged with green, incurved to present a globular shape; anthers dark purple, numerous; styles short, stout with about 20 stigmas radiating outward like wheel spokes
Fruits: ellipsoidal, about 1 in (25 mm) long, greenish yellow; edible
Twigs: glabrous, brownish; lenticels many; pith brown, plated (lamellate); leaf scars raised with single bundle trace
Buds: terminal missing, lateral enclosed in petiole base

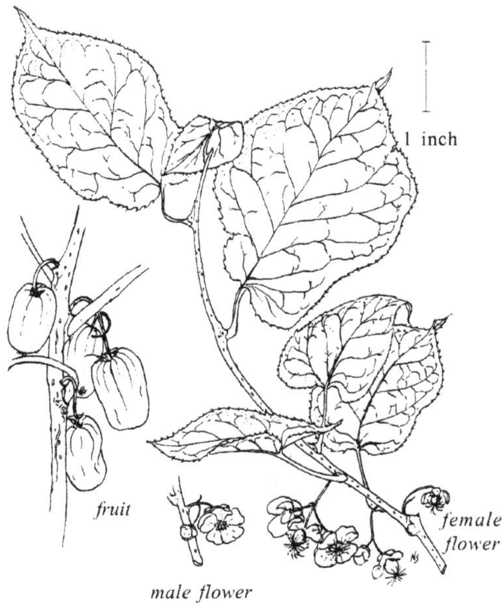

Actinidia kolomikta

Actinidia kolomikta (Rupr.) Maxim. - kolomitka actinidia

Kolomitka actinidia is often planted for its showy, variegated foliage. It is adapted to the Omaha, Nebraska, area where the male plants are usually planted because their foliage is more colorful than the foliage of the female (Scott 1998).

Planning

Height: up to 20 ft (6 m)
Habit: twining vine; requires support
Site: sun for best foliage coloration, part shade; pH and soil adaptable, tolerates infertile soil; zone 5
Use: arbor

Recognition

Leaves: egg-shaped to oblong; long pointed (acuminate), base rounded or heart-shaped; margin unequally toothed, bristly; hairy on veins beneath when young; white to pink blotch at the apex extending to the middle or beyond, especially on male plants
Flowers: 1-3, white, fragrant; anthers yellow
Fruits: oblong-egg-shaped, about 1 in (25 mm) long; greenish to yellowish; edible, sweet
Twigs: similar to *A. arguta*
Buds: similar to *A. arguta*

Clusiaceae Lindl. (Guttiferae Choisy) - St. Johnswort family

The family Clusiaceae as defined by Cronquist (1981) consists of about 50 genera and 1,200 species. The two subfamilies, Clusioideae and Hypericoideae, have been treated as distinct families by other authorities. Plants in the former subfamily are woody, tropical species, while those of Hypericoideae are adapted to temperate climates.

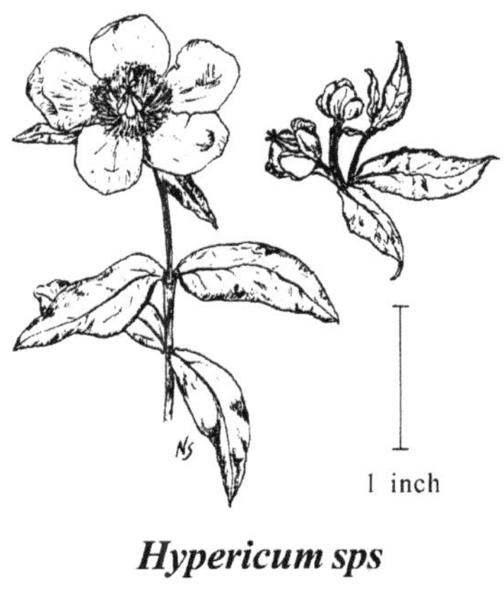

Hypericum sps

Subfamily Hypericoideae

Recognition

Habit: trees, shrubs, herbs; deciduous or evergreen
Leaves: opposite or whorled, simple; margins mostly entire; surface covered with resin cavities (glandular-punctate); stipules usually wanting
Flowers: clusters (cymes) or seldom solitary, bisexual, regular; sepals 4-5, overlapping; petals distinct or fused at base; stamens many, often clustered
Fruits: capsules, drupes or berry-like

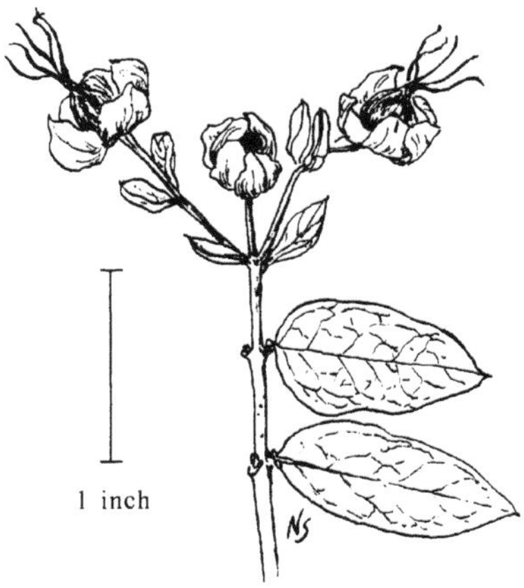

Hypericum patulum

Hypericum L. - St. Johnswort

About 40 of the 300 species of *Hypericum* grow in the U.S. Some species are grown for their handsome summer flowers, some are considered weeds, and a few are poisonous to livestock.

In addition to the two species described herein, several others may be adapted to the southern prairies. They include: *H. calycinum* L., a procumbent semi-evergreen shrub that grows well on poor sandy soils; and *H. frondosum* Michx., an upright shrub bearing large bright yellow flowers (Dirr 1990). *H. patulum* Thunb. var. *henryi* Bean is a small semi-evergreen shrub that tolerates dry soils and has no insect or disease problems but can suffer dieback in severe winters (Spangler and Riperda 1977).

Dirr (1990) believes that the *Hypericum* species have not been adequately explored and developed as landscape plants in the U.S.

Recognition

Habit: shrubs, herbs; deciduous or evergreen
Leaves: opposite, sometimes whorled; margins entire; short-petioled or sessile
Flowers: solitary, or terminal or axillary clusters; 5-parted, petals with margins overlapping, yellow; stamens distinct or in bundles
Fruits: dry, splitting (capsules), rarely berry-like

Hypericum kalmianum L. - Kalm St. Johnswort

Kalm St. Johnswort grows in open woodlands in the northeastern U.S. including the Great Lakes Region. Trial plantings have survived extremely low temperatures (-30 degrees F or -34 degrees C), but drouth and high pH soils are detrimental to the plant (North Central Plant Introduction Station 1975). Richley (1988) reports that Kalm St. Johnswort suffers winter dieback in southwestern Nebraska.

Hypericum kalmianum

Planning

Height: up to 3 ft (1 m)
Habit: shrub; deciduous, but leaves persisting into winter; dense, spreading; July flowers on new growth; autumn foliage colorful
Site: sun, part shade; pH and soil adaptable; zone 4
Use: border, specimen

Recognition

Leaves: linear oblong, sometimes wider above middle, about 1 in (25 mm) long; surface bluish green above, whitish waxy coating beneath (glaucous)
Flowers: clusters of 3, about 1 in (25 mm) across, bright yellow; sepals leaf-like; stamens free, half as long as petals; styles 5, united below, slightly longer than stamens
Fruits: egg-shaped, beaked, about 0.3 in (8 mm) in diameter
Twigs: 4-angled

Hypericum prolificum L. - shrubby St. Johnswort

Shrubby St. Johnswort, native to the eastern U.S., has little or no insect and disease problems, but severe drought and cold have caused injury in test plantings (North Central Plant Introduction Station 1959). This shrub grows extremely well in dry, rocky and calcareous soils (Dirr 1990).

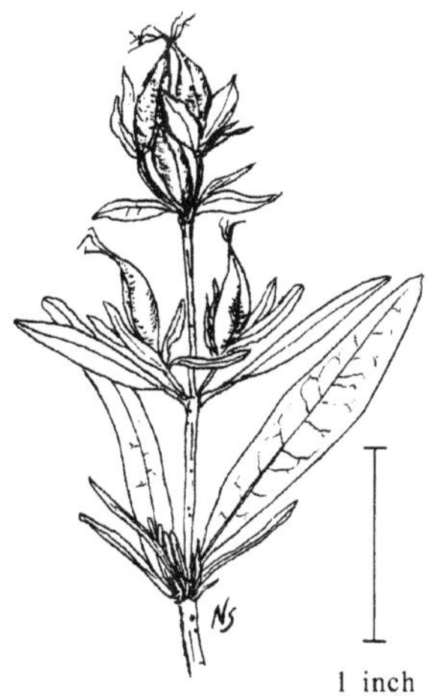

Hypericum prolificum

Planning

Height: up to 5 ft (1.5 m)
Habit: shrub; deciduous; upright, rounded; summer bloom on new growth
Site: sun, part shade; pH and soil adaptable; drouth tolerant; zone 5
Use: border, specimen

Recognition

Leaves: linear oblong, sometimes wider above middle, about 2 in (5 cm) long; surface above dark green to blue-green
Flowers: terminal and axillary clusters, few to several-flowered; each about 1 in (25 mm) across; bright yellow; styles 3, united at base
Fruits: capsules up to 0.6 in (15 mm) long, not furrowed; persisting through winter
Twigs: 2-edged, glabrous
Bark: exfoliating

Order Malvales

Order Malvales, as defined by Cronquist (1981), consists of five families and over 3,000 widely distributed species, mostly found in the tropics.

Recognition

Habit: trees, shrubs, vines, herbs
Leaves: alternate, rarely opposite, simple or sometimes palmately compound; often palmately veined
Flowers: bisexual, less often unisexual, usually regular; sepals distinct or fused at base (connate), usually not overlapping; petals distinct, sometimes wanting; stamens usually numerous, distinct
Fruits: variable

Tiliaceae Juss. - linden family

The linden family, consisting of over 40 genera and 450 species, is most abundant in the tropics. The trees of this family provide shade, ornamentation and valuable lumber. Jute, used for making burlap and cord, is obtained from herbaceous linden species native to Asia.

Tiliaceae *Tilia*

Recognition

Habit: trees, shrubs, herbs; usually deciduous

Leaves: simple, alternate, rarely opposite; margins entire, toothed, or lobed; stipules often deciduous

Flowers: variously clustered, sometimes paired or solitary; bisexual, sometimes unisexual; symmetrical; sepals usually 5, petals 5, rarely wanting; stamens 10 or more, sterile stamens (staminodes) sometimes present

Fruits: capsules, berries or nuts

Tilia L. - linden or basswood

Many *Tilia* species have been named, but at present only about 30 are widely recognized as valid. They are found naturally in eastern North America, Mexico, Europe, central China and Japan (Harlow and Hararr 1968). Four species, native to the U.S., are commonly called basswoods by the timber industry and foresters. Horticulturists prefer linden for the common name.

The soft, light, straight-grained wood of basswood is easy to work, does not warp or check, and does not impart taste or odor to foodstuff. The wood is used in the honey industry, in furniture and cabinetry, woodenware, and interior finishing, and is a favorite of many wood carvers. The tough, fibrous inner bark can be used for cords, coarse cloth and many other purposes. These bast fibers may account for the common name, "basswood," a corruption of "bastwood."

The basswood's sweet-scented flowers are attractive to bees and provide nectar that produces a mild, light-colored honey. A mix of *Tilia* species in a "honey orchard" will provide a pasture for bees throughout the month of June. However, flowers of *T. tomentosa* and *T. ×euchlora* are reputed to incapacitate bees with some sort of narcotic dose that sometimes proves fatal to the bees (Johnson 1973).

Recognition

Habit: trees; deciduous

Leaves: broadly egg-shaped or heart-shaped, toothed; tip pointed, base uneven; petioles slender; emerging seed leaves (cotyledons) palmately 5-lobed

Flowers: long-stalked clusters (cymes or corymbs) attached to twigs by strap-shaped, pale green, membranous bracts; pale yellow, fragrant

Fruits: globose to egg-shaped, nut-like, woody; seeds 1 to 3

Twigs: somewhat zigzag, glabrous or hairy; lenticels numerous, pale, inconspicuous; leaf scars half-round to crescent-shaped, bundle scars 3 or in 3 groups or scattered; stipule scars prominent, unequal length; pith homogenous

Buds: terminal wanting; lateral diverge from stem, lopsided, broadly egg-shaped; scales red to green, 2-3 visible, thick, obtuse

Bark: dark gray and smooth becoming thick and tough, sometimes breaking into scales or furrows; inner bark tough and fibrous

Wood: light, soft, nearly white, straight-grained

Tilia americana L. - American basswood

American basswood grows throughout the native woodlands of eastern Nebraska, west to Cherry County along the Niobrara River, and west to Morse Bluff on the Platte River. It is also native from the eastern portions of the Dakotas, Kansas and Oklahoma to the Atlantic Coast from North Carolina to Maine.

Although American basswood prefers deep, moist, fertile soils, it can be found growing on dry rocky ridges, except in the West where it survives on north slopes only. Plantings of this species will be most successful on well-watered, partially shaded sites protected from desiccating winds.

American basswood matures into a broad-crowned tree suitable for large areas. Dirr (1990) rates it as a "good and handsome native tree which should be left in the woods." However, Richley (1988) says it is a good tree for southwestern Nebraska although it is slower growing than some of the lindens. This observation is reinforced by Pool (1951) who says, "It is easily moved and readily becomes established in a great variety of conditions, doing very well even in western Nebraska if watered frequently during periods of drought." Pool further opines that the American basswood is superior to the introduced European lindens.

'Redmond', introduced in 1927 by Plumfield Nurseries of Fremont, Nebraska, is a selection of *T. americana*, according to Dirr (1990). Others believe that 'Redmond' is a hybrid (Fairchild and Klett 1993; Sutton 1998). The original tree stands in Fremont, maintaining a distinctive pyramidal form. Other cultivars of similar form are 'Fastigiata' and 'Legend'.

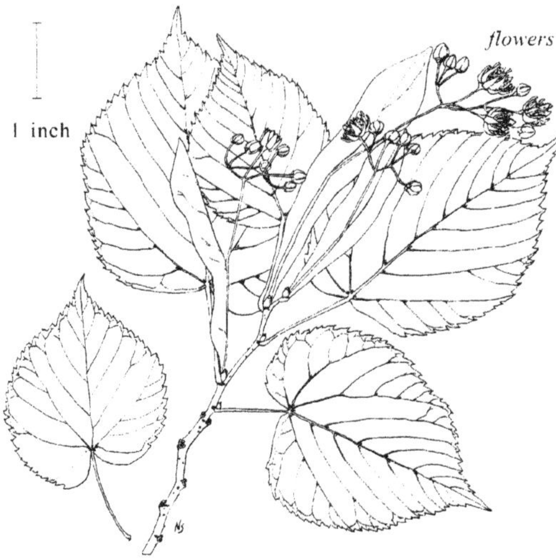

Tilia americana

Planning

Height: up to 70 ft (21 m); Nebraska champion, 1987, 100 ft (30.5 m) tall, 13 ft (3.9 m) trunk circumference, Hastings; National champion, 1987, 94 ft (29 m) tall, 24 ft (7.3 m) trunk circumference, Pennsylvania
Habit: tree; deciduous; pyramidal becoming broad, rounded; flowers after leaves fully grown
Site: sun, part shade; prefers deep fertile moist soil, pH adaptable; drouth tolerant; zone 3
Use: shade, bee pasture, lumber

Recognition

Leaves: 5-6 in (13-15 cm) long, 3-4 in (8-10 cm) wide; coarse teeth turning inward; surfaces glabrous except sometimes hairy below, tufts in vein axils often buffy; petioles glabrous, 2-3 in (5-7 cm) long, stipules soon deciduous (caducous)
Flowers: calyx lobes 5, hairy, pale yellow, curved inward toward petals; petals 5, glabrous, creamy white, erect, lance-shaped, longer than stamens; sterile stamens (staminodes) 5, petal-like, white, smooth, opposite petals; fertile stamens many, often in 5 clusters opposite petals; ovary white, hairy, style white, stigma greenish, 5-lobed
Fruits: globose, about 0.4 in (9 mm) diameter, gray hairy, without ribs; 2 to 7 per bract; persisting into winter; seeds usually 1
Twigs: glabrous, red-gray
Buds: red, egg-shaped, glossy, glabrous, sometimes tipped with hairs

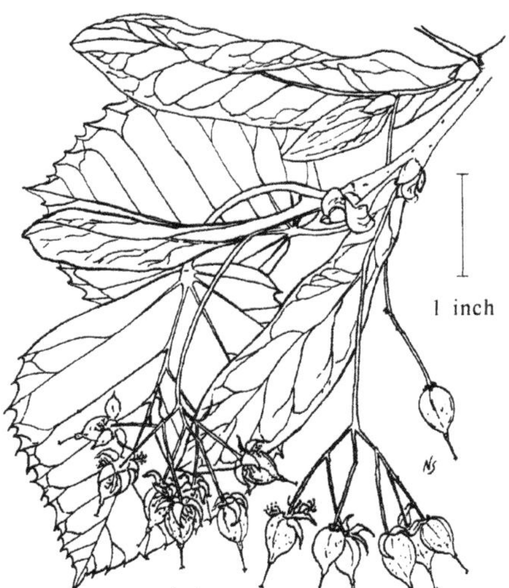

Tilia cordata

Tiliaceae *Tilia*

Tilia cordata Mill. - littleleaf linden

Littleleaf linden—the most commonly planted of all lindens—is native to Europe. It is one of the parents of a hybrid widely planted in Europe, *T. ×europea* L., a cross with *T. platyphyllos*—dubbed the "common lime" in Europe. Dirr (1990) states that this hybrid is inferior to *T. cordata* and *T. ×euchlora* for planting in the U.S.

Littleleaf linden is an excellent shade tree for a variety of sites, including streets and malls (if soil is fertile and water is adequate). It is quite commonly shaped into hedges in Europe. Many cultivars have been selected, including 'Greenspire', which is used widely as a street tree; 'Morden', from the Canadian research station of that name; and 'XP110', rated high in Ohio tests.

Richley (1988) states that normal precipitation in southwestern Nebraska is inadequate to maintain littleleaf linden in good vigor. Littleleaf linden trees, including the cultivars 'Fairview' and 'June Bride', are well adapted to the Fort Collins area (Fairchild and Klett 1993). Littleleaf linden seedlings of unknown ancestry at age 20 years growing at Prairie Pines Arboretum are remarkably vigorous (Bagley 1998). Basal sprouts were somewhat troublesome during their first few years of growth.

Planning

Height: up to 60 ft (18 m)
Habit: tree; deciduous; pyramidal becoming oval; bloom in late June after leaves full size
Site: sun, part shade; prefers fertile, moist, well-drained soil, pH adaptable; zone 3
Use: shade, hedge, windbreak, street, bee pasture

Recognition

Leaves: about 2 in (5 cm) long, sometimes broader than long; finely toothed; surfaces glabrous except for brown tufts in vein axils and a waxy bloom (glaucous) beneath; petioles glabrous, about 1 in (25 mm) long
Flowers: sterile stamens (staminodes) wanting, fertile stamens about as long as petals
Fruits: globose, gray hairy, slightly or not ribbed
Twigs and buds: similar to *T. americana*

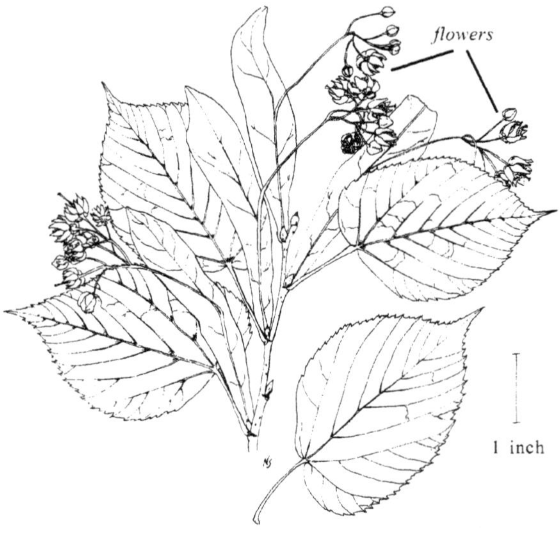

Tilia ×euchlora

Tilia ×euchlora K. Koch (*T. cordata* × *T. dasystyla* Stev.) - Crimean linden

Crimean linden, a handsome tree with smooth, glossy leaves, is reputed to be tolerant of air pollution, heat and drouth. In Europe, where aphids are a problem for lindens, this hybrid is the most "aphid-proof" linden. According to Dirr (1990), suckering that is often prevalent on grafted trees can be reduced by propagating with rooted cuttings. Crimean lindens usually bloom in early June in Lincoln (Bagley 1998).

Planning

Height: up to 60 ft (18 m)
Habit: tree; deciduous; crown oval; prolific basal sprouts
Site: sun; pH and soil adaptable; zone 4
Use: shade
Caution: bees may be injured by nectar

Recognition

Leaves: as in *T. cordata* except 2 to 4 in (5-10 cm) long; surface pale green below; petioles 1-2 in (3-5 cm) long
Flowers: as in *T. cordata*
Fruits: egg-shaped, about 0.25 in (6 mm) long; woolly, slightly 5-ribbed
Twigs and buds: similar to *T. americana*

Tilia heterophylla Vent. - white basswood

White basswood, native to the Appalachian region, is occasionally found as far west as eastern Missouri (Settergren and McDermott 1968). Its preference for moist, well-drained sites probably limits the successful planting of white basswood to the eastern prairies. White basswood growing along the street south of Bancroft Hall, UNL City Campus, are healthy (Bagley 1998). They were in full bloom June 20, 1989.

Planning

Height: up to 60 ft (18 m); National champion, 1986, 101 ft (31 m), North Carolina
Habit: tree; deciduous; crown irregular; late June bloom
Site: sun; prefers fertile, moist, well-drained soils; zone 5
Use: shade

Recognition

Leaves: margins finely toothed (serrate), teeth appear to be glandular tipped; glabrous above, white, sometimes brownish hairy mat beneath, hairs fascicled; petioles glabrous, about 1.6 in (4 cm) long
Flowers: as in *T. americana*
Fruits: slightly elongate, tip short-pointed; rusty hairy
Twigs and buds: similar to *T. americana*

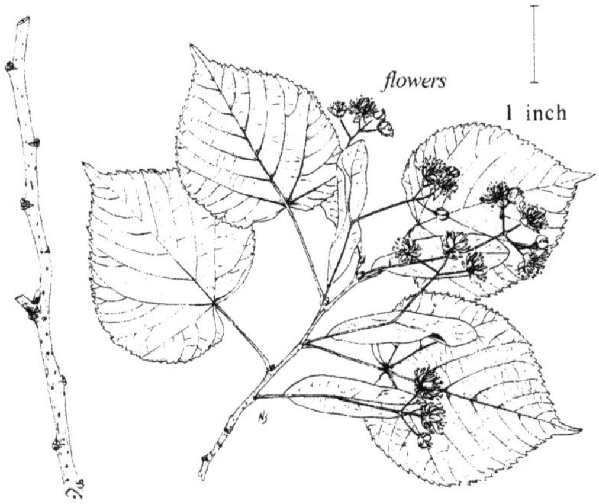

Tilia platyphyllos

Tilia platyphyllos Scop. - bigleaf linden

Though the tree is named bigleaf linden, the leaves of this native of Europe are not nearly as large as those of the American basswood. Because leaf pubescence can be so variable, this linden's ribbed nutlets are its most consistent identifying characteristic. Bigleaf linden has been widely planted in Europe for centuries.

Cultivars of this highly variable species include: 'Fastigiata', with strongly ascending branches; 'Laciniata', a small tree with irregularly lobed leaves; and 'Rubra', with red winter twigs.

Several healthy bigleaf linden trees on the UNL campus were in full bloom June 1, 1989, more than two weeks before those of nearby littleleaf lindens (Bagley 1998).

Planning

Height: up to 70 ft (21 m)
Habit: tree; deciduous; crown variable; early June bloom
Site: sun; pH and soil adaptable; zone 4
Use: shade

Recognition

Leaves: as in *T. americana* except 3-5 in (8-13 cm) long; short hairy, sometimes glabrous above, hairy beneath, especially on veins, rarely nearly glabrous; petioles hairy to nearly glabrous, 0.5-2 in long (15-50 mm) long
Flowers: similar to *T. cordata* except stamens longer than petals
Fruits: pear-shaped, long-hairy, 5-ribbed; about 0.5 in (13 mm) long
Twigs: long-hairy to glabrous
Buds: similar to *T. americana*

Tilia tomentosa Moench - silver linden

Silver linden, native to southeastern Europe, is the most tolerant to heat and drouth of all the lindens (Dirr 1990). A young specimen growing at Maxwell Arboretum is healthy.

Cultivars include: 'Green Mountain', a rapidly growing heat- and drouth-tolerant tree with a dense crown; and 'Sterling', a tree with a broad pyramidal crown and lustrous dark green leaves with silvery underside.

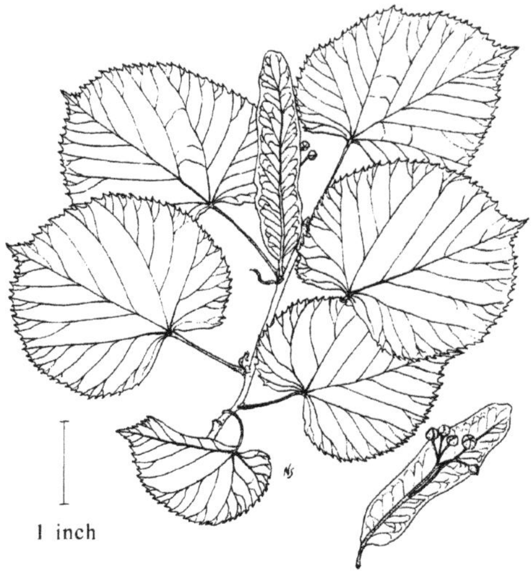

Tilia tomentosa

T. petiolaris DC., very similar to silver linden but with pendulous branches and longer petioles, is generally considered a distinct species, but Dirr (1990) states that it might better be considered as a cultivar of silver linden.

Planning

Height: up to 65 ft (20 m)
Habit: tree; deciduous; pyramidal becoming oval
Site: sun; pH adaptable, prefers moist, fertile soil; heat and drouth tolerant; zone 5
Use: street, shade, specimen, windbreak
Caution: nectar may be harmful to bees

Recognition

Leaves: similar to *T. americana*, except 2-5 in (5-10 cm) long; slightly hairy becoming glabrous above, white, stellate-hairy beneath; petiole hairy (tomentose), about 1.3 in (3 cm) long
Flowers: sterile stamens (staminodes) present
Fruits: egg-shaped, whitish hairy, faintly 5-angled
Twigs: soft short-hairy
Buds: similar to buds of *T. americana* except soft short-hairy

Malvaceae Juss. - mallow family

The mallow family comprises 75 genera and over 1,000 species of herbs, shrubs and small trees native to tropical and temperate regions. The North American species of mallow are herbs or shrubs largely confined to mild climates. Included in this family are hollyhock, okra and cotton. In warmer climates, wild mallow species are often shrubby perennials and some can attain tree size.

Recognition

Habit: herbs, shrubs, trees; deciduous or evergreen
Leaves: alternate, simple; usually palmately veined, more or less lobed; stipules usually present
Flowers: solitary or in various clusters, sometimes large and showy; bisexual, rarely unisexual; usually regular; often enclosed by leaf-like bracts; perianth 5-parted; stamens numerous, filaments fused into a tube
Fruits: dry (capsules), splitting lengthwise into 5 valves, seldom berries or samaras; seeds glabrous to hairy

Hibiscus L. - hibiscus

Over 200 hibiscus species, largely of the tropics, are noted for their showy flowers. Species native to North America are all herbs or shrubs.

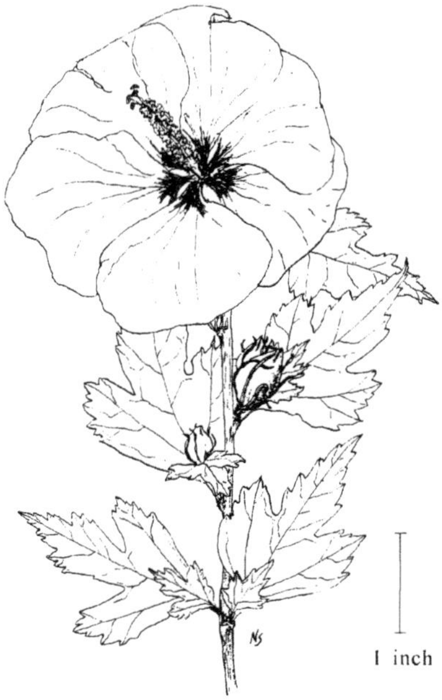

Hibiscus syriacus

Hibiscus syriacus L. - shrubalthea

Shrubalthea, a native of China and India that has been cultivated for many years in the U.S., has escaped cultivation in the east of our country. Numerous cultivars, many having double or triple flowers include: 'Diana', with its long-lasting, pure white flowers; 'Helene', with a purple blush at the base of white flowers; 'Aphrodite', that has dark pink flowers; and 'Minerva', with lavender flowers (Dirr 1990). These cultivars are all triploid and set little fruit. Trial plantings of 'Helene' and 'Rubrum' failed at PERC, Fort Collins (Fairchild and Klett 1993).

Planning

Height: up to 10 ft (3 m), occasionally 20 ft (6 m)
Habit: shrub, tree; deciduous; upright branches; flowers in summer and early autuomn on new growth; fruits persist into winter
Site: sun, part shade; pH adaptable; zone 5-6
Use: specimen, screen, hedge, grouping
Caution: avoid extremely wet or dry sites

Recognition

Leaves: egg-shaped, up to 4 in (10 cm) long; 3-lobed, margin coarsely toothed; base broad, wedge-shaped, or rounded; glabrous above, hairs on veins beneath; petiole up to 1 in (25 mm) long
Flowers: axillary, solitary on new growth; petals distinct, white, red, purple, or a combination thereof, bell-shaped, 2 to 4 in (5 to 10 cm) across; calyx with unequal lobes, subtended by secondary involucre of linear bractlets shorter than calyx; style 5-cleft at apex
Fruits: capsule, splitting, 5-valved; broad at base tapering to a beak, up to 0.8 in (2 cm) long; seeds flattened, margins hairy
Twigs: hairy when young, fluted near enlarged tip
Buds: not evident, covered by leaf base

Order Violales

The order Violales includes 24 families and nearly 5,000 species (Cronquist 1981).

Recognition

Habit: trees, shrubs, herbs; deciduous or evergreen

Leaves: alternate or opposite, simple, occasionally compound; with or without stipules

Flowers: bisexual or unisexual, mostly regular; sepals distinct, sometimes fused below, seldom forming a tube with terminal teeth; petals distinct, less often united, or wanting; stamens 5 to many

Fruits: dry or fleshy, splitting or not

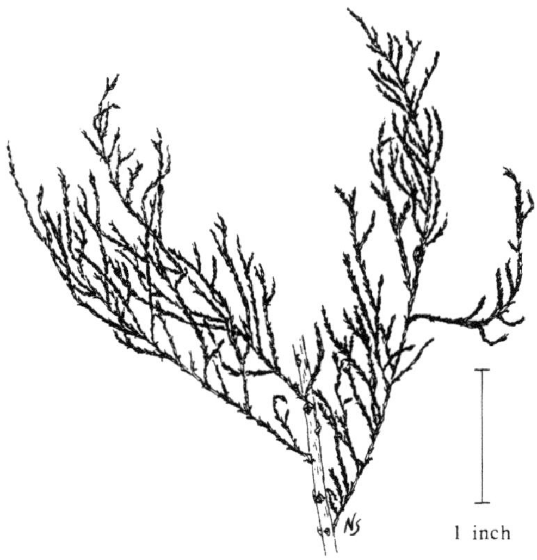

***Tamarix* sps.**

Tamariaceae Link - tamarisk family

The tamarisk family of four or five genera and 100 species is native to the tropical and temperate regions of Europe, Africa, the Middle East and Asia.

Recognition

Habit: shrubs, trees; deciduous, often evergreen

Leaves: alternate, commonly scale-like; mostly sessile; stipules wanting

Flowers: solitary or more often in clusters; sepals persistent, overlapping (imbricate); petals alternate with sepals, sometimes persistent; stamens as many or more often twice as many as the petals, or numerous and fused at the base into 5 bundles; styles distinct or connate at base

Fruits: dry, splitting into a cavity (loculicidal capsules)

Tamarix L. - tamarisk

An estimated dozen of the 75 species of tamarisk have been introduced into the U.S. as ornamentals, or for erosion control and sand dune stabilization. These plants, also known as salt-cedars because of their scale-like leaves and ability to grow on saline soils, have become naturalized in warm regions of the U.S. They thrive in hot, dry weather.

Tamarix species are very similar, making their positive identification difficult. Descriptions differ among taxonomists which suggests that interspecific hybridization is common. *T. gallica*, French tamarisk, is one of the hardier tamarisks that may grow in some parts of the central prairies.

Recognition

Habit: shrubs, trees; evergreen, deciduous if frozen

Leaves: usually scale-like, clasping twigs

Flowers: clusters dense, showy; usually terminal on short stems or sessile; perianth 4-5 parted, petals pinkish; stamens usually 4 to 5, distinct, or fused at base

Fruits: capsules; seeds bearded, rarely winged

Twigs: very slender, smaller ones falling with leaves

Tamarix parviflora DC. - small flower tamarisk

Small flower tamarisk, an import from southeastern Europe, has escaped cultivation and become a weedy pest on riverbeds in the U.S. Southwest where it thrives, competing for water with domestic use and irrigation. It is widely scattered throughout Kansas, usually mixed with five-stamen tamarisk (Stephens 1973).

Planning

Height: up to 15 ft (4.5 m); National champion, 1986, 14 ft (4.2 m), Virginia
Habit: shrub, tree; spreading, open, arching; late spring bloom on previous year's growth
Site: sun; pH and soil adaptable, avoid high fertility; drouth and heat tolerant; zone 5
Use: naturalizing, wildlife habitat

Recognition

Leaves: egg-shaped, partly encircling twig, thin, dry (scarious); tip pointed
Flowers: slender clusters up to 1.5 in (4 cm) long; perianth 4-parted; petals egg-shaped, upright, persistent, light pink, calyx lobes obtuse; anthers end abruptly in sharp-pointed tip (apiculate); styles 3, half as long as ovary
Fruits: capsule
Twigs: dark purple

Tamarix chinensis Lour. (*T. ramosissima* Ledeb., *T. pentandra* Pall.) - five-stamen tamarisk

Five-stamen tamarisk, a native of southeastern Europe and central Asia, has escaped cultivation in the U.S. It can be found growing in some southern states, north to Massachusetts and west through Kansas, where it is quite commonly found on the Arkansas River flood plain. It is also commonly found in New Mexico, Arizona, California and Oregon. Scattered "wild" five-stamen tamarisk plants have been found in Nebraska and North Dakota where it is occasionally grown in yards and gardens.

Five-stamen tamarisk can be effectively managed as a herbaceous perennial by removing all old stems each winter (flowers are borne on new wood only).

Planning

Height: up to 20 ft (6 m); National champion, 1981, 34 ft (10.3 m), New Mexico
Habit: shrub, tree-like; many stemmed, dense but loose and open on fertile soils; pink summer bloom on new growth
Site: sun; pH and soil adaptable, prefers sandy, moist, alkaline soils; drouth and heat tolerant; salt-spray tolerant; zone 3
Use: naturalizing, windbreak, wildlife habitat
Caution: aggressive, spreads quickly on amenable sites

Recognition

Leaves: scale-like, trough-shaped, appressed; tip pointed, base clasping; covered with whitish, waxy covering
Flowers: clusters spike-like, up to 2 in (5 cm) long; perianth 5-parted; petals elliptic, often cupped, pink; calyx lobes egg-shaped, acute to rounded, green, margins slightly toothed; stamens 5, anthers pink; styles spoon-shaped, half as long as ovary
Fruits: capsules egg-shaped, yellowish or pinkish, smooth; petals and calyx lobes usually present
Twigs: purple to red becoming grayish, roughened by raised leaf scars; upper twigs often deciduous with leaves
Buds: reddish, nearly globular, often 3 at each node
Bark: red, becoming dark brown, exfoliating into thin strips

Order Salicales

Salicaceae is the only family in the order Salicales.

Salicaceae Mirbel - willow family

The two genera in the willow family, *Populus* and *Salix*, are comprised of over 340 tree and shrub species that are native throughout temperate and north-temperate regions of the world. The wind-pollinated flowers appear ahead of or in conjunction with the leaves. Capsules release large numbers of tiny, cottony, winged seeds early in the summer. These seeds must germinate and take root in a matter of hours or perish because they do not contain endosperm, a built-in food supply.

These fluffy seeds of the willow family, though picturesque in flight, can become a nuisance when they lodge in air conditioner screens or other places where they are not welcome. Hence, those plants used for ornamental purposes are generally "cottonless" male clones propagated by cuttings.

The willow family species are valuable for use in ornamental, windbreak and forest plantings on sites to which they are adapted. They provide habitat and food for many animal species. The light, pale wood is an important source of biomass for fuel and paper production. These species are naturally short-lived as compared to many woody plants. They become increasingly so if located on inhospitable sites where they are easy prey to insects and diseases.

Recognition

Habit: trees, shrubs; deciduous

Leaves: simple, alternate; stipules present, often conspicuous, frequently early deciduous (caducous)

Flowers: clusters (catkins) unisexual; usually on separate trees (dioecious), sometimes on the same cluster (androgynous); calyx vestigial, petals wanting; stamens 1 to many; pistil 1 with many ovules; wind pollinated

Fruits: dry, splitting (capsules), 2 to 4 valved; seeds small, numerous, long silky hairs attached

Populus L. - cottonwoods, poplars

The wood of *Populus* species—among the fastest growing trees of the Temperate Zone—is widely used commercially for paper pulp, lumber and, in recent years, as an energy source. Easily propagated, fast-growing clones of the species and their hybrids—both natural and those derived from breeding projects—have created a good source of the wood used by paper mills, energy producers and others.

Optimum growth of *Populus* species and hybrids can only be achieved on fertile soils with adequate water. Site differences within the prairies are great enough that cultivars cannot be expected to thrive, or even survive, at all locations.

Regional research plantations containing native provenances and cultivars, hybrid or otherwise, will provide information allowing us to determine the most desirable clones for that region. Selected varieties or cultivars should be maintained by a certification program to insure identity.

The genus *Populus* is a complex of species that is difficult to classify because of their predisposition to hybridize. Approximately forty species are found throughout the Northern Hemisphere. An excellent reference for poplar nomenclature, culture and performance is **The Culture of Poplars in Eastern North America** (Dickman and Stuart 1983).

Recognition

Habit: trees
Leaves: usually broadest near base and about as long as broad; petioles long
Flowers: arching or drooping clusters appearing in advance of leaves, each flower solitary, inserted on a disc and subtended by a ragged (laciniate) bract
Fruits: capsules; seeds sumerous with long silky hairs
Twigs: circular to angled; pores (lenticels) conspicuous; pith 5-angled in section, homogenous
Buds: terminal present; resinous or nonresinous; scales many, overlapping, lowermost bud scale directly above leaf scar; flower buds separate and wider than leaf buds
Bark: smooth, pale becoming darker and furrowed with age

Grouping Key for *Populus* (cottonwoods, poplars)

Leaves woolly or hairless, petioles compressed or round; flowers with 5-20 stamens; capsules oblong, usually 2-valved; buds hairy or glabrous and sticky; bark smooth, green to gray, becoming pitted and rough at base: **group one - aspens and white poplars**

Leaves large and leathery, usually hairy when young; base heart-shaped (cordate), petiole compressed at apex; flowers with 12-40 stamens; capsules 2- to 3-valved; buds cone-shaped, glabrous, barely sticky; bark rough, scaly: **group two - Leucoides**

Leaves usually whitish below, base wedge-shaped to heart-shaped, margin translucent, petiole rounded or 4-sided, usually grooved above; flowers with 18-60 stamens; capsules 2-4 valved; buds sticky, fragrant; bark furrowed: **group three - balsam poplars**

Leaves usually triangular and broad at base, border translucent, coarsely toothed; flowers with 15-30 stamens; capsules 2-4 valved; buds sticky; bark furrowed: **group four - cottonwoods and poplars**

Group one - aspens and white poplars

Populus alba L. - white poplar

White poplar, a tree with silvery maple-like leaves, is commonly mistaken for *Acer saccharinum*, silver maple. It has been used as a "gully stopper" because of its prolific root suckers. White poplar is normally a large tree with a spreading crown, but 'Pyramidalis'— possibly a cross between *P. alba* and *P. nigra* 'Italica'—is columnar, similar in form and growth rate to Lombardy poplar. 'Pyramidalis', sometimes sold as bolleana white poplar, is susceptible to stem canker diseases that often shorten its life. Fairchild and Klett (1993) reported that a 10-year-old bolleana white poplar growing at PERC, Fort Collins, was healthy, possibly because the dry Colorado climate may reduce incidence of disease. The bolleana poplar is extremely hardy, growing at 7000-8000 feet above sea level in the San Luis Valley of Colorado (Sutton 1998).

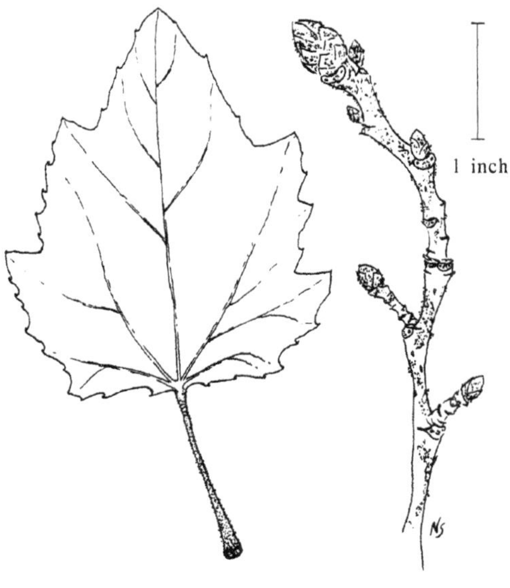

Populus alba

Hybrids resulting from crosses of white poplar with bigtooth and European aspen are of unknown value for prairie plantings. 'Tower', a cultivar of *P. canescens* (Ait.) Sm., gray poplar, a native of Europe, is similar to bolleana poplar. Some authorities consider 'Tower' to be a hybrid between *P. alba* and *P. tremula*. At PERC it is fast growing, winter hardy and adapted to alkaline soil (Fairchild and Klett 1993). Scientists there cite the incidence of cytospera canker and vigorous sucker growth as potential problems.

Planning

Height: up to 95 ft (29 m); Nebraska champion, 1981, 102 ft (31.2 m) tall, 14.5 ft (4.4 m) trunk circumference, Fremont; National champion, 1989, 96 ft (29 m) tall, 21 ft (6.3 m) trunk circumference, Wisconsin
Habit: tree; crown irregular, round-topped; prolific root suckers
Site: sun, part shade; pH and soil adaptable; drouth and air pollution tolerant; zone 4
Use: erosion control, naturalizing in large landscapes

Recognition

Leaves: oval in outline, often 3- to 5-lobed or with a few coarse teeth; white felt-like covering on lower surface; petiole hairy (tomentose)
Flowers: catkins–male pendulous, anthers purplish
Fruits: capsules egg-shaped, glabrous
Twigs: greenish gray covered with whitish, dense, cottony wool
Buds: appressed, conical; scales overlapping, brown, covered with white matted hairs
Bark: thin, smooth, greenish white, becoming dark and furrowed at base of large trees

Populus grandidentata Michx. - bigtooth aspen

Bigtooth aspen, a pioneer species native to the northeastern U.S. including eastern Iowa and Minnesota, grows on uplands but reaches its best development on flood plains. Seeds carried by wind travel great distances allowing seedlings to colonize burned-over forests and abandoned farmlands. Root suckers rapidly increase the population. Like other members of this family, bigtooth aspen will not succeed itself in a stand.

Bigtooth aspen is not widely planted as an ornamental because of its short life and suckering habit. These aspens in Lincoln have reached a diameter of about 12 inches (30 cm) and a height of 30 to 40 feet (9-12 m); they prefer loamy soils. Several groves of P. grandidentata can be found thriving on old farmsteads in the sandhills south of O'Neill, Nebraska (Sutton 1998).

Planning

Height: up to 50 ft (15 m); National champion, 1984, 132 ft (40 m), Michigan
Habit: tree; crown oval, irregular
Site: sun; pH and soil adaptable, prefers moist, fertile soil, tolerates dry and sandy soils; zone 3
Use: screen, windbreak, fiber production

Recognition

Leaves: nearly round to ovate; dark green above, lower surface fine-hairy, becoming glabrous when mature; margin coarse toothed, abrupt tips sharp to blunt; petiole slender, over 1 in (25 mm) long
Flowers: catkins
Fruits: capsules flask-shaped, about 0.2 in (6 mm) long; glabrous; ripening late spring
Twigs: moderately stout, brownish with fine, gray hairs or glabrous
Buds: egg-shaped, point outward; scales brown, overlapping, partly covered with fine gray hairs
Bark: smooth, olive green, becoming dark and furrowed

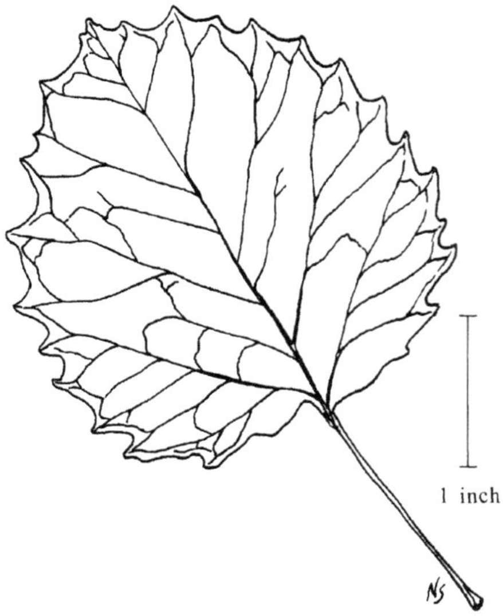

Populus grandidentata

Populus tremula L. - European aspen

European aspen, a native of Europe, Asia and North Africa, served up delicious meals for beaver before its demise (Johnson 1973). Richley (1988) rates European aspen as fair in southwestern Nebraska growing on loam soils but not adapted to clays.

The European aspen's habit is variable. 'Pendula' is a weeping form that bears grayish purple catkins (Dirr 1990). 'Erecta', a narrow, fastigiate tree, could be a substitute for Lombardy poplar in Colorado (Fairchild and Klett 1993). Fairchild and Klett report that a young tree growing at PERC, Fort Collins, has performed well when given adequate moisture, and it did not produce root suckers. Dirr (1990) reports excellent specimens at the Minnesota Landscape Arboretum and speculates that it may be best adapted to cold climates.

Planning

Height: up to 60 ft (18 m)
Habit: tree; crown variable, round, open to narrow
Site: sun; pH adaptable, prefers moist loams; zone 3
Use: windbreak, screen

Recognition

Leaves: nearly round to oval; hairy (tomentose) when unfolding becoming glabrous; thin; teeth coarse, rounded; petiole flattened, glabrous, often as long as blade
Flowers: catkins deeply lobed, fringed
Fruits: capsules
Twigs: glabrous
Buds: egg-shaped, pointed, slightly sticky
Bark: smooth, light colored

Populus tremuloides Michx. - quaking aspen

Quaking aspen, the most widely distributed poplar in North America, is an aggressive pioneer tree of the northlands from Newfoundland and Labrador west to northwest Alaska, south through the Rockies and western states and into Mexico. It is found over much of the northeastern and north-central states of the U.S., with isolated natural stands found in northern Nebraska along the Niobrara River.

Quaking aspen is famous for its smooth white bark and tremulous leaves becoming golden in autumn. However, admirers may be disappointed if they plant it with hopes for autumn gold unless they choose trees from a local or near-local provenance. Trees originating in the Rockies of Colorado are plagued by pests such as scale, and succumb to stem canker disease at a young age when planted in eastern Nebraska (Bagley 1998).

Quaking aspen trees propagated by the late Allen Wilke—from what he believed to be a native aspen grove north of Columbus, Nebraska—continue healthy growth after 15 years at Prairie Pines Arboretum but do not exhibit a bright fall color.

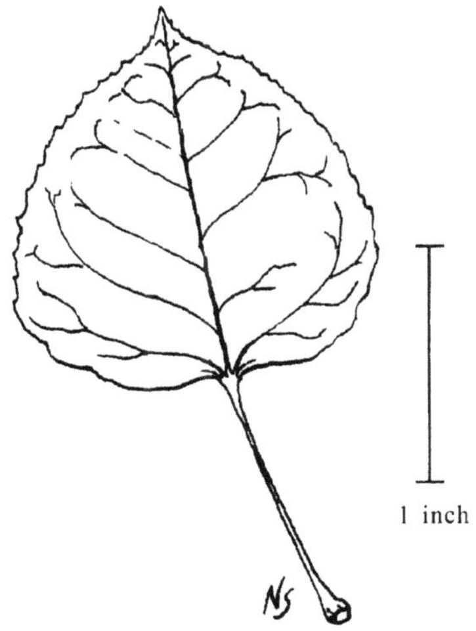

Populus tremuloides

Genetic information of quaking aspen is limited, but variability within the species is undoubtedly great due to its wide natural distribution. It crosses readily with bigtooth and European aspens. Seed crops from quaking aspen are few and far between and stem cuttings do not root readily. It reproduces mainly by its phenomenal ability to produce root suckers. It may best be enjoyed in its natural habitat unless one is prepared to cope with an ever-expanding population of suckers.

Planning

Height: up to 60 ft (18 m); National champion, 1982, 109 ft (33 m), Michigan
Habit: tree; crown pyramidal becoming narrow and rounded; prolific suckering
Site: sun; pH and soil adaptable, tolerant of low fertility and drouth; zone 1
Use: naturalizing

Recognition

Leaves: nearly round; glabrous, somewhat lustrous; margin finely toothed, hairy (ciliate) until midsummer; petiole flattened from the sides, glabrous; stipules long narrow, immediately deciduous (caducous)
Flowers: catkins—male drooping, anthers purple; female erect or drooping, many flowered
Fruits: capsules flask shaped, about 0.3 in (6 mm) long; seeds over 3 million per lb (6 million per kg)
Twigs: smooth, brown; lenticels conspicuous, pith 5-angled in cross-section
Buds: terminal cone-shaped, sharp-pointed; lateral appressed to the twig; scales 6 to 7, red-brown, glabrous, may be slightly sticky
Bark: smooth, greenish white to cream-colored with dark, thickened spots, becoming dark and furrowed in old age

Group two - Leucoides

Populus heterophylla L. - swamp cottonwood

Swamp cottonwood is a native of the swamps of the eastern coastal plain and of scattered locations along the Mississippi and Ohio River basins, growing mostly on waterlogged clay soils. It is rarely planted. Two Asian species belonging to this group, *P. lasiocarpa* Oliv. and *P. wilsonii* Schneid., might be hardy in the southeastern prairies.

Planning

Height: up to 50 ft (15 m); National champion, 1991, 140 ft (42.8 m), Ohio
Habit: tree; crown round-topped, open
Site: sun; moist, fertile soil, tolerates long periods of flooding; zone 5
Use: shade

Recognition

Leaves: egg-shaped, long-hairy when unfolding, becoming glabrous; margin toothed, base heart-shaped or sometimes rounded; petiole slender, about 3 in (7.5 cm) long
Flowers: catkins; female few-flowered
Fruits: capsules red-brown, splitting into 2 to 3 parts
Twigs: hairy, becoming shiny dark brown; pith orange
Buds: egg-shaped, barely sticky; scales red-brown
Bark: dark red-brown, deeply fissured

Group three - balsam poplars

Populus angustifolia James - narrowleaf cottonwood

Narrowleaf cottonwood, native to the Rocky Mountains from northern Mexico to southern Canada, grows along mountain streams and other moist sites.

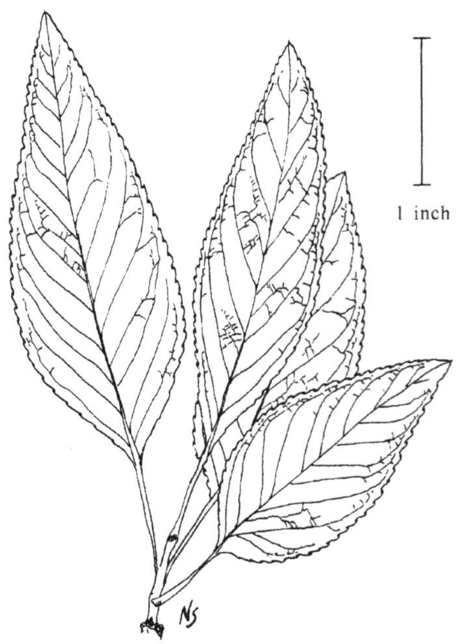

Populus angustifolia

It also resides in the Black Hills and in extreme northwestern Nebraska as small scattered stands growing along streams of the Pine Ridge area of Sioux and Dawes counties.

Although narrowleaf cottonwood is planted in the Rocky Mountain region, we have no knowledge of prairie plantings. Progeny of this cottonwood growing in the natural stands of western Nebraska or nearby Colorado and Wyoming should succeed in the shortgrass prairie region if water is adequate. However, they might be severely injured by diseases at a young age in the humid east.

Planning

Height: up to 50 ft (15 m); National champion, 1973, 79 ft (23.0 m), Oregon
Habit: tree; crown narrow, rounded top
Site: sun; pH and soil adaptable on moist soil; zone 3
Use: shade
Caution: short-lived in a humid climate

Recognition

Leaves: narrow (lanceolate ovate-lanceolate), up to 4 in (10 cm) long; gradually narowed to tip, base wedge-shaped; margin finely toothed, tips inturned (crenate-serrulate); surfaces glabrous; petiole slender, less than one-third length of blade
Flowers: catkins—male, stamens red
Fruits: capsule 2-valved; seeds greenish, silky-hairy
Twigs: glabrous, lenticels vertical, inconspicuous; leaf scars elliptical, bundle scars 3
Buds: slender, long-pointed; scales overlapping, chestnut brown, resinous, somewhat aromatic
Bark: light gray, smooth, becoming furrowed near base of old trees

Populus ×acuminata Rydb. - lanceleaf cottonwood

Although Pool (1951) and Stephens (1973) describe this Rydberg discovery as a species, Preston (1947) and others suggest that lanceleaf cottonwood may be a natural hybrid (*P. deltoides* var. *occidentalis* × *P. angustifolia*). It grows throughout the eastern foothills of the Rocky Mountains and is found to a limited extent in northwest Nebraska.

Seedling progeny of two young lanceleaf cottonwoods from Carter Canyon (southwest of Scottsbluff, Nebraska) succumbed to disease before the end of their second summer in a Lincoln plantation (Bagley 1998).

Planning

Height: up to 40 ft (12 m)
Habit: tree; rounded crown
Site: sun; pH adaptable on moist soil; zone 3
Use: shade
Caution: short-lived in a humid climate

Recognition

Leaves: egg-shaped or somewhat triangular, thick and leatherly; apex abruptly narrowed (acuminate), base wedge-shaped; margin coarsely toothed (crenate-serrate); surfaces glabrous: petioles slender, round, over one-half as long as blade
Flowers: catkins—male, stamens red, 15 to 20
Fruits: 2 to 3 valved
Twigs: slender, yellow-brown
Buds: long, narrow, glabrous; scales overlapping, light brown, shiny, resinous, nonaromatic
Bark: nearly white, smooth, becoming dark, ridged

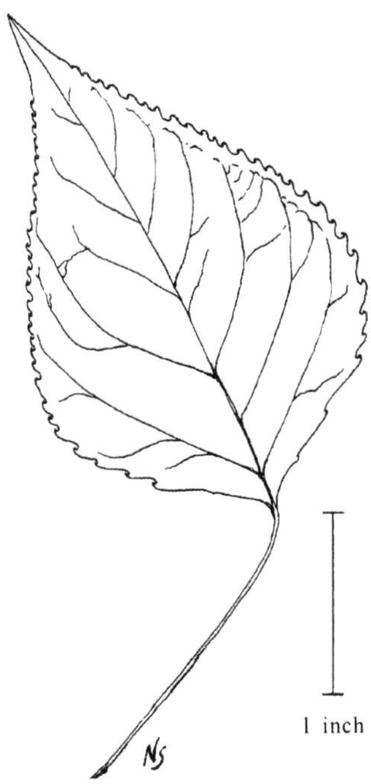

Populus ×acuminata

Salicaceae *Populus*

Populus balsamifera L. - balsam poplar

Balsam poplar, a tree of the northland, ranges over a wide territory from Newfoundland to Northwest Alaska, including the Great Lakes Region, the northern Rockies and the Black Hills with scattered outliers to extreme northwestern Nebraska. However, since it is easily confused with black poplar (*P. trichocarpa*)—similar in appearance except for the seed capsules—some of these reported locations at the edge of its range may be black poplar or natural hybrids of the two species. Balsam poplar also crosses with *P. deltoides*. One such hybrid cultivar, widely recommended for the northern Great Plains, is 'Northwest' (Hoag 1965).

Balsam poplar is a shallow-rooted tree prefering moist bottomland sites. A young balsam poplar in the Bessey Arboretum, Halsey, was healthy in 1985 (Bagley 1998). Research is needed to determine provenaces best adapted to prairie sites.

Planning

Height: up to 50 ft (15 m); National champion, 1986, 138 ft (41.8 m), Michigan
Habit: tree; crown pyramidal, irregular
Site: sun, part shade; pH adaptable, prefers fertile loams, abundant moisture; zone 1
Use: shade
Caution: protect from excessive wind

Recognition

Leaves: broadly egg-shaped (ovate); base rounded or heart-shaped, tip tapered to a point (acuminate), margin fine-toothed (crenate-serrate), sometimes hairy (ciliate); dark green, glabrous above, pale with rusty brown blotches below, glabrous or with a few hairs on midrib; petioles shorter than blade, hairy, abruptly broadened at base
Flowers: catkins—male, stamens 25 to 30, anthers yellow
Fruits: glabrous, 2 to occasionally 3 valved
Twigs: brown, often with waxy coating
Buds: cone-shaped, sharp; scales red-brown, resinous, fragrant, glabrous, fine hairs along edges (ciliate)
Bark: olive, becoming gray-brown to black, ridged

Populus trichocarpa Torr. & Gr. - black cottonwood

Black cottonwood, the largest of the American cottonwoods, is native to the West Coast from California to Alaska and to the forests of Idaho and Montana. 'Androscoggin' is a disease prone hybrid resulting from a cross of black cottonwood with Japanese poplar, *P. maximowiczii* (Dickman and Stewart 1983). Richley (1988) reports vigorous young 'Androscoggin' trees growing in southwestern Nebraska.

Planning

Height: up to 60 ft (18 m); National champion, 1984, 155 ft (47 m), Oregon
Habit: tree; crown narrow, rounded
Site: sun; pH 6.0-7.0 best, prefers moist, fertile loams in a humid climate; zone 3
Use: shade

Recognition

Leaves: egg-shaped (ovate), up to 4 in (10 cm) long; tip pointed, base rounded or heart-shaped (cordate); margin finely toothed with fine hairs (ciliate); green, glabrous above; rusty, silvery or pale green below
Flowers: catkins—male densely flowered, female loose Fruits: capsules rounded, 3 valved, 0.2 in (4 mm) long; short-hairy or scurfy
Buds: long conical, glabrous, resinous, fragrant
Bark: grayish, smooth, becoming dark gray and furrowed
Wood: light, weak

Group four - cottonwood and poplars

Populus deltoides var. *deltoides* Bartr. - eastern cottonwood

Eastern cottonwood, one of the largest native trees of the Midwest, grows naturally in the eastern U.S. and into the foothills of the Rocky Mountains. The westernmost trees of the eastern cottonwood have been named *P. sargentii* Dode, but recently a varietial designation, *P. deltoides* var. *occidentalis* Rydb., is in favor.

Morphological differences within eastern cottonwood provenances are nearly as great as differences among the various provenances. However, eastern cottonwood provenances differ significantly in disease resistance and vigor (Ying and Bagley 1976a). The population of the eastern cottonwood trees of the central Great Plains is a mixture of trees with characteristics of easterly and westerly provenances. Much of this heterogeneity is caused by man's planting activities. It is difficlt to be certain that a cottonwood tree is native to the area in which it is growing.

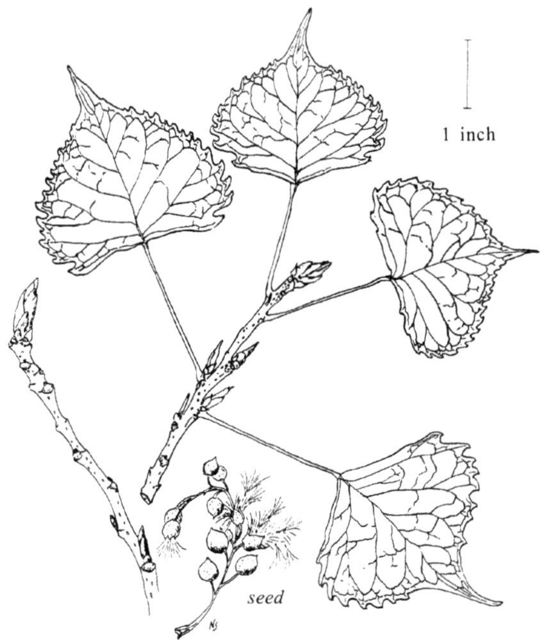

Populus deltoides

Leaves from trees that are believed to be native eastern cottonwood vary in shape and size across the continent, becoming progressively smaller and wider in proportion to length toward the West until they are wider than long in the western provenances. The bark becomes progressively smoother and thinner toward the West for stems of similar diameter (Ying and Bagley 1976a). We believe there is no clear-cut demarcation between variety *deltoides* and variety *occidentalis* and will not attempt to describe the latter.

City dwellers seldom have space for mature cottonwoods. Their best uses are for shade, wind protection and wood production in rural settings, where soils are fertile and water is plentiful. The wood can be used as an energy source thereby qualifying cottonwood as an alternative agricultural crop (Bagley 1988). Paper mills use it to produce high quality glossy paper. Cottonwood lumber is used for pallets and light construction.

'Platte', 'Mighty Mo' and 'Ohio Red' are selections from a 1961 provenance research plantation of eastern cottonwood growing at UNARDC, Ithaca, Nebraska. These male (and therefore "cottonless") trees are resistant to stem canker disease and are among the largest in the plantation. Although leaf diseases are not serious for these cultivars, 'Ohio Red' is notably less prone to contract these ailments than 'Platte' and 'Mighty Mo'.

'Siouxland', a South Dakota Agricultural Experiment Station introduction, is highly resistant to leaf rust diseases and recommended for shelter plantings in the northern Great Plains (Hoag 1965). However, in the more southerly plains states, especially Nebraska, it is short-lived due to stem canker disease (Bagley 1973).

Morphological differences among cultivars are barely perceptible. Care by a certification system must be exercised to maintain the integrity of each.

Eastern cottonwood is easily propagated by hardwood cuttings. Trees originating from the west and northwest part of the cottonwood's natural range can be successfully propagated by cuttings more easily than natives of the eastern U.S. (Ying and Bagley 1977). Eastern provenances root predominantly at the base of the cutting and should not be planted more than 2-3 in (5-8 cm) deep. Consequently, the practice of using long cuttings to facilitate survival in nonirrigated line-out beds or field plantings places the rooting zone too deep for optimum growth. Western provenances of eastern cottonwood, *P. nigra*, and many of the hybrids initiate adventitious roots along the entire stem making the depth of planting less critical for success.

Planing

Height: up to 80 (25 m); National co-champions, 1991; 96 ft (29.4 m) tall, 420 in (10.67 m) trunk circumference, Gosper County, Nebraska, and 85 ft (26 m) tall, 433 in (11 m) trunk circumference, Cassia County, Idaho

Habit: tree; crown pyramidal, becoming broad, open and irregular; autumn leaves brilliant yellow

Site: sun; pH and soil adaptable, prefers moist fertile soils; zone 3

Use: shade, windbreak, fiber production, lumber

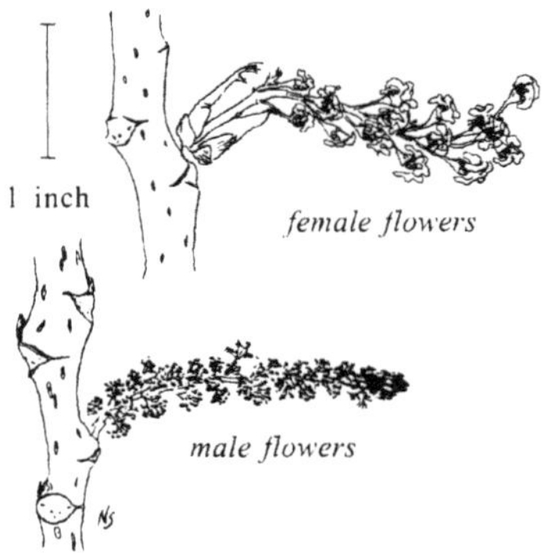

Populus deltoides

Recognition

Leaves: broadly triangular (deltoid-ovate), up to 5 in (13 cm) long; tip long-pointed, base squared-off (truncate) to heart-shaped (cordate); teeth coarse and incurved (crenate-dentate) except at base and apex; surfaces shiny, glabrous; stipules narrow, about 0.2 in (4 mm) long, promptly deciduous (caducous); petioles glabrous, flattened from the sides, often with glands at junction with blade

Flowers: catkins appearing before leaves; male densely flowerd, stamens yellow or reddish; female with 25 to 50 flowers

Fruits: flask-shaped, about 0.3 in (7-8mm) long, ripening in early summer

Twigs: stout, glabrous, gray-green to yellowish; lenticels prominent; leaf scars broad, shield-shaped, 3 bundle scars; pith star-shaped

Buds: terminal—about 0.7 in (19 mm) long, conical, sharp pointed; scales 6 to 7, lustrous brown, resinous; lateral—two sizes, flower buds larger; lowermost scale directly above leaf scar

Bark: thin, smooth, gray-green becoming thick, deeply furrowed, ash-gray

Populus nigra L. - black poplar

Black poplar, a species of European and Asian origir is well known as one parent of many hybrids resulting from crosses with eastern cottonwood, *P. deltoides*. This all began with natural hybridization after the introduction of our native cottonwood to southern Europ over 250 years ago.

Black poplar has also been crossed with *P. laurifolic* Ledeb. to produce Berlin poplar; and with *P. angulata* Ait., a close relative of *P. deltoides*, to produce *P. ×robusta* Schneid. Both are adapted to the northern plains, but their use has not been widespread (Hoag 1965). 'Robusta' appears to be well adapted to the Foı Collins area with only minor infestation by cytospera canker (Fairchild and Klett 1993). Canker disease is a cause for the short life of 'Robusta' in eastern Nebraska (Bagley 1998).

Black poplar is best known for the fast-growing, cottonless columnar *P. nigra* var. *italica* Muenchh., better known as Lombardy poplar, named after its plac of origin, a province in Italy. It is an ideal tree for planting borders, screens and windbreaks. Lombardy poplar is a stately, long-lived tree in the dry climates oı Utah and Idaho, along irrigation canals and in other places where soil water is adequate.

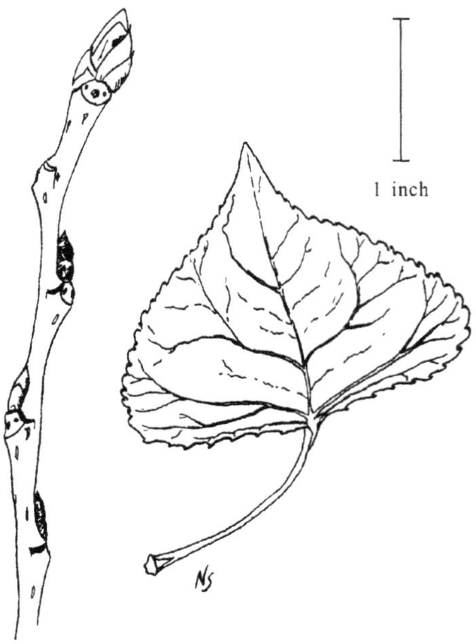

Populus nigra

Unfortunately, Lombardy poplar is short lived on our eastern prairies because of stem canker disease. Serious dieback often occurs eight to ten years after planting. 'Thevestina', or more correctly 'Afghanica', is a female clone supposedly more resistant to canker though no good research evidence supports this claim (Dirr 1990).

Planning

Height: up to 70 ft (21 m)
Habit: tree; crown wide-spreading to columnar
Site: sun; pH and soil adaptable; zone 3
Use: screen, windbreak, border
Pests: cytospera canker

Recognition

Leaves: triangular to broadly diamond-shaped (rhombic-ovate), up to 4 in (10 cm) long; tip long-pointed (acuminate), base variously wedge-shaped (cuneate); margin toothed (crenate-serrate); surfaces glabrous; petioles slender, up to 2 in (5 cm) long
Fruits: capsule 2-valved
Twigs: slender, glabrous, yellow-green with reddish patches on young twigs, becoming yellow-green to gray-green
Buds: cone-shaped; terminal up to 0.4 in (1 cm) long; lateral closely pressed to stem; scales brown, glabrous, slightly sticky
Bark: thin, light gray, becoming furrowed on lower trunk

Populus ×canadensis Moench. - **hybrid poplar**

Hybrids resulting from crosses between *P. deltoides* and *P. nigra* have produced many cultivars. The first were natural ones resulting from the export of eastern cottonwood to Europe in the seventeenth century. Among those several European hybrids, the best-known import to the U.S. is 'Eugenei'.

Dr. Ernst J. Schreiner of the Northeast Forest Experiment Station, U.S. Forest Service, developed many poplar hybrids during the 1950s. Most of them that were planted in Lincoln by the U.S. Forest Service Research Center succumbed to stem diseases within a few years. However, one of these hybrids, NE-237, later named 'Nor'easter', is a vigorous, fast-growing clone that is resistant to stem canker and leaf rust diseases (Bagley 1973). This female clone produces pods that drop when partially open, releasing very little "cotton" and sterile seeds. 'Nor'easter' is very easily propagated by hardwood cuttings and root sprouts; it has been released to local nurseries.

Several other cultivars of *P. ×canadensis* are sold by nurseries with glowing claims of "instant shade." All are vigorous and fast growing, but often deteriorate and die from stem diseases at an early age. They are excellent trees for wood harvest on a short rotation and are easily propagated by hardwood cuttings and sprout profusely from young stumps after harvest.

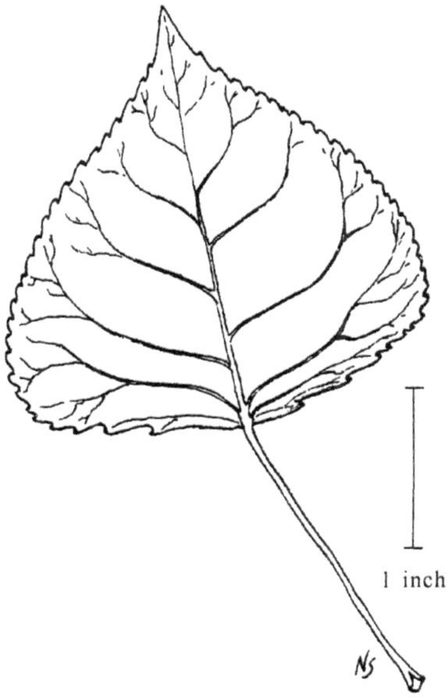

Populus ×canadensis

Planning

See *P. deltoides* and *P. nigra*.

Recognition

See *P. deltoides* and *P. nigra*. Clones are variable. Characteristics may be similar to either parent or an intermediate. Flowers may be sterile.

Salix L. - willow

The willows are a large, variable group of plants that defy separation into neat compartments but have been divided into about 300 species. Natural interbreeding among willows that are considered different species produces offspring that differ greatly in appearance. (Seedling offspring of a known species will often differ from published descriptions of the parent.)

About 80 willow species, mostly shrubs, are native to North America. They are the only woody plants capable of growing in the tundras of the high altitudes and northerly latitudes. Willows are found on beach dunes at sea level and along streams almost everywhere else. Because they are prolific, they survive even though they are prone to storm breakage and stem canker diseases and are attractive to many insect pests and browsing animals.

These fast-growing, easily propagated plants add interest and beauty to our landscapes with their colorful stems, attractive catkins and unusual forms. They provide habitat for wildlife and are used to control streambank erosion. Though the wood is lightweight and not durable when in contact with moisture, it is tough and strong. It is used for core stock in veneers and for making crates and slack cooperage. Pliable shoots of some willow species are used for making baskets and furniture. Willow bark was the original source of aspirin.

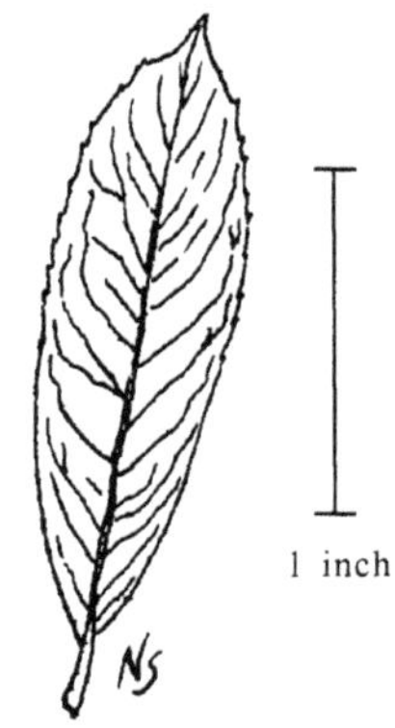

Salix purpurea

The famous Babylon weeping willow, *S. babylonica* L.—a rare species adapted only to southerly climes—has been crossed with hardier species to produce weeping willows for planting in more northerly landscapes. One such hybrid, a cross with *S. fragilis*, is *S.* ×*blanda* Anders, Wisconsin or 'Niobe' weeping willow, a pendulous tree with glossy green leaves. Another hybrid believed to be of the same parentage is the 'Thurlow' weeping willow (*S.* ×*elegantissima* K. Koch). *S.* ×*chrysocoma* Dode., 'Tristis', golden weeping willow, is now considered to be a cross between *S. alba* L. and *S. babylonica* (Dirr 1990). Unfortunately, this weeping willow has also been sold as 'Niobe'. A Nebraska champion willow identified as *S. babylonica* growing on the Lydick farm northwest of Tekamah is undoubtedly one of these hybrids. It was 108 feet (33 m) tall in 1987 (Kuhns 1988).

S. elaeagnos Scop., Rosemary willow, a shrub with long linear leaves, is recommended for wet areas. *S. sachalinensis* Fr. Schmidt, 'Sekka', Japanese fantail willow, is a large shrub with twisted and, sometimes, flattened branchlets (Dirr 1990). *S. gracilistyla* Miq., rosegold pussywillow, is grown for its early, showy pinkish catkins that portend the coming of spring.

S. lucida Muhl., shining willow—with its dark orange twigs and shiny yellow-green leaves—is native to wet soils of the northeastern U.S. *S. purpurea* L., purpleosier willow, a shrub with purplish leaves dies back during winter in southeast Nebraska. *S. matsudana* var. *tortuosa* Rehd., corkscrew willow—a small tree planted for its contorted and twisted branches—has little else to recommend it because of its short life.

S. myrsinifolia Salisb. is an extremely variable species of little importance (Rehder 1940). However, it has been crossed with *S. retusa* L. to produce a hybrid, *S.* ×*cotteti* Kern. The latter suffered severe cold injury in Maxwell Arboretum during the winter of 1991-92.

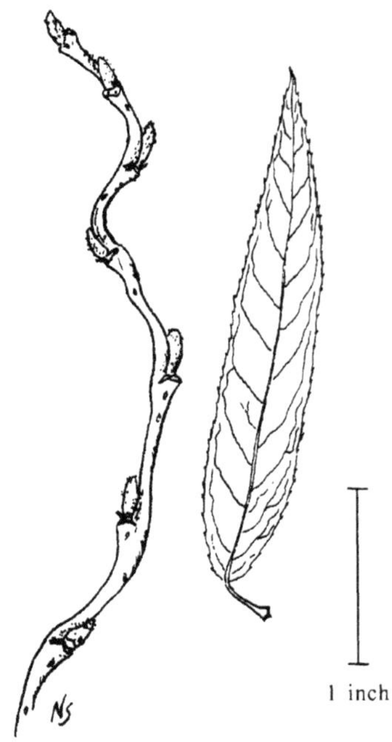

Salix matsudana tortuosa

Many more willow species could be mentioned here but we have included only those that our readers are most likely to encounter in the prairie region. We apologize if we have omitted your favorite willow.

We have also omitted much detail about flowers and fruits because they are available for observation for only a short time. Recommended references for the serious student of willows are **Woody Plants of the North Central Plains** (Stephens 1973) and **The Genus *Salix* in the Black Hills of South Dakota** (Froiland 1962).

Recognition

Habit: shrubs, trees
Leaves: usually several times longer than broad (lanceolate), margins smooth (entire) or toothed
Flowers: male and female usually on different trees (dioecious); clusters (catkins) densely flowered, elongated, ascending
Fruits: capsules, 2-valved
Twigs: often colorful—red, yellow, green, purple—frequently brittle and drooping; lenticels inconspicuous, leaf scars v-shaped, bundle scars 3; pith uniform, usually round in cross-section
Buds: terminal absent; lateral covered with a single cap-like scale; flower buds separate and larger than leaf buds

Salix alba L. - white willow

White willow, a native of Europe and a source of pliable lumber used for basketry, is also used in England for cricket bats and balls. White willow was imported to the prairies as a possible hedging plant for livestock enclosures. It has escaped cultivation and become naturalized, growing along streams on moist sites.

A hybrid, *S. matsudana* Koidz. × *S. alba*, is a recent import from Australia. This vigorous, insect- and disease-resistant tree is widely used as windbreaks in crop protection there and appears to be worthy of extensive trials here.

White willow twigs often provide winter color. Several cultivars have been selected for this trait including: 'Vitellina', with bright yellow stems; 'Britzensis', with bright red first-year stems; and the golden weeping willow, 'Tristis', considered to be the hardiest and most beautiful of the weeping types (Dirr 1990).

Planning

Height: up to 60 ft (18 m); National champion, 1991, 133 ft (40.7 m), Michigan
Habit: tree; crown open, round-topped; flowers appear with the leaves
Site: sun; pH adaptable, tolerates wet soils; zone 2
Use: screen, shade, naturalizing, windbreak

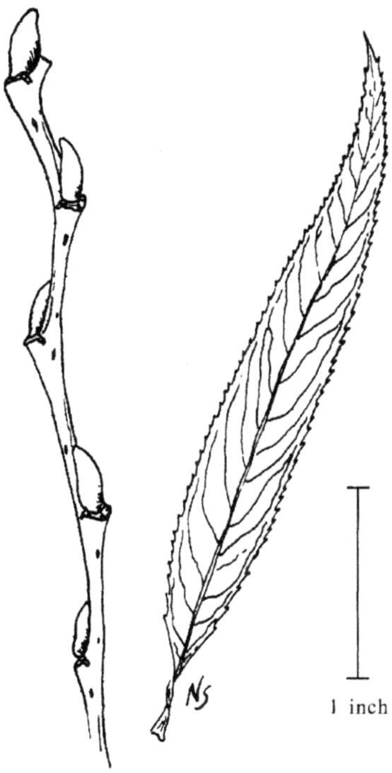

Salix alba

Recognition

Leaves: narrow (lanceolate) with widest part about the middle, up to 4 in (10 cm) long; tip long-pointed, base wedge-shaped; margin fine-toothed (serrulate); silky hairy becoming glabrous above, bloomy and silky beneath; stipules narrow, early deciduous; petioles up to 0.5 in (12 mm) long, grooved and hairy above, usually glandular near the blade

Fruits: cone shaped capsules, 0.2 in (5 mm) long, glabrous, yellowish

Twigs: flexible, greenish brown to light golden yellow, sometimes reddish

Buds: brown to golden yellow, sparsely hairy; tips usually turned toward stem

Bark: light brown with shallow furrows

Salix amygdaloides Anders. - peachleaf willow

Peachleaf willow, native to much of the central and northern prairies, is found abundantly in the Rocky Mountains. It is a pioneer growing naturally along our perennial and intermittent streams.

Peachleaf willow bears close resemblance to black willow (*S. nigra*) with which it mingles in the eastern prairies. Peachleaf willow can usually be distinguished from black willow by its wider leaves (which are whitish on the underside rather than light green), its longer petioles, a lack of persistent stipules, and the fact that green twigs will not break cleanly at the base with a sharp bend.

Peachleaf willow wood is light, soft and weak. Disease and wind take their toll resulting in a tangle of broken willow tops. The bark tends to cling to the trunk after death, keeping the wood moist and thus becoming a good medium for fungi, including the edible oyster shell mushroom. The dead wood also provides homes and food for woodpeckers and other birds.

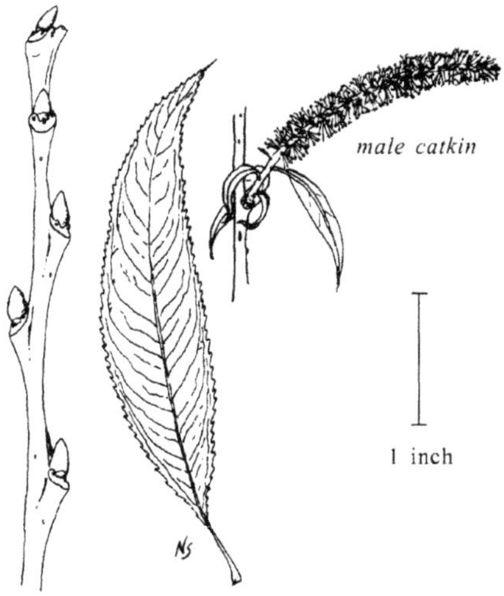

Salix amygdaloides

Although peachleaf willow is easy to propagate and useful for streambank erosion control and animal browse, it is seldom planted. Bulldozers have taken a heavy toll on the native prairie population of the peachleaf willow to make way for grassed waterways and center pivot irrigation systems.

Planning

Height: up to 60 ft (18 m); National champion, 1989, 58 ft (17.6 m), Wisconsin

Habit: tree; crown round-topped, branches somewhat drooping

Site: pH adaptable, prefers moist, fertile soil, drouth tolerant; zone 2

Use: wildlife habitat, streambank erosion control, naturalizing

Recognition

Leaves: much longer than wide (lanceolate); surfaces glabrous, pale blue-green to whitish (glaucous) beneath; petioles 0.2-1.0 in (5-25 mm) long, glaucous; stipules minute and early deciduous or absent

Flowers: catkins on leafy branches, axis hairy

Fruits: capsule yellow at maturity; 2 valves spreading, glabrous

Twigs: light yellow to brownish, glabrous; do not break cleanly with sharp bend at base; leaf scars narrow, crescent to U-shaped

Buds: conical to egg-shaped (ovoid), swollen on the side away from the stem (gibbous); single scale glabrous

Bark: deeply fissured, shaggy

Salix bebbiana Sarg. var. *bebbiana* Richards. - **Bebb willow**

Bebb willow is native of an area that stretches from Alaska across Canada to Newfoundland, south across the northern half of the U.S., and down through the Rockies to New Mexico and Arizona. It is the most abundant willow found in the Black Hills (Froiland 1962).

Bebb willow, *S. bebbiana* Sarg., is known as "diamond" willow in some localities. However, it is only one of several species with diamond-shaped dimples from which novelty walking sticks, lamps, and furniture are fashioned (Little 1980a). Interesting patterns are created by these stem indentations that reveal brown heartwood in contrast to the white of sapwood. Fungi are the probable cause of these pits that occur at points of twig attachment. Stems often die at approximately 2 inches in diameter (Straight 1993).

Variety *perrostrata* Schneid. is not widespread in the U.S. but is common in the Rocky Mountain region and the Black Hills. Leaves are entire, sparsely hairy, becoming smooth (glabrate). This variety grows in a drier habitat and is more vigorous in the Black Hills than variety *bebbiana* (Froiland 1962). Intergradation of the two varieties often makes positive identification impossible.

Planning

Height: up to 20 ft (6 m)
Habit: shrub; much branched, irregular, gnarled; stems several, thicket-forming
Site: sun; moist, fertile soil; zone 2
Use: naturalizing, wildlife habitat

Recognition

Leaves: elliptic to narrow egg-shaped, wider above the middle (obovate), up to 2 in (5 cm) long; somewhat leathery; tip short-pointed, base wedge-shaped; dark green with long silky hairs (villous) above, lower surface gray to blue-green, usually hairy; margin entire to shallow-toothed; petioles 0.2 in (6 mm) long, hairy, no glands at summit; stipules more or less ovate, early deciduous
Flowers: catkins expanding with leaves; very short-stalked, usually 2 small leaves at base; densely flowered
Fruits: hairy; long tapering from base to long slender beak
Twigs: golden to reddish brown, becoming brown; hairy; lenticels vertical; leaf scars raised
Buds: flattened, flange on side, broadly egg-shaped; variably hairy
Bark: gray, shallow furrowed

Salix caroliniana Michx. - **Carolina willow**

Carolina willow, a small tree native to the coastal plains of the southeastern U.S., also grows extensively in Missouri. Stephens (1973) reports it is also in southeast Nebraska. Carolina willow leaves resemble those of the peachleaf willow and its bark is similar to that of the black willow. It may be confused with *S. eriocephala* where their ranges overlap.

Planning

Height: up to 30 ft (9 m); National champion, 1984, 57 ft (17.4 m), Virginia
Habit: tree; crown spreading, branches slightly drooping
Site: sun; prefers sandbars, streambanks; zone 5
Use: erosion control, naturalizing

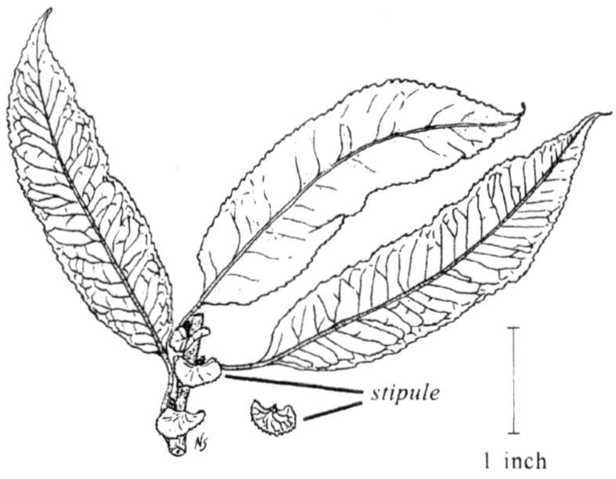

Salix caroliniana

Salicaceae *Salix*

Recognition

Leaves: much longer than wide (lanceolate); fine teeth (serrate), tips light-colored, calloused; young leaves hairy (tomentose); dark green, glossy above, silvery white, glaucoous beneath; petioles hairy (tomentose), 0.2-0.5 in (5 to 10 mm) long; stipules egg-shaped to kidney-shaped (reniform), persistent
Flowers: catkins terminal, appearing with leaves
Fruits: capsules egg-shaped, glabrous
Twigs: red-brown to gray-brown, hairy or glabrous; leaf scars narrow crescent-shaped
Buds: dark brown, sparsely to densely hairy
Bark: brownish gray, furrowed, ridges firm

Salix discolor Muhl. - pussywillow

Pussywillow, a well-known small tree native to wet sites throughout much of the northeastern U.S., including North and South Dakota, is widely planted for its silky, soft, silver-gray male catkins that at maturity turn golden with pollen. This shrubby willow is a harbinger of spring and furnishes welcome food for bees at a critical time of year. Many a new tree is started for a friend when roots are discovered on twigs that have been collected for a spring bouquet.

Pussywillows are short-lived, especially on unfavorable sites because of stem canker diseases. They can easily be confused with *S. bebbiana*, Bebb willow. Goat willow (*S. caprea* L.), native to Europe and Asia, is also very similar in appearance to pussywillow. 'Pendula' ('Kilmarnock') is a weeping form of goat willow used as groundcover or trained as a small weeping tree.

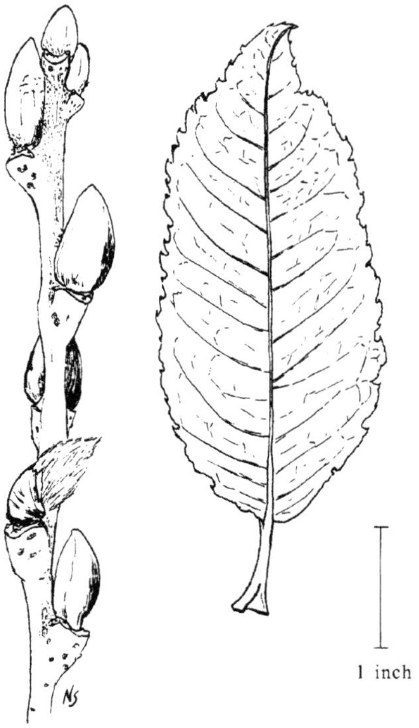

Salix discolor

Planning

Height: up to 20 ft (6 m); National champion, 1983, 47 ft (14.2 m), Michigan
Habit: tree; single or multiple trunks, rounded top
Site: sun, part shade; prefers moist soils; zone 3
Use: naturalizing, streambank erosion control

Recognition

Leaves: elliptic, widest at middle or just above the middle (obovate), up to 3 in (8 cm) long; tip pointed, base wedge-shaped (cuneate); margin wavy (undulate), sparsely toothed; surface hairy at first becoming smooth, dark green above, whitened below; petioles up to 0.4 in (1 cm) long; stipules up to 0.1 in (1 mm) long, promptly deciduous
Flowers: before leaves, catkins densely flowered; male showy, anthers yellow
Fruits: wide spreading at maturity, hairy
Twigs: red-purple to red-brown, glabrous; lenticels small, oval, yellow; leaf scars hairy on upper edge
Buds: egg-shaped (ovoid) somewhat appressed to stem, scale covered with short fine hairs (puberulent)
Bark: gray-brown with tight, fine ridges, becoming fissured

Salix eriocephala Michx. - **Missouri River willow**

A local common name for Missouri River willow is "diamond willow." This willow grows along the Missouri and Mississippi rivers and their main tributaries from southern Missouri to Minnesota and North Dakota. The Misssouri willow is not found abundantly. Some authorities consider *S. eriocephala* to be a variety of *S. rigida* (Van Der Linden and Farrar 1984).

Planning

Height: up to 20 ft (6 m)
Habit: tree; single stem, sometimes multiple stems
Site: sun; prefers sandbars, wet streambanks; zone 3
Use: streambank erosion control, naturalizing

Recognition

Leaves: elliptic to much longer than wide (lanceolate), about 3 in (8 cm) long; tip long-pointed (acuminate), base rounded; margin glandular toothed; upper surface dark green, lustrous, pale beneath, glabrous or with a few hairs; midrib hairy both surfaces, veins usually yellowish, often united near margin; petioles hairy, about 0.5 in (12 mm) long; stipules leaf-like, persistent or deciduous
Flowers: before the leaves
Fruits: capsules egg-shaped with long neck, glabrous; valves recurling sharply at maturity
Twigs: green-brown, covered with white or gray hairs first year, becoming red-brown, glabrous
Buds: red-brown, flattened, sharp; scale covered with grayish hairs; male flower buds often showing white hairs in winter
Bark: smooth, light gray with slight red tinge, becoming shallowly fissured, tightly appressed; lenticels large, horizontal, elliptic or diamond-shaped
Wood: dark brown, moderately durable

Salix humilis Marsh. - **dwarf gray willow**

The dwarf gray willow, an inhabitant of sandy upland prairies throughout much of the eastern U.S., is usually found only along roadsides or as fence rows because of livestock browsing. It differs greatly from its bottomland cousins that prefer moist soils. The dwarf gray's often hairy twigs and leaves with their margins rolled under may be adaptations that have allowed its dry-site survival.

Hoary willow, *S. candida* Fluegge ex Willd., an inhabitant of the swamps and bogs of northern North America, can be confused with dwarf gray willow. However, their ranges barely overlap and site requirements are quite different.

Dwarf gray willow deserves an opportunity for more trials in our landscape plantings. A specimen planted at Horning State Farm in 1980 is a thriving, handsome shrub (Bagley 1998).

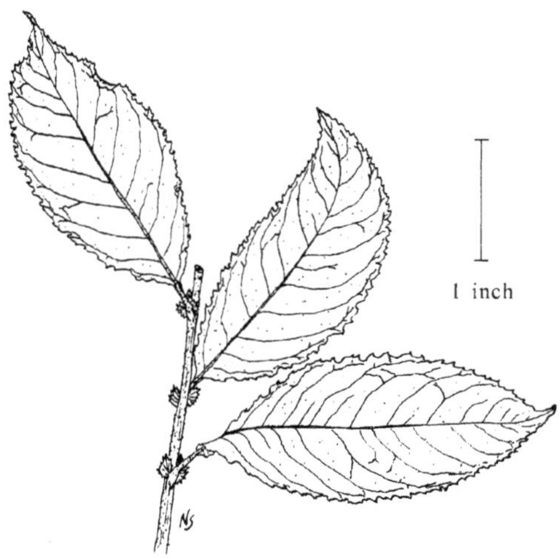

Salix humilis

Planning

Height: up to 5 ft (1.5 m)
Habit: shrub; stems single or clustered; colonial; leaves usually crowded near twig tips
Site: sun; pH and soil adaptable, drouth tolerant; zone 3
Use: naturalizing, wildlife habitat, grouping, border

Salicaceae *Salix*

Recognition

Leaves: long and narrow, wider above the middle (oblanceolate) to elliptical, about 3 in (8 cm) long; tip rounded or pointed, base tapered; margin smooth, rolled under (revolute); upper surface slightly hairy, lower silvery white, hairy; petioles up to 0.4 in (1 cm) long, hairy; stipules leaf-like, hairy
Flowers: catkins axillary, before the leaves; male—stamens 2, anthers red-orange
Fruits: capsules hairy, strongly recurved at maturity
Twigs: yellow, gray hairy
Buds: egg-shaped, scale reddish with white hairs
Bark: gray-brown, rough

Salix interior Rowlee - sandbar willow

This shrub (or occasionally small tree) is also known as *S. exigua* Nutt. ssp. *interior* (Rowlee) Cronq. and coyote willow is its sobriquet (Stephens 1973). It grows throughout the prairies, northwest to Alaska and northeast to the New England states and southeastern Canada.

Sandbar willow is one of the first plants to become established on sandbars where it forms dense thickets that provide shelter and nesting sites for birds, and food for deer, beaver and rabbits. A thick, matted root system provides excellent erosion control. Sandbar willow can be easily recognized by the very narrow leaves with wide spaced marginal teeth and very short petioles.

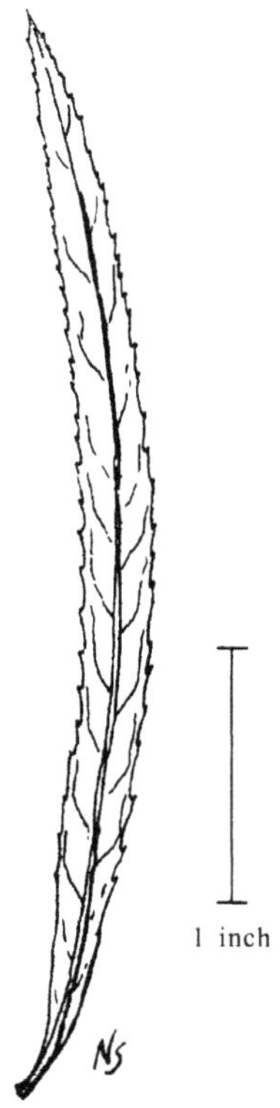

Salix interior

Planning

Height: up to 15 ft (4.5 m)
Habit: shrub, tree
Site: sun; pH adaptable, prefers sandbars and wet streambanks; zone 1
Use: streambank erosion control, naturalizing

Recognition

Leaves: linear, sharp pointed, about 4 in (10 cm) long, less than 0.4 in (1 cm) wide (lanceolate); margins with shallow, wide-spaced, glandular teeth (remotely dentate); surfaces at first soft and silky, becoming glabrous; yellow-green above, paler below, sometimes slightly hairy; lateral veins curve and run nearly parallel to leaf margin, ending in a tooth; petioles up to 0.2 in (6 mm) long, grooved; stipules early deciduous
Flowers: catkins solitary or paired at the end of new leafy shoots
Fruits: capsules narrowly egg-shaped, glabrous, 2 valves spreading at maturity
Twigs: slender, glabrous or finely hairy; leaf scars narrow, nearly straight
Buds: narrow egg-shaped (ovoid), about 0.1 in (2-3 mm) long, sharp pointed
Bark: thin, smooth, becoming shallow fissured

Salix nigra Marsh. - **black willow**

The largest of the willows, black willow grows naturally on lowlands throughout the eastern U.S. west to Minnesota and extreme eastern Nebraska, and south through eastern Kansas and most of Oklahoma and Texas.

Black willow is often confused with peachleaf willow, a species more commonly found in the westen prairies. Natural crosses have occurred between the two species (Fowells 1965). Clues to the identity of black willow are its leaf characteristics and the clean break a bent twig makes at the juncture of a larger stem.

The fast-growing black willow is easily propagated by root and stem cuttings. In fact, fence post size cuttings have been rooted in gully control projects. The lightweight wood is used in furniture and baskets. Black willow, short-lived and susceptible to breakage by wind and ice, is rarely used in landscape plantings.

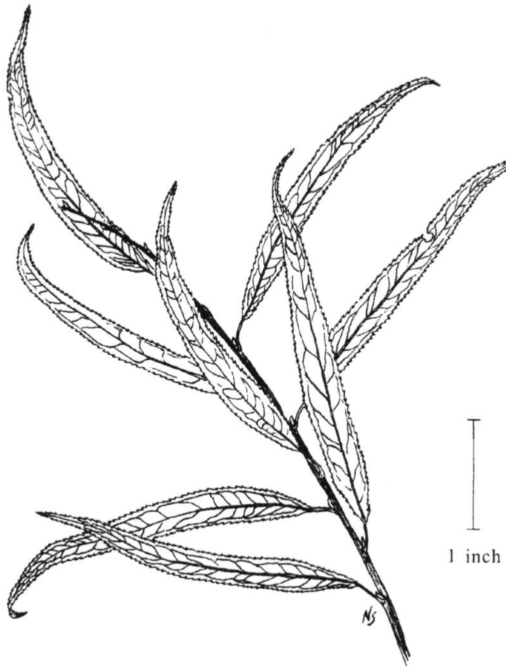

Salix nigra

Planning

Height: up to 60 ft (18 m); Nebraska champion, 1988, 52 ft (15.9 m), Orchard (NOTE: this tree may really be *S. amygdaloides*); National champion, 1973, 114 ft (35 m), Michigan
Habit: tree; broadly irregular
Site: sun; pH and soil adaptable as long as water is plentiful; zone 4
Use: naturalizing, shade, wood

Recognition

Leaves: much longer than wide (lanceolate), up to 6 in (15 cm) long; margin finely toothed with incurved callous-tipped teeth; glabrous, lustrous above, light green beneath, hairy, becoming glabrous; petioles glabrous, 0.1-0.5 in (3-12 mm) long; stipules conspicuous and persistent or small and deciduous
Flowers: catkins terminal on short, new leafy shoots; male noticeable with yellow bracts and anthers
Fruits: capsules short-stalked, glabrous, 0.2 in (6 mm) long; seeds 2-3 million per pound
Twigs: slender, reddish brown becoming gray; often brittle, sharp bend at base produces clean break; leaf scars narrow, linear to crescent-shaped
Buds: narrowly conical, less than 0.1 in (2 mm) long; scale shiny, red-brown
Bark: thick, gray-brown, fissured with shaggy scaly ridges

Salix pentandra L. - laurel willow

Laurel willow, a handsome native of Europe, is noted for its polished, dark green foliage. However, Dirr (1990) notes that the leaves of some laurel willows located in central Illinois are usually destroyed by disease by August. These willows at Maxwell Arboretum and at the South Dakota Arboretum, Brookings, usually remain relatively free of disease. Apparently leaf diseases that plague laurel willow in Illinois are less serious in a drier climate.

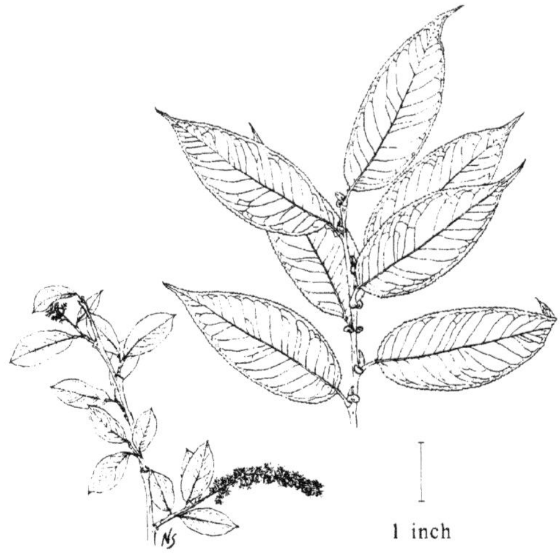

Salix pentandra

Planning

Height: up to 60 ft (18 m)
Habit: tree, sometimes shrubby; crown rounded
Site: sun, part shade; pH and soil adaptable; zone 4
Use: specimen, grouping, screen, windbreak

Recognition

Leaves: elliptic to much longer than wide, up to 5 in (13 cm) long; tip pointed, base rounded; margin teeth with glandular tip; dark green, shiny above, pale, glabrous beneath; petioles up to 0.4 in (1 cm) long, glandular at leaf base; stipules persistent, oblong to kidney-shaped (reniform) with gland-tipped teeth
Flowers: catkins on leafy stalks; male—stamens 5-12; female—ovary glabrous, style short, stigma 2-lobed
Twigs: lustrous, brownish green, sticky when young
Buds: yellow-green, curved inward
Bark: gray, fissured

Salix petiolaris Sm. - meadow willow

Meadow willow grows along streambanks and in wet meadows in central and western Nebraska, in the Black Hills and in east central Colorado. It is also native to a broad belt that stretches from New Brunswick to Saskatchewan, Canada, and south to Tennesee and Wisconsin. Meadow willow is often dwarfed to dense, stubby shrubs by browsing deer and other animals (Stephens 1973).

Planning

Height: up to 7 ft (2 m)
Habit: shrub; rounded
Site: sun; soil and pH adaptable with plentiful water; zone 2
Use: naturalizing, wildlife habitat

Recognition

Leaves: much longer than wide (lanceolate), up to 3 in (7.5 cm) long; tip long-pointed, base wedge-shaped to slightly rounded; margin finely toothed, teeth pointing forward; surface silky when young, becoming glabrous above and nearly so below; petioles up to 0.4 in (1 cm) long, glabrous, broadly grooved above; stipules narrow, hairy, early deciduous
Flowers: catkins appearing with leaves; male—stamens 2, anthers yellow, female—ovary silky hairy
Fruits: capsules conical, short, silky, appressed hairy
Twigs: red to yellow brown, glabrous but sometimes hairy; lenticels not noticeable
Buds: red-brown, glabrous; cap-like scale with fine longitudinal streaks (striate), base often light-colored
Bark: gray-green or red-brown, often becoming purplish; older trunks brown, smooth or scaly

Salix rigida Muhl. **var. watsonii** (Bebb) Cronq. - **yellow willow**

Yellow willow is a variable species found naturally throughout most of the U.S. and southern Canada. Descriptions vary greatly; consequently, the list of synonyms is long. The name and description used here is favored by Stephens (1973) and Elias (1980). They also list *S. lutea* Nutt. as a synonym. Rehder (1940) does not mention *S. lutea*, and states that *S. rigida* is a synonym for the preferred *S. cordata* Muhl. *S. lutea* is described by the Great Plains Flora Association (1986), but the Association states that this plant is very similar to *S. eriocephala* Michx. and would perhaps be "better treated as a variety of it."

Rich Straight collected a branch that bore a "pine-cone" gall from a small tree on the Loup River near St. Paul, Nebraska, which closely resembles *S. lutea* (Froiland 1962). The willow cone gall midge that causes this gall infests only *lutea*, according to Froiland. Stephens (1973) reports that "pine-cone" galls are often present on *S. rigida*.

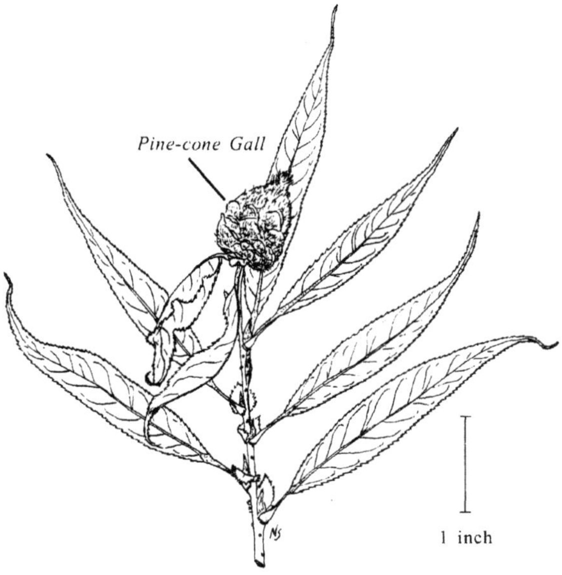

Salix rigida

Diamond-shaped dimples are common on willow stems bearing "pine-cone" galls throughout east-central Nebraska, especially along the Loup River (Straight 1993). Straight reports that the heartwood of these willows is brown, contrary to the statement by Stephens that there is no distinction between the sapwood and heartwood of yellow willow.

Planning

Height: up to 20 ft (6 m)
Habit: shrub, tree
Site: sun; soil and pH adaptable, prefers moist soil; zone 2
Use: streambank erosion control, naturalizing

Recognition

Leaves: elliptic to much longer than wide (lanceolate), up to 2.5 in (6 cm) long; tip short-pointed, base rounded to almost heart-shaped; margin finely toothed (crenate-serrate), teeth tipped with glands; surfaces hairy at first, becoming glabrous; petioles up to 0.3 in (8 mm) long, usually red, flattened above, glabrous; stipules kidney-shaped (reniform), glandular toothed
Flowers: capsules appearing before or simultaneously with leaves
Fruits: egg-shaped, abruptly narrowed to long beak, glabrous; 2 valves splitting and curling at maturity
Twigs: red-brown to greenish red, glabrous, sometimes with gray waxy coating (pruinose)
Buds: appressed with tip turned outward; cap-like scale greenish or purplish red
Bark: gray-brown, tight, finely fissured; lenticels horizontal, orange, prominent

Order Capparales

Capparales is comprised of five families and about 4,000 species (Cronquist 1981); the largest is Brassicaceae with 3,000 species.

Recognition

Habit: trees, shrubs, vines, herbs; deciduous or evergreen
Leaves: alternate, rarely opposite, simple or compound, with or without stipules
Flowers: terminal clusters (racemes, panicles); bisexual, rarely unisexual; regular or irregular; sepals distinct, commonly 4, petals distinct, commonly 4; stamens 4 to many
Fruits: variable

Brassicaceae Burnett **- mustard or crucifer family**

This family of about 350 genera is composed mostly of herbaceous plants. A few are considered to be subshrubs or shrubs.

Recognition

Habit: herbs, subshrubs, shrubs; deciduous or evergreen
Leaves: alternate, rarely opposite; simple, sometimes dissected
Flowers: bisexual, regular, perianth 4-parted; stamens commonly 6, the outer 2 shorter than the inner 4
Fruits: dry, usually splitting (dehiscent)

Iberis L. **- candytuft**

About 30 species of candytuft are native to southern Europe, western Asia and northern Africa.

Iberis sempervirens L. **- candytuft**

Candytuft provides an evergreen groundcover adapted to the southern prairies. In Lincoln it becomes a bronze color during average winters and winter injury is minimal (J. Morgenson, 1992).

Planning

Height: up to 1 ft (30 cm)
Habit: subshrub; sprawling, evergreen, early spring bloom
Site: sun, part shade; pH and soil adaptable; zone 5
Use: groundcover

Recognition

Leaves: alternate, linear-oblong, up to 2 in (5 cm) long; tip rounded, margin entire; dark green
Flowers: terminal clusters (racemes) about 1.3 in (3 cm) across; white
Fruits: siliques, about 0.25 in (6 mm) long

1 inch

Iberis sempervirens

Order Ericales

Eight families and about 4,000 species are included in order Ericales (Cronquist 1981).

Recognition

Habit: shrubs, trees, few herbs; evergreen or deciduous
Leaves: alternate, sometimes opposite or whorled, simple; stipules wanting
Flowers: clusters (racemes); bisexual, sometimes unisexual; regular or slightly irregular; perianth 5-parted, sometimes 3-7, petals distinct or connate to form a sympetalous corolla, rarely wanting; stamens as many as or twice as many as petals, rarely 2 or many
Fruits: various

Clethraceae Klotzch - **pepperbush family**

The pepperbush family, sometimes dubbed the white-alder family, is comprised of one genus, *Clethra* (Cronquist 1981).

Clethra L. - pepperbush

Species of this native genus of the Americas, Asia and the Madeira Islands of the Pacific are often planted as ornamentals because of their fragrant, showy flowers. Two pepperbush species grow in the eastern U.S.

Recognition

Habit: shrubs, trees; evergreen or deciduous
Leaves: alternate; simple, usually toothed (serrate); without stipules
Flowers: terminal clusters (racemes or panicles); bisexual, regular; sepals 5, fused to form tube below, overlapping, persistent in fruit; petals 5, distinct or nearly so, overlapping; stamens 10, usually free and distinct; ovary superior, 3-celled; style cylindric, slender, 3-parted at apex
Fruits: dry, splitting, dehiscent, 3-valved (capsules); seeds many
Buds: hairy

Clethra acuminata Michx. - cinnamon clethra

Cinnamon clethra, native to the southern Appalachian Mountains, grows in fertile, moist, deciduous woodlands (Elias 1980). It may have limited application in the prairies, however, because of its acidic soil requirement and questionable ability to withstand cold.

Planning

Height: up to 20 ft (6 m); National champion, 1981, 27 ft (8.2 m), South Carolina
Habit: shrub, tree; deciduous; erect, often suckering; early summer fragrant bloom
Site: sun, part shade; acidic, moist soil; zone 6
Use: naturalizing

Recognition

Leaves: egg-shaped (ovate), elliptic or longer, up to 8 in (20 cm) long; long-pointed; finely toothed except near base; glabrous above, slightly hairy beneath
Flowers: elongated, 1-sided clusters (racemes), white; sepals pointed (acute)
Fruits: capsules narrowly egg-shaped (ovate), about 0.2 in (5 mm) long, nodding

Clethraceae *Clethra*

Clethra alnifolia L. - summersweet clethra

Summersweet clethra, native to the Atlantic Coastal Plain from Maine to Florida, bears fragrant summer flowers. Cultivars to consider are 'Hummingbird' and 'Paniculata'. Two other cultivars, 'Pink Spires' and 'Rosea', both display flowers in shades of pink. Variety *tomentosa*, woolly summersweet, bears matted hairs on its leaves and twigs.

Planning

Height: up to 8 ft (2.5 m)
Habit: shrub; deciduous; erect, round-topped, often suckering; summer blooms on new growth
Site: sun, shade; moist, acidic soil; tolerates salt spray; zone 4
Use: border, specimen, bee pasture

Recognition

Leaves: egg-shaped, wider above middle (obovate); up to 4 in (10 cm) long; tip pointed, base tapered; sharply toothed except near base; glabrous, or nearly so; petioles up to 0.7 in (18 mm) long, hairy
Flowers: compound clusters (panicled racemes), white; sepals obtuse
Fruits: capsules nearly round, 0.1 in (3 mm) across
Twigs: brown, hairy; pith light brown
Buds: sessile, egg-shaped; scales loose

Clethra barbinervis Sieb. & Zucc. - Japanese clethra

Japenese clethra has not been extensively planted in the U.S., though it may possibly be adapted to the southeastern portion of the central prairie. Dirr (1940) describes Japanese clethra as a beautiful, large shrub or small tree with a bark that is striking in appearance.

Planning

Height: up to 20 ft (6 m)
Habit: shrub, tree; deciduous; branches spreading; fragrant summer bloom; bark striking
Site: sun, part shade; moist, fertile soil; zone 6
Use: specimen

Recognition

Leaves: egg-shaped to wider above the middle (obovate), up to 5 in (13 cm) long; tip pointed, base wedge-shaped; sharply toothed; shiny, dark green and glabrous above, hairy on veins beneath

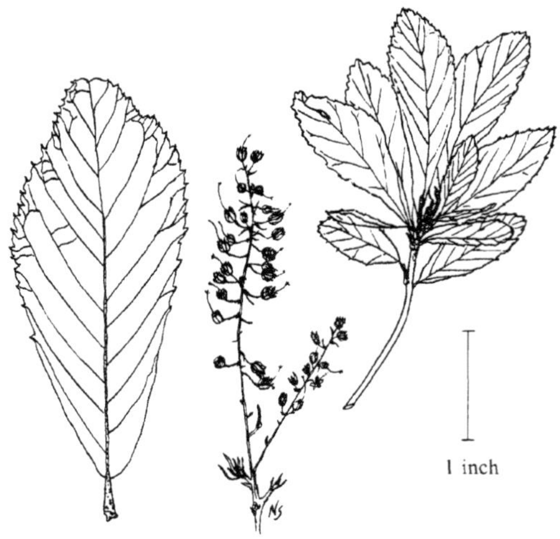

Clethra barbinervis

Flowers: compound clusters (panicles, racemes); flowers white, 0.3 in (8 mm) across; sepals broad egg-shaped (ovate) with a minute abrupt tip (mucronulate); petals slightly shorter than stamens
Fruits: capsules subglobose, about 0.2 in (5 mm) across
Twigs: nearly glabrous
Bark: smooth, polished, gray to rich brown, sometimes exfoliating

Ericaceae Juss. - **heath family**

The heath family, numbering about 125 genera and 3,500 species, grows in cool regions, especially in Asia and South Africa, as well as in the high elevations of the tropics (Cronquist 1981). Many species flourish on highly acidic soils in the coniferous forests of rainy climates.

Species of the heath family are dependent upon soil fungi for uptake of necessary minerals. A symbiotic relationship exists between these plants and soil fungi that infect the roots to form mycorrhiza ("fungus roots"). However, even with soil amendments and protection from climatic extremes, growing members of the heath family is difficult in soils that are high in lime, in climates with highly fluctuating temperatures, or in areas where soils are often dry.

Some heath family species prized as ornamentals are the azaleas, rhododendrons, and mountain-laurel. The blueberries and cranberries on the grocer's shelves are fruits of *Vaccinium* species and the huckleberries are fruits of the genus *Gaylussacia*. Wintergreen oil is distilled from the leaves and fruits of checkerberry (*Gaultheria procumbens* L.) found in the southern Appalachian Region.

Chimaphila umbellata (L.) Bart. var. *occidentalis* (Rybd.) Blake, wintergreen, is native to the moist, wooded hillsides of the Black Hills. It is also found in the western U.S., much of Canada and around the Great Lakes (Stephens 1973). Rehder (1940) places it in the shinleaf family (Pyrolaceae). Wintergreen is a low, evergreen subshrub that might serve as a groundcover under a pine canopy on northern prairie sites.

A few of the genera and species described here have been grown successfully in the prairies by acidifying the soil and providing protection from temperature extremes and low humidity.

Recognition

Habit: shrubs, trees, rarely herbs; deciduous, often evergreen
Leaves: simple; alternate, rarely opposite or whorled; entire or toothed; stipules wanting; often leathery
Flowers: solitary or clustered; terminal or axillary; bisexual; regular or nearly so; sepals 3-7, usually 5, persistent; petals 3-7, usually 5, fused, at least below, to form a corolla tube, often urn-shaped; stamens twice as many as corolla-lobes, sometimes up to 20, distinct, rarely fused
Fruits: dry and splitting (capsules), rarely berries or drupes; seeds minute, usually many

Arctostaphyllos Adans. - **manzanita**

Most of the 40-odd species of *Arctostaphyllos* are known as manzanitas, Spanish for "little apple," because the mature fruits of the plants resemble diminutive apples (U.S. Forest Service 1937). With one exception (mentioned below) these plants are natives of western North America and are adapted to coarse, stony soils that are low in fertility.

Recognition

Habit: shrubs, sometimes prostrate, rarely small trees; evergreen
Leaves: alternate; entire, rarely toothed; petioled or sessile
Flowers: clusters, terminal, nodding; corolla urn- or bell-shaped, 4 to 5-lobed; calyx 4 to 5-parted, persistent; stamens 8-10, included within the corolla
Fruits: drupes, smooth or warty; pulp mealy, granular or fleshy; or dry with 4 to 10 nutlets

Arctostaphyllos patula Greene - **greenleaf manzanita**

Greenleaf manzanita is a broadleaved evergreen native of the western U.S. mountains. It forms dense and extensive brush fields on old burns of coniferous forests (U.S. Forest Service 1937). Greenleaf manzanita thrives on sandy, acidic soils, especially if also protected from winter cold by snow.

Ericaceae *Arctostaphylos*

Planning

Height: up to 7 ft (2 m)
Habit: shrub; upright, spreading by root shoots; spring bloom, fruits persist into winter
Site: sun; drouth tolerant; low nutrient, well-drained soil; pH 6-7; zone 5
Use: soil stabilization, border, foundation

Recognition

Leaves: elliptic to orbicular, up to 2 in (5 cm) long; leathery; tip variable, base rounded; surfaces bright green, glabrous
Flowers: clusters dense, up to 2 in (5 cm) long, stalks minutely glandular; corolla usually pink, sometimes white, about 0.3 in (7 mm) long; calyx 5-lobed, pinkish with white papery tips, persistent on fruits; stamens 10, anthers with pair of bent-back bristles on back
Fruits: globose, about 0.4 in (9 mm) across; glossy, dark brown to black; persistent
Twigs: minutely hairy, glandular
Stems: rigid, very crooked
Bark: smooth, dark red

Arctostaphyllos uva-ursi (L.) Spreng. - bearberry

Bearberry, native to the subarctic regions of all continents, also grows as far south as northern California in the Sierras and Mexico in the Rocky Mountain chain. It is native as far south as the Black Hills, Illinois and Virginia. Bearberry is usually associated with coniferous forests.

It is also commonly known as kinnikinick, a word for a tobacco substitute used by Native Americans. Bearberry leaves have been used as a source of tannin for curing leather and its mature fruits are relished by wildlife.

Bearberry is a pest-free plant on sites that suit its requirements. Dirr (1990) calls it an "outstanding ground cover for that different effect" and the "emerald carpet" of groundcovers. It is worthy of trial on sandy soils of the northern prairies. Experimental plantings should include plants from a wide range of origins.

Sutton (1998) has observed the closely related pine-mat manzanita (*A. nevadensis* var. *coloradensis* Gray) thriving at the Denver Botanical Garden.

Planning

Height: up to 1 ft (30 cm)
Habit: shrub; spreading to diameter of 15 ft (4.5 m); rooting at nodes; summer foliage glossy green, bronze to reddish in autumn
Site: light shade; drouth tolerant; prefers infertile, sandy soil; pH 4.5-6.0; zone 2
Use: groundcover, erosion control, wildlife habitat

Recognition

Leaves: egg-shaped, wider above middle; width variable, up to 1 in (25 mm) long; leathery, tip rounded, base long-tapered; margin entire, sometimes fringed (ciliate); surfaces glabrous, dark green and glossy above; petioles about 0.1 in (3 mm) long
Flowers: clusters (racemes) of 3 to 6; corolla white to pinkish, about 0.2 in (6 mm) long, 5-lobed, glabrous outside, hairy inside; calyx 5-lobed; stamens 10
Fruits: globose about 0.7 in (8 mm) across; pink to bright red, shiny, glabrous; dry, mealy interior; seeds usually 5, hard, often remain united
Twigs: glabrous to minutely hairy
Buds: egg-shaped, slightly hairy, light red, about 0.05 in (1 mm) long; flower buds globular in axils of bracts at stem tip
Bark: exfoliating, shreddy

Calluna Salisb. - Scotch heather

Scotch heather, the single species of *Calluna*, is native to Europe and Asia Minor and naturalized in northeastern North America.

Calluna vulgaris (L.) Hull - Scotch heather

Over 600 cultivars of Scotch heather are recognized. Although its soil requirements are fastidious and the plant must be protected from excessive wind (especially in summer), some cultivars can be grown successfully in the northeastern prairie.

Dirr (1990) recommends the following based upon his observations in southern Illinois: variety *alba*, bearing white flowers; 'County Wicklow', bearing double pink flowers and dark green foliage; 'H. E. Beale', a silvery double pink form that flowers from August to October; and 'Robert Chapman', bearing greenish yellow summer foliage that changes to reddish in winter and rose-purple flowers in August and September.

Erica, a genus of 200 species, is closely related to *Calluna*. *Erica* species are "not as cold hardy and do not appear to be as happy in American gardens as *Calluna*," according to Dirr (1990). Leaves of these species are arranged in whorls, and the calyx is shorter than the corolla and not colored. *Erica* species bloom in winter or spring.

Planning

Height: up to 2 ft (60 cm)
Habit: shrub; upright; evergreen; summer bloom, winter foliage green to bronze
Site: sun, part shade; infertile, sandy, well-drained soil; pH 4.5-6.0; zone 5
Use: groundcover, edging, rock garden

Recognition

Leaves: opposite, simple, 4-ranked, scale-like, keeled, about 0.1 in (2 mm) long; minutely hairy; without petioles (sessile)
Flowers: clusters terminal (racemes), rosy pink; corolla urn-shaped, 4-parted; calyx 4-parted, colored, longer than the corolla; stamens 8, shorter than corolla
Fruits: dry, 4-valved (capsules); included in persistent perianth
Twigs: circular (terete) but appearing squarish
Buds: solitary, minute; scales 3

Enkianthus Lour. - enkianthus

Only one of the 10 species of *Enkianthus*, natives of eastern Asia and the Himalayas, is likely to be successfully grown in our prairies.

Enkianthus campanulatus (Miq.) Nichols. - redvein enkianthus

Redvein enkianthus is noted for its brilliant autumn foliage. This pest-resistant plant should grow successfully in the south of the central prairie if soil and protection requirements are met. Selections include 'Albiflorus' (white flowered) and variety *palibinii* (red flowered).

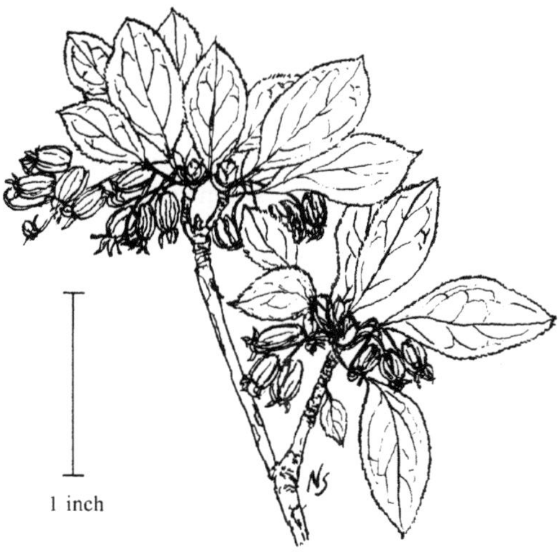

Enkianthus campanulatus

Planning

Height: up to 10 ft (3 m)
Habit: shrub, tree; deciduous; upright with layered branches, foliage tufted; spring flowers, autumn foliage brilliant
Site: sun, part shade; moist soil; pH 4.5-6.5; zone 5
Use: specimen, border
Caution: protect from excessive wind

Recognition

Leaves: alternate, mostly crowded at branch ends; elliptic, up to 3 in (8 cm) long; tip pointed, teeth on margin bearing stiff awns; scattered bristly hairs above and on veins beneath; petioles about 0.5 in (12 mm) long
Flowers: clusters terminal, pendulous (umbel-like racemes); yellowish, veined with red; perianth 5-parted, sepals much longer than wide; corolla bell-shaped with 5 rounded lobes, deciduous; stamens 10, included within corolla, anthers with upright awns
Fruits: dry (capsules), 5-valved, egg-shaped; about 0.2 in (5 mm) across
Twigs: slender, glabrous, often reddish

Kalmia L. - kalmia

Kalmia, a North American genus of eight species, grows from Alaska south to the northwestern U.S., across Canada to Newfoundland, and south along the East Coast. These shrubs, usually evergreen, occasionally reach tree size. Some plant parts are apparently poisonous to people and livestock if ingested in large quantities.

K. angustifolia, sheep laurel, is a low-growing evergreen native to Newfoundland and Hudson Bay and south into Michigan and Georgia. It might be grown in the northern prairies if soil and protection requirements are met. *K. microphylla* (Hook.) Heller is a dwarf subalpine shrub of western North America that is found growing in wet meadows and along pond and stream borders from Alaska south to Colorado and California (U.S. Forest Service 1937).

Kalmia latifolia L. - mountain-laurel

Mountain-laurel, native to the Appalachians, is "one of the most beautiful American flowering shrubs" (Rehder 1940). People have attempted and will undoubtedly continue to attempt to meet the rather exacting site requirements for growing this ornamental plant.

Mountain-laurel leaves contain andromedotoxin, a substance very poisonous to livestock, particularly sheep (Harlow and Harrar 1968). **The Laurel Book** (Jaynes 1975) is an excellent treatise about mountain-laurel and its many cultivars.

Planning

Height: up to 15 ft (4.5 m); National champion, 1991, 25 ft (7.6 m), North Carolina
Habit: shrub, tree; evergreen; symmetrical, becoming open and straggly with age; spring bloom; fruits persist through winter
Site: part shade; cool, moist soil; pH 4.5-6.0; zone 5
Use: grouping, border, naturalizing, specimen
Caution: protect from wind, winter sun

Recognition

Leaves: alternate or irregularly whorled; elliptic, up to 4 in (10 cm) long; leathery; margin entire, tip pointed, base tapered; surfaces glabrous, dark green above; petioles about 0.7 in (2 cm) long

Flowers: clusters terminal (corymbs); corolla saucer-shaped, white to deep rose with purple markings, 0.9 in (22 mm) across; stamens 10, anthers held in pouches, springing out when corolla is touched

Fruits: dry (capsules), 5-valved; about 0.2 in (6 mm) across; persist through winter

Twigs: glabrous, red-green to red-brown; leaf scars often whorled and clustered near twig end, embedded, half-round to shield-shaped

Buds: leaf smaller than flower; egg-shaped (ovate), outer scales usually 2, green; flower buds clustered at twig ends, about 1 in (2.5 cm) long, scales glandular, hairy

Bark: thin, red-brown, becoming divided into long, narrow scales

Leucothoe D. Don - **leucothoe**

Leucothoe, a genus of about 35 species, is native to the American continents and parts of Asia and Africa.

Leucothoe fontanesiana (Teud.) Sleum. [*L. catesbaei* (Walt.) Gray] - fetterbush

Fetterbush, a native of the mountains of Virginia, Georgia and Tennessee, is a plant for the avid gardener looking for a shade-loving evergreen shrub. However, Dirr (1990) cautions that it should only be placed in ideal sites because stress predisposes the plant to a leaf spot that can be fatal.

'Girard's Rainbow' and 'Scarletta' display showy new growth, and 'Lovita' and 'Nana' are compact shrubs. A fetterbush plant of unknown pedigree grows in Lincoln (Sutton 1998). Dirr (1990) recommends pruning to the ground after flowering.

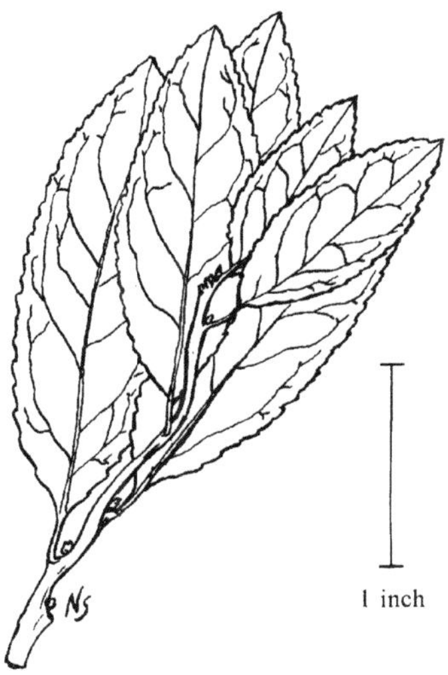

Leucothoe fontanesiana

Planning

Height: up to 5 ft (1.5 m)

Habit: shrub; evergreen; spreading, arching branches; spring bloom; emerging foliage colorful, becoming bronze to purplish in winter

Site: part to full shade; fertile, cool, moist soil, pH 4.5-6.0; zone 5

Use: grouping, border, underplanting

Pests: leaf spot caused by several fungi serious for stressed plants

Caution: protect from excessive wind and sun

Recognition

Leaves: alternate, much longer than wide (ovate-lanceolate), up to 5 in (13 cm) long; tip long-pointed, base rounded or wide tapered; margin fine-toothed with hairy fringe (appressed ciliate-serrulate); lustrous dark green above, glabrous or with scattered hairs beneath; petioles about 0.5 in (12 mm) long

Flowers: clusters (racemes) axillary, long, hanging; corolla slender pitcher-shaped (cylindric-ovoid), about 0.2 in (6 mm) long, deciduous, white, usually reddish in bud; sepals oblong, scarcely overlapping

Fruits: dry, splitting (capsules)

Twigs: long, slender, reddish, smooth, shiny

Buds: round, several outer scales

Oxydendrum DC. - sourwood

Oxydendrum, a genus of one species, is native to the southeastern U.S. south of Pennsylvania and mostly east of the Mississippi River.

Oxydendrum arboreum (L.) DC. - sourwood

Sourwood is an all-season tree displaying dark green foliage that, when found in its native habitat, becomes a mixture of yellows, reds and purples in autumn. This tree attracts bees by the thousands with its many tiny white flowers that resemble those of lily of the valley. Sourwood is seldom planted in the prairies because it requires acidic soils, but it is nevertheless a tree of season-long beauty worth a little extra effort to grow. A specimen on the UNL East campus is not thriving (Sutton 1998).

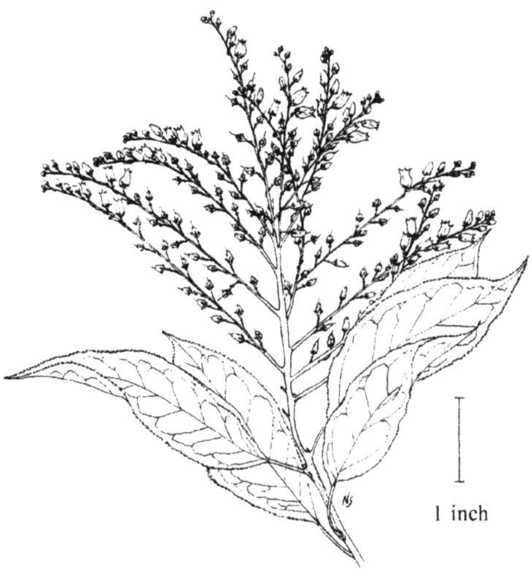

Oxydendrum arboreum

Planning

Height: up to 30 ft (9 m); National champion, 1975, 118 ft (35.8 m), North Carolina
Habit: tree; deciduous; crown narrow oblong, summer bloom, autumn foliage colorful; fruits persist into winter
Site: sun, part shade; moist, well-drained soil; pH 5.0-6.5; zone 5
Use: specimen, wood for paneling, specialty items

Recognition

Leaves: elliptic-oblong, up to 8 in (20 cm) long; tip long-pointed, base widely tapered; finely toothed (serrulate); lustrous, dark green and glabrous above, sparsely hairy on veins beneath; petioles about 0.7 in (2 cm) long
Flowers: clusters terminal, drooping (wide panicles of 1-sided racemes), up to 10 in (25 cm) long; corolla deciduous, urn-shaped, about 0.3 in (7 mm) long, lobes 5, short, white; sepals 5; stamens 10, anthers not awned
Fruits: dry, splitting (capsules), 5-valved; yellowish becoming brown
Twigs: glabrous or sparingly hairy, olive or bright red, zigzag, lenticels conspicuous; leaf scars half-round, bundle scars c-shaped
Buds: terminal absent; lateral globose, partly embedded in bark; scales several, overlapping
Bark: grayish, furrowed and ridged
Wood: hard, attractive grain

Pieris D. Don - pieris

About eight species of *Pieris* are native to North America and eastern Asia.

Recognition

Habit: shrubs, trees; evergreen
Leaves: alternate, toothed, rarely entire, petioled
Flowers: clusters (panicles) terminal; corolla urn-shaped, lobes 5, short, deciduous; calyx lobes separate; stamens 10, included within corolla, anthers with 2 reflexed awns on back
Fruits: dry and splitting (capsules); seeds small, similar to sawdust

Pieris floribunda (Pursh) Benth. & Hook. - **mountain pieris**

Mountain pieris, native to the Appalachians of Virginia and Georgia, is slow growing but tolerates soils of higher pH than most species of *Ericaceae* (Dirr 1990). 'Brower's Beauty' is a hybrid resulting from crossing Japanese pieris and mountain pieris. 'Millstream', a slow-growing cultivar, is suitable for rock gardens (Dirr 1990).

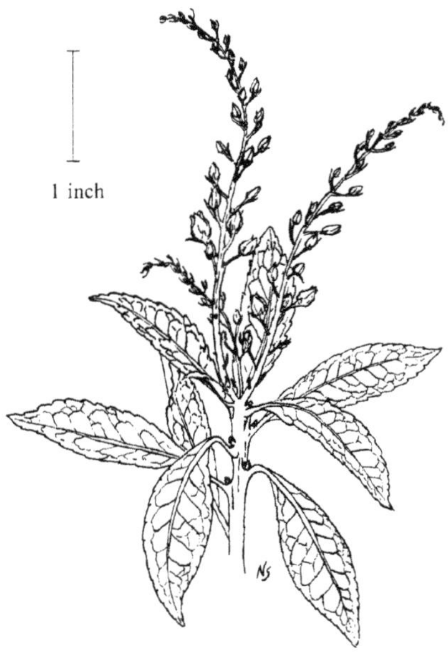

Pieris japonica

Planning

Height: up to 7 ft (2 m)
Habit: shrub; low, rounded; April bloom
Site: sun, part shade; moist, well-drained soil; pH 5.0-6.5; zone 5
Use: border, grouping, rock garden
Caution: protect from wind, winter sun

Recognition

Leaves: elliptic to oblong, up to 3 in (8 cm) long; tip pointed, base obtuse; minutely toothed with hairs along margin (ciliate); surfaces glabrous
Flowers: clusters (panicles) dense, upright, up to 4 in (10 cm) long; corolla angled, up to 0.2 in (6 mm) long, white; calyx lobes egg-shaped (ovate)
Fruits: capsules globose to egg-shaped, about 0.2 in (5 mm) long
Twigs: covered with straight, stiff hairs (strigose)

Pieris japonica (Thunb.) D. Don - **Japanese pieris**

Japanese pieris, a beautiful, broadleaf evergreen, requires shelter from wind and an acidic soil having plenty of organic matter. Japanese pieris plants on the UNL campus suffered cold injury to their flower buds, grew very slowly, and succumbed after 6 years (Sutton 1998).

Dirr (1990) points out that large numbers of Japanese pieris cultivars are flooding the market, but "how far they will travel in the nursery trade is questionable." Japanese pieris cultivars range in form from compact to tall and slender, with flowers and foliage of many sizes and colors.

Planning

Height: up to 10 ft (3 m)
Habit: shrub; upright with spreading branches to dense and compact; early spring flowers
Site: sun, part shade; moist, well-drained soil; pH 5.0-6.0; zone 5
Use: specimen, border, grouping
Caution: shelter from wind

Recognition

Leaves: oblong, wider above middle, up to 3.5 in (9 cm) long; tip pointed, base tapered, minutely toothed; surfaces glabrous
Flowers: clusters (panicles) drooping, up to 4.7 in (12 cm) long; corolla egg-shaped, up to 0.3 in (8 mm) long; calyx lobes oblong
Fruits: capsules depressed globose, about about 0.2 in (6 mm) across
Twigs: glabrous

Rhododendron L. - rhododendron

About 800 rhododendron species—flowering shrubs and small trees—are native to the temperate and colder regions of North America, Europe and Asia, but can be found in areas that extend south as far as New Guinea and Australia. Over 40 species have been found in the Himalayas, including *R. aboreum*, which has been crossed with European and North American species to yield brightly colored cultivars. An untold number of rhododendron cultivars are available, many of which are hardy enough to withstand the rigors of much of the prairie region.

We do not have the space nor expertise to describe the hundreds of rhododendron hybrids that have been developed. Dirr (1990) describes many of the hardier and more common species and hybrids. Bowers (1960), Leach (1961), and others have published books about these extraordinary plants. Membership in the **American Rhododendron Society** provides an excellent way to become acquainted with the plants and successful growers.

Although the genus *Rhododendron* has been subdivided into several subgenera and sections, two major groups are commonly recognized: azaleas and true rhododendrons. We limit our descriptions here to a few hardy species in each—a sampling to whet your appetite.

Recognition

Habit: shrubs, rarely trees; evergreen or deciduous
Leaves: alternate, entire; rarely minutely toothed (serrulate) with marginal hairs (ciliate); petioled
Flowers: clusters usually terminal (umbel-like racemes), sometimes solitary or few, rarely from lateral buds; corolla variable from wheel-shaped (rotate) to tubular, slightly irregular; calyx 5-parted, often very small; stamens 5-10, rarely more
Fruits: dry and splitting (capsules); seeds minute, numerous
Buds: scales several to many, overlapping
Wood: dense, hard, fine-grained

Grouping Key for *Rhododendron*

Shrubs; leaves mostly deciduous, never scaly, frequently hairy; flowers largely funnelform, stamens mostly 5; twigs usually hairy: **azaleas**

Shrubs, trees; leaves mostly evergreen, often scaly or with small dots on undersurface; flowers often bell-shaped, stamens 10 or more: **true rhododendrons**

Azaleas

'Northern Lights', a deciduous hybrid bearing dainty pink flowers that was bred and released at the University of Minnesota, withstands cold temperatures (Flemmer 1985). Dirr (1990) says that the book, **Azaleas** (Galle 1987), is a classic tome on these plants.

Rhododendron arborescens (Pursh.) Torr. - sweet azalea

Sweet azalea, native to the Appalachians of southern Pennsylvania and south to Georgia and Alabama, is one of the hardier native species (Dirr 1990).

Planning

Height: up to 20 ft (6 M)
Habit: shrub; deciduous; erect, loosely branched; spring bloom fragrant, autumn foliage colorful
Site: sun, part shade; moist, fertile soil; pH 5.0-6.5; zone 5
Use: border, mass, naturalizing

Recognition

Leaves: elliptic to wider above the middle, up to 3 in (8 cm) long; tip pointed to obtuse, margin hairy (ciliate); bright green and glabrous above, sometimes hairy on midrib below
Flowers: clusters of 3 to 6, terminal, appearing after leaves; corolla funnel-shaped (funnelform), tube up to 1.2 in (3 cm) long, longer than lobes, hairy inside, lobes egg-shaped (ovate), pointed, white or pinkish; sepals egg-shaped or narrow-oblong, margins hairy; stamens 5, twice as long as tube
Fruits: capsules oblong to egg-shaped, about 0.6 in (15 mm) long
Twigs: glabrous, rarely with few scattered hairs, often slightly bloomy

Rhododendron schlippenbachii Maxim. - royal azalea

Royal azalea, a native of Korea and Manchuria, is "one of the most delicate and beautiful of the azaleas for northern gardens" (Dirr 1990). It tolerates soils of higher pH than other azaleas.

Planning

Height: up to 10 ft (3 m)
Habit: shrub; deciduous; upright, rounded; spring bloom, autumn foliage colorful
Site: sun, part shade; moist, well-drained soil; pH 6.5-7.0; zone 5
Use: border, naturalizing, grouping

Recognition

Leaves: usually 5 at branch ends, appearing as if whorled; egg-shaped but wider above middle (obovate), up to 5.5 in (14 cm) long; tip usually blunt, base tapered; sparingly hairy, becoming glabrous except on veins beneath
Flowers: clusters of 3-6 with the leaves, terminal; corolla somewhat wheel-shaped (rotate-funnelform), pale to rosy pink; sepals egg-shaped; stamens 10, unequal, longest as long as petals
Fruits: capsules oblong to egg-shaped, about 0.6 in (15 mm) long, glandular
Twigs: glandular hairy, becoming glabrous second year

True Rhododendrons

Cultivars adapted to the southern portion of the central prairie, based on trials at the Wichita Horticultural Research Center of Kansas State University, include hybrids 'Boudoir', 'Herbert', 'Snowball' and 'Purple Splendor' (Pair 1988).

Rhododendron carolinianum Rehd. - Carolina rhododendron

Carolina rhododendron, native to the Blue Ridge Mountains of the Carolinas and Tennessee, is one of the hardiest of the species (Spangler and Ripperda 1977). It "flowered nicely" in an Illinois garden (Dirr 1990). The 'P.J.M.' hybrid is derived from crossing *R. carolinianum* with *R. dauricum* var. *sempervirens* Sims. These evergreen shrubs bear lavender pink flowers in early spring. 'P.J.M.' is the hardiest broadleaf rhododendron at the Minnesota Landscape Arboretum (Dirr 1990).

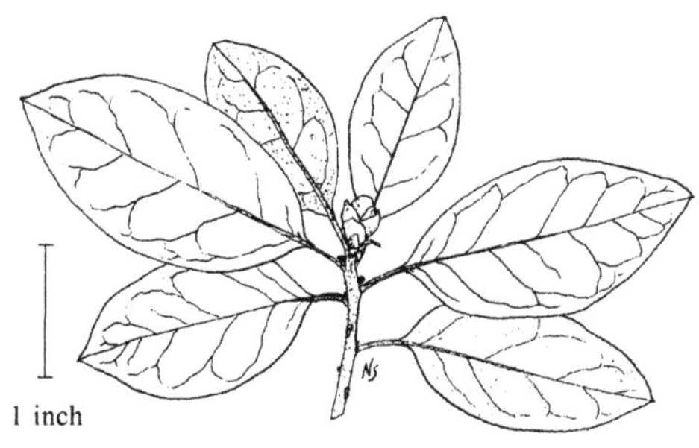

***Rhododendron* 'P.J.M.'**

Ericaceae *Rhododendron*

Planning

Height: up to 7 ft (2 m)
Habit: shrub; evergreen; rounded; spring bloom, winter foliage purplish
Site: sun, part shade; moist, well-drained soil; pH 4.5-6; zone 5
Use: foundation, border, grouping
Caution: protect from wind, winter sun

Recognition

Leaves: elliptic, up to 3 in (8 cm) long; tip pointed, base tapered; glabrous above, usually rust-colored, scaly beneath, aromatic when bruised
Flowers: clusters of 4 to 10 (umbel-like racemes) terminal; corolla broad funnel-shaped, tube about as long as the 5 lobes or shorter, about 1.5 in (4 cm) across, white to pale rose to lilac-rose; stamens 10, shorter than corolla, long-hairy at base; style glabrous, shorter than stamens
Fruits: capsule 5-valved, splitting (dehiscent)
Twigs: glabrous, reddish

Rhododendron catawbiense Michx. - catawba rhododendron

Catawba rhododendron, native to the Piedmont Plateau and the nearby Appalachian Mountains from Virginia to Georgia and Alabama, grows on ridges and bluffs and along streams associated with white pine, Fraser fir and various hardwoods. It occasionally forms impenetrable thickets (Elias 1980). Usually shrubby, it sometimes reaches the stature of a small tree in its native habitat.

Described by Dirr (1990) as a very handsome broadleaf evergreen, catawba rhododendron is also one of our hardier native species. It is a parent of many hybrid, cold-hardy cultivars including 'Purple Splendor' which is adapted to southern Kansas (Pair 1988) .

Planning

Height: up to 15 ft (4.5 m); National champion, 1991, 26 ft (7.9 m), North Carolina
Habit: shrub, tree; evergreen; upright, leggy; spring bloom; fruits persist into winter
Site: sun, part shade; moist, well-drained soil; pH 4.5-6.0; zone 5
Use: grouping, naturalizing

Recognition

Leaves: elliptic to oblong, up to 6 in (15 cm) long; tip obtuse, coming to a short, sharp point (mucronulate), base rounded; glabrous, leathery; lustrous above, whitish beneath
Flowers: clusters many-flowered, terminal; corolla bell-shaped, about 2.4 in (6 cm) across, 1.5 in (4 cm) long, lilac purple to purplish rose with olive-green spots, lobes broad, rounded; sepals triangular; stamens 10, filaments hairy at base; style red, ovary rusty-hairy
Fruits: capsules egg-shaped (ovate), 0.5 in (12 mm) long, rusty-hairy
Twigs: stout, circular, glabrous, yellowish green becoming red-brown; leaf scars somewhat heart-shaped to linear, bundle scars 1 or broken into several in u-shape
Buds: leaf—usually lateral, conical, many-scaled, solitary, dark green, scales deciduous; flower—terminal, conical, 1.5 in (4 cm) long
Bark: thin, red-brown, smooth, later peeling into thin scales

Rhododendron maximum L. - rosebay rhododendron

Rosebay rhododendron, native to Nova Scotia and Ontario south to Georgia and Alabama, grows along mountain streams. The flowers attract bees, but the honey produced from the nectar is poisonous. The hard, fine-grained wood is used for tool handles and the bowls of briar pipes (Elias 1980). "Requires cool, moist, well drained root run and is not ideally suited to 'normal' landscape situations," notes Dirr (1990).

Planning

Height: up to 30 ft (9 m); National champion, 1981, 40 ft (12.2 m), South Carolina
Habit: shrub, tree; evergreen; loose, open, thicket-forming; June bloom
Site: shade; cool, moist, well-drained soil; pH 6.5-7.0; zone 4
Use: naturalizing

Recognition

Leaves: oblong, up to 8 in (20 cm) long; tip pointed, base wedge-shaped (cuneate); dark green above, matted hairs below; petioles hairy when young
Flowers: clusters rounded, umbrella-shaped; corolla 5-lobed, rose to white, spotted olive green to orange, about 1.5 in (4 cm) across; sepals egg-shaped, as long as ovary
Twigs: hairy, becoming glabrous

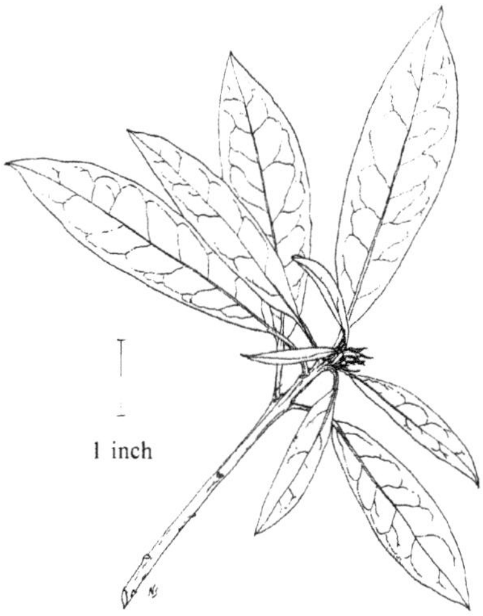

Rhododendron maximum

Rhododendron mucronulatum Turcz. - **Korean rhododendron**

Although Korean rhododendron, a native of Korea, Manchuria and Japan, is deciduous, it is more closely related to the true rhododendrons than to the azaleas. This rhododendron blooms very early in the spring, often subjecting its flowers to freeze injury. Dirr (1990) recommends planting it where the sun's rays do not strike it during February and March.

Planning

Height: up to 8 ft (2.5 m)
Habit: shrub; deciduous; upright oval; early spring bloom, ahead of leaves, autumn foliage colorful
Site: part shade; moist well-drained soil, ph 4.5-6.0; zone 5
Use: foundation, border

Recognition

Leaves: elliptic to much longer than wide; up to 3 in (8 cm) long, tip pointed to rounded; thin textured, surfaces scaly, very aromatic when crushed
Flowers: clusters of 3 to 6 from lateral buds at twig ends; corolla funnel- to bell-shaped, about 1.5 in (4 cm) across, rosy purple; sepals small, scaly; stamens 10, long-hairy (villous) below middle; style glabrous, longer than stamens
Twigs: sparingly scaly, light brown
Buds: flower—egg-shaped (ovate), glabrous, light brown

Vaccinium L. - **blueberry**

Five species of *Vaccinium* are native to the prairie fringes (Stephens 1973): *V. membranaceum* Torr., bilberry, a native of southern Canada, grows on rocky hillsides in the Black Hills; *V. scoparium* Leib., grouseberry, a native of the Rocky Mountains from Canada to Colorado, also grows in the open woods on the mountaintops of the Black Hills; *V. arboreum* Marsh., farkleberry—a tree-like shrub native to the southeastern U.S.—is found in extreme southeastern Kansas; and *V. stamineum* L., buckberry, and *V. vacillans* Torr., lowbush blueberry, both natives of the eastern U.S., are found on dry, wooded hillsides in extreme southeastern Kansas and eastern Oklahoma.

Although most of these blueberries will probably withstand extremes of temperature and are tolerant of dry sites, they require an acidic soil for survival. Dirr (1990) recommends that gardeners with calcareous soils who wish to grow *Vaccinium* species construct raised beds and mix in pine bark, peat moss or other acidic organic materials. It is possible to harvest berries in eastern Nebraska from *V. corymbosum*, highbush blueberry, and *V. angustifolium*, lowbush blueberry, if grown in partial shade with special soil treatments and protected from extreme winds (Rassmussen 1991).

Recognition

Habit: shrubs, rarely small trees; deciduous or evergreen
Leaves: alternate, entire or toothed, not resinous dotted
Flowers: solitary or clusters, axillary or terminal; corolla urn-shaped to bell-shaped, 4 to 5 lobes; calyx 4 to 5 lobes, persistent on fruit; stamens 8 or 10; ovary inferior
Fruits: berries many-seeded, crowned by persistent calyx
Twigs: red-brown, glabrous or few hairs; leaf scar with 1 bundle scar
Buds: terminal absent; lateral egg-shaped (ovate), scaly, red-brown

Order Ebenales

Cronquist (1981) lists five familes and about 1,750 species in Ebenales, most of which grow in tropical and subtropical regions.

Recognition

Habit: trees, shrubs; deciduous or evergreen
Leaves: alternate, rarely opposite; simple, often firm and leathery; margin entire, seldom toothed; stipules usually wanting
Flowers: variously clustered, often axillary or at nodes of old wood; bisexual or unisexual, regular; sepals 4 to 6, more or less distinct; petals jointed (sympetalous), 4 to 6 lobes, rarely distinct; stamens mostly 2 to 3 times as many as corolla lobes
Fruits: various, dry or fleshy, splitting or non-splitting

Sapotaceae Juss. - sapodilla family

Most Sapotaceae species—70 genera and about 800 species—are tropical and subtropical plants (Cronquist 1981). Many are important timber trees and some bear edible fruits. One species yields an inelastic rubber—gutta-percha—used for insulating sea cables. The sapodilla tree, *Manilkara zapota* (L.) Van Royen, is one of the most important trees of Central America (Elias 1980). The dark red sapodilla wood is hard, heavy and durable, its fruit is edible, and chicle—an important ingredient of chewing gum—is obtained from the plant's milky sap.

Six genera, both native and naturalized, grow in the extreme southern U.S., and one species is native in the south of the central prairie.

Recognition

Habit: trees, shrubs; deciduous or evergreen
Leaves: alternate, seldom opposite, simple, mostly entire, often leathery
Flowers: clusters or single, axillary; usually bisexual, regular; sepals distinct or nearly so, usually 5 (4-8), persistent on fruit; petals united (sympetalous), 4 to 8 overlapping lobes, usually same number as sepals; stamens 5-8, joined to base of petals (epipetalous), often alternating with sterile stamens (staminodes); stigma and style 1
Fruits: berries, fleshy or dry; seeds one to several, large, seed coat shiny, usually thick and hard

Bumelia Sw.

One of about 30 species of *Bumelia* is native in the south of the central prairie.

Bumelia lanuginosa (Michx.) Pers. - gum bumelia

Gum bumelia is a native that grows across the southern U.S., in northern Mexico, and as far north as southeastern Kansas and southern Missouri. It grows on dry, rocky sites in open woodlands and on floodplains (Stephens 1973).

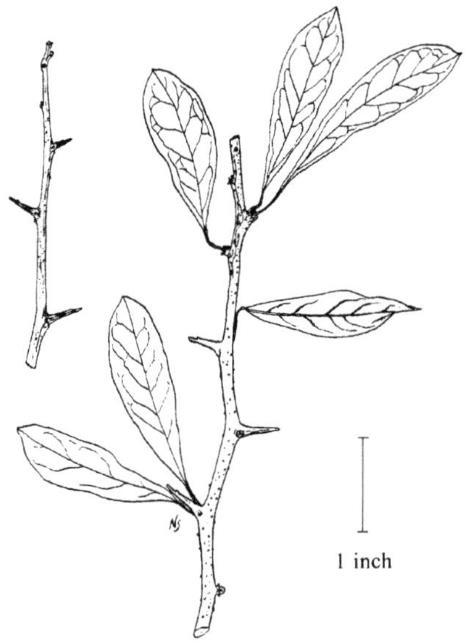

Bumelia lanuginosa

Planning

Height: up to 40 ft (12 m)

Habit: tree, shrub; deciduous; rounded, irregular crown, usually spiny; early summer flowers, leaves persist into winter

Site: sun; pH and soil adaptable; tolerates drouth; zone 6

Use: naturalizing, wildlife habitat

Recognition

Leaves: egg-shaped, wider above middle (obovate) to spoon-shaped; leathery; up to 3 in (8 cm) long; tip rounded, base tapered; upper surface dark green, glossy with hairs on midrib, densely white to rusty woolly beneath; petioles hairy

Flowers: clusters of 20 to 30, axillary, dense: bisexual; petals 5, white, each 3 lobed; calyx lobes 5, distinct, rusty hairy outside; stamens 5, alternating with 5 petal-like, sterile stamens (staminodes)

Fruits: egg-shaped (ovate), about 0.5 in (12 mm) long with persistent calyx lobes; glabrous, glossy black; seeds 1

Twigs: rigid, tough, rusty hairy at first, often with spine at leaf base

Buds: broadly egg-shaped, red-brown, densely hairy

Bark: nearly black with blocky ridges

Wood: hard, heavy

Ebenaceae Gurke - ebony family

Five genera of Ebenaceae—consisting of about 450 species—are native to the tropical and subtropical forested regions of the world (Cronquist 1981). Two species of one genus (*Diospyros*) are native to the U.S.

The ebony family, noted for the valuable wood of many of its tropical species, produces black or brownish black lumber that is sold as ebony. However, true ebony is obtained from *Diospyros ebenum* Konig. Many species are cultivated for their delicious fruits enjoyed by both humans and wildlife.

Recognition

Habit: trees, shrubs; mostly deciduous

Leaves: alternate, simple; stipules wanting

Flowers: mostly unisexual, commonly dioecious; regular, borne as small clusters in the axils; female usually single with sterile stamens; male with vestigial ovary; calyx 3- to 7-lobed, persistent, sometimes becoming a part of the fruit (accrescent); corolla 3- to 7-lobed; stamens usually twice the number of corolla lobes, often paired, filaments of pairs sometimes connate

Fruits: berries, juicy to somewhat leathery, sometimes capsules

Diospyros L. - persimmon, ebony

About 400 species of *Diospyros*, the most important genus of the ebony family, are native to eastern and southwestern Asia, the Malaya Peninsula and the Mediterranean Region. Two species are natives of the U.S. One of the two, *D. texana* Scheele, grows in southwestern Texas and northeastern Mexico.

Japanese persimmon (*D. kaki* L. f.), a native of China that has long been cultivated for its fruits in China and Japan, is now grown for that purpose in California and the southeastern U.S. A Japanese persimmon tree in the Department of Horticulture orchard at UNL dies back each winter (Gustafson 1992).

Diospyros virginiana L. - persimmon

What mischievous person has not enticed a friend to taste a smooth, orange persimmon fruit and then laughed at the pained expression as lips pucker? What a difference when the fruit becomes mushy and wrinkled—and edible for humans and wild creatures.

Native fruit fanciers have selected many persimmon cultivars bearing fruit to their liking. The wood has been used for such special purposes as shuttles for textile weaving, golf club heads and parquet flooring. Persimmon is seldom used for landscaping and protection plantings. Primarily a native tree of the southeastern U.S., persimmon grows as far north as Pennsylvania and southeastern Iowa, and west to eastern Kansas, Oklahoma and Texas.

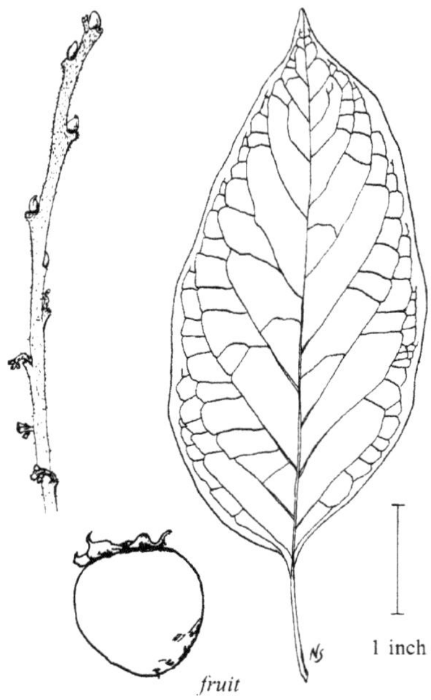

Diospyros virginiana

Planning

Height: up to 60 ft (18 m); Nebraska champion, 1987, 57 ft (17 m), Falls City; National co-champion, 1970, 131 ft (40 m), Missouri (5 other co-champions are shorter but with larger diameters and wider crown spreads)

Habit: tree; deciduous; crown oval-rounded; colony forming; autumn foliage colorful in the south, seldom so on trees in the central and northern prairies

Site: sun, part shade; pH and soil adaptable; drouth tolerant; zone 5

Use: fruit, naturalizing, wildlife habitat, wood

Recognition

Leaves: elliptical to oval, up to 5 in (13 cm) long; entire; tip pointed, base rounded; lustrous, dark green above, somewhat hairy beneath, glabrous at maturity; petioles hairy or smooth

Flowers: clusters axillary on new growth; unisexual, usually dioecious but sometimes both on same tree; perianth 4-lobed; corolla urn-shaped, yellow; calyx connate at base; male—solitary or in clusters of 2 to 3; stamens 16, vestigial pistil in center, without style; female—solitary, vestigial stamens 6 to 10, ovary broadly ovoid, styles 4, joined half their length

Fruits: berries subglobose, about 1.5 in (4 cm) across, orange to orange-purple, wrinkled when ripe; calyx and base of styles persistent

Twigs: zigzag, hairy, gray-brown; pith white, sometimes chambered; leaf scars half-round, raised (shelf-like), bundle scar 1

Buds: terminal absent; lateral egg-shaped, closely appressed; scales, 2 visible, shiny, sometimes slightly hairy, first scale as long as bud

Bark: nearly black, broken into squarish blocks

Wood: sapwood creamy white, heartwood blackish brown, very hard, strong and tough

Styracaceae Dumort. - **storax family**

The storax family—10 genera and approximately 150 species—is mostly of tropical origin, but species of *Halesia* and *Styrax* are also native to North America. *Styrax benzoin* Dryand. is the source of gum benzoin. A gummy resin, storax, is obtained from *S. officinalis* L. (Cronquist 1981).

Recognition

Habit: shrubs, trees; deciduous or evergreen

Leaves: simple, alternate, entire or toothed; usually covered with star-shaped hairs beneath

Flowers: clusters, seldom solitary; bisexual, seldom unisexual; regular; calyx tubular, usually 4-5 lobes, persistent on fruit; corolla usually 2 to 5 petals united below to form a short tube; stamens commonly twice the petal number, filaments usually attached to corolla tube; pistil 1, style slender

Fruits: dry or fleshy

Halesia L. - silverbell-tree

Four *Halesia* species—three found in eastern and southeastern U.S. and one found in China—are noted for their showy, white flowers. They are characterized by scaly, red-brown bark, the few-flowered clusters of bell-shaped flowers along the branchlets, and their dry, elongated fruits with 2 or 4 lengthwise wings.

Two-winged silverbell, *H. diptera* Ellis, and little silverbell, *H. parviflora* Michx., both natives of the southeastern U.S., might be hardy in the southeastern prairie.

Halesia carolina L. - Carolina silverbell

Carolina silverbell, a tree of wooded slopes and streambanks in the mountains of Virginia, the Carolinas and Georgia, also grows as far west as eastern Oklahoma. A specimen in Maxwell Arboretum, UNL, is usually injured by cold and has yet to bloom.

Variety *monticola* Rehd., listed as a species by some authorities, tends to grow larger than the species with larger flowers and fruit.

Planning

Height: up to 30 ft (9 m); National champion, 1987, 86 ft (26 m), Tennessee

Habit: shrub, tree; deciduous; crown broad, rounded; spring bloom before leaves

Site: sun, part shade; moist soil, high organic matter; pH 5.0-6.5; zone 5

Use: border, specimen

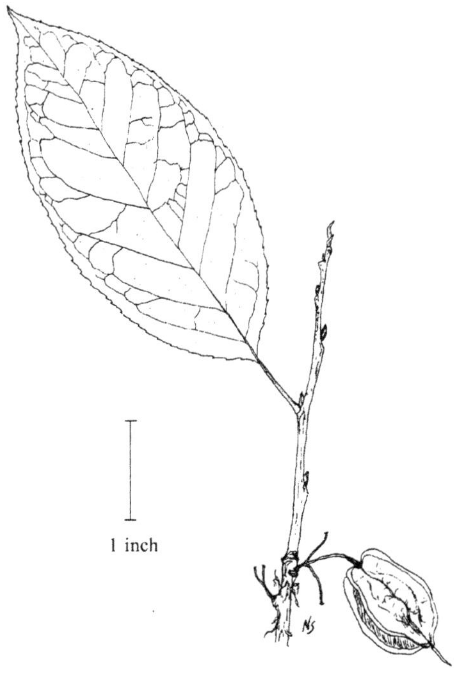

Halesia carolina

Styracaceae *Pterostyrax, Styrax*

Recognition

Leaves: elliptic to egg-shaped, up to 5 in (13 cm) long; tip pointed, base tapered to rounded; fine-toothed; densely hairy above, becoming glabrous; petioles about 0.3 in (8 mm) long
Flowers: clusters of 2 to 5 axillary on branches of previous season; calyx 0.2 in (5 mm) long, glabrous, lobes sharp-pointed; corolla white, rarely pale rose, bell-shaped, about 0.6 in (15 mm) long, lobes shallow; stamens 10-16
Fruits: drupes, oblong, dry, 4-winged, about 1.5 in (4 cm) long, green, becoming light brown to reddish; persistent into winter
Twigs: slender, somewhat zigzag, hairy becoming glabrous, lenticels becoming horizontally elongated; pith white, becoming spongy or chambered on old twigs; leaf scars raised, heart-shaped, 1 crescent-shaped bundle scar
Buds: terminal absent; lateral sometimes stalked, egg-shaped, scales 2 to 4, reddish, often one above the other at a scar (superposed)
Bark: reddish brown with chalky white streaks, becoming thin and scaly on old trunks

Pterostyrax Sieb. & Zucc. - epaulettetree

Four *Pterostyrax* species are native to eastern Asia.

Pterostyrax hispida Sieb. & Zucc. - fragrant epaulettetree

Fragrant epaulettetree—a handsome and unusual pest-free tree from Japan—is difficult to locate in commerce (Dirr 1990). Trial plantings are warranted.

Planning

Height: up to 30 ft (9 m)
Habit: tree; deciduous; crown round, branches slender; fragrant bloom in early June
Site: sun; prefers moist, well-drained, acidic soil; zone 5
Use: specimen, border, grouping
Caution: shelter from strong winds

Recognition

Leaves: oblong, up to 8 in (20 cm) long; tip pointed, base rounded or wedge-shaped, margin fine-toothed (denticulate); glabrous above, sparingly hairy beneath; petioles about 0.5 in (12 mm) long
Flowers: clusters (panicles) pendulous, hairy; calyx 5-toothed, petals distinct or slightly cohering at base, white; stamens 10, nearly free, exserted; style slender, slightly longer than the stamens
Fruits: drupes oblong-cylindric, dry; about 0.4 in (1 cm) long; 10-ribbed
Twigs: sparingly hairy, fetid odor when bruised; pith continuous, off-white
Buds: terminal hairy and naked, about 0.5 in (12 mm) long; lateral about 0.3 in (8 mm) long, egg-shaped; scales 2 visible

Styrax L. - snowbell

About 120 species of snowbell inhabit the tropical and warm temperate regions of Europe, Asia and the Americas. Five-parted white flowers, usually with 10 stamens, are found on unbranched, elongated spikes at twig ends. Fruits are rounded, dry to fleshy, and split along three or four valves.

Two of the four North American species, *S. grandifolius* Ait. and *S. americanus* Lam., might be adapted to the southeastern prairies.

Styrax japonicus Sieb. & Zucc. - Japanese snowbell

Japenese snowbell, native to China and Japan, is described as "a very dainty tree" that is relatively unknown but worthy of extensive landscape use (Dirr 1990). Flowers are very visible despite the fact that they appear at the same time as the leaves. "The leaves perch in pairs like butterflies along the tops of the branches while the dainty little white flowers hang beneath" (Johnson 1973).

Unpublished records of the North Central Plant Introduction Station (1954-1984) recommend planting Japanese snowbell on protected, moist sites in the southeast of the central prairie.

Styrax japonicus

Planning

Height: up to 30 ft (9 m)
Habit: shrub, tree; deciduous; wide-spreading branches; fragrant spring bloom appearing with leaves
Site: light shade; moist soil, high organic matter; pH 5.5-7.0; zone 5
Use: specimen, border
Caution: protect from strong winds

Recognition

Leaves: broad elliptic, up to 3.5 in (9 cm) long; tip pointed, base tapered; margin slightly notched (denticulate); star-shaped hairs quickly disappear except for axillary tufts beneath; petioles 0.3 in (8 mm) long
Flowers: clusters of 3 to 6 pendulous, terminal, each on slender stalks about 1.2 in (3 cm) long; corolla 5-lobed, bell-shaped, lobes spreading, elliptic-oblong; calyx glabrous, 5-toothed
Fruits: egg-shaped (ovoid) drupes, about 0.5 in (8 mm) long, grayish; seeds 1, shiny brown
Twigs: star-shaped hairs disappear quickly, slender, zigzag; pith green
Buds: naked, scurfy; one above another at a leaf scar (superposed)
Bark: smooth, gray-brown with orange-brown irregular, interlacing fissures

Symplocaceae Desf. - sweetleaf family

The sweetleaf family is comprised of one genus and over 300 species, mostly occuring in tropical and subtropical regions of Asia and the American continents (Cronquist 1981).

Symplocos Jacq. - sweetleaf

Symplocos tinctoria (L.) L'Her., common sweetleaf, is native to the U.S. from Delaware to southeastern Oklahoma. It is evergreen in the southern part of its range. Attractive clusters of yellow flowers appear before the leaves. A yellow dye is extracted from the leaves and bark (Cronquist 1981). Although seldom planted as an ornamental, provenance trials may prove worthworthwhile.

Symplocaceae *Symplocos*

Recognition

Habit: trees, shrubs; evergreen, rarely deciduous

Leaves: alternate, simple, often sweet-tasting; bladder-like water cells exserted from epidermis often present; stipules none

Flowers: clusters or solitary, axillary to a bract; bisexual, sometimes unisexual (polygamous), regular; calyx usually 5-lobed; corolla usually of 5 petals, joined at base to form a short tube; stamems 12 or more, attached to base of corolla tube; ovary usually inferior

Fruits: usually drupes, crowned by persistent calyx lobes

Symplocos paniculata (Thunb.) Miq. - sapphireberry

Sapphireberry, a native of Asia from the Himalayas through China and in Japan, is a small tree or large shrub that deserves trial planting in our prairies. Colorful fruits are delicacies to many birds. According to Dirr (1990), sapphireberry prefers full sun and is not seriously affected by diseases and insects.

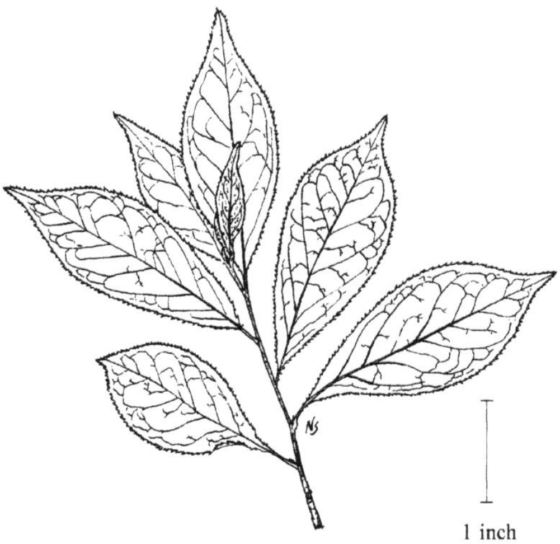

Symplocos paniculata

Planning

Height: up to 20 ft (6 m)

Habit: tree, shrub; deciduous; crown spreading; fragrant flowers in early June, bright turquoise blue fruits in autumn

Site: sun; pH 5 to 7; zone 5

Use: border, specimen, wildlife habitat

Recognition

Leaves: elliptic or egg-shaped, wider above the middle (obovate); tip pointed, base wedge-shaped; glabrous above, usually hairy below with conspicuous veins; petioles hairy, about 0.2 in (5 mm) long

Flowers: clusters (panicles), 2 to 3 in (5-8 cm) long on previous season's growth; bisexual or unisexual; corolla creamy white, about 0.5 in (12 mm) across, divided nearly to base; stamens distinct, numerous

Fruits: drupes ellipsoidal, about 0.3 in (8 mm) long, bright blue; calyx persists on fruits

Twigs: gray, hairy, becoming glabrous; pith plated

Buds: cone-shaped, pointed, often more than one (superposed), gray-brown, scales overlapping; flower buds larger than leaf buds, globose

Bark: gray with flat ridges separated by dark fissures, becoming stringy

Subclass Rosidae

The subclass Rosidae consists of 18 orders and 114 familes (Cronquist 1981).

Recognition

Habit: woody, herbaceous; deciduous or evergreen

Leaves: simple or pinnately compound; less often, palmately compound

Flowers: petals distinct or fused at the base, or sometimes much reduced or wanting

Order Rosales

Cronquist (1981) defines Rosales as a highly diverse group of plants consisting of 24 families and about 6,600 species.

Recognition

Habit: trees, shrubs, vines, herbs; deciduous or evergreen
Recognition: leaves alternate, opposite or whorled, simple or compound, with or without stipules
Flowers: bisexual, seldom unisexual; regular or somewhat irregular; petals distinct, seldom fused near the base, or wanting; stamens numerous

Hydrangaceae Dumort. - hydrangea family

The hydrangea family includes approximately 17 genera and about 170 species (Cronquist 1981). Species of the hydrangea family are found mainly in North America and eastern Asia.

Schizophragma hydrangeoides Sieb & Zucc., Japanese hydrangea vine, is a climbing shrub with aerial rootlets very similar to those of the *Hydrangea anomala*. Japanese hydrangea vine is easily identified by its solitary sepal of the sterile flowers, its longitudinally splitting bark, and buds with 2 to 4 pairs of hairy chestnut-brown or hairy fringed scales (Rehder 1940). It deserves trials to determine its range of adaptation on the prairies.

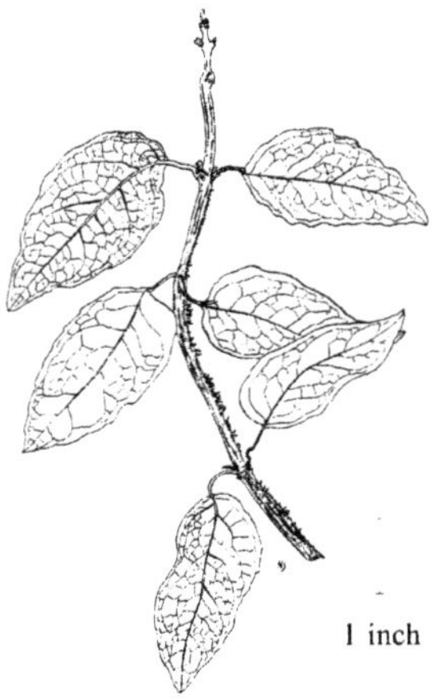

Decumaria barbara

Decumaria barbara L., another climbing shrub with aerial rootlets, has recently been added to Maxwell Arboretum, UNL. The leaves are entire or nearly so, the bark is exfoliating, and the white flowers are bisexual on terminal clusters. *Decumaria barbara* is native to the southeastern U.S.

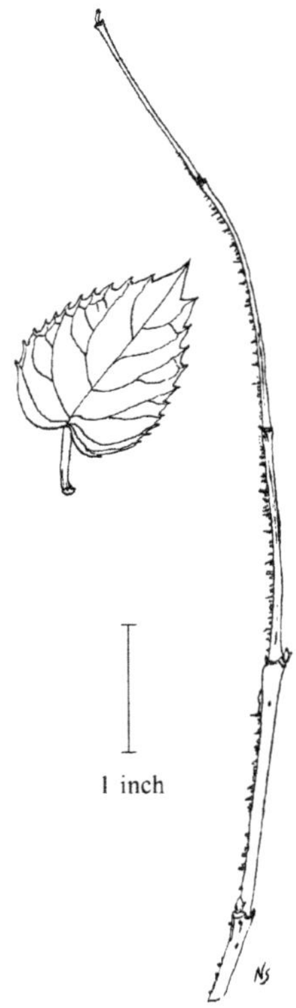

Schizophragma hydrangeoides

Recognition

Habit: shrubs, trees, vines, herbs; deciduous or evergreen
Leaves: simple; opposite, seldom whorled; stipules none
Flowers: clusters, usually small; usually bisexual and regular, sometimes marginal flowers are sterile with enlarged sepals; calyx, corolla and stamens arising from cup-like hypanthium attached to lower part of ovary (perigynous or half-inferior); petals distinct, 4- to 5- parted; stamens distinct, pistil 1
Fruits: usually dry, splitting, many-seeded capsules, seldom berries

Deutzia Thunb. - **deutzia**

About 50 species of *Deutzia* grow in eastern Asia and the Himalayas and two other species of it grow in Mexico. Many varieties have been recognized and several hybrids have been created. We describe *D. gracilis*, the species most likely to succeed in the prairies.

Twig dieback is often a problem for *Deutzia* species growing in cold climates. Varieties with pink or rose-colored flowers are especially susceptible to dieback (Dirr 1990). The beautiful white-flowered *D. ×lemoinei*, created by French horticulturist Andre Lemoine by crossing *D. parviflora* Bge. with *D. gracilis*, is one of the hardier varieties.

Field tests indicate that Lemoine deutzia can be successfully grown in the southeastern portion of the central prairie (North Central Plant Introduction Station 1977), but winter dieback and chlorosis are evident on plants growing in Fort Collins (Fairchild and Klett 1993). Dirr (1990) states that Lemoine deutzia is grown successfully in Madison, Wisconsin, and Chaska, Minnesota.

Cultivars include 'Avalanche', a compact shrub bearing white flowers in small clusters on arching branches, and 'Compacta', a dwarf bearing pure white, large flowers in dense clusters (Dirr 1990). Lemoine deutzia is characterized by scattered star-shaped hairs on its leaf undersides and brown exfoliating bark.

Recognition

Habit: shrubs; deciduous, rarely evergreen
Leaves: opposite, toothed; stipules wanting
Flowers: clusters, rarely solitary, usually terminal on lateral branches; bisexual; petals 5; calyx 5-toothed; stamens 10, rarely more, shorter than petals, filaments usually winged and toothed at apex; styles distinct, 3-5
Fruits: capsules, 3- to 5-valved; seeds minute, numerous
Twigs: hollow
Buds: scales many, overlapping, glabrous or nearly so
Bark: often shreddy (exfoliating)

Deutzia gracilis Sieb. & Zucc. - **slender deutzia**

Slender deutzia, native to Japan, has been successfully grown in Lincoln with diligent care on fertile soil. Twig dieback over winter can be severe and chlorosis is often evident on plants growing at PERC in Fort Collins (Fairchild and Klett 1993). Although slender deutzia provides a beautiful spring flower display, visually it is very ordinary for the remainder of the year.

D. scabra Thunb., fuzzy deutzia, is a taller, rather coarse shrub that blooms later then slender deutzia does. Flowers are often tinged with pink. Star-shaped hairs (stellate-pubescence) are present on both leaf surfaces.

Deutzia gracilis

Planning

Height: up to 5 ft (1.5 m)
Habit: shrub; deciduous; broad rounded; spring bloom
Site: sun, part shade; pH adaptable; fertile, moist soil; zone 5
Use: hedge, border, grouping

Recognition

Leaves: much longer than wide (oblong-lanceolate), up to 2.5 in (6 cm) long; toothed; tip long-pointed (acuminate), base broadly tapered to rounded; scattered star-shaped (stellate) hairs above, nearly glabrous beneath

Flowers: clusters erect, each flower about 0.6 in (16 mm) long, all fertile; petals valvate in bud, white, elliptic to oblong; sepals with greenish white lobes, deciduous on fruit; stamens 10, filaments winged with short spreading teeth

Twigs: pith white, becoming hollow

Buds: egg-shaped

Bark: yellowish gray, glabrous

Hydrangea L. - hydrangea

The hydrangeas, native to North and South America and eastern Asia, are planted in our landscapes for their summer floral displays. They grow best on fertile, moist soils in mild climates. Most species are hardy only in the most southerly prairies.

Nursery catalog pictures of bigleaf hydrangea (*Hydrangea macrophylla* [Thunb.] DC.), with its bountiful blue or pink flowers, lure us each January to include it in our spring planting. In our enthusiasm we fail to read the fine print, which states that bigleaf hydrangea is subject to cold injury if planted north of zone 7 and, as a consequence, may not bloom because most of the previous year's growth has died. 'Nikko Blue', an old standard, blue-flowering form, is grown with short-term expectations—enjoy it while you may (Finke 1996).

If soils are not acidic, aluminum sulphate must be added to the soil to induce bright flower color on bigleaf hydrangeas. Blue coloration requires a pH range of 5.0-5.5; pink, a pH range of 6.0-6.5 (Dirr 1990).

Recognition

Habit: shrubs, sometimes climbing; deciduous

Leaves: usually toothed, rarely lobed; petioled

Flowers: clusters terminal (usually corymbs), 4- to 5-parted; often with sterile marginal flowers; bisexual; 3-4 sepals, 4- to 5-parted; stamens 8-20, usually 10; ovary inferior or half-superior, styles 2-5, free

Fruits: capsules, splitting at apex; seeds minute, winged or wingless

Twigs: pith white

Bark: exfoliating

Hydrangea anomala Don **ssp.** *petiolaris* Sieb. & Zucc. - **climbing hydrangea**

Climbing hydrangea is primarily grown for its glossy green foliage. The branches provide depth to walls because they protrude out from the structures (Dirr 1990). Climbing hydrangea grows best on east and north exposures. The species (*H. anomala*) differs from the subspecies *petiolaris* in its display of longer, more narrow leaves (elliptic-ovate) and fewer stamens (9-15).

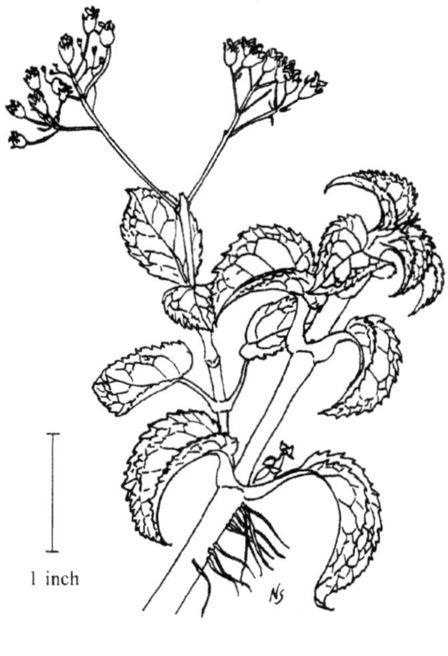

Hydrangea anomala petiolaris

Hydrangeaceae *Hydrangea*

Planning

Height: up to 80 ft (24 m)
Habit: shrub, vine; clings to trees, structures; early summer bloom, foliage green until leaf drop, exfoliating bark effective in winter
Site: shade to light shade; pH adaptable; fertile, moist soil; zone 5
Use: cover for arbor or masonry wall, free-standing bush

Recognition

Leaves: broad egg-shaped to oval, up to 4 in (10 cm) long; tip pointed, base rounded or heart-shaped; toothed; nearly glabrous, dark green and shiny above, often tufts on vein axils beneath; petioles up to 4 in (10 cm) long
Flowers: clusters flat-topped (corymbs), up to 10 in (25 cm) across; outer flowers sterile, about 1.2 in (3 cm) across, sepals entire; inner flowers fertile, dull white, petals cohering at apex; stamens 15-20; styles usually 2
Fruits: capsules; seeds winged all around
Twigs: older stems brown with exfoliating, shaggy bark; root-like holdfasts along internodes

Hydrangea arborescens L. - smooth hydrangea

Smooth hydrangea, native to the eastern U.S. as far west as southeastern Kansas and eastern Oklahoma, grows among rocks in shady woodlands, on flood plains, and in open areas. Because this hydrangea tends to become unkempt and blooms on new wood only, Dirr (1990) suggests mowing it in late fall. Smooth hydrangea often winterkills to near the ground in Kansas (Stephens 1973).

H. arborescens var. *grandiflora*, a wilding discovered in Ohio, bears sterile flowers only on subglobose heads up to 7 inches (18 cm) across (Rehder 1940). 'Annabelle', which bears flower clusters up to one foot (30 cm) across, was discovered by J. C. McDaniel, University of Illinois. 'Annabelle' is a very "tough" landscape plant (Finke 1996) and is hardy in southeastern North Dakota (North Central Plant Introduction Station 1975).

Hydrangea arborescens

Planning

Height: up to 5 ft (1.5 m)
Habit: shrub; wide, suckering; summer bloom on new wood, autumn foliage nondescript
Site: light to full shade; pH adaptable; moist, fertile soil; zone 4
Use: border, grouping

Recognition

Leaves: egg-shaped to elliptic, up to 8 in (20 cm) long; tip abruptly pointed, base rounded to heart-shaped; margin toothed; glabrous to variously hairy; petioles up to 2.5 in (6 cm) long, hairy, broadened to a semi-clasping base

Flowers: clusters flattish (corymbs), up to 6 in (15 cm) across on long, hairy stalks; sterile flowers up to 0.8 in (2 cm) across, green, becoming white then brown; fertile flowers dull white, styles persistent on fruits

Fruits: capsules 8- to 10-ribbed, about 0.1 in (2 mm) across

Twigs: stout, glabrous; pith about one-half of stem, white; leaf scars broad U-shaped, bundle scars 3

Buds: egg-shaped, divergent, about 0.1 in (3 mm) long; scales 4-6, greenish brown, glabrous

Bark: brownish, exfoliating into thin sheets

Hydrangea paniculata Sieb. - panicle hydrangea

Panicle hydrangea, the hardiest and largest of the hydrangeas, is native to Japan and eastern China. It can be trained into a small tree. 'Grandiflora', commonly known as peegee hydrangea, produces nearly all sterile flowers in panicles up to one foot (30 cm) long. Flowers turn pinkish purple with age, finally brown, and persist into winter. Other cultivars include 'Praecox', that blooms earlier than 'Grandiflora' and is hardier, and 'Tardiva', that flowers quite late in summer (Dirr 1990). Fairchild and Klett (1993) report minor winter dieback on 'Grandiflora' at Fort Collins.

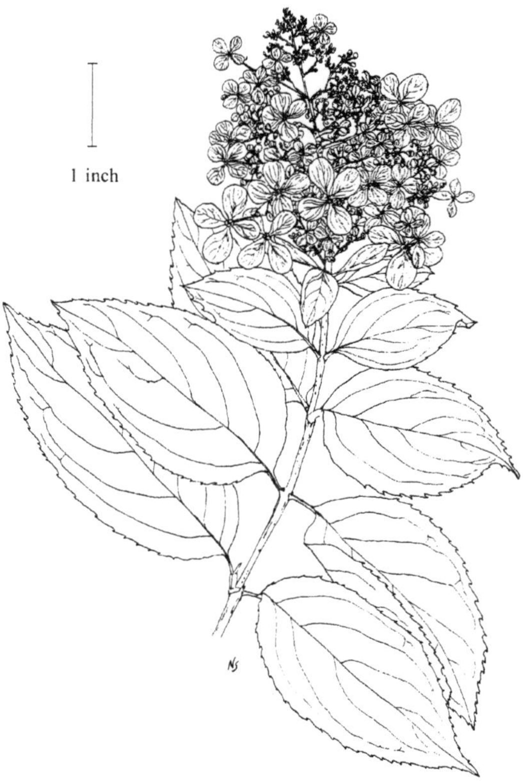

Hydrangea paniculata

Planning

Height: up to 15 ft (4.5 m)

Habit: shrub, tree; upright, spreading, often becoming unkempt with age; blooms in mid-July and later on new wood, flowers persist into winter; autumn foliage green, eventually yellow to red, not spectacular

Site: sun, part shade; pH adaptable; fertile, moist soil; zone 4

Use: border, grouping, specimen

Recognition

Leaves: sometimes whorled, especially on flowering stems; elliptic to egg-shaped (ovate), up to 6 in (15 cm) long; tip pointed, base tapered to rounded; margin toothed; scattered hairs above, bristly hairy beneath, especially on veins; petioles up to 1 in (25 mm) long

Flowers: clusters pyramidal (panicles), up to 8 in (20 cm) long, flowers mostly fertile to mostly sterile; yellowish white with petals free and expanding; stamens 10, styles 2-3; sterile flowers with 4 entire, elliptic sepals, white changing to purplish pink

Twigs: stout, hairy, reddish brown with gray, vertical streaks

Buds: globose; scales 4-6, overlapping, glabrous, brownish

Bark: ridged and furrowed on large stems

Hydrangea quercifolia Bartr. - **oakleaf hydrangea**

Oakleaf hydrangea, native to Georgia and Florida and west to Mississippi, is noted for its excellent summer bloom, colorful autumn foliage, and interesting bark that is most apparent in winter. Unfortunately, residents of the more northerly prairies are likely to be disappointed with the oakleaf's uninspiring autumn foliage coloration because of the large difference in latitude between our prairies and its native habitat.

Oakleaf hydrangea is rated high in pest resistance (Dirr 1990; Spangler and Ripperda 1977). Several cultivars have been selected for their large, mostly sterile flower clusters. 'Snowflake' is one of the most beautiful, with panicles over one foot long (Dirr 1990). Others include: 'Harmony', with mostly sterile large panicles; 'Roanoke', with loose, open clusters; and 'Snow Queen', with dense, upright flower clusters turning pink as they mature.

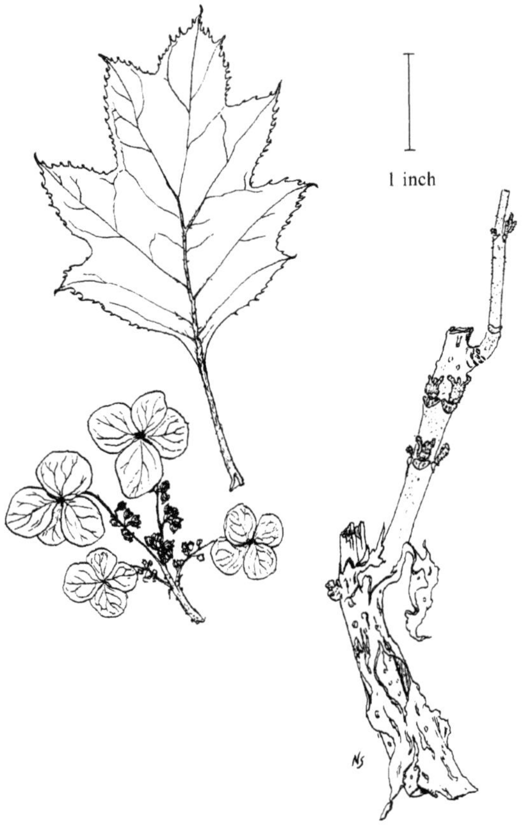

Hydrangea quercifolia

Planning

Height: up to 7 ft (2 m)
Habit: shrub; coarse textured, upright, stoloniferous; summer bloom, autumn foliage colorful, decorative bark
Site: sun, prefers part shade; pH adaptable; moist, fertile soil; zone 5
Use: border, grouping
Caution: protect from strong winds

Recognition

Leaves: somewhat egg-shaped (ovate) in outline; up to 8 in (20 cm) long; 3 to 7 broad lobes, margin toothed; glabrous above, whitish to light brown hairy (tomentose) beneath; petioles up to 2.5 in (6 cm) long, often exude milky sap when broken
Flowers: clusters erect (panicles), up to 12 in (30 cm) long; flowers white, changing to purplish pink, then brown, outer flowers sterile, about 1.2 in (3 cm) across
Fruits: about 0.1 in (3 mm) long; seeds wingless
Twigs: stout, hairy when young; leaf scars triangular with prominent bundle scars; pith light brown
Buds: divergent, terminal much larger than lateral; scales 4-6, overlapping, brownish, hairy (tomentose)
Bark: cinnamon brown, exfoliating

Jamesia Torr. & Gr. - **waxflower**

One species is native to North America.

Jamesia americana Torr. & Gr. - **waxflower**

Waxflower grows on warm southern exposures from the canyons of the foothills up to the subalpine zone in the Rocky Mountains of Wyoming and Colorado and parts of Utah and Arizona. It has been introduced into Scotland with success. Weber (1990) reported discovering a beautiful bush in full flower in Glasgow in 1957. Waxflower may prove to be useful in western prairie landscapes. Some exploration and testing seems to be in order.

Planning

Height: up to 3 ft (1 m)
Habit: shrub; deciduous; upright; autumn foliage colorful
Site: sun; experimental in the prairies
Use: specimen

Recognition

Leaves: ovate to elliptic, up to 2.5 in (6 cm) long; penniveined, margin toothed (serrate), tip pointed; wrinkled and slighty hairy above, copiously soft-hairy beneath; petioles short
Flowers: clusters terminal (cymes); fertile; each flower about 0.6 in (15 mm) across; petals 5, white, sometimes pinkish outside, sepals much shorter than petals; stamens 10, shorter than petals; ovary nearly superior, styles 3-5, distinct, persistent
Fruits: capsules about 0.2 in (4 mm) long, 3-5 valved; seeds numerous
Twigs: hairy when young, pith solid
Buds: densely hairy, terminal with 2 pairs of outer scales, lateral with 1 pair
Bark: exfoliating

Philadelphus L. - mockorange

The sweetly scented flowers of many of the 40 species of *Philadelphus* account for the common name mockorange. Many hybrids have been developed that produce white or creamy white blooms in profusion, scented and unscented. Most mockoranges are large, leggy shrubs with coarse, dull foliage that do little for the landscape once the blooms are gone. Hoag (1965) advises using restraint in selecting mockorange cultivars for the landscape. Dirr (1990) provides minimal treatment to *Philadelphus* species "since the only redeeming character is fragrance and many cultivars do not offer this."

Littleleaf mockorange (*P. microphyllus* Gray), a native of rocky slopes of southern Colorado, New Mexico and Arizona, has played an important part in the development of hybrids and many cultivars. Victor Lemoine of France did the crossing and Arnold Arboretum brought them to this country. Perhaps this small shrub with exquisitely fragrant flowers can occupy a niche in our landscapes, especially in the western prairie. It is worth a try.

Lemoine mockorange, *P. ×lemoinei* Lemoine (*P. microphyllys × P. coronarius*), is the parent of many cultivars with single and double fragrant flowers. This mockorange has been used to develop another common hybrid, virginal mockorange.

Hybrids are generally less hardy than *P. coronarius* in North Dakota (Hoag 1965). Three hybrids and several cultivars growing in PERC suffer winter dieback and varying degrees of chlorosis (Fairchild and Klett 1993). Lemoine mockorange appears more hardy than the others on the Fort Collins site, with only minor chlorosis and no pest problems.

Recognition

Habit: shrubs; deciduous
Leaves: 3- to 5-nerved at base
Flowers: terminal on lateral branches, solitary or in clusters; petals 4, white or rarely purple near base, sepals 4; stamens numerous; ovary inferior or half-superior; styles 4, more or less fused together
Fruits: capsules 4-valved
Twigs: stout, pith solid; leaf scars triangular
Buds: terminal lacking; lateral usually hidden by petiole base; scales overlapping

Hydrangaceae *Philadelphus*

Philadelphus coronarius L. - sweet mockorange

Sweet mockorange, a native of southeastern Europe and Asia Minor, was once commonly planted in the U.S. and continues to survive in the yards of many older homes. When in bloom these shrubs provide a backdrop for many graduation photographs. Hybrids with larger or double-form flowers, even though often not as fragrant, have become more popular.

Hoag (1965) recommends planting sweet mockorange in the northern Great Plains because it is better adapted to that climate and more tolerant of dry soils than its hybrid descendants are. Pruning out old wood or cutting back the entire plant after blooming reduces the tendency for the shrub to become leggy and unkempt.

Sweet mockorange is a long-lived shrub with no serious pest problems. With proper care, it and some related hybrids can serve a useful purpose where many other more colorful species may fail.

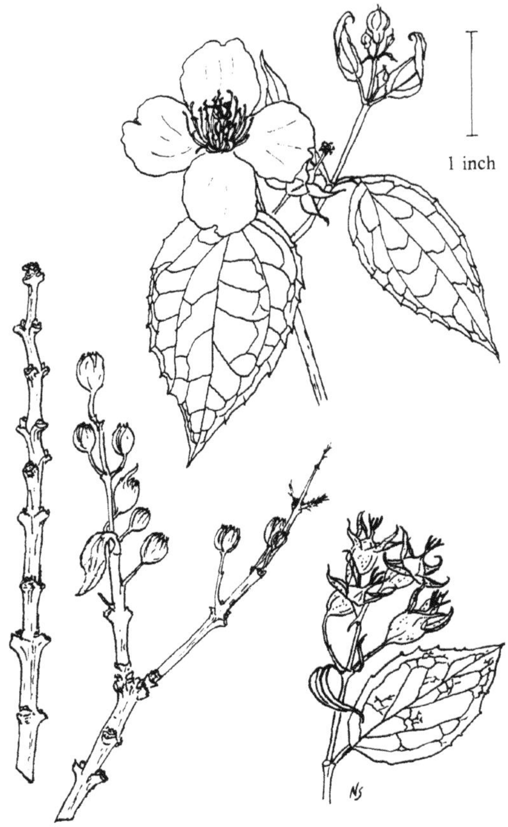

Philadelphus sp.

Planning

Height: up to 10 ft (3 m)
Habit: shrub; upright, rounded, becoming leggy; late spring bloom fragrant
Site: sun, part shade; moist, fertile soil, pH 5.0-7.5; zone 4
Use: border

Recognition

Leaves: egg-shaped (ovate), up to 4 in (10 cm) long: tip pointed, base broad, tapering to rounded; teeth widely spaced; glabrous, except tufts in vein axils beneath, veins sometimes hairy; petioles hairy, about 0.2 in (5 mm) long
Flowers: clusters 5- to 7-flowered (racemes), individual flowers about 1.2 in (3 cm) across; calyx usually glabrous, petals white; styles shorter than stamens and usually divided about half its length
Twigs: glabrous or slightly hairy, pith pale or white
Buds: scales 2, hairy
Bark: reddish brown, exfoliating on second year stems

Grossulariaceae A. DC. - **currant family**

The currant family is comprised of about 25 genera and 350 species (Cronquist 1981). It is considered by some authorities to be subfamily Ribesioideae Engl. within Saxifragaceae Dum., which includes the single genus *Ribes*, composed of currants and gooseberries. Cronquist places *Itea* in the currant family.

Recognition

Habit: shrubs, sometimes trees; deciduous or evergreen
Leaves: alternate, simple; often glandular toothed, sometimes deeply cleft; veins pinnate or palmate
Flowers: terminal or axillary clusters or solitary; bisexual or seldom unisexual (sometimes dioecious); sepals lobed or forming a tube, sometimes petal-like and showy; petals alternate with sepals, sometimes wanting; stamens opposite sepals, staminodes sometimes present, alternating with sepals; styles distinct to variously fused
Fruits: capsules or berries

Itea L. - **sweetspire**

About ten species of *Itea* are native to southeast Asia. *Itea japonica* 'Beppu', an Asian species, is not as hardy as the North American native (Dirr 1990). However, Gibson (1992) states that 'Beppu' plants have survived two winters with minimal injury at the Lincoln Regional Center Arboretum. 'Beppu' bears smaller flowers than *I. virginica*.

Itea virginica L. - **Virginia sweetspire**

Virginia sweetspire is native to the eastern U.S., from New Jersey south to the Gulf states and west to Missouri. 'Henry's Garnet' provides brilliant, reddish purple fall color and its flowers are superior to those of the species (Dirr 1990).

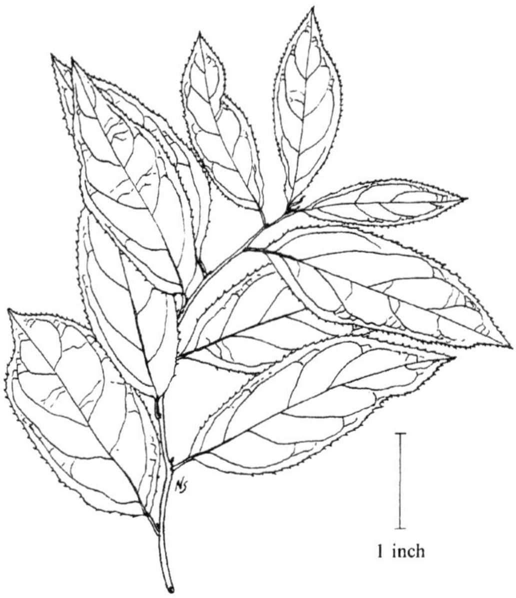

Itea virginica

Planning

Height: up to 6 ft (2 m)
Habit: shrub; deciduous in cold climates, evergreen in the south; arching with rounded top, suckering; bright autumn foliage, early summer flowers
Site: sun; prefers moist soils but withstands dry conditions; zone 5
Use: naturalizing

Recognition

Leaves: elliptic to oblong; tip pointed, base wedge-shaped; margin fine-toothed (serrulate); glaucous above, sparingly hairy beneath; petioles hairy, grooved above, about 0.2 in (5 mm) long
Flowers: clusters upright (racemes), dense, hairy, 2 to 6 in (5-15 cm) long; bisexual; sepals 5, persistent, petals 5, about 0.2 in (4 mm) long; stamens 5, as long as petals
Fruits: capsules, about 0.3 in (8 mm) long, narrow, hairy, persistent; seeds flattened
Twigs: long, straight, slender, hairy while young; pith plated (lamellate)
Buds: accessory above true lateral (superposed); scales 3-4

Ribes L. - currant or gooseberry

Ribas, the Persian and Arabic name for a species of rhubarb (*Rheum*), inspired the generic name *Ribes* for this group of plants because a fruit syrup of European red currant resembles a famous acidulous Arabic syrup made from the young stems of rhubarb (U.S. Forest Service 1937). No wonder rhubarb and gooseberries are a compatible mixture in jams and pies. About half of the 150 *Ribes* species—natives of the cool temperate regions of the Northern Hemisphere and Andes Mountains—are cultivated for their edible fruits and ornamental value.

Many currants and gooseberries are alternate hosts for white pine blister rust, *Cronartium rubicola*, a deadly disease of the soft pines (subgenus *Haploxylon*). *Ribes* species have been subjected to eradication programs in forest areas of the U.S. where white pines are important timber trees. This disease is not likely to be a problem in the prairies because white pine populations are small and scattered.

Currants and gooseberries have been divided into two, sometimes even four, genera. We place them all in one genus, *Ribes,* and four subgenera as in Rehder (1940).

Recognition

Habit: shrubs; usually deciduous
Leaves: simple, alternate; palmately veined and often palmately lobed, with or without stipules
Flowers: commonly in clusters (racemes), bisexual or unisexual (dioecious); usually 5-parted, regular; ovary inferior (hypanthium attached atop the ovary); calyx usually petal-like (petaloid), cylindric to wheel-shaped (rotate), colored, larger than petals; stamens 5, alternate with petals; pistil 1, styles 1-2
Fruits: berries, several seeded
Twigs: unarmed or prickly; leaf scars with 3 bundle scars
Buds: scales several, thin, dry (scarious), or leaf-like

Grouping Key for *Ribes* (currants, gooseberries)

(Rehder 1940)

Flowers dioecious, clustered (usually racemes); stalks (pedicels) jointed; twigs unarmed or rarely with small paired nodal prickles; bud scales thin, dry (scarious) or leaf-like (herbaceous): **subgenus *Berisia* (currants)**

Flowers bisexual, usually in racemes; stalks jointed; twigs unarmed; bud scales usually leaf-like: **subgenus *Ribesia* (currants)**

Flowers bisexual, saucer-shaped, clusters (racemes) pendulous, 6 to 20 flowers; stalks jointed beneath ovary; fruits separating from stalks; twigs armed with spines and prickles; bud scales thin, dry: **subgenus *Grossularioides* (currants)**

Flowers bisexual, never saucer-shaped, solitary to clusters of 4; stalks not jointed; fruits remain attached to stalks; twigs with nodal spines, often with bristles, rarely nearly wanting; bud scales thin, dry: **subgenus *Grossularia* (gooseberries)**

Subgenus *Berisia* - currants

Ribes alpinum L. - alpine currant

Alpine currant is a hardy European import adapted to most prairie sites. Early emerging leaves provide a long green season and the scarlet berries of the female plants add colorful contrast to the foliage.

Alpine currant and the dwarf 'Green Mound' are winter hardy and grow on alkaline soils in Fort Collins (Fairchild and Klett 1993). Other compact or dwarf forms include 'Aureum', with its yellowish leaves, 'Pumila', 'Compacta' and 'Nana' (Dirr 1990).

Planning

Height: up to 7 ft (2 m)
Habit: shrub; upright becoming rounded; April bloom unobtrusive, early green leaves; scarlet berries on female plants
Site: sun, part shade; drouth tolerant, pH and soil adaptable; zone 3
Use: hedge (tolerates pruning), screen, grouping, naturalizing

Recognition

Leaves: oval or egg-shaped (ovate), up to 2 in (5 cm) long; 3, rarely 5 lobes, margin toothed (dentate); base squared off (truncate) or with rounded lobes (subcordate); petioles about 0.5 in (12 mm) long, glandular hairy
Flowers: dioecious (bisexual sometimes present), greenish yellow; male in upright clusters of 20-30, up to 2 in (5 cm) long; female clusters half as long; calyx tube wheel-shaped (rotate), sepals egg-shaped
Fruits: berries scarlet, glabrous, subglobose, about 0.3 in (8 mm) across, not edible
Twigs: unarmed, light brown, often shiny, ridges extend down from leaf scar edges
Buds: stalked; scales thin, dry, overlapping (imbricate), pointed

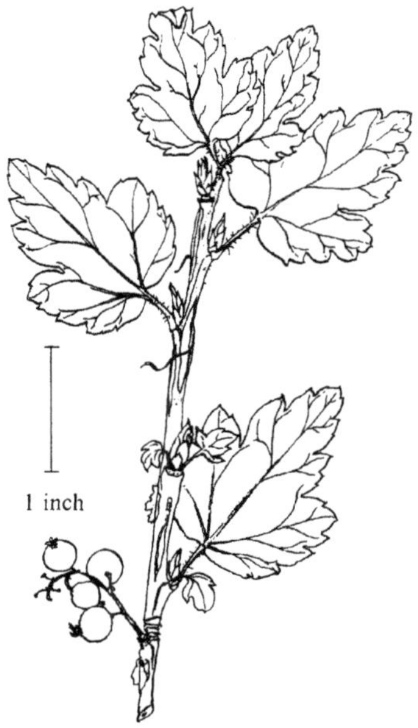

Ribes alpinum

Ribes diacanthum Pall. - Siberian red currant

Siberian red currant, native of northern Asia, is a true currant with small prickles. This shrub is slightly smaller and has more lustrous leaves than alpine currant. Slight chlorosis has been noted on Siberian red currant plants on alkaline soils, but it is less susceptible to foliar diseases than most other currants (North Central Plant Introduction Station 1960). One of the few specimens observed in Nebraska is at the Hans Burchard Arboretum at Branched Oak Lake near Lincoln (Sutton 1998).

Planning

Height: up to 5 ft (1.5 m)
Habit: shrub; upright, dense; ornamental fruit in summer (female)
Site: sun, part shade; pH and soil adaptable; zone 3
Use: hedge, screen, wildlife habitat

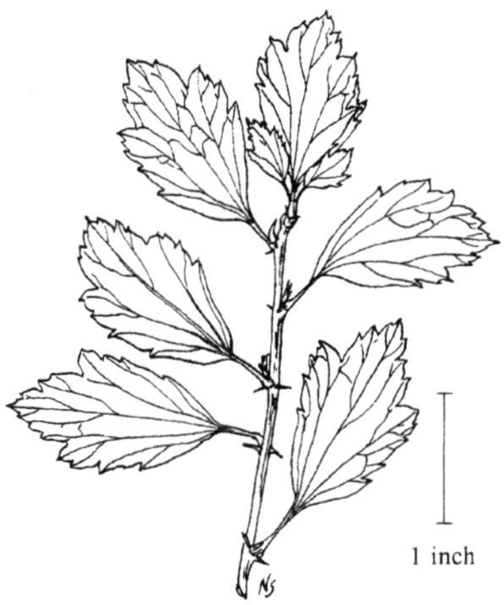

Ribes diacanthum

Recognition

Leaves: egg-shaped (ovate) to wider above middle (obovate), up to 1.4 in (35 mm) long; slightly 3-lobed, lobes rounded and sparingly toothed, base widely tapering to rounded; surfaces lustrous with rudimentary veins
Flowers: dioecious, greenish yellow; male clusters up to 1.2 in (3 cm) long, female one-half as long; calyx wheel-shaped (rotate), sepals ovate
Fruits: berries globose, scarlet, smooth, bitter
Twigs: 2 small prickles at nodes, often smaller scattered prickles

Subgenus *Ribesia* - currants

Ribes triste Pall., swamp red currant, is native to the prairie state of North Dakota. This inhabitant of shaded, wet, brushy areas is a low decumbent shrub that roots at the nodes (Stephens 1973). Flowers are inconspicuous.

Ribes americanum Mill. - American black currant

American black currant, a native of the northeastern U.S., southern Canada and the Rocky Mountains, also grows in the northern Great Plains on moist, wooded hillsides and brushy areas. The fruits of these shrubs, dubbed "beaver berries" by the Dakota Sioux, were used to make jelly by the early settlers. The roots provided a medicine for the Omaha Indians (Gilmore 1977). The extended period of fruit ripening provides an excellent, lasting food source for wildlife.

Planning

Height: up to 3 ft (1 m)
Habit: shrub; few branches; May bloom, late summer fruit
Site: sun, part shade; pH and soil adaptable; zone 2
Use: wildlife habitat, hedge, naturalizing, fruit

Recognition

Leaves: broadly egg-shaped (ovate), up to 3 in (8 cm) long and wide; 3-5 pointed lobes with deep, sharp notch (sinus) between; base squared off or with rounded lobes; margin toothed and hairy (ciliate); upper surface glabrous, sparsely hairy beneath, golden yellow, glandular; petioles about 2 in (5 cm) long, curly-hairy
Flowers: bisexual; clusters axillary, drooping (racemes), up to 2 in (5 cm) long, 2 to 12 flowers; calyx tube cylindric, lobes oblong, pale yellow, spreading or recurved; petals white, about 70% of sepal length; stamens 5 opposite calyx lobes, shorter than petals
Fruits: berries globular, about 0.4 in (1 cm) across; red-purple to nearly black, smooth, glossy, sweet, not tasty; many-seeded; floral parts persistent
Twigs: unarmed; brownish, minutely curly-hairy (puberulent), orange glands; ridges extending down from center and both sides of v-shaped leaf scar; pith white

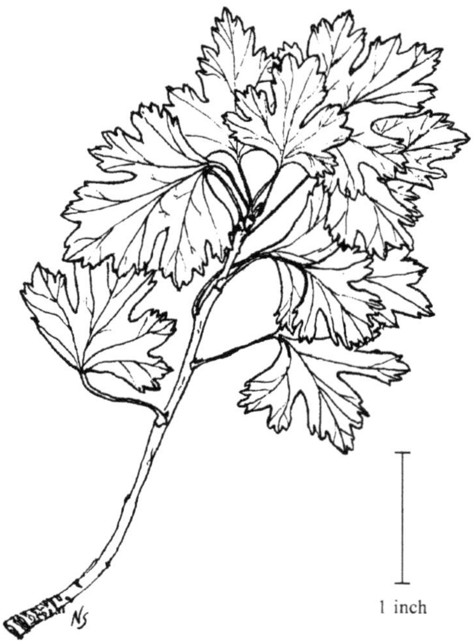

Ribes americanum

Buds: stalked, egg-shaped (ovoid), minutely hairy, gray-brown tinged with red; outer scales short-pointed (apiculate)
Bark: red-brown, smooth, tight

Ribes cereum Dougl. var. *inebrians* (Lindl.) Hitchc. - squaw currant

The native range of squaw currant, also known as wax currant, extends from northwest Nebraska west and north to include the western U.S. and southwestern Canada. Variety *inebrians*, considered by some authorities to be a separate species, is native to the Rockies and Nebraska.

Squaw currant is a hardy, drouth-resistant shrub growing naturally at elevations from 4,000 to 9,000 feet on dry rocky hillsides, but it lacks the refinement necessary for ornamental gardens (Sutton and Johnson 1974). The squaw currant is a good shrub for dry, minimum maintenance gardens. Its gnarled branching pattern, short internodes and small leaves make it a candidate for bonsai culture.

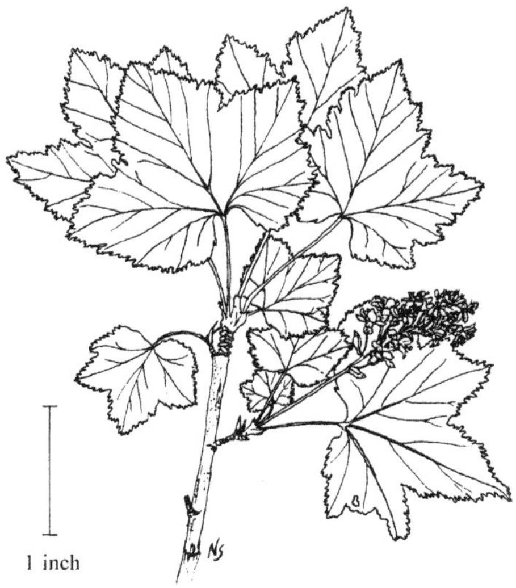

Ribes cereum

Planning

Height: up to 5 ft (1.5)
Habit: shrub; crooked, coarse branches; late spring bloom, summer fruit
Site: sun; pH and soil adaptable; drouth tolerant; zone 3
Use: wildlife habitat, naturalizing

Recognition

Leaves: often clustered at end of short spur; rounded to kidney-shaped (reniform), about 1 in (25 mm) across; 3 rounded lobes with shallow acute notches (sinuses), margin irregularly lobed and toothed with short hairs (ciliate), some glandular; surfaces hairy and glandular hairy; petioles hairy, grooved above, up to 0.6 in (15 mm) long
Flowers: bisexual; clusters drooping (racemes), 1 to 6 flowers; calyx tube cylindric, pinkish, purplish, pale green or nearly white, lobes triangular, nearly white, about 0.1. in (2 mm) long, spreading or reflexed, hairy and glandular outside; petals erect, white; stamens 5, opposite calyx lobes
Fruits: berries globular, about 0.4 in (1 cm) across, red, covered with glands on short stalks (stipitate); floral parts persistent
Twigs: unarmed, brownish, minutely hairy (puberulent), splitting second year to show reddish inner bark; leaf scars narrow extending about halfway around stem
Buds: egg-shaped (ovoid), very short-stalked; outer scales brown, often 3-ribbed with marginal hairs (ciliate)

Ribes odoratum Wendl. - clove currant

Clove currant, native to much of the Great Plains region, grows on a variety of sites, from sandy to chalk bluffs among prairie plants and along the edges of open woodlands. In the western part of its range the clove currant intergrades with the closely related golden currant, *R. aureum* Pursh. The leaves of the latter are less deeply lobed, its calyx lobes are longer, and its fruits more often yellow (Stephens 1973, U.S. Forest Service 1937).

Wild clove currant was a source of food for Native Americans and early settlers. 'Crandall', a large-berried cultivar, has been cultivated for food (U.S. Forest Service 1937). Hoag (1965) recommends the species for planting as an ornamental and in windbreaks wherever a hardy shrub of its size is needed. Clove currant is adapted to the Fort Collins area (Fairchild and Klett 1993)

Planning

Height: up to 5 ft (1.5 m)
Habit: shrub; upright arching, open, colony forming, fragrant; April bloom; edible fruit; autumn foliage ornamental
Site: sun, part shade; pH and soil adaptable; prefers moist, well-drained soil; drouth tolerant; zone 3
Use: wildlife habitat, fruit, naturalizing, erosion control, shelter, hedge

Recognition

Leaves: broadly egg-shaped (ovate), up to 2.5 in (6 cm) across; 3-5 lobes, tips variable, margins coarsely toothed, hairy (ciliate), base tapering (cuneate) to squared-off (truncate); surfaces green with scattered hairs (glabrate), veins hairy; petioles hairy, up to 2 in (5 cm) long

Flowers: clusters short (racemes), 5 to 10 flowers, bisexual; tube cylindric, calyx lobes oblong, spreading or reflexed; petals pale yellow often with red tips, one-half length of sepals, recurved; stamens alternate with petals

Fruits: berries globose, about 0.4 in (1 cm) across, purple-black, smooth; veins show as light lines

Twigs: reddish brown, hairy at first; fruit spurs gray-rough; pith greenish, leaf scars narrowly crescent-shaped

Buds: narrowly egg-shaped, pointed; outer scales keeled, margins hairy (ciliate)

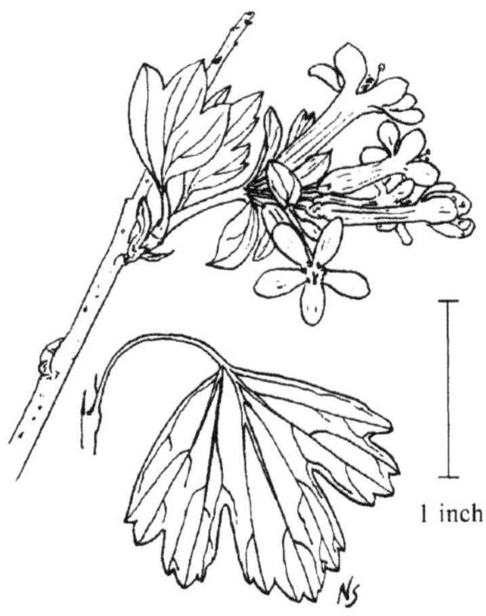

Ribes odoratum

Ribes sativum Syme. - red currant

Many currant hybrids and cultivars developed for fruit production have originated from red currant, a native of western Europe. Some have escaped and gone wild in the eastern U.S. 'Red Lake' is popular for its fruits, either fresh or in preserves.

Planning

Height: up to 5 ft (1.5 m)
Habit: shrub; upright; summer fruit
Site: sun; pH and soil adaptable; zone 4
Use: fruit, wildlife habitat

Recognition

Leaves: suborbicular, up to 2.8 in (7 cm) wide; 3-5 lobes, short-pointed, margin toothed; hairy on veins beneath

Flowers: clusters drooping, many-flowered (racemes); bisexual; receptacle saucer-shaped, green or slightly purple

Fruits: berries usually red, edible

Twigs: hairy when young

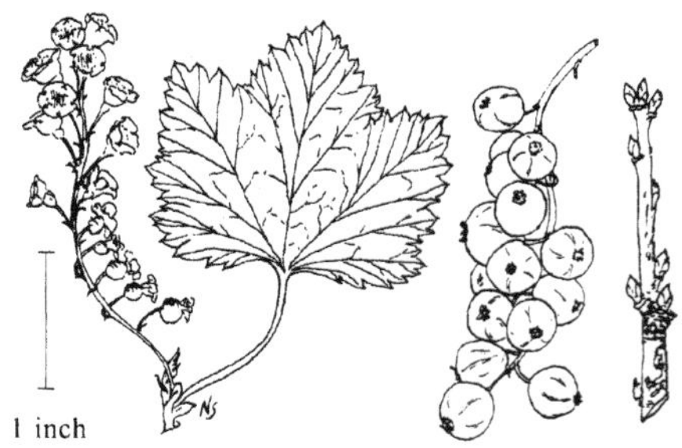

Ribes sativum

Subgenus *Grossularioides* - currants

Ribes lacustre (Pers.) Poir. - prickly currant

Prickly currant, a native of many of the cooler regions of North America, is an exception to this mostly spine-free group of plants. Prickly currant, a spreading, often decumbent inhabitant of moist woods and swamps, grows in the Black Hills.

Planning

Height: up to 3 ft (90 cm)
Habit: shrub; sparse; stems slender, weak, often decumbent
Site: shade; moist, fertile soils; zone 3
Use: naturalizing

Recognition

Leaves: broadly egg-shaped to circular in outline, up to 2 in (5 cm) across; 3-5 lobes, deeply cleft, narrowed at base; margins toothed (crenate-dentate), often tipped with stipitate glands; surfaces with scattered stipitate glands; petioles about as long as blades

Flowers: bisexual; clusters of 2 to 10, pendulous; flowers greenish or purplish; calyx tube saucer-shaped; petals much broader than long

Fruits: berries globose, about 0.3 in (8 mm) across; dark purple, stipitate glandular

Twigs: clustered nodal spines, densely covered with soft prickles

Buds: narrow egg-shaped, yellowish; glabrous

Subgenus *Grossularia* - gooseberries

Gooseberries are also sometimes considered to be in the genus *Grossularia* by some authorities. The acidic fruits of many gooseberry species are used for sauces and pies and are relished by birds and small animals. Fruit pickers note that a small section of the fruit stalk clings to the fruit, a characteristic that distinguishes the gooseberry from the currant.

A gooseberry bush of an unknown variety was one of the few woody plants that survived and produced copious fruit during the late 1920s on a Sandhills homestead in eastern Colorado (Bagley 1998).

Ribes cynosbati L - prickly gooseberry

Prickly gooseberry, a native of eastern North America, is found in wooded areas on the fringes of the prairies, from Manitoba and North Dakota south to Oklahoma. Its spiny fruits distinguish prickly gooseberry from other species.

Planning

Height: up to 3 ft (1 m) high

Habit: shrub; broad, open; armed with spines and prickles

Site: part shade; moist, fertile soil; zone 3

Use: naturalizing

Recognition

Leaves: broadly egg-shaped in outline; deeply 3- to 5-lobed, base somewhat square (truncate) but variable; margins toothed (crenate-dentate), surfaces hairy (pubescent) and stipitate glandular; petioles shorter than blade, hairy and glandular

Flowers: clusters of 2 to 4, hairy and glandular; calyx tube glabrous inside, sepals shorter than bell-shaped receptacle; stamens shorter than sepals; ovary bristly (setose)

Fruits: berries globose, hairy (pubescent) and spiny, dark purple; old floral parts persistent

Twigs: 2-3 dark red spines at each node; prickles on internodes pointed to base (retrorse)

Buds: narrowly egg-shaped (ovoid), sharp pointed (acute), pale brown; scales thin, crinkly, margins hairy (ciliate)

Ribes grossularia L. - European gooseberry

European gooseberry, a native of Europe, northern Africa and the Caucasus Mountains, has been cultivated for its fruit since the sixteenth Century. It is the parent of a large number of hybrids from which many cultivars have been selected. Variety *uva-crispa* (L.) Sm. has hairy branches and leaves and bears small, yellow, hairy fruits (Rehder 1940).

Planning

Height: up to 3 ft (1 m)

Habit: shrub; many-stemmed, rounded; summer fruit

Site: sun; pH and soil adaptable; zone 4

Use: fruit, wildlife habitat

Grossulariaceae *Ribes*

Recognition

Leaves: roundish, up to 2.5 in (6 cm) across, base heart-shaped to broadly tapered, 3-5 lobes; margin irregularly lobed and toothed; surfaces glabrous or hairy
Flowers: solitary or in twos; calyx tube hairy inside; sepals about as long as the bell-shaped receptacle, greenish, usually hairy, recurved during anthesis; stamens about half as long as sepals; ovary hairy, often glandular
Fruits: berries globose to egg-shaped, hairy or glabrous, green, yellow or red
Twigs: variously armed, bristles and spines mostly 3-parted

Ribes hirtellum Michx. - gooseberry

'Pale Red', one of the better native edible gooseberries, is a selection of *R. hirtellum*. This species is native to southeastern Canada, the northeastern U.S. and Iowa, North Dakota and the Black Hills.

R. inerme Rydb., whitestem gooseberry, ranges from British Columbia to Montana south to New Mexico and California. The shrub's whitish bark is the reason for the common name. Whitestem gooseberry is so similar to *R. hirtellum* that Stephens (1973) combines the two species in his description. Both are characterized by few or no spines or prickles.

Planning

Height: up to 3 ft (1 m)
Habit: shrub; branches stiff, few or no spines or bristles
Site: sun, part shade; pH and soil adaptable, drouth tolerant; zone 3
Use: fruit, wildlife habitat, naturalizing

Recognition

Leaves: somewhat rounded, 3-5 lobes; square (truncate) to heart-shaped (cordate) at base, lobes toothed (crenate-serrate); upper surface glabrous to slightly hairy, variably hairy beneath; petioles variably hairy, sometimes with long, branched hairs
Flowers: clusters of 2 to 4 (racemes), greenish or purplish; calyx tube cylindric to bell-shaped, hairy inside, less so outside, lobes same length as tube, bent back or spreading; petals inserted on rim of calyx tube, thin, white to pinkish, half as long as calyx lobes; stamens twice as long as petals; ovary glabrous, green
Fruits: berries globular, smooth, dark red-purple; edible
Twigs: nodal spines few or none, internodal bristles few or none
Buds: outer scales hairy, ciliate

Ribes missouriense Nutt. - Missouri gooseberry

Misssouri gooseberry, a spiny, unkempt little shrub, is native to much of the eastern U.S. including the eastern prairies. It appears in new tree plantings soon after trees become large enough to provide bird perches—proof of its value as food for wildlife. Berries have been gathered from the wild plants for pies and jams, but the Missouri gooseberry has seldom been cultivated.

Missouri gooseberry has served as a parent for hybrids. 'Pixwell', a popular garden cultivar, and 'Dakota Dwarf', a compact clone with ornamental qualities, are products of breeding by Dr. A. F. Yeager of the North Dakota Agricultural Experiment Station (Hoag 1965).

Planning

Height: up to 5 ft (1.5 m)
Habit: shrub; sparse, spreading; May bloom, summer fruit
Site: sun, shade; pH and soil adaptable; zone 4
Use: fruit, wildlife habitat, naturalizing

Recognition

Leaves: roundish, about 2 in (6 cm) across, 3 to 5 lobes separated by shallow narrow notches (sinuses); margins hairy (ciliate), base tapering to shallow heart-shaped (subcordate); glabrous above, hairy on veins only below; petioles about 0.1 in (2 mm) long, hairy

Flowers: on old wood fruiting spurs; calyx tube cylindric, white, lobes 5, oblong, strongly reflexed; petals 5, white, less than half as long as sepals; stamens 5, alternate with petals and twice as long as sepals; ovary green, glabrous

Fruits: berries globose, about 0.5 in (12 mm) across, capped by persistent style, whitish lines from base to tip, surface granular, black

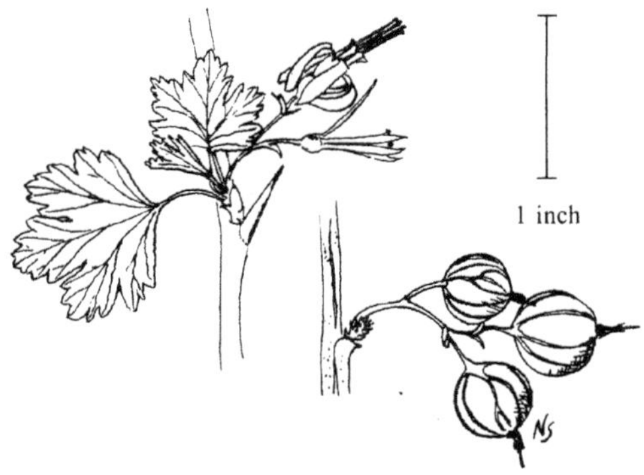

Ribes missouriense

Twigs: glabrous, brown, often tinted reddish; weak internode prickles, 1-3 rigid prickles at nodes below leaf scars, those on fruiting stems dark red-brown, up to 0.4 in (1 cm) long, nearly straight; leaf scars crescent-shaped; pith white, becoming brown and pitted

Buds: narrowly egg-shaped, long pointed; scales red-brown, keeled, tip spreading

Bark: exfoliating into thin, papery scales curling laterally

Ribes oxycanthoides L. - northern gooseberry

Northern gooseberry is a native of southern Canada and the northern U. S. from Michigan and the Dakotas south to northwest Nebraska. *Ribes setosum* Lindl., another gooseberry with a similar natural range, is often confused with northern gooseberry. See Stephens (1973) for detailed differences between the two species.

Planning

Height: up to 3 ft (1 m)

Habit: shrub; much branched, spreading; May flowers, summer fruit

Site: sun; pH and soil adaptable; drouth tolerant; zone 3

Use: naturalizing, wildlife habitat

Recognition

Leaves: broader than long, up to 2 in (5 cm) wide; 3 to 5 lobes, widely cleft about halfway to midrib; margin coarsely toothed, hairy (ciliate); base square (truncate) to heart-shaped (cordate); surfaces sparsely hairy above, hairy and glandular beneath; petioles hairy, grooved above, about 0.7 in (18 mm) long

Flowers: 1 to 3 at nodes; calyx tube bell-shaped, greenish or pinkish, glabrous outside, hairy inside, lobes 5, about twice petal length, white, spreading; petals erect, white; stamens 5, equaling or exceeding petals, attached to calyx tube rim and extending downward on tube

Fruits: berries globular, smooth, dark red-purple

Twigs: brownish to purplish, exfoliating into papery flakes, glabrous; weak internodal prickles, stiff nodal spines 0.2-0.3 in (4-8 mm) long; leaf scars raised, narrow crescent-shaped; pith greenish

Buds: narrow egg-shaped, pointed; scales thin, dry, glabrous, slightly keeled, margins hairy (ciliate), turned out

Bark: tight, with large horizontal lenticels; becoming loose in thin strips on very old stems

Rosaceae Juss. - **rose family**

A bite from a peach or an apple may have been one of your first encounters with the rose family. Or you may remember most vividly the white spring blooms on the cherry tree or your pricked finger as you plucked a yellow rose from grandma's arbor.

The many and varied members of the rose family provide us with edible fruits and wood for fine furniture. They beautify our gardens, provide habitat for wildlife, and protect our farmsteads from wind. They also provide a pharmacopia of herbal medicines. For instance, a spirea extract is the source of a key ingredient of aspirin, poultices of blackberry and potentilla have been used as folk remedies for toothache, and cherrybark extract is a cough remedy.

Authorities differ on the number of species belonging to the rose family. Elias (1980) places approximately 120 genera and 3,600 species of herbaceous and woody plants in the family. Cronquist (1981) includes about 100 genera and 3,000 species, most of which are native to the Northern Hemisphere. Forty-five woody species of the rose family are native to the North Central Plains (Stephens 1973).

Natural interbreeding among closely related rose species has resulted in new combinations of stem, leaf and flower characteristics to confound the taxonomist. However, this compatibility within and among the rose family species is a boon to man because plant breeders can easily interchange genetic traits among plants to produce hybrids with a great variety of flowers, fruits and forms, as well as resistance to pests and adaptation to a wide range of site conditions.

At first glance, flowers of the rose family species are as remarkably similar as the fruits are different. Both sexes are usually present in the same flower (bisexual or perfect), and the petals are in fives except for some "double-flowered" varieties. Fruits are dry or fleshy and differ greatly in appearance and taste.

Recognition

Habit: trees, shrubs, herbs; deciduous, rarely evergreen

Leaves: alternate, rarely opposite; simple, variously compound or dissected; stipules usually present and often prominent, sometimes attached to petioles

Flowers: solitary, petals often showy, distinct and equal or often in clusters (more or less cymose); usually bisexual, regular and 5-parted; stamens numerous, often in sets of 10 or 20; pistils 1 to many, separate or united, often enclosed by a hypanthium to which sepals, petals and stamens are attached (ovary inferior or perigynous)

Fruits: dry (akenes or follicles) or fleshy (pomes, drupes or aggregates of drupelets)

Grouping Key for Rosaceae

(Rehder 1940)

Shrubs, rarely trees or herbs; leaves simple or compound, with or without stipules; flowers usually small, in large clusters, white to pink or purple; pistils 2 or more, each with several ovules; fruits dry, splitting (follicles), seeds several, rarely non-splitting (achenes) with 1 seed; twigs unarmed: **subfamily Spiraeoideae**

Trees, shrubs; leaves alternate, simple or compound, stipules small, early deciduous; flowers solitary or in clusters, pistils 1, ovary inferior, fused with calyx tube (hypanthium), carpels 2-5; fruits more or less fleshy (pomes), flesh derived from enlarged hypanthium enclosing bony, leathery or papery carpels; twigs sometimes armed: **subfamily Maloideae**

Herbs, shrubs, rarely trees; leaves deciduous, sometimes evergreen; simple or often compound, stipules present; flower clusters variable; pistils more than 2, each with single ovule, ovaries superior; fruits dry, non-splitting (akenes) or coherent drupelets; twigs often armed: **subfamily Rosoideae**

Trees, shrubs; leaves simple, stipules present, usually early deciduous, petioles often glandular; flowers solitary or in clusters, petals white to pink, rarely wanting; hypanthium usually cup-shaped, free from ovary, deciduous; pistil 1 with single ovule; stamens 10 to many; fruits drupes, 1, rarely 2-seeded, sometimes splitting; twigs sometimes spiny: **subfamily Prunoideae**

Subfamily Spiraeoideae

Chamaebatiaria Maxim. - tansybush

The sole *Chamaebatiaria* species is native to the southwestern U.S.

Chamaebatiaria millefolium (Torr.) Maxim. - tansybush

Tansybush plants grown from seed collected in northern Arizona and southern Utah have been evaluated on Great Plains sites by the North Central Plant Introduction Station (1966). These plants failed in the north central prairie, but several specimens thrive at Colby, Kansas, and the Denver Botanical Gardens. Further trials of the drouth-tolerant tansybush for use in the southwestern prairie should be rewarding.

Planning

Height: up to 5 ft (1.5 m)
Habit: shrub; deciduous, but persist into winter; spreading, flowers June-August
Site: sun; prefers well-drained soils and low humidity; zone 5
Use: specimen, border

Recognition

Leaves: alternate, simple; ovate-oblong to much longer than wide, 1-3 in (25-75 mm) long; deeply divided (pinnatifid) with numerous small segments; gray-green; stipules much longer than wide, entire; petioles short
Flowers: clusters terminal, leafy, hairy, 3-6 in (8-15 cm) long; flowers about 0.5 in (12 mm) across; petals 5, white; calyx tube inversely conical, sepals 5, erect; stamens about 60
Fruits: follicles, splitting, hairy to nearly glabrous, about 0.2 in (5 mm) long, style upright; seeds few, terete
Twigs: star-shaped hairy (stellate), glandular
Bark: cherry brown

Exochorda Lindl. - pearlbush

The four *Exochorda* species, natives to Asia, are handsome spring-flowering shrubs with large, white flowers in terminal clusters (racemes). *E. giraldii* Hesse and *E. korolkowii* Lav., very similar in apearance to common pearlbush, are adapted to the southeastern prairies (Dirr 1990).

Exochorda racemosa (Lindl.) Rehd. - common pearlbush

Common pearlbush, a native of eastern China, is seldom grown in today's landscapes, despite its freedom from pests and wide site adaptability. Pearlbush derives its common name from the tight, white, spherical flower buds. Its appearance will be improved and future blooming will be aided by removing some of the largest canes after flowering.

Exochorda ×*macrantha* Schneid., a hybrid (*E. racemosa* × *E. korolkowii*), blooms more profusely than common pearlbush. 'The Bride' is a bushy dwarf growing to about three feet (1 m) tall (Dirr 1990).

Exochorda racemosa

Planning

Height: up to 15 ft (4.5 m)
Habit: shrub; loose, irregular; deciduous
Site: sun, light shade; adaptable, tolerates heat, drouth; zone 5
Use: border, windbreak

Recognition

Leaves: alternate, simple; elliptic to elliptic-oblong, 1-2 in (25-50 mm) long; margins smooth (entire) or toothed (serrate) above the middle on vigorous shoots; underside whitish
Flowers: clusters (racemes) of 6 to 10 at the end of short, lateral stems of the previous year; flowers white, 1.5 in (37 cm) across, odorless
Fruits: capsules broad, top-shaped, 5-valved, persist into winter
Twigs: slender, brown, roughened by lenticels and longitudinal fissures
Buds: egg-shaped (ovoid); scales about 10, fringed

Neillia D. Don

About 10 *Neillia* species—graceful shrubs with spreading, slender branches—are natives of the Himalayas and China.

Neillia sinensis Oliv. - Chinese neillia

Chinese neillia, a native of central China, is little troubled by insects and disease (Dirr 1990).

Planning

Height: up to 6 ft (2 m)
Habit: shrub; deciduous; rounded with slender, spreading branches
Site: sun, part shade; soil adaptable; zone 6
Use: grouping, border

Recognition

Leaves: ovate to oblong, about 2.5 in (6 cm) long; doubly toothed (serrate), slightly lobed; tip pointed, base squared (truncate) to rounded; nearly glabrous at maturity, bright green; stipules long, narrow, entire, early deciduous (caducous)

Flowers: clusters of 12 to 20 (racemes), nodding, about 2 in (5 cm) long; petals pink; calyx tube cylindric with glandular bristles, lobes much shorter than tube, erect, about as long as petals

Fruits: pods enclosed by calyx tube, splitting along inner suture; seeds lustrous

Twigs: glabrous, brown, peeling with age

Buds: egg-shaped (ovoid), often one above another (superposed)

Petrophytum Rydb.

Five species of *Petrophytum*—low-growing evergreen, prostrate subshrubs—grow in western North America.

Petrophytum caespitosum (Nutt.) Rydb. - **rock-spirea**

Rock-spirea, native to the Black Hills and the mountains of the western U.S., grows on rocky, wooded hillsides. Its flowers and fruits are very similar to those of the *Spiraea* species.

Planning

Height: up to 2 in (5 cm)

Habit: subshrub; evergreen, retains brown leaves; tuft- or mat-forming (caespitose), up to 2 ft (60 cm) across

Site: part shade; rock crevices, prefers moist, limestone soil; zone 4

Use: rock garden

Recognition

Leaves: simple, 1-nerved; alternate, rosette-like; narrow, wider at outer end (spatulate), about 0.4 in (1 cm) long; margin entire, surfaces silky-hairy (sericeous); stipules and petioles wanting

Flowers: clusters spike-like (racemes), upright, crowded, about 0.8 in (2 cm) long; stalks about 1.2 in (3 cm) long; calyx tube top-shaped (turbinate), lobes hairy, green, often red-tinged; petals white, narrow, wider near tip (spatulate); stamens about 20, exserted, anthers pale yellow; pistils 5, tightly clustered, distinct

Fruits: follicles, usually 5; calyx persistent

Twigs: coarse; retain brown leaves for several years

Buds: terminal only, naked

Physocarpus Maxim. - **ninebark**

The ninebarks are shrubby, multistemmed plants of spreading habit, seldom growing over ten feet tall. Most are natives of North America. The common name "ninebark" refers to their multicolored, shreddy and exfoliating bark. Generally, the plants possess clusters of few to many flowers that are effective but not showy. Slender, arching branches tend to flatten from snow loads. Winter injury is often the cause of an unkempt appearance.

Recognition

Habit: shrubs; deciduous

Leaves: alternate; fine toothed (serrate); usually 3-lobed with 3 veins joining at base

Flowers: clusters (corymb-like), bisexual; petals spreading, white or rarely pinkish; stamens 20-40; pistils 1-5, united at base

Fruits: dry (follicles), inflated, opening along two sutures; seeds 2-5, yellowish, shiny

Twigs: bark peeling in thin strips

Buds: small; scales 5, brown

Physocarpus malvaceus (Greene) Ktze. - **mallow ninebark**

Mallow ninebark, the most common species of *Physocarpus* found in western North America, grows on a variety of sites: dry rocky slopes, streambanks in the open and in timber, and across a wide altitudinal range. It is a native of an area that extends from British Columbia to Oregon, Nevada, Wyoming and Montana (U.S. Forest Service 1937). Fairchild and Klett (1993) say that mallow ninebark is a shrub for use on difficult sites; it is winter hardy and displays only minor chlorosis on new growth and has shown no insect or disease problems at PERC.

Planning

Height: up to 8 ft (2.5 m)
Habit: shrub; dense, branches arching; May-June flowers; fruits persist into winter; bark exfoliating
Site: sun, part shade; site adaptable; drouth tolerant; zone 3
Use: border, screen, grouping, natural areas

Recognition

Leaves: rounded to egg-shaped (ovate), often broader above the middle (obovate), 3 to 5 lobes; doubly toothed (serrate); base often heart-shaped (cordate); upper surface glabrous or hairy (stellate-pubescent), stellate pubescence beneath
Flowers: clusters terminal (corymbs), tops rounded; flowers about 0.4 in (1 cm) across; petals white; calyx 5-lobed, persistent
Fruits: follicles, densely hairy (stellate pubescence); usually 2 united for about half the length; calyx folded over; flattened with erect beaks; splitting when falling; seeds 1-several, shiny
Twigs: hairy (stellate pubescence)

Physocarpus monogynus (Torr.) Coult. - **mountain ninebark**

Mountain ninebark grows on the dry, rocky soils of generally north-facing slopes in the Black Hills and southern Rocky Mountains. It is very similar in size and form to *P. opulifolius* var. *intermedius*, but it sends out root suckers and is characterized by star-shaped hairs (stellate pubescence) on foliage and flowers.

Mountain ninebark is probably best adapted to the western prairies because, like many other plants native to our arid West, it may be injured by leaf diseases in humid climates (North Central Plant Introduction Station 1975).

Planning

Height: up to 3 ft (1 m)
Habit: shrub; erect, compact, suckering, June flowers
Site: sun; soil adaptable; drouth tolerant; zone 3
Use: grouping, low hedge

Recognition

Leaves: distinctly 3-5 lobed, nearly round overall, 1-2 in (25-50 mm) long; lobed, margin toothed (biserrate); upper surface dark green, glabrous or nearly so, pale below with sparse, star-shaped hairs (stellate); petioles grooved above, sparsely hairy
Flowers: clusters terminal (corymbs), tops rounded, 1-1.5 in (25-40 mm) across; hypanthium cup-shaped, calyx lobes triangular, margins whitish; petals white, nearly round, spreading; stamens 30-40; ovaries 2-4
Fruits: follicles, densely stellate-hairy; usually 2, united half the length, flattened with spreading beaks; calyx lobes persistent
Twigs: red-brown, glabrous or nearly so; ridge extends downward on each side of leaf scar
Buds: egg-shaped (ovoid); outer scales loose, marginal hairs (ciliate)

Physocarpus opulifolius (L.) Maxim. - **common ninebark**

Common ninebark grows naturally in scattered locations—generally along streambanks in the Black Hills, north-central Nebraska, parts of Colorado, and eastward to New York and Virginia (Stephens 1973). Although adapted to a wide range of site conditions, common ninebark may develop chlorosis on alkaline soils (Richley 1988).

A rather coarse, unkempt shrub, common ninebark is primarily suited to natural areas and mass plantings on large landscapes. The common ninebark's tangled twigs often hide its attractive, brown inner bark that is exposed by the shredding, papery outer bark.

Stephens (1973) states that variety *intermedius* (Rydb.) Robins. is a native of the northern Great Plains. Its most obvious characteristic is permanent star-shaped hairs on the fruits. Dirr (1990) adds that this variety is a low-growing, fine-textured form. Leaves of 'Luteus' and 'Dart's Golden' are initially yellow, becoming green with age. Variety *nanus* (Kirchn.) Zab. is a dwarf form that transplants easily and is adapted to a wide variety of site conditions (North Central Plant Introduction Station 1960).

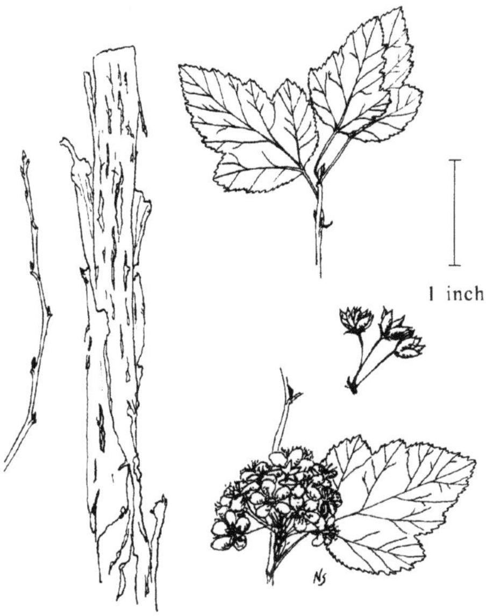

Physocarpus opulifolius

Planning

Height: up to 9 ft (3 m)
Habit: shrub; coarse; May flowers on new growth
Site: sun, partial shade; site adaptable except for high pH; zone 3
Use: border, screen

Recognition

Leaves: round to oval in outline, 3 to 5 small lobes, irregularly toothed (crenate-serrate); upper surface glabrous, sometimes star-shaped (stellate) hairs on veins beneath; petioles grooved above, glabrous, 0.25-1 in (6-25 mm) long
Flowers: clusters (corymbs) of 10 to 30, 1-2 in (25-50 mm) in diameter; calyx a shallow cup, lobes triangular, hairy; petals white or pink, orbicular, spreading; stamens 25-35; pistils 4-5, woolly
Fruits: follicles 3-5, inflated, hairless (except var. *intermedius*), reddish, 0.25-0.5 in (6-12 mm) long; seeds yellow, glossy, smooth
Twigs: shiny, brownish, glabrous, a narrow ridge extending downward from each side of the leaf scar
Buds: egg-shaped (ovoid), appressed to stem, slightly hairy

Sorbaria A. Br. - falsespirea

Sorbaria, a genus of seven species that is native to Asia, derives its name from *Sorbus* (mountain ash) because of its leaf similiarity. Showy terminal clusters of white flowers resemble those of the spireas, hence the common name.

S. arborea Schneid., tree falsespirea, bears loose, spreading flower clusters and sharply double-toothed leaflets, sometimes hairy beneath (stellate-pubescent). Fruit stalks are recurved. Tree falsespirea is adapted to zone 6 (Rehder 1940).

S. aitchisonii Hemsl., Kashmir falsespirea, is usually red-stemmed with hairless, single-toothed leaflets. Flowers are borne on upright clusters. It is worthy of trial plantings, especially in the southern portion of the central prairie.

Recognition

Habit: shrubs; deciduous
Leaves: alternate, compound (pinnate), toothed (serrate or doubly serrate), stipulate
Flowers: clusters terminal (panicles); calyx tube cup-shaped, sepals reflexed; petals white, oval to orbicular, overlapping (imbricate); stamens 20-50
Fruits: follicles splitting on inner face (ventral); seeds numerous
Twigs: cylindrical (terete) in cross-section
Buds: egg-shaped (ovoid); scales several

Sorbaria sorbifolia (L.) A. Br. - Ural falsespirea

Ural falsespirea is adapted to much of the prairie region where water is ample. Plants at PERC in Fort Collins suffer dieback during severe winters but grow back vigorously and bloom reliably (Fairchild and Klett 1993).

This large, coarse-textured suckering shrub with lush foliage produces a thicket-like growth that may be undesirable in formal landscapes but useful for erosion control on steep slopes where space is available.

Planning

Height: up to 15 ft (4.5 m)
Habit: shrub; coarse, multistemmed, suckering; flowers June-July on new wood
Site: sun, light shade; pH and soil adaptable; zone 3
Use: border, grouping, bank cover, naturalizing

Recognition

Leaves: leaflets lance-shaped, 13-25, 2-4 in (5-10 cm) long, margins sharply toothed (doubly serrate) from base to tip; surfaces usually glabrous
Flowers: clusters terminal (panicles), upright, 4-10 in (10-25 cm) long; calyx glabrous; stamens twice as long as petals
Fruits: follicles glabrous; style terminal, recurved
Twigs: stout, greenish pinkish downy, becoming gray-brown, glabrous; pith large, brown and continuous

Sorbaria sorbifolia

Spiraea L. - spirea

Although several spireas are native to North America, most of the cultivated species are originally from temperate areas of Asia. More than 80 species and a large number of natural and man-induced hybrids are recognized within this genus. It is difficult to distinguish among the species without all plant parts available for examination.

Individual spirea flowers are small, but flower clusters are generally large and showy. The spirea's dry fruit clusters simply fade into the brownish twigs.

Most spireas are relatively free of insects and diseases, though soils with a high lime content, a limited water supply, or overcrowding and shading can cause unsatisfactory growth. Spireas require timely and regular pruning for best flowering. Ascertain whether flowers are borne on new or old growth to prevent destruction of flower-bearing stems before they are able to bloom.

Recognition

Habit: shrubs; deciduous
Leaves: alternate, simple; toothed, or rarely entire, sometimes lobed
Flowers: clusters (racemes, corymbs, or panicles); bisexual, rarely otherwise; sepals and petals 5; stamens 15-60, pistils usually 5, distinct
Fruits: follicles split along inner suture (ventral); seeds several, minute
Twigs: slender
Buds: small; scales several

Grouping Key for *Spiraea*

(Rehder 1940)

Flower clusters (fascicles or umbrella-like racemes) sessile, axillary on branches of the previous season; borne in spring or early summer; flowers white: **section Chamaedryon**

Flower clusters (corymbs) compound, flat or hemispheric, broader than high; usually appearing in midsummer; flowers white, often pink to purple: **section Calospira**

Flower clusters (panicles) dense, pyramidal, longer than broad, on twig tips; appearing in midsummer; flowers white to pink: **section Spiraria**

Section Chamaedryon

Spiraea ×*cinerea* Zab. (*S. cana* × *S. hypericifolia*) - cinerea spirea

'Grefsheim', a selection of cinerea spirea, is early flowering.

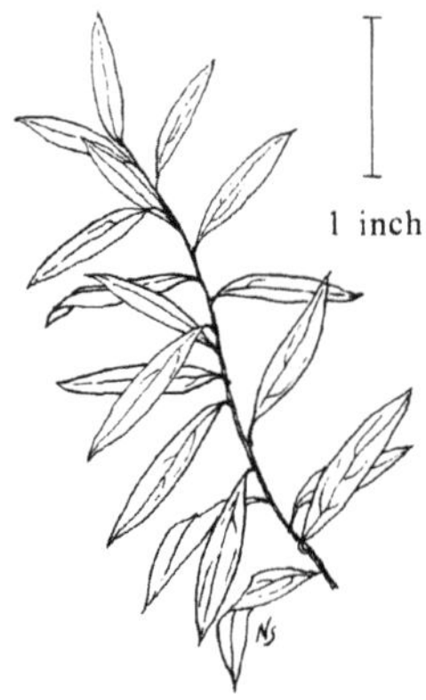

Spiraea ×*cinerea* **'Grefsheim'**

Planning

Height: up to 5 ft (1.5 m)
Habit: shrub; dense, arching stems; blooms before leaves
Site: sun; moist, fertile soils; zone 5
Use: grouping, border

Recognition

Leaves: much longer than wide (narrow-elliptic to lanceolate), about 1 in (25 mm) long; tip pointed, base tapered; margin entire or fine-toothed at apex; surface glabrous above, appressed silky hairs beneath
Flowers: clusters stalkless along the twigs; petals white, longer than stamens

Spiraea nipponica Maxim. - nippon spirea

Nippon spirea, a multistemmed shrub from Japan, may suffer chlorosis on alkaline soils and winter injury in the north and west of the central Great Plains, according to field tests (North Central Plant Introduction Station 1976). 'Snowmound' and 'Halward's Silver' suffer minor to moderate dieback at PERC in Fort Collins, but flower reliably; chlorosis is minor, mainly on new growth (Fairchild and Klett 1993).

'Snowmound', a dwarf selection, is often confused with variety *tosaensis* (Yatabe) Mak., but the leaves of the cultivar are wider, approaching an oval shape (Dirr 1990).

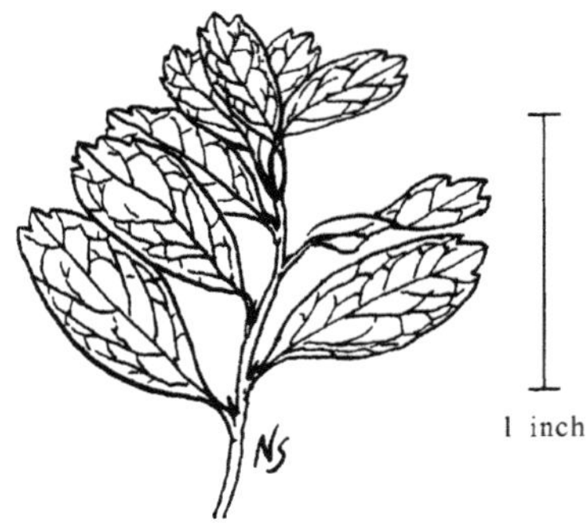

Spiraea nipponica

Planning

Height: up to 7 ft (2 m)
Habit: shrub; upright, stiff, spreading; blooms late May-early June
Site: sun; moist soil, pH adaptable; zone 4
Use: border, grouping, specimen, hedge

Recognition

Leaves: wider above the middle (obovate) to oval, about 1 in (25 mm) long; rounded teeth at apex (crenate), rarely entire; surfaces glabrous or nearly so, dark green above, blue-green below; texture papery
Flowers: clusters many-flowered, rounded, on leafy stalks; petals nearly round, longer than stamens
Fruits: follicles upright, slightly hairy with spreading style
Twigs: angled, glabrous
Buds: short; scales 2 visible

Spiraea prunifolia Sieb. & Zucc. - **bridalwreath spirea**

The white flowers of bridalwreath spirea are harbingers of spring in many yards, and the colorful, orange autumn foliage is a consistent occurrence. However, if not properly pruned the bridalwreath becomes straggly and stingy with its flowers. Bridalwreath spirea is often replaced by more floriferous spirea hybrids or other shrubs of more pleasing form.

Double-petaled variety *plena* Schneid. is more commonly cultivated than variety *simpliciflora* Nakai, the single-flowered variety.

Spiraea prunifolia

Planning

Height: up to 9 ft (3 m)
Habit: shrub; open, coarse, often leggy; blooms late April-early May on old wood
Site: sun; soil adaptable; zone 5
Use: border

Recognition

Leaves: elliptic to elliptic-oblong, about 1.5 in (4 cm) long; tip pointed, base narrowed, minutely toothed (denticulate); glabrous or sparsely hairy below; petioles about 0.1 in (2 mm) long
Flowers: clusters (umbels) sessile, 3 to 6 flowers, leafy bracts at base; petals inverted egg-shape (obovate), longer than stamens
Fruits: follicles spreading, glabrous
Twigs: angled, slender, zigzag; hairy or nearly glabrous

Spiraea thunbergii Sieb. - **Thunberg spirea**

Thunberg spirea, an early flowering shrub with tiny leaves on slender branches, is a native of China and Japan. Garland spirea, *S.* ×*arguta* Zab. (*S. thunbergii* × *multiflora* Zab.) is similar to Thunberg spirea except that it is taller and more vigorous, and blooms later than the Thunberg.

Although Thunberg spirea is often sheared, more flowers will be obtained on unpruned plants. However, without corrective pruning, both Thunberg and garland spirea will become open and straggly.

Planning

Height: up to 5 ft (1.5 m)
Habit: shrub; spreading, arching, fine textured; blooms before leaves in April on old wood; colorful autumn foliage
Site: sun; soil adaptable; zone 5
Use: border, grouping

Spiraea thunbergii

Recognition

Leaves: much longer than wide (linear-lanceolate), about 1 in (25 mm) long, 0.2 in (5 mm) wide; sharply toothed (serrate); surfaces glabrous, light green
Flowers: clusters (umbels) sessile, 3 to 5 flowers; petals inverted egg-shaped (obovate), much longer than stamens
Fruits: follicles diverging, terminal style spreading
Twigs: slender, zigzag, light brown

Spiraea trilobata L. - threelobe spirea

Threelobe spirea is native to the harsh climates of northern China, Siberia and Turkestan. This hardy, profuse bloomer, even under shady conditions, fills the need for a compact shrub on small properties (Hoag 1965).

'Swan Lake' is compact and floriferous and 'Fairy Queen' is distinguished by its dark green lobed leaves and frothy masses of white flowers in May and June (Dirr 1990).

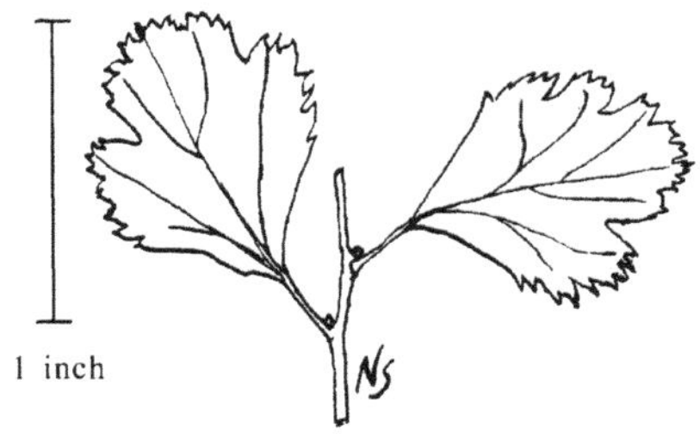

Spiraea trilobata

Planning

Height: up to 5 ft (1.5 m)
Habit: shrub; dense, compact, arching, bloom April-May
Site: sun, partial shade; soil adaptable; zone 4
Use: border, grouping

Recognition

Leaves: round in general outline, about 0.8 in (2 cm) across; usually obscurely 3- or sometimes 5-lobed, 3- to 5-nerved at base: coarsely toothed (crenate-dentate); surfaces glabrous, pale blue-green beneath
Flowers: clusters (umbels) many-flowered; petals longer than stamens
Fruits: follicles slightly diverging, obtuse, style ascending; sepals persistent, upright
Twigs: slender, zigzag, glabrous

Spiraea ×*vanhouttei* (Briot) Zab. (*S. cantonensis* Lour. × *S. trilobata*) - Vanhoutte spirea

The popularity and hardiness of Vanhoutte spirea is attested to by its continued presence throughout the prairies around abandoned farmsteads. These neglected shrubs often receive a helpful, albeit haphazard, pruning by hungry rabbits during snowy winters. Insects and diseases pose no problems and it is adapted to alkaline soils (Fairchild and Klett 1993).

Abundant, white spring flowers, borne in tight clusters along arching branches, are the trademark of Vanhoutte spirea. Autumn sometimes colors the foliage a reddish purple.

Planning

Height: up to 8 ft (2.5 m)
Habit: shrub; rounded, fine textured; blooms April-May-June
Site: sun; pH and soil adaptable; zone 3
Use: border, specimen, grouping

Recognition

Leaves: egg-shaped, sometimes wider above middle (obovate), about 1.2 in (3 cm) long; often slightly 3- to 5-lobed; tip pointed, base tapering; irregularly coarsely toothed above the middle (incised-serrate); surfaces glabrous, bluish green beneath
Flowers: clusters (umbels) many-flowered; flowers 0.3 inch (8 mm.) across; petals twice as long as the sometimes sterile stamens
Fruits: follicles slightly diverging, styles half upright; persistent sepals spreading
Twigs: glabrous

Spiraea ×vanhouttei

Section Calospira

Spiraea albiflora (Miq.) Zab. - Japanese white spirea

Japanese white spirea, a compact shrub with stiff, upright branches, produces an abundance of white flowers (Rehder 1940). Dirr (1990) states that it has also been listed as *S. japonica* var. *alba* or 'Alba', but if *japonica* is preferred, the cultivar name should be 'Albiflora'.

Planning

Height: up to 1.5 ft (50 cm)
Habit: shrub; dense, upright; blooms June-July on new wood
Site: sun; requires acidic soil; zone 5
Use: facer, border, grouping

Recognition

Leaves: much longer than wide, broader below middle (lanceolate); about 2.5 in (64 mm) long; margin coarsely, sometimes doubly toothed (serrate); tip pointed; base wedge-shaped, not toothed; surfaces glabrous, green above, blue-green below; petioles short
Flowers: clusters (corymbs) dense, minutely hairy (puberulous), terminal large, many smaller ones below bloom later; petals white, shorter than stamens
Fruits: follicles glabrous, slightly or not diverging; styles persistent, spreading; sepals persistent, reflexed
Twigs: angled, glabrous

Spiraea betulifolia Pall. var. *lucida* (Dougl.) C. L. Hitchc. - western birchleaf spirea

Western birchleaf spirea, an Asian-North American species, is quite variable, but all forms of the plant have a common trademark—large flat-topped clusters (corymbs) of white flowers. Variety *betulifolia* has been used by Dr. F. L. Skinner of Manitoba, Canada, as one parent of hybrids 'Rosabella' and 'Summersnow' (Hoag 1965).

The North American native, variety *lucida*, grows in open woods or bushy areas from South Dakota and Wyoming west to Oregon and British Columbia (Stephens 1973). It is not as showy as its Asian cousin, but it has potential for informal shrub masses and could be useful for erosion control. Subalpine spirea, *S. densiflora* Nutt., a native of the same region as *lucida*, bears rose-colored flowers and fruits with upright or spreading sepals.

Planning

Height: up to 1.5 ft (50 cm)
Habit: shrub; many branches, suckering, blooms July-August on new wood
Site: sun; soil adaptable, drouth tolerant; zone 4
Use: border, erosion control, grouping

Recognition

Leaves: oval, about 1.5 in (4 cm) long; coarsely toothed (crenate-serrate), mostly above middle; tip pointed to narrowly rounded, base broadly wedge-shaped (cuneate); surfaces glabrous; petioles glabrous, flat or trough-shaped, about 0.2 in (5 mm) long
Flowers: clusters (corymbs) single, flat-topped; calyx glabrous, green, lobes triangluar, reflexed; petals white, spreading, nearly round, often notched at tip; stamens 15-20
Fruits: follicles glabrous, about 0.1 in (2 mm) long, splitting; calyx persistent, erect or reflexed
Twigs: glabrous, red-brown, lenticels inconspicuous
Buds: egg-shaped (ovoid), glabrous; scales sometimes with a few marginal hairs (ciliate)

Spiraea bullata Maxim. **- crispleaf spirea**

Crispleaf spirea, a native of Japan, is a low-growing shrub with unusual foliage and rosy pink flowers. It should be on the list for trial plantings.

Planning

Height: up to 1.5 ft (45 cm)
Habit: shrub; upright, blooms June-July
Site: sun; experimental; zone 5
Use: groundcover, grouping

Recognition

Leaves: round to egg-shaped, about 1 in (25 mm) long; toothed (incisely serrate), tip pointed; bluish green above, grayish green below, glabrous, appearing blistered between the veins (bullate); petioles short
Flowers: clusters (corymbs) terminal, dense, compound, 1.5-3 in (40-75 mm) across; lateral clusters smaller, blooming later; flowers deep rosy pink; stamens slightly longer than petals, reddish pink
Twigs: hairy (villous) when young

Spiraea ×bumalda Burvenich **(S. japonica × S. albiflora) - bumald spirea**

This hybrid, bumald spirea, is widely planted. Its best-known cultivars, 'Anthony Waterer' and 'Froebelii', are similar in appearance. Their flowers are pinkish, but the former has narrower leaves and blooms later than 'Froebelii'. Other bumald spirea cultivars are: 'Crispa' with its twisted leaves, 'Gold Flame' with mottled yellowish foliage, and 'Norman' and 'Gumball', both short and compact.

Bumald spirea and its cultivars are subject to chlorosis on high lime soils and have been injured by drouth and cold in western Nebraska (North Central Plant Introduction Station 1957) and at Fort Collins (Fairchild and Klett 1993).

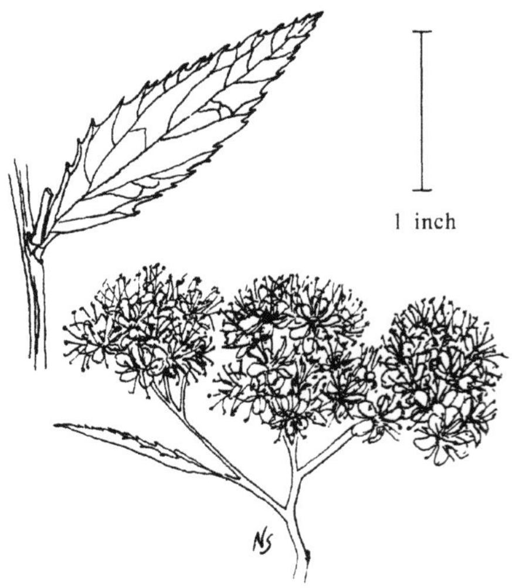

Spiraea ×bumalda

Planning

Height: up to 3 ft (1 m)
Habit: shrub; dense, erect, flat-topped, suckering; blooms June-August on new wood
Site: sun; prefers acidic, moist soil; zone 4
Use: facer, specimen, foundation, erosion control, grouping
Suggestion: treat as herbaceous perennial in harsh climates

Recognition

Leaves: much longer than wide (ovate lanceolate), doubly toothed (serrate); tip pointed, base narrowed; glabrous, pinkish red-purple when young, becoming dark green above, light gray-green below (varies with cultivar)
Flowers: clusters (corymbs) flat-topped, many-flowered, 4-6 in (10-15 cm) across; flowers white to deep pink
Fruits: follicles diverging
Twigs: glabrous, vertical lines

Spiraea japonica L. - **Japanese spirea**

Some of the many cultivars of Japanese spirea are: 'Nana' (also referred to as variety *alpina*), a fine-textured and low-growing plant with pink flowers; 'Atrosanguinea', with large deep, rose-red flowers; 'Ovalifolia', with white flowers; and 'Ruberrima', with deep pink flowers.

Herman (1985) states that 'Little Princess', a cultivar similar to 'Nana' in appearance, is hardier than the latter and easy to transplant. 'Little Princess' suffers winter dieback at Fort Collins, but dead twigs are well hidden by the dense foliage except during severe winters if it has been killed back to the ground (Fairchild and Klett 1993). Fairchild and Klett report that 'Little Princess' suffers minor chlorosis but flowers reliably.

Planning

Height: up to 5 ft (1.5 m)
Habit: shrub; rounded, but quite variable; foliage red, becoming green with bluish cast; blooms June-July on new wood
Site: sun; moist, fertile soil, low alkalinity; zone 4
Uses: facer, border, grouping

Recognition

Leaves: egg-shaped (ovate) to ovate-oblong, 1-3 in (2-8 cm) long; doubly toothed (incised-serrate), tip pointed, base narrowed (cuneate); pale beneath, usually hairy on the veins; petioles about 0.1 inch (3 mm) long
Flowers: clusters (corymbs) terminal, loose, compound, about 4 in (10 cm) across; lateral clusters smaller, blooming later; petals pale to deep pink, sometimes white; stamens much longer than petals
Fruits: follicles glabrous, diverging
Twigs: glabrous, round (terete) in cross-section

Spiraea ×superba (Froebel) Zab. **(*S. albiflora* × *S. corymbosa*** Raf.**) - striped spirea**

Striped spirea is little known, but worthy of trial.

Planning

Height: up to 1.5 ft (45 cm)
Habit: shrub; upright; flowers on new wood
Site: experimental
Use: grouping, border

Recognition

Leaves: oblong, about 2 in (5 cm) long; tip pointed, base narrowed, toothed (serrate)
Flowers: clusters (corymbs) terminal, solitary; petals light rose or whitish pink; stamens twice the petal length
Fruits: follicles parallel; sepals persistent, reflexed
Twigs: striped, dark brown

Spiraea wilsonii Duthie - Wilson spirea

Wilson spirea, a native of central and western China, is an attractive plant with a dense growth habit that is similar in appearance to Vanhoutte spirea. Fairchild and Klett (1993) state that Wilson spirea is adapted to the soils and climate of the Fort Collins area and blooming is consistent.

Planning

Height: up to 9 ft (3 m)
Habit: shrub; dense, rounded with arching branches, fine-textured; bloom May-June on wood of previous year
Site: sun; pH and soil adaptable, moist; zone 4
Use: border, hedge, specimen, grouping

Recognition

Leaves: variable, oval to oblong or wider above the middle; apex variable, base wedge-shaped; margin toothed (dentate) toward apex or entire; dull green and hairy above, grayish and long-hairy beneath; petioles very short
Flowers: clusters (corymbs) dense, glabrous, on short lateral branches, about 1.5 in (4 cm) across; petals white; stamens about as long as petals
Fruits: follicles spreading, sparingly hairy on the inner suture
Twigs: hairy, becoming glabrous
Buds: egg-shaped; scales several

Section Spiraria

Spiraea alba Du Roi - meadowsweet

Meadowsweet, a low-growing, suckering shrub native to the eastern U.S., also grows east of the Missouri River in the Dakotas where it frequents sandy soils in and around shallow potholes and thrives in the dappled shade of open aspen groves (Stephens 1973). Meadowsweet is seldom, if ever, listed in nursery catalogs.

Planning

Height: up to 3 ft (1 m)
Habit: shrub; upright, dense, suckering; midsummer flowers on new wood
Site: sun, light shade; experimental; zone 3
Use: grouping, erosion control

Recognition

Leaves: oblong, wider above middle, 1-2 in (25-50 mm) long; sharply toothed (serrate) except on lower portion; tip pointed, base long tapered; surfaces glabrous or minutely hairy (puberulous); petioles trough-shaped, about 0.1 in (3 mm) long
Flowers: clusters (panicles) woolly, pyramidal; flowers white; stamens white, longer than petals
Fruits: follicles glabrous, sepals persistent, upright
Twigs: glabrous to minutely hairy with fine longitudinal lines (striated), lenticels conspicuous, 3 ridges extending down from each leaf scar
Buds: egg-shaped; scales glabrous, margins hairy (ciliate)

Spiraea alba

Spiraea ×*billiardii* (*S. douglasii* × *S. salicifolia*) - billiard spirea

'Triumphans', a selection of the hybrid billiard spirea, bears dense, conical clusters of purplish rose flowers and is considered to be among the "finest of late-flowered spireas" (Dirr 1990).

Spiraea ×billiardii

Planning

Height: up to 6 ft (2 m)
Habit: shrub; upright, suckering; blooms June-August on new wood
Site: sun; prefers acidic soil; zone 4
Use: border, grouping

Recognition

Leaves: oblong, 1-3 in (25-75 mm) long; sharply and often doubly toothed (serrate) above lower third of margin; tip pointed, base wedge-shaped; gray hairs beneath when young
Flowers: clusters (panicles) narrow, dense, 4-8 in (10-20 cm) long; petals bright rose, stamens twice as long
Fruits: follicles parallel, glabrous; styles spreading
Twigs: hairy, brown

Stephanandra Sieb. & Zucc. - stephanandra

The four species of *Stephandra* are all natives to eastern Asia. Tanaka stephanandra, *S. tanakae* Franch., is a vigorous shrub but is not preferable to *S. incisa* for landscape planting (Dirr 1940). Tanaka stephandra is characterized by shallow-lobed leaves with doubly serrate teeth, flower clusters subtended by long oblong bracts, and white flowers with 15 to 20 stamens.

Stephanandra incisa (Thunb.) Zab. - cutleaf stephanandra

The zigzag winter branch pattern of cutleaf stephanandra, a compact, rounded shrub, is distinctive. It should withstand the rigors of winter in the central prairies but it is not recommended for planting in the northern Great Plains (Hoag 1965). 'Crispa', a more compact plant, can serve as a woody groundcover, especially on steep banks (Dirr 1990).

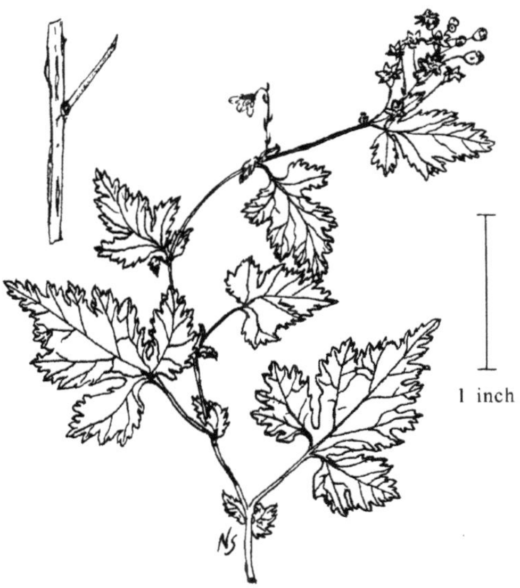

Stephanandra incisa

Planning

Height: up to 7 ft (2 m)
Habit: shrub; deciduous; dense, fine-textured, arching, suckering; fall foliage reddish, May-June flowers not showy
Site: sun, light shade; prefers moist, well-drained, acidic soil; zone 5
Use: hedge, screen, border, groundcover, erosion control

Recognition

Leaves: simple, alternate, 2-ranked, egg-shaped in outline; about 1.5 in (4 cm) long; lobes sharply and deeply cut, margins notched, lower incisions halfway to midrib; bright green above, veins hairy beneath; stipules linear, sparingly toothed; petioles up 0.5 in (12 mm) long

Flowers: clusters (panicles) terminal, loose, about 2 in (5 cm) long; bracts awl-shaped, shorter than stalks; flowers bisexual, petals 5, greenish or yellowish white, inconspicuous; stamens 10

Fruits: follicles split open at base; seeds 1-2, globose, lustrous

Twigs: slender, round (terete), zigzag, brown

Buds: egg-shaped; 2 at each leaf scar, one above the other (superposed); scales 4

Subfamily Maloideae

Amelanchier Med. - serviceberry

The genus *Amelanchier* consists of about 25 species of small trees and shrubs that inhabit the continents of the Northern Hemisphere. Their delicate white flowers, which appear ahead of the leaves, herald the approach of spring.

The small, fleshy fruits of serviceberry mature in early summer and are attractive and useful to many small animals, birds and people. Native Americans worked serviceberry fruits into their dried pemican. Canadian scientists are selecting large-fruited cultivars for commerce.

These hardy plants, primarily natives of cool climates, are relatively free of disease except for leaf rusts that may prevail where temperatures are warm and humidity is high.

Utah serviceberry, *A. utahensis* Koehne, is a native of the foothills and medium elevations of Colorado, New Mexico and Nevada. This shrub or small tree is very similar to *A. alnifolia*. The ranges of these two species overlap in the central and southern Rockies. It would be useful to observe Utah serviceberry in test plantings along with provenances of the other *Amelanchier* species, especially those native to the prairies.

Running serviceberry, *A. stolonifera* Wieg., is a small, dense shrub that spreads by stolons. It is suitable for natural areas and mass plantings. Running serviceberry specimens at PERC in Fort Collins exhibit slight chlorosis on new growth and dieback occurs during severe winters (Fairchild and Klett 1993).

Serviceberries are closely related to mountainashes (*Sorbus*) and chokeberries (*Aronia*). An intergeneric hybrid, ×*Amelasorbus* Rehd., has been recognized by Rehder (1940).

Authorities do not agree on the names or descriptions of several *Amelanchier* species. We describe in detail three species that are native to the Great Plains with Stephens (1973) as the authority for nomenclature.

Recognition

Habit: trees, shrubs; deciduous

Leaves: alternate, simple; egg-shaped to oblong, usually widest at or below the middle

Flowers: clusters terminal, flowers rarely solitary; bisexual; petals 5, white; calyx tube cup-shaped with 5 short lobes, adheres to ovary; stamens 10-20

Fruits: pomes, small, berry-like, blue-black, crowned by calyx lobes

Buds: conspicuous, pointed

Amelanchier alnifolia Nutt. - Saskatoon serviceberry

Saskatoon serviceberry is a native of an area that stretches from Alaska to northern California, east to Wisconsin and south into northern Nebraska, Colorado and New Mexico. Closely related *A. florida* Lindl. is considered a synonym of *A. alnifolia* by some authorities. Saskatoon serviceberry grows on streambanks, hillsides and mountain slopes that have a wide range of climatic conditions.

Site variability, enhanced by great differences in altitude and latitude throughout its geographic range, contributes to a potential for great genetic diversity among plants in the serviceberry's population. Cultivars 'Dusky' and 'Honeywood' have been selected by Canadian scientists for high quality fruit production. 'Regent', which is considered to be hybrid, is a dense, suckering shrub that is winter hardy in Fort Collins with only minor chlorosis apparent on new leaves (Fairchild and Klett 1993). 'Success' is abundantly fruitful (Dirr 1990).

Amelanchier alnifolia

Planning

Height: up to 15 ft (5 m), National champion, 1975, 27 ft (8.2 m), Oregon
Habit: shrub, small tree; upright, spreading by stolons; white flowers before leaves
Site: sun, partial shade; site adaptable; zone 3
Use: fruit, grouping, border, wildlife habitat, erosion control, naturalizing

Recognition

Leaves: oval to oblong, about 1.5 in (4 cm) long; sharply and coarsely toothed along upper margin, lateral veins entering the teeth; tip rounded to slightly indented, base rounded to heart-shaped; emerging leaves folded lengthwise, long-hairy becoming nearly glabrous; petioles about 0.6 in (15 mm) long, slender, more or less hairy, grooved above
Flowers: clusters (racemes) of 2 to 10 flowers; petals oblong; calyx lobes triangular, reflexed, long-hairy on inside, glabrous on outer surface
Fruits: pomes rounded, about 0.4 in (1 cm) long, near black, sweet; seeds about 6
Twigs: sparsely white-hairy, becoming glabrous
Buds: egg-shaped, pointed, about 0.2 in (5 mm) long; scales overlapping (imbricate), chestnut brown, glabrous or with a few hairs

Amelanchier arborea (Michx.) Fern. - shadblow serviceberry

Shadblow serviceberry grows in the understory of our eastern North American hardwood forests from New Brunswick to Florida, and west to the eastern edge of the Great Plains including the Missouri River valley in Nebraska.

Shadblow serviceberry is known by a confusing array of scientific and common names. Stephens (1973) considers *A. arborea* to be synonymous with *A. canadensis* (L.) Med. Dirr (1990) indicates that the two species are used interchangeably in the nursery trade. Little (1953) states that Fernald has shown that the epithet *canadensis* should be applied to the shrubby species known as *A. oblongifolia* Roem. Rehder (1940) does not recognize *arborea* but uses the specific epithet *canadensis* for a tree with tomentose leaves and *laevis* for a very similar tree with glabrous leaves. Elias (1980) considers *A. arborea* to be a synonym of *A. laevis* Wieg.

Since all of the above species have a similar natural range, we are grouping them together under *A. arborea*. We hope that other taxonomists will some day agree with us, or come to some other reasonable conclusion.

Obviously, this population provides a wonderful opportunity for plant breeders to develop hybrids of great variance in habit, flower and fruit characteristics, and adaptability to a wide variety of sites. *A.* ×*grandiflora* Rehd. is a dependable small tree for much of our region. 'Robin Hill', a pink-flowered selection of this hybrid, grows at PERC in Fort Collins with no problems (Fairchild and Klett 1993).

Several cultivars have been selected for their exceptional autumn color but brilliance is affected by the latitude of the planting site. A partial list from Dirr (1990) includes: 'Autumn Brilliance', selected in Illinois; 'Autumn Sunset', selected in Georgia; 'Ballerina' from the Netherlands; 'Cole' from Ohio; and 'Princess Diana' from Wisconsin.

Amelanchier arborea

Planning

Height: up to 25 ft (7.5 m); Nebraska champion, 1988, 25 ft (7.5 m), Indian Cave State Park; national champion, 1986, 60 ft (18.3 m), Virginia

Habit: shrub, tree; crown rounded, often multistemmed; blooms early spring, autumn foliage reddish

Site: sun, partial shade; soil and pH adaptable, prefers moist, well-drained sites but tolerates poor drainage; zone 3

Use: specimen, wildlife habitat, naturalizing, street, border, fruit

Recognition

Leaves: egg-shaped to oval, about 3 in (75 mm) long; margin finely toothed, tip abruptly pointed, base rounded or slightly heart-shaped, veins branching to form network along margin; dark green and variably hairy above, paler and variably hairy below; petioles glabrous

Flowers: clusters (racemes) of 3 to 15 flowers, variably hairy; petals oblong, straight or twisted; calyx lobes hairy, sharply recurved; stamens 20

Fruits: pomes rounded, about 0.3 in (8 mm) in diameter; red, becoming dark purple, variably tasty

Twigs: glabrous to slightly hairy

Buds: terminal about 0.4 in (1 cm) long; lateral shorter, appressed to stem; scales overlapping, red-brown to yellow-green, glabrous, fringed with hairs (ciliate)

Amelanchier sanguinea (Pursh) DC. - dwarf juneberry

The natural range of dwarf juneberry extends from northern Nebraska into the Dakotas, eastward to Maine. Stephens considers *A. humilis* Wieg. to be a synonym. It is not common in the nursery trade. This red-twigged shrub deserves trial plantings in the northern prairies.

Planning

Height: up to 10 ft (3 m)
Habit: shrub; colony forming, red twigs
Site: sun, partial shade; pH and soil adaptable; zone 3
Use: wildlife habitat, naturalizing

Recognition

Leaves: oval to oblong, about 2 in (5 cm) long; margin variable, entire to variably serrate, veins entering teeth; densely long-hairy at first becoming nearly glabrous; petioles hairy or glabrous
Flowers: clusters (racemes) of 10 to 15 flowers, hairy (tomentose); petals much longer than wide, broadest above middle; calyx lobes hairy, becoming reflexed at the middle; stamens about 20
Fruits: pomes rounded, about 0.4 in (1 cm) in diameter, dark purple with a slight waxy bloom, sweet
Twigs: red, sometimes with orange cast; glabrous except at nodes
Buds: broadly egg-shaped (ovoid), about 0.2 in (5 mm) long; scales overlapping, wine red, nearly glabrous, fringed with hairs (ciliate), inner scales densely silky-hairy on outside

Aronia Med. - chokeberry

Aronia is comprised of three closely related shrubby, deciduous species, native only to North America. Chokeberries are closely related to the mountainashes (*Sorbus*) and serviceberries (*Amelanchier*). An intergeneric hybrid between *Sorbus* and *Aronia*, ×*Sorbaronia* Schneid., has been identified (Rehder 1940).

Chokeberries provide fruit for birds, dense thickets for wildlife cover, and bright colored autumn foliage. They are successfully grown in much of the prairie region, even as far west as Fort Collins, where the only problem has been minor chlorosis (Sutton 1998).

Black chokeberry, *A. melanocarpa* (Michx.) Ell., is a shrub shorter than red chokeberry, its fruits are blackish purple, and its twigs, leaves and flower clusters are glabrous. Otherwise, it is very similar to red chokeberry described herein. Black chokeberry variety *elata* Rehd. is a shrub with leaves, flowers and fruits that are larger than the species. Dirr (1990) states that this variety grows well in Minnesota and rates it better than the species for landscape use.

A. prunifolia (Marsh.) Rehd., purple-fruited chokeberry, is now considered to be a hybrid between red and black chokeberry (Dirr 1990). Flowers are white, fruits are purplish black persisting into winter, and leaves, twigs and flower clusters are hairy, similar to red chokeberry. Although chlorosis recurs on the plant at PERC in Fort Collins, purple-fruited chokeberry is winter hardy and consistently fruitful (Fairchild and Klett 1993).

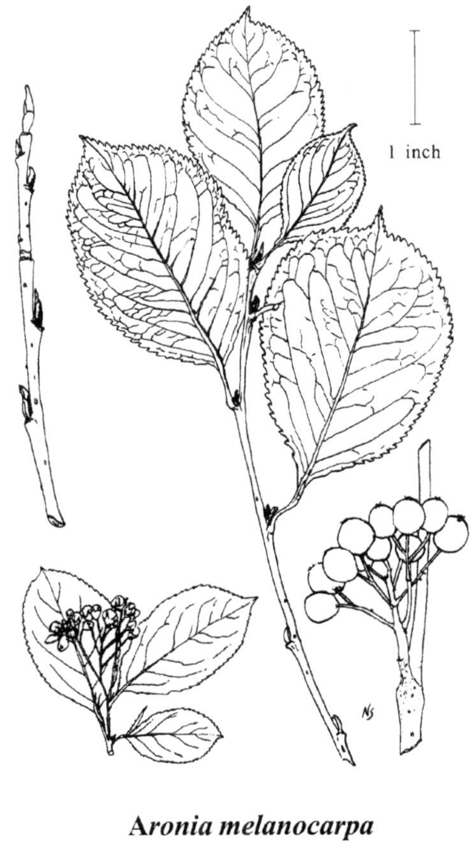

Aronia melanocarpa

Aronia arbutifolia (L.) Ell. - red chokeberry

Red chokeberry, a native of much of the eastern U.S., is an upright multistemmed shrub. An eight-year-old compact cultivar of red chokeberry, 'Brilliantissima', is 3 feet (1 m) tall in Lincoln; it blooms abundantly and brightens autumn days with its red fruit and scarlet foliage.

Planning

Height: up to 10 ft (3 m)
Habit: shrub; deciduous; upright, suckering; autumn foliage and fruits showy
Site: sun, partial shade; soil adaptable, prefers moist, well-drained sites; zone 5
Use: border, massing, naturalizing, specimen
Caution: develops chlorosis at pH 8

Recognition

Leaves: alternate, simple; oblong, sometimes wider above the middle; tip sharp-pointed, base wedge-shaped, margin with black-tipped teeth; upper surface lustrous dark green with dark glands along the midrib, gray woolly (tomentose) beneath; petioles hairy; stipules deciduous
Flowers: clusters (corymbs) dense, hairy, 9 to 20 flowers, about 1.2 in (3 cm) across; petals white, spreading; anthers reddish to purple, styles 5, joined at base (connate)
Fruits: pomes, fleshy, persistent into winter, bright red, 0.25 in (6 mm) diameter; calyx lobes persistent
Twigs: woolly (tomentose)
Buds: appressed; scales, 3 visible

Chaenomeles Lindl. - floweringquince

Three *Chaenomeles* species, all natives of Asia, are planted for flower effect and for hedges because they adapt to shearing.

Floweringquince is sometimes confused with quince (*Cydonia oblonga* Mill.), a shrub or small tree with similar flowers and fruits that is a native of Iran grown chiefly for food. Quince leaves are entire and styles are free.

C. ×*superba* (Frahm) Rehd. displays a wide range of flower colors and is intermediate between the two parents, the Japanese and the common flowering quinces. *C. sinensis* (Thouin) Koehne, which can reach the stature of a small tree, bears solitary pink flowers on leafy shoots. Its twigs are spineless and its stipules long and narrow (lanceolate).

Recognition

Habit: shrubs, small trees; deciduous or half-evergreen
Leaves: simple, alternate; toothed (serrate or crenate); with stipules
Flowers: solitary or fascicled; 5-parted; stamens 20 or more, styles 5, joined at base (connate)
Fruits: large, many-seeded
Twigs: sometimes spiny
Buds: small; scales, 2 visible

Chaenomeles japonica (Thunb.) Lindl. - Japanese floweringquince

Japanese floweringquince, a thorny, diminutive shrub that blooms earlier in the season than *C. speciosa* does, has proved to be reasonably hardy at PERC in Fort Collins where it experiences some injury during extremely cold winters and minor chlorosis on the region's highly alkaline soil (Fairchild and Klett 1993).

Planning

Height: up to 6 ft (2 m)
Habit: shrub; rounded; dense by suckering; spiny twig ends; April-May flowers on one-year-old wood; new foliage reddish becoming glossy green
Site: sun; fertile, moist soils, prefers pH less than 8; zone 5
Use: specimen, border

Recognition

Leaves: broadly egg-shaped (ovate to obovate); glabrous; coarsely toothed (crenate-serrate); apex obtuse to pointed; stipules egg-shaped to heart-shaped, persistent
Flowers: clusters leafless; flowers about 1.5 in (4 cm) across, various shades of red; calyx lobes upright, entire
Fruits: pomes yellow, 1.5 in (4 cm) in diameter; tart, bitter
Twigs: rough to touch (scabrous), sometimes spiny
Buds: solitary; brown, egg-shaped

Chaenomeles speciosa (Sweet) Nakai - common floweringquince

Common floweringquince, formerly known as *C. lagenaria* (Loisel.) Koidz., is the hardiest and most common floweringquince planted. However, 'Texas Scarlet' died of winter injury at Fort Collins (Fairchild and Klett 1993).

Differences among the many cultivars of floweringquince are so minute as to be meaningless (Dirr 1990). Among the few that Dirr recommends are: 'Texas Scarlet', bearing profuse red flowers on a spreading, compact plant; 'Cameo', with apricot pink double flowers; and 'Jet Trail' and 'Nivalis', both bearing white flowers.

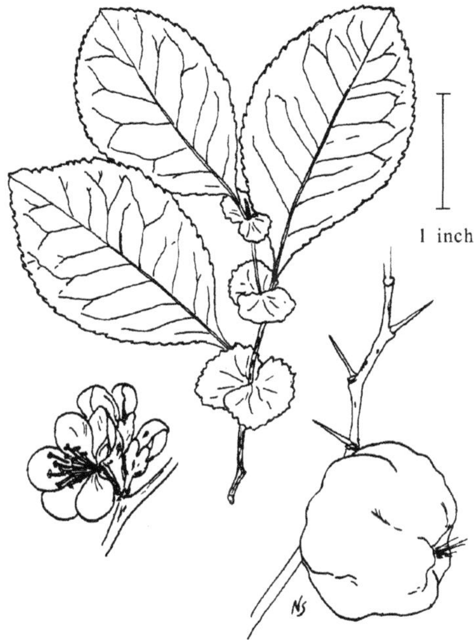

Chaenomeles speciosa

Planning

Height: up to 10 ft (3 m)
Habit: shrub; upright to rambling; April-May bloom on old wood ahead of leaves
Site: sun, partial shade; acidic to neutral soils; drouth tolerant; zone 5
Use: border, grouping, hedge
Note: fruits can be used in preserves and jellies

Recognition

Leaves: egg-shaped to oblong, about 2 in (5 cm) long; tip pointed, base wedge-shaped, margins sharply toothed (serrate); bronze red becoming lustrous dark green above, glabrous; stipules about 1 in (25 mm) in diameter, rounded, toothed, persistent
Flowers: clusters of 2-4 or solitary, leafless; flowers about 1.7 in (4 cm) across, scarlet red, varying to pink and white; calyx lobes upright, entire
Fruits: pomes many seeded, about 2 in (5 cm) in diameter, fragrant
Twigs: brown, often hairy; somewhat thorny, especially tips
Buds: terminal lacking; lateral egg-shaped, scales overlapping

Cotoneaster Ehrh. - cotoneaster

The cotoneasters, deciduous and evergreen shrubs that number about 50 species, are native to moist, temperate zones of Europe and Asia. Deciduous species usually flush leaves very early in the spring; evergreen species are not likely to be hardy enough to survive the winters of the northern and central prairies.

Generally, cotoneasters are grown for their attractive fruits and dark green foliage that often changes to red in autumn. The cotoneasters' flowers are attractive to bees and their fleshy fruits (pomes) are relished by birds and small mammals. Fireblight, a bacterial disease, is their greatest liability.

C. ignavus Wolf and *C. moupinensis* Franch. are adapted to the Fort Collins area (Fairchild and Klett 1993). These species (growing at PERC) are winter hardy and do not suffer chlorosis on alkaline soil. The former is a large dense shrub that bears white flowers and purplish red fruit, but fall color is not important. The latter is a dense medium-sized shrub that displays pink flowers in May and black fruit that persists through winter, and autumn foliage is reddish. These species, not readily available commercially, were obtained from the Cheyenne Horticulture Research Station.

Recognition

Habit: shrubs; deciduous or evergreen
Leaves: alternate, simple, margins entire; stipules awl-shaped, early deciduous (caducous); petioles short
Flowers: clusters (corymbs) few to many-flowered, sometimes solitary; bisexual; petals white or pinkish, upright or spreading; sepals 5, short; stamens about 20, styles 2-5, distinct
Fruits: pomes fleshy, red or black; crowned by persistent calyx; nutlets 2-5
Buds: small; scales several, overlapping

Grouping Key for *Cotoneaster*

(Rehder 1940)

Petals upright, egg-shaped, wider above middle (obovate), usually pinkish; fruits red or black: **section Orthopetalum**

Petals spreading, nearly circular, white; fruits red: **section Chaenopetalum**

Section Orthopetalum

Cotoneaster apiculatus Rehd. & Wils. - cranberry cotoneaster

The trademark of cranberry cotoneaster is its showy, red fruits that hang from arching branches during late summer and early fall. Also, its glossy green foliage that becomes bronze red in autumn enhances the landscape.

Cranberry cotoneaster is generally hardy in the central prairies (North Central Plant Introduction Station 1960). Dirr (1990) reports it performs well in the Salt Lake City–Provo, Utah, area.

Planning

Height: up to 5 ft (1.5 m)
Habit: shrub; deciduous; wide-spreading, dense; foliage glossy green in summer to bronze red in autumn; red fruits August-September
Site: sun, partial shade; soil and pH adaptable; drouth tolerant; zone 5
Use: foundation, groundcover

Recognition

Leaves: nearly round to oval, less than 1 in (25 mm) across; contracted to a minute point (apiculate), margin undulating; dark lustrous green above, slightly hairy below

Flowers: solitary; petals upright, obovate, pinkish; sepals short-acuminate

Fruits: red, about 0.25 in (1 cm) in diameter, almost sessile

Twigs: reddish purple with appressed hairs, becoming gray-brown, ragged

Buds: 2 outer scales parted, exposing hairy interior

Cotoneaster divaricatus Rehd. & Wils. - spreading cotoneaster

"Spreading" is an apt description of the habit of this handsome shrub of central and western China. The finely textured, glossy, green foliage becomes bright red in late October in eastern Nebraska.

Dirr (1990) rates spreading cotoneaster as one of the most desirable and trouble-free ornamental species of cotoneaster (though it may be subject to winter injury on the northern prairies). A planting of spreading cotoneaster failed to establish at PERC in Fort Collins (Fairchild and Klett 1993).

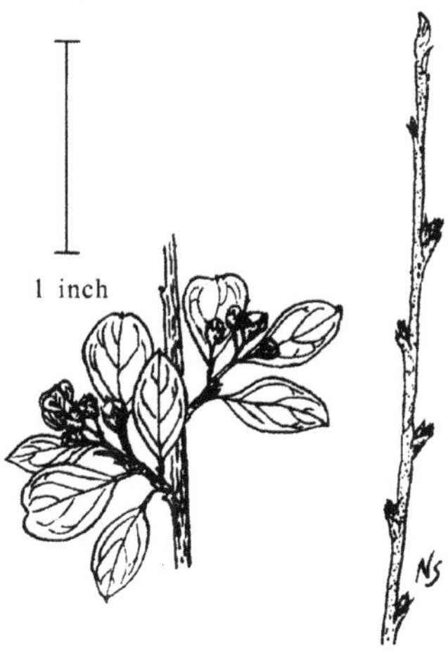

Cotoneaster divaricatus

Planning

Height: up to 7 ft (2 m)

Habit: shrub; deciduous; wide-spreading, fine textured; summer foliage glossy green, autumn red-purple; red fruit September-November

Site: sun, light shade; soil and pH adaptable; zone 5

Use: border, hedge, grouping, specimen

Recognition

Leaves: oval to elliptic, less than 0.7 in (2 cm) long; glabrous, dark green above, lighter green, sometimes slightly hairy below

Flowers: clusters of 3 or solitary; petals upright, obovate, light pink

Fruits: fleshy, red, 0.3 in (8 mm) across

Twigs: slender, appressed hairy becoming glabrous and brown

Buds: scales part to reveal hairy interior

Cotoneaster horizontalis Decne. - rockspray cotoneaster

Rockspray cotoneaster, an interesting, low-growing dense shrub with glossy green foliage and branches that form a fishbone pattern, may be borderline in hardiness for much of the prairie region.

Planning

Height: up to 2 ft (60 cm)

Habit: shrub; deciduous north to evergreen south; flat with horizontal branches, layered

Site: sun, light shade; soil and pH adaptable; zone 5

Use: rock garden, border, specimen

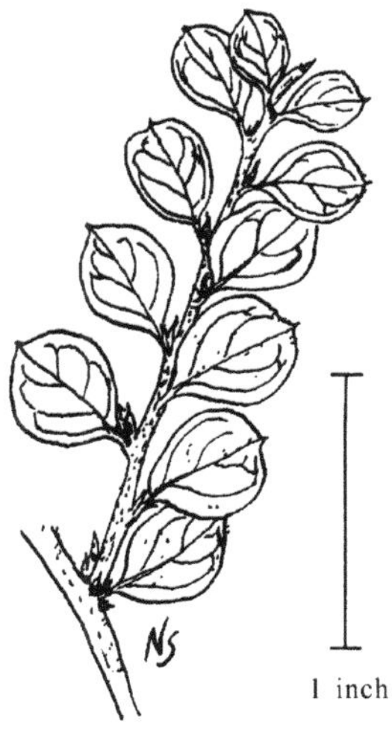

Cotoneaster horizontalis

Recognition

Leaves: nearly round, about 0.5 in (12 mm) long; flat (not undulating); dark green, glabrous above, sparsely hairy below
Flowers: single or in pairs, petals upright, obovate, pink
Fruits: fleshy, bright red, 0.2 in (5 mm) across
Twigs: arranged in alternating pattern, like a fishbone

Cotoneaster integerrimus Med. - European cotoneaster

Rehder (1940) describes European cotoneaster as an upright, medium shrub that reaches a height of seven feet (2 m) and is hardy in zone 6. Morgenson (1988) notes that 'Centennial', a selection from seed purported to be collected from European cotoneaster plants in China and released by the Natural Resources Conservation Service Plant Materials Center in Bismarck, N.D., is an open, fireblight-resistant shrub that can ultimately attain a height of 12 feet (3.5 m).

'Centennial' is presently marketed by Lincoln-Oakes Nurseries of Bismarck for inclusion in conservation plantings. It is a vigorous, fruitful shrub growing in an eastern Nebraska windbreak (Bagley 1998). Dr. Gabor Schmidt (1994), University of Horticulture and Food Industry, Budapest, Hungary, opines that this cultivar closely resembles and may actually be *C. hupehensis* Rehd. & Wils. or *C. multiflorus*. His opinion is based on leaf and fruit characteristics only; flowers were not available.

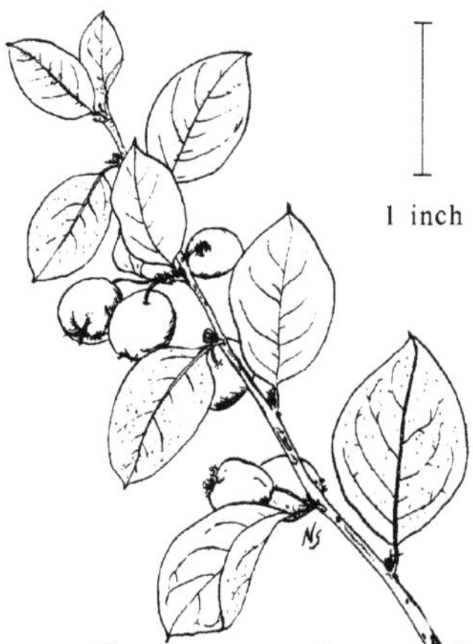

Cotoneaster integerrimus

Planning

Height: up to 12 ft (3.5 m)
Habit: shrub; deciduous; upright, spreading; showy, red fruit in late summer and autumn
Site: sun; soil and pH adaptable; drouth tolerant; zone 3
Use: border, wildlife habitat, specimen, windbreak

Recognition

Leaves: round to egg-shaped, up to 2 in (5 cm) long; dark green, glabrous above, hairy beneath
Flowers: clusters of 2 to 4 flowers, nodding; petals upright, obovate, pinkish
Fruits: red, 0.2 in (6 mm) across

Cotoneaster lucidus Schlect. - hedge cotoneaster

The common name provides a clue to the typical use of hedge cotoneaster, a shrub of narrow, upright habit. But its ability to adapt to a wide range of site conditions extends its usefulness beyond the realm of urban landscapes to rural wind protection and wildlife plantations.

Hedge cotoneaster is often confused with *C. acutifolius* Turcz., Peking cotoneaster, which is similar except for its dull green hairy leaves. The shrub that is presently planted widely throughout the prairies in conservation plantings as Peking should more properly be labeled hedge cotoneaster (North Central Plant Introduction Station 1959).

Cotoneaster lucidus

Planning

Height: up to 12 ft (3.5 m)
Habit: shrub; deciduous; upright, round-topped; usually taller than broad; branches slender; black fruit persists; lustrous foliage in very early spring becomes colorful autumn foliage
Site: sun, light shade; pH and soil adaptable; drouth tolerant; zone 4
Use: screen, hedge, grouping, wildlife habitat, windbreaks

Recognition

Leaves: egg-shaped to oblong, 1-2 in (25-50 mm) long; dark green, glabrous above, lighter below, with hairy veins, becoming nearly glabrous; petioles hairy
Flowers: clusters (cymes) of 3 to 8 flowers; petals upright, obovate, pinkish white
Fruits: fleshy, black; 0.2 in (5 mm) across
Twigs: slender, hairy, becoming glabrous; sometimes peeling, light brown
Buds: scales parted showing hairy interior

Section Chaenopetalum

Cotoneaster congestus Baker - Pyrenees cotoneaster

Pyrenees cotoneaster rates high marks for compact growth and showy red fruits. Dirr (1990) describes it as a handsome, small evergreen suitable for rock gardens. Decurved branches form a compact rounded mass resembling a small haystack.

Low-growing Pyrenees cotoneaster is hardy but not totally evergreen at South Dakota State University, Brookings, and Colorado State University, Fort Collins (Klett 1975).

Planning

Height: up to 2 ft (60 cm)
Habit: shrub; evergreen in mild climates; spreading, mounding, compact; red fruits
Site: sun, light shade; soil and pH adaptable; zone 5
Use: border, rock garden, groundcover, foundation

Recognition

Leaves: oval to wider above middle (obovate); green above, whitish below, hairy, becoming glabrous
Flowers: solitary; petals spreading, nearly round, pinkish white
Fruits: bright red, 0.25 in (6 mm) across
Twigs: hairy, soon glabrous

Cotoneaster dammeri Schneid. - bearberry cotoneaster

Bearberry cotoneaster, a fast-growing native of Central China, is considered by Dirr (1990) to be one of the best evergreen groundcovers. 'Skogholm' was exceptionally vigorous in Dirr's Illinois garden in Champaign-Urbana.

Bearberry cotoneaster and two other evergreen cotoneaster species of similar habit, *C. conspicuus* Marquand, wintergreen cotoneaster, and *C. microphyllus* Lindl., little-leaf cotoneaster, should be considered for trial plantings in the central and southern prairies.

Planning

Height: up to 1.5 ft (45 cm)
Habit: shrub; evergreen in mild climates; prostrate with creeping stems, bright red fruits in late summer
Site: sun, light shade; zone 6
Use: groundcover, border, foundation, rock garden

Recognition

Leaves: elliptic to oblong, about 1 in (25 mm) long; tip pointed to obtuse, sometimes with short abrupt tip (mucronulate); dark green, glabrous above; hairy below, becoming glabrous
Flowers: solitary or in pairs, about 0.4 in (1 cm) across; petals white, spreading, nearly round; anthers purple
Fruits: red, 0.25 in (6 mm) across; nutlets usually 5
Twigs: hairy, becoming glabrous, red-brown
Buds: scales parted showing hairy interior

Cotoneaster multiflorus Bge. - manyflower cotoneaster

The flowers and fruits of manyflower cotoneaster, a native of western China, make it the most spectacular of all the cotoneasters (Dirr 1990). Dirr also reports that manyflower cotoneaster is the most vigorous and trouble-free of the cotoneasters on the campus of the University of Illinois.

Cotoneaster multiflorus

Planning

Height: up to 15 ft (4.5 m)
Habit: shrub; upright and arching; May flowers, fruits August-October
Site: sun; pH and soil adaptable; zone 4
Use: border, grouping, specimen, windbreak, wildlife habitat

Recognition

Leaves: egg-shaped (ovate), up to 2 in (5 cm) long; bluish green above, densely hairy beneath, soon glabrous
Flowers: clusters (corymbs) many-flowered; petals spreading, nearly circular, white; 0.5 in (12 mm) across; malodorous
Fruits: fleshy, red, 0.3 in (8 mm) across
Twigs: purple, becoming reddish green and finally gray; sometimes slightly hairy

Crataegus L. - hawthorn

The hawthorns, found as natives throughout the temperate regions of the Northern Hemisphere, vary in habit from suckering shrubs to small trees. Dense, often thorny branches bearing fleshy fruits provide cover and food for wildlife. Although deer and domestic animals feed on their leaves and twigs, the thorns protect many hawthorn species from excessive browsing and rubbing, allowing them to increase and persist in eastern prairie pastures, much to the dismay of stockmen. Cedar-hawthorn rust, fireblight, aphids and leaf blotch miner are sometimes serious problems for hawthorns, but pest-resistant cultivars have been discovered.

Interbreeding among *Crataegus* species has produced a population so variable that taxonomists have named over 1,000 species. Freeman (1992) states that apomixis, a method of vegetative reproduction, is common in *Crataegus*.

Most woody plant books only describe representatives of hawthorn groups or series that differ greatly from each other. Detailed descriptions of many hawthorn species can be obtained from the **Manual of the Trees of North America** (Sargent 1933).

Representative hawthorn species important to the prairies are described herein. Readers will discover plants that do not quite match our descriptions nor the descriptions of other authors.

Recognition

Habit: shrubs, small trees; deciduous, rarely half-evergreen
Leaves: alternate, toothed and often shallow lobed
Flowers: clusters (corymbs) or rarely solitary; bisexual; petals 5, white, rarely pale yellow, pink or red, calyx 5-lobed; stamens 5-25, styles 1-5
Fruits: pomes fleshy; seeds nut-like, 1 to 5
Twigs: rigid, more or less zigzag, often armed with axillary, unbranched, sharp slender thorns, usually over 1 in (25 mm) long; lenticels oblong, pale; bundle scars 3
Buds: terminal usually present; globose, reddish; scales fleshy, 4 to 8
Bark: scaly to shallow furrowed
Wood: heavy, hard, tough, close-grained

Grouping Key for *Crataegus*

Leaves slightly or not lobed, veins extending only to lobe and teeth points: **Group One**

Leaves usually conspicuously lobed, veins extending to points of lobes and sinuses, at least on larger leaves: **Group Two**

Group One

Crataegus arnoldiana Sarg. - Arnold hawthorn

Arnold hawthorn, a small, spiny tree of dense habit that is native to the New England states, is a close relative of *C. mollis*. It is adapted to most prairie sites, including alkaline soils such as those in the Fort Collins area (Hoag 1965, Fairchild and Klett 1993).

'Homestead', a selection released by the Natural Resources Conservation Service Plant Materials Center in Bismarck, N. D., is propagated by certified seed that is available from approved growers in North Dakota, South Dakota and Minnesota.

Planning

Height: up to 20 ft (6 m)
Habit: small tree; stout branches form an oval crown, thorny; flowers April-May
Site: sun; pH and soil adaptable; zone 3
Use: wildlife habitat, screen, windbreak, fruit
Caution: susceptible to cedar-hawthorn rust

Recognition

Leaves: egg-shaped, sharply toothed (serrate), base squared off (truncate); hairy below
Flowers: clustered, stalks slender; flowers 0.7 in (2 cm) across; petals white, pale yellow on 'Homestead'
Fruits: nearly round, 0.6 in (15 mm) in diameter, red; edible
Twigs: hairy, zigzag; spines chestnut brown, straight to slightly curved, 2-3 in (5-8 cm) long

Crataegus crusgalli L. - cockspur hawthorn

Authors differ in their descriptions of the cockspur hawthorn's habit, from rather large and tree-like to short and shrubby. However, they are unanimous in their evaluation of the thorns—they are sharp and can be very dangerous. These weapons on a rounded crown of wide-spreading branches present a formidable barrier. Persistent fruits provide wildlife food.

Cockspur hawthorn is a native of and commonly found in the states adjoining Nebraska to the east and south, as well as throughout most of the eastern U.S. Successful plantings can be found in Nebraska along Interstate 80 as far west as Grand Island.

Thornless variety *inermis* performed well in six trial plantings across Kansas with a minimal incidence of leaf rust (Hensley et al. 1991); the authors concluded that it is a desirable plant for most Kansas landscapes. Variety *inermis* is one of the best of the hawthorn species in the PERC collection at Fort Collins (Fairchild and Klett 1993).

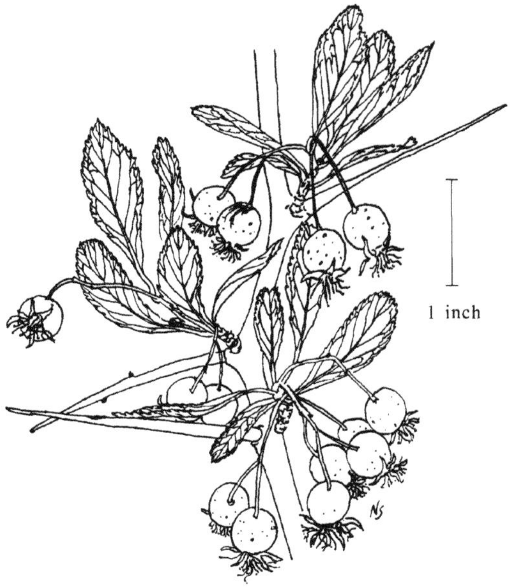

Crataegus crusgalli

C. ×*lavallei* Herincq, (*C. crusgalli* × *C. pubescens*) is a small, oval-headed and usually thornless tree that has toothed oblong leaves unlike those of most other hawthorns. Red fruits are persistent into winter. Lavalle hawthorn is rust resistant and widely adaptable (Dirr 1990).

Planning

Height: up to 20 ft (6 m); National champion, 1987, 40 ft (12.2 m), Virginia
Habit: shrub, tree; crown rounded, wide-spreading, dense, thorny or thornless; May flowers white; autumn foliage maroon red; fruits red, persistent
Site: sun; soil and pH adaptable; drouth tolerant; zone 3
Use: specimen, grouping, screen, barrier, street

Recognition

Leaves: leathery, egg-shaped, wider above the middle (obovate), not lobed, base wedge-shaped (cuneate), apex rounded or small-pointed; margins sharply toothed except near the base; upper surface dark green, lustrous, lower surface pale, glabrous; petioles stout, about 0.3 in (8 mm) long
Flowers: clusters many-flowered, about 2.5 in (6 cm) across; flowers white, malodorous; calyx glabrous, lobes linear; stamens 10, anthers pink
Fruits: nearly round, ripening red, hard, 0.4 in (1 cm) across; calyx lobes persistent
Twigs: rigid, glabrous with waxy covering; thorns numerous, very sharp, up to 3 in (8 cm) long (except *inermis*)

Crataegus mollis Scheele - downy hawthorn

Downy hawthorn, a native of eastern Nebraska along the Missouri River, also is native south to eastern Texas and northeast to Maine. It is a picturesque though small tree, except in localities where cedar-hawthorn rust causes excessive leaf and fruit discoloration in late summer.

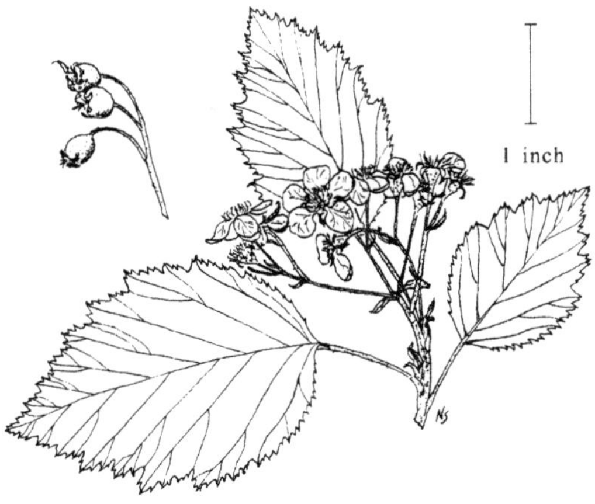

Crataegus mollis

Planning

Height: up to 30 ft (9 m); Nebraska champion, 1977, 35 ft (10.7 m), Blair; National champion, 1972, 52 ft (15.9 m), Michigan
Habit: tree; crown wide-spreading, variably thorny; white flowers early May; showy red fruits drop when ripe
Site: sun; pH and soil adaptable, zone 4
Use: wildlife habitat, specimen
Caution: susceptible to cedar-hawthorn rust

Recognition

Leaves: broadly egg-shaped, doubly toothed, 4-5 pairs of short lobes; hairy below at first, later chiefly on the veins; petioles hairy, about 1.5 in (4 cm) long; stipules curved, toothed, hairy
Flowers: clusters of 5-7, about 3 in (8 cm) across, long soft-hairy (tomentulose); flowers white, malodorous; stamens 20, styles 4-5
Fruits: globose, red, about 0.6 in (15 mm) across; thick, juicy; nutlets 3 to 5
Twigs: long-hairy first season; thorns to 2 in (5 cm), straight

Crataegus pedicellata Sarg. - scarlet hawthorn

Scarlet hawthorn, a native of the northeastern U.S., is winter hardy and adapted to alkaline soils (Fairchild and Klett 1993).

Planning

Height: up to 20 ft (6 m)
Habit: tree, large shrub; broad oval, thorny; May flowers, showy red fruit not persistent; autumn foliage golden
Site: sun; pH and soil adaptable; zone 3
Use: barrier, naturalizing
Caution: susceptible to cedar-hawthorn rust

Recognition

Leaves: broad egg-shaped, about 3 in (8 cm) long; coarsely toothed (often doubly serrate), base squared (truncate) to broadly wedge-shaped (cuneate); 4-5 short, acute lobes above middle; rough to touch above, nearly glabrous below; petioles long, slender
Flowers: clusters slightly hairy, loose; flowers white; sepals coarsely toothed, glandular; stamens 5 to 10, anthers rose-colored
Fruits: pear-shaped to ellipsoidal, lustrous, bright scarlet; calyx conspicuous; flesh thin, mealy; nutlets 4 to 5
Twigs: slender; thorns slightly curved, about 1.5 in (4 cm) long

Crataegus punctata Jacq. - dotted hawthorn

Dotted hawthorn is a native of a broad area across Iowa, eastern Kansas and Oklahoma, east to Maine. It seems to be reasonably adapted to southwestern Nebraska (Richley 1988). A nearly thornless cultivar, 'Ohio Pioneer', selected at the Secrest Arboretum in Wooster, Ohio, has exhibited good growth and fruiting characteristics (Dirr 1990).

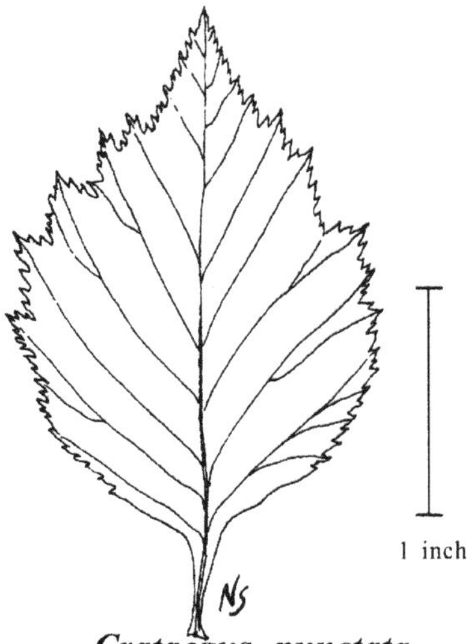

Crataegus punctata

Planning

Height: up to 25 ft (8 m); National champion, 1979, 38 ft (11.6 m), West Virginia
Habit: tree, shrub; crown flat-topped, horizontal branching; thorny or thornless (var. *inermis*); summer foliage grayish green; autumn fruits dark red, not persistent
Site: sun; pH and soil adaptable; zone 5
Use: specimen, screen
Caution: susceptible to cedar-hawthorn rust

Recognition

Leaves: egg-shaped, broadest above the middle (obovate), unlobed or sometimes obscurely lobed on upper half, base wedge-shaped; hairy beneath especially along the veins; petioles margined
Flowers: clusters hairy, many flowered; flowers white, not fragrant; sepals nearly entire; stamens 20
Fruits: globe to pear-shaped, about 0.6 in (15 mm) in diameter; flesh thick, juicy; red or yellow; nutlets 3-5
Twigs: unarmed, or armed with straight thorns, 1-3 in (3-7 cm) long

Crataegus rivularis Nutt. - river hawthorn

River hawthorn, a native of Idaho, Nevada and New Mexico, is adapted to the alkaline soils and winter climate of Fort Collins (Fairchild and Klett 1993).

Planning

Height: up to 15 ft (4.5 m)
Habit: shrub, tree; oval, ascending branches, thorny, suckering; May flowers, red in midsummer; fruit dropping when ripe; autumn foliage yellow
Site: sun; soil and pH adaptable, prefers moist, well-drained site; zone 3
Use: specimen, screen, naturalizing
Caution: susceptible to cedar-hawthorn rust
Pests: aphids

Recognition

Leaves: egg-shaped, about 2 in (5 cm) long; slightly lobed, toothed (serrate), base wedge-shaped (cuneate); hairy on veins, becoming glabrous; petioles about 0.2 in (5 mm) long
Flowers: clusters glabrous; flowers white, about 0.4 in (1 cm) across; sepals entire; stamens 20
Fruits: red
Twigs: thorns about 0.7 in (2 cm) long

Crataegus succulenta Link - fleshy hawthorn

Fleshy hawthorn, a native of southern Canada and much of the northern U.S. including the Dakotas, northwest Nebraska, Iowa and Missouri, is hardy and drouth tolerant but susceptible to cedar-apple rust (North Central Plant Introduction Station 1979). Charles Sargent considered it to be one of the six best native hawthorns in the U.S. (Dirr 1990).

Variety *macracantha* Eggl. (*C. macracantha* Lodd.) usually differs from fleshy hawthorne by its ten stamens with yellow anthers (Rehder 1940). However, Stephens (1973) attributes these characteristics to *succulenta*. Variety *macracantha* grows at Fort Collins where aphids are the only serious problem (Fairchild and Klett 1993).

C. ×*mordenensis* (*succulenta* × *laevigata*), 'Toba', produced at the Morden Experiment Station in Canada, is an attractive large shrub or small tree that bears white to pinkish double flowers and red fruits (Hoag 1965). This hybrid is winter hardy and adapted to the alkaline soil at PERC where aphids are the only problem (Fairchild and Klett 1993).

Red haw (*C. chrysocarpa* Ashe), a closely related species, has a similar natural range but is not found as commonly in Nebraska as fleshy hawthorn. However, it grows in Jones Canyon, Garfield County, Nebraska (Bagley 1998). According to Stephens (1973), red haw is smaller than fleshy hawthorn, even shrub-like, with red glands appearing on the petioles and leaf teeth tips. The fruit is hard.

Planning

Height: up to 20 ft (6 m)

Habit: shrub, tree; dense, thicket-forming, thorny; autumn foliage purple-red, fruits bright red in September

Site: sun; pH and soil adaptable; drouth tolerant; zone 3

Use: naturalizing, conservation plantings

Recognition

Leaves: leathery, broadly egg-shaped, about 2.5 in (6 cm) long; shallow lobes on upper half; margins toothed, without obvious glands; surfaces usually hairy, especially along veins below; petioles variable, up to 1 in (25 mm)

Flowers: clusters flat-topped, hairy, many-flowered; petals white, nearly circular, calyx lobes reflexed, hairy, gland-toothed; stamens 10 to 20, anthers rose-colored or yellow

Fruits: nearly round, bright red; about 0.5 in (12 mm) in diameter; flesh thick, mealy, sometimes juicy; surface smooth, sometimes with small light-colored spots; nutlets 3 to 5

Twigs: dark red, glabrous; thorns slender, slightly curved, about 2 in (5 cm) long, dark red, becoming gray

Buds: broadly egg-shaped to nearly round, dark red, glossy

Crataegus viridus L. - green hawthorn

Creek banks and flood plains are favorite habitats for green hawthorn in the southern and eastern seaboard states and the Mississippi Valley, north to Illinois. Small outliers occur in eastern Kansas and Oklahoma. Stephens (1973) describes it as more delicate than the other hawthorns—having fine branches and a graceful appearance.

'Winter King'—a tree with green, bloomy twigs, colorful fall foliage, vase-shaped form and resistance to cedar-apple rust—is one of the outstanding green hawthorn cultivars appropriate for landscape use (Dirr 1990). *C. nitida* Sarg., glossy hawthorn, is a closely related species bearing dull red fruits that persist into winter.

Planning

Height: up to 25 ft (8 m); National champion, 1981, 40 ft (12.2 m), West Virginia
Habit: tree; crown dense, rounded, moderately thorny; white May flowers, red autumn fruits persisting into winter
Site: sun; soil adaptable; zone 5
Use: specimen, border, wildlife habitat

Recognition

Leaves: egg-shaped, the broadest part often above the middle; with or without small shallow lobes near the tip, margins finely toothed (serrate), sometimes only above the middle; lustrous above, hair tufts often at vein intersections beneath; petioles about 1 in (25 mm) long
Flowers: clusters of 6 to 12, erect, glabrous; petals white, circular; calyx lobes broadly triangular, tip sharp-pointed, fine hairs along margin (ciliate); stamens 20, styles 5
Fruits: nearly round, about 0.3 in (8 mm) in diameter; red, juicy; nutlets 4 to 6
Twigs: glabrous, gray waxy coating; thorns few, at right angles to stem, about 1.5 in (4 cm) long
Buds: broadly egg-shaped, red, glabrous, usually waxy

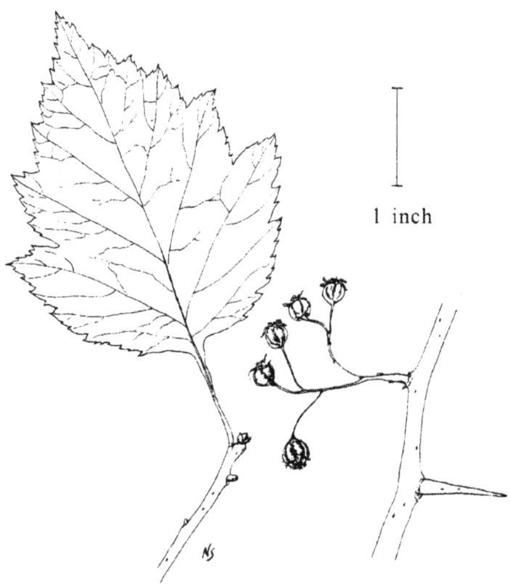

Crataegus viridus

Group Two

Crataegus laevigata (Poir.) DC. - English hawthorn

English hawthorn, a tree long grown in Europe and North Africa, has spawned a large number of cultivars. It is one of a few hawthorns that have produced plants with flowers that are white as in 'Autumn Glory', red as in 'Crimson Cloud' and double-red as in 'Paul's Scarlet'. 'Crimson Cloud' and 'Paul's Scarlet' are winter hardy and adapted to the alkaline soils at PERC in Fort Collins, but are plagued with aphids. 'Autumn Glory' failed to become established at PERC (Fairchild and Klett 1993).

'Toba', a hybrid selection produced at Morden Experiment Station, Manitoba, Canada, from a cross between 'Paul's Scarlet' and *C. succulenta*, bears double white flowers that fade to light pink (Dirr 1990). 'Toba' blooms and produces fruit reliably, is winter hardy and is adapted to the alkaline, clay soil at Fort Collins, but the showy red fruits do not persist (Fairchild and Klett 1993). 'Toba' is frequently planted in Manitoba (Hoag 1965).

Crataegus laevigata

Planning

Height: up to 25 ft (7.5 m)
Habit: tree; round-topped, dense, thorny; flower colors variable, May bloom; fruits showy red, do not persist into winter
Site: sun, part shade; pH and soil adaptable; zone 4
Use: specimen, screen, street

Recognition

Leaves: broad egg-shaped, 3- to 5-lobed; lobe margins minutely toothed (serrulate), base wedge-shaped (cuneate); glabrous, dark green above; petioles about 0.4 in (1 cm) long
Flowers: clusters glabrous, 5 to 12 flowers; petals white to red, stamens 20, styles 2-3
Fruits: subglobose, scarlet; nutlets 2
Twigs: stout spines to 1 in (25 mm)

Crataegus monogyna Jacq. - singleseed hawthorn

Singleseed hawthorn, a long-cultivated native of Europe and adjoining continents, shares many characteristics with *C. laevigata*. Many cultivars have been selected including 'Biflora' (Glastonbury Thorn) that in mild climates will flower once in mid-winter and again in May (Dirr 1990).

'Stricta', a dense, columnar tree, grows at PERC in Fort Collins with no serious problems (Fairchild and Klett 1993). A closely related European species, *C. ambigua* C. A. May., grows successfully at the City Parks Arboretum in Blair, Nebraska, in southwestern Nebraska, and in eastern Colorado.

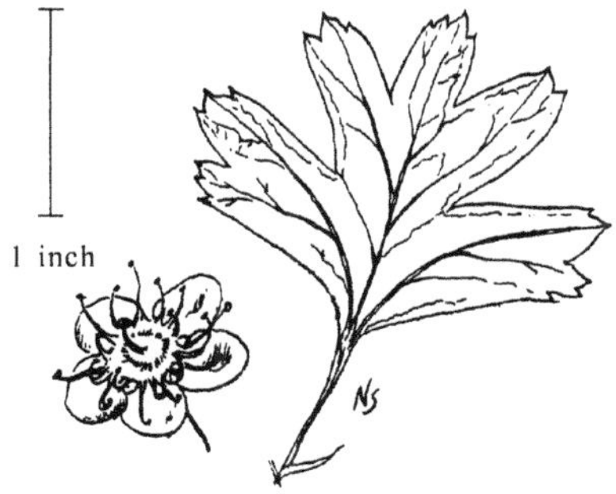

Crataegus monogyna

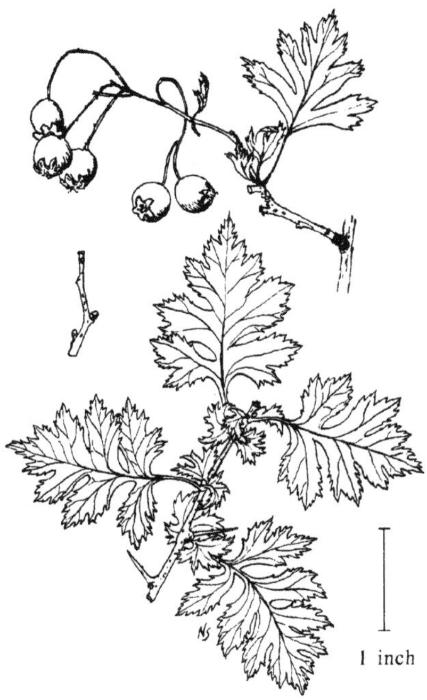

Crataegus ambigua

Planning

Height: up to 30 ft (9 m); National champion, 1989, 64 ft (19.6 m), Washington
Habit: tree; crown rounded, dense, very thorny; May flowers; red autumn fruits obscured by dense foliage
Site: sun, part shade; pH and soil tolerant, prefers moist, well-drained sites; zone 4
Use: border, specimen, street
Caution: susceptible to fireblight

Recognition

Leaves: broad egg-shaped, deeply 3- to 7-lobed with few teeth at apex; shiny dark green
Flowers: white; style 1 (usually)
Fruits: red, ellipsoid; nutlet 1
Twigs: spines stout

Crataegus phaenopyrum (L.f) Med. - Washington hawthorn

Washington hawthorn, a native of the southeastern U.S. including central and southern Missouri, is widely planted as an ornamental throughout North America and Europe. It has proved effective in conservation and wildlife plantings. Richley (1988) rates Washington hawthorn a "durable" tree for southwestern Nebraska.

Attractive flowers, bright orange to scarlet autumn leaves, persistent red fruits and disease resistance are attributes of Washington hawthorn. Cultivars include: 'Fastigiata', a narrow columnar form; 'Clark', a prolific fruiter; and 'Vaughn', a selection of *C. crusgalli* × *C. phaenopyrum* (Dirr 1990).

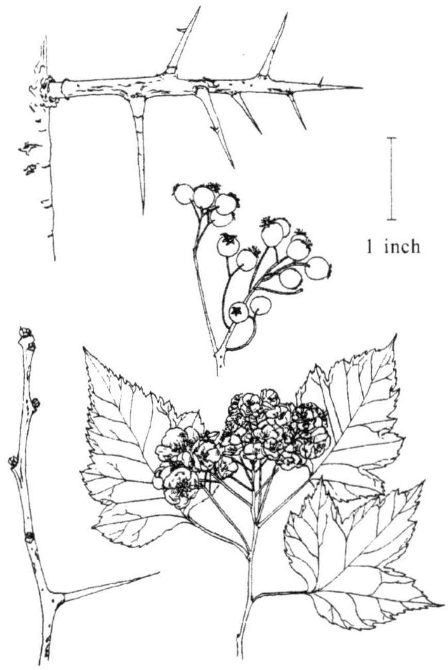

Crataegus phaenopyrum

Planning

Height: up to 30 ft (9 m); National champion, 1989, 36 ft (11 m), Virginia
Habit: tree; crown oval, upright branching, dense, thorny; white flowers in early June; red fruits persist into winter
Site: sun; soil and pH adaptable; zone 4
Use: specimen, screen, wildlife habitat, windbreak

Recognition

Leaves: triangular egg-shaped, about 1 in (25 mm) long, broadest near the base; base rounded to slightly heart-shaped; 1 to 4 pairs of shallow lobes; tip pointed, margin toothed; dark green and glabrous; petioles slender, about 1 in (25 mm) long
Flowers: clusters along the twigs, many-flowered; flowers 0.5 in (12 mm) across; sepals entire, deltoid; stamens 20, anthers yellow or pink
Fruits: nearly round, red, 0.3 in (8 mm) in diameter; sepals deciduous, leaving scar on fruit; flesh thin, dry; nutlets 3 to 5
Twigs: glabrous; spines slender, up to 3 in (8 cm) long, may occasionally be branched

Crataegus pinnatifida Bge. - Chinese hawthorn

This is the most attractive hawthorn growing in the northern plains (Hoag 1965).

Planning

Height: up to 20 ft (6 m)
Habit: tree; short thorns or none; red fruit showy, edible
Site: experimental; zone 3
Use: border, specimen, fruit

Recognition

Leaves: broadly egg-shaped (triangular to rhombic ovate), about 3 in (8 cm) long; base broadly wedge-shaped (cuneate) or squared off (truncate); lobes 5-9, lower pair divided nearly to midrib, sharply toothed (serrate); shiny, hairy on veins on both sides; stipules toothed; petioles about 1.5 in (4 cm) long
Flowers: clusters hairy (villous), about 2.5 in (65 mm) across; flowers white; stamens 20, anthers pink
Fruits: fruits nearly round to pear-shaped, up to 1 in (25 mm) across; red, dotted; nutlets 3-4
Twigs: spines wanting or short

Malus Mill. - apple

Are apples really apples? Or are they pears? A statement about differing opinions among taxonomists is in order here. We place apples in the genus *Malus* as do Rehder (1941), Little (1953), Porter (1967), Bailey Hortarium (1976) and Dirr (1990). Pool (1951), Stephens (1973) and Elias (1980) include apples and pears in the genus *Pyrus*.

About 25 apple species are native to temperate regions. From the time that man began to manage plants and, later, to cultivate them, apples have been an important food. Selection and breeding have resulted in many cultivars, some long forgotten. Others are now staples of commerce and there are surely more to come.

Edible species provided early man a source of a vinegar-like liquid, brandy and cider. Other, less-edible species, are pleasing in bloom, their fruits are colorful, and they are a source of food for animal and bird life. They fulfill the human desire for beauty in the landscape.

The introduction into the U.S. of Siberian and Chinese species in the early 1800s encouraged such a plethora of hybrids with complex pedigrees that we can mention only a small fraction of the large number of hybrids and cultivars recognized today.

Crabapples, among the most popular of flowering trees, are long-lived and adapted to a wide variety of soils, but grow best on moderately heavy, fertile, well-drained acidic soil. Full sun and freedom from drouth stress are required for the best flowers and fruits. Many crabapples make up for drab autumn foliage by displaying attractive fruits that sometimes persist into winter.

Crabapple cultivars are being developed that are resistant to cedar-apple rust, fireblight and scab. Coordinated testing programs at the University of Nebraska and other universities will continue to evaluate a steady stream of new selections.

Among the best cultivars at UNL from 1986 to 1889 are: 'Bob White', 'Centurion', 'Christmas Holly', 'Harvest Gold', 'Mary Potter', 'Molten Lava', 'Red Jade', 'Red Jewel', 'Royalty', 'Ruby Luster', 'Sugar Tyme', 'White Cascade', 'Indian Summer', 'Prairifire' and 'Profusion'. The latter three plus 'Centzam', 'Coral Burst', 'David', 'Flame', and 'Pink Spirea' are well adapted to the Fort Collins area (Fairchild and Klett 1993).

'Magenta', a hybrid crabapple, is an attractive compact tree that reaches up to 30 ft (9 m) in height. The leaves are red in the spring and retain a reddish hue for most of the year. May flowers are pink and the one-half inch (14 cm) fruits are dark red. 'Magenta' is a release from the Rose Lake Plant Materials Center of the Natural Resources Conservation Service and is propagated by seed. It is recommended for use in creating shelterbelts and roadside beautification in the eastern prairies.

Recognition

Habit: trees, rarely shrubs; deciduous, rarely half-evergreen
Leaves: variably egg-shaped or elliptic to oblong; toothed, sometimes lobed; folded or rolled up longitudinally in the bud (convolute)
Flowers: clusters at ends of spur-like twigs; petals white to rose-colored, rounded at tip; anthers usually yellow, styles 2-5, united at base (connate)
Fruits: rounded, fleshy with or without grit-cells, calyx persistent or deciduous
Twigs: rarely with spines
Buds: egg-shaped; scales several, overlapping

Malus baccata (L.) Borkh. - Siberian crabapple

Selections of Siberian crabapple, a native of central Asia, include variety *mandschurica* (Maxim.) Schneid., that bears bright red fruits, and 'Midwest', a release by the Natural Resources Conservation Service Plant Materials Center in Bismarck that is adapted to the northern Great Plains as far south as Kansas. Variety *columnaris* Rehd. reaches a height of over 20 feet with a crown spread of four feet. This crabapple is one of the parents of many cultivars that have produced trees valued for their flowers and fruits, both ornamental and edible.

Crosses between Siberian crabapple and redvein crab, *M. pumila* var. *niedzwetzkyana*, produced 'Rosybloom', an internationally known race of hardy, flowering crabapples. 'Rosybloom' is characterized by the presence of reddish purple pigment on the veins, petioles, flowers and fruits. Of similar breeding are 'Hopa', released by N. E. Hansen of South Dakota, and 'Almey' , introduced by the Morden Experiment Station in Manitoba. These cultivars did much to increase the popularity of pink-flowered crabapples. 'Red Splendor' and 'Radiant' are among the many other popular cultivars selected from the same parentage (Hoag 1965).

Planning

Height: up to 45 ft (14 m)
Habit: tree; round top; white bloom, red or yellow fruit
Site: sun; soil adaptable; zone 3
Use: specimen, naturalizing, border

Recognition

Leaves: elliptic to egg-shaped; tip pointed, base wedge-shaped to rounded, finely toothed; glabrous or slightly hairy beneath
Flowers: white, about 1.2 in (3 cm) across; calyx glabrous, lobes long-pointed; styles usually longer than stamens
Fruits: nearly round, about 0.4 in (1 cm) across, red or yellow, calyx deciduous
Twigs: glabrous

Malus coronaria (L.) Mill. - sweet crabapple

Sweet crabapple, a native of Ohio and its neighboring states, with outliers as far away as Georgia and Missouri, bursts into bloom after most cultivated apples have dropped their petals. Flowers are at first pink, then they become white and release a pervading fragrance. Peattie (1950) describes it as "an odor like that of apple blossoms but more tart and wild and more penetrating." The tangy, fragrant fruit provides food for animals but is never quite mellow enough for human consumption except for an occasional jar of jelly. 'Charlottae' is a selection with double petals.

Planning

Height: up to 30 ft (9 m); National champion, 1976, 37 ft (11.2 m), Virginia
Habit: tree; crown broad, twigs spiny
Site: sun; soil adaptable; zone 5
Use: specimen, naturalizing

Recognition

Leaves: egg-shaped to much longer than wide (lanceolate); tip pointed, base usually rounded, irregularly toothed (those on vegetative shoots lobed); hairy when young, becoming glabrous
Flowers: pinkish, maturing white, about 1.2 in (3 cm) across; calyx hairy, lobes long-pointed
Fruits: usually wider than long, about 1 in (25 mm) long; yellow-green with waxy covering; flesh firm, fragrant, very acidic
Twigs: hairy at first, becoming glabrous, red-brown

Malus ionensis (Wood) Brit. - prairie crabapple

The natural habitat for prairie crabapple is the prairie states of Minnesota, Illinois, Missouri and Iowa, with scattered populations in adjoining states including eastern Kansas and Nebraska.

Peattie (1950) describes prairie crabapple as a tree that is "in flower the fairest and the most sweetly scented, in fruit, if not the largest, at least the most fragrant and handsome in color and tempting to the eye."

'Plena Nova', a double pink-petaled form, is a selection from the old standby, Bechtel crabapple. Unfortunately, all relatives of prairie crabapple are very susceptible to cedar-apple rust, a leaf disfiguring disease associated with humid areas where the alternate host, eastern redcedar, abounds.

Malus ionensis

Planning

Height: up to 20 ft (6 m); National champion, 1991, 46 ft (14 m), Michigan
Habit: tree; crown broad, trunk sprout-forming
Site: sun; soil adaptable; zone 3
Use: specimen, naturalizing
Caution: susceptible to cedar-apple rust

Recognition

Leaves: egg-shaped to oblong; tip pointed, base wedge-shaped to rounded: coarsely toothed and somewhat lobed; densely hairy when young becoming dark green, shiny above, paler and hairy beneath; petioles hairy
Flowers: white to pink, about 1.6 in (4 cm) across; calyx densely hairy, lobes long-pointed
Fruits: usually wider than long, about 1 in (25 mm) long; depressed at tip and base, yellow-green, fragrant, very acidic
Twigs: somewhat spiny; densely hairy, becoming glabrous, red-brown

Malus pumila Mill. - common apple

The common apple, a native of Europe and western Asia that has been cultivated since ancient times, is an important species of the apple lineage grown primarily for their fruits. Also, variety *niedzwetzkyana* (Dieck) Schneid., redvein crab, played an important part in the creation of hybrid ornamental crabs when it was crossed with Siberian crabapple and other crabaple species (Hoag 1965).

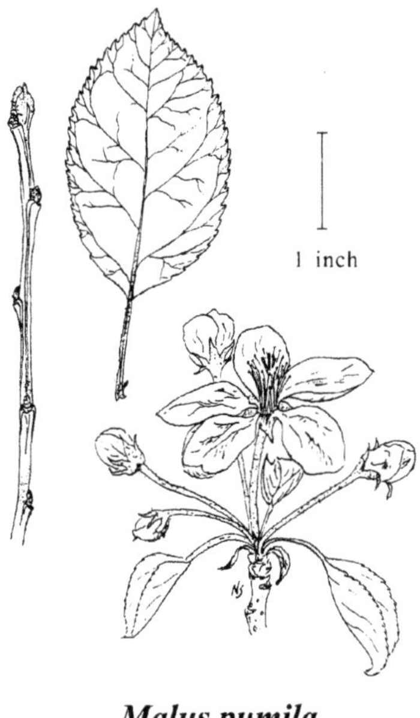

Malus pumila

Planning

Height: up to 40 ft (12 m)
Habit: tree; crown wide-spreading
Site: sun; soil adaptable; zone 3
Use: orchard, specimen
Caution: varieties variably resistant to cedar-apple rust

Recognition

Leaves: broad-elliptic to egg-shaped; tip pointed, base wedge-shaped; toothed (crenate-serrate); hairy on both sides becoming nearly glabrous above; petioles hairy
Flowers: white with touch of pink; calyx hairy, lobes longer than tube, pointed
Fruits: nearly round, depressed at tip and base; calyx persistent, impressed
Twigs: hairy becoming glabrous
Buds: hairy

Malus sargentii Rehd. - Sargent crabapple

Sargent crabapple, a hardy, diminutive Japanese species, is often propagated on dwarfing rootstock or sometimes on six-foot standards for special landscape effects. White-flowered Sargent crabapple has failed to achieve popularity because of the competing pink-flowered hybrids (Hoag 1965).

'Roselow' is a short, compact, seed-propagated shrub released by the Rose Lake Plant Materials Center, Natural Resources Conservation Service, East Lansing, Michigan. It is recommended for planting in windbreak and wildlife plantings in areas from the eastern Great Plains to the East Coast.

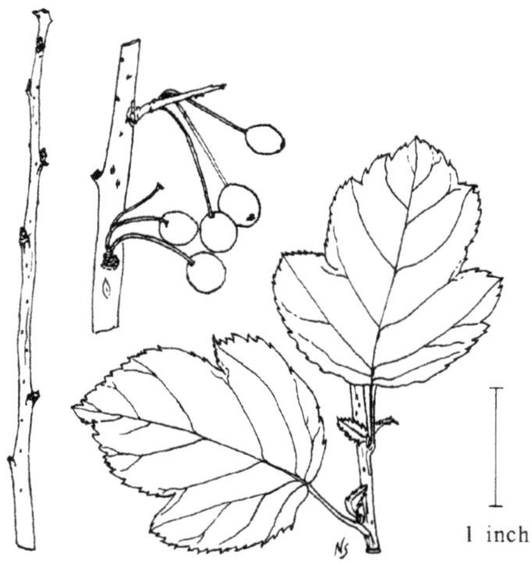

Malus sargentii

Planning

Height: up to 8 ft (2.5 m)
Habit: shrub; crown dense, wide-spreading, thorny; mid-May blooms; fruits persist into winter
Site: sun; soil adaptable up to pH 7; zone 3
Use: border, naturalizing

Recognition

Leaves: egg-shaped, mostly 3-lobed on shoots; tip pointed, base variable; margin sharp-toothed; long hairy (villous) when young, becoming nearly glabrous; petioles hairy
Flowers: white, about 1 in (25 mm) across; calyx glabrous, petals oval, rounded at base, overlapping; styles usually 4
Fruits: nearly round, about 0.4 in (1 cm) across; dark red, slightly bloomy
Twigs: hairy, becoming glabrous

Malus sieboldii (Reg.) Rehd. - Toringo crabapple

Toringo crabapple is an Asian species of shrub and tree form (the latter form is variety *arborescens* Rehd). Toringo crabapple is a parent of the purpurea hybrids—the result of crosses with redvein crab, *M. pumila* var. *niedzwetzkyana* (Hoag 1965).

Purple-leafed 'Royalty' and 'Purple Wave' are two of these purpurea hybrids. Merging the purpurea hybrids with the 'Rosybloom' hybrid race (See *M. baccata*) produced an aggregate of hybrids with pink, rose or reddish purple blooms, with green to purplish foliage (Hoag 1965).

Planning

Height: up to 15 ft (5 m), tree form up to 30 ft (10 m)
Habit: shrub, tree; spreading crown; fruits persist
Site: sun; soil adaptable up to pH 7; zone 5
Use: border, naturalizing

Recognition

Leaves: elliptic to egg-shaped (broader on shoots), 3 to 5 lobes; tip pointed, margin sharply toothed; hairy, becoming glabrous above; petioles hairy
Flowers: pink in bud becoming nearly white, about 0.8 in (2 cm) across; calyx hairy, petals oblong to wider above middle (obovate), wedge-shaped at base; styles 3-4
Fruits: nearly round, about 0.3 in (7 mm) across, red or brownish yellow
Twigs: hairy, becoming glabrous

Photinia Lindl. - photinia

One of the 40 photinia tree and shrub species, natives to southern and eastern Asia, is potentially useful in prairie landscapes.

Photinia villosa (Thunb.) DC. - Oriental photinia

Persistent bright red fruits of Oriental photinia brighten landscapes and feed birds. Its preference for well-drained, acidic soils limits its extensive use on the prairies. However, we believe trial plantings are desirable to determine its potential value.

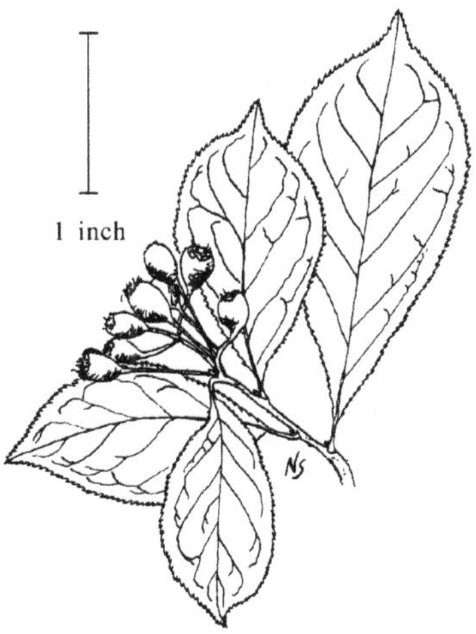

Photinia villosa

Planning

Height: up to 15 ft (5 m)
Habit: shrub, tree; deciduous; taller than broad; autumn foliage reddish; red fruits persistent into winter
Site: sun, light shade; pH less than 7; zone 5
Use: specimen, border, wildlife habitat
Caution: susceptible to fireblight

Recognition

Leaves: alternate, simple; egg-shaped, about 2.5 in (6 cm) long; wider above middle (obovate); tip pointed, base wedge-shaped; sharp teeth fine, gland-tipped; dark green, glabrous above, long soft hairs below (villous); petioles less than 0.2 in (5 mm) long
Flowers: clusters hairy, many-flowered, warty, about 1.5 in (4 cm) across; petals 5, white; stamens about 20, styles usually 2, at least partially joined (connate)
Fruits: fleshy, bright red, 0.3 inch (8 mm) across; calyx lobes persistent, converging
Twigs: hairy when young; lenticels prominent; leaf scars 2-ranked, crescent-shaped, bundle traces 3
Buds: terminal lacking, egg-shaped, pointed; scales 4, keeled, tip sharp (mucronate)

Pyracantha Roem. - firethorn

The six species of firethorn, natives of southeastern Europe and central Asia, are so named because of their bright red fruits and spiny branches. The strongly right-angled branching patterns are compatible to the creation of espaliers. They also can be sheared to form topiaries.

Pyracantha coccinea Roem. - scarlet firethorn

The leaves of scarlet firethorn, a native of Italy and western Asia, gradually become purple and eventually drop during most winters in southeastern Nebraska. Prolonged severe cold, such as the winter of 1983-84, may kill the entire plant. The thorny branches of scarlet firethorn provide "cat proof" nesting sites and food for cardinals, robins and other feathered fruit eaters.

'Lalandei' is one of the most commonly planted and hardier cultivars among the many mentioned by Dirr (1990). 'Walker's Pride', selected in Lincoln for its hardiness and compact size, is marketed locally, but further trial plantings are desirable to confirm its attributes.

Planning

Height: up to 15 ft (5 m)
Habit: shrub; evergreen, semi-evergreen; upright, stiff, thorny branches; flowers along stems of previous year's growth; orange-red fruit fades with freezing
Site: sun, partial shade; soil adaptable; zone 6
Use: hedge, barrier, espalier, trellis, wildlife habitat

Recognition

Leaves: alternate, simple; longer than wide; tip usually sharp, base wedge-shaped; teeth shallow, rounded (crenulate-serrulate), sometimes tipped with a small gland; lustrous dark green above, sometimes slightly hairy beneath at first; stipules minute, early deciduous (caducous); petioles hairy
Flowers: clusters (corymbs) many-flowered, about 2.5 in (6 cm) long, hairy; bisexual; petals white, nearly round (suborbicular), spreading; stamens 20, anthers yellow
Fruits: nearly round, about 0.2 in (6 mm) across, fleshy, red to orange-red; calyx persistent; nutlets 5
Twigs: woolly at first, becoming glossy; spines about 0.6 in (15 mm) long

Pyracantha coccinea

Pyrus - pear

Ancestors of our common fruit-bearing pear were probably selected and cultivated in ancient Persia. Some historians say that the apple in the Garden of Eden was really a pear. So what are the real differences between apples and pears? Not many, according to taxonomists, some of whom lump apples and pears together under *Pyrus* (See *Malus*).

A distinctive flavor, a granular flesh with plentiful grit cells, and a stalk attached to a bump instead of a dip on the fruit characterize the pear species. Most of the 20 species are grown as ornamentals.

'Angel Wing', a cultivar of *P. regelii* Rehd. with gray-green, deeply dissected leaves and drooping form, grows in Fremont, Nebraska, and Fort Collins, Colorado (Sutton 1998). Dirr (1990) notes that *P. pyrifolia* (Burm.) Nakai, Chinese sand pear, a native of central and western China, blooms profusely at the Arnold Arboretum and has orange-red autumn foliage.

Recognition

Habit: trees, rarely shrubs; thorny or not; deciduous or half-evergreen
Leaves: simple; toothed or entire, rarely lobed; rolled inward in the bud (involute)
Flowers: clusters (umbel-like racemes), white, rarely pinkish; sepals reflexed or spreading; anthers usually red; styles free, 2 to 5
Fruits: pear-shaped, flesh with many grit cells; calyx persistent or deciduous
Twigs: sometimes thorny
Buds: scales overlapping (imbricate)

Pyrus calleryana Dcne. - callery pear

Callery pear, a native of China, was once only used as grafting understock before being recognized for its ornamental qualities. Cultivars include: 'Bradford', the most common, and 'Aristocrat', 'Autumn Blaze', 'Capitol', 'Chanticleer' and many others. Most are thornless and vary in their form, branch angle and date of autumn coloration. Cold hardiness can probably be predicted by observing the date of the callery pear's foliage color change at a given latitude.

Although most of the callery pears planted in Lincoln are purported to be 'Bradford', we have noted significant differences in dates of change to autumn foliage coloration among these trees. Injury to a considerable number of these trees occured during two winters of little snowfall and quite variable temperatures (1986-87, 1988-89). The outer trunk bark popped loose on many callery pear trees during or after the first such winter, but no mortality was noted. Flower buds and twig tips were injured during the winter of 1988-89.

Major twig dieback on callery pears from the Morden Experiment Station, Manitoba, has occurred at Prairie Pines Arboretum. However, they recover quickly from the injury as vigorous "water sprouts" soon hide the injured twigs. The Morden trees bloom and fruit consistently and profusely, but require considerable maintenance pruning to remove dead twigs and surplus sprouts.

Dirr (1990) tempered his enthusiasm for using 'Bradford' as a street and general purpose tree by cautioning that splitting trunks and graft incompatabilities are beginning to appear on older trees. It is obvious from our experience in Lincoln that provenance testing may be the best solution for finding the best callery pear cultivars for prairie sites.

Pyrus calleryana

Planning

Height: up to 30 ft (9 m)
Habit: tree; deciduous; crown variably pyramidal; white blooms before or with leaves; fruits persistent but productivity variable
Site: sun; soil adaptable; drouth tolerant; zone 5
Use: street, specimen, wildlife habitat
Note: fireblight resistant

Recognition

Leaves: egg-shaped to broadly so, leathery; tip pointed, base usually rounded; teeth rounded (crenate); surfaces usually glabrous, lustrous dark green; petioles about 1 in (3 cm) long
Flowers: clusters glabrous, about 3 in (8 cm) in diameter; flowers malodorous, about 0.9 in (23 mm) across; petals white; stamens 20, styles 2 rarely 3
Fruits: nearly round about 0.4 in (1 cm) across; brown, dotted; calyx deciduous
Twigs: woolly, becoming glabrous
Buds: hairy, elongated egg-shaped

Rosaceae *Pyrus*

Pyrus communis L. - common pear

The common pear, a native of southern Europe and western Asia, is planted in orchards and backyards throughout the temperate regions. The strong, right-angled branch patterns and ability to respond positively to pruning combine to make it an excellent subject for espaliers.

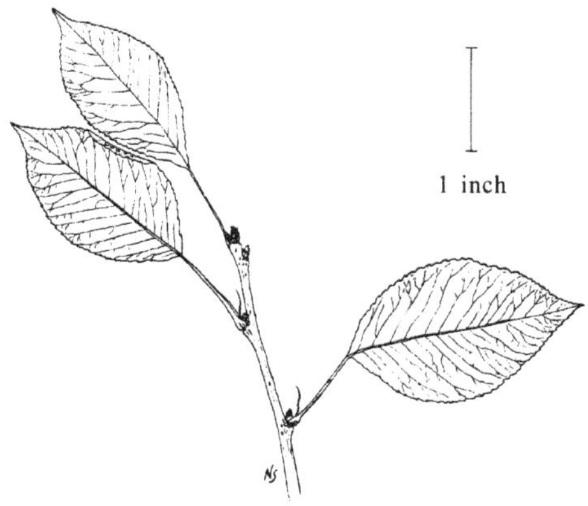

Pyrus communis

Planning

Height: up to 30 ft (9 m); National champion, 1991, 59 ft (18 m), Washington
Habit: tree; crown round-topped, broad; deciduous, sometimes spiny
Site: sun; soil adaptable; zone 5
Use: orchard, espalier
Caution: susceptible to fireblight

Recognition

Leaves: egg-shaped to elliptic; tip pointed, base often rounded; margins fine-toothed (crenate-serrulate); surfaces glabrous or long hairy (villous) when young; petioles about 1.2 in (3 cm) long
Flowers: clusters variably hairy; flowers white, about 1.2 in (3 cm) across
Fruits: inversely conical (turbinate) to globular; calyx persistent; edible
Twigs: slightly hairy to glabrous, sometimes spiny

Pyrus salicifolia Pall. - willowleaf pear

The Eurasian willowleaf, a tree that bears silvery foliage, produces a copious amount of white flower clusters. A healthy young tree grows in Maxwell Arboretum, UNL. 'Pendula', a tree with graceful drooping branches, is illustrated in many mail order catalogs. Dirr (1990) suspects that the very similar 'Silver Frost' in reality is 'Pendula' "with a fancy name."

Planning

Height: up to 25 ft (8 m)
Habit: tree; deciduous; round-topped to pendulous, not thorny; foliage silvery
Site: sun; soil adaptable; zone 5
Use: specimen
Caution: susceptible to fireblight

Recognition

Leaves: much longer than wide, about 2.5 in (6 cm) long; tapering gradually to each end; margin entire; surfaces white, densely woolly (tomentose), becoming glabrous above
Flowers: clusters dense, hairy (tomentose), many-flowered; flowers white, about 0.8 in (2 cm) across
Fruits: pear-shaped, green, about 1 in (25 mm) across; calyx persistent; short-stalked; not edible
Twigs: first year densely woolly (villous)

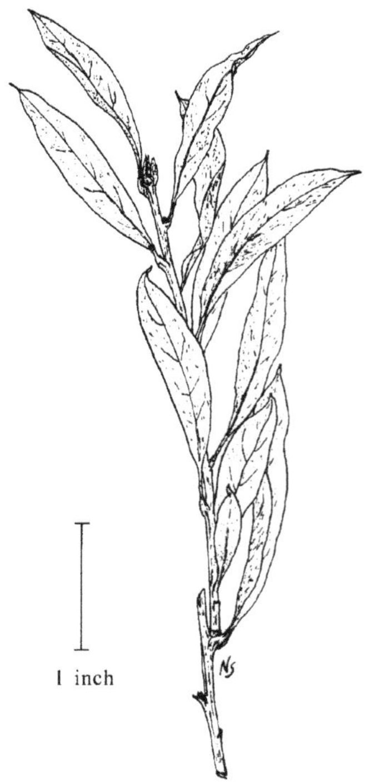

Pyrus salicifolia

Pyrus ussuriensis Maxim. - Ussurian pear

Ussarian pear, a native of northeastern Asia, is well adapted to the northern prairie climate. Although Hoag (1965) rates this pear as questionable for use in the northern plains because of its slow growth, it is often planted for wildlife habitat and windbreaks.

Ussarian pear is well adapted to the Fort Collins area (Fairchild and Klett 1993). 'McDermand', a fireblight-resistant cultivar released by the Natural Resources Conservation Service Plant Materials Center, Bismarck, N.D., is a selection from trees established by Dr. N. E. Hansen of South Dakota State University, from seed he collected near Harbin, Manchuria, in 1924.

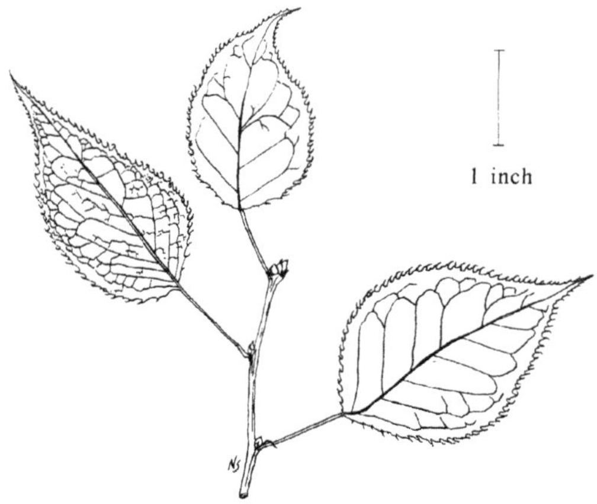

Pyrus ussuriensis

Planning

Height: up to 30 ft (9.1 m)
Habit: tree; deciduous; pyramidal to oval; autumn foliage reddish; April flowers white
Site: sun; pH and soil adaptable, drouth tolerant; zone 3
Use: wildlife and conservation plantings, screen, specimen

Recognition

Leaves: egg-shaped to orbicular, about 3 in (8 cm) long; tip pointed, base rounded; margin bristly toothed (setosely serrate); surfaces glabrous or nearly so; petioles about 1.5 in in (4 cm) long
Flowers: clusters dense, hemispherical, glabrous; flowers 1.2 in (3 cm) across, pink in bud, finally white; petals obovate, gradually narrowed to base; styles hairy (pilose) near base
Fruits: nearly round, greenish yellow, becoming light brown, about 1.3 in (33 mm) across, calyx persistent; not edible
Twigs: green-brown, surface glabrous; lenticels light brown, prominent
Buds: egg-shaped, glabrous

Sorbus L. - mountainash

The mountainashes are native to the northern latitudes and higher elevations of North America, Europe and Asia. They grow best in cool, moist habitats along swamp borders and on north slopes. Extreme heat and drouth combined with low humidity and strong winds predispose mountainash species to fireblight and borers.

Decorative leaves, large white flower clusters and bright red fruits that persist into winter are distinctive characteristics of the mountainash species. Their fruits are valuable wildlife food—especially for birds, squirrels and other rodents.

Intraspecific hybrids of *Sorbus* present an array of plants that are confusing, even to professional horticulturists. Intergeneric hybrids have been identified between *Sorbus* and *Aronia* (×*Sorbaronia* Schneid.), *Cotoneaster* (×*Sorbocotoneaster*), *Amelanchier* (×*Amelasorbus* Rehd.) and *Pyrus* (×*Sorbopyrus* Schneid.). Apomictic reproduction further confounds the taxonomy of the genus (Dirr 1990).

Recognition

Habit: trees, shrubs; deciduous
Leaves: alternate, simple or compound (odd-pinnate)
Flowers: clusters large (compound terminal corymbs); petals 5, white, rarely pink; sepals 5, tiny; stamens 15-20, styles free or united at base (connate)
Fruits: fleshy, nearly round, orange to red, bitter; nutlets 1-2
Buds: large, pointed, often gummy; scales overlapping (imbricate)
Wood: light, not strong

Grouping Key for *Sorbus* (mountainash)

Leaves compound (pinnate): **Group One - mountainashes**

Leaves simple, lobed or toothed (serrate); sometimes white hairy (tomentose) beneath: **Group Two - whitebeams**

Group One - mountainashes

Several European and Asian species of the mountainash that are worth noting by tree planters that desire a challenge are: *Sorbus tianshanica* Rupr., a handsome shrub or small tree reportedly resistant to disease; and *Sorbus cashmiriana*, Kashmir mountainash, with pinkish white flowers and white fruit that has been used in breeding programs and may be worthy of trial (Dirr 1990).

Sorbus aucuparia (Sieb. & Zucc.) K. Koch - European mountainash

European mountainash, widely cultivated in North America, reproduces naturally in cool, moist climates. This species can be successfully grown in a protected prairie landscape that promotes the vigor necessary for it to resist insect and disease attacks.

European mountainash exhibits few problems at PERC in Fort Collins (Fairchild and Klett 1993). These researchers report that 'Michred', a narrow tree bearing bright red fruit, is attractive and produces flowers and fruit consistently, and that 'Xanthocarpa', selected for its yellow fruit, does not produce flowers and fruit reliably and is occasionally chlorotic.

Many other European mountainash cultivars have been selected for tree form, fruit size and color, and leaf characteristics.

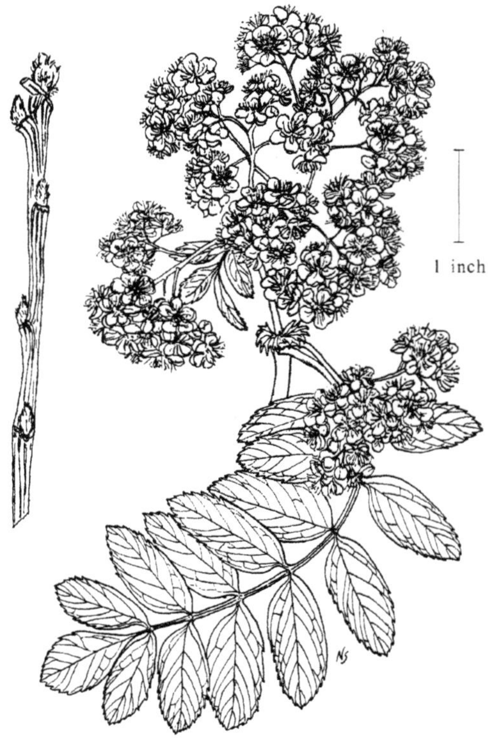

Sorbus aucuparia

Planning

Height: up to 40 ft (12 m); National champion, 1984, 43 ft (13 m), Washington
Habit: tree; crown oval; May flowers, fruit persists into winter; autumn foliage color variable
Site: sun; pH and soil adaptable; prefers cool, moist site; zone 4
Use: specimen, shade
Caution: protect from excessive wind

Recognition

Leaves: leaflets oblong, 9 to 15; toothed on the upper two-thirds of margin; tip pointed or rounded; dull, dark green above; hairy beneath changing to a thin, waxy bloom (glaucescent); stipules small, early deciduous (caducous)
Flowers: clusters (corymbs) flat-topped, dense, many-flowered, 3 to 6 in (7-15 cm) across, usually hairy; flowers about 0.3 in (8 mm) across, petals white; stamens about as long as petals
Fruits: orange-red, about 0.3 in (9 mm) across
Twigs: hairy at first, becoming glabrous and shiny
Buds: woolly, not gummy, broadly cone-shaped, about 0.4 in (1 cm) long
Bark: light grayish brown, smooth, becoming roughened

Sorbus pohuashanensis (Hance) Hedl. - **mountainash**

A specimen of this mountainash in the Fort Collins arboretum is winter hardy, has no pest problems, and is only occasionally chlorotic (Fairchild and Klett 1993). Its flower and fruit production varies from year to year.

Planning

Height: up to 20 ft (6 m)
Habit: tree; broad oval, upright branches; May flowers, orange-red fruits; bark shiny, bronze colored
Site: sun; moist soil, pH adaptable; zone 4
Use: specimen, street

Recognition

Leaves: leaflets oblong, 11 to 15; pointed, toothed from below the middle; hairy and waxy (glaucescent) beneath; stipules large, toothed, persistent
Flowers: clusters (corymbs) woolly-hairy (floccose-tomentose), about 4.5 in (11 cm) across
Fruits: subglobose, about 0.3 in (7 mm) across; orange-red
Twigs: hairy (villous)
Buds: white-villous

Sorbus scopulina **Greene - Rocky mountainash**

Rocky mountainash, a native of the Black Hills, the Rocky Mountains and the Pacific Northwest (including Alaska), has been successful in trial plantings in the northern prairies on well-drained, moist soils (North Central Plant Introduction Station 1973).

Other North American species, *S. americana* Marsh., *S. decora* (Sarg.) Schneid., and *S. sitchensis* will tolerate cold temperatures, but cannot cope with drouth, excessive hot wind or low humidity.

Planning

Height: up to 20 ft (6 m)
Habit: tree, shrub; June flowers; September fruits orange-red
Site: sun, part shade; cool, moist, well-drained soils; zone 3
Use: specimen

Recognition

Leaves: leaflets 9 to 13, sessile, elliptic, about 1.7 in (45 mm) long; sharply toothed (serrate); glabrous; petioles widened at base, hairy along shallow groove above; stipules slender, egg-shaped (ovate-lanceolate), long-pointed, sometimes persistent

Flowers: clusters (corymbs) terminal, flat-topped, 4 to 6 in (10-15 cm) across; sparingly hairy (pilose); petals white, egg-shaped, margin irregular, slightly reflexed, hairy at base inside; calyx lobes triangular, hairy, ascending between petals; stamens many, anthers white

Fruits: globose, 0.3 in (8 mm) across; orange-red, smooth; seeds 1-5

Twigs: glabrous with thin, waxy coat or hairy; lenticels numerous, light-colored; leaf scars U-shaped, raised, bundle scars 5, hidden by petiole stump that is later deciduous

Buds: narrowly egg-shaped, pointed; scales with hairy edges (ciliate)

Group Two - whitebeams

Sorbus alnifolia (Sieb. & Zucc.) K. Koch - Korean mountainash

Korean mountainash is not well known, but Dirr (1990) rates it as one of his favorite all-around trees because of its beech-like bark, beautiful foliage and flowers, and spectacular fruits. It is sometimes referred to as Japanese whitebeam, but it lacks the white woolly hairs that give whitebeams their common name.

This native of central China, Korea and Japan, is growing successfully at the Arnold and Minnesota arboretums. Richley (1988) reports that it appears to be better adapted to southwestern Nebraska soils and climate than any of the other mountainashes. Trial plantings are needed throughout the prairies to assess its further potential.

Planning

Height: up to 50 ft (15 m)

Habit: tree; pyramidal, becoming rounded; May flowers; persistent orange to scarlet fruits September-October; autumn foliage colorful

Site: sun; pH and soil adaptable; zone 4

Use: specimen

Caution: avoid high stress sites

Recognition

Leaves: simple, egg-shaped to elliptic, 2-4 in (5-10 cm) long; tip short-pointed, base rounded; margins unequally toothed; surfaces lustrous dark green, glabrous above, slightly hairy beneath; petioles about 0.6 in (15 mm) long

Flowers: clusters (corymbs) of 6 to 10, flat-topped, about 3 in (8 cm) across; nearly glabrous; flowers white, 0.4 in (1 cm) across; styles 2

Fruits: nearly round, about 0.3 in (8 mm) across, orange to scarlet, bloomy, speckled with dark lenticels; calyx deciduous

Twigs: lustrous, reddish brown; lenticels prominent, gray, becoming diamond-shaped

Buds: oblong; scales several, dark margined, inner hairy, often matted with gum

Bark: gray, smooth

Sorbus aria (L.) Crantz - **whitebeam mountainash**

The woolly white undersides of simple leaves should leave no doubt that this native of Europe belongs among the whitebeams. A whitebeam mountainash in the Fort Collins arboretum is well adapted to the soil and climate without major pest problems (Fairchild and Klett 1993).

Planning

Height: up to 35 ft (11 M)
Habit: tree; egg-shaped or rounded; May flowers; red fruits September-October; autumn foliage reddish purple; bark golden bronze
Site: sun; pH adaptable, soil moist, well-drained; zone 4
Use: specimen, shade

Recognition

Leaves: leaves egg-shaped to elliptic, about 3 in (8 cm) long, leathery; tip variably pointed, base usually wedge-shaped (cuneate); sharple toothed (doubly serrate); glabrous, bright green above except on nerves, white woolly beneath (tomentose); veins 8-12 pairs; petioles woolly, about 0.6 in (15 mm) long
Flowers: clusters (corymbs) about 3 in (8 cm) across, woolly; flowers about 0.6 in (15 mm) across, white; styles usually 2, joined at base
Fruits: subglobose, about 0.6 in (12 mm) across; orange-red to scarlet, mealy; calyx persistent
Twigs: woolly (tomentose)
Buds: greenish, sticky (viscid)
Bark: bronze

Sorbus mougeotii Soyer-Willemt and Godr. - **lobed whitebeam mountainash**

A lobed whitebeam mountainash growing at PERC in Fort Collins is adapted to the soil and climate of that site (Fairchild and Klett 1993). However, flowering is not consistent and fruit set is not abundant.

Planning

Height: up to 30 ft (9 m)
Habit: tree, shrub; oval to round, suckering; May flowers, red-orange fruits; autumn foliage reddish purple; bark shiny bronze
Site: sun; pH adaptable, prefers moist, well-drained soil; zone 4
Use: Specimen

Recognition

Leaves: simple, pinnately lobed; tip pointed, margins toothed; veins 8-12 pairs; grayish white hairy (tomentose) beneath
Flowers: clusters hairy (tomentose); flowers white, 0.6 in (12 mm) across; styles usually 2
Fruits: subglobose, about 0.3 in (8 mm) across, red; calyx persistent; edible
Twigs: hairy (tomentose)
Buds: sticky; scale margins hairy
Bark: bronze

Subfamily Rosoideae

Cercocarpus H.B.K. - mountain-mahogany

About 19 species of genus *Cercocarpus*, natives mainly of the Southwest U.S. and Mexico, inhabit dry, mountainous regions. Foliage and twig tips provide browse for cattle, deer, sheep and goats.

The autumn display of feathery, long-tailed seeds is the trademark of mountain-mahogany. Its natural habitat and reddish brown, hard, heavy wood account for the common name. These slow-growing shrubs adapt to hedging and are likely candidates for bonsai culture.

Recognition

Habit: shrubs, small trees; evergreen, deciduous

Leaves: simple, alternate, mostly in clusters; margins toothed or entire

Flowers: clusters of 1 to 10 in leaf axils; petals none; calyx tube elongated, cylindric, expands into a cup shape with 5 petal-like deciduous lobes, greenish white or reddish; ovary 1-celled, enclosed in calyx tube, style exserted

Fruits: dry, one-seeded, not splitting (akenes); enclosed in calyx tube and terminating with long curved or twisted style

Twigs: spur-like, often spine-like; roughened by leaf scars

Cercocarpus ledifolius Nutt. - curlleaf mountain-mahogany

Slow-growing curlleaf mountain-mahogany is a native of an area that extends from eastern Washington to the Grand Canyon, and in the Rockies from western Colorado to Montana at elevations from 4500 ft (1360 m) in the north to 9000 ft (2700 m) in the south. This most common mountain-mahogany species grows on dry, rocky ridges and on south and west slopes where it serves as deer and elk habitat. The prized wood is extremely hard, and, though brittle, has been used by Native Americans for bows.

Excellent specimens of curlleaf mountain-mahogany can be found growing at PERC in Fort Collins and at the West Central Research and Extension Center at North Platte, Nebraska. Fairchild and Klett (1993) report that it is slow to establish, and that it adapts to sites on the Front Range of the Rockies.

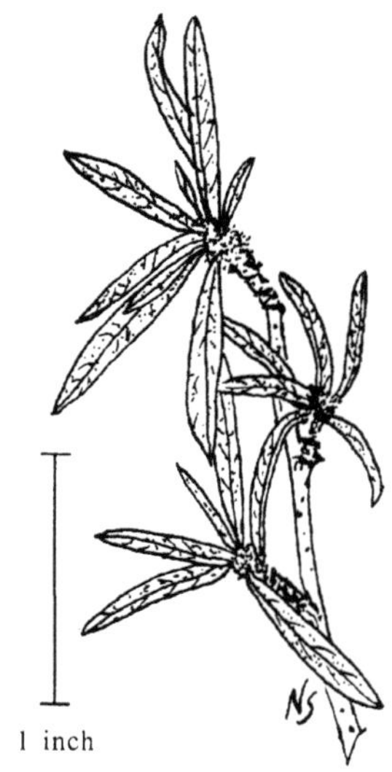

Cercocarpus ledifolius

Planning

Height: up to 15 ft (5 m), rarely 40 ft (12 m)

Habit: shrub, tree; evergreen; upright, open; silvery white fruits persist into winter

Site: sun; soil and pH adaptable, prefers well-drained sites; drouth tolerant; zone 4

Use: naturalizing, low maintenance, specimen, border

Recognition

Leaves: entire, leathery; edges rolled under, elliptic to much longer than wide, abruptly pointed; hairy above, becoming glabrous, hairy beneath, midrib prominent

Flowers: solitary or clusters of 2 to 3; calyx top-shaped, hairy, deciduous except lower cylindric portion

Fruits: densely soft-hairy, tipped by feathery tail 2-3 in (50-75 mm) long

Twigs: red-brown, hairy, becoming glabrous

Cercocarpus montanus Raf. - mountain-mahogany

Mountain-mahogany, a shrubby plant bearing striking, plumose fruits, grows in the Wildcat Hills and Pine Ridge of Nebraska, and from there north to Montana, south to New Mexico, and west to eastern California and Washington. It is found on a wide variety of sites, ranging from moist valleys to dry slopes and ridges (U.S. Forest Service 1937).

Arboretum specimens of mountain-mahogany at Fort Collins are variable in growth habit, and seed set is not abundant or consistent (Fairchild and Klett 1993). Mountain-mahogany is a good candidate for low-maintenance landscapes. The high atmospheric humidity found in the eastern prairies may cause foliar diseases. However, a specimen at Prairie Pines Arboretum near Lincoln is healthy, but very slow growing.

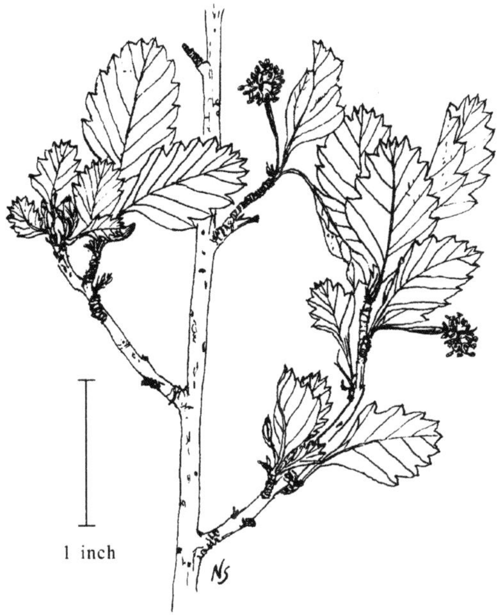

Cercocarpus montanus

Planning

Height: up to 10 ft (3 m), rarely 20 (6 m)
Habit: shrub; deciduous; upright, open; plumose fruits persist into winter
Site: sun; soil and pH adaptable, prefers well-drained sites; drouth tolerant; zone 4
Use: naturalizing

Recognition

Leaves: broadly egg-shaped or reverse egg-shaped (obovate); toothed above middle (crenate-serrate), tip rounded or pointed, base wedge-shaped; prominent lateral veins ending in teeth; green and sparse silky hairs above, densely hairy (tomentose) beneath; blade decurrent on half of petiole; stipules early deciduous
Flowers: solitary in leaf axils at end of short spur; calyx cylindric, hairy outside, expanding to shallow cup shape, lobes (sepals) hairy, triangular; stamens 20-40, inserted on cup rim; ovary cylindric, densely hairy; style hairy
Fruits: soft-hairy, tipped by silvery, feathery tail (elongated style) up to 4 in (10 cm) long
Twigs: red, hairy, becoming gray-waxy; developing short, spur-like branches
Buds: egg-shaped, pointed; scales brown, hairy

Cowania Don. - cliffrose

Four species of cliffrose are natives of the southwestern U.S. and Mexico.

Cowania stansburiana Gray var. *stansburiana* Torr. - cliffrose

Cliffrose is named for its fragrant, rose-shaped flowers and its frequent cliff habitat. Bitter twigs have earned it another common name—"quinine-bush." This native of southern Colorado and Utah grows on dry, rocky sites on a variety of soil formations (Sutton and Johnson 1974).

Cliffrose is a browse plant for livestock and game animals. Although it is seldom, if ever, planted in landscapes, we believe that it has potential for use in the semi-arid prairies. This slow-growing plant is a good candidate for bonsai culture.

Planning

Height: up to 10 ft (3 m), occasionally taller
Habit: shrub; evergreen, much branched, spreading, sometimes approaching tree-like; aromatic; flowers fragrant
Site: sun; soil adaptable, prefers well-drained sites; drouth tolerant; zone 6
Use: naturalizing, border

Recognition

Leaves: alternate, simple, at end of short, spurlike twigs; lobes 3 to 7, margins curled under; dark green, gland-dotted above, white-hairy beneath, leathery
Flowers: solitary at end of spur-like twigs, stalked; about 0.7 in (18 mm) across; petals white to sulfur yellow, sometimes purple, rounded, spreading; calyx united into top-shaped tube without bracts, densely white-hairy, persistent; pistils few on convex receptacle, styles exserted; stamens numerous, in 2 rows
Fruits: akenes hairy, dry, 1-seeded, not splitting; tipped by feathery tails (styles) up to 2 in (5 cm); 4 to 10 seeds borne on a flat disc in the hypanthium
Twigs: hairy, bitter

Fallugia Endl. - Apache plume

Fallugia is a genus of one species.

Fallugia paradoxa Endl. - Apache plume

Apache plume is a native of the home of the Apache Tribe—western Texas and southwestern Colorado to southern Nevada, south into Mexico. This common and, occasionally, locally abundant shrub grows on a variety of soils, from fertile, moist canyon bottoms to dry, rocky ridges, and provides browse for livestock and game animals (U.S. Forest Service 1937). It is unobtrusive in its native habitat until delicate, white, raspberry-like flowers appear in early summer, followed by clusters of feathery-plumed seed clusters with a fancied resemblance to eagle-plumed Apache warbonnets.

Richley (1988) reports that Apache plume grows well on dry, exposed sites in southwestern Nebraska. It is marketed by western nurseries, but it is planted infrequently outside the western states.

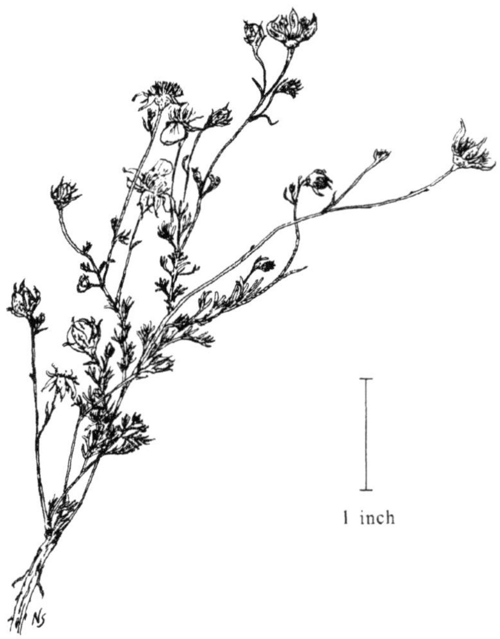

Fallugia paradoxa

Planning

Height: up to 6 ft (2 m)
Habit: shrub; deciduous, evergreen in warm climates; upright, many-branched; autumn fruits striking
Site: sun; soil adaptable; drouth tolerant; zone 6
Use: border, naturalizing

Recognition

Leaves: alternate, simple; clustered or borne singly; wedge-shaped, pinnately divided into 3-7 narrow-oblong blunt-tipped lobes, margins rolled under; soft hairy, thick, glandular; stalkless

Flowers: solitary or sometimes clustered at end of nearly leafless stalks, about 1.2 in (3 cm) across; petals white, calyx hairy with long, tapering, sometimes 3-cleft tips with bracts at base; pistils many on convex receptacle; stamens numerous in 3 rows

Fruits: dry, non-splitting one-seeded (akenes); narrow-oblong; tipped by feathery tails (styles) up to 2 in (5 cm) long, turning reddish with age

Twigs: delicately woolly, becoming shreddy

Kerria DC. - Japanese kerria

The only species of Japanese kerria is a native of central and western China.

Kerria japonica (L.) DC. - Japanese kerria

Although Japenese kerria requires considerable pruning to keep it neat, Dirr (1990) believes that this China native should be used more extensively because of its shade tolerance and relative freedom from serious pests. Its yellow-green twigs are effective winter color that can complement yellow-twig dogwood in the landscape. Cultivars include 'Picta', with variegated leaves, and 'Pleniflora', with double flowers.

Planning

Height: up to 10 ft (3 m)

Habit: shrub; deciduous; upright, arching, becoming a rounded, twiggy mass; texture fine; early spring blooms bright yellow

Site: shade, protected location; zone 5

Use: border, massing

Caution: avoid over-fertilizing for most effective flowering

Recognition

Leaves: alternate, simple; oblong egg-shaped; tip long-pointed, doubly toothed (serrate); bright green, glabrous above, slightly hairy beneath on veins; petioles about 0.4 in (1 cm) long

Flowers: solitary at end of previous year's growth; about 1.5 in (4 cm) across; petals bright yellow, oval; pistils 2-5 on convex receptacle; stamens and styles half as long as petals

Fruits: dry, 1-seeded, not splitting (akenes); not showy

Twigs: green, glabrous, glossy, zigzag

Buds: scales overlapping, greenish brown, glabrous

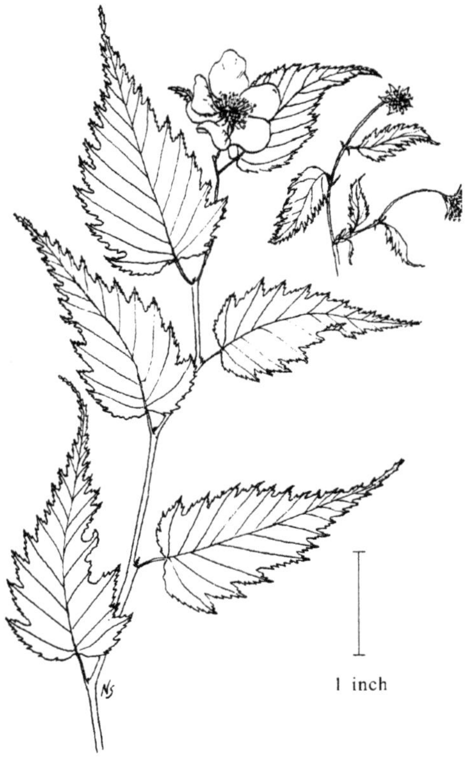

Kerria japonica

Neviusia Gray - snow-wreath

The only species of snow-wreath is a native of Alabama.

Neviusia alabamensis - snow-wreath

Snow-wreath is a novelty for persons who want a different kind of plant (Dirr 1990). It grows at the Morton Arboretum in Chicago, but we are not aware of the extent of plantings in the prairies. However, it is worthy of trial.

Planning

Height: up to 6 ft (2 m)
Habit: shrub; deciduous; upright, arching, becoming rounded; feathery flowers in May
Site: sun, partial shade; soil adaptable; zone 5
Use: border
Caution: experimental

Recognition

Leaves: alternate, simple; egg-shaped to oblong; tip pointed, doubly toothed; nearly glabrous
Flowers: clusters (cymes) of 3 to 8, rarely solitary; flowers about 1 in (25 mm) across; petals wanting; sepals petal-like, white, toothed, persistent, spreading; stamens numerous, longer than sepals, showy; pistils borne on convex receptacle, styles shorter than stamens
Fruits: dry, 1-seeded, not splitting (akenes), not showy
Twigs: golden brown, slightly hairy (puberulent)
Buds: egg-shaped; scales overlapping, short-hairy, becoming nearly devoid of hairs (glabrate)

Potentilla L. - cinquefoil

Some 300 *Potentilla* species inhabit the north temperate and frigid zones, the high mountains of South America and some tropical areas (U.S. Forest Service 1937). Most cinquefoils are herbaceous, some are woody only at the base of the plant (suffrutescent), and a very few are shrubs.

The name cinquefoil describes the five fingerlike leaflets characteristic of many *Potentilla* species. Many species are astringent and high in tannic acid content. They are believed by some to possess powerful medicinal properties.

Recognition

Habit: herbs, shrubs; deciduous
Leaves: palmately or pinnately compound; stipules attached to and sheathing the petiole
Flowers: clusters (cymes) terminal or solitary; white, yellow or red; sepals 5, bracts 5, alternating with sepals; petals 5, nearly round (orbicular); stamens 10-30; pistils many, styles deciduous
Fruits: dry, non-splitting (akenes)

Potentilla fruticosa L. - shrubby cinquefoil

Shrubby cinquefoil occupies a wide variety of sites—mountains and humid forests to arid plains—throughout the Northern Hemisphere. The numerous varieties have contributed to the breeding and selection of an astounding array of cultivars, many so similar that labels are mandatory for identification. Dirr (1990) summarizes the characteristics of some 57 cultivars. We make no recommendations and attempt only to describe an "average" plant, if such can be found.

Shrubby cinquefoil becomes ragged with time—some of its cultivars more quickly than others. As twigs die back, renewal pruning is required. Eleven cultivars growing at PERC in Fort Collins are adapted to alkaline soil and are generally winter hardy (Fairchild and Klett 1993). Near the end of a hot, dry growing season mites can cause the foliage to turn bronze colored and sometimes die. Among the cultivars that consistently provide good floral displays at the Fort Collins site are the white-flowered 'Abbotswood', and yellow-flowered 'Coronation Triumph', 'Klondike' and 'Longacre'.

Planning

Height: up to 3 ft (1 m), but varies with cultivar
Habit: shrub; upright stems, rounded outline, but quite variable; flowers June to frost
Site: sun, tolerates partial shade; soil and pH adaptable; zone 3
Use: border, massing, hedge, foundation

Recognition

Leaves: compound (pinnate); leaflets usually 5 to 7, stalkless, elliptic to linear oblong, about 0.6 in (15 mm) long; margin entire, somewhat turned under (revolute); silvery-hairy, more dense below; stipules 0.3 in (8 mm) long, lower half joined to petiole, hairy, papery, persistent into winter
Flowers: solitary or in few-flowered clusters on new twig ends; green parts hairy; about 1 in (25 mm) across; petals round, entire, spreading, usually bright yellow, shades of yellow, white and shades of red, twice as long as calyx tubes; calyx saucer-shaped with 5 lobes alternating with 5 bractlets, green, silky-hairy; stamens 10-20, anthers yellow
Fruits: dry, 1-seeded, not splitting (akenes); hairy
Twigs: silky hairy becoming shiny brown, shreddy with age
Buds: egg-shaped, concealed beneath stipules; scales thin, fringes hairy (ciliate)

Potentilla fruticosa

Rhodotypos Sieb. & Zucc. - black jetbead

The only species of black jetbead is a native of central China and Japan.

Rhodotypos scandens (Thunb.) Mak. - black jetbead

Black jetbead, though not commonly found in our landscapes, is rated as a tough, durable shrub for shady areas (Dirr 1990). It survives in Lincoln with litle evidence of injury from either climatic extremes or pests.

Planning

Height: up to 6 ft (2 m)
Habit: shrub; deciduous; ascending, loosely branched; spring bloom; fruits persist through spring into early summer
Site: sun, shade; soil and pH adaptable; zone 5
Use: border, massing

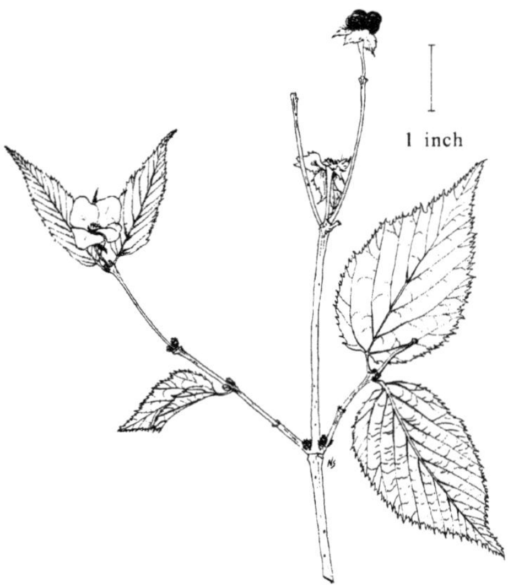

Rhodotypos scandens

Recognition

Leaves: opposite, simple, egg-shaped to oblong; long-pointed, base rounded; sharply doubly toothed (serrate); bright green, glabrous above, silky hairy beneath when young; petioles about 0.2 in (5 mm) long

Flowers: solitary at end of short twigs, about 1.5 in (4 cm) across; petals 4, white; calyx tube nearly flat; sepals 4, egg-shaped, toothed, 4 alternate bracts at base; stamens many, half as long as petals; carpels usually 4

Fruits: drupes dry, black, 3 to 4 in a group, about 0.3 in (8 mm) long; surrounded by persistent calyx

Twigs: green, shiny, becoming brown, streaked with gray; lenticels orange

Buds: diverging from stem; egg-shaped, glabrous; scales overlapping (imbricate)

Rosa L. - rose

Roses have been admired and cultivated throughout the ages—the object of festivals from the days of ancient Persia to the modern day California Rose Parade. Public gardens and parks all over the world take pride in their rose displays. Roses occupy a special place in the hearts of many home gardeners as well.

Nursery catalogs and garden magazines heighten our expectations for a colorful spring and summer as we peer out at a bleak midwinter landscape. Unfortunately, we discover that we must adopt a rigorous program of disease and insect control, fertilization, watering, protection and pruning to maintain many of the most prized rose plants from year to year. If deer are part of the environment, rose flower buds may disappear before blooms have a chance to appear.

Highly prized hybrid tea roses are the most tender, requiring mulching over winter for protection from temperature extremes and desiccation. Floribundas and grandifloras are somewhat more hardy. Fortunately, among this vast rose population there are vigorous and faithful bloomers—wild roses and their hybrids—that can provide beauty in our garden and bouquets for our tables even though we have not the time, resources or knowldege to grow hybrid teas. Some of these roses are suited for protection and wildlife habitat plantations. Several have beem important contributors to rose breeding programs.

Rehder (1940) conservatively estimates the number of species of *Rosa* to be between 100 and 200 throughout the temperate and subtropical regions of the Northern Hemisphere. He cites authors that have recognized as few as 30 species to one who enumerates over 4,000 species in Europe and Asia. Many articles, monographs and books have been written specifically about American roses, European roses, Asiatic roses and cultivated roses.

A concise history of rose development is presented by Hoag (1965). He also describes species and cultivars that are hardy in the northern prairies. **Roses** by Peter Beales (1992) is an illustrated encyclopedia and grower's manual which describes almost 2,000 old and modern roses.

Rose species are difficult to describe because of the high variability of many of the diagnostic characters and their tendency to hybridize freely. Minor characteristics of a section or a group of species may occur in various combinations in other sections of groups and are therefore not sufficiently constant to be trustworthy in identifying a particular species. This is especially true of stem prickles; glands, hairiness, and margins of leaflets and sepals; and the form of the fruits (hips).

Wild roses grow on sites ranging from exposed, hot and dry to shaded, cool and moist throughout the prairies and the western mountain ranges of North America. They are normally scattered, providing a small part of the groundcover and only forming patches or thickets under especially favorable conditions.

Roses are browsed by domestic and wild animals. The fruits, a source of vitamin C, provide sustenance to many birds and mammals. Native Americans have used the roots, stems, leaves, flowers and fruits for food and medicine.

Our text is limited to only those roses native to the Great Plains and a few hardy exotics. We have relied greatly on descriptions by Stephens (1973) for those that are native to the north central plains.

Recognition

Habit: shrubs (upright, trailing or climbing); deciduous or evergreen
Leaves: odd pinnately compound, rarely simple; leaflets commonly toothed, end pair often the largest; stipules, 2 partly fused appear as wings at petiole base
Flowers: solitary or in clusters (corymbs) at tips of short twigs; petals 5 or more, variously colored, rounded or reverse heart-shaped; sepals 5, sometimes variously lobed, erect or spreading, often persisting after fruit maturity; stamens numerous; pistils numerous
Fruits: dry, 1-seeded (akenes) enclosed in a fleshy false fruit (hip); usually red
Twigs: upright, climbing or trailing; prickly, sometimes bristly; rarely unarmed

Grouping Key for *Rosa* Subgenus *Eurosa*

(Rehder 1940)

Stems arching or trailing, prickles and bristles straight or slightly curved; leaflets 7-9; flowers solitary; sepals entire, erect, persistent: **section Pimppinellifoliae**

Stems upright or arching, prickles hooked, rarely straight; leaflets 5-9; flower clusters (corymbs) usually many-flowered; sepals, outer lobed, reflexed after anthesis and deciduous or erect and persistent: **section Caninae**

Stems upright, often with bristles; prickles paired, straight, sometimes scattered short prickles; leaflets 5-9; flowers solitary or clusters (corymbs) few-flowered, pedicels glandular hispid, rarely smooth; sepals, outer entire or with few erect lobes, spreading after flowering, soon deciduous: **section Carolinae**

Stems upright, often bristly; prickles in pairs below stipules or scattered, usually straight; stems often bristly; leaflets 5-11; flower clusters usually many-flowered; sepals usually entire, upright after flowering and persistent, rarely deciduous: **section Cinnamomeae**

Stems climbing, trailing, rarely upright, prickles hooked; leaflets 3-9; flower clusters (corymbs) few to many flowered; sepals, outer pinnate, rarely entire, early deciduous: **section Synstylae**

Section Pimpinellifoliae

Rosa foetida Herrm. - Austrian brier

Austrian brier, a dependable Eurasian rose, can be found growing in abandoned farmyards, thriving in spite of neglect. Because it is somewhat tender in cold climes and subject to black spot disease, Hoag (1965) instead recommends 'Harison's Yellow Rose' (*R. foetida* var. *persiana* (Lem.) Rehd. × *R. spinosissima* L.) as one of the more reliable shrub roses for much of the Great Plains.

'Persiana' bears double yellow flowers and 'Austrian Copper' exhibits petals that are copper-red above and yellow below. Dirr (1990) describes the latter as a magnificant rose that "is a tremendous experience to see in full flower, perhaps no more handsome rose."

Planning

Height: up to 15 ft (5 m)
Habit: shrub; deciduous; arching; yellow May-June flowers; fruits rarely seen
Site: sun; drouth tolerant; soil adaptable; zone 5
Use: specimen
Caution: highly susceptible to black spot disease

Recognition

Leaves: leaflets 5-9, oval or wider above the middle, about 1.2 in (3 cm) long, teeth glandular; glabrous above, glandular and slightly hairy below; stipules glandular toothed

Flowers: solitary or small clusters, odor unpleasant, about 2.5 in (6 cm) across; petals deep yellow to red-orange, sepals leafy at tip

Fruits: nearly round, about 0.5 in (12 mm) across, red

Twigs: gray, prickles many, slender, abruptly widened at base

Section Caninae

Rosa eglanteria L. - sweet-brier

Sweet-brier is aptly named for its fragrant foliage. This pink-flowered, vigorous rose produces bounteous large hips that persist into winter.

R. pomifera Herrm., a rose of similar habit, bears slender, straight spines and fruits nearly 1 inch (25 mm) across. *R. glutinosa* sibth. & Sm., a smaller rose with many prickles on both stems and fruits, is aptly named pine-scented rose (Scott 1998)

Planning

Height: up to 8 ft (2.5 m)

Habit: shrub; deciduous; much-branched, upright to arching; June bloom; fragrant foliage; conspicuous red fruits persist through winter

Site: sun; soil adaptable; zone 3

Use: naturalizing, border, wildlife habitat

Recognition

Leaves: leaflets 5-7, oval to nearly round; margins doubly toothed, gland tipped; glabrous above, glandular hairy beneath; petiole and rachis glandular hairy; stipule margin glandular hairy

Flowers: clusters small, each about 1.5 in (4 cm) across; stalks (pedicels) glandular hispid; petals pink; sepals long, lobed, pointed, margins glandular toothed, reflexed in bloom, tardily deciduous

Fruits: hips nearly round to egg-shaped; red, conspicuous

Twigs: green with reddish tinge, glabrous; prickles strong, widened at base, curved

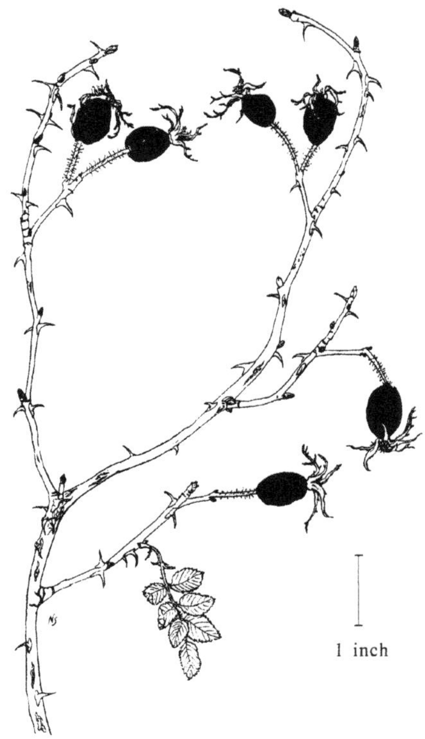

Rosa eglanteria

Rosa rubrifolia Vill. - redleaf rose

Redleaf rose, named for the red foliage of this hardy European native, provides contrast in the landscape, but the foliage mutes the effect of its pink flowers. Redleaf rose is only occasionally planted, mostly in the northern prairies. Hoag (1965) rates it questionable for landscape planting, but Dirr (1990) says it is a fine addition to a shrub border.

Planning

Height: up to 7 ft (2 m)
Habit: shrub; deciduous; erect, dense; reddish foliage; flowers pink to red
Site: sun for best foliage color; soil adaptable; zone 3
Use: border

Recognition

Leaves: leaflets 5-7, egg-shaped to elliptic, up to 1.5 in (4 cm) long, toothed (serrate); red-purple, glabrous, not glandular
Flowers: clusters few-flowered; flowers up to 1.5 in (35 mm) across, petals pink to red; sepals usually entire, longer than petals
Fruits: hips nearly round, about 0.5 in (12 mm) across, red; calyx deciduous
Twigs: purplish bloom, prickles few, hooked or straight

Section Carolinae

Rosa carolina L. - Carolina rose

Carolina rose, a native of the eastern U.S., grows most commonly on moist sites. Stephens (1973) states that though it is a native in eastern Kansas, the plant's tops die back during cold winters there.

Planning

Height: up to 3 ft (1 m)
Habit: shrub; deciduous; thicket-forming, prickly; pink or white flowers June-July
Site: sun; moist soil; zone 5
Use: naturalizing

Recognition

Leaves: leaflets 5-7, elliptic, about 1.3 in (3 cm) long; tip pointed, base wedge-shaped; margin toothed, with or without glands, hairy (ciliate) or not; upper surface glabrous, lower with few hairs on veins; petioles about 0.4 in (1 cm) long; stipules united nearly whole length of petiole (adnate), margin hairy glandular; central axis (rachis) may have a few prickles, with or without glands
Flowers: solitary, sometimes small clusters; flowers about 2 in (5 cm) across; stalks (pedicels) glandular; calyx lobes about 0.7 in (18 mm) long, tapering to a point, margin with narrow projections; petals pink, notched at apex
Fruits: hips nearly round, about 0.4 in (1 cm) across, orange-red with scattered glands on short stalks (stipitate); calyx lobes persistent or deciduous
Twigs: green to green-brown, glabrous; prickles, paired long, scattered prickles shorter
Buds: egg-shaped, red

Rosa virginiana Mill. - Virginia rose

Virginia rose is one of the more handsome North American roses (Dirr 1990). Foliage of this native of the eastern U.S. is glossy dark green in summer, and crimson and yellow in autumn.

Planning

Height: up to 6 ft (2 m)
Habit: shrub; deciduous; erect, dense; few or no suckers; June bloom; autumn foliage colorful; red fruit persistent
Site: sun, light shade; tolerates sandy soils; zone 4
Use: border, hedge

Recognition

Leaves: leaflets 7-9; elliptic to wider above middle (obovate), up to 2 in (5 cm) long; margin toothed, ascending (serrate); tip pointed, base wedge-shaped (cuneate); lustrous above, hairy on veins below

Flowers: clusters usually few-flowered, sometimes solitary; flowers about 2.5 in (6 cm) across, pink; stalks and receptacle glandular-hispid

Fruits: hips nearly round, about 0.5 in (12 mm) across, red

Twigs: reddish, bristly when young; paired prickles below stipules, thick-based, usually curved, or variously prickly

Section Cinnamomeae

Rosa acicularis Lindl. - prickly rose

Prickly rose (var. *bourgeauiana* Crep.), a native of northern North America stretching south into Colorado and South Dakota, thrives on streambanks and wooded hillsides on moist soil (Stephens 1973). Varieties *acicularis* and *nipponensis* Koehne are natives of Europe and Asia. Although many prickly rose varieties and hybrids have been described, few have been grown commercially.

Planning

Height: up to 3 ft (1 m)

Habit: shrub; deciduous; often quite prickly; flowers early June

Site: sun, part shade; soil adaptable; zone 2

Use: naturalizing, wildlife habitat

Recognition

Leaves: leaflets 5-9; central axis (rachis) grooved above, minutely hairy (puberulent), glandular, prickles few; leaflets elliptic, about 1 in (25 mm) long; tip pointed to rounded, base rounded to wedge-shaped; margins toothed (serrate) and hairy (ciliate); upper surface usually glabrous, soft hairy below; petioles grooved above, hairy and glandular; stipule surface and margins similar to leaflets

Flowers: usually solitary on new lateral shoots, about 1.7 in (45 mm) across; pedicels glabrous; calyx lobes usually narrowed in middle, wider at tip, 0.8 in (22 mm) long; petals pink or red

Fruits: hips pear-shaped, about 0.7 in (18 mm) long; bright, orange-red, smooth; calyx lobes persistent, erect

Twigs: reddish; prickles needle-like, weak

Buds: broadly egg-shaped to rounded; brown-red; hairy at tip

Rosa arkansana Porter var. *suffulta* (Greene) Cockerell - Arkansas rose

Arkansas rose, a native of much of the U.S. from Mexico to southern Canada, is so much a part of the prairies that it is also called prairie rose. It withstands mowing but succumbs to repeated herbicide applications.

Planning

Height: up to 2 ft (50 cm)

Habit: shrub; deciduous; loose colonies; numerous prickles; June flowers

Site: sun; pH and soil adaptable; drouth tolerant; zone 2

Use: naturalizing

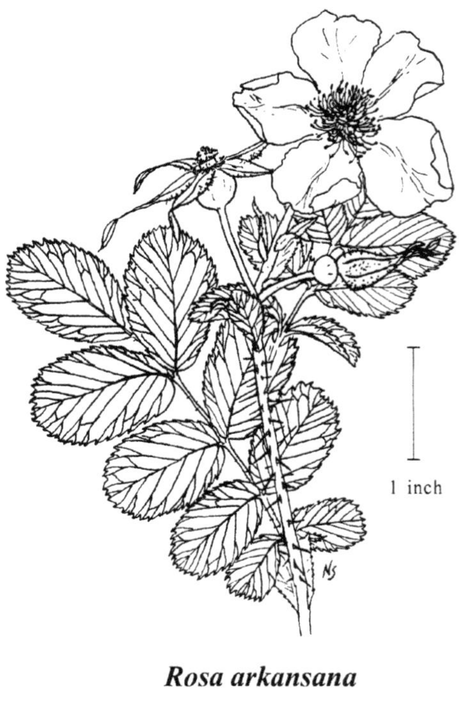

Rosa arkansana

Recognition

Leaves: leaflets 5-9, central axis (rachis) hairy; leaflets oval to wider above the middle, about 0.8 in (2 cm) long; tip rounded to pointed, base wedge-shaped, margins sharply serrate; upper surface usually glabrate, soft hairy to nearly glabrous below; petiole and rachis hairy; stipules variaaably hairy, tips spreading, margin entire, minutely toothed, or glandular dentate.

Flowers: solitary or clusters of 2 to 4, terminal on new growth; about 1.5 in (4 cm) across; calyx lobes long, narrow tapered to sharp point, about 0.3 in (8 mm) long; petals pink, notched at outer end

Fruits: hips nearly round, about 0.5 in (12 mm) across, bright red; calyx lobes persistent, erect or spreading

Twigs: reddish, sometimes bristly; prickles numerous, round in cross-section (terete)

Buds: egg-shaped, tip rounded

Rosa blanda Ait. - meadow rose

Meadow rose, a native of the northern prairies and the northeastern U.S. as well as Ontario and Quebec, grows on hillsides and along roads. Meadow rose is not as common as *R. arkansana*, perhaps because it is not as well armed as its prairie cousin. Although it is occasionally marketed by nurseries, meadow rose is not popular because of its suckering habit.

Planning

Height: up to 3 ft (1 m)

Habit: shrub; deciduous; colony forming, prickles only at stem base; early June flowers, bright red stems

Site: sun; soil adaptable, drouth tolerant, zone 2

Use: naturalizing

Recognition

Leaves: leaflets 5-7, central axis (rachis) hairy; leaflets oval to wider above the middle (obovate), about 0.8 in (2 cm) long; tip rounded, base rounded to wedge-shaped; margins hairy (ciliate); upper surface usually glabrous, soft hairy to nearly glabrous below; petioles hairy, stipules hairy below

Flowers: solitary or small clusters (corymbs) on new growth from lateral buds; flowers about 2 in (5 cm) across; calyx lobes tapered to long slender point, about 0.6 in (15 mm) long; petals pink, often streaked with red, notched at outer end

Fruits: hips nearly round, 0.5 in (13 mm) across, bright red; calyx lobes spreading or erect

Twigs: red; prickles none, or few

Buds: egg-shaped (ovoid); scales red, somewhat keeled

Rosa rugosa Thunb. - rugosa rose

The ability to grow on difficult sites is a desirable trait of rugosa rose, a native of northern China, Korea and Japan. Dirr (1990) rates it as one of the most trouble-free roses. Rugosa rose that is growing on a severely eroded site in a wildlife habitat planting near Lincoln produces copious flowers and fruits despite climatic extremes and infertile soil.

Based on flower, fruit and foliage characteristics, insect and disease resistance, cold hardiness, form, and color, rugosa rose and related hybrids comprise the most promising group of landscape roses (Epping and Hasselkus 1989). Hoag (1965) states that rugosa rose is subject to chlorosis on high lime soils.

'Frau Dagmar Hastrup', a rugosa seedling, was the best all-around rose among many cultivars in Wisconsin trials (Epping and Hasselkus 1989). This cultivar produces abundant fragrant, light pink, single flowers, and large, red hips are very evident from July to November.

Other recommended rugosa cultivars include: 'Albo-plena' which produces fragrant, double white flowers, but no fruit; 'Blanc Double de Coubert' which differs from 'Albo-plena' only in height and flower form; and 'Belle Poitvine', a hybrid with large, semidouble light mauve-pink flowers that drops the hips shortly after they form (Dirr 1990).

Rosaceae *Rosa*

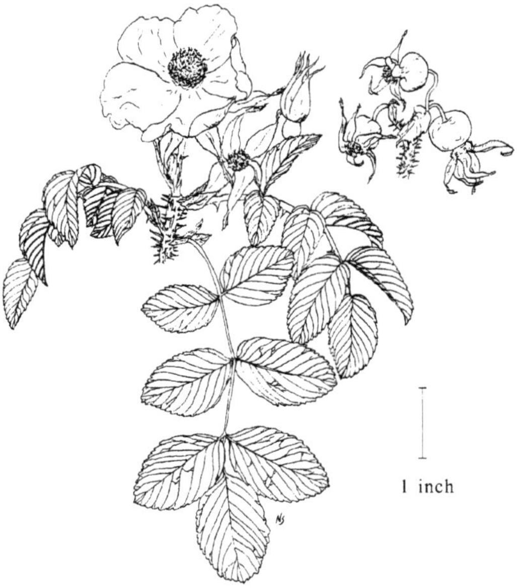

Rosa rugosa

Planning

Height: up to 6 ft (2 m)

Habit: shrub; deciduous; upright, forming dense low thicket; foliage dark green; large, colorful blooms, summer through autumn; orange-red fruits, autumn until early winter

Site: sun; soil adaptable except high pH; drouth tolerant; zone 3

Use: border, erosion control, wildlife habitat

Recognition

Leaves: leaflets 5-9, up to 2 in (5 cm) long, elliptic to wider above middle (obovate); tip blunt to pointed; wrinkled (rugose), thick and firm; lustrous and glabrous above; hairy, slightly bloomy, veins netted (reticulate) beneath; petioles long-hairy (tomentose), bristly

Flowers: usually solitary, about 2.7 in (7 cm) across, purple to white; stalks bristly; calyx lobes tapering to a sharp point, margin entire, surfaces hairy, glandular beneath

Fruits: hips nearly round, about 0.8 in (20 mm) across, red; calyx lobes spreading, persistent

Twigs: hairy, densely bristly and prickly

Rosa woodsii Lindl. - Woods rose

Woods rose is a western species found over a large section of the western U.S. and parts of Canada, including much of the Great Plains from Kansas to Minnesota and north into Canada. Woods rose can be distinguished from the Arkansas rose (*R. arkansana*), often a site companion, by the prickles found only at the nodes and glandular tipped leaf serrations (Stephens 1973). However, Rehder (1940) states that the Woods rose's prickles may be numerous and scattered along the stem.

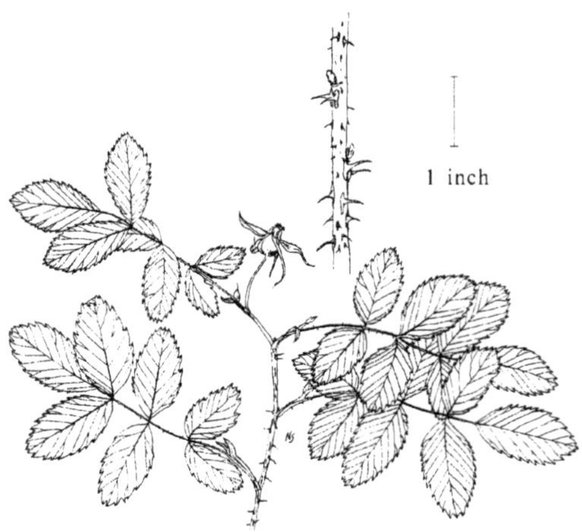

Rosa woodsii

Planning

Height: up to 5 ft (1.5 m)

Habit: shrub; deciduous; thicket forming; spring bloom; purplish twigs

Site: sun; pH and soil adaptable; drouth tolerant; zone 3

Use: erosion control, wildlife habitat

Recognition

Leaves: leaflets 7-9, short prickles on central axis (rachis); leaflets oval to wider above the middle (obovate), up to 1 in (25 mm) long; tip rounded, base wedge-shaped; glabrous and glaucous above, hairy to nearly glabrous below; margins toothed (coarsely serrate), glandular; stipules hairy, free tips spreading, glandular toothed or glands wanting

Flowers: solitary, sometimes 2 to 3, about 1.5 in (4 cm) across, stalks and receptacle glabrous; calyx lobes long-linear, hairy; petals pink, notched at apex

Fruits: hips nearly round, about 0.4 in (1 cm) across; orange-red, smooth; calyx persistent, spreading or erect

Twigs: glabrous, purplish; 2 straight or slightly recurved prickles below each node

Buds: egg-shaped (ovoid), glabrous, glossy, red

Section Synstylae

Rosa multiflora Thunb. - multiflora rose

Multiflora rose, a native of Korea and Japan, has been widely planted in the U.S. as a living fence, often around newly constructed farm ponds. It is also used as a grafting understock for specimen roses.

The agressive multiflora rose can be a troublesome pest in pastures and waste areas in the southeastern U.S. including the southeastern prairies.

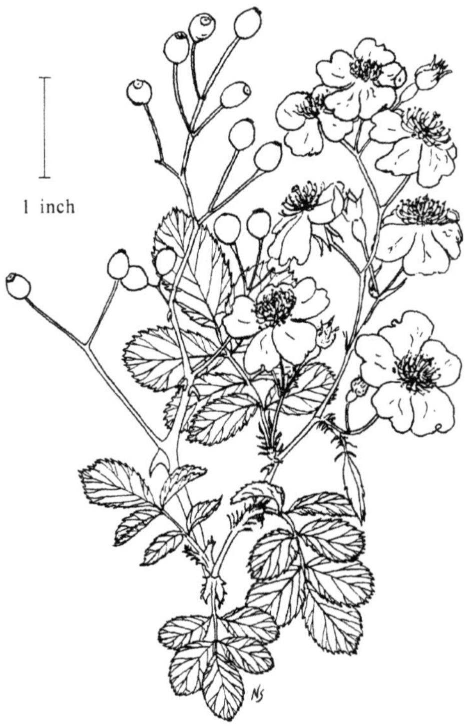

Rosa multiflora

Planning

Height: up to 10 ft (3 m)

Habit: shrub; deciduous; arching to trailing, thicket forming; spring flowers

Site: sun, part shade; soil adaptable; drouth tolerant; zone 5

Use: border, living fence, wildlife habitat

Caution: can escape to become troublesome on uncultivated sites

Recognition

Leaves: leaflets 7-9; central axis (rachis) hairy, often glandular; leaflets elliptic to egg-shaped, sometimes wider above middle (obovate), up to 1.3 in (3 cm) long; tip pointed to rounded, base wedge-shaped; margin sharply toothed, not glandular; upper surface glabrous, lower hairy; petioles hairy; stipules divided pinnately into narrow segments with glands on short stalks (stipitate)

Flowers: clusters (panicles) many-flowered, about 5 in (12 cm) long, stalks hairy, glandular; calyx lobes variable in shape, hairy, reflexed and appressed to hypanthium; petals pink to white, fragrant

Fruits: hips round to egg-shaped, about 0.2 in (5 mm) across, red; calyx deciduous

Twigs: red to green, glabrous; armed with stout recurved prickles

Buds: egg-shaped, at right angles to stem; scales red, often toothed

Rosa setigera Michx. **var. *tomentosa*** T. & G. - **prairie rose**

The climbing or trailing prairie rose—a native of the eastern U.S., including the southeastern prairies—may grow canes up to 15 feet (4.5 m) long in a single season, but it seldom reaches five feet (1.5 m) tall without support. The leaflets of variety *setigera* are nearly glabrous on the under surface.

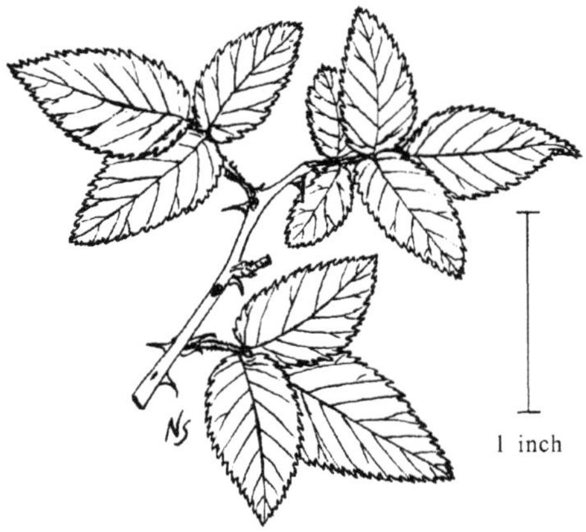

Rosa setigera

Planning

Height: up to 15 ft (5 m) on a trellis
Habit: shrub; deciduous; climbing, trailing; spring bloom; autumn foliage colorful
Site: sun; soil adaptable; zone 5
Use: wildlife habitat

Recognition

Leaves: leaflets 3, sometimes 5; terminal up to 2 in (5 cm) long, narrow to broadly egg-shaped, stalk (petiolule) about 0.3 in (7 mm) long, red stipitate glandular; lateral nearly stalkless (sessile); margin sharply toothed; tip pointed, base rounded; glabrous above, hairy on veins below; stipules narrow, slightly hairy, margin short-stalked stipitate glandular; petioles red, stipitate glandular
Flowers: clusters (corymbs) many-flowered, terminal on new growth; flowers about 2 in (5 cm) across; stalks stipitate glandular; petals pink, deeply notched; calyx lobes egg-shaped, entire, hairy, reflexed in flower
Fruits: hips nearly round, about 0.4 in (1 cm) across; smooth, yellow to red; calyx lobes deciduous
Twigs: green to reddish; prickles scattered or paired, straw-colored, recurved, enlarged and flattened at base
Buds: egg-shaped, red, glabrous

Rosa wichuraiana Crep. - **memorial rose**

Memorial rose, a trouble-free climber imported from Asia, is evergreen in warm climes but deciduous in the northern prairies. It is a parent of many other rose climbers. Dirr (1990) reports that memorial rose effectively controls erosion on highway embankments.

Planning

Height: canes to 15 ft (4.5 m)
Habit: procumbent shrub; semi-evergreen, deciduous; early summer bloom
Site: sun; soil adaptable; zone 6
Use: groundcover, trellis, erosion control

Recognition

Leaves: leaflets 7-9, nearly round to broad egg-shaped, up to 1 in (25 mm) long; tips blunt; lustrous, glabrous; stipules toothed
Flowers: clusters (corymbs) pyramidal, few to many-flowered, about 2 in (5 cm) across; petals white, fragrant
Fruits: hips egg-shaped, about 0.5 in (12 mm) long, red
Twigs: prickles scattered, strong, curved

Rubus L. - raspberry, blackberry

Most of the 400-plus species of genus *Rubus* grow throughout the Northern Hemisphere as shrubs, subshrubs or herbs. The woody species are cultivated for their ornamental value and edible fruits.

Differences among the species are not clearly defined because many of the species hybridize freely. Apomixis also accounts for difficulties in describing and identifying species (Freeman 1992).

R. pubescens Raf., creeping blackberry, has been found growing in northwest Nebraska and in the Black Hills, but it dies back so completely each winter that it can hardly be called a woody plant (Stephens 1973).

Recognition

Habit: shrubs, herbs; deciduous in the central and northern prairies
Leaves: alternate, simple or compound
Flowers: solitary or clustered; terminal or rarely in leaf axils; bisexual, rarely unisexual; calyx usually 5-parted, lobes persistent; petals 5, sometimes wanting, white to pink
Fruits: aggregates of drupelets, sometimes dry

Grouping Key for *Rubus*

(Rehder 1940)

Leaves simple, palmately lobed; stipules long, narrow (lanceolate), partially attached to petioles (adnate); flowers with spreading petals, style club-shaped, torus flat; stems unarmed, usually upright; bark exfoliating: **subgenus *Anoploratus***

Leaves compound, ternate or pinanate, sometimes simple; stipules linear, sometimes adnate; fruit separating from receptacle (torus) at maturity; stems upright, climbing (scandent) or procumbent, usually prickly, bristly (setose) or grandular; usually biennial: **subgenus *Idaeobatus* (raspberries)**

Leaves compound, ternate or 5 foliate (digitate), rarely 7; petals always present; stipules adnate to petiole; fruit usually black or dark purple, receptacle (torus) remains together at maturity; stems prickly, usually angled in cross-section; usually biennial: **subgenus *Eubatus* (blackberries)**

Subgenus *Anoploratus*

Rubus deliciosus Torr. - boulder raspberry

Boulder raspberry, a native of moist, well-drained sites on the eastern slopes of the Rocky Mountains, has thrived in trials in the northern Great Plains (North Central Plant Introduction Station 1959). Wyman (1969) states that 'Plena', a double-flowered cultivar, is adapted to partially shaded sites where it can serve as a specimen or as an accent in naturalistic landscapes.

R. parviflorus Nutt., thimbleberry, found in Ontario, Minnesota, the Black Hills and much of the western U.S., has simple leaves and unarmed stems (Stephens 1973). It grows along streambanks, generally where shaded.

Planning

Height: up to 5 ft (2.5 m)
Habit: shrub; arching, unarmed, suckering
Site: sun, partial shade; zone 5
Use: naturalizing, specimen

Recognition

Leaves: simple, nearly round, 3- sometimes 5-lobed; tips blunt to pointed, base squared to heart-shaped; margin irregularly toothed; glandular below, sparingly hairy on veins
Flowers: mostly solitary, about 2 in (5 cm) across, white
Fruits: hemispheric, dark purple, about 0.4 in (1 cm) across
Twigs: unarmed, minutely hairy when young
Bark: flaky

Subgenus *Idaeobatus* - raspberries

Rubus ideus L. ssp. *sachalinensis* (Levl.) Focke - red raspberry

Many forms of red raspberry, natives of Eurasia and in North America from Alaska to New Mexico and from Labrador to North Carolina, have been selected for cultivation. Several cultivars are fruitful in much of the central and northern prairie if they are protected from extreme winds and provided with adequate water.

Excellent fruit can be picked in the fall from the everbearers 'Heritage' and 'September'. Best results are obtained by removing all old canes after harvest to allow all fruits to be produced on new growth.

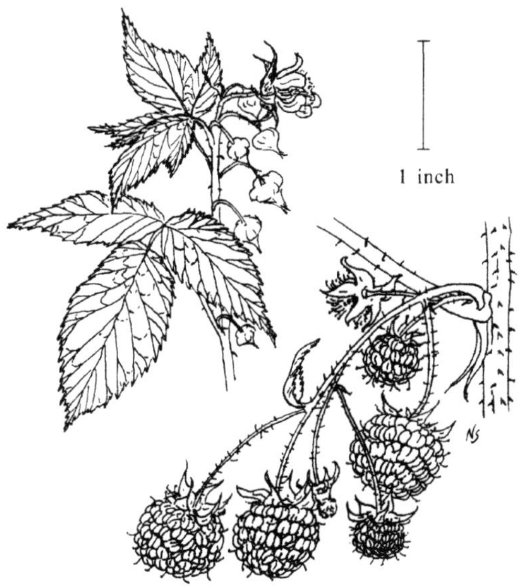

Rubus ideus L. ssp. *sachalinensis*

Planning

Height: up to 6 ft (2 m)
Habit: shrub; erect to arching, suckering, armed with prickles
Site: sun; soil adaptable; zone 4
Use: naturalizing, fruit

Recognition

Leaves: leaflets 5 (3 on flowering canes), egg-shaped; tip pointed, base usually rounded, irregularly toothed; petioles variously hairy, bristly glandular; stipules long, narrow, hairy, glandular, fused to petiole for about 0.2 in (4 mm)
Flowers: clusters, usually terminal on short branches of new growth; petals 5, white, ascending; calyx lobes 5 with short slender tips, glandular, hairy, often bristly, reflexed or spreading
Fruits: hemispherical, about 0.6 in (15 mm) across; separates easily from receptacle (torus), hollow; finely hairy, red, edible
Twigs: nearly glabrous or covered with soft slender yellowish prickles
Buds: egg-shaped, red-brown; scales minutely hairy, margins fringed (ciliate)

Rubus occidentalis L. - black raspberry

The hardy black raspberry is a native of much of the U.S., but if it is not yet native in your neighborhood, it probably soon will be "native" if it has been planted in any nearby gardens. New plants will start from cane tips where they touch moist soil, and seedlings will grow where birds drop seeds after a berry feast.

Several black raspberry cultivars are available for fruit production. Native Americans dried its fruits for winter consumption and brewed tea from the leaves.

Planning

Height: up to 5 ft (1.5 m)
Habit: shrub; arching, trailing
Site: sun, partial shade; soil adaptable; zone 3
Use: naturalizing, wildlife habitat, fruit
Note: roots will develop where branch tips touch moist soil

Recognition

Leaves: leaflets 3-5 (usually 3 on flowering canes), egg-shaped; tip pointed, base rounded to square; dark green above with few hairs, densely white woolly-hairy (tomentose) below; petioles glabrous with 1 or 2 prickles; stipules long narrow, hairy
Flowers: clusters few-flowered, prickly; petals 5, white, half as long as sepals; calyx flat, 5-lobed, sharp-pointed, hairy both sides, green, spreading
Fruits: hemispherical, about 0.5 in (13 mm) across; separate from receptacle (torus) at maturity, hollow; purple-black, edible
Twigs: glabrous, prickles broad-based, recurved
Buds: narrowly egg-shaped, red-brown, hairy; partially hidden by old leaf base

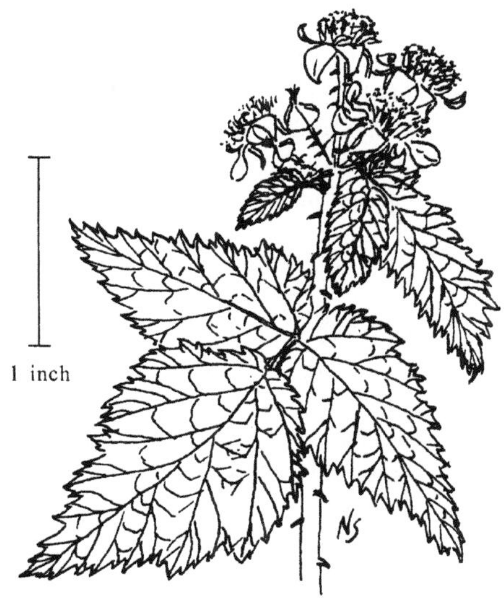

Rubus occidentalis

Subgenus Eubatus (blackberries)

Rubus flagellaris Willd. - dewberry

Large, tasty fruits of this native of the eastern U.S.—including Minnesota and south through eastern Kansas to Texas—provide a flavorful dessert for any feast, be it for man, bird or small mammal (Stephens 1973). Cultivars bearing tasty and bountiful fruits have been selected.

Planning

Height: up to 10 ft (3 m) on trellis
Habit: shrub; trailing, vine-like; mid-June fruits
Site: sun, part shade; soil adaptable; zone 3
Use: trellis, fruit production

Recognition

Leaves: leaflets 3-5, egg-shaped; margins toothed, hairy (ciliate); surface sparse-hairy above, soft hairs below; petioles with a few prickles, stipules hairy
Flowers: clusters (corymbs) few-flowered or solitary; petals 5 or more, white, often slightly pinkish, cleft on ends; calyx lobes 5, divided to near base, densely woolly, tips awl-shaped
Fruits: oval, about 0.7 in (18 mm) long; red, becoming black at maturity; drop from calyx with receptacle (not hollow); sweet, juicy, edible
Twigs: prostrate, glabrous; armed with recurved, wide-based prickles
Buds: ovoid, appressed hairy, brown

Rubus frondosus Bigel. (***R. pensilvanicus*** Poir.) - **highbush blackberry**

Highbush blackberry, a native of much of the eastern U.S., also grows in eastern Kansas and extreme southeastern Nebraska. Other blackberries with similar ranges and overlapping characteristics are *R. allegheniensis* Porter and *R. ostryifolius* Rybd. Several cultivars with desirable characteristics for fruit production have been selected.

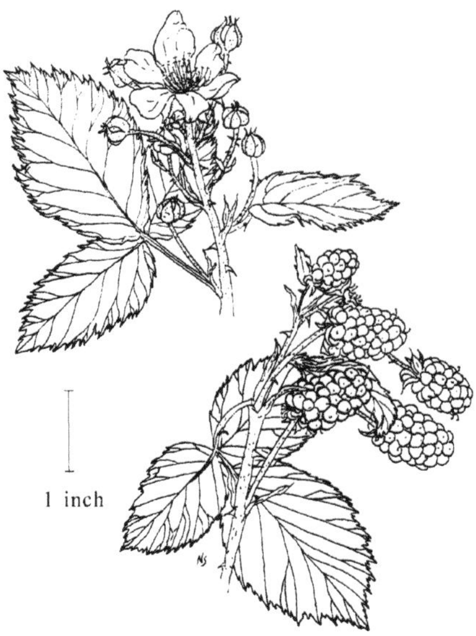

Rubus frondosus

Planning

Height: up to 7 ft (2 m)
Habit: shrub; arching, thicket-forming; early July fruits
Site: sun, part shade; soil adaptable; zone 4
Use: naturalizing, wildlife habitat, fruit production

Recognition

Leaves: leaflets palmately compound (5-foliate, 3 on flowering canes); egg-shaped, tip pointed, base usually rounded, irregularly toothed; petioles grooved above, densely hairy with a few prickles; stipules long, narrow, minutely toothed, hairy
Flowers: clusters (racemes) terminal; petals 5, white; calyx lobes 5, divided to near base, abruptly sharp-pointed, hairy, spreading or reflexed
Fruits: oval, about 0.5 in (12 mm) long; red, black at maturity, glossy smooth, sweet; drop from calyx with receptacle (not hollow)
Twigs: hairy, glandular, armed with recurved broad-based prickles
Buds: egg-shaped; scales, outer red-brown, appressed-hairy, inner densely white-hairy (pubescent)

Subfamily Prunoideae

Prinsepia Royle - **prinsepia**

The *Prinsepia* species—three or four of which are natives of eastern Asia and northwestern China south to the Himalayas—are not commonly found in our landscapes. They are pest resistant, require little maintenance, and adapt to difficult sites (Dirr 1990).

Recognition

Habit: shrubs; deciduous
Leaves: alternate, mostly in clusters, entire or toothed; stipules longer than wide, persistent
Flowers: 1 to 4 in axils of previous year's branches or in axillary racemes; bisexual; petals 5, spreading, nearly round, clawed; calyx tube cup-shaped, sepals 5; stamens 10 or more
Fruits: drupes; stones slightly sculptured
Twigs: spiny; pith plated (lamellate)
Buds: hairy, enclosed by stipules

Prinsepia sinensis (Oliv.) Oliv. - **cherry prinsepia**

Cherry prinsepia, a native of Manchuria, would undoubtedly be more widely planted were it not for its thorns. It can grow in shade and compete with other plants (Hoag 1965). The ornamental fruits are relished by birds.

Although minor chlorosis has occurred on cherry prinsepia at PERC in Fort Collins, it is a good shrub for eastern Colorado (Fairchild and Klett 1993).

Planning

Height: up to 10 ft (3 m)
Habit: shrub; rounded, dense, spiny; very early spring leaves, fragrant flowers; red summer fruits persist into winter
Site: sun, shade; soil and pH adaptable; zone 3
Use: hedge, screen, wildlife habitat
Note: tolerates pruning

Recognition

Leaves: oval to much longer than wide, sometimes wider above middle; entire or shallow-toothed on the upper two-thirds, margins finely hairy (ciliate); otherwise glabrous, bright green
Flowers: fascicles of 1 to 4; light yellow, about 0.6 in (15 mm) across; stamens 10
Fruits: drupes nearly round, cherry-like, about 0.6 in (15 mm) across; orange-red to purple; stone egg-shaped, compressed
Twigs: slender, spines about 0.3 in (8 mm) long; pith plated
Buds: concealed by brown hairs; scales indistinct
Bark: shreddy with age

Prinsepia sinensis

Prinsepia uniflora Batal. - **hedge prinsepia**

Hedge prinsepia, a native of northwest China, is rarely planted in our prairies but deserves trials. Foliage color is especially good and fruits are quite handsome (Dirr 1990).

Planning

Height: up to 5 ft (1.5 m)
Habit: shrub; rounded, dense; very dark green foliage, late summer fruits
Site: sun, part shade; soil adaptable; zone 3
Use: hedge, screen, wildlife habitat

Recognition

Leaves: narrow-oblong; entire or finely toothed; dark green, glabrous
Flowers: fascicles of 1 to 3; white, 0.6 in (15 mm) across; stamens 10
Fruits: drupes purplish red, bloomy, about 0.5 in (12 mm) across; stone egg-shaped, compressed
Twigs: slender, spines about 0.5 in (12 mm) long; pith finely plated
Buds: slender, scales indistinct

Prunus L. - apricot, cherry, peach, plum

We conjure up visions of cherry pie and peaches and cream when *Prunus* species come to mind. Their fruits have long been an important food source for people and animals. Some species are large enough to also produce fine hardwood lumber. Many are valued ornamentals and serve as wildlife habitat.

Between 125 and 400 species are natives of the north temperate zone. A few species grow in the Andes of South America, and 30 are natives of North America.

Recognition

Habit: trees, shrubs; usually deciduous
Leaves: alternate, simple, toothed, rarely entire
Flowers: solitary or clustered; petals usually white, sometimes shades of red; pistil 1, stamens many
Fruits: rounded, composed of a hard stone containing a seed, surrounded by dry to juicy thick pulp
Buds: scales many, overlapping (imbricate)

Grouping Key for *Prunus*

(Rehder 1940)

Leaves deciduous; flowers solitary or in clusters, before or rarely with leaves, stalked; fruits grooved (sulcate), glabrous, usually bloomy, stone compressed, usually longer than broad, smooth or nearly so, stalk usually remaining attached when falling; terminal buds wanting, axillary buds solitary: **subgenus *Prunophora* (plums, apricots)**

Leaves deciduous; flowers solitary or few, before leaves, short-stalked or sessile; fruits matted-hairy (tomentose), rarely glabrous, usually dry and splitting, stone pitted or smooth; terminal buds present; axillary buds 3, flower buds on outside: **subgenus *Amygdalus* (peaches, almonds)**

Leaves deciduous; flowers white or rose-colored, solitary, fascicled or clustered (racemes), usually stalked; calyx tube cup-shaped or tubular; fruits not grooved, not bloomy, stone turgid or nearly round, smooth or furrowed and pitted; terminal buds present: **subgenus *Cerasus* (cherries)**

Leaves deciduous; flowers in many-flowered clusters (racemes), leafy at base, rarely leafless, calyx tube short, bell-shaped (campanulate); fruits not grooved, not bloomy, stone turgid or nearly round; terminal buds present: **subgenus *Padus* (cherries)**

Leaves evergreen, lustrous dark green above; flowers white, fragrant, in 12- to many-flowered clusters, leafless at base; fruits purple to black: **subgenus *Laurocerasus* (laurel)**

Subgenus *Prunophora* - plums, apricots

Grouping Key for Subgenus *Prunophora*
(Rehder 1940)

Flowers solitary or 2s, rarely 3s, stalked, usually appear with the leaves; fruits glabrous, stones often sculptured: **section Euprunus**

Flowers clustered, 2-5, stalked; fruits glabrous, stones usually smooth: **section Prunocerasus**

Flowers usually solitary, usually sessile or short-stalked, appear before leaves; fruits hairy (pubescent), stones smooth or pitted: **section Armeniaca**

Section Euprunus

Prunus cerasifera Ehrh. - cherry plum

Cherry plum, a native of western Asia, would be little known or planted as an ornamental were it not for the purple-leaved variety *atropurpurea* Jaeg. M. Pissard introduced this variety to Europe as 'Pissardii' more than 100 years ago. Its purple leaves are larger than those of the species and the flowers are pink. Many other purple-leaved cultivars have been developed by crossbreeding and selection. Consequently, they have now become so well known that they are overused (Johnson 1973, Dirr 1983).

Hybrids include: *P.* ×*cistena*, a cross between 'Atropurpurea' and *P. pumila*; and *P.* × *blireiana*, a cross between 'Atropurpurea' and *P. mume* (Sieb.) Sieb. & Zucc.

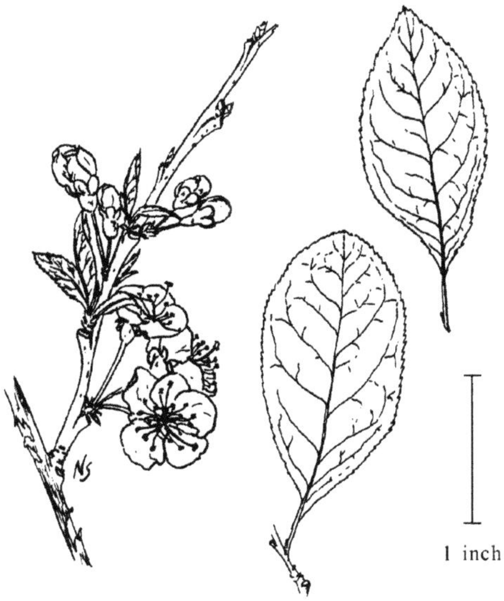

Prunus cerasifera

Planning

Height: up to 25 ft (8 m)
Habit: rounded top; flowers short-lived, appear ahead of leaves; cultivars are purple-leaved
Site: sun for best purple color, partial shade; pH and soil adaptable, prefers moist, well-drained sites; zone 5
Use: specimen, grouping

Recognition

Leaves: egg-shaped to elliptic, about 2 in (5 cm) long; fine-toothed; apex pointed, base tapered to rounded; hairy on midrib below
Flowers: usually solitary, about 0.9 in (13 mm) across; white
Fruits: nearly round, about 1 in (25 mm) across; red, slightly bloomy
Twigs: glabrous, sometimes thorny

Prunus domestica L. - garden plum

Garden plum, a native of Europe and Asia, has long been cultivated in many forms for its fruits. It escapes from orchards and reproduces on its own in many parts of the U.S. A double-flowered variety, *plantierii*, is occasionally used in landscape plantings.

Planning

Height: up to 40 ft (12 m); National champion, 1976, 45 ft (13.7 m), Oregon
Habit: tree, sometimes shrubby; round-topped
Site: sun; pH and soil adaptable; zone 5
Use: specimen, fruit

Recognition

Leaves: elliptic, sometimes wider above the middle, about 3 in (8 cm) long; tip pointed, base tapering; coarsely toothed (crenate-serrate); hairy at first, becoming glabrous above, net-veined (reticulate) below; petioles about 0.8 in (2 cm) long
Flowers: solitary or clusters of 2-3, about 0.7 in (18 mm) across; greenish white, sepals hairy inside
Fruits: usually egg-shaped to oblong, about 1 in (25 mm) long, blue to blue-black; stone free, smooth
Twigs: usually glabrous

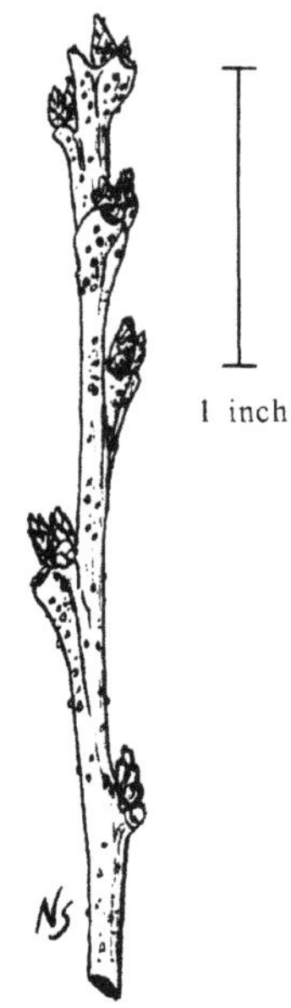

Prunus domestica

Prunus spinosa L. - blackthorn

The name of this Eurasian native shrub is very descriptive. Blackthorn—more functional than ornamental—provides child and pet-proof barriers. The astringent fruits have been used to flavor gin. Variety *purpurea* Andre, which may be a hybrid, has purple leaves and pink flowers. Variety *plena* West. produces double-white flowers. It is a tough plant and has been used in conservation plantings in S.W. Nebraska. (Sutton 1998).

Planning

Height: up to 12 ft (4 m)
Habit: shrub; densely thorny, suckering; spring bloom before leaves
Site: sun; soil adaptable; zone 5
Use: barrier, wildlife habitat

Recognition

Leaves: elliptic to oblong, widest above middle; apex rounded; finely toothed; surfaces hairy, sometimes becoming glabrous
Flowers: usually solitary, about 0.5 in (13 mm) across; white
Fruits: upright, nearly round about 0.5 in (12 mm) across; blue-black, bloomy, astringent; stone more or less pitted
Twigs: hairy, spiny
Buds: nearly round to egg-shaped

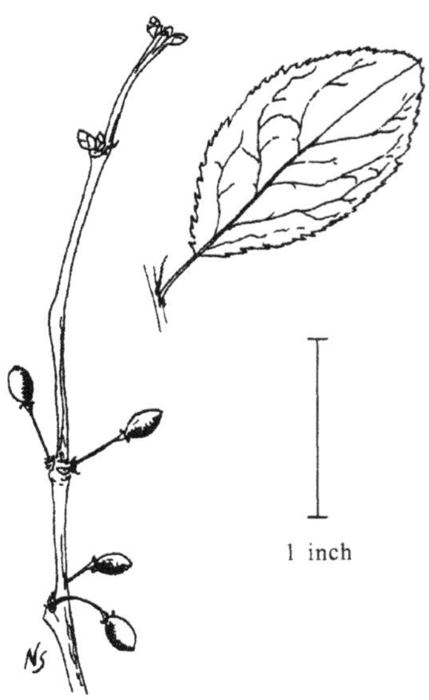

Prunus spinosa

Section Prunocerasus

Prunus americana Marsh. - American plum

We nominate American plum to the woody plant "All-Star Team for the Prairies." This native of much of the eastern U.S., including the Great Plains and a localized area in the Rocky Mountain states, is found in niches along woodland edges, roadsides and craggy bluffs.

The American plum is a workhorse in our windbreak and wildlife habitat plantings throughout the prairies. This plum was planted in most of the Great Plains shelterbelts during the 1930s and it has survived almost all hardships here except the crush of bulldozers (Read 1958).

During much of the year this nondescript shrub receives scant attention, except when its impenetrable spiny thicket spares a cottontail from a redtail hawk's sharp talons, or impedes our passage to an adjoining field. Suddenly, as if by magic, mounds of snowwhite flowers burst out in early May on these bushes to then become larders for hungry bees and a reminder that corn planting time is here. Connoisseurs of fine foods note the locations of American plum flowers, and barring untimely spring freezes, they hope to reap a tangy August fruit harvest for the makings of delightful biscuit and pancake toppings.

Many cultivars that bear large, tasty fruits have been selected from wild American plum bushes and from crosses with other plum species. Some bushes, often those with the largest, most tasty fruits, are less inclined to form thickets, making them more desirable as backyard shrubs.

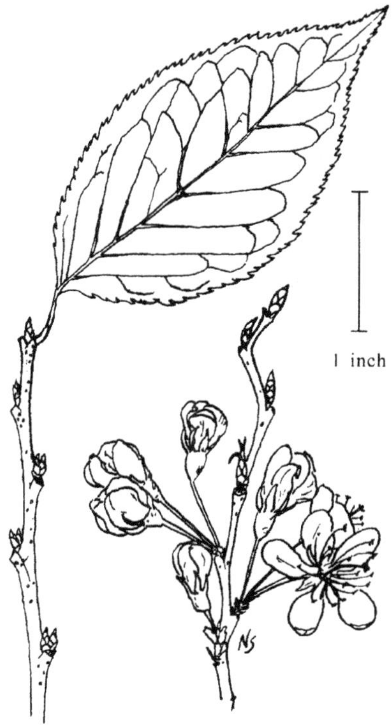

Prunus americana

The native range of the closely related Canada plum, *P. nigra* Ait., is more northerly than that of the American plum. 'Princess Kay', a selection of Canada plum, bears double flowers and few fruits. Canada plum is the parent of the hybrid 'Muckle plum' (see *P. tenella*, dwarf Russian almond).

Planning

Height: up to 16 ft (5 m); Nebraska champion, 1980, 22 ft (6.7 m), Tecumseh; National champion, 1972, 35 ft (10.7 m), Michigan

Habit: tree, shrub; broad-topped, thicket forming; early spring bloom, late summer fruits

Site: sun, partial shade; soil and pH adaptable; drouth tolerant; zone 3

Use: naturalizing, wildlife habitat, windbreak, fruit

Recognition

Leaves: oval to oblong-oval, about 3 in (8 cm) long, widest at middle or above the middle, double-toothed (serrate); apex pointed, base rounded to tapered; surfaces glabrous or sparsely hairy below; petioles hairy on upper surface only, glands occasionally at summit; stipules linear, divided into 2-3 linear segments, toothed, hairy, early deciduous (caducous)
Flowers: clusters of 2-5, with or before leaves, stalks glabrous; petals white; calyx bell-shaped, lobes over 0.1 in (3 mm) long, oblong, margin usually entire, fringed with hairs (ciliate)
Fruits: slightly elongated, about 1 in (25 cm) in diameter, ripening orange-red with slight bloom, juicy, edible; stone somewhat wrinkled (rugose)
Twigs: rigid with small spine-like side twigs, reddish brown, glabrous
Buds: egg-shaped, pointed; scales brown, glabrous
Bark: brown, scaly; lenticels horizontal, prominent
Wood: hard, heavy, strong; heartwood red-brown, sapwood narrow

Prunus angustifolia Marsh. - chickasaw plum

Chickasaw plum is a native of the southeastern U.S., including Indiana, Kansas, Oklahoma and Texas. It is commonly found in sandy regions where it is also known as sandhill plum (Stephens 1973). Chickasaw plum is seldom planted, but it deserves trials, especially on sandy sites.

Planning

Height: up to 25 ft (8 m); National champion, 1988, 32 ft (9.7 m), North Carolina
Habit: shrub, tree; broad, thicket-forming; April flowers, July fruits
Site: sun; sandy and loam soils preferred; zone 5
Use: protection, wildlife habitat, naturalizing, fruit

Recognition

Leaves: elliptic to lance-shaped, about 2 in (5 cm) long; fine-toothed (serrate), gland tipped; apex pointed, base rounded to tapered; surfaces glabrous; petioles with 2 glands near leaf base, hairy when young
Flowers: clusters of 2-4, with or before leaves; petals white; calyx lobes less than 0.1 in (1 mm) long, margin hairy (ciliate)
Fruits: nearly round, about 0.8 in (2 cm) across; ripening yellow to red with slight bloom, juicy, edible
Twigs: red, shiny, becoming dull reddish brown, often spine-tipped
Buds: pointed, mostly grouped at twig or spur ends; scales red-brown, pointed
Bark: brown, splitting and curling with age
Wood: hard, heavy; heartwood red-brown, sapwood wide

Prunus hortulana Bailey - hortulan plum

The natural range of the hortulan plum centers on Missouri and extends into eastern Kansas, into Illinois as far north as southern Wisconsin, and east up the Ohio River Valley to West Virginia. Several large-fruited cultivars have been selected, dating back to the campaigns of General Andrew Jackson (Peattie 1950). One selection, first named 'Old Hickory' by Captain William Dodd, was handed from neighbor to neighbor for over 50 years before being renamed 'Miner' and discovered by rofessional nurserymen. 'Golden Beauty', which bears yellow-skinned fruits, was introduced by a Texas nursery in 1874.

September visitors to Horning State Farm near Plattsmouth marvel at the copious display of bright red fruits on the hortulan plum trees. This plum deserves further trials throughout the prairies.

Hortulan plum is often confused with *P. munsoniana* Wight & Hedr., wild goose plum, a small thicket-forming tree with a similar natural range. The latter is most easily distinguished by leaf margins with incurved teeth which bear glands on the infacing side and by the tendency of the leaves to fold lengthwise (Stephens 1973).

Planning

Height: up to 20 ft (6 m); National champion, 1972, 27 ft (8.2 m), Missouri
Habit: tree, shrub; round-topped, not suckering; spring bloom with the leaves; colorful September fruits
Site: sun, partial shade; soil adaptable; zone 5
Use: specimen, windbreak, wildlife habitat, fruit

Recognition

Leaves: egg-shaped to ovate-lanceolate; apex long-pointed, base tapering to rounded; fine toothed (serrate), gland tips pointing outward; upper surface lustrous, glabrous, hairy on veins below; stipules linear, toothed, glandular, early deciduous
Flowers: clustered on short spurs, each about 0.6 in (15 mm) across; white; hypanthium cone-shaped, glabrous; calyx lobes egg-shaped, glandular toothed, hairy at base on inner side
Fruits: nearly round, about 1 in (25 mm) across; bright red, flesh firm, sweet, juicy; stone egg-shaped
Twigs: red-brown, lenticels not obvious
Buds: egg-shaped; scales chestnut brown, margins hairy fringed (ciliate)
Wood: hard, heavy, reddish brown; sapwood white, wide

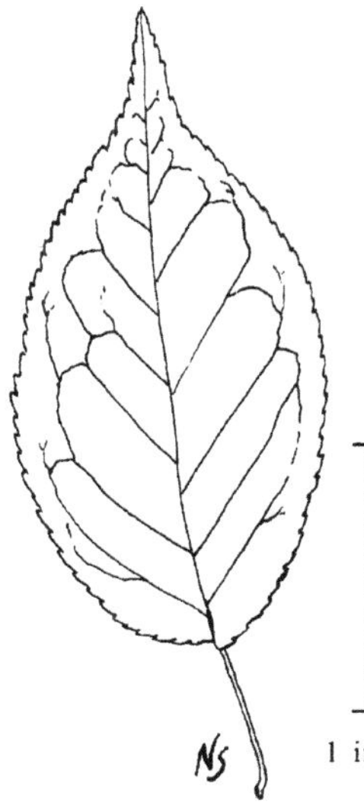

1 inch

Prunus hortulana

Prunus mexicana S. Wats. - big tree plum

Big tree plum is a native of open woods, bottomlands, and prairie hillsides in much of the Midwest, the south central U.S. and northeastern Mexico. It is often overlooked in its native habitat because of its similarity to American plum, but big tree plum can be distinguished from the latter by its hairy twigs, petioles, and flower stalks, and by its larger purple fruits with a whitish bloom.

Planning

Height: up to 12 ft (4 m); National champion, 1991, 15 ft (4.6 m), Texas
Habit: tree, shrub; round-topped, single or in thickets; bloom appears with leaves, summer fruits
Site: sun, partial shade; soil adaptable; zone 5
Use: wildlife habitat, windbreak

Recognition

Leaves: egg-shaped to elliptic; apex abruptly short-pointed, base rounded; fine-toothed; glabrous above, somewhat wrinkled (rugose), hairy below, especially along veins; petioles hairy on all sides; stipules toothed, hairy, early deciduous
Flowers: clusters of 3-4, stalks finely hairy; flowers about 0.8 in (2 cm) across, petals white; calyx lobes oblong, hairy, reflexed after anthesis
Fruits: slightly elongated, about 1 in (25 mm) long; red, becoming purple when ripe with waxy bloom; flesh clings to stone, thick, juicy, sweet; stone nearly round, smooth
Twigs: minutely hairy (puberulent), red-brown; leaf scars often hairy on upper margin
Buds: egg-shaped, hairy, red-brown
Bark: peeling into thin scales; lenticels horizontal on young branches; old trunks black, rough, deeply furrowed
Wood: hard, heavy, brown; sapwood narrow

Section Armeniaca

Prunus armeniaca L. - apricot

Apricot, a native of China, has long been cultivated for its fruit; it is now a major California commercial crop. Varieties *mandshurica* Maxim. and *sibirica* L. are hardy, with potential ornamental value in the northern Great Plains (Hoag 1965, Klett 1975). Several varieties are adapted to the Fort Collins area, but are vulnerable to spring frosts and cannot be relied upon for fruit production (Fairchild and Klett 1993). 'Scout', a release from Minnesota, is hardy for northern areas and provides an occasional fruit crop.

Planning

Height: up to 30 ft (9 m)
Habit: tree; round-headed; blooms before leaves; autumn foliage golden
Site: sun; soil and pH adaptable; zone 4
Use: specimen, fruit, screen

Recognition

Leaves: broadly egg-shaped, about 3 in (8 cm) long; fine-toothed; apex abruptly pointed, base rounded; sometimes tufted at vein axils below
Flowers: solitary, about 1 in (25 mm) across, nearly stalkless; white or pinkish
Fruits: round, about 1.2 in (3 cm) across; ripening yellow with a red cheek, juicy, edible; stone smooth

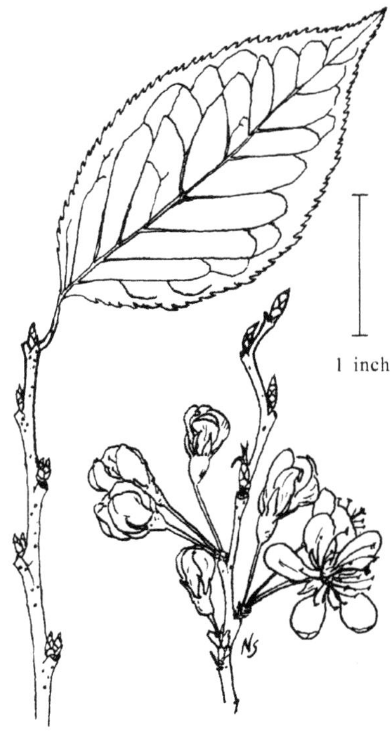

Prunus armeniaca

Subgenus *Amygdalus* - peaches, almonds

The almond, *P. amygdalus* Batsch, which lends its name to this subgenus, has been cultivated for thousands of years for its edible nut. This native of western Asia and northern Africa, now grown in vast California orchards, is not adapted to the central and northern prairies.

Prunus persica (L.) Batsch - peach

The peach, a native of China, has been cultivated for its fruit since the beginning of agriculture. The ancestral species, if still in existence, probably bears only slight resemblance to the many cultivars now grown primarily for their fruits.

Peaches to be grown as ornamentals are selected for flower characteristics, foliage color and form. Many cultivars are double-flowered, with red or rose-pink blooms. One of the best known is 'Klara Meyer' (Johnson 1973).

Special care is necessary to allow peaches to grow for a reasonable life span. Fruit crops are uncertain in areas where unseasonable frosts are common. Because the peach is so prone to insect and disease injury, Dirr (1990) advises the average homeowner that "the best approach is to avoid peaches altogether."

Planning

Height: up to 15 ft (5 m); National champion, 1986, 18 ft (5.5 m), Virginia
Habit: tree; round-topped, spreading; blooms before the leaves appear
Site: sun; soil adaptable; zone 6
Use: specimen, fruit
Pests: very prone to damage by stem borers

Recognition

Leaves: elliptic, much longer than wide, broadest at about the middle, about 4 in (10 cm) long; apex long-pointed, base rounded; sharply toothed; shiny above, glabrous; petioles glandular
Flowers: usually solitary, about 1.2 in (3 cm) across, very short-stalked; petals pink; sepals hairy on outside surface
Fruits: nearly round, about 2.5 in (6 cm) across, yellow with reddish blush, hairy (tomentose); stone deeply pitted and furrowed
Twigs: glabrous, red-green
Buds: hairy
Bark: reddish brown; smooth, becoming scaly

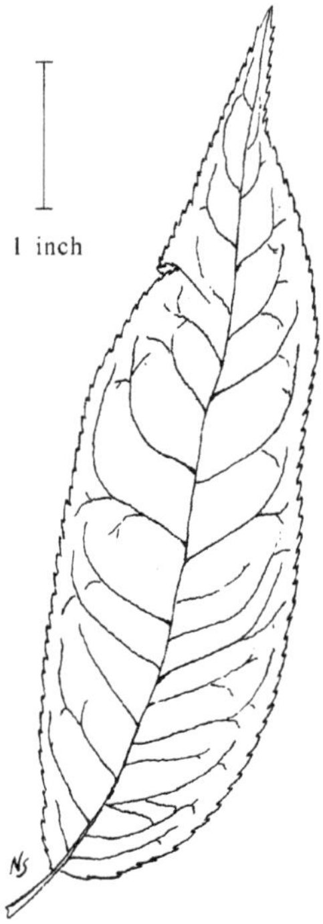

Prunus persica

Prunus tenella Batsch - dwarf Russian almond

Although its flowers are colorful, dwarf Russian almond is seldom cultivated. An upright almond form that was crossed with *P. nigra*, Canada plum, has produced the 'Muckle Plum', which has promise as an interesting pink-flowered, small tree-sized ornamental (Hoag 1965).

A specimen of the white-flowered variety *alba* (Schneid.) Rehd., growing at the Cheyenne Horticultural Research Station, is only three feet (1 m) tall at age 50 years (Sutton 1998). ND-283, a selection of this variety, is being tested in the northern Great Plains by the Natural Resources Conservation Service Plant Materials Center, Bismarck, N.D.

Planning

Height: up to 5 ft (1.5 m)
Habit: shrub; thicket-forming by suckering
Site: sun; soil and pH adaptable; pest resistant; zone 3
Use: border, protection planting, wildlife habitat

Recognition

Leaves: egg-shaped to oblong, sometimes broadest above middle, about 2.5 in (6 cm) long; apex pointed, base tapering; sharply toothed; surfaces glabrous, dark green
Flowers: stalkless, 1-3, about 0.6 in (15 mm) across; petals various shades of red, sometimes white; calyx tubular, tube much longer than the toothed lobes
Fruits: egg-shaped, about 0.8 in (2 cm) long, hairy; stone wrinkled (rugose)
Twigs: glabrous

Prunus triloba Lindl. var. *multiplex* (Bge.) Rehd. - flowering almond

A mature specimen of this double-flowered almond, a native of China, is a spectacular plant in full bloom (Hoag 1965). Hoag further recommends that although very early blooming, flowering almond has a place in the landscapes of the northern Great Plains. It should not be propagated on American plum rootstalks because of the tendency for plum suckers to proliferate.

Flowering almond crossed with *P. pedunculata* (Pall.) Maxim. has produced 'Prairie Almond', which Hoag (1965) believes may prove to be valuable in most of the northern plains. Dwarf flowering almond, *P. glandulosa* Thunb., is less hardy than flowering almond.

Planning

Height: up to 10 ft (3 m)
Habit: shrub; April bloom, yellowish autumn foliage
Site: sun; soil and pH adaptable; zone 4
Use: border, grouping

Recognition

Leaves: broad elliptic, sometimes wider above middle, about 2 in (5 cm) long; sometimes 3-lobed at apex, pointed, base tapering; coarsely toothed; glabrous above, slightly hairy below
Flowers: 1-2, about 1 in (25 mm) across; petals double, pinkish; calyx tube (hypanthium) broad bell-shaped, as long as lobes and shorter than stalks (pedicels)
Fruits: nearly round, about 0.5 in (12 mm) across, hairy; stone with thick bony shell
Twigs: glabrous or slightly hairy

Subgenus *Cerasus* - cherries

Grouping Key for Subgenus *Cerasus*
(Rehder 1940)

Shrubs; flowers sessile or stalked, calyx tube tubular or cup-shaped; 2-3 buds at nodes on flowering twigs: **section Microcerasus**

Shrubs, trees; flowers in fascicles or short few-flowered racemes, calyx tube usually cylindric, sepals upright or spreading: **section Pseudocerasus**

Shrubs, trees; leaves toothed (dentate), obtuse teeth; flowers in sessile umbels, rarely short racemes, surrounded at base by the persistent bud scales: **section Eucerasus**

Shrubs, trees; leaves fine-toothed (crenulate or serrulate); flowers in umbels or few- to 12-flowered racemes, their basal bud scales deciduous before anthesis; bracts deciduous: **section Mahaleb**

Section Microcerasus

Prunus besseyi Bailey - western sandcherry

Western sandcherry, a hardy shrub native to the Great Plains, is found in an area that extends from central Kansas north to Minnesota and Manitoba, and west to Wyoming and Colorado. Those of us fortunate enough to live in the sandhills of Nebraska and eastern Colorado during the 1920s recall our delight when, on a "thirsty" summer day far from the homestead we stumbled upon a sprawling sandcherry bush growing among the grass and sagebrush, loaded with ripe, juicy fruits. Unfortunately, the wild sandcherry population has dwindled—a victim of livestock overgrazing and herbicide misuse.

Sunday supplement advertisements occasionally proclaim the sandcherry as a garden shrub that produces a bountiful fruit harvest at an early age. However, western sandcherry does not thrive for long outside its native habitat—sandy, prairie hillsides. It has been unfruitful and short-lived on the clay soils at Prairie Pines Arboretum near Lincoln. Flowering and fruit production is reliable at Fort Collins, but western sandcherry exhibits chlorosis on the high pH clay soil there, possibly because of irrigation at the arboretum (Fairchild and Klett 1993).

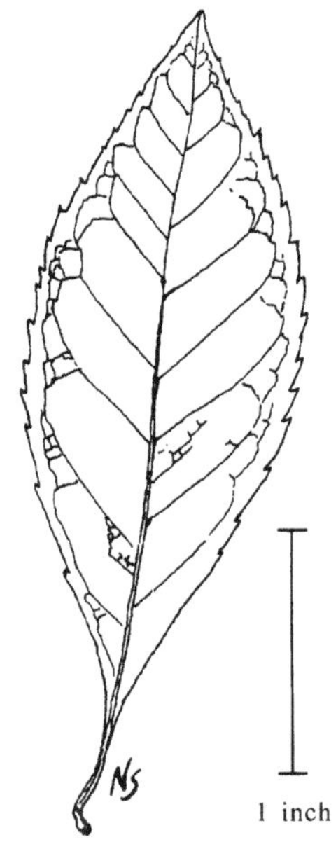

Prunus besseyi

Planning

Height: up to 6 ft (2 m)
Habit: shrub; sprawling; early spring blooms; midsummer fruits; autumn foliage purplish
Site: sun; prefers well-drained loams and sandy soils; drouth tolerant; zone 4
Use: border, fruit, wildlife habitat, naturalizing, foundation

Recognition

Leaves: often clustered near end of short twig; elliptical, sometimes broader above middle, about 2 in (5 cm) long; apex pointed to rounded, base long slender tapered (attenuate); surfaces glabrous; margin shallow toothed on the upper two-thirds; petioles grooved; stipules variably long and narrow, early deciduous
Flowers: 2 from a single bud, each about 0.5 in (12 mm) across; petals white, oval, base abruptly narrowed, tip rounded; calyx tube bell-shaped, lobes oblong, about 0.1 in (1.5 mm) long, glabrous; stamens 25-30; ovary egg-shaped, glabrous
Fruits: round to egg-shaped, about 0.5 in (13 mm) across, minute point at tip; deep purple, flesh purple, edible with distinctive flavor; stone oval, flattened, shallow pitted
Twigs: red, becoming grayish with waxy coating (pruinose)
Buds: leaf buds rounded, glossy; flower buds twice as long

Prunus ×*cistena* N. E. Hansen - purpleleaf sandcherry

This cross between the purple-leaved cherry plum (*P. cerasifera* var. *atropurpurea*) and the eastern sandcherry (*P. pumila*)—introduced in 1910 by Dr. N. E. Hansen of South Dakota State University—is one of the hardiest purple leaf plants (Dirr 1990). Minor chlorosis is often present on these plants in the Fort Collins area, but they are winter hardy (Fairchild and Klett 1993).

This purple leaf hybrid resembles eastern sandcherry except for leaf and flower color. Proper pruning, to control a tendency to become leggy, makes purpleleaf sandcherry a very satisfactory landscape plant. Cultivars include 'Minnesota Red', with its deep reddish purple leaves, and 'Big Cis', a fast-growing, small tree (Dirr 1990).

Planning

Height: up to 10 ft (3 m)
Habit: shrub, small tree; upright, suckering; spring bloom after leaves appear; purple foliage
Site: sun (partial shade reduces flowering); pH and soil adaptable; zone 3
Use: border, screen, specimen

Recognition

Leaves: longer than wide, widest above the middle, about 1.5 in (4 cm) long; red-purple, hairy beneath along lower midrib; petioles minutely hairy above
Flowers: solitary or in 2s; white to pinkish; stalks and calyx reddish; anthers purple
Fruits: round, not grooved, purple-black, few

Prunus glandulosa Thunb. - dwarf flowering almond

Dirr (1990) describes dwarf flowering almond, a native of China and Japan, as "single season quality, appearing distraught and alone in summer, fall and winter." It is sometimes confused with the shrub flowering almond (*P. triloba*). Dwarf flowering almond cultivars 'Alboplena' and 'Sinensis' bear double white and double pink flowers, respectively.

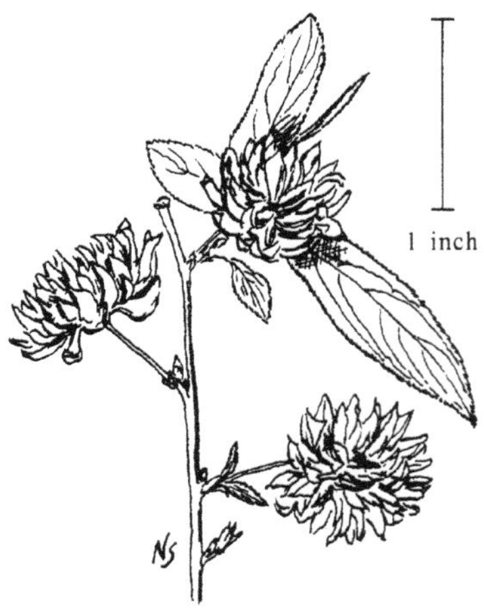

Prunus glandulosa

Planning

Height: up to 5 ft (1.5 m)
Habit: shrub; weakly multistemmed, spreading, straggly; early May bloom; fruits rare
Site: sun; zone 5
Use: border, grouping

Recognition

Leaves: much longer than wide, widest below the middle, about 2.5 in (6 cm) long; apex pointed, base tapering; margin toothed; glabrous or slightly hairy below on midrib; petioles about 0.2 in (6 mm) long
Flowers: 1-2, stalks (pedicels) about 0.4 in (1 cm) long; petals white or pink; calyx tube (hypanthium) cup-shaped, lobes (sepals) fine-toothed
Fruits: nearly round, about 0.5 in (12 mm) across; red
Twigs: glabrous, rarely minutely hairy

Prunus tomentosa Thunb. - **Manchu cherry**

Manchu cherry, often known as Nanking cherry, is a native of a wide area of China and Japan. It is an attractive shrub despite minor chlorosis, and it produces flowers and fruits reliably with no pest problems at PERC in Fort Collins (Fairchild and Klett 1993). The early ripening fruits are tasty when eaten fresh or in jams and jellies, but birds are strong competitors for the fruit harvest. Stem diseases cause dieback on plants located on shady sites at Prairie Pines Arboretum near Lincoln, probably due to the greater humidity caused by shading.

Korean cherry, *P. japonica* Thunb. is another small shrub that produces edible fruit. Hoag (1965) rates it as questionable in value without further testing and selection. Korean cherry bears white stalked flowers and fruits that ripen to a purplish color.

Prunus tomentosa

Planning

Height: up to 10 ft (3 m)
Habit: shrub; broad, dense; early spring blooms; fruits ripen in early summer
Site: sun; pH and soil adaptable; drouth tolerant; zone 3
Use: border, wildlife habitat, windbreak, hedge, naturalizing
Tip: full sun promotes fruiting and long life

Recognition

Leaves: broad elliptic to wider above the middle, about 2.5 in (6 cm) long; apex abruptly pointed, unequally toothed; wrinkled (rugose), dull dark green and hairy above, densely long hairy below (villous); petioles about 0.1 in (3 mm) long, glandular
Flowers: 1-2, nearly stalkless; pinkish in bud, becoming white; hypanthium (calyx tube) tubular, hairy outside, lobes toothed
Fruits: nearly round, about 0.4 in (1 cm) across; scarlet, edible
Twigs: long-hairy (tomentose)

Section Pseudocerasus

Prunus sargentii Rehd. - **Sargent cherry**

Sargent cherry, native to Japan, is a tree for all seasons: rose-pink flowers; emerging reddish-tinted leaves for spring that turn dark green in summer; bronze-red foliage in autumn; and polished red-brown bark in winter.

The species, including a column-shaped cultivar, 'Columnaris', are rated as excellent in the Wooster, Ohio, Shade Tree Evaluation Plots (Dirr 1983). Borer injury has been reported by Marshall's Nursery, Arlington, Nebraska (Sutton 1998).

Planning

Height: up to 50 ft (15 m)
Habit: tree; vase-shaped to columnar; flowers before leaves; June-July fruit; colorful autumn foliage
Site: sun; soil adaptable; zone 5
Use: shade, specimen

Recognition

Leaves: broad elliptic to oblong, wider above the middle, about 4 in (10 cm) long; apex long-pointed, base usually rounded; sharply toothed; glabrous, slight waxy bloom beneath; petioles about 0.7 in (18 mm) long, 2-4 glands near blade
Flowers: clusters (umbels) of 2 to 4, sessile; flowers on pedicels about 1 in (25 mm) long; petals pink, about 1.3 in (3 cm) across; hypanthium (calyx tube) narrow bell-shaped, glabrous, lobes upright or spreading, glabrous, entire
Fruits: round to egg-shaped, about 0.4 in (1 cm) long; purple-black
Twigs: glabrous
Bark: smooth, chestnut brown; lenticels conspicuous

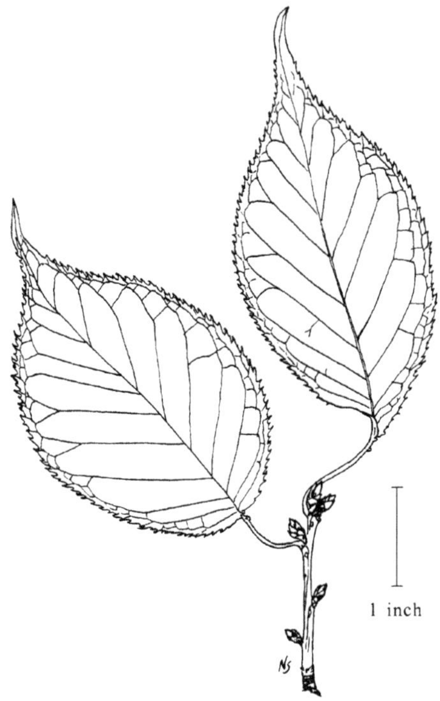

Prunus sargentii

Prunus subhirtella Miq. - rosebud cherry

Rosebud cherry from Japan is aptly named. Its several cultivars provide a bountiful display of pink to reddish pink blossoms before the leaves emerge. Cultivars of the weeping variety, *pendula* (Maxim.) Tanaka, include: 'Pendula Rosea', 'Pink Cloud' and 'Pendula Rubra', each with different shades of double pink flowers.

Dirr (1990) states that this fast-growing, graceful tree would be "garden worthy" even if it never flowered. He proclaims rosebud cherry to be tolerant of cold and heat extremes. However, Richley (1988) reports that rosebud cherry does not survive in southwestern Nebraska. This cherry deserves widespread trials to determine its worth throughout the central prairie.

Planning

Height: up to 40 ft (12 m)
Habit: tree; variable, upright to weeping; spring blooms appear before leaves
Site: sun; soil adaptable; zone 5
Use: specimen

Recognition

Leaves: egg-shaped to elongated, about 3 in (8 cm) long; apex pointed, sharply toothed; hairy on veins below; petioles hairy, glandular, about 0.2 in (6 mm) long
Flowers: clusters (umbels) of 2 to 5, sessile (no peduncles); petals pink; hypanthium (calyx tube) cylindric, narrowed at base, hairy; bracts (involucre) deciduous
Fruits: round, about 0.3 in (8 mm) across; shiny black
Twigs: hairy

Section Eucerasus

Prunus avium L. - mazzard cherry

This Eurasian cherry, known for its use in fine cabinets in Europe, is sometimes planted in our prairies for its sweet fruits. Mazzard cherry produces a significant commercial fruit crop in the western coastal states of the U.S. It has become naturalized in parts of the eastern U.S. where it can be found along roadsides and in open woods (Elias 1980).

Mazzard cherry fruit-producing cultivars include 'Black Tartarian' and 'Yellow Glass'. 'Plena', which bears large double white flowers, could be more widely used as an ornamental (Dirr 1990).

P. ×fontanesiana (*P. avium* × *P. mahaleb*) is not commercially available but should be considered for propagation and use in the Fort Collins area (Fairchild and Klett 1993). These authors state that a specimen of this hybrid, acquired from the Cheyenne Horticulture Research Station, is well adapted to the PERC site. It produces excellent spring flower displays and colorful autumn foliage.

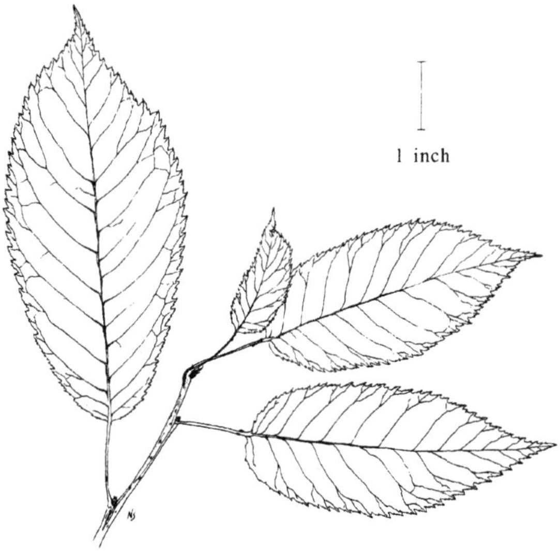

Prunus avium

Planning

Height: up to 80 ft (24 m); National champion, 1989, 82 ft (24.8 m), Pennsylvania
Habit: tree; pyramidal; spring bloom; autumn foliage yellowish
Site: sun; soil adaptable; zone 4
Use: specimen, fruit, shade

Recognition

Leaves: oblong-ovate, about 4 in (10 cm) long; apex pointed, base tapering; margin unequally toothed; slightly wrinkled (rugose) above, hairy below; petioles up to 1.5 in (4 cm) long, glandular near blade
Flowers: solitary or in clusters of 3-5, without leafy bracts; petals white; calyx tube (hypanthium) urn-shaped, about as long as lobes (sepals), lobes entire
Fruits: nearly round to heart-shaped, about 1 in (25 mm) across; yellow to dark red or nearly black, sweet, juicy; stone rounded to slightly flattened
Buds: inner scales reflexed

Prunus cerasus L. - sour cherry

Sour cherry, native to southeastern Europe and Asia, has escaped from cultivation in the eastern U.S. where it can be found along roadsides and in open woodlands. Commercial orchards in the prairies and elsewhere in the U.S. produce large quantities of sour cherry fruits.

The sour cherry is one of the most reliable fruit-producing trees for use by the prairie backyard farmer. Many homesteaders of the western plains planted sour cherry trees and reaped bountiful fruit harvests unless the trees were destroyed by a chance hailstorm or prolonged drouth. Although such events discourage large orchards, an occasional sour cherry tree can still be found in western prairie backyards.

'Northstar' and 'Meteor' are hardy sour cherry cultivars introduced by the University of Minnesota. Two other cultivars, 'Montmorency' and 'Early Richmond' are reliable fruit producers of long standing.

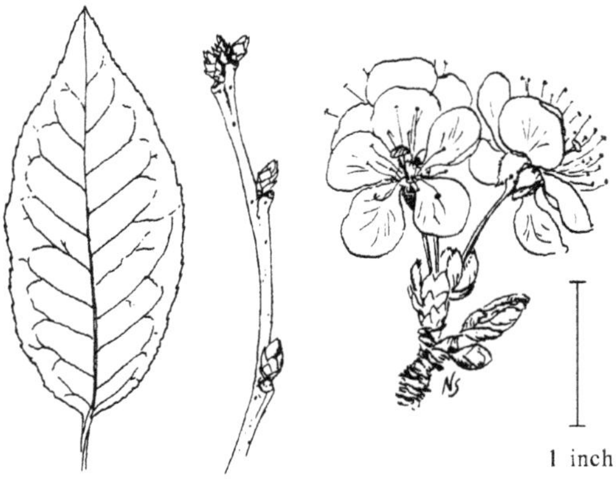

Prunus cerasus

Planning

Height: up to 50 ft (15 m); National champion, 1972, 68 ft (20.6 m), Michigan
Habit: tree; round-topped; early spring bloom, summer fruits
Site: sun; pH and soil adaptable; zone 4
Use: specimen, fruit

Recognition

Leaves: elliptic to egg-shaped, sometimes broader above the middle, about 2.5 in (6 cm) long; apex pointed, base rounded; doubly toothed; dark green and glabrous above, slightly hairy beneath; petioles about 0.6 in (15 mm) long
Flowers: clusters dense, usually subtended with a few leafy bracts; petals white; hypanthium urn-shaped, about as long as lobes (sepals) which are usually minutely toothed
Fruits: nearly round, depressed at apex, about 0.5 in (12 mm) across; red, juicy, sour, edible
Twigs: glabrous
Buds: inner scales upright

Prunus fruticosa (Pall.) Maxim. - **ground cherry**

Ground cherry, a native of Europe and Siberia, mimics the native western sandcherry in form, but its foliage is an attractive, dark glossy green. The ground cherry is often planted in the northern plains, especially in Canada (Hoag 1965).

'Scarlet', a hardy ground cherry cultivar selected from a Mongolian population by the Natural Resources Conservation Service Plant Materials Center of Bismarck, forms a dense thicket. It is recommended for planting in the northern prairies.

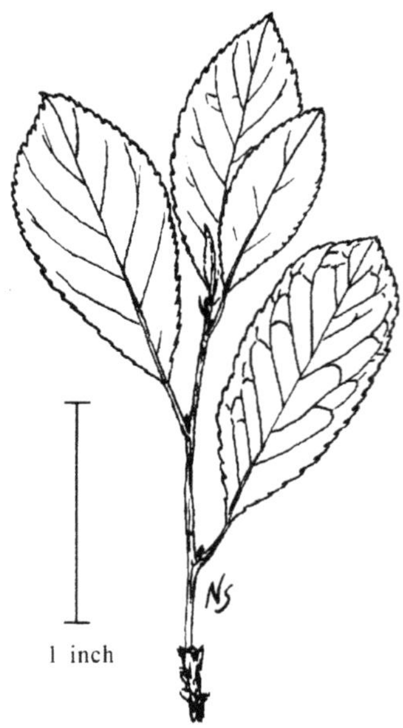

Prunus fruticosa

Planning

Height: up to 6 ft (1.8 m)
Habit: shrub; strongly suckering; July fruits
Site: sun; soil adaptable; zone 3
Use: wildlife habitat, fruit, windbreak

Recognition

Leaves: elliptic to oblong, wider above the middle, about 1.5 in (4 cm) long; apex rounded to short-pointed; glabrous and dark glossy green; petioles about 0.3 in (8 mm) long
Flowers: clusters of 2 to 4 subtended by leafy bracts; petals white; hypanthium twice as long as calyx lobes (sepals) which are broad and rounded
Fruits: nearly round, about 0.4 in (1 cm) across; dark red
Twigs: glabrous

Section Mahaleb

Prunus mahaleb L. - mahaleb cherry

Mahaleb cherry has found its way across most of the eastern U.S. from its original home in Europe by outliving the garden cherries grafted to its roots. It is most frequently found near abandoned farmsteads and orchards. Mahaleb cherry, an attractive, small tree, is adapted to the Fort Collins area (Fairchild and Klett 1993). Birds harvest its ripe fruits, and jelly can be made from the fruits if enough can be found, even though they are small and bitter.

Planning

Height: up to 20 ft (6 m)
Habit: tree; rounded, broad; suckering; April flowers; yellow autumn foliage
Site: sun; pH and soil adaptable; zone 5
Use: wildlife habitat, windbreak, grafting understock, specimen, shade

Recognition

Leaves: broad-oval to egg-shaped, about 1.5 in (4 cm) long; apex pointed, base rounded to heart-shaped; dark green, lustrous above, midrib hairy below; petioles flattened or grooved on upper side, glabrous, often 1-2 glands at summit; stipules egg-shaped, red-glandular, early deciduous
Flowers: clusters of 6 to 8, elongated, terminating a new leafy shoot; petals white; hypanthium (calyx tube) largest at apex, glabrous, slightly longer than the lobes; sepals (lobes) reflexed sharply when flowers are fully open
Fruits: egg-shaped, about 0.3 in (8 mm) long: glossy, bitter, red-purple
Twigs: brownish, hairy at first, later coated with gray wax
Buds: pointed; scales red-brown, edges pale gray
Bark: smooth with large horizontal lenticels, becoming dark gray and scaly

Prunus pensylvanica L. - pin cherry

Pin cherry grows naturally in an area that extends across Canada from Newfoundland to British Columbia, south in the Appalachians to Georgia, and in the Rocky Mountains, south to Colorado. It grows in the wooded areas of the Great Lake States, North Dakota and the Black Hills. Pin cherry is often associated with aspen and birch trees.

Known as a "fire species," pin cherry is a valuable aid to natural reforestation after fires because it quickly forms pure stands which provide protection for seedlings of other species of greater value. Pin cherry trees soon die, allowing long-lived trees to emerge from the understory. The sharply bitter fruits provide food for birds.

Pin cherry is not often planted as an ornamental. 'Stockton Double Pincherry', which bears small, white double flowers, has been available from nurseries in the northern plains but is not widely tested (Hoag 1965).

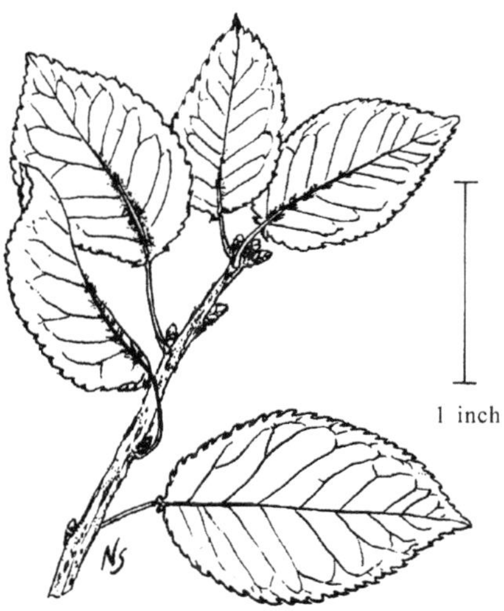

Prunus pensylvanica

Planning

Height: up to 50 ft (15 m); National champion, 1982, 85 ft (25.9 m), Tennessee
Habit: tree; rounded, broad
Site: sun; soil adaptable; zone 2
Use: wildlife habitat, naturalizing

Recognition

Leaves: elliptic to much longer than wide, broadest at base or middle, about 3.5 in (9 cm) long; apex long-pointed, base rounded; margin with fine, rounded, irregular teeth, large triangular glands on incurved side, sometimes overlapping the next tooth; glabrous, shiny above; petioles glabrous or slightly hairy, glands often near blade; stipules glandular toothed, early deciduous
Flowers: clusters of 4 to 5 usually short; petals creamy white; hypanthium (calyx tube) largest at apex, glabrous, about same length as lobes; sepals (lobes) triangular with rounded tips, glabrous
Fruits: nearly round, about 0.2 in (6 mm) across; glossy red, sharply bitter; stone surface pebbled, tip pointed, ridged on inner side
Twigs: red-brown, glabrous, wax-coated; lenticels conspicuous, orange
Buds: clustered near twig end, dull pointed; scales difficult to distinguish
Bark: purple-brown, becoming dark, split, with edges curling outward; horizontal lenticels obvious
Wood: lightweight, soft

Subgenus *Padus* - cherries

Prunus maackii Rupr. - Amur chokecherry

This Asian relative of our native chokecherry is noted for its coppery, exfoliating bark. Visitors marvel at the beauty and size of a young Amur chokecherry tree at Prairie Pines Arboretum near Lincoln. A specimen at PERC in Fort Collins is winter hardy, a reliable bloomer, exhibits only minor chlorosis on new growth, and grows few suckers (Fairchild and Klett 1993).

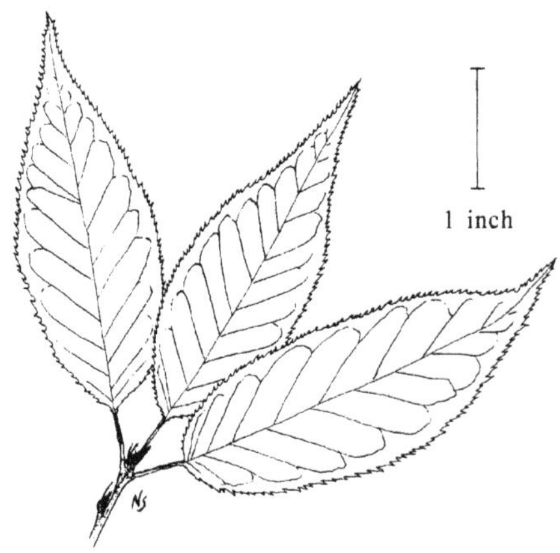

Prunus maackii

Planning

Height: up to 40 ft (12 m)
Habit: tree; round-topped; May bloom; fruit ripens in early summer; shiny copper-colored bark; autumn foliage yellow
Site: sun; soil and pH adaptable; prefers moist, well-drained site; zone 3
Use: specimen, shade, wildlife habitat, fruit for birds

Recognition

Leaves: elliptic to oblong-egg-shaped, about 2.5 in (8 cm) long; apex pointed, base rounded; finely toothed; gland dotted below, veins slightly hairy; petioles about 0.5 in (12 mm) long; stipules prominent and present through midsummer
Flowers: clusters (racemes) dense, hairy, about 2.5 in (6 cm) long, stalk (peduncle) glabrous; petals white; hypanthium narrowly bell-shaped, longer than the glandular-toothed lobes (sepals); style about as long as stamens, shorter than petals
Fruits: round, about 0.2 in (5 mm) across; black; calyx deciduous
Twigs: hairy
Bark: brownish yellow to reddish brown, flaking or exfoliating

Prunus padus L. - European birdcherry

European birdcherry, a native of Europe and northern Asia, is a quite variable species with many recognized varieties. Although closely related to our native chokecherry, it is tree-like. The best known variety is *P. padus* var. *commutata* Dipp., commonly know as mayday tree because of its early blooms. Large clusters of white fragrant flowers on mayday trees brighten early May days on the UNL campus. Mayday trees growing at PERC in Fort Collins produce abundant suckers, but they do not suffer from cold or chlorosis (Fairchild and Klett 1993). Hoag (1965) considers birdcherry to be non-suckering.

Cultivars, some of which are considered to be varieties by Rehder (1940) include: 'Watereri', noted for its large pendulous flower clusters; 'Plena', with double flowers; 'Summer Glow', with red-purple summer leaves; and 'Berg', a tight globe-headed form with crimson summer foliage (Dirr 1940). Trial plantings of selection SD-131 are currently being evaluated for the Northern Great Plains by the Natural Resources Conservation Service Plant Materials Center in Bismarck.

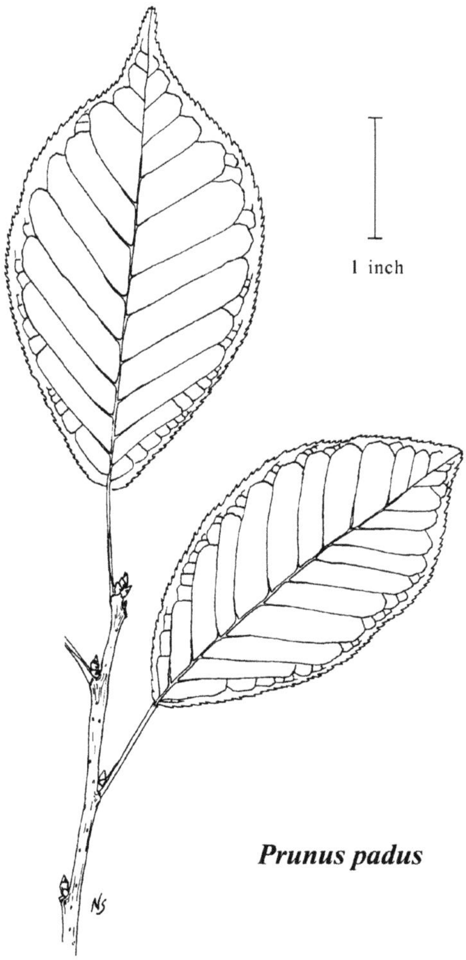

Prunus padus

Planning

Height: up to 30 ft (9 m)
Habit: tree; round-topped; spring blooms; July fruits; yellow autumn foliage
Site: sun; pH and soil adaptable, prefers moist, well-drained sites; zone 3
Use: grouping, specimen, wildlife habitat, fruit for birds

Recognition

Leaves: elliptic to wider above the middle, about 3.5 in (9 cm) long; apex abruptly pointed, base usually rounded; sharply fine-toothed; dull, dark green above, grayish and glabrous below, sometimes tufts on vein axils; petioles about 0.6 in (15 mm) long, glabrous
Flowers: clusters (racemes) loose, drooping, about 5 in (13 cm) long, stalk (peduncle) leafy, glabrous; petals white, fragrant, twice as long as stamens; hypanthium hairy inside; stamens longer than style
Fruits: round, about 0.3 in (7 cm) across, black; stone wrinkled (rugose)
Twigs: finely hairy or glabrous

Prunus serotina Ehrh. - black cherry

Black cherry, the largest of our native cherries, is best known for its attractive wood that is used for many purposes where smooth grain, stability, and ability to take a fine finish are important. When large black cherry trees were plentiful, their wood was often used as a mahogany imitation, but now the reverse is often true. These trees are seldom used in landscaping. However, the black cherry's white flower clusters and glossy, dark green leaves are attractive, and its fruits are a delicacy for birds, bears and small animals. We use the fruit for jams, jellies, and beverages and the bark for an extract to quiet coughs.

Capulin cherry, *P. serotina* var. *salicifolia* (Kunth) Koehne, a native of the highlands of Central America and northern South America, has been imported to California where it is valued for its fruits.

The natural range of black cherry includes every state in the U.S. east of Nebraska and south to the Gulf of Mexico through eastern Kansas and Texas. Pool (1951) states that this cherry is found in Nebraska as far west as Franklin County in the Republican River Valley and near Weeping Water in Cass County. Black cherry also grows naturally in south central Texas, west to Arizona, and south through the highlands of Mexico to Guatemala. It has become naturalized in northwestern South America. This wide natural range of black cherry, representing great differences in site characteristics, has produced a species of great genetic diversity.

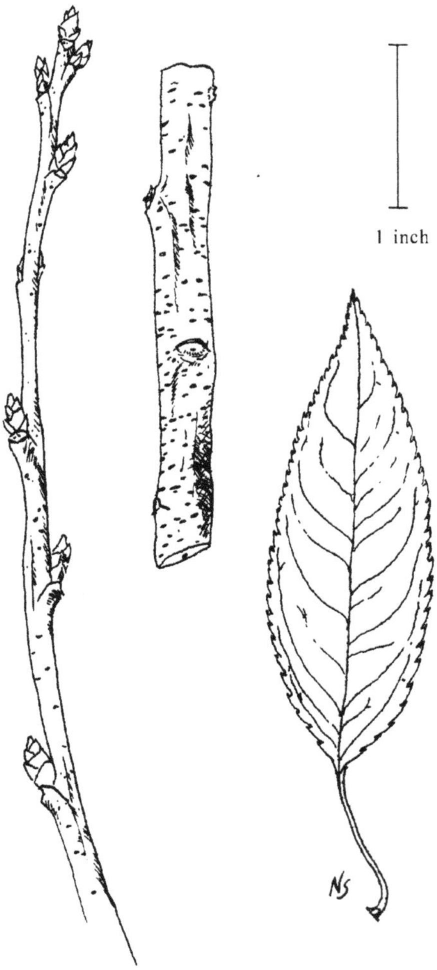

Prunus serotina

In 1961, Dr. E. M. Meader, University of New Hampshire, sent seed derived from crossing Capulin cherry with native New Hampshire black cherry trees to Dr. Fred Gowen in the Department of Horticulture and Forestry at UNL. Trees grown from this seed were planted at Horning State Farm near Plattsmouth. The hybrid trees in the Horning plantation and their progeny, planted elsewhere, are quite variable, ranging from short and multistemmed to tall and straight. Small, shrubby trees generally produce large, tasty fruits that ripen earlier than the fruits of the tall, straight-stemmed trees. The astringent fruits of the tall trees are small, but they provide food for birds and animals.

Very early leaf flushing and blooming is a common trait of these hybrid trees. Untimely, late spring freezes that injure American plum blossoms and destroy newly emerging leaves of other deciduous trees do not injure these hybrids. In Nebraska their foliage generally remains green until late November. Seedlings covered by snow in late November have retained green leaves throughout the winter. Winter dieback has never been detected (Bagley 1998).

This hybrid population presents an opportunity for selection of cultivars with a variety of forms and fruit characteristics. These hybrids deserve further trial plantings throughout our prairies.

Planning

Height: up to 60 ft (18 m); Nebraska champion, 1986, 63 ft (19 m), Neligh; National champion, 1986, 138 ft (41.8 m), Michigan
Habit: tree; round-topped; flowers with leaves
Site: sun; pH and soil adaptable, prefers moist, fertile sites; zone 3
Use: specimen, shade, lumber, wildlife habitat

Recognition

Leaves: narrowly egg-shaped to long and narrow (oblong-lanceolate), about 3 in (8 cm) long; apex pointed, base tapering to rounded; margin with fine, incurved, callus-tipped teeth; glabrous, dark green, lustrous above, paler beneath; petioles about 0.5 in (12 mm) long, glabrous, often red on upper side, 2 glands at leaf base; stipules glandular toothed, early deciduous
Flowers: clusters (racemes) many-flowered, loose, about 3 in (8 cm) long, stalk (peduncle) leafy, glabrous: petals white, slightly longer than stamens; hypanthium bell-shaped, glabrous, lobes pointed, often toothed
Fruits: round, about 0.4 in (1 cm) across; black, juicy, edible; calyx persistent
Twigs: glabrous, red-brown; lenticels minute, pale, becoming prominent, elongated horizontally
Buds: egg-shaped; scales red-brown, glabrous
Bark: becoming black with recurved scaly plates
Wood: moderately strong, heavy, hard

Prunus virginiana L. - common chokecherry

Common chokecherry fruits, although somewhat astringent and acidic to the taste, provided food and beverage for Native Americans and early settlers across our nation and southern Canada. The Dakota Tribe named the period of fruit ripening "The Moon of the Black Cherry." A few hardy people still comb the countryside in August looking for this fruit of distinctive flavor from which they make jelly and wine.

Common chokecherry grows along woodland borders, in niches among steep hills, and in canyons along intermittent and flowing streams on the prairies and in the mountains. Though less common today in its original habitat, common chokecherry manages to survive along roadsides and fencerows.

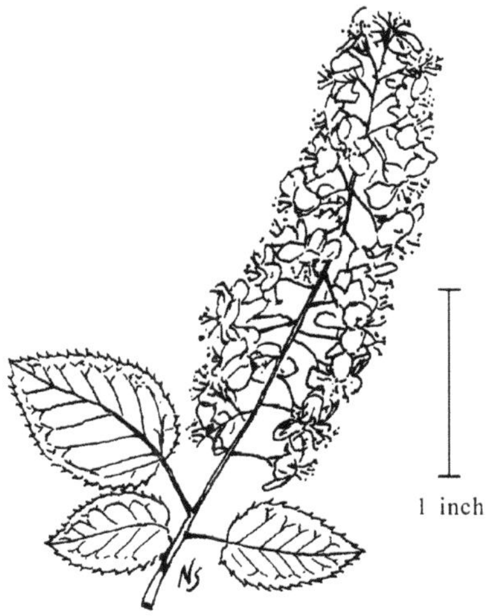

Prunus virginiana

P. melanocarpa Rydb., western chokecherry, described by Poole (1951) as native to western Nebraska and the Rocky Mountains, is now considered to be *P. virginiana* var. *melanocarpa* Sarg. It sometimes attains tree-like form (Sutton 1998). Darker fruits, smaller leaves and glandless petioles characterize the western variety, but these features are so variable and vague that positive identification is seldom possible. Western cchokecherry in its western habitat reveals a larger more tree-like form (Sutton 1999). However, there are, no doubt, significant differences in hereditary traits to be discovered within the wide-ranging chokecherry population.

A North Dakota nursery has introduced a purple-leafed form. This cultivar, named 'Schubert' or 'Shubert', may be grown as a very large shrub or small tree; it can also be grafted onto non-suckering mayday tree rootstock (Hoag 1965). 'Schubert' could be an asset on difficult sites where color contrast is desired. This cultivar, also named Canadian red cherry, is adapted to the PERC site at Fort Collins (Fairchild and Klett 1993).

Planning

Height: up to 25 ft (8 m); Nebraska champion, 1980, 26 ft (7.9 m), Lincoln; National champion, 1982, 67 ft (20.3 m), Michigan
Habit: shrub, tree; crown round-topped; suckering; flowers appear with the leaves
Site: sun; pH and soil adaptable; drouth tolerant; zone 2
Use: wildlife habitat, protection planting, border, specimen, fruit

Recognition

Leaves: broad elliptic to egg-shaped, some wider above the middle, about 3 in (8 cm) long; apex pointed, base tapering to rounded; fine sharp teeth pointing outward or toward apex, without glands when full grown; upper surface dark green, glossy, sparsely hairy beneath, veins often very hairy; petioles about 0.6 in (15 mm) long, glabrous, grooved above, glandular near leaf blade; stipules linear, toothed, promptly deciduous
Flowers: clusters (racemes) many-flowered, dense, about 2.5 in (6 cm) long, stalk (peduncle) leafy, hairy; petals white, almost twice stamen length; hypanthium bell-shaped, glabrous, calyx lobes rounded, glandular toothed
Fruits: round, about 0.4 in (1 cm) across; usually dark purple (sometimes yellow), juicy, edible; calyx deciduous
Twigs: glabrous, dark brown; lenticels prominent, becoming horizontally elongated on large branches
Buds: conical, light brown; scales with grayish edges
Bark: red-brown becoming fissured and scaly

Subgenus *Laurocerasus* - laurels

Cherrylaurel, *P. laurocerasus* L., is an evergreen shrub commonly planted as a hedge in the southeastern U.S. and Europe.

Order Fabales

The order Fabales is comprised of three families and about 14,000 species distributed worldwide (Cronquist 1981). These plants, commonly known as legumes, are considered by some authorities to be in the order Rosales. All are placed in a single family, Leguminosae or Fabaceae.

Most plants of Fabales are herbaceous. Whether woody or herbaceous, annual or perennial, most Fabales members have somewhat similar flowers and fruits and finely divided leaves. Bacteria-containing nodules—present on the roots of many Fabales species—convert atmospheric nitrogen to nitrates that are usable by other plants.

Recognition

Habit: trees, shrubs, vines, herbs; deciduous or evergreen

Leaves: mostly alternate; usually pinnately compound (once or twice), less often palmately compound or trifoliate (3 leaflets), seldom simple; stipules usually present, sometimes as prickles or spines

Flowers: mostly clusters (racemes, corymbs, spikes or heads) bisexual, rarely unisexual (dioecious or monoecious); symmetrical to irregular; calyx usually 5 sepals, distinct or united to form a lobed tube, seldom much-reduced or obsolete; corolla of usually 5 petals, distinct or united (connate) to form a lobed tube, or 2 lower petals often connate and the others distinct; stamens 5, 10, to many, often joined by their filaments to form a sheath around the ovary; pistil 1

Fruits: commonly dry and splitting along both sutures (legumes), sometimes not splitting or breaking into 1-seeded joints; seeds hard, embryo large, cotyledons 2

Grouping Key for Fabales

(Cronquist 1981)

Trees, shrubs, rarely herbs, sometimes spiny; leaves mostly alternate, compound (bipinnate); flowers symmetrical, in dense clusters (heads, spikes, racemes); usually bisexual; calyx of usually 5 sepals, united to form a tube with valvate lobes, seldom reduced or obsolete; petals distinct or united below to form a tube; stamens usually 10, commonly long-exserted, colored, conspicuous; fruit a legume: **Mimosaceae (mimosa family)**

Trees, shrubs, rarely herbs; leaves alternate, usually pinnately compound, stipules present; flowers small and inconspicuous to showy, in clusters (racemes, spikes or cymes), bisexual, usually irregular, not pea-like (papilionaceous); sepals 5, usually distinct or the 2 upper connate or all united into a 5-toothed cup; petals 5, distinct, imbricate, irregular, the uppermost generally internal to the 2 adjacent and often smaller; stamens rarely extending beyond petals, not showy; fruit commonly a legume: **Caesalpiniaceae (caesalpinia family)**

Herbs, less often trees, shrubs, vines; leaves alternate, rarely opposite, usually compound (pinnate, digitate or trifoliate); flowers in clusters (racemes, spikes or heads); sepals mostly 5, more-or-less joined into a lobed tube; petals usually 5, irregular, pea-like, upper petal (standard) enclosing the others in bud, 2 lateral petals (wings) similar, mostly distinct; calyx 5-toothed or 2-lipped; stamens mostly 10, 9 filaments joined in open sheath around the pistil, the 10th usually separate (diadelphous); fruit commonly a legume: **Fabaceae (pea family)**

Mimosaceae R. Brown - **mimosa family**

Woody plants of the mimosa family venture as far north as extreme southern Kansas in the prairie region. Most of the 40 genera and 2,000 or more species are found in arid and semi-arid climates of tropical and subtropical regions. In addition to their ornamental and protective uses, they are a source of valuable products including gum arabic and lumber.

Acacia Mill. - **acacia**

The acacias, mostly trees, are found in nearly all tropical and subtropical regions. They are unsurpassed in their ability to thrive upon and improve arid and sterile soils (U.S. Forest Service 1937).

In addition to providing valuable wood, dyes, gums and tannins, many acacias are used as ornamentals. Those with low tannin content are valuable browse for livestock. About 15 acacia species are native to North America, mostly Mexico and the southwestern U.S.

Acacia angustissima (Mill.) Ktz. **var. *hirta*** (Nutt.) Robins. - **prairie acacia**

Prairie acacia—a native of Mexico and the southwestern U.S.—grows as far north as extreme southern Kansas where its top usually winterkills to the ground (Stephens 1973).

Planning

Height: up to 2 ft (60 cm)
Habit: shrub; dies back to ground in winter (suffrutescent) if growing north of Oklahoma
Site: sun; soil and pH adaptable; drouth tolerant; zone 6
Use: forage, wildlife habitat

Recognition

Leaves: alternate; leaflets nearly stalkless (sessile), oblong; pale yellow-green; petioles with 2 lateral ridges, upper surface flat or grooved, hairy; stipules narrow, minutely toothed
Flowers: clusters (racemes), nearly globose; short, axillary, about 0.6 in (15 mm) across; corolla cylindric, greenish white, 5-lobed; stamens many, long exserted, filaments white or pinkish
Fruits: legumes constricted between seeds, 3 to 6 per pod
Twigs: brown, ridged
Buds: nearly round; scales 2-3, red-green

Mimosa L. - **mimosa**

Leaves of the mimosas—natives of tropical America—fold together in reponse to a touch. Over 400 species of trees, shrubs, vines and herbs have been recognized in the genus *Mimosa.*

Mimosa borealis Gray - **fragrant mimosa**

Fragrant mimosa is found in extreme southwestern Kansas, Oklahoma, Texas and New Mexico.

Planning

Height: up to 3 ft (1 m)
Habit: shrub; deciduous; crown open, crooked branches prickly; May flowers on old wood
Site: sun; drouth tolerant, leaves fold together during hot dry weather; zone 6
Use: wildlife habitat

Recognition

Leaves: alternate; leaflets oblong or egg-shaped, about 0.2 in (4 mm) long; margin entire; tip abruptly pointed, short (mucronate), base rounded; surfaces green, glabrous, dotted with minute holes (punctate-dotted); petioles flattened or grooved above, dotted; leaf stalk (rachis) extends beyond last pair of leaflets; stipules awl-shaped
Flowers: clusters, globose or dome-shaped, axillary: petals rose-colored; stamens 10, long-exserted, filaments purple, anthers yellow
Fruits: legumes curved, about 2 in (5 cm) long, constricted between seeds, tip sharp; seeds with U-shaped marking on each surface
Twigs: ridged; prickles numerous, tips recurved; persistent stipules rigid
Buds: white, woolly

Prosopis L. - mesquite

About 30 mesquite species grow in the warmer regions of the Americas, Asia and Africa. Three species are native to the U.S., all found in the Southwest. Mesquite leaves and their nutritious fruits, rich in protein, are sources of food for wildlife and livestock (U.S. Forest Service 1937). The hard, durable wood makes excellent fence posts and fuel, but mesquite trees are usually too small to yield lumber.

Prosopis glandulosa Torr. - honey mesquite

The natural range of honey mesquite extends from southwest Kansas into Texas and Mexico and west to California. It can become a dominant rangeland plant, especially if the range is subjected to overgrazing.

Planning

Height: up to 14 ft (4 m)
Habit: shrub, tree; deciduous; trunks often clustered, thorny; spring flowers on old wood
Site: sun; soil and pH adaptable; drouth enduring; zone 6
Use: wildlife habitat, forage, fuel

Recognition

Leaves: alternate; leaflets linear, about 1.2 in (3 cm) long, end pair usually curved; margin entire, tip abruptly pointed (mucronate); surfaces yellow-green, glabrous; petioles grooved above, about 3 in (8 cm) long
Flowers: clusters (spikes) pendulous; petals yellow-green; stamens 10, exserted, filaments white, becoming yellow as anthers open; anthers yellow
Fruits: legumes solitary or clusters of 2-3, about 5 in (13 cm) long, not splitting, constricted between seeds, abruptly pointed; seeds embedded in pithy, glutinous material
Twigs: armed, spines rigid, nearly straight, up to 1 in (25 mm) long, usually in pairs at leaf axils
Buds: egg-shaped; scales abruptly constricted, points spreading

Caesalpiniaceae R. Brown - caesalpinia family

Fossil records have revealed that the caesalpinia family of about 150 genera and 2,200 species is the oldest of the Fabales (Cronquist 1981).

Cercis L. - redbud

Two of the eight redbud species are native to North America, one is native to Europe and western Asia, and five are native to China. They are too small for their durable wood to be used commercially.

The Eurasian species, *C. siliquastrum* L., is known as the Judas-tree. *C. chinensis* Bge., Chinese redbud, is widely planted in the more southerly regions of the U.S. as an ornamental.

Cercis canadensis L. - eastern redbud

The eastern redbud—a native of southeastern Nebraska, south into Mexico and east to the Atlantic as far north as New Jersey—grows along streambanks, woodland edges and rocky hillsides.

Flowers of this redbud are among the first to appear in spring. Unfortunately, eastern redbuds are sometimes inadvertently planted so far north of their native habitat that unseasonably cold temperatures injure flower buds and sometimes kill part or all of the crown.

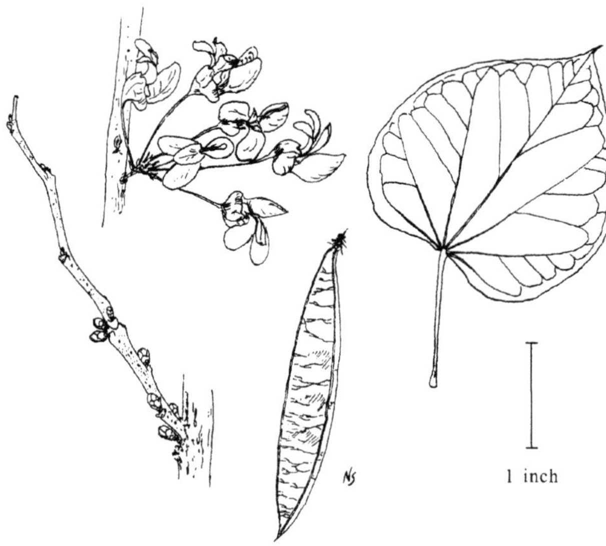

Cercis canadensis

Fairchild and Klett (1993) report that redbud trees (of unspecified origin) growing at PERC in Fort Collins are attractive with dense growth habits even though partially shaded, but that good flower and fruit displays have occurred only after mild winters. They also state that lack of flower bud hardiness is the reason for unreliable blooming, but that the trees have not suffered winter dieback. Fairchild and Klett recommend that eastern redbud trees for planting in northeastern Colorado be grown from a northern seed source. The Minnesota Landscape Arboretum is successfully growing seedlings from an eastern redbud tree at the University of Minnesota Horticulture Research Center (Dirr 1990).

A row of eastern redbud trees growing along the driveway south of the old orchard shed on East Campus, UNL, is a source of seed that might provide hardy trees capable of blooming reliably in northerly latitudes. They have been thriving and providing a magnificent floral display annually for the past 30 years. These redbuds were grown from seed collected by Victor Miller (1988) from a tree planted by Earl Maxwell in what is now the Maxwell Arboretum. We suggest that these eastern redbuds be designated by an appropriate name, such as 'Miller-Maxwell'.

Variety *alba* Rehd. bears white flowers. Eastern redbud cultivars include 'Royal White', a white-flowered tree hardy in central Illinois, and 'Pinkbud', a pure, true pink-flowered tree discovered near Kansas City (Dirr 1990).

Planning

Height: up to 30 ft (9 m); Nebraska champion, 1985, 35 ft (10.7 m), Lyons; National champion, 1991, 39 ft (11.9 m), Virginia

Habit: tree; deciduous; crown rounded to spreading; April flowers appearing before leaves; fruits persist into winter

Site: sun, part shade; pH and soil adaptable, tolerates drouth; zone 5

Use: specimen, border, naturalizing

Caution: very sensitive to the herbicide 2-4,D

Recognition

Leaves: simple; heart-shaped, palmately veined; margin entire, short-pointed at tip; glabrous above, hairs on veins below; petioles enlarged at base and summit; stipules minute

Flowers: clusters of 4 to 8 on old wood; calyx lobes 5, broad, rounded, purplish; petals 5, purplish to pinkish (or white), irregular; stamens 10, distinct, filaments white, anthers rose-colored; ovary green, style and stigma red

Fruits: legumes flat, about 2.5 in (6 cm) long; seeds oval, 8-12

Twigs: glabrous, dark gray-brown, often zigzag; leaf scars hairy along margin, bundle scars 3

Buds: terminal absent; leaf buds often 2 at a node (superposed), scales 2; flower buds rounded, stalked, scales many

Bark: smooth, becoming ridged and scaly

Wood: durable, tough, hard

Gleditsia L. - honeylocust

Gleditsia, a genus of 12 species native to tropical and temperate climes, is represented by two species in the U.S. One of these, *G. aquatica* Marsh., waterlocust, is native to the southeastern Coastal Plain of the U.S. Its fruits are much shorter than those of the honeylocust.

Gleditsia triacanthos L. - honeylocust

Honeylocust—a native of the eastern prairies from southeastern South Dakota south to the Gulf of Mexico, and east to the Appalachians—is most commonly found on moist, fertile bottomlands, but it is a survivor on drouthy sites. Though often armed with formidable spines, variety *inermis* Willd. (forma *inermis* (L.) Zabel) is unarmed or nearly so.

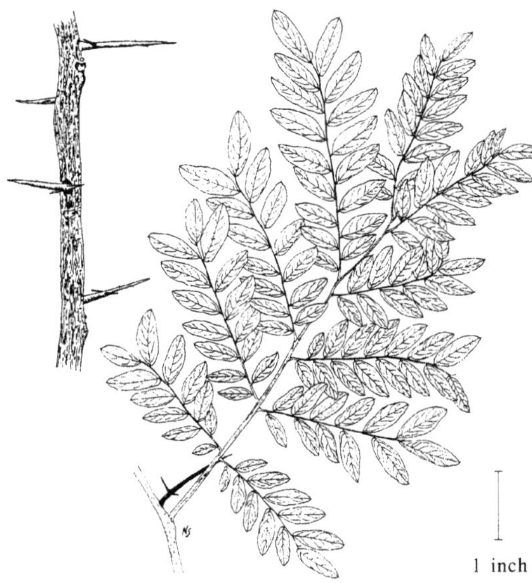

Gleditsia triacanthos

Honeylocusts were planted in many of the Great Plains shelterbelts during the 1930s by the Prairie States Forestry Project. Their survival in these plantations was excellent when verified in 1954 except on sites in southern Kansas and Oklahoma where severe drouth prevailed from 1952 to 1954 (Read 1958). Several honeylocust trees planted in 1917 on a sandy rangeland site at a Yuma County, Colorado, homestead survive today (Bagley 1998).

Honeylocusts serve admirably as shade in pastures and yards because grasses thrive under their airy canopy. Unfortunately, these trees can become a nuisance in heavily grazed pastures. The nutritious pods are relished by cattle and sheep. The hard seeds pass through digestive tracts unharmed and prepared for rapid germination. The trees protected by thorns proliferate while thornless trees succumb to browsing.

Seedlings grown from seeds that have been collected from thornless honeylocusts are mostly without spines. They are commonly planted in windbreaks or other large plantations. Many thornless and more-or-less fruitless cultivars have been selected and are propagated for ornamental plantings. The podless 'Moraine' and 'Fairview' and pod-bearing 'Imperial', 'Shademaster', and 'Skyline' are all adapted to the soil and climate of the Fort Collins area, but 'Green Glory' failed to establish at PERC (Fairchild and Klett 1993). 'Maxwell', reputably hardy at low temperatures, was introduced to the nursery trade by Earl Maxwell, extension forester at UNL during the 1930s.

A few of the many honeylocust cultivars of interesting habit include: 'Bujotii', elegant and pendulous; 'Elegantissima', dense, shrubby, with elegant foliage; 'Pin Cushion', whose foliage is bunched along the stem; and 'Summergold' and 'Sunburst' that have golden leaves on new growth that eventually change to green (Dirr 1990).

Honeylocust trees are a potential agricultural crop because of the nutrituous qualities of their palatable pods (Smith 1953, Bagley 1988). 'Calhoun'—discovered in Georgia by the Tennessee Valley Authority in the late 1930s during a search for honeylocust trees bearing pods high in sugar—bears wide, thick, fleshy pods. Twenty-five-year-old 'Calhoun' trees have produced fruits approximately every third year since age eight at Horning State Farm near Plattsmouth and at Prairie Pines Arboretum near Lincoln. The open-pollinated, mostly thornless seedling progeny of 'Calhoun' became fruitful at about five years of age, bearing pods intermediate in size to those of the parent trees. The 'Calhoun' honeylocust and its progeny have been free from noticeable cold injury at these eastern Nebraska sites (Bagley 1998).

Planning

Height: up to 65 ft (20 m); Nebraska champion, 1977, 69 ft (20.9 m), Bennet; National champion, 1972, 115 ft (34.8 m), Michigan

Habit: tree; thorny or thornless; crown usually spreading; autumn foliage yellow, soon dropping; spring flowers inconspicuous but fragrant

Site: sun; pH and soil adaptable, drouth tolerant; zone 4

Use: shade, street, windbreaks, wildlife habitat, forage for livestock, firewood, lumber

Pests: mimosa webworm and leaf hoppers can be troublesome in ornamental plantings

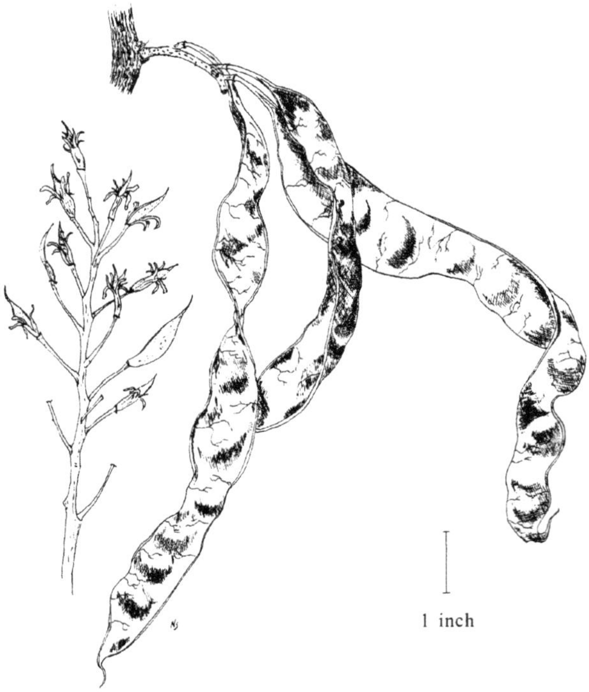

Gleditsia triacanthos

Recognition

Leaves: once or twice compound (bipinnate), leaflets broadest near base, margin obscurely toothed or entire; glabrous and shiny above, hairy on midrib below, about 1 in (25 mm) long

Flowers: clustered; unisexual and bisexual may be on the same tree (polygamo-dioecious), not showy; calyx 4-lobed, linear, green, hairy; petals 4, egg-shaped (obovate) to oblong, green with red-brown marks (male) or yellowish (female), hairy; stamens 4, opposite calyx lobes; ovary linear, hairy, green, style sharply curved, stigma 2-lobed

Fruits: legumes twisted, over 8 in (20 cm) long, brown, glabrous or velvety hairy, sometimes persisting into late winter; seeds oval, brown, hard, about 0.4 in (1 cm) long, about 12 per pod

Twigs: glabrous, greenish to brown, zigzag, thickened at nodes; sometimes armed with sharp spines, single or clustered, often 3-forked

Buds: terminal wanting; lateral minute, sunken, often covered with leaf scars, more than one per node (superposed)

Bark: gray to black, smooth, hard, becoming roughened into broad plates, often bearing branched thorns

Wood: very heavy and hard, light red to reddish brown, easily split

Gymnocladus L. - coffeetree

Fossil discoveries show that the coffeetree once occupied a much greater natural range than it does today. Two species are known—one in China, *G. chinensis* Baill., and the other in the U.S. Its hard, attractive wood is used for wall panels and cabinets.

Gymnocladus dioicus (L.) K. Koch. - Kentucky coffeetree

A native of eastern Nebraska and Kansas, Kentucky coffeetree's natural range also extends south through Oklahoma and Arkansas, northeasterly through western Pennsylvania and New York and west across southern Michigan through Iowa. It is most commonly found along streams on rich moist soils. However, coffeetree adapts to a wide range of soils including the compacted, topsy-turvy soils found on city building lots.

Richley (1988) describes Kentucky coffeetree as an outstanding tree for planting in southwestern Nebraska. It is well adapted to the site at PERC in Fort Collins where no problems have been noted (Fairchild and Klett 1993).

On a bleak winter day, Kentucky coffeetree presents a gaunt yet picturesque sight—sparse, open-crowned with thick, almost club-like twigs that offer little hindrance to wind and sun. In summer its very large compound leaves cast a welcome shade. Coffeetree is seldom bothered by insects and diseases.

Propagation is simple. Pick a pod from a likely coffeetree; scarify, and plant a hard, marble-sized seed. A hardy seedling will emerge; protect it from hungry cottontails and deer. In 15 to 20 years the tree, if a female, will be decorated with reddish brown pods during winter—a bane to those who dislike yard chores, but a potential cup of brew to those who love experimentation. There is no guarantee that the "coffee" will be as aromatic and flavorful as that brewed from the beans of another plant that grows in the mountains of Colombia.

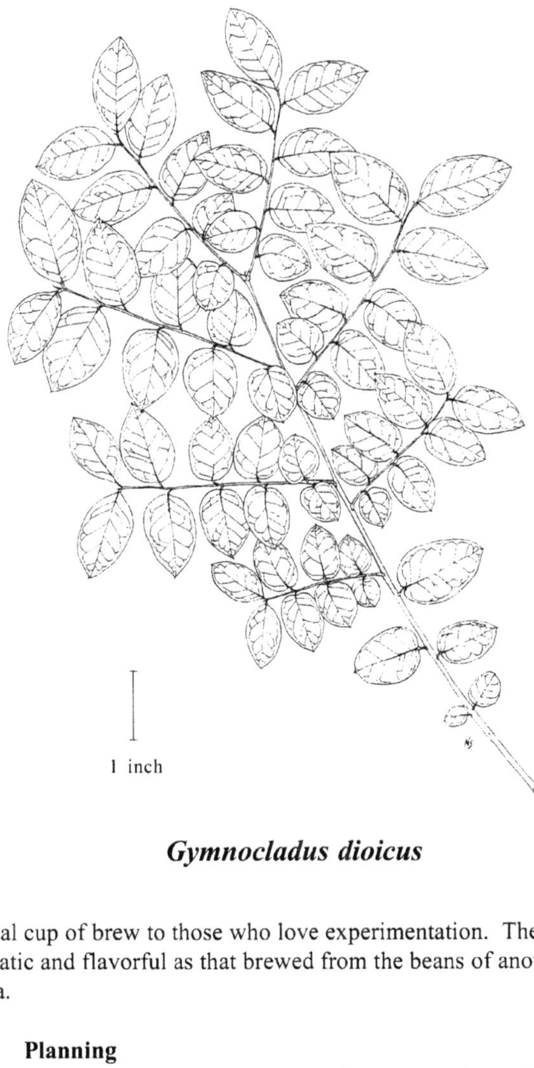

1 inch

Gymnocladus dioicus

Planning

Height: up to 60 ft (18 m); Nebraska champion, 1986, 72 ft (21.8 m), Leigh; National champion, 1986, 112 ft (33.9 m), Michigan

Habit: tree; crown oval; autumn foliage yellow, soon dropping; flowers inconspicuous after leaves; fruits persist into winter

Site: sun; pH and soil adaptable, drouth tolerant; zone 4

Use: shade for large areas, specimen, lumber

Caution: highly sensitive to the herbicide 2-4,D

Recognition

Leaves: twice compound (bipinnate); up to 3 ft (90 cm) long and 2 ft (60 cm) broad; leaflets about 2 in (5 cm) long, broadest near base, tip pointed; light green; hairy when young, becoming less so; margin smooth (entire)

Flowers: clusters (racemes) elongated; unisexual (dioecious), occasionally bisexual (perfect); calyx tube (hypanthium) urn-shaped, hairy, ridges from the base to the 5 calyx lobes and petals; petals 5, whitish, hairy; stamens 10; ovary green with white hairs, style curved near end

Fruits: legumes stalked, leathery, about 5 in (13 cm) long; persistent and closed most of winter; seeds surrounded by green glutinous pulp; nearly round, about 0.5 in (12 mm) in diameter; smooth, hard, dark brown

Twigs: stout, hairy at first, becoming glabrous, light brown, often mottled; lenticels orange, conspicuous; pith large, homogenous, pinkish brown

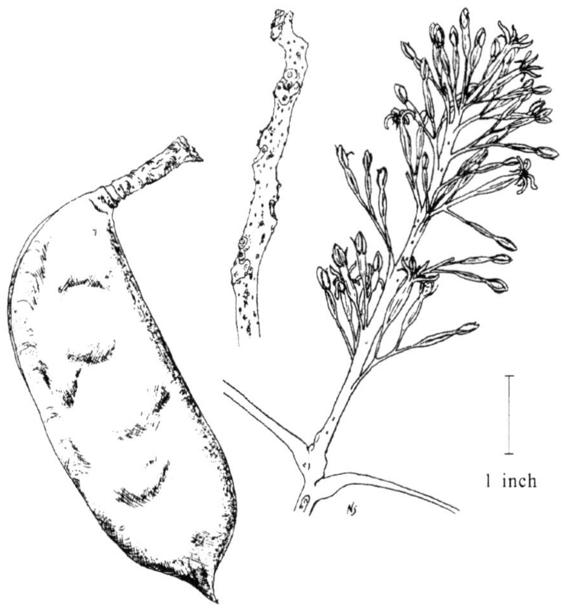

Gymnocladus dioicus

Buds: terminal wanting; lateral partly concealed, one above another (superposed); scales indistinct

Bark: smooth, gray-brown, becoming shallow-fissured with distinctive recurved scale-like ridges

Wood: light red to reddish brown; heavy, hard, moderately durable

Fabaceae Lindl. - pea family

This largest legume family includes many valuable species—wild and domesticated, herbaceous and woody. Four hundred genera and 10,000 species—mostly herbaceous natives growing worldwide—are included in Fabaceae (Cronquist 1981). This family of the clovers, beans, peas and alfalfa is second only to Poaceae (grasses) in agricultural importance.

Salt tree, *Halimodendron halodendron* (L.) Voss, an unusual deciduous small tree or shrub, is native to the saline soils of Turkestan and vicinity. Its blue-green leaves are borne on spiny branches that are silky-hairy when young. The pea-shaped June flowers are purplish pink, and the brownish yellow mature pods—about 0.7 in (2 cm) long—persist through winter. Fairchild and Klett (1993) report that salt tree has a tendency to sucker and suffers twig dieback and a reduced floral display each winter at PERC in Fort Collins. It could be useful for planting on sunny and dry, alkaline sites.

Amorpha L. - amorpha

About 15 species of these shrubby, rarely herbaceous plants, are native to North America.

Recognition

Habit: shrubs, subshrubs, rarely herbs

Leaves: alternate, pinnately compound, leaflets odd-numbered, margins usually smooth (entire); stipules awl-shaped, early deciduous (caducous)

Flowers: clusters dense, terminal (panicled spikes or racemes); calyx bell-shaped, gland-dotted, teeth 5, equal or unequal; corolla a single upright standard, other petals wanting

Fruits: legumes usually gland dotted; seeds 1

Amorpha canescens Nutt. - leadplant amorpha

Leadplant amorpha, native to much of the prairie region, grows among the bluestem and Indian grasses except where it is depleted by overgrazing. Cattle relish this shrub decked with purple flowers and gray, woolly foliage.

Leadplant amorpha can be sown along with native grasses and forbes for prairie re-establishment. An impermeable seedcoat and dormant embryo conspire to try one's patience; but lo and behold, in about three to five years the new prairie will abound with these woolly shrubs. Seedlings or rooted plants are planted in sunny locations where contrast is desired in the landscape. Leadplant amorpha that is transplanted from native meadows seldom survives.

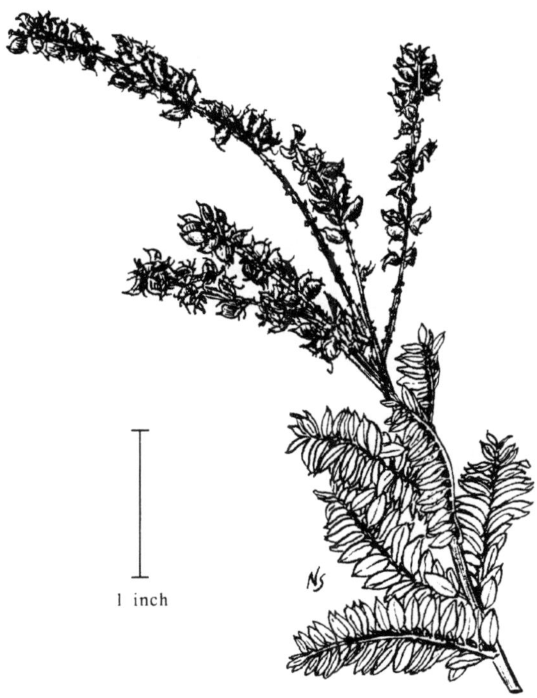

Amorpha canescens

Planning

Height: up to 3 ft (1 m)
Habit: shrub, subshrub; rounded, single to many-stemmed; early June bloom; twigs die back in winter
Site: sun; pH and soil adaptable, drouth tolerant; zone 3
Use: rock garden, prairie re-establishment, wildlife habitat

Recognition

Leaves: curved; leaflets, 15-20 pairs plus small terminal, each about 0.5 in (12 mm) long; tip short pointed (mucronate) or rounded, base rounded; surfaces covered with dense, long gray hairs (tomentose); petioles woolly
Flowers: clusters spike-like (racemes), multiple, terminal, hairy, densely flowered; calyx tube obconical, woolly, purple, lobes 5; petal (standard), purple, enclosing 10 stamens and pistil
Fruits: legumes about 0.2 in (4 mm) long; calyx persistent; seeds oval, beaked, smooth, olive-brown
Twigs: flexible, hairy, gray-brown
Buds: egg-shaped, hairy
Bark: gray, smooth, peeling on old stems (exfoliating)

Amorpha fruticosa L. - indigobush

Indigobush is a leggy, rather scraggly shrub that grows in ungrazed prairies and open woodlands throughout most of the U.S. It is often found along wet roadside ditches and prairie creeks where it grows on a wide range of soil types. Winter pruning will keep indigobush reasonably presentable in a landscape planting.

Indigobush specimens growing at PERC in Fort Collins seldom suffer winter dieback and are not plagued by chlorosis, insects or diseases (Fairchild and Klett 1993).

Planning

Height: up to 10 ft (3 m)
Habit: shrub; multistemmed, irregular, open; midsummer bloom
Site: sun; pH and soil adaptable; tolerates infertile, sandy soils and drouth; zone 3
Use: wildlife habitat, naturalizing

Recognition

Leaves: leaflets 15-24, about 1 in (25 mm) long; hairy (appressed pubescent); tip rounded with minute prickle-like point (mucronate); petioles hairy, about 0.8 in (2 cm) long, lower end abruptly enlarged
Flowers: clusters terminal, spike-like (racemes), usually in 3's; calyx tube obconical, becoming nearly hairless (glabrate), lobes green, hairy, red-streaked, glandular; petal (standard) purplish blue, encloses stamens and pistil; filaments white, purplish toward tip, anthers orange, extending beyond petal
Fruits: legumes about 0.3 in (8 mm) long, curved, warty, persist into winter; seeds surrounded by sticky fluid, oval, beaked, smooth, tan-brown
Twigs: rigid, glabrous
Buds: ovoid, red-brown, hairy
Bark: smooth, old trunks slightly fissured

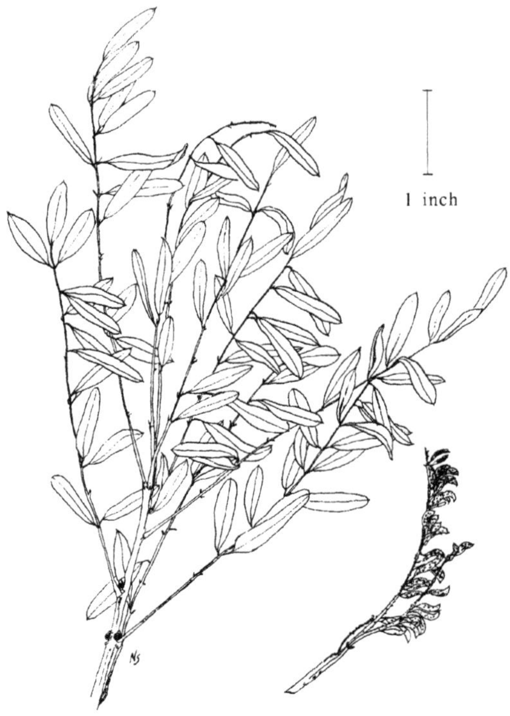

Amorpha fruticosa

Amorpha nana Nutt. - dwarf indigo amorpha

Dwarf indigo amorpha—often single-stemmed, about one foot (30 cm) tall—grows naturally in the northern part of Nebraska with a range extending from Colorado to Minnesota, north into Manitoba and Saskatchewan in Canada. Stephens (1973) recorded an isolated stand in southwestern Kansas that grew to three feet (1 m) in height. Freeman (1992) states that a number of populations are known in Kansas.

A closely related low shrub, *A. brachycarpa* E. J. Palmer, that bears gray-green foliage and purple flowers, is described by Hoag (1965) as hardy and well adapted to the northern plains. July flowers are borne in short, cylindrical terminal clusters. Grooved twigs are covered with fine gray hairs. Dirr (1983) considers this amorpha to be one of the best for garden use.

Planning

Height: up to 1.5 ft (45 cm)
Habit: shrub; often single-stemmed; late May bloom
Site: sun; pH and soil adaptable, tolerates infertile soil and drouth; zone 3
Use: rock garden, prairie revegetation, wildlife habitat

Recognition

Leaves: leaflets about 0.5 in (11 mm) long; glabrous above, sparingly hairy below and glandular; margin slightly notched (crenate), tip rounded, shallow notched (emarginate) or minutely pointed (nucronate); petioles hairy, about 0.2 in (5 mm) long

Flowers: clusters solitary, spike-like (racemes); calyx tube obconical, green with reddish cast, blistery, lobes 5, pointed with marginal hairs (ciliate); petals purple, enclose stamens and pistil; stamens 10, filaments purple, joined (connate) at base, anthers purple, exserted

Fruits: legumes about 0.2 in (4 mm) long, red-brown, glandular, hairy at tip; seeds oval, beaked, smooth

Twigs: flexible, glabrous

Buds: conical; outer scales with hairy margin, inner scales densely hairy, brown

Bark: split lengthwise, slightly rough

Caragana Lam. - peashrub

The 50-odd species of *Caragana*, widespread natives of Asia, are primarily planted for their abundant bright yellow flowers. These hardy, drouth-resistant shrubs are adapted to a wide range of soil conditions including high lime, but they are susceptible to a host of insects that are especially injurious in warm climes.

Hoag (1965) and Dirr (1990) summarize characteristics of several species. Most of them are tall, narrow shrubs, best suited for screens and windbreaks. Dirr (1990) states that two of the prettiest peashrubs—Pygmy peashrub (*C. pygmaea* DC.) and dwarf peashrub (*C. aurantiaca* Koehne)—mature at about 4 ft (1.2 m) tall and tolerate dry, alkaline sites.

Caragana arborescens Lam. - Siberian peashrub

Siberian peashrub was planted extensively as a shrub row in shelterbelts of the Prairie States Forestry Project in the Dakotas and to a lesser extent in Nebraska during the 1930s. Read (1958) reports that 10-year survival and growth was good in the shelterbelts except on sites with poor drainage and high salt content. He found that Siberian peashrub produced "exceptionally uniform and dense barriers" and that it "can withstand defoliation during the growing season without undue loss of vigor."

Siberian peashrub is a useful, nitrogen-fixing shrub that will grow on windswept, drouthy, infertile sites despite attacks by defoliating insects. Other than minor chlorosis, Fairchild and Klett (1993) report no injury or dieback on Siberian peashrub at PERC in Fort Collins.

Vegetatively propagated cultivars include: 'Pendula', stiffly weeping; 'Nana', a dwarf; 'Sutherland', upright; and 'Lorbergii' with narrow leaves and flowers (Dirr 1990).

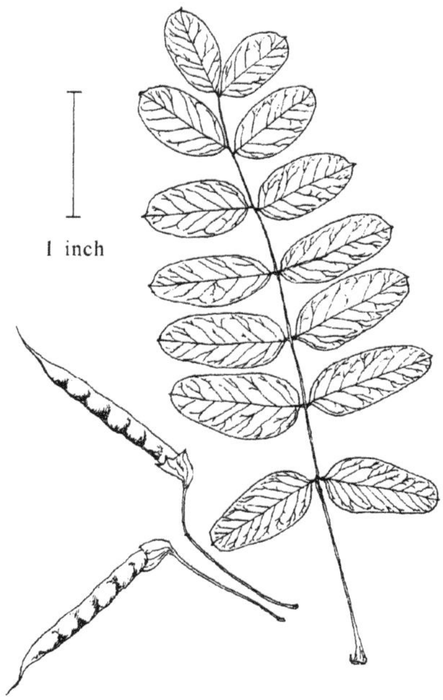

Caragana arborescens

Planning

Height: up to 15 ft (4.5 m)

Habit: shrub or cultivated as a tree; stems erect, arching with age; suckering; April-May yellow blooms on old wood; sometimes spiny

Site: sun, part shade; pH and soil adaptable, tolerates infertile soil and drouth; zone 3

Use: hedge, screen, windbreak, wildlife habitat, naturalizing

Pests: defoliating insects

Recognition

Leaves: leaflets 8-14, about 0.7 in (18 mm) long; elliptic to wider above the middle, rounded at apex with abrupt sharp point (mucronate); hairy when young, becoming less so with age; stipules developing into a pair of short spines on some plants, shorter than the petioles

Flowers: clusters of 1 to 4, pea flower shape; calyx tubular, short-toothed; petals bright yellow

Fruits: legumes up to 2 in (5 cm) long, linear, round in cross-section, becoming brown; seeds 3-5

Twigs: hairy at first, green for several years, scurfy, lenticels horizontal

Caragana frutex (L.) K. Koch - **Russian peashrub**

Russian peashrub 'Globosa'—slow-growing, sun-loving and drouth-tolerant—is adapted to the climate and soils of Fort Collins where it maintains a formal compact growth habit without pruning (Fairchild and Klett 1993). The authors state that the sparse flowers are hidden by the foliage.

Planning

Height: species up to 10 ft (3 m); 'Globosa' up to 2 ft (60 cm)

Habit: upright; suckering; unarmed; 'Globosa' flowers unimportant, hidden by dark green foliage

Site: sun; pH and soil adaptable, tolerates drouth; zone 3

Use: hedge, border ('Globosa': foundation, specimen)

Recognition

Leaves: leaflets 4, up to 1 in (25 mm) long; obovate to oblong-obovate; tip rounded or notched (emarginate), base wedge-shaped (cuneate)

Flowers: clusters of 1 to 3; bright yellow, about 1 in (25mm) long; stalks nearly as long; calyx bell-shaped

Fruits: legumes cylindric; about 1.3 in (35 mm) long; glabrous

Twigs: slender, yellowish

Caragana maximowicziana Komar. - **maximowic peashrub**

A low-growing peashrub, the spiny, wide-spreading maximowic peashrub produces a good display of bright yellow pea-shaped flowers at PERC (Fairchild and Klett 1993). It is also drouth tolerant and adaptable to infertile sites.

Planning

Height: up to 6 ft (2 m)

Habit: rounded, spreading to twice its height; spiny; April-May flowers

Site: sun; pH and soil adaptable, drouth tolerant; zone 3

Use: barrier, hedge, naturalizing

Recognition

Leaves: rachis spiny; leaflets 4-6, up to 0.4 in (1 cm) long; much longer than wide (oblong-lanceolate); abruptly short tip (mucronate); stipules becoming spiny (spinescent)

Flowers: solitary, short-stalked; bright yellow, about 1 in (25 mm) long; calyx cylindric, glabrous, short-toothed

Cladrastis Raf. - **yellowwood**

Only one of the four yellowwood species is native to the U.S. A nice specimen of one of them, *C. platycarpa* (Maxim.) Mak., Japanese yellowwood, grows in the Arnold Arboretum.

Cladrastris kentukea (Dum.-Cours.) Rudd - **yellowwood**

This tree—also known as *C. lutea* (Michx.) K. Koch—is native to the southeastern U.S. where it grows on fertile, well-drained soils. Its fragrant flowers provide nectar for bees. The wood takes an excellent finish and is prized for gunstocks. It has also been used as a source of yellow dye.

The most northerly natural stands of yellowwood adjacent to the prairies occur in southern Missouri and Illinois, but it is hardy in the Minnesota Landscape Arboretum (Dirr 1990). Several planted yellowwood trees are growing in southeastern Nebraska communities, the largest of which is growing at the corner of 20th and Euclid in Lincoln. The performance of young trees indicates that yellowwood is adapted to the soils and climate of Fort Collins (Fairchild and Klett 1993). 'Rosea', a beautiful, pink-flowered cultivar, has been distributed by the Arnold Arboretum (Dirr 1990).

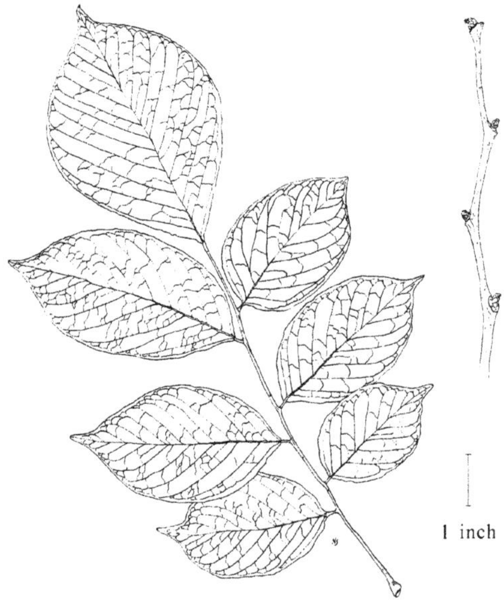

Cladrastris kentukea

Planning

Height: up to 40 ft (12 m); Nebraska champion, 1979, 53 ft (16.1 m), Lincoln; National champion, 1987, 72 ft (21.8 m), Ohio
Habit: tree; deciduous; crown rounded; early June bloom; autumn foliage yellow
Site: sun, part shade; pH and soil adaptable; zone 4
Use: specimen, shade
Caution: prune in summer to avoid "bleeding"

Recognition

Leaves: pinnately compound; leaflets odd numbered, often alternate on the rachis, about 3 in (8 cm) long; elliptic, margin smooth (entire); surfaces glabrous; petioles enlarged at base, enclosing buds
Flowers: clusters (panicles) elongated, hanging; calyx bell-shaped; petals 5, white, standard circular, wings and keels straight; stamens 10, distinct
Fruits: legumes about 3 in (8 cm) long, flattened, glabrous, papery; seeds 4-6, flat, brown
Twigs: zigzag, glabrous, red-brown
Buds: tiny, hidden by leaves or ringed by leaf scars
Bark: smooth, becoming somewhat roughened, silvery gray
Wood: yellow, heavy, hard

Colutea L. - **bladder-senna**

About 10 *Colutea* species are natives of an area that extends from southern Europe to the Himalayas.

Colutea arborescens L. - **bladder-senna**

Bladder-senna, a coarse, vigorous shrub, bears beautiful pea-shaped yellow flowers in late June and early July. Richley (1988) reports that bladder-senna grows in southwestern Nebraska. Excellent specimens are also growing at the Land Institute in Salina, Kansas (Sutton 1998).

Cultivars include 'Bullata', a dwarf form, and 'Crispa' with wavy leaf margins (Dirr 1990). *C. orientalis* Mill. is a shorter shrub with reddish brown flowers that has been crossed with *C. arborescens* to produce *C. ×media* Willd. which bears blue-green foliage and flowers that are usually a red-brown hue.

Planning

Height: up to 7 ft (2 m)
Habit: shrub; deciduous; leggy; foliage bright green; June-July blooms on new growth, yellow
Site: sun; tolerates infertile soil; zone 5
Use: naturalizing, screen

Recognition

Leaves: odd pinnately compound; leaflets elliptic, sometimes wider above middle, less than 1 in (25 mm) long, apex notched or with sharp abrupt point (mucronate); stipules persistent
Flowers: clusters of 6 to 8 (racemes); petals yellow, standard with red markings; calyx bell-shaped, 5-toothed; the upper stamen free
Fruits: legumes inflated and bladder-like, non-splitting, closed at apex; hairy becoming glabrous, greenish, often slightly reddish near base; about 0.5 in (12 mm) long
Twigs: hairy at first, decurrent lines extending from nodes; leaf scars elevated, broadly crescent shaped, 1-3 bundle traces
Buds: globose, often 2 at node (superposed); scales 2-4, appressed-hairy

Colutea arborescens

Cytisus L. - broom

The genus *Cytisus*—consisting of about 50 species native to central and southern Europe—is known for its showy flowers. Dirr (1990) recommends a close look at *C. decumbens* (Durand) Spach, prostrate broom, as a groundcover or in a sunny rock garden. *C. hirsutus* L., the hardiest of the brooms, is a dwarf that has been successfully grown in Minnesota. The hybrid *C. ×praecox* Bean, Warminster broom, produces an abundance of yellow flowers.

Recognition

Habit: shrubs, rarely trees; deciduous or evergreen
Leaves: compound; leaflets usually 3, sometimes minute
Flowers: axillary or in terminal clusters or heads; petals yellow, white, or rarely purple
Fruits: legumes flat, splitting open (dehiscent); 2- to many-seeded

Cytisus scoparius (L.) Lk. - Scotch broom

Scotch broom is a plant for poor soils, especially sands, where it can be a stabilizer. Though considered a plant for sunny locations, Richley (1988) states that it grows much better in light shade in southwestern Nebraska.

Blooming Scotch broom in a mass planting is a spectular sight. Although normally yellow-flowered, cultivars have been selected with flowers of rosy pink to purple.

Planning

Height: up to 7 ft (2 m)
Habit: shrub; deciduous; rounded; wider than tall; May-June bloom on old wood
Site: sun, part shade; pH adaptable, prefers sandy, infertile soils, drouth tolerant; zone 5
Use: erosion control, naturalizing, wildlife habitat
Caution: temperatures below 0 degrees F (18 degrees C) may injure

Recognition

Leaves: leaflets in 3's, much longer than wide, sometimes wider above middle (obovate), about 0.5 in (12 mm) long; on upper stem reduced to 1 leaflet; sparingly hairy

Flowers: axillary, usually 1-2; calyx bell-shaped, few hairs, slightly longer than wide; petals usually bright yellow; style longer than keel, spirally incurved

Fruits: legumes less than 2 in (5 cm) long, margins long-hairy (villous)

Twigs: slender, grooved or angled, green, hairy when young

Genista L. - genista

About 100 species comprise *Genista*, natives of Europe, North Africa and western Asia. Though quite common in European gardens, the genistas have not been widely planted in America. They prefer dry, sandy soils and lots of sun. Those of more northerly origin should be at home on many prairie sites and certainly deserve trial plantings.

In addition to the one described here, Dirr (1990) mentions two species, *G. pilosa* L., a procumbant shrub, and *G. hispanica* L. with outstanding foliage color.

Recognition

Habit: shrubs; deciduous or half-evergreen, sometimes nearly leafless

Leaves: alternate, rarely opposite, simple or sometimes 3-foliate; margins smooth (entire); stipules small or wanting

Flowers: clusters (racemes) or heads terminal, rarely axillary; petals yellow, rarely white

Fruits: legumes linear to egg-shaped; 1- to many-seeded

Twigs: spiny or unarmed

Genista tinctoria L. - woadwaxen

Woadwaxen, a native of southern Europe and western Asia, is quite variable with numerous varieties from which several cultivars have been selected. Given these characteristics and its preference for sunny sites on sandy soils, varieties may be selected that are adapted to much of the prairies. Richley (1988) reports fair performance for woadwaxen growing in southwestern Nebraska.

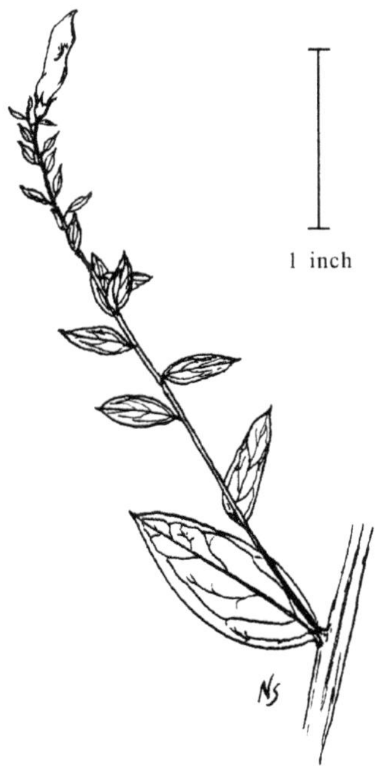

Genista tinctoria

Planning

Height: up to 3 ft (1 m)

Habit: shrub; deciduous; vertical, spiky stems; branching sparse, not spiny; June-September blooms on new growth

Site: sun; pH adaptable, prefers infertile, sandy soils; drouth tolerant; zone 4

Use: mass, naturalizing, border, wildlife habitat

Recognition

Leaves: simple, alternate, elliptic-oblong, 0.5-1.0 in (12-25 mm) long; surfaces nearly glabrous, margin hairy (ciliate); stipules persistent

Flowers: clusters (racemes) many-flowered, sometimes branched, up to 3 in (8 cm) long; petals yellow

Fruits: legumes compressed, narrow oblong, usually glabrous; seeds 6-10

Twigs: slender, green, stripe-grooved; leaf scars raised, minute, bundle trace 1

Buds: solitary, egg-shaped

Indigofera L. - indigofera

Indigofera—a genus of 300 species, many of whch are native to subtropical and tropical regions—is widely planted for ornament even though many of its species may die back to the ground in cold climates. An excellent specimen of *I. amblyantha* Craib var. *purdomii* Rehd., pink indigo, survives in Colby, Kansas (Sutton 1998).

Recognition

Habit: shrubs, subshrubs, herbs; deciduous
Leaves: alternate, odd-pinnately compound, rarely 3-foliate; leaflets usually opposite, margins smooth (entire); stipules awl-shaped adhering at base to petioles
Flowers: clusters (racemes) axillary on current growth; standard orbicular to oblong, wings oblong, keel with swelling on either side, pink to purple; calyx teeth 5, nearly equal
Fruits: legumes nearly round to linear-oblong, usually round or angular in cross section, splitting open (dehiscent)

Indigofera kirilowii Maxim. - Kirilow indigo

Kirilow indigo is native to northern China, Korea and Japan. It will grow on calcareous soils (Dirr 1990). Kirilow indigo may be treated as an herbaceous perennial in cold climates because new shoots will bear flowers in June and July even if the plant is killed to the ground the previous winter.

Planning

Height: up to 3 ft (1 m)
Habit: shrub; stems erect; dense; suckering; June-July blooms on new growth
Site: sun; pH and soil adaptable; zone 5
Use: groundcover, erosion control, naturalizing

Recognition

Leaves: leaflets 5-13, nearly round to elliptic, about 1 in (25 mm) long; tip rounded with abrupt prickle (mucronate), base rounded; surfaces with sparse appressed hairs
Flowers: clusters (racemes) dense, erect, up to 4.5 in (12 cm) long; flower stalks (peduncles) longer than petioles; petals rose-colored; calyx teeth unequal, slightly hairy
Fruits: legumes about 1.5 in (4 cm) long
Twigs: slightly hairy, becoming glabrous

Laburnum Med. - golden chaintree

Three *Laburnum* species, all natives of southern Europe and western Asia, are shrubs or small trees. *L. anagyroides* Med., goldenchain laburnum, and *L. alpinum* Bercht. & Prsl., Scotch laburnum (parents of the hybrid described below) are noted for their lavish display of yellow flowers in late spring.

The laburnums should be protected from wind and bright sun, but they can grow on high pH soils. Dirr (1990) considers them to be short-lived. These plants, especially the seeds, are poisonous if ingested.

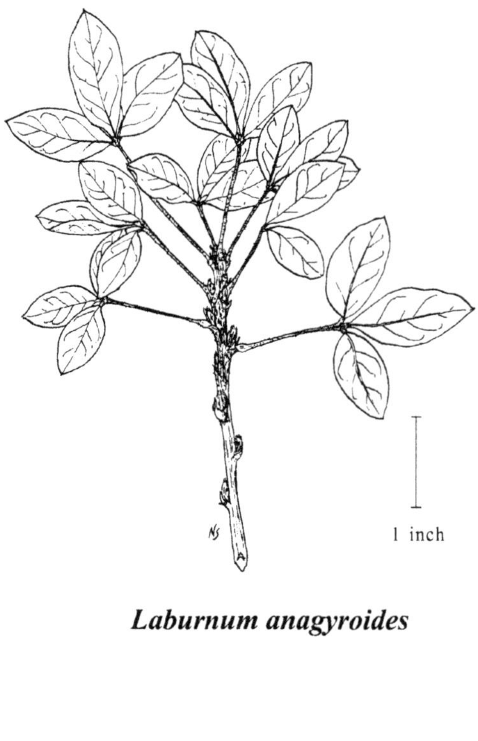

Laburnum anagyroides

Recognition

Habit: trees, shrubs; deciduous
Leaves: alternate; 3-foliate; petioled; stipules wanting; leaflets subsessile
Flowers: clusters (racemes) terminal, pendulous; calyx 2-lipped; petals yellow, all free; stamens 10, joined (connate)
Fruits: legumes linear, compressed, tardily splitting; seeds several
Buds: round to egg-shaped (globose-ovoid); scales 2-3

Laburnum ×watereri Dipp. - Waterer laburnum

This hardy hybrid blooms more profusely than its parents do. Waterer laburnum can be used as a border or grouped to provide an effective color display.

Waterer laburnum grows at 5900 feet (1800 m) in southwestern Colorado and in Santa Fe, New Mexico (Sutton 1998). Although a young plant is growing at PERC in Fort Collins, the cultivar 'Vossii' failed to overwinter (Fairchild and Klett 1993).

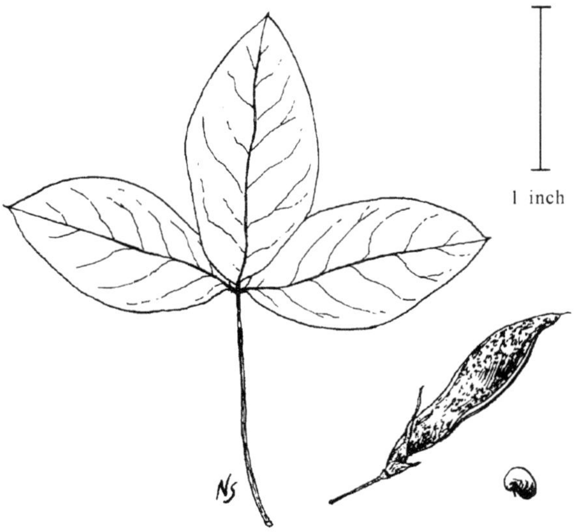

Laburnum ×watereri

Planning

Height: up to 15 ft (4.5 m)
Habit: shrub, small tree; upright, top round; May blooms yellow
Site: light shade; pH adaptable, prefers moist, well-drained, fertile site; zone 5
Use: border, specimen, mass
Caution: protect from excessive wind; poisonous, avoid ingestion

Recognition

Leaves: leaflets sparingly hairy, becoming glabrous, bright green with bluish tinge
Flowers: clusters (racemes) pendulous, 6-10 in (15-25 cm) long; petals bright yellow
Fruits: legumes about 2 in (5 cm) long with narrow wing; slightly hairy; seeds few, rarely developed
Twigs: green
Buds: egg-shaped; scales 2-4, silvery-hairy

Maackia Rupr. & Maxim. - maackia

The genus *Maackia,* a native of eastern Asia, is represented by six species of small deciduous trees. The maackias are closely related to the yellowwoods except that their leaflets are opposite on the rachis and their solitary leaf buds are not hidden by the petiole base. Dirr (1990) states that *M. chinensis* Takeda, a small shrubby tree with leaves that are silvery gray pubescent at emergence, is less hardy than *M. amurensis* (described below).

Maackia amurensis Rupr. - Amur maackia

Dirr (1990) states that Amur maackia is surviving at the Minnesota Landscape Arboretum. Two young trees are growing in Prairie Pines Arboretum near Lincoln without any problems.

Planning

Height: up to 20 ft (6 m)
Habit: tree; deciduous; top round; summer blooms white
Site: sun; pH and soil adaptable; zone 4
Use: street tree, border
Caution: wind protection may be necessary

Recognition

Leaves: alternate, odd-pinnately compound; leaflets 7-11, usually opposite on the rachis, long egg-shaped to elliptic, about 3 in (8 cm) long; margins smooth (entire); surfaces glabrous
Flowers: clusters (racemes), dense, upright; petals white; calyx bell-shaped (campanulate), 5-toothed; stamens distinct except near base
Fruits: legumes compressed, 2-3 in (5-8 cm) long, splitting; seeds 1-5
Twigs: dark brown to nearly black, usually glabrous
Buds: egg-shaped; scales several, overlapping (imbricate), dark brown

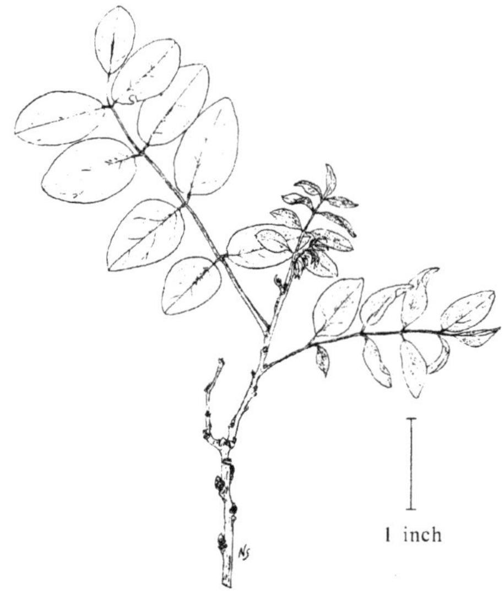

Maackia amurensis

Robinia L. - locust

Ten to 20 *Robinia* species are native to the U.S. and Mexico, four of which reach tree size. All are characterized by sharp spines along the twigs; pea-shaped, showy, often fragrant, flowers; and a tendency to produce thickets by root suckers. Nitrogen-fixing bacteria that are associated with locust root nodules add nitrates to the soil.

R. neomexicana Gray—a prickly shrub or small tree native to southern Colorado, Arizona, New Mexico, and small areas of adjoining Mexico, Nevada, and Texas—is a valuable erosion control plant adapted to the extreme southern prairies.

R. fertilis Ashe and *R. hispida* L., very similar low shrubs beset with rigid bristles, are used to control erosion on steep slopes. The latter species thrives in the Sandhills Region of Nebraska. Cultivars include: 'Arnot', a good sandy soil stabilizer; 'Monument', small and compact with rose-colored flowers; 'Flowering Globe', with dark pink flowers; and 'Casque Rouge', with purplish red flowers.

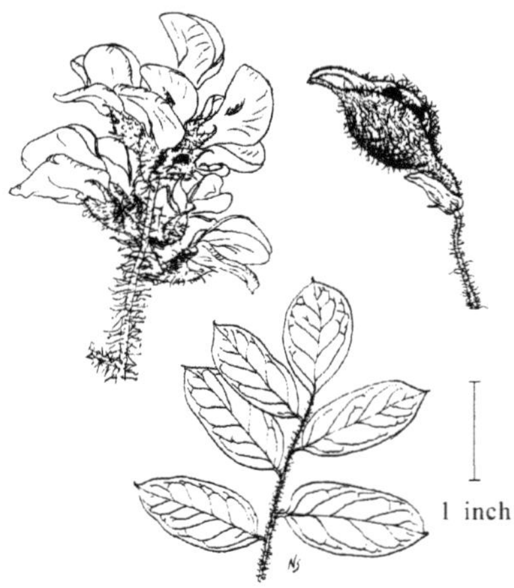

Robinia fertilis

Recognition

Habit: trees, shrubs; deciduous; spiny
Leaves: alternate, odd-pinnately compound; leaflets 7-19, opposite on the rachis
Flowers: clusters (racemes) pendulous, elongated, branched; calyx bell-shaped, 5-toothed, slightly 2-lipped; petals white to pink or pale purple; stamens 10 (9 united, 1 free)
Fruits: legumes oblong, flat, 2-valved, glabrous or hairy; seeds several
Twigs: often zigzag, usually with stipular spines; leaf scars raised, triangular or 3-lobed, decurrent ridges from the base; bundle scars 3, indistinct
Buds: terminal absent; lateral downy, naked (no scales), 3-4 superposed, hidden by petiole base

Robinia pseudoacacia L. - black locust

Black locust, originally native to the Appalachians from Georgia to Pennsylvania and in the Ozarks of Missouri and Arkansas, has been so widely planted that it is now naturalized throughout much of the U.S.

The name "locust" was originally attached to the carob, *Ceratonia siliqua*, during biblical times because John the Baptist was said to have survived in the wilderness by eating locusts (insects). Later it was divulged to be the nutritious pods of carob, sometimes called St. John's Bread. Locust became commonly used as the name for several other legume family species.

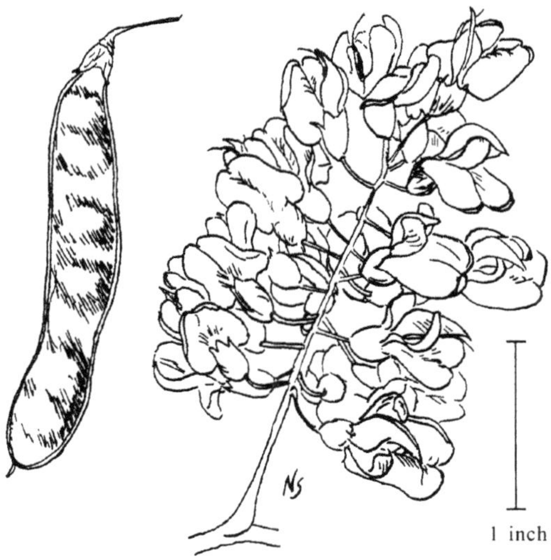

Robinia pseudoacacia

In the U.S., black locust is often severely damaged by a borer insect (*Megacyllene robiniae*) that can thoroughly "honeycomb" both sapwood and heartwood, weakening the tree and causing easy breakage. Borer tunnels also provide easy access for a deadly heart rot fungus (*Fomes rimosus*). Fast-growing trees on fertile soils often escape or outgrow serious borer injury. Magnificent black locust trees can be found throughout the prairies on streambanks and other fertile, well-watered sites.

Black locust was one of the first tree species sent by early colonists to Europe where it is now used in landscape and erosion control plantings and is treasured for its lumber, firewood, foliage for forage, and flowers for bee "pasture." It has since been planted around the world, and, fortunately, the wood boring insect has not followed.

Black locust wood is among the hardest and strongest woods obtained from our native trees, and ranks high, along with osage orange, in decay resistance. The looming scarcity of fossil fuels may stimulate research to control locust borer damage. Black locust wood could replace chemically treated woods where decay resistance is important. Energy savings would be significant; an added benefit would be increased supplies of delicious honey from its fragrant flowers.

Several black locust cultivars that vary in form and flower character—some of which are thornless (var. *inermis* DC)—have been selected. 'Idaho', supposedly a black locust-*R. hispida* cross, is better adapted than the species is to extremes of heat and cold (Dirr 1990). *R.* ×*ambigua* Poir. 'Idahoensis'—a cross between black locust and *R. viscosa* (Rehder 1940)—is being evaluated at PERC in Fort Collins (Fairchild and Klett 1993).

Planning

Height: up to 65 ft (20 m); Nebraska champion, 1980, 63 ft (19.1 m), Fort Atkinson State Historical Park; National champion, 1974, 96 ft (29.1 m), New York

Habit: tree; crown oval; produces root sprouts; early June bloom white, fragrant; pods persist into winter

Site: sun, part shade; pH and soil adaptable, prefers fertile, moist, loamy soil; tolerates drouth and infertile soil; zone 4

Use: shade, erosion control, windbreak, lumber, firewood, fence posts

Pests: locust borer (*Megacyllene robiniae*)

Recognition

Leaves: leaflets egg-shaped to elliptic, 1-2 in (25-50 mm) long; rounded at apex, abruptly terminated with sharp tip (mucronate); upper surface glabrous, lower slightly hairy; petioles about 0.2 in (5 mm) long, base enlarged; stipules tiny, becoming spines

Flowers: clusters (racemes) of 10 to 30; petals white; calyx lobes 5, lower sharp-pointed, 2 upper fused, notched at apex

Fruits: legumes 2-4 in (5-10 cm) long, straight, ridged on both sides, abruptly pointed; slightly hairy or glabrous; splitting on both sides; seeds 3-10, mottled brown

Twigs: glabrous, paired stipular spines usually present

Buds: nearly hidden by rusty hairs

Bark: rough, deeply furrowed

Wood: greenish yellow to brown, very hard, heavy and strong; heartwood resists decay

Robinia viscosa Vent. - clammy locust

Clammy locust, a native of the ridge tops at lower altitudes in the Appalachians from Pennsylvania to Alabama, is planted in landscapes for its rose-colored flowers and soil stabilizing capability. It suckers profusely.

Planning

Height: up to 30 ft (9 m)
Habit: tree; thicket forming
Site: sun; pH and soil adaptable; zone 4
Use: naturalizing, erosion control

Recognition

Leaves: leaflets egg-shaped, 13-25, apex variable, base broadly tapered, 1-1.5 in (25-40 mm) long; usually hairy beneath; petioles sticky-glandular

Flowers: clusters (racemes) up to 3 in (8 cm) long, stalks hairy; petals pink with yellow blotch on standard; calyx dark red, hairy

Fruits: legumes about 3 in (8 cm) long; hairs erect, glandular, sticky

Twigs: hairs stiff, glandular, sticky (viscid)

Sophora L. - sophora

This genus of many species, native to warm regions, is composed of evergreen and deciduous trees, shrubs and perennial herbs. Two species, *S. secundiflora* (Ortega) Lag., coralbean, and *S. affinis* Torr. & Gray, Texas sophora, are natives of Texas and Mexico. They are probably not hardy very far north of their native range.

Sophora japonica L. - Japanese pagodatree

Japanese pagodatree, a native of China and Korea, is a tree suitable for infertile soils. Though it suffers twig dieback during most winters in the Lincoln vicinity, it produces a beautiful flower display in mid-summer, a time when few trees and shrubs are in bloom.

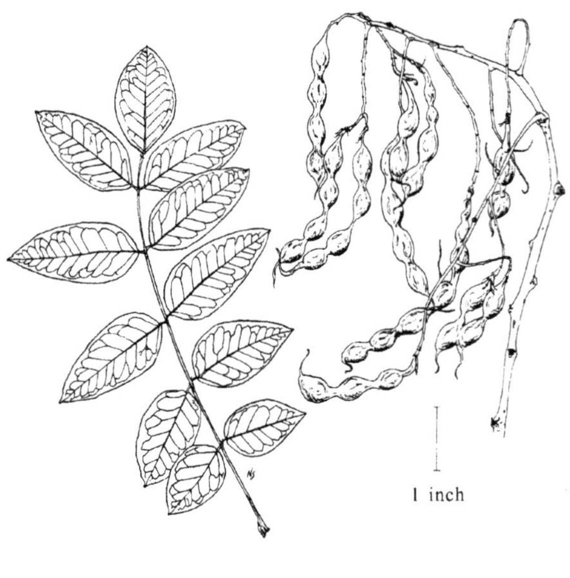

Sophora japonica

Cultivars that exhibit variations in form and flower color include: 'Pendula', a weeping form; 'Princeton Upright', with upright branches; 'Regent', straight and fast growing; and 'Variegata', with white speckled leaves (Dirr 1990).

Planning

Height: up to 50 ft (15 m); Nebraska champion, 1981, 64 ft (19.4 m), Arbor Lodge Historical Park, Nebraska City

Habit: tree; crown broad, rounded; summer bloom creamy white, fragrant; pods persist into winter

Site: sun, part shade; pH and soil adaptable, tolerates heat, drouth, pollution, infertile soils; zone 5

Use: shade, specimen

Recognition

Leaves: alternate, odd-pinnately compound; leaflets opposite on rachis; egg-shaped, 1-2 in (25-50 mm) long, pointed; shiny above with waxy bloom (glaucous), appressed hairs below

Flowers: clusters (panicles) loose, terminal, up to 1 ft (30 cm) long and wide; calyx bell-shaped, 5-toothed, green; petals creamy-white; stamens 10, distinct, fused only at base

Fruits: legumes round in cross-section (terete), 2-4 in (5-10 cm) long; glabrous, green, becoming yellow and then brown; seeds 1-6

Twigs: green up to 4-5 years of age, glabrous; nodes prominent; pith greenish

Bark: gray-brown, ridged

Wisteria Nutt. - wisteria

Nine *Wisteria* species, all deciduous twining shrubs, are natives of North America and East Asia. *W. sinensis* (Sims) Sweet, Chinese wisteria, is similar to *W. floribunda* except that its flowers are blue-violet, its leaflets number 7 to 13, and its pods are densely hairy. *W. venusta* Rehd. & Wils., silky wisteria, is distinguished by hairy leaflets, white flowers in 4 to 6 inch (10-15 cm) clusters, and compressed pods covered with dense, firm hairs (velutinous). These two species are recommended for zone 6 (Dirr 1990).

W. frutescens (L.) Poir., American wisteria, a native of the southeastern U.S. from Virginia to Texas, is adapted to zone 6. Its pale lilac-colored flowers, each with a single yellow spot, are borne on a compact 5-inch (13 cm) cluster (raceme) on new growth from June to August and its pods are glabrous. 'Nivea' is white-flowered. Another native species, *W. macrostachya* Nutt., is adapted to the Madison, Wisconsin, area (Hasselkuss 1993). Its pods are glabrous and flower clusters are about 10 inches (25 cm) long and glandular.

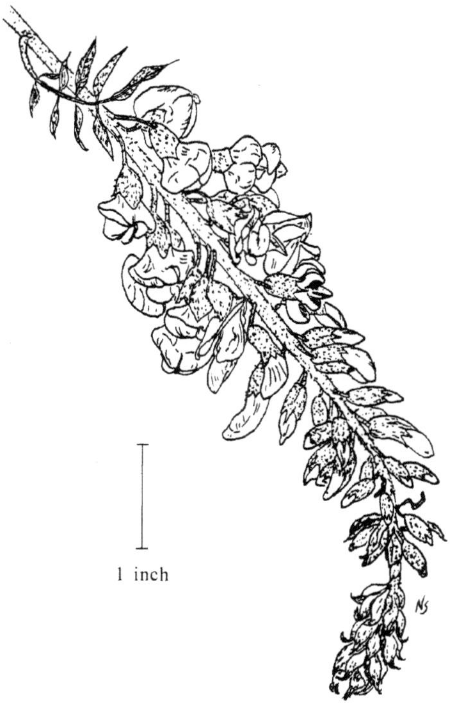

Wisteria floribunda

Wisteria floribunda (Willd.) DC. - Japanese wisteria

This vigorous twiner from Japan should be supported by a strong structure. Protective measures will be required on sites inhabited by cottontail rabbits and deer. The many Japanese wisteria cultivars vary in flower color, fragrance and cluster size.

Planning

Height: up to 20 ft (6 m)
Habit: twining shrub; deciduous; late May bloom on old wood; pods persist into winter
Site: sun, part shade; pH and soil adaptable; zone 5
Use: trellis
Pests: resistant
Caution: support with sturdy structure; protect from excessive wind

Recognition

Leaves: alternate, odd-pinnately compound; leaflets long egg-shaped (ovate-oblong), 13-19, about 2 in (5 cm) long; apex pointed (acuminate), base rounded; surfaces appressed hairy becoming glabrous; petioles flattened at base
Flowers: clusters (racemes) slender, pendulous; flowers opening first at base and then toward apex; petals violet (cultivars white, blue, various), standard reflexed, wings sickle-shaped (falcate) with ear-like appendages at base (auriculate), keel obtuse with short abrupt tip (mucronate); calyx bell-shaped (campanulate), hairy, 5-toothed, 2 upper teeth short and broad; upper stamen free
Fruits: flattened, usually slightly constricted between seeds, 4-6 in (10-15 cm) long, narrowed toward base, velvety; splitting tardily
Twigs: somewhat angled, 2 spine-like projections at leaf scars
Buds: appressed, oblong, tip pointed; one scale usually surrounds most of bud, reddish brown, hairy

Order Proteales

Two families have been placed in Proteales (Cronquist 1981). One family, Proteaceae Juss., consists of evergreen shrubs and trees native to tropical and subtropical regions. The other family, Elaeagnaceae, is described below.

Elaeagnaceae Juss. - oleaster family

The oleaster family, represented by three genera and about 50 species, is native to temperate and subtropical regions of much of the Northern Hemisphere and Australia.

Recognition

Habit: trees, shrubs; deciduous or evergreen
Leaves: alternate or opposite; simple; margins smooth (entire), surfaces covered with silvery or brown peltate scales or stellate hairs; stipules none
Flowers: solitary or clustered in leaf axils; bisexual, sometimes unisexual (dioecious or polygamo-dioecious); perianth regular, mostly 4-parted, petals none; calyx lobes 4, rarely 2 or 6, often petal-like; stamens as many or twice as many as sepals
Fruits: dry or fleshy; seeds one

Elaeagnus L. - oleaster

An estimated 45 *Elaeagnus* species are native to southern Europe and Asia, and one species is native to North America in a region that extends from Alaska to the northern Great Plains. The oleasters are planted for ornament, windbreaks and wildlife habitat. Some, if not all, are soil builders that harbor nitrogen-fixing microorganisms on their roots.

Recognition

Habit: trees, shrubs; deciduous or evergreen; flowers in spring or fall
Leaves: alternate; dotted with silvery or brown scales
Flowers: single or clustered, axillary; bisexual or unisexual (polygamous); sepals 4, stamens 4
Fruits: drupe-like, fleshy to leathery; seeds 1, hard
Twigs: often spiny, dotted with silvery or brown scales

Elaeagnus angustifolia L. - Russianolive

Russianolive—a native of Central Asia and southern Europe—was introduced into the U.S. in colonial times, but made its greatest impact on the Great Plains as a component of many of the shelterbelts planted by the Prairie States Forestry Project during the 1930s. Russianolive is planted for windbreaks, wildlife habitat, and landscape beautification. It is well adapted to much of the western prairie which has allowed it to become naturalized along many of our rivers and streams, notably the Platte River, thanks to birds and animals that partake of its fruit.

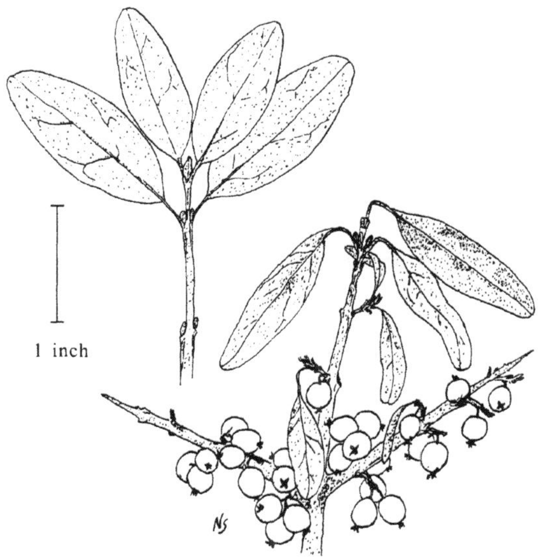

Elaeagnus angustifolia

Russianolive tolerates drouth, extreme cold, polluted air and highly saline soils. Transplant survival is high. It tolerates shearing and forms an impenetrable hedge (Sutton 1998). Peterson (1976) reports a disease caused by *Botryodiplodia theobromae* is responsible for the widespread mortality of Russianolive in the Great Plains, especially in the eastern part of the region. Consequently, many of these trees become straggly at an early age because of dead twigs and branches.

Russianolive maintains its greatest vigor in the dry climate of the western Great Plains where soil moisture is adequate. A well-adapted specimen at PERC in Fort Collins is not affected by insects or disease (Fairchild and Klett 1993).

Three Russianolive selections planted in 1962 at Horning State Farm near Plattsmouth demonstrated some of the genetic variability of the species. One selection was very poorly adapted and was removed after a few years. The second selection from trees that were widely planted in Great Plains plantings was adequate, but not as outstanding as the third, known as Colorado 143.

Colorado 143, informally named 'Kinghorn' after the Colorado nursery from which it originated, was very vigorous and healthy and grew larger than the others. Its large fruits that ripen with a red blush is decorative and persists into winter. Russianolive trees growing in the Denver area that bear similar fruits are also exceptionally healthy.

'King Red', a Russianolive cultivar released by the Natural Resources Conservation Service Plant Materials Center of Los Lunas, New Mexico, deserves trial plantings throughout the Great Plains. It is very similar to 'Kinghorn' in appearance; they may be of similar origin.

Planning

Height: up to 30 ft (9 m); Nebraska champion, 1980, 55 ft (16.8 m), circumference 112 in (2.8 m), Omaha; National champion, 1991, 52 ft (15.8 m), circumference 149 in (3.8 m), South Dakota

Habit: shrubby tree; deciduous; crown rounded, occasional thorns; May-June yellow flowers fragrant, not showy; fruits persist into winter

Site: sun; pH and soil adaptable, tolerates drouth and high alkalinity; zone 3

Use: windbreak, wildlife habitat, hedge, screen

Elaeagnaceae *Elaeagnus*

Recognition

Leaves: oblong to linear lance-shaped, widest near the base; margin often undulating, tip variable, base wedge-shaped (cuneate); dull green above, flecked with star-shaped (stellate) hairs, silvery white below, densely covered with scale-like stellate hairs; petioles silvery white
Flowers: clusters of 2 to 3; usually bisexual; yellow inside
Fruits: densely covered with silvery scales, yellow; developing a red blush on some trees; flesh mealy
Twigs: silvery scaly first season, becoming dark brown, often bearing thorns 1 in (25 mm) long
Buds: egg-shaped, gray, scaly
Bark: gray-brown, shallow fissured, often exfoliating into long shreddy strips
Wood: light weight, heartwood dark brown

Elaeagnus commutata Bernh. - silverberry

Silverberry, a native of Alaska, western Canada and the northern Great Plains, grows naturally on exposed hillsides, often on sandy soils. This hardy shrub with showy foliage and fragrant flowers might be more widely planted except for its "take over" attitude. One plant soon becomes a colony because of its stoloniferous habit. However, where space is adequate and the site is difficult for plant extablishment and growth, silverberry is worth a try, especially in the northern Great Plains.

Silverberry shrubs planted in Prairie Pines Arboretum near Lincoln formed vigorous colonies and prospered for about five years before suffering a sudden decline in vigor and eventual death during a year of above average rainfall (Bagley 1998).

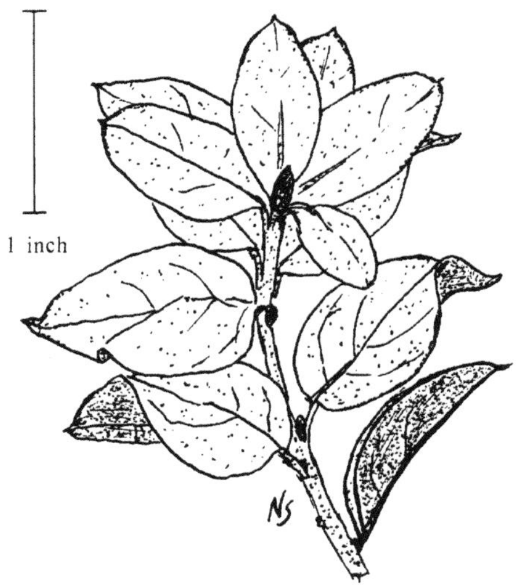

Elaeagnus commutata

Planning

Height: up to 15 ft (4.5 m)
Habit: shrub; deciduous; stoloniferous; fragrant flowers on new growth
Site: sun; pH and soil adaptable, tolerates infertile soil and drouth; zone 3
Use: wildlife habitat, erosion control, naturalizing

Recognition

Leaves: oval; margin entire and undulating, often folded lengthwise; surfaces covered with silvery scales, less dense above; scattering reddish brown scales below, more evident on midrib and petiole
Flowers: 1 to 3 in leaf axils; bisexual
Fruits: covered with silvery scales, scattering reddish scales; flesh dry
Twigs: densely covered with reddish brown scales

Elaeagnus umbellata Thunb. - autumnolive

Its nitrogen-fixing capability enables the autumnolive, an attractive shrub with contrasting green and silvery foliage, to make good growth on infertile soils. Nearby plants, including trees, also benefit from the extra nitrogen. Its flowers attract bees, and birds eat the small, red, tasty fruits and scatter seeds far and wide, resulting in unwanted autumnolives. Occasional plants bear fruits that persist into winter (Bagley 1998).

Autumnolives at PERC in Fort Collins become large, dense shrubs that are winter hardy and free of insects and diseases, but 'Cardinal' failed to establish after being planted there (Fairchild and Klett 1993). 'Redwing', released by the Natural Resources Conservation Service's Rose Lake Plant Materials Center of East Lansing, Michigan, is more winter hardy than 'Cardinal' and 'Elsberry', another release by the NRCS from the Plant Materials Center at Elsberry, Missouri. 'Redwing' fruits mature earlier and are larger and sweeter than 'Cardinal' fruits.

A shrub of similar habit, *E. multiflora* Thunb., cherry elaeagnus—a native of China and Japan—is not hardy at the Minnesota Landscape Arboretum (Dirr 1990).

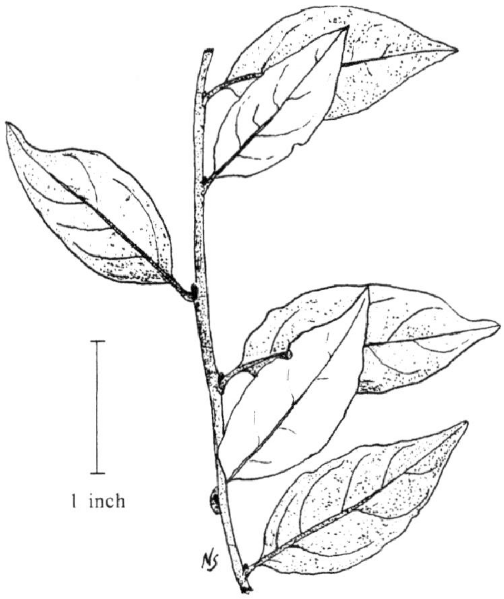

Elaeagnus umbellata

Planning

Height: up to 15 ft (4.5 m)

Habit: shrub; deciduous; spreading becoming arching; suckering, self-seeding; fragrant May flowers; edible fruits variably persistent

Site: sun, part shade; pH and soil adaptable, tolerates infertile soils and drouth; zone 3

Use: grouping, wildlife habitat, windbreak, border, hedge

Recognition

Leaves: elliptic to oblong, wider near base; tip obtuse to short-pointed, base rounded to wide wedge-shaped, margin crinkled; upper surface bright green with obscure silvery scales becoming less visible with age; dense silvery scales below with reddish brown flecks; petioles about 0.2 in (6 mm) long

Flowers: bisexual, yellowish white; calyx tube much longer than limb, scaly outside

Fruits: nearly round, juicy, about 0.25 in (7 mm) across, covered with silvery and brown scales, becoming red

Twigs: silvery, speckled with brown, often spiny

Shepherdia Nutt. - buffaloberry

The three species of this genus are native to North America. Although these species are similar to those in other genera of the oleaster family in their fruit and foliage characteristics, buffaloberries can easily be distinguished by their opposite leaves. They tolerate soils of low fertility because of their ability to utilize atmospheric nitrogen.

Recognition

Habit: shrubs, trees; deciduous or evergreen

Leaves: opposite, stipules none

Flowers: unisexual, male and female on separate plants (dioecious); axillary in clusters (spikes) or female sometimes solitary; calyx 4-parted, stamens 8; style slender with 1-sided stigma

Fruits: drupe-like, juicy; seeds 1

Buds: 1 or 2 pairs of outer scales

Shepherdia argentea Nutt. - silver buffaloberry

Silver buffaloberry grows along streams throughout much of Nebraska, north and northeast into southern Canada and Minnesota and westward to Arizona and New Mexico (Stephens 1973). The ability to grow at varied sites probably contributes to a species with great genetic diversity, a premise which we should take advantage of in our search for plants adapted to various prairie sites.

Several silver buffaloberries of unknown origin planted at Prairie Pines Arboretum in 1970 died soon after becoming densely shaded by faster growing trees. A later planting in a sunny, more exposed nearby location has fared much better (Bagley 1998).

Shepherdia argentea

This well-armed plant protects copious juicy, bright red fruits with sharp daggers masquerading as leafy spur twigs. The drupe-like fruits are tart treats for those with patience who can endure sharp pricks during the harvest.

Planning

Height: up to 15 ft (4.5 m); National champion, 1975, 22 ft (6.7 m), Oregon
Habit: shrub, small tree; deciduous; crown round-topped; thicket-forming; foliage silvery
Site: sun; pH and soil adaptable, drouth tolerant; zone 3
Use: wildlife habitat, windbreak, naturalizing

Recognition

Leaves: elliptic to oblong; tip rounded, base tapered; surfaces and petioles covered with silvery scales, often with reddish centers, more dense beneath; petioles about 0.2 in (5 mm) long
Flowers: male usually in clusters of 4, sepals spreading or reflexed; female often solitary with erect sepals
Fruits: nearly round, usually bright red, sometimes yellow; glossy, smooth; about 0.3 in (8 mm) across; calyx and style often persist; edible
Twigs: densely covered with silvery white scales, becoming red-brown and smooth; many sharp-pointed; leaf scars crescent-shaped, bundle scars 1
Buds: vegetative oblong, scales 2, valvate; flower nearly round on pedicels, male clustered, covered with silvery white scales
Bark: gray-brown in loose, long, thin strips

Shepherdia canadensis (L.) Nutt. - russet buffaloberry

Russet buffaloberry, a shrubby relative of the larger silver buffaloberry, is a native of more northerly climes and higher altitudes than the silver. It is native to the Black Hills, North Dakota, much of the Rocky Mountains and the West Coast, Alaska, and across Canada into the northeastern U.S.

A predominance of brown or red-brown scales on the twigs with lesser amounts sprinkled on the silvery underside of the leaves distinguishes it from other plants. 'Xanthocarpa' has yellow fruits.

Planning

Height: up to 7 ft (2 m)
Habit: shrub; rounded; foliage silver-green
Site: sun; pH and soil adaptable, tolerates drouth and infertile soils; zone 3
Use: wildlife habitat, erosion control, naturalizing

Recognition

Leaves: elliptic to egg-shaped, rounded at tip and base; yellow-green above with scattered star-shaped (stellate) hairs, silvery below speckled with red-brown, lobed, peltate scales; petioles covered likewise, about 0.25 in (7 mm) long
Flowers: dioecious or monoecious; sepals red scaly outside, yellowish inside, reflexed and spreading
Fruits: nearly round, ranging from bright red to yellow, glossy, smooth, with a few red-brown peltate scales, about 0.3 in (8 mm) across; juicy, insipid
Twigs: green beneath dense overlapping red scales, not spiny; leaf scars half-round, bundle scar 1
Buds: vegetative egg-shaped covered with silvery and red scales; flower buds nearly round, clustered on short spur branches

Hippophae L. - seabuckthorn

The two *Hippophae* species are natives of Europe and Asia. *H. salicifolia* D. Don is very similar to the species described below. Its leaves are dull green above with a brown midrib and its fruits are yellow. It is seldom planted in the U.S.

Hippophae rhamnoides L. - seabuckthorn

This seabuckthorn species, like other members of the oleaster family, is silvery gray and adapted to infertile soils because of its nitrogen-fixing capability. Orange fruits persist through winter, providing color but apparently not food for birds. Dirr (1990) suggests that the fruits may be too acidic for their taste. Resistance to salt spray makes seabuckthorn a candidate for roadside plantings.

Seabuckthorn shrubs growing in Colby, Kansas, and Brookings, South Dakota, present an interesting, rugged form with healthy foliage (Sutton 1998). It is winter hardy at PERC in Fort Collins and pests are not a problem (Fairchild and Klett 1993). Richley (1988) rates its performance good in southwestern Nebraska. Seabuckthorn is certainly a plant deserving greater use in the Great Plains.

Planning

Height: up to 25 ft (7.5 m)
Habit: shrub; deciduous; loose and open; often spiny; flowers appear before leaves; foliage silver-green; orange fruit persists through winter
Site: sun; pH and soil adaptable, tolerates drouth and infertile soils including sands; zone 4
Use: naturalizing, erosion control, border, windbreak

Hippophae rhamnoides

Recognition

Leaves: alternate; narrow, 1-3 in (25-60 mm) long, about 0.3 in (8 mm) wide; silvery white scales above are more prominent below; petioles scale covered, 0.1 in (2 mm) long

Flowers: clusters (racemes) axillary; unisexual (dioecious); yellowish; calyx tube with 2 minute lobes

Fruits: nearly round, about 0.3 in (8 mm) across, bright orange

Twigs: covered with silvery brownish scales and star-shaped hairs, often with terminal and axillary spines

Buds: terminal absent; lateral with 2 loose, thin, brownish scales

Order Myrtales

The order Myrtales is comprised of 12 families and more than 9,000 species (Cronquist 1981).

Recognition

Habit: woody, herbaceous; deciduous or evergreen

Leaves: simple, most commonly entire, mostly opposite; stipules rarely well-developed or none

Flowers: clustered, bisexual or seldom unisexual, usually regular; sepals often appearing as lobes on the hypanthium; petals distinct, alternating with sepals, or sometimes wanting; stamens numerous; ovary compound

Thymelaeaceae Juss. - mezereum family

The mezereums, a family of 50 genera and 500 species, is widely distributed throughout the world.

Recognition

Habit: shrubs, trees, rarely herbs; deciduous or evergreen

Leaves: alternate, rarely opposite; margin entire

Flowers: 4-5 parted; petals small, sometimes scale-like or wanting

Fruits: nuts, drupes, berries, rarely dry and splitting

Daphne L. - daphne

Only two of the 50 species of this Eurasian genus and a hybrid are likely candidates for prairie landscape plantings. However, Hoag (1965) adds to the list *D. tangutica*, a low evergreen shrub, stating that it survives and blooms in the northern plains with no protection other than snow cover. Fragrant, showy flowers are rewards for providing the extra care required for its survival.

Recognition

Habit: shrubs; deciduous or evergreen

Leaves: usually alternate

Flowers: clustered, usually with bracts; bisexual; petals lacking; calyx tube (hypanthium) bell-shaped, corolla-like, usually hairy outside, 4- or rarely 5-lobed; stamens 8 or 10 in 2 rows, not exserted, style short

Fruits: drupes, leathery or fleshy; seeds 1

Daphne cneorum L. - rose daphne

Rose daphne forms a dense, evergreen mat that requires protection from wind and winter sun. If snow is unreliable, pine boughs are good substitutes for protection of rose daphne. Dirr (1990) recommends pruning annually immediately after blooming.

D. ×burkwoodii Burkwood (*Daphne caucasica* Pall. × *D. cneorum*) is a low shrub that in the northern plains remains green until late winter when the leaves drop (Hoag 1965). Cultivars include 'Arthur Burkwood', 'Somerset' and 'Carol Mackie'. Herman (1965) reports the latter cultivar "performing well in early trials" in South Dakota.

Planning

Height: up to 1 ft (30 cm)
Habit: shrub; prostrate; evergreen; spring bloom, often blooms again in late summer, fragrant
Site: sun, part shade in winter; neutral pH; moist; zone 5
Use: groundcover, specimen

Recognition

Leaves: alternate, much longer than wide (oblanceolate), about 0.8 in (2 cm) long; obtuse with a sharp, abrupt tip; lustrous dark green above, slight waxy bloom beneath
Flowers: clusters of 6 to 8 terminal; bright rosy pink, each about 0.4 in (1 cm) across
Fruits: drupes yellowish brown
Twigs: slender, hairy when young

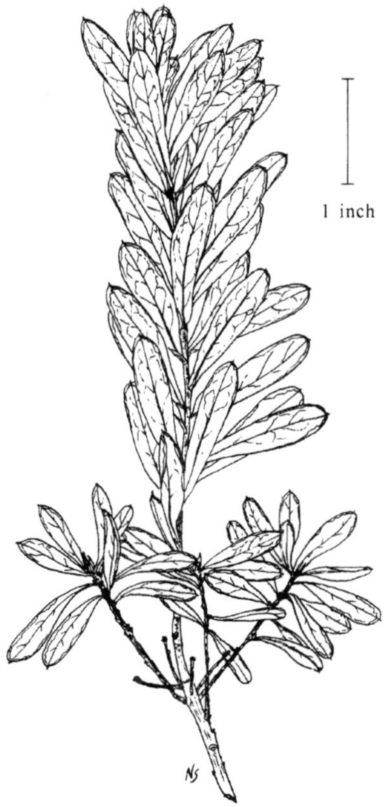

Daphne ×burkwoodii

Daphne mezereum L. - February daphne

February daphne, native to Europe and Siberia, is hardy in the northern Great Plains where heavy, high-lime soils have contributed most to midsummer loss (Hoag 1965). Dirr (1990) states that it "is a most enchanting daphne because of the deliciously fragrant flowers."

Variety *alba* West. bears dull white flowers and yellow fruits. 'Paul's White' and 'Bowles White' bear pure white flowers and 'Autumnalis' blooms in autumn and is usually not fruitful.

Planning

Height: up to 3 ft (1 m)
Habit: shrub; deciduous; erect; early spring blooms appear ahead of leaves, fragrant
Site: sun, part shade; neutral pH; moist; zone 4
Use: specimen, border

Recognition

Leaves: alternate, oblong, up to 3 in (8 cm) long; obtuse to pointed
Flowers: clusters axillary, stalkless, usually 3-flowered; lilac to rosy purple, each about 0.5 in (12 mm) across
Fruits: drupes scarlet
Twigs: stout, glabrous

Dirca L. - leatherwood

Dirca, a native of North America, is represented by two species, one of which is described below. Neither is commonly sold in the nursery trade.

Dirca palustris L. - leatherwood

Leatherwood is a native of southeastern Canada and the eastern U.S. south to Florida and west to Missouri. Native Americans and early settlers used its tough, leathery bark to make baskets, bow strings, fish lines and other necessities.

Planning

Height: up to 6 ft (2 m)
Habit: shrub; deciduous; many-branched, rounded; spring bloom appears before leaves on old wood; autumn foliage yellow
Site: shade, part shade; neutral pH; moist soil; zone 4
Use: naturalizing

Recognition

Leaves: alternate, elliptic to wider above middle, up to 3 in (8 cm) long; tip obtuse, base wedge-shaped: glabrous above, hairy beneath when young; petioles about 0.1 in (2 mm) long
Flowers: clusters axillary, 2-3 flowered; bisexual; petals lacking, sepals 4-5, yellow; stamens 8, exserted, style slender, as long as stamens
Fruits: drupes oval, about 0.3 in (8 mm) long; pale green or reddish; seeds shiny, brown
Twigs: brownish to olive; lenticels white, conspicuous
Buds: terminal lacking; lateral conical, enclosed by petiole base before leaf-fall; scales 4, silky, indistinct
Bark: tough, leathery

Onagraceae Juss. - evening primrose family

Most species of the evening primrose family are herbs found in temperate and subtropical regions. Shrubs and small trees of *Fuschia* are widely planted ornamentals in subtropical and warm temperate regions.

Oenothera L. - evening primrose

One *Oenothera* species, a native of the prairies and the Intermountain Region of the U.S., is considered to be a subshrub (suffrutescent) by some authorities (Stephens 1973).

Oenothera serrulata Nutt. - evening primrose

Evening primrose, herbaceous or with a woody base, is found in open prairies and brushy areas on a variety of soils (often dry ones).

Planning

Height: up to 16 in (40 cm)
Habit: subshrub; June blooms
Site: sun; pH and soil adaptable; drouth tolerant; zone 3
Use: grouping, naturalizing

Recognition

Leaves: alternate, simple, much longer than wide (lanceolate to linear), up to 2 in (5 cm) long; tip pointed, base tapered, shallow-toothed; surfaces variable from densely white hairy (strigose) to hairless above and slightly hairy beneath; sessile
Flowers: solitary, axillary near the end of the stem; bisexual; petals 4, rounded, about 0.5 in (15 mm) across, bright yellow; calyx 4-lobed, strongly ribbed, reflexed; stamens 8
Fruits: dry (capsules), cylindric, 4-angled, about 0.8 in (2 cm) long
Twigs: gray-brown, flexible; pith brown; die back to below the lowest flower
Buds: naked
Bark: red-brown, exfoliating into thin papery strips

Order Cornales

Four families and about 150 species are placed in Cornales (Cronquist 1981).

Recognition

Habit: woody plants, rarely herbs
Leaves: opposite, seldom alternate; simple; seldom with stipules
Flowers: mostly bisexual, often 4-parted, regular; petals distinct, sometimes wanting; ovary inferior to other flower parts (epigynous)
Fruits: drupes or berries

Nyssaceae Dum. - tupelo family

Five of the eight species in the three genera of Nyssaceae are natives of China and southeastern Asia. The other three species, members of genus *Nyssa*, are natives of North America.

Nyssa L. - tupelo

Tupelos, noted for their brilliant scarlet leaves of autumn, are in demand for ornamental plantings. The forest tupelos are harvested for their lumber, veneer and woodpulp.

Nyssa sylvatica Marsh. - black tupelo

Black tupelo grows on uplands and along stream bottoms from Maine to Oklahoma and east Texas, including Indiana and southern Missouri. It is also found in southern and central Mexico. Variety *biflora*, swamp tupelo, grows in coves and swamps from Maryland to Florida, west to the Mississippi River and north to Tennessee.

Provenance research should be rewarding because of the likelihood of considerable genetic variability in the black tupelo population. Research would identify which trees are most likely to survive throughout the prairie region. Young black tupelo trees are flourishing in southestern Nebraska, many of which provide brilliant foliage in mid-October.

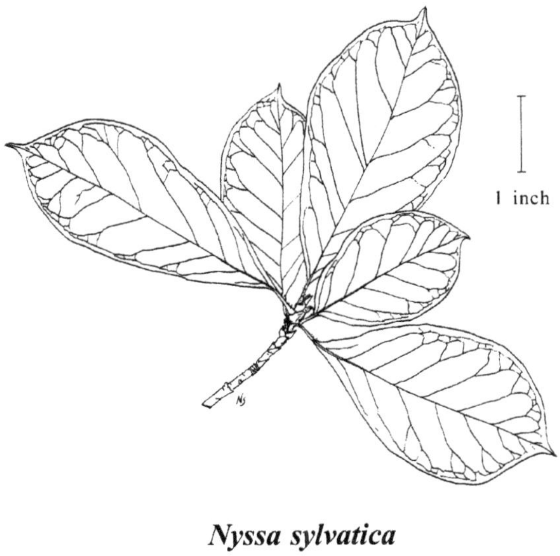

Nyssa sylvatica

Planning

Height: up to 40 ft (12 m); National champion, 1991, 132 ft (40.4 m), West Virginia
Habit: tree; deciduous; pyramidal becoming rounded to irregular; autumn foliage colorful
Site: sun, part shade; prefers moist soils, neutral to acidic; tolerates low fertility and drouth; zone 4
Use: shade, naturalizing, specimen, bee pasture, wildlife habitat, lumber

Recognition

Leaves: alternate; simple; often crowded near twig tips; elliptic, up to 6 in (15 cm) long; tip pointed to rounded, base broadly tapered to rounded; margin entire or rarely coarsely toothed; dark green and shiny above, paler and somewhat hairy below; petioles somewhat reddish, up to 1 in (25 mm) long; stipules wanting

Flowers: minute, greenish white; male or female and bisexual on same tree (polygamo-dioecious); male in dense, many-flowered heads on slender hairy stalks, calyx 5-toothed, petals 5, stamens 5-12, exserted; female and bisexual in clusters of 1 to several on long slender stalks; calyx tube bell-shaped, stamens 5-10, often functioning

Fruits: drupes oblong to egg-shaped, 0.5 in (12 mm) long; pendent on long slender stalks; dark blue; flesh thin, bitter; stone hard, indistinctly ribbed

Twigs: greenish brown to red-brown; lenticels small; leaf scars half-round to heart-shaped, bundle scars 3; pith diaphrammed; spur shoots common

Buds: terminal egg-shaped to globose; visible scales 3-5, yellow-brown to red-brown, glabrous or slightly hairy at tip; lateral smaller than terminal, divergent from twig

Bark: dark gray to black, ridged to blocky

Wood: moderately hard, tough, not easily split

Cornaceae Dum. - **dogwood family**

Most species of Cornaceae—11 genera and about 100 species—occupy temperate regions of the Northern Hemisphere. A few species are native to subtropical and tropical regions of Africa and South America. *Cornus* is the only genus relevant to the prairie region.

Cornus L. - **dogwood**

Except for one Peruvian species, the forty-odd species of *Cornus* are natives of the temperate regions of the Northern Hemisphere. All but two are trees or shrubs.

Dogwoods are widely grown for their attractive flowers, fruits and colorful stems. The fruits are valuable wildlife food. Unfortunately, their stems and foliage are relished by mammals, notably cottontail rabbits and deer—which sometimes creates a problem in the establishment of habitat plantations. On the other hand, the dogwood's dense, fibrous root systems insure high initial survival of transplants.

The low spreading bunchberry (*C. canadensis* L.) is a beautiful deciduous groundcover adapted to acidic soils. It grows among cool, moist woodlands in northern latitudes and high altitudes.

Dogwoods are easily recognized in summer by their leaves and in winter by their leaf scars and buds. However, differences among many of the species are difficult to detect.

Recognition

Habit: trees, shrubs, rarely herbs; deciduous, rarely evergreen

Leaves: opposite, rarely alternate; margins entire; surfaces usually covered with 2-armed appressed (malpighiaceous) hairs; veins curve along margins toward tip (arcuate); stipules wanting

Flowers: clusters (cymes) or heads, terminal; some surrounded by involucral bracts; bisexual, 4-parted; calyx teeth minute, petals egg-shaped to oblong

Fruits: drupes

Twigs: green, purple, yellow or red; pith white or brown; leaf scars narrow v- to crescent-shaped, raised, often meeting the opposite leaf scar to encircle the twig; bundle scars 3, often indistinct

Buds: terminal present, lateral often enclosed by a persistent petiole base; leaf buds narrow, solitary, often covered by 2 or 3 valvate or overlapping scales

Grouping Key for *Cornus*
(Rehder 1940)

Flowers white; clusters (cymes or panicles) without involucre; stones subglobose: **section Thelycrania**

Flowers yellow; clusters in dense umbels with yellowish involucre not exceeding the flowers, deciduous at anthesis; stones ellipsoidal to oblong: **section Macrocarpium**

Flowers greenish yellow; clusters in dense umbels with persistent involucre of white or pink bracts; fruits in dense clusters, individually distinct; stones ellipsoidal to oblong: **section Benthamidia**

Flowers greenish yellow; clusters in dense umbels with persistent involucre of white or pink bracts; fruits connate (globose fleshy head); stones ellipsoidal to oblong: **section Benthamia**

Section Thelycrania

Cornus alba L. - tatarian dogwood

Tatarian dogwood—red-stemmed native of Siberia and Manchuria—and two similar species native to North America, *C. baileyi* and *C. sericea*, are often sold as "red-stemmed dogwood" without reference to the species. Differences in health and growth rate among the plants sold under the "red stem" label have been observed. Provenance research plantations that include representatives of these three species would identify which plants could be planted on specific sites for best performance.

Cultivars with variegated leaves include 'Spaethii', 'Elegantissima' and 'Gouchaultii'. The stems of 'Sibirica' are bright coral red and fruits are bluish.

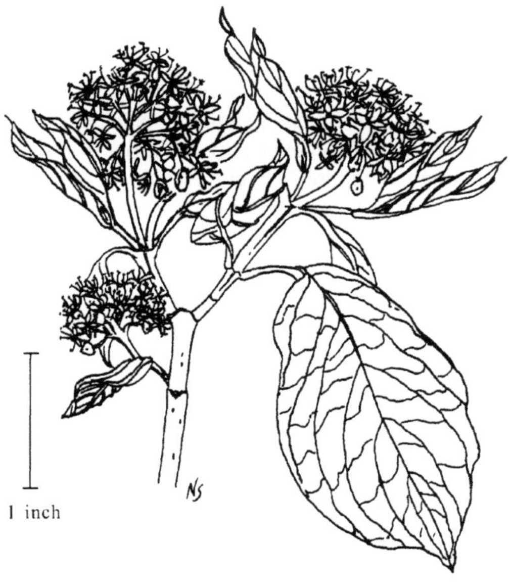

Cornus alba

Planning

Height: up to 10 ft (3 m)

Habit: shrub; deciduous; stems erect, arching with age; rich red bark on young stems is effective in winter; flowers May-June

Site: sun, part shade; pH and soil adaptable; zone 3

Use: border, grouping, wildlife habitat, windbreak

Pests: leaf diseases may cause unsightly late summer foliage

Recognition

Leaves: opposite; egg-shaped to elliptic; tip pointed, base usually rounded; 5-6 pairs lateral veins; wrinkled (rugose) above, whitened below (glaucous); petioles up to 1 in (25 mm) long
Flowers: clusters (cymes) flat-topped, variably hairy; flowers yellowish white
Fruits: drupes round, about 0.3 in (7 mm) across, white or slightly bluish; stone somewhat pointed at ends
Twigs: red, hairy, becoming glabrous; lenticels long oval, vertical; pith white
Buds: terminal swollen near base; scales not overlapping (valvate), hairy; flower buds larger than leaf buds

Cornus alternifolia L. F. - pagoda dogwood

Pagoda dogwood, with its distinctive branch pattern and leaf arrangement, is native to southeastern Canada and the eastern U.S. as far west as Minnesota and south into Arkansas and Georgia. Its tier-like branch arrangement is responsible for its common name.

A healthy specimen planted over 30 years ago by Professor Ronald Taven grows north of the former biochemistry building on the East Campus of UNL. Pagoda dogwoods at PERC in Fort Collins are well adapted with no insect or disease problems (Fairchild and Klett 1993). Hoag (1965) states that this dogwood deserves more plantings in the northern plains.

'Argentea' is a shrubby form with variegated leaves (Dirr 1990).

Giant dogwood (*C. controversa* Hemsl.), a native of Japan and China, is similar to pagoda dogwood in its leaf and branching habit, but is capable of growing much larger than the pagoda. Dirr (1990) states that giant dogwood may be more urban tolerant than other dogwoods, and therefore has been used as a street tree.

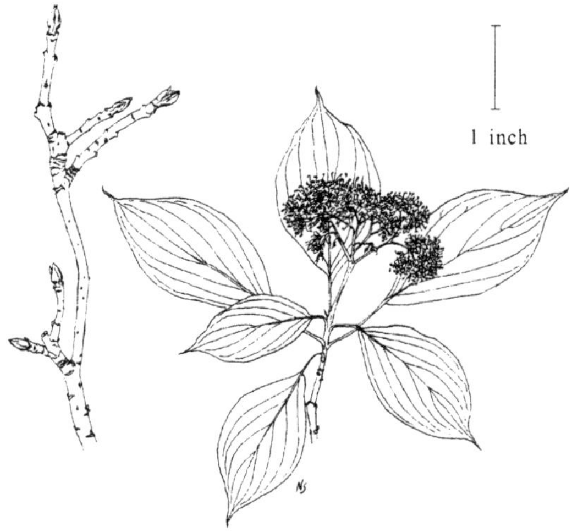

Cornus alternifolia

Planning

Height: up to 25 ft (7.5 m)
Habit: tree, shrub; deciduous; tiered horizontal branching; late May blooms; red fruit stalks; autumn foliage reddish purple
Site: sun, part shade; pH and soil adaptable, prefers cool, moist, fertile sites; zone 3
Use: border, naturalizing, wildlife habitat, specimen

Recognition

Leaves: alternate to nearly opposite, clustered near branch ends; elliptic, margin somewhat wavy; tip pointed, base wedge-shaped; dark green and glabrous above, appressed sparse hairs below, slightly bloomy (glaucescent); petioles grooved above, up to 2 in (5 cm) long
Flowers: clusters flat-topped; flowers fragrant, creamy white
Fruits: nearly round, about 0.2 in (5 mm) across; green to red, blue-black at maturity; flesh bitter; stalks red
Twigs: glabrous, shiny, green becoming brown, odoriferous when bruised; pith white
Buds: slightly hairy, shiny, slender; scales 2-3, overlapping
Bark: thin, smooth; becoming furrowed and ridged

Cornus amomum Mill. - silky dogwood

Silky dogwood—a native of southeastern Nebraska and eastern Kansas, as well as an area that extends southeast to Georgia, and north to New England—grows along streams and swamp edges. The plants native to the central prairies have been identified as ssp. *obliqua* (Raf.) Wils.J.S. (Stephens 1973), but some authorities consider them to be *C. obliqua* Raf. (Rehder 1940). Silky dogwood is very similar to roughleaf dogwood (*C. drummondii*), differing only slightly in fruit and leaf characteristics.

Silky dogwood suffers mild chlorosis and variable winter dieback at the PERC site in Fort Collins (Fairchild and Klett). 'Indigo', propagated by seed, was released by the Rose Lake Plant Materials Center of the Natural Resources Conservation Service. It is recommended for conservation plantings in most states east of the Mississippi River, but trial plantings in the western prairies are warranted.

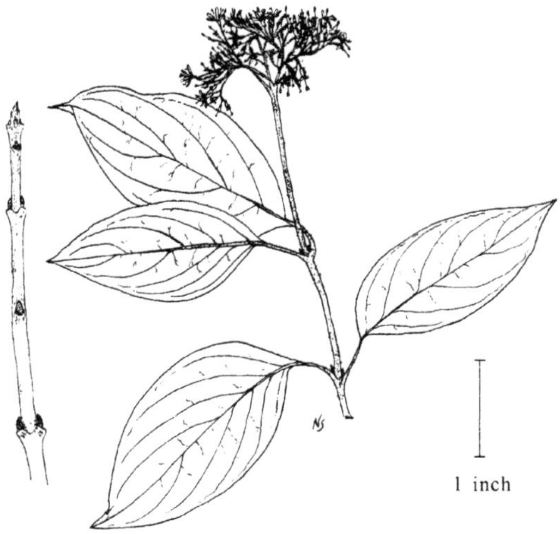

Cornus amomum

Planning

Height: up to 8 ft (2.5 m)
Habit: shrub; deciduous; many stems, round-topped becoming straggly; June flowers; late summer fruits; autumn foliage reddish purple
Site: partial shade, tolerates full sun; prefers moist soils, pH 5.0 to 7.0; zone 4
Use: border, grouping, naturalizing, wildlife habitat, windbreak

Recognition

Leaves: opposite; narrow egg-shaped to elliptic; tip pointed, base rounded; dark green with appressed, long hairs above; pale green below, hairs more prevalent on veins; petioles grooved or flattened above, appressed hairy, up to 0.8 in (2 cm) long
Flowers: clusters (cymes) many-flowered, round-topped, appressed hairs on all parts; yellowish white, not fragrant; style green, widened just below stigma
Fruits: nearly round, about 0.4 in (1 cm) in diameter; blue, juicy; calyx lobes persistent; seeds irregularly round or egg-shaped, rough, with 5-7 irregular ridges
Twigs: appressed hairy, becoming glabrous; reddish green; pith light brown
Buds: hairy, scales not overlapping (valvate); terminal flower bud egg-shaped, lateral vegetative buds flattened, appressed to twig

Cornus drummondii C.A. Mey. - roughleaf dogwood

Roughleaf dogwood—a native of southeastern South Dakota, eastern Nebraska and Kansas, south through eastern Texas and Mississippi, and east to Ohio—grows on dry uplands and moist lowlands. Some authorities consider it to be a variety of *C. asperifolia* Michx.

Roughleaf dogwood becomes efficiently established by bird activity if it is not heavily browsed by grazing animals. Though seldom planted, it should not be overlooked as part of erosion control and wildlife habitat plantings.

Planning

Height: up to 15 ft (4.5 m); National champion, 1987, 43 ft (13 m), Mississippi

Habit: shrub, small tree; deciduous; multistemmed, thicket-forming; May-June flowers; autumn fruits

Site: sun, partial shade; pH and soil adaptable, drouth tolerant; zone 4

Use: naturalizing, wildlife habitat, windbreak

Recognition

Leaves: opposite; egg-shaped to elliptic; tip pointed, base rounded; 3-5 pairs veins; stiff short hairs roughen upper surface; paler beneath with longer hairs, somewhat woolly in appearance; petioles with slight groove near blade above, hairy, up to 0.4 in (1 cm) long

Flowers: clusters (cymes) flat-topped; calyx white, petals creamy white; styles white, often gradually enlarged upward

Fruits: round, white with appressed hairs, milky juice; seeds round, smooth, with longitudinal vein lines

Twigs: reddish brown, hairy, pith white; second year becoming gray, glabrous; pith brown

Buds: reddish to dark brown, hairy; scales not overlapping (valvate), terminal swollen below middle, lateral egg-shaped

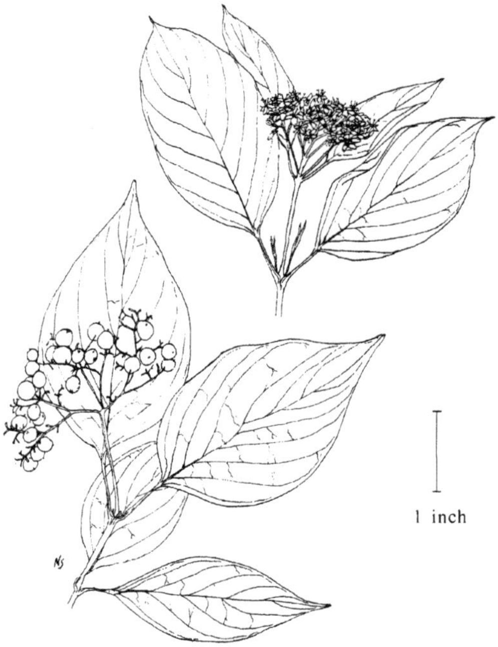

Cornus drummondii

Cornus foemina Mill. ssp. *racemosa* (Lam.) J.S. Wils. - gray dogwood

Gray dogwood grows naturally in northern Nebraska and the Dakotas and eastward to Maine in the area north of the Ohio River. Subspecies *foemina*, found in the southeastern U.S. from Georgia to east Texas, is characterized by its dark red twigs and blue fruits (Stephens 1973). Some authorities consider gray dogwood to be *C. racemosa* Lam. (Rehder 1940; Dirr 1990).

Gray dogwood at PERC in Fort Collins suffers moderate to severe chlorosis and variable winter dieback (Fairchild and Klett 1993).

The similarity in general appearance—color, form and thicket-forming ability—among the three dogwood species, gray, roughleaf (*C. drummondii*) and redosier (*C. sericea*), is a source of confusion.

The best winter identification character for gray dogwood is its persistent bright red peduncles and pedicels (Sutton 1998).

Planning

Height: up to 10 ft (3 m); National champion, 1975, 38 ft (11.5 m), Michigan

Habit: shrub, tree; deciduous; multistemmed, thicket-forming; early June flowers; contrasting winter colors, red to gray

Site: sun, shade; prefers moist, well-drained soil, pH less than 8.0; zone 3

Use: border, grouping, naturalizing, wildlife habitat, windbreak

Recognition

Leaves: opposite; elliptic; tip pointed, base wedge-shaped to rounded, 3-4 pairs veins; appressed hairs, sparse above, many beneath; petioles hairy, slightly flattened or grooved above, about 0.4 in (1 cm) long

Flowers: clusters pyramidal (paniculate) with central axis; flowers creamy white; style abruptly enlarged to the stigma

Fruits: nearly round, white, minutely appressed hairs; style often persistent; seed egg-shaped, sometimes unevenly so, with 8 longitudinal veins

Twigs: tan to gray-red; hairy near tip, quickly becoming glabrous; pith white, becoming brown by second year

Buds: egg-shaped, pointed, red-brown; scales not overlapping (valvate); a small bud often on each side of terminal

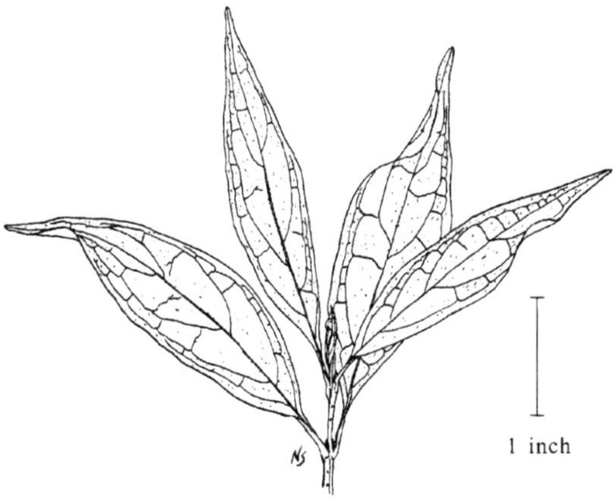

Cornus foemina

Cornus paucinerva Hance - few-nerved dogwood

Few-nerved dogwood, a native of central China, is seldom planted. It is worthy of trial plantings to determine its value on prairie sites, even though it has no outstanding visible attributes to set it apart from other dogwoods. A young plant grows slowly in full shade at Prairie Pines Arboretum near Lincoln.

Planning

Height: up to 10 ft (3 m)
Habit: shrub; deciduous
Site: experimental; zone 5
Use: experimental

Recognition

Leaves: opposite; broad oval to elliptic; tip abruptly pointed, base rounded to wedge-shaped; 2-4 pairs veins; dark green above, light green below; sparse, long, appressed white hairs on both surfaces, dense on prominent veins below with dark red hairs interspersed, at least early in the season; petioles grooved above, hairy, about 0.2 in (5 mm) long

Flowers: clusters (cymes) dense, long-stalked; calyx teeth longer than disc; style thickened at apex

Fruits: black, round, about 0.2 in (5 mm) across

Twigs: greenish red, short appressed-hairy; pith white

Buds: narrow; scales appressed-hairy, dark brown

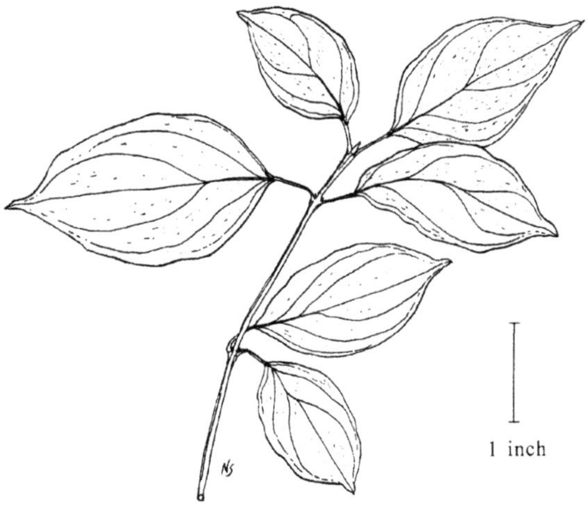

Cornus paucinerva

Cornaceae *Cornus*

Cornus sanguinea L. - red dogwood

A 40-year-old leggy specimen of red dogwood grows in Maxwell Arboretum at UNL. Dirr (1940) rates it "coarse in all seasons," somewhat of a "sore thumb" plant that is difficult to blend into the landscape.

Cornus sanguinea

Planning

Height: up to 15 ft (4.5 m)
Habit: shrub; deciduous; round-topped, multistemmed, thicket-forming; early June flowers; autumn foliage red, greenish purple in most of the prairie
Site: sun, partial shade; pH and soil adaptable; zone 5
Use: border, grouping, naturalizing, wildlife habitat

Recognition

Leaves: opposite; broad elliptic to egg-shaped; tip pointed, base wedge-shaped, 3-5 pairs veins; dark green above, very sparse long hairs on both surfaces; petioles hairy, slightly flattened or grooved above
Flowers: clusters (cymes) flat-topped, hairy; flowers dull white; style gradually thickened upward
Fruits: nearly round, about 0.2 in (5 mm) across; purplish black
Twigs: dark red above, green on lower side, appressed hairs sparse to almost woolly
Buds: terminal egg-shaped, lateral slender; scales hairy, not overlapping (valvate)

Cornus sericea L. ssp. *stolonifera* Fosberg (*C. stolonifera* Michx.) - redosier dogwood

Redosier dogwood grows on streambanks and moist uplands across Canada from New Foundland to Alaska and across the northern U.S. from New England to the West Coast. *C. baileyi* Coult. & Evans, a native of the northeastern U.S. and Ontario, is considered by some authorities to be a form of *C. sericea.* These two shrubs along with *C. alba*, tatarian dogwood, are often sold as "red-stemmed" dogwood without regard to species.

Dirr (1983) states that redosier dogwood is stoloniferous, spreading by underground stems. However, Sutton (1998) avers that it forms a cluster by rooting along the stems of the lower branches where they touch the soil.

Redosier dogwood cultivars include: yellow-stemmed 'Flaviramea'; green-stemmed 'Nitida'; and low-growing and compact forms 'Kelseyi' and 'Isanti'. The latter is winter hardy and suffers no chlorosis or pest problems at PERC in Fort Collins (Fairchild and Klett 1993). Variety *coloradensis* 'Cheyenne' is red-stemmed and shorter than the species.

The widespread natural range of redosier dogwood indicates a great potential for fruitful provenance research. Undoubtedly, plants of many habits can be selected that are adapted to most prairie sites.

Planning

Height: up to 15 ft (4.5 m); National champion, 1983, 26 ft (7.9 m), Idaho
Habit: shrub, small tree; deciduous; usually multistemmed, rounded; early June blooms, August-September fruits; autumn foliage reddish purple, winter stem color variable
Site: sun, partial shade; pH and soil adaptable including poor drainage; drouth tolerant; zone 3
Use: border, grouping, naturalizing, wildlife habitat, erosion control, windbreak
Pests: resistant

Recognition

Leaves: opposite; egg-shaped to elliptic; tip pointed, base wedge-shaped to rounded, 5-6 pairs lateral veins; upper surface with appressed hairs, not rough; lower surface whitened, variably hairy with occasional red hairs; petioles flattend above, about 0.6 in (15 mm) long, hairy, often with red hairs

Flowers: clusters (cymes) flat-topped to slightly rounded, stalks hairy; flowers dull white; calyx lobes with white and red hairs; style white, hairy, enlarged below stigma

Fruits: round, about 0.3 in (7 mm) across; dull white with few appressed hairs, style persistent; seed obliquely egg-shaped, ends pointed or base rounded, 8 white longitudinal veins

Twigs: green, yellow or red, hairy when young; lenticels elliptic, light-colored; pith white; lenticels horizontal

Buds: terminal leaf short-stalked, swollen near base, long-pointed, white and red appressed hairs; scales not overlapping (valvate); terminal flower buds large; lateral leaf buds appressed to twigs

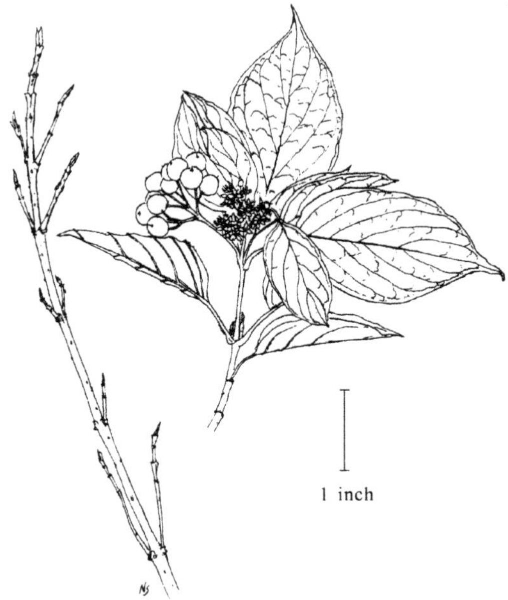

Cornus sericea

Section Macrocarpium

Cornus mas L. - Corneliancherry dogwood

The yellow flowers of Corneliancherry dogwood—a native of southern Europe and western Asia—are among the earliest blooms to appear in Nebraska's springs. At Prairie Pines Arboretum near Lincoln the late March blooms are invariably subjected to temperatures in the teens (F.) within a month after anthesis. Yet fruit crops have never been reduced by these freezing temperatures. However, fruits were scarce for the first three years that the young trees bloomed because most of the flowers were sterile (without pistils). The fourth year most flowers were fertile, the bees played their part, and the fruit crop was bountiful. The 20-year-old Corneliancherry dogwoods at Prairie Pines remain free of insect, disease and cold injuries (Bagley 1998).

The cherry-like fruits of this dogwood are edible to man and beast, though a bit tart for some palates. Corneliancherry dogwoods have provided fresh, dried and pickled fruit for thousands of years in their native lands of Europe and Asia. There the fruit is used in soft drinks and alcoholic beverages.

Although the autumn foliage of Corneliancherry dogwood is colorful in its native lands, the best we have seen in the Lincoln area is a tinge of reddish purple on the leaf tips in November. Our sharp freezes usually come too early for the color change to be completed before leaf fall. Despite this shortcoming, Corneliancherry dogwood is a hardy plant that can add variety to landscapes on much of the prairie. Richley (1988) believes that it will grow well in southwestern Nebraska.

'Redstone', a recent release by the Natural Resources Conservation Service Plant Materials Center at Elsberry, Missouri, is recommended for windbreak and wildlife plantings. Other cultivars include: 'Alba', a white-fruited form; 'Golden Glory', a hardy upright form with abundant flowers; 'Nana', a low-growing, small-leaved form; and 'Variegata' with leaf margins irregularly colored creamy white (Dirr 1990).

'Golden Glory' suffered minor winter dieback and bud kill during severe winters at the PERC site in Fort Collins, but it was adapted to the site's alkaline clay soil (Fairchild and Klett 1993).

The hard-coated seeds require a long period of stratification for germination. Hundreds of volunteer seedlings appeared under Corneliancherry dogwoods at Prairie Pines the second year after wood chip mulch was applied.

Planning

Height: up to 20 ft (6 m)

Habit: shrub with many upright stems, small tree by pruning; deciduous; crown oval to rounded; late March flowers, August fruits; foliage remains green into late fall on northern prairies

Site: sun, part shade; pH and soil adaptable, drouth tolerant; zone 5

Use: border, hedge, specimen, wildlife habitat, windbreak, fruit

Pests: none

Recognition

Leaves: opposite; egg-shaped to elliptic; tip pointed, base wedge-shaped, 3-5 pairs veins; dark green above with sparse, long, appressed hairs; light green beneath with long appressed hairs, more dense on prominent veins; petioles up to 0.4 in (1 cm) long, hairy

Flowers: clusters (umbels) short-stalked, dense; 4, yellowish, broad-ovate bracts at base of cluster, deciduous at anthesis or shortly thereafter; petals yellow

Cornus mas

Fruits: oblong, about 0.6 in (15 mm) long, bright cherry red

Twigs: angled, green with reddish streaks; short, appressed hairs, becoming gray and glabrous

Buds: flower stalked, axillary or terminal on short twigs, nearly round, about 0.2 in (5 mm) in diameter, enclosed by downy boat-shaped bracts; leaf buds downy, inconspicuous

Bark: exfoliating, flaky

Section Benthamidia

Cornus florida L. - flowering dogwood

Flowering dogwood, the aristocrat of flowering trees, is native to the eastern hardwood forests from Maine and Michigan, south to northern Florida, and to near the Gulf of Mexico in Texas. It grows on a variety of soils from moist stream bottoms to uplands. Flowering dogwood is found near or under taller trees, such as the oaks and hickories, and also in the company of small trees like redbuds and hawthorns. Bracts surrounding the flowers are responsible for its showy floral display.

Flower buds are susceptible to extreme cold and may become a casualty even though the tree survives without other apparent injury. Consequently, it is very important that flowering dogwoods planted in the central and northern prairies be from seed of northerly origin. Trial plantings of flowering dogwood at

PERC in Fort Collins have not been successful (Fairchild and Klett 1993).

The long list of flowering dogwood cultivars—most selected for appearance in their native habitat without regard to origin—becomes confusing to a landscaper looking for a hardy specimen. Provenance research will provide the most effective method for determining the proper plants for each prairie site. The extensive natural latitudinal range of eastern dogwood provides good assurance that plants adapted to a sizeable portion of the prairies can be discovered.

Two seedling flowering dogwoods planted at Prairie Pines Arboretum near Lincoln have thrived for many years, have seldom failed to bloom, and have displayed colorful foliage in early October. These dogwoods grow in sheltered locations on fertile sites, but receive no supplemental water. They have survived several seasons of severe drouth with no apparent ill effects (Bagley 1998).

Flowering dogwood fruits, though poisonous to humans, are relished by birds and small mammals (Elias 1980). Leaves and twigs are browsed heavily by deer and cottontail rabbits, necessitating protective measures during establishment. Its hard, tough wood has been used for tool handles, wheel hubs and other specialty items.

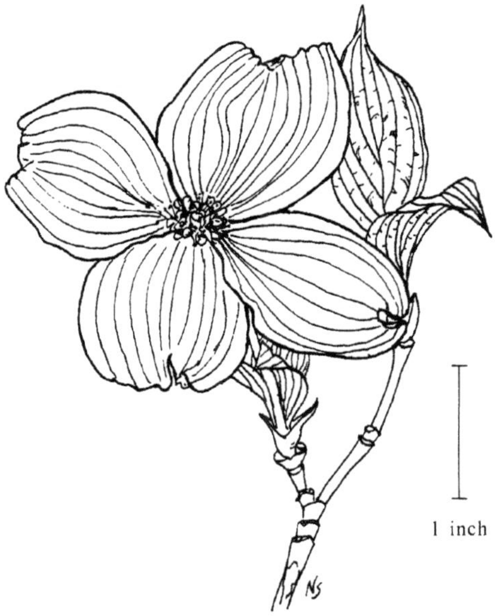

Cornus florida

Planning

Height: up to 30 ft (9 m); National champion, 1989, 33 ft (10 m), Virginia
Habit: tree; deciduous; crown flat-topped, horizontal branching; April flowers, September-December fruits; autumn foliage colorful
Site: partial shade, sun; prefers pH neutral to acidic, moist, well-drained soil; zone 5
Use: specimen, grouping, naturalizing, wildlife habitat
Caution: protect from excessive wind
Pests: susceptible to many in its native range; less so on prairie sites

Recognition

Leaves: opposite; egg-shaped to elliptic; tip pointed, base wedge-shaped (cuneate); dark green with short appressed hairs above, pale green beneath with appressed hairs, veins woolly; petioles sparsely hairy, grooved above, about 0.4 in (1 cm) long
Flowers: clusters (heads) dense, crowded, surrounded by 4 white, sometimes pink, squarish or rounded petal-like structures (bracts)
Fruits: glossy red in stemless clusters of 2-6; ellipsoid, about 0.4 in (1 cm) long, flesh dry; stone ellipsoid, about 0.3 in (8 mm) long with longitudinal grooves
Twigs: hairy, pale green, becoming reddish, glabrous, encircled by v-shaped leaf scars; lenticels small, dark; pith greenish, spongy
Buds: flower terminal, globose, enclosed by 4 bracts; peduncle about 0.4 in (1 cm) long, 2 bracts at base; vegetative buds inconspicuous, egg-shaped
Bark: dark gray-brown to black, shallow-fissured with thin, squarish plates
Wood: fine textured, very heavy and hard

Section Benthamia

Cornus kousa Hance - Japanese dogwood

This handsome tree—native to Japan, Korea, and China—develops exfoliating multicolored bark which, along with its other virtues, prompts Dirr (1990) to proclaim that if he could have only one dogwood tree in his garden, he would choose Japanese dogwood. Flowers of the Japanese appear two or three weeks later in the season than those of the flowering dogwood. Ten-year-old Japanese dogwoods at Prairie Pines Arboretum near Lincoln remain healthy, but they have not produced flower buds.

The many cultivars of Japanese dogwood display a great variety of form, bract color and foliage characteristic. The population includes: large-fruited 'Big Apple'; low, spreading 'Dwarf Pink'; weeping 'Elizabeth Lustgarten'; fastigiate 'Fanfare'; variegated 'Gold Star'; and floriferous 'Madame Butterfly'. Cultivars reputed to be most cold hardy are 'Silver Star', 'Moonbeam' and 'Triple Crown' (Dirr 1990).

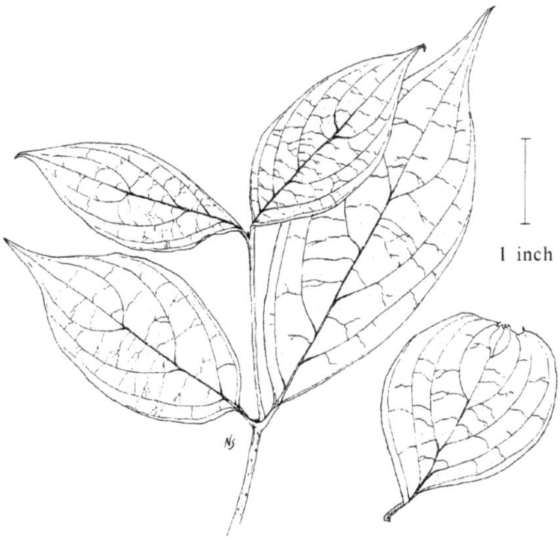

Cornus kousa

Planning

Height: up to 15 ft (4.5 m)
Habit: tree, shrub; deciduous; crown rounded; mid-June flowers; August-October fruits; exfoliating bark; autumn foliage may be colorful (but is not at Lincoln)
Site: sun, partial shade; prefers pH neutral to acidic, moist, well-drained soils; zone 5
Use: specimen, border, foundation
Caution: protect from excessive winds

Recognition

Leaves: opposite; elliptic to egg-shaped; tip pointed, base wedge-shaped to rounded; appressed hairs above and below; petioles about 0.2 in (5 mm) long
Flowers: clusters (heads) dense, globose (connate), surrounded by 4 white, petal-like pointed bracts; stalks 0.2 in (6 mm) long
Fruits: pinkish to red, borne in rounded syncarp resembling a raspberry, edible though somewhat mealy; stone ellipsoid, about 0.2 in (6 mm) long
Twigs: gray-red, hairy, becoming maroon and less hairy in autumn; pith greenish
Buds: flower terminal, sharp-pointed with globose base, 2 silky bud scales not overlapping (valvate); vegetative buds partially enclosed by petioles, inconspicuous
Bark: exfoliating, forming mosaic of gray, tan, and brown

Order Celastrales

Celastrales is comprised of 11 families and about 2,000 species (Cronquist 1981).

Recognition

Habit: woody plants, seldom herbs; deciduous, evergreen
Leaves: opposite or alternate; simple; entire or toothed

Celastraceae R. Brown - bittersweet family

The bittersweet family is represented worldwide by about 50 genera and 800 species. Seven genera are native to North America. In addition to their ornamental value, Celastraceae species yield oils, extracts for medicines and yellow dyes. The fine-grained wood is used for specialty products.

Recognition

Habit: trees, shrubs, vines; deciduous, evergreen
Leaves: alternate or opposite; simple; stipules mostly small and early deciduous or none
Flowers: clusters (cymes) axillary or terminal, seldom solitary; bisexual or sometimes unisexual; commonly greenish or white, regular, 4- to 5-parted; sepals overlapping (imbricate), seldom valvate; petals distinct, or rarely wanting; stamens 5, rarely 10; pistil 1, ovary superior, style short or nearly wanting, stigma 2- to 5-lobed
Fruits: dry, splitting (capsules), rarely not splitting and winged (samaras) or fleshy (drupes)

Celastrus L. - bittersweet

Over 30 species of *Celastrus* are natives of Asia, Australia and America. These twining shrubs are cultivated chiefly for their colorful winter fruits. Bittersweet quickly covers fences and trellises and soon gets out of hand without diligent attention. Small trees can be strangled by twining stems.

A tangle of bittersweet vines provides an excellent bird nesting site and persistent fruits for food. Birds return the favor by depositing seeds under favorite perches where, if they are undisturbed and moisture is favorable, new plants emerge to scramble up and over all objects in the vicinity.

Recognition

Habit: shrubs, usually twining; deciduous, rarely evergreen
Leaves: alternate; petioled, toothed
Flowers: clusters axillary or terminal; unisexual and bisexual on the same plant (polygamo-dioecious), or dioecious; 5-parted
Fruits: capsules splitting into 3 valves each containing 1-2 seeds within a crimson fleshy appendage (aril); about 0.3 in (8 mm) across
Buds: scales several, overlapping

Celastrus orbiculatus Thunb. - Oriental bittersweet

Oriental bittersweet, a rampant climbing native of Japan and China, has escaped cultivation in the northeastern U.S. It resembles our native American bittersweet, but apparently is not as well adapted to cold climes. Fruits are dropped soon after ripening, making it undesirable for decorative purposes (Sutton 1998).

Another Asian bittersweet, *C. loeseneri* Rehd. & Wils., dies back during winter in southeastern Nebraska (North Central Plant Introduction Station 1973). A chambered pith and more circular leaves distinguish this plant from Oriental bittersweet.

Celastraceae *Celastrus*

Planning

Height: up to 40 ft (12 m) long
Habit: shrub; deciduous; climbing by twining; rhizomatous; colorful autumn fruit, not long persistent
Site: sun, part shade; pH and soil adaptable; drouth tolerant; zone 5
Use: arbor, fence cover, naturalizing

Recognition

Leaves: nearly round to wider above the middle and somewhat elongated (suborbicular to oblong-obovate), up to 4 in (10 cm) long; abruptly pointed, base broadly tapering; toothed (crenate-serrate)
Flowers: clusters (cymes) axillary; stalk (peduncle) about the same length as its branches (pedicels)
Fruits: capsules orange-yellow; arils scarlet
Twigs: round; pith solid, white

Celastrus scandens L. - American bittersweet

Twigs of American bittersweet bearing orange and crimson fruits have long been popular for winter holiday decorations. This native to the eastern hardwood forests of the U.S. and southeastern Canada is widely scattered throughout the prairies along creek banks and fence rows. The Sioux Indians avoided it, calling the fruit "snake food" in the belief that it was poisonous. European settlers had no such qualms. Widespread fruit harvesting, along with indiscriminate herbicide use and animal browsing, has decimated the wild bittersweet population.

Birds are effective in increasing the bittersweet population where fruits are available and seedlings are protected from browsing and herbicides. The vigorous twining vines can quickly engulf fences and climb trees, sometimes becoming detrimental to the latter. Although some flowers may bear both male and female parts, most are either one or the other (dioecious). Obviously, fruiting is dependent upon vines of both sexes growing in reasonable proximity.

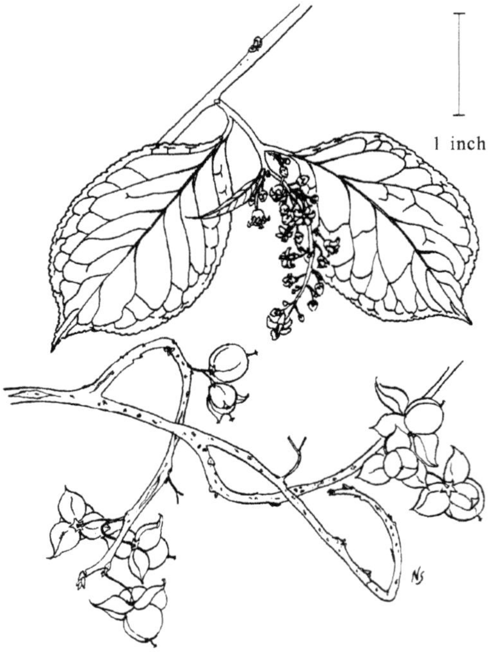

Celastrus scandens

Planning

Height: up to 40 ft (12 m) long
Habit: twining shrub; deciduous; rhizomatous; autumn foliage yellow, short-lived; colorful fruits September-December
Site: partial shade, full sun for good fruit production; pH and soil adaptable; drouth tolerant; zone 3
Use: arbor, fence, naturalizing, wildlife habitat
Caution: male and female plants required for abundant fruit production

Recognition

Leaves: egg-shaped to elliptic, up to 4 in (10 cm) long; tip long-pointed, base broadly tapered; margin toothed, surfaces glabrous; petioles about 0.5 in (12 mm) long

Flowers: unisexual (dioecious), sometimes bisexual (polygamo-dioecious); male in terminal or axillary clusters (panicles) on new growth, many-flowered; petals 5, yellowish; stamens 5, pistil rudimentary; female terminal only, 5-25 flowers; petals 6, usually reflexed; vestigial stamens 5; ovary egg-shaped, green, glabrous, stigma 3-lobed

Fruits: capsules globular, about 0.4 in (1 cm) across, orange, surface granular, wide-spreading after splitting; arils crimson, slightly creased, appearing to be in 6 sections

Twigs: twining, glabrous; leaf scars half-round; bundle scar 1, elongated; pith white, solid

Buds: often at right angles to stem, egg-shaped, pointed; outer scales pointed, red-brown

Bark: light brown, smooth with prominent lenticels, peeling into thin flakes

Euonymus L. - euonymus

Attractive foliage and colorful fruits are the attributes of *Euonymus*, a genus of 175 species, most of which are native to the Himalayas, China and Japan. Four species are native to North America.

Recognition

Habit: trees, shrubs, sometimes creeping or climbing by rootlets; deciduous, evergreen

Leaves: opposite, rarely alternate or whorled; simple; usually glabrous; petioled

Flowers: clusters (cymes) axillary; flowers 4- to 5-parted; usually bisexual

Fruits: capsules, usually lobed, sometimes winged; 4- to 5-celled, each cell containing 1-2 seeds with a fleshy appendage (aril), usually orange

Twigs: usually 4-angled, often greenish

Buds: small to conspicuous; scales overlapping (imbricate)

Leaves Deciduous

Euonymus alata (Thunb.) Sieb. - winged euonymus

Winged euonymus, a native of Asia, is a pleasantly greer shrub perfectly suited to remain in the landscape background. However, by autumn a burst of crimson glory thrusts it to center stage. Hoag (1965) notes that foliage color of winged euonymus frequently does not develop in the northern plains.

Timing of foliage coloration is hereditary—determined by the latitude of origin of the parent population. After the first sharp freeze the leaves gradually succumb to autumn winds and a more subtle color appears as the opening capsules expose their orange-red arils. The fruits cling well into the winter before they begin dotting the snow below. The quantity of fruits and the prominence of the corky wings on branches vary among plants.

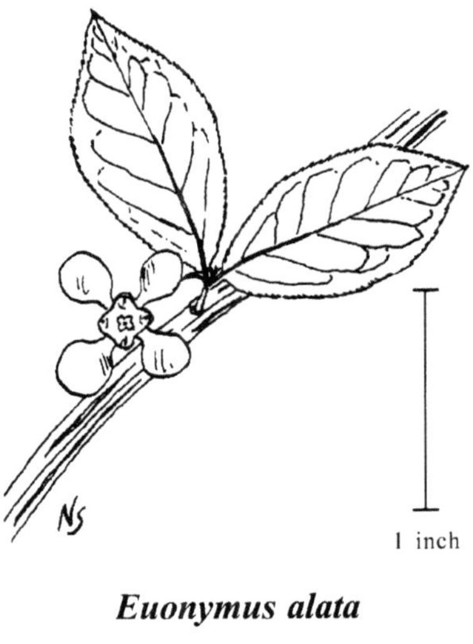

Euonymus alata

'Compactus' attains a height of about 6 feet (2 m); it is not as winter hardy as other forms of the species; and it is more likely to be troubled with chlorosis in high lime soils (North Central Plant Introduction Station 1971). Test plantings of variety *koreanus* indicate more resistance to chlorosis than 'Compactus' and its leaves persist near the ground (North Central Plant Introduction Station 1963, 1971). Other compact cultivars of winged euonmyus include 'Nordine Strain', 'Rudy Haag', 'October Glory' and 'Angelica' (Dirr 1990).

Celastraceae *Euonymus*

Planning

Height: up to 15 ft (4.5 m)
Habit: shrub; rounded to spreading, dense; brilliant red autumn foliage; fruits persist into winter
Site: sun, partial shade; pH and soil adaptable, prefers moist, fertile, well-drained sites; zone 4
Use: specimen, screening, grouping, wildlife habitat, hedging
Note: mulching is desirable on sunny sites

Recognition

Leaves: elliptic to sometimes broader above middle (obovate), up to 2 in (5 cm) long; tip pointed, base tapered; finely and sharply toothed (serrate); surfaces glabrous; petioles 0.1 in (2 mm) long
Flowers: clusters, usually 3-flowered cymes; bisexual; yellow-green, each about 0.2 in (6 mm) across
Fruits: capsules purplish, divided nearly to base into 4 pods, sometimes 1-3; opening to expose orange-red aril
Twigs: glabrous, greenish; 2-4 corky wings, inconspicuous to prominent; leaf scars half-round, bundle scar 1, appearing as a curved line; pith greenish, spongy or excavated
Buds: brownish, glabrous, strongly diverging from stem

Euonymus atropurpurea Jacq.- eastern wahoo

Eastern wahoo, often referred to as burning bush, is native to much of the eastern U.S., including the eastern part of the central prairie. Although not an abundant prairie plant, it was important to Native Americans.

Another Indian name for eastern wahoo, "Walking-Ghost-Stick," probably relates to its open growth habit (Gilmore 1977). The Sioux used the hard, close-grained wood for arrows, and it is believed that the name "wahoo" is derived from an Indian word meaning "wood of the arrow" (Van Der Linden and Farrar 1984). A Nebraska creek and a town nearby borrowed its name. The wahoo's bitter root bark, when pulverized, served as a remedy for whatever ailed early settlers (though the seeds are reputedly poisonous).

A sparse branching habit and the tendency to form colonies has limited the use of eastern wahoo in decorative plantings. However, it serves in natural settings and larger landscapes as a spot of bright red for a week or so in mid-September, followed by a myriad of persistent scarlet fruits that fade gradually as winter deepens.

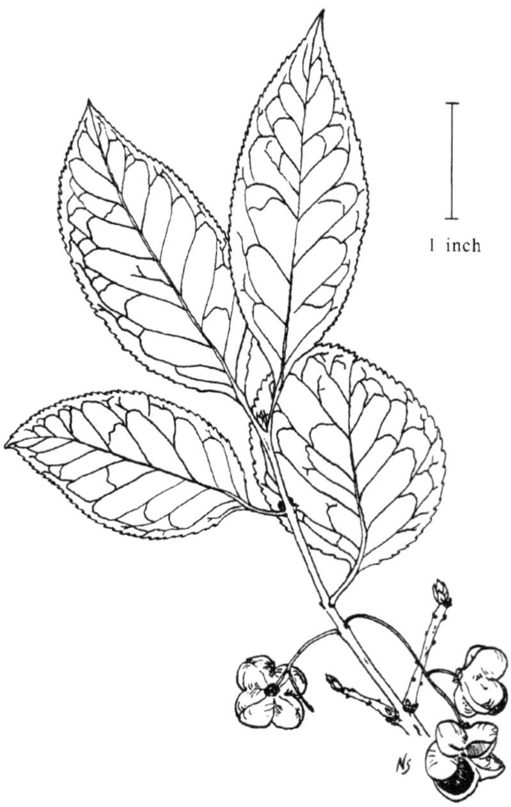

Euonymus atropurpurea

Planning

Height: up to 20 ft (6 m)
Habit: shrub, tree; crown wide, irregular; autumn foliage colorful, short-lived; scarlet fruits persist into winter
Site: sun, part shade; pH and soil adaptable, prefers deep fertile soils; zone 3
Use: naturalizing, wildlife habitat, grouping

Recognition

Leaves: egg-shaped to elliptic, up to 3.5 in (9 cm) long; tip long-pointed, base tapered, margin with fine glandular-tipped teeth; glabrous above, hairy on veins beneath; petioles glabrous, up to 0.6 in (15 mm) long

Flowers: clusters (cymes) of 7-15 on new growth; bisexual; flowers 4-parted, purple, each about 0.4 in (1 cm) across; stamens 4, anthers purple or yellow; ovary mostly imbedded in disc, style wanting, stigma green, 4-lobed

Fruits: capsules yellow-brown, tinged with purple; deeply 4-lobed, 1-2 often underdeveloped, opening to expose scarlet aril

Twigs: circular to somewhat 4-sided in cross-section; purplish green, nodes slightly enlarged; leaf scars half-round, bundle scar 1, large; pith greenish white

Buds: terminal egg-shaped, about 0.3 in (6 mm) long; lateral much smaller, somewhat appressed or not strongly divergent; scales sharp-pointed (apiculate), 4-ranked, glabrous

Bark: thin, ash-gray, becoming scaly

Euonymus bungeana Maxim. - winterberry euonymus

Late autumn days at Prairie Pines Arboretum near Lincoln are brightened by masses of colorful fruits clothing 'Pink Lady', one of several cultivars of winterberry euonymus. A native of China and Manchuria, this release by the Natural Resources Conservation Service Plant Materials Center at Manhattan, Kansas, is adapted to most sites in Kansas, Colorado, New Mexico and Oklahoma, based on trial plantations made in cooperation with agricultural experiment stations of those several states.

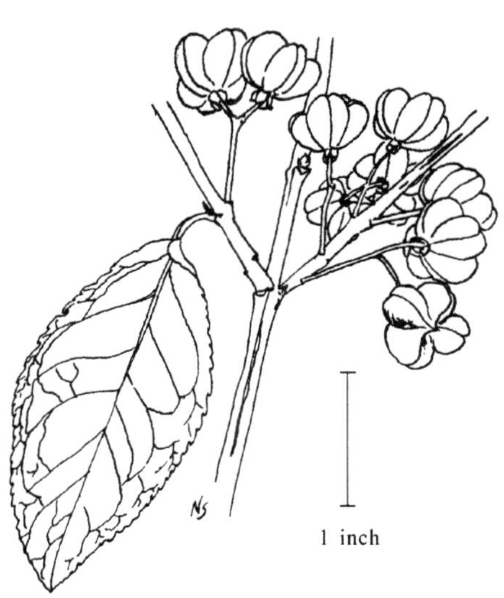

Euonymus bungeana

Hoag (1965) and Herman (1985) report that winterberry euonymus is adapted to the northern Great Plains. Varieties *pendula* Rehd. and *semipersistens* (Rehd.) Schneid. are characterized by pendulous branches and persistent, half-evergreen leaves, respectively.

Winterberry euonymus has not been widely planted, but this small tree has a place in our prairie landscapes. Although Hasselkuss (1993) and Fairchild and Klett (1993) state that suckering does occur, no root sprouts have been noticed on trees at Prairie Pines Arboretum. Fairchild and Klett also note that 'Pink Lady' and 'North Platte' suffer no chlorosis at PERC in Fort Collins, but minor twig dieback has occured during winter. They state that 'Pink Lady' is more fruitful than the latter.

Planning

Height: up to 20 ft (6 m)

Habit: shrub, tree; crown rounded, branches pendulous; autumn foliage yellow to reddish; fruits pink October-December; suckering variable

Site: sun, partial shade; pH and soil adaptable; zone 4

Use: specimen, screen, naturalizing, wildlife habitat

Note: fruit production is enhanced by sunshine

Recognition

Leaves: elliptic to oblong up to 4 in (10 cm) long; tip abruptly long-pointed, base wedge-shaped; margins finely toothed, wavy; surfaces glabrous, light green; petioles up to 0.8 in (2 cm) long

Flowers: clusters (cymes) of 3-7; bisexual, flowers 4-parted, yellowish; anthers purple

Fruits: capsules greenish yellow shading to reddish purple, opening to expose orange-red arils; sepals persistent

Twigs: circular in cross-section; greenish with purplish blotches, glabrous; leaf scars half-round, bundle scar 1; pith white

Buds: terminal nearly globose; scales sharp-pointed (apiculate); lateral buds globose

Euonymus europaea L. - European burningbush

Varieties and cultivars of European burningbush—a native of Europe and western Asia—present an array of forms, fruitfulness and foliage color. Variety *aldenhamensis* Gibbs and 'Red Cascade' are noted for their bountiful and colorful fruits. Hoag (1965) describes the former as sufficiently hardy in the northern plains with shelter.

'Red Cap', a shrub with dense, rounded growth habit, displays good fall color and showy red fruits that persists into winter. It is adapted to the PERC site at Fort Collins (Fairchild and Klett 1993).

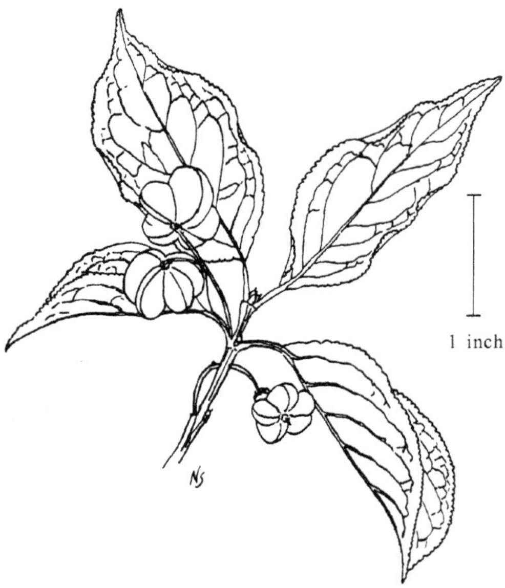

Euonymus europaea

Planning

Height: up to 30 ft (9 m)

Habit: shrub, tree; columnar becoming broad; colorful autumn foliage and fruits

Site: sun, partial shade; pH and soil adaptable, drouth tolerant; zone 4

Use: screening, grouping, border, naturalizing

Recognition

Leaves: elliptic to oblong to wider above the middle (obovate), up to 3.5 in (9 cm) long; long-pointed, base broad tapered to rounded, margin with rounded teeth (crenate); surfaces glabrous; petioles up to 0.5 (12 mm) long

Flowers: clusters of 3-5; bisexual; flowers 4-parted, yellow-green; anthers yellow

Fruits: capsules pink to red, 4-lobed, smooth, opens to expose orange aril

Twigs: purplish, glabrous, becoming dark gray; leaf scars half-round, bundle scar 1; pith white, angular in cross-section, spongy

Buds: nearly globular, plump, greenish brown, glabrous, about 0.1 in (3 mm) long, slightly divergent

Euonymus hamiltoniana Wall. - Hamilton euonymus

Hamilton euonymus, an Asian native closely related to *E. europaea*, is adapted to the soils and climate of Fort Collins with no pest problems (Fairchild and Klett 1993). Variety *sieboldianus* (var. *yedoensis*), known as Yeddo euonymus, is very susceptible to scale insects (Dirr 1990). Trial plantings of Yeddo euonymus at Ithaca, Nebraska (UNARDC), have been successful. Variety *maackii* is a pleasing ornamental in the northern Great Plains (Herman 1985).

The reports above present results of experiments with similar, closely related plants, but the confused nomenclature will cause difficulty in locating a plant that is identical to one of the above varieties. Yeddo euonymus is also considered to be a separate species, *E. yedoensis* (Rehder 1940).

Planning

Height: up to 15 ft (4.5 m)
Habit: shrub, tree; coarse textured, upright, thicket-forming; pink fruits persist into winter
Site: sun, part shade; pH and soil adaptable; zone 4
Use: screen, wildlife habitat, naturalizing

Recognition

Leaves: elliptic to wider above middle (obovate-oblong), up to 5 in (13 cm) long; abruptly sharp-pointed, base broadly tapered, fine-toothed margin (crenate-serrulate); petioles up to 0.5 in (12 mm) long
Flowers: clusters of many; bisexual; flowers 4-parted, yellowish; anthers purple
Fruits: capsules pink to red, deeply 4-lobed, usually closed or slightly open; aril orange
Twigs: stout, greenish red, glabrous
Buds: plump, greenish in summer, becoming purplish

Euonymus nana Bieb. - dwarf euonymus

Dwarf euonymus, a native of the Caucasian Mountains and western China, is adapted to much of the Great Plains (North Central Plant Introduction Station 1960, Hoag 1965). Variety *turkestanica* Dieck is more upright than the species. Plants of the species and variety are growing at Colby, Kansas (Sutton 1998). Dwarf euonymus specimens growing at PERC in Fort Collins are well adapted to the site, but fruit set is inconsistent (Fairchild and Klett 1993).

Planning

Height: up to 3 ft (1 m)
Habit: shrub; deciduous, but evergreen to semi-evergreen in mild climates; irregular, straggly; autumn foliage red; midsummer fruits pink and orange
Site: sun, partial shade; pH and soil adaptable, drouth tolerant; zone 3
Use: border, naturalizing, wildlife habitat, foundation, groundcover

Recognition

Leaves: usually alternate or whorled; linear, about 2 in (5 cm) long; tip rounded, margin entire or nearly so; bluish green; petioles less than 0.1 in (1 mm) long
Flowers: 1-3, on slender stalks; brownish purple
Fruits: capsules 4-lobed, pink; seed brown, partially covered by orange aril
Twigs: angled, glabrous

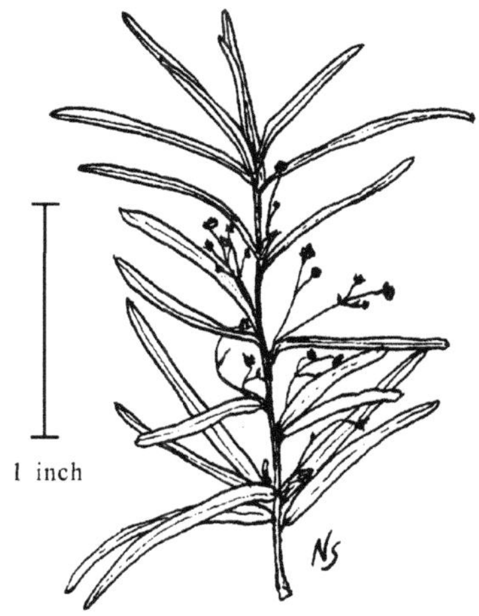

Euonymus nana turkestanica

Leaves Evergreen

Euonymus fortunei (Turc Hand.—Mazz. - wintercreeper

Wintercreeper, a native of C is a taxonomist's and landscaper's delight (or night) with its seemingly never-ending list of cultivars re either fruiting or nonfruiting and differing in fo and leaf size, shape and color. Dirr (1990) suggests stopping new introductions and consolidating these plants into a few reasonably well-defined, useful cultivars.

Rehder (1940) recognizes 7 varieties of *E. fortunei*, each quite variable in form and leaf character. Two varieties common in the nursery trade are: var. *colorata* with its deep glossy green foliage that becomes plum purple in winter, usually nonfruiting; and var. *radicans* with its medium to dark green leaves, wavy appearance, and sporadic flowering and fruiting (Dirr 1990). The non-flowering juvenile form of the species is quite different in leaf character from the fruiting adult form.

Wintercreeper is a popular landscape plant in spite of sometimes serious insect and disease problems which may be caused by stressful site conditions. For example, three unidentified and quite different cultivars have been thriving with no pests at Prairie Pines Arboretum near Lincoln for over 30 years as ground, wall and trellis cover. Occasionally, the leaves become brown due to severe cold, but new leaves appear as if by magic with the advent of spring.

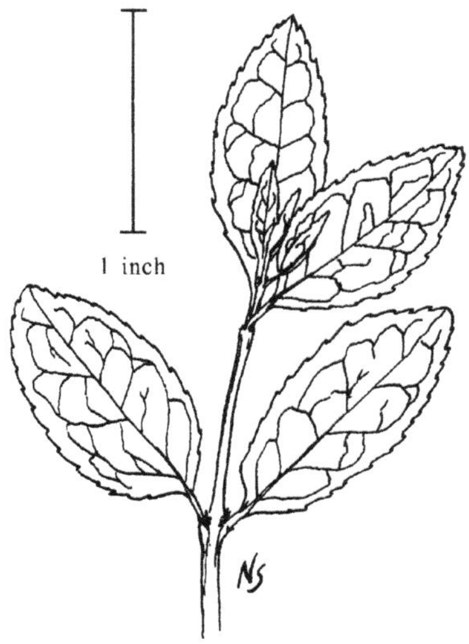

Euonymus fortunei

Planning

Height: up to 40 ft (12 m) long; 6 in (15 cm) high as groundcover
Habit: shrub; climbing or trailing; dark green foliage year-round, some reddish purple in winter, browning in sub-zero temperatures; colorful fruits on some varieties, October into winter
Site: shade, partial sun; pH and soil adaptable, thrives on moist fertile soils, drouth tolerant on protected sites; zone 5
Use: wall cover, groundcover, low hedge, grouping
Caution: aerial rootlets will leave scars on wooden walls when removed

Recognition

Leaves: opposite; elliptic to elliptic-ovate up to 2 in (5 cm) long; tip acute to obtuse, base variably rounded, margin toothed; surfaces glabrous, dull to glossy green, sometimes becoming purple in winter, veins often silvery; petioles about 0.2 in (5 mm) long
Flowers: clusters (cymes) of 5-10; greenish white
Fruits: capsules pinkish, opening to expose orange aril
Twigs: green, sometimes with purplish blotches, glabrous, somewhat warty, rootlets hold fast to walls or take root in soil
Buds: narrow elliptic, pointed, appressed to slightly divergent, green, brown, or purplish red

Euonymus kiautschovica Loes. - spreading euonymus

'Manhattan', a hybrid cultivar of spreading euonymus developed at Kansas State University, failed to establish at PERC in Fort Collins because of winter cold (Fairchild and Klett 1993). Other cultivars include: 'Dupont', a vigorous, hardy, hybrid form; 'Hobbs', a dense mounded form; and 'Paulii', a hardy, upright form with leathery leaves.

The useful range of spreading euonymus can be extended somewhat to the north by providing protection from winter wind and sun. Plants on the north sides of buildings on the UNL campus survived for many years, sprouting back after tops were killed by severe cold.

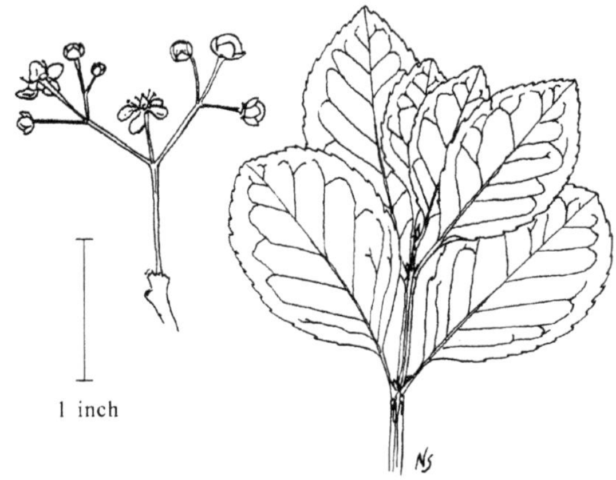

Euonymus kiautschovica

Planning

Height: up to 10 ft (3 m)
Habit: shrub; rounded; evergreen, semi-evergreen or deciduous; autumn fruit colorful
Site: sun, partial shade; pH and soil adaptable; zone 6
Use: hedge, screen, grouping

Recognition

Leaves: opposite; broad elliptic to oblong, sometimes wider above the middle, up to 3 in (8 cm) long; tip pointed to obtuse, base tapering, shallow-toothed; surfaces shiny dark green, veins not evident beneath; petioles up to 0.3 in (8 mm) long
Flowers: clusters loose, many-flowered; greenish white
Fruits: capsules subglobose, 4-lobed, smooth, pink; opening to expose orange-red aril
Twigs: green, rounded in cross-section
Buds: conical, sharp-pointed, often red-tinged to brown in winter

Paxistima Raf. - paxistima

Two species are native to North America. *Pachistima* is an older spelling of the generic name.

Recognition

Habit: shrubs; evergreen
Leaves: opposite; simple; entire or finely toothed; short-petioled
Flowers: clusters few-flowered, axillary; bisexual; flowers green or reddish, 4-parted
Fruits: capsules leathery, 2-valved; seeds 1-2, thin lacerated appendages (arils)
Twigs: quadrangular, warty
Buds: egg-shaped; scales, 2 pairs

Paxistima canbyi Gray - Canby paxistima

Canby paxistima grows on calcareous soil in the mountains of Virginia and West Virginia. Although Dirr (1990) reports that it tolerates high pH soils, plants of this species at North Platte, Nebraska, are chlorotic (Sutton 1998).

Celastraceae *Paxistima*

Spangler and Ripperda (1977) recommend planting Canby paxistima on an acidic, moist, well-drained site with protection from winter sun and winds. According to Hoag (1965), it seldom survives on heavy clay soils. Hoag also states that Canby paxistima is well established in the International Peace Garden on the North Dakota—Manitoba border where snow cover preserves its evergreen characteristic.

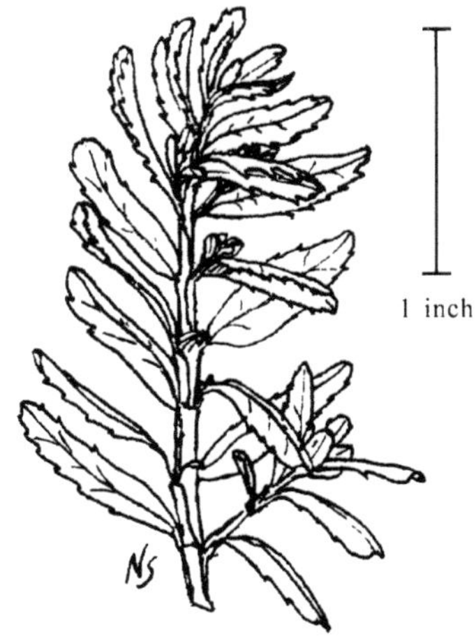

Paxistima canbyi

Planning

Height: up to 1 ft (30 cm)
Habit: shrub; decumbent, branches root on contact with moist soil
Site: sun, part shade; pH and soil adaptable, best on a fertile, moist but well-drained site; zone 4, zone 3 with snow cover
Use: groundcover, low hedge

Recognition

Leaves: linear to narrow-oblong, up to 1 in (25 mm) long; finely toothed above the middle, margins rolled toward the back (revolute); surfaces glabrous, shiny, dark green; petioles less than 0.1 in (2 mm) long
Flowers: about 0.2 in (5 mm) across, greenish red, on pedicels about 0.2 in (5 mm) long
Fruits: capsules about 0.2 in (4 mm) long
Twigs: slightly 4-sided, becoming rough and corky; pith very small, brownish and spongy; leaf scars minute, crescent-shaped, slightly raised; bundle scars indistinct
Buds: minute, egg-shaped, sessile, appressed to twig

Paxistima myrsinites Raf. - **mountain-lover**

Mountain-lover, a native of mountains from British Columbia to California and New Mexico, grows slowly on moist, acidic soils on forested sites. This is a plant for the dedicated collector of unusual species. The foliage resembles that of the boxwood.

Planning

Height: up to 1.5 ft (45 cm)
Habit: shrub; spreading, branches stiff
Site: partial sun, shade; moist acidic soil; zone 4, zone 3 with snow cover
Use: groundcover, low hedge

Recognition

Leaves: elliptic to oblong, wider above middle, up to 1.2 in (3 cm) long; margins slightly rolled back, finely toothed or nearly entire, glossy green
Flowers: short-stalked, red
Fruits: capsules about 0.3 in (8 mm) long, whitish
Twigs: slightly 4-angled

Trypterygium Hook. F. - **threewingnut**

Three species of *Trypterygium* are native to eastern Asia.

Trypterygium regelii Sprague & Takeda - **Regels threewingnut**

Regels threewingnut is a plant for adventurous gardeners.

Planning

Height: up to 7 ft (2 m) or climbing taller with support
Habit: shrub; deciduous; summer blooms
Site: soil and pH adaptable; zone 5
Use: experimental

Recognition

Leaves: alternate; broad-elliptic to egg-shaped, up to 6 in (15 cm) long; tip pointed, base tapered or rounded, margin with incurved teeth; surfaces glabrous; petioles about 0.6 in (15 mm) long
Flowers: clusters (panicles) terminal, up to 8 in (20 cm) long, also in axils of terminal leaves; unisexual and bisexual on same plant (polygamous); flowers yellowish white, 5-parted
Fruits: nutlets, 3-winged, 0.6 in (15 mm) long; seeds 1
Twigs: angular, warty; without tendrils
Buds: conical; scales 2 pairs

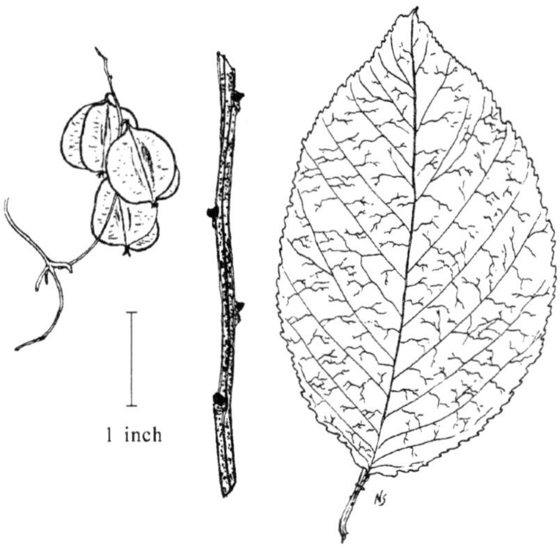

Trypterygium regelii

Aquifoliaceae Bartl. - **holly family**

About 400 species of the holly family are widely distributed throughout tropical and temperate regions of the world.

Recognition

Habit: trees, shrubs; evergreen, few deciduous
Leaves: alternate, rarely otherwise; simple; stipules minute, soon falling, or wanting
Flowers: clusters, sometimes solitary, axillary or terminal: unisexual (dioecious) or unisexual and bisexual (polygamous); flowers regular, usually 4-parted; sepals imbricate, more or less united at the base, but distinct, sometimes early deciduous; petals (united at the base in *Ilex*), imbricate, sometimes valvate, or rarely wanting; stamens usually alternate with petals, sometimes more numerous; ovary superior (hypogynous), compound, style short, stigma lobed
Fruits: fleshy, berry-like (drupes)

Ilex L. - **holly**

Decking the halls with boughs of holly and wreaths of fir celebrate the birth of Christ, rituals which date back to the 16th century and perhaps earlier. At that time fresh, green boughs were meant to brighten homes and lighten hearts during the bleak winter months in the north of Europe. The tradition continues today in Christian homes throughout the world.

Aquifoliaceae *Ilex*

English holly *(I. aquifolium* L.), a native of Europe and Asia, is the tree with "varnished" leaves that has been cultivated for so long that its cultivars and hybrids outnumber the *Ilex* species itself (Johnson 1973). A 40-foot silver variegated variety growing in Tillamook, Oregon, is the largest known English holly in the U.S. (American Forestsry Association 1992).

English holly bears flowers and fruits clustered on the previous year's branches. Leaves of Japanese holly, *I. crenata* Thunb., are spine-free and its fruits are black. Japanese holly has fostered as many as 380 cultivars, sixty of which are commonly offered by the nursery trade in the U. S. (Dirr 1990). Many of these cultivars are difficult to identify without labels.

Not all of the 380 species of holly are evergreen. Male and female flowers are usually borne on separate plants. Thus, not all plants bear the highly decorative fruit which is valuable wildlife food but is somewhat poisonous to humans. Male plants must be included in holly plantations to assure fruit production.

Hollies, especially the evergreens, present a challenge to those who wish to grow them in the Prairie Region. For some species the prairie's winters are too cold, summers are too hot, and winds are too persistent. Many prairie soils are too calcareous as well. The hollies prefer cool, moist, fertile, acidic soils, and partial winter shade is necessary for the evergreen species.

Recognition

Habit: trees, shrubs; evergreen or deciduous
Leaves: margins entire or toothed, often spiny and wavy; petioled
Flowers: solitary or clustered (fascicles or cymes); usually unisexual (dioecious); petals broad, united at base, sepals persistent
Fruits: berry-like drupes with 2-8 bony 1-seeded nutlets
Twigs: circular in cross-section; lenticels numerous, inconspicuous; leaf scars half-round to crescent-shaped, bundle scars a curved line
Buds: terminal egg-shaped with 3 outer scales; lateral smaller
Bark: thin with warty lenticels, becoming scaly

Grouping Key for *Ilex*

(Rehder 1940)

Leaves deciduous; flowers solitary in leaf axils or fascicled with the leaves; fruits red: **subgenus *Prinos***

Leaves evergreen; flowers solitary, clustered or fascicled in leaf axils; fruit colors vary: **subgenus *Aquifolium***

Subgenus *Prinos*

Ilex decidua Walt. - deciduous holly

Deciduous holly, a native of most of the eastern U.S. south of the Nebraska–Kansas border, grows in lowland areas. The western boundary of its range extends from extreme southeastern Kansas south throughout most of eastern Texas. Fruit-bearing winter twigs of deciduous holly are often gathered for Christmas decorations.

Healthy deciduous holly plants grow in Lincoln landscapes and to the southeast. Cultivars include: 'Byers Golden', yellow-fruited; 'Council Fire', orange-fruited; 'Pocahontas', red-fruited; and 'Warren Red', also red-fruited and one of the hardier cultivars (Dirr 1990).

Planning

Height: up to 30 ft (9 m)
Habit: shrub, tree; upright, many-branched; colorful fruits September-April
Site: sun, partial shade; soil and pH adaptable, prefers acidic, moist, well-drained sites; zone 5
Use: wildlife habitat, border, holiday decorations

Recognition

Leaves: alternate, often clustered on short spurs; thickish, egg-shaped, usually broadest above middle, up to 2.7 in (7 cm) long; margin obtusely toothed; tip rounded to notched, tapering to narrow base; surfaces glabrous except hairs on midrib beneath, veins impressed; petioles grooved, sometimes hairy, 0.3 in (7 mm) long
Flowers: clusters of 6-11 at end of short spur; calyx lobes 4, egg-shaped, green with red tip, often toothed or with hairy margin; petals 4, yellow to white, joined at base, 3 times longer than sepals, somewhat recurved; stamens 4; ovary green, glabrous, stigma with minute, erect lobes
Fruits: drupes globose, 0.3 in (8 mm) across, orange to red; calyx persistent, stigma persistent as a flat knob; nutlets ribbed on all surfaces
Twigs: glabrous, variably gray; spurs numerous on second-year's growth; pith greenish, leaf scars half-round; stipules often persistent
Buds: globular, glabrous; scales light gray, overlapping, pointed
Bark: light gray, smooth except for small, warty lenticels

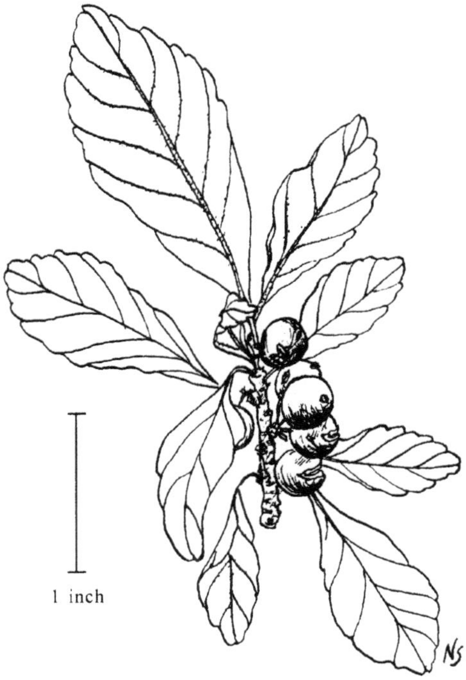

Ilex decidua

Ilex verticillata (L.) Gray - common winterberry

Common winterberry, the most northerly of the native hollies, grows in swamps and along streambanks from Nova Scotia to Minnesota, south to Missouri and Florida. Chlorosis has been noted on plants of this species growing at South Dakota University at Brookings, at UNL, and at Desoto Bend Wildlife Refuge near Blair, Nebraska. However, common winterberries growing on fertile soil with a pH of 7.0 at Prairie Pines Arboretum near Lincoln are healthy and fruitful with no soil amendments (Bagley 1998).

Fruitful and colorful common winterberry cultivars include: 'Afterglow', a slow-growing compact globe with orange to orange-red fruits; 'Chrysocarpa', a yellow-fruited form that drops its leaves earlier than red-fruited plants; and 'Winter Red' which bears a profusion of bright red fruits (Dirr 1990). Hybrid (*I. verticallata* × *I. serrata*) cultivar 'Sparkleberry' (female), developed at the National Arboretum, bears persistent brilliant red fruits when paired with 'Apollo' (male).

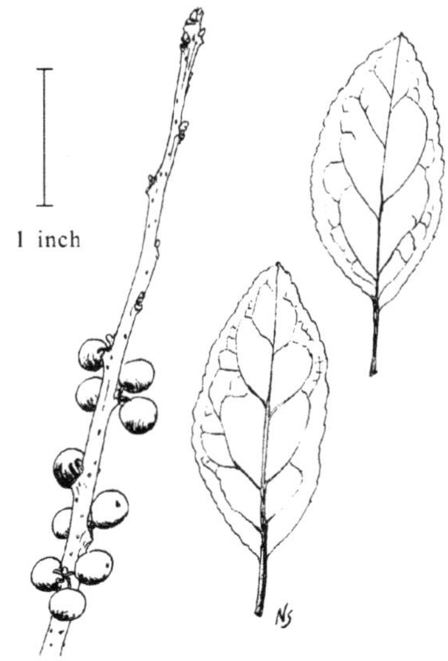

Ilex verticillata

Ilex laevigata (Dum.-Cours.) Gray., smooth winterberry, is very similar to common winterberry except that its slightly larger fruits are borne singly and its leaf petioles are shorter. This native of the East Coast is not commonly planted.

Planning

Height: up to 10 ft (3 m)
Habit: shrub; rounded, rhizomatous; bright red fruits September-December
Site: sun, partial shade; fertile, moist, slightly acidic soil; zone 4
Use: grouping, border, wildlife habitat

Recognition

Leaves: alternate; elliptic to sometimes wider above middle, up to 3 in (8 cm) long; tip pointed, base tapered, toothed; hairy beneath, especially on veins; petioles about 0.4 in (1 cm) long
Flowers: clusters short-stalked, male about 6, female fewer; flowers 5-8 parted
Fruits: drupes often in pairs, globose, 0.3 in (8 mm) across, bright red, persistent; nutlets 4-5, smooth
Twigs: olive to purplish brown, glabrous or with fine hairs; peeling second year; pith white to greenish, somewhat excavated; leaf scars half-elliptical with 2 small, black projections on either side
Buds: globose, in pairs (superposed), lower flower, upper leaf

Subgenus *Aquifolium*

Occasionally, evergreen plants with holly-like leaves masquerade as holly and the name "Holly" is sometimes attached to them. These are *Mahonia aquifolium*, Oregongrape, and *M. repens*, creeping mahonia—both small shrubs with compound leaves and bluish black grape-like fruit.

Ilex glabra (L.) Gray - **inkberry**

Inkberry, so named for its black-purple fruit, grows naturally in swampy areas along the east coast of North America from Nova Scotia south to Florida and west to Mississippi. Temperatures below freezing may cause leaf browning of the inkberry, especially if it is exposed to sun and wind. Cultivars include 'Compacta', a dwarf female clone, and 'Ivory Queen', a white-fruited form.

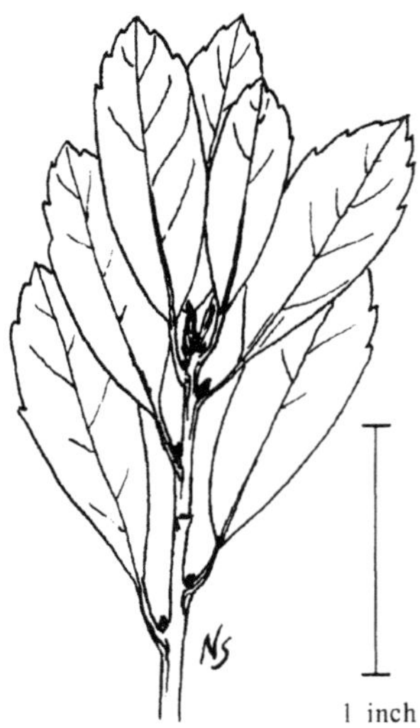

Ilex glabra

Planning

Height: up to 10 ft (3 m)
Habit: shrub; half-evergreen in cold climates; upright, rhizomatous; flowers on current season's growth; fruits September through May
Site: sun, partial shade; moist, fertile, slightly acidic soils, tolerates wet sites; zone 3
Use: grouping, foundation, hedge, wildlife habitat

Recognition

Leaves: alternate; egg-shaped to elongated, wider above middle (obovate), up to 2 in (5 cm) long; tip pointed to rounded, base tapered, margin entire or a few obtuse teeth near apex; surfaces glabrous; petioles 0.2 in (5 mm) long
Flowers: male flowers 3 or more on a single stalk, female often solitary; petals 6 to 8, creamy
Fruits: drupes globose, 0.2 in (6 mm) across, black, persistent
Twigs: green, hairy, becoming glabrous

Ilex ×meservae S. Y. Hu - Meserve holly

Mrs. Kathleen Meserve of St. James, New York, made the initial crosses between the hardy but unspectacular Asian holly, *I. rugosa* Fr. Schmidt, and the ornamental English holly. 'Blue Princess' and 'Blue Prince' are the most cold hardy of the five clones.

Mrs. Meserve has also released 'China Girl' and 'China Boy', hybrids from crossing Asian holly with *I. cornuta* Lindl., Chinese holly.

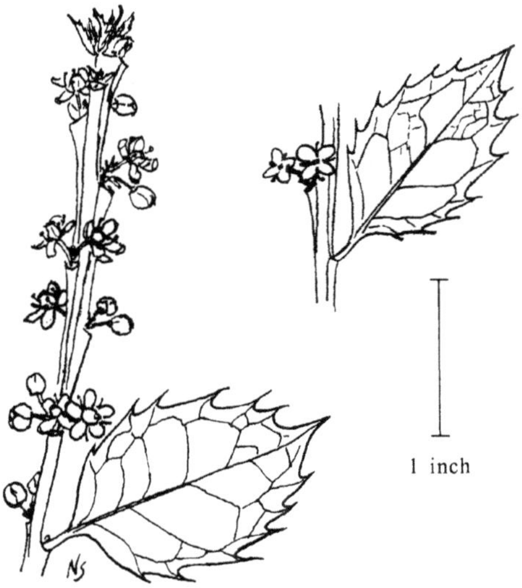

Ilex ×meservae

Planning

Height: up to 5 ft (1.5 m)
Habit: shrub; spreading; colorful fruits, fall into winter
Site: shade, partial shade; fertile, moist, slightly acidic soil; zone 5
Use: specimen, border

Recognition

Leaves: alternate; egg-shaped, up to 2 in (5 cm) long; margins wavy, spiny-toothed; leathery; surfaces glossy blue-green; petioles 0.2 in (5 mm) long
Flowers: clusters of 2-3
Fruits: drupes globose, bright red

Ilex opaca Ait. - American holly

American holly—a native of the southeastern U.S. south of a line from southeastern Oklahoma to Connecticut—grows best on fertile, moist bottomlands. A tree of this species grows well on the north side of a tall office building in Lincoln, protected from midday sun and south winds, and another tree of southeastern Missouri origin grows well in an unshaded residential Lincoln yard (Sutton 1998).

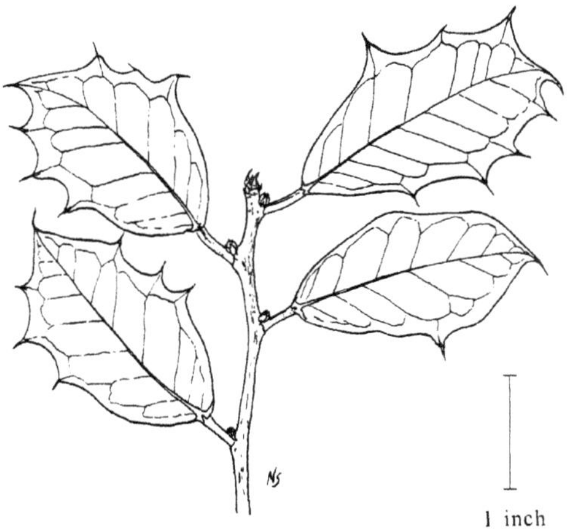

Ilex opaca

Planning

Height: up to 30 ft (9 m); National champion, 1987, 74 ft (22.4 m), Alabama
Habit: tree; pyramidal, dense; flowers on current season's growth; red fruits October into winter
Site: sun, shade; fertile, moist, slightly acidic soil; zone 5
Use: specimen, grouping, Christmas boughs
Caution: protect from excessive winds

Recognition

Leaves: alternate; elliptic, broadest near base or middle, up to 4 in (10 cm) long; tip pointed, base rounded; margin coarsely spiny-toothed, rarely nearly entire; stiff, leathery; surfaces glabrous; petioles up to 0.5 in (1 cm) long
Flowers: clusters, male 3-12, female 2-3, or solitary; flowers dull white
Fruits: drupes globose, dull red, 0.4 in (1 cm) across, on stalks 0.2 in (5 mm) long
Twigs: fine, rust-colored hairs (puberulous), becoming glabrous, pale brown
Buds: scales narrow, overlapping
Bark: thin, gray with warty projections
Wood: tough, dense, ivory white

Nemopanthus Raf. - mountainholly

This genus of one species is native to the northeastern U.S.

Nemopanthus mucronatus (L.) Trel. - mountainholly

Mountainholly bears attractive red fruit during late summer. This shrub is appropriate for persons looking for unusual new plants.

Planning

Height: up to 10 ft (3 m)
Habit: shrub; deciduous; slender branhes
Site: experimental; zone 4
Use: specimen, border, foundation

Recognition

Leaves: alternate; elliptic to oblong, up to 1.5 in (35 mm) long; margin entire or slightly toothed; tip with short, abrupt point (mucronate); petioles up to 0.5 in (12 mm) long
Flowers: male clusters of 1-4, female solitary, on slender axillary stalks; unisexual, polygamo-dioecious; each flower about 0.2 in (5 mm) across; petals white, linear, free; sepals minute, deciduous
Fruits: drupes, subglobose, 0.3 in (8 mm) across, dull red; nutlets 4-5
Twigs: slender, glabrous, purplish, becoming ashy gray
Buds: egg-shaped; scales 3

Order Euphorbiales

Four families are included in Euphorbiales (Cronquist 1981).

Recognition

Habit: woody, less often herbs; evergreen, deciduous
Leaves: simple, sometimes compound
Flowers: unisexual, perianth seldom showy or wanting; stamens variable, ovary superior (hypogynous)
Fruits: dry, splitting into 2 halves (schizocarps), less often drupaceous

Buxaceae Dum. - **boxwood family**

Five genera and some 60 species of the boxwood family are widely distributed in the tropics and temperate regions.

Recognition

Habit: subshrubs, shrubs, trees; evergreen, seldom deciduous
Leaves: opposite, seldom alternate; simple; stipules wanting
Flowers: clusters (spikes, racemes or heads); unisexual (usually monoecious), regular; perianth not differentiated, tepals not petal-like or wanting; stamens opposite tepals or numerous, ovary superior
Fruits: dry and splitting (capsules) or fleshy

Buxus L. - **boxwood**

Toparius, a Roman god, was the inspiration for the practice of shearing plants into strong, geometrical shapes. Boxwood, a traditional plant for forming low hedges around flower beds in formal gardens, was a favorite subject of the topiary art during the height of the Roman Empire and in Elizabethan England from whence it was broughtto the North American colonies. Topiary is featured today in the reconstruction of the gardens of colonial Williamsburg, Virginia.

Recognition

Habit: shrubs, trees; evergreen
Leaves: opposite; simple; entire; tip rounded or with shallow notch (emarginate), base wedge-shaped; surfaces glabrous, veins arising along central midrib; petioles short
Flowers: clusters axillary or terminal, consisting of a terminal female flower and several male flowers, fragrant; female with 6 tepals, ovary 3-celled, styles 3; male with 4 tepals, stamens 4, much longer than tepals, pistil rudimentary
Fruits: capsules subglobose, 3-horned
Twigs: green, variously angled in cross-section
Buds: solitary, egg-shaped, sessile; scales 1-2 pairs visible

Buxus microphylla Sieb. & Zucc. - **littleleaf boxwood**

Littleleaf boxwood, a native of the Orient, has been widely planted in climates where common boxwood is likely to be injured by cold temperatures. Patience is a requisite for cultivating the extremely slow-growing littleleaf boxwood. Leaves tend to become yellowish brown in winter. 'Green Gem', a compact green mound, requires little pruning and retains its green color throughout winter.

Variety *koreana* Nakai, a plant with somewhat smaller leaves than the species, will grow in much of the Prairie Region on fertile, moist, well-drained soils if protected from wind and winter sun. A peat or leaf mold mulch is desirable.

Forms of variety *koreana*, 'Winter Gem', 'Winter Beauty' and 'Tide Hill' retain green foliage all winter, and 'Pincushion' and 'Tall Boy' become bronze-brown in winter (Dirr 1990). An extremely hardy, unnamed cultivar of littleleaf boxwood is grown by Clayton Berg, a Montana nurseryman (Sutton 1998).

Buxus microphylla

Buxaceae *Buxus, Pachysandra*

Planning

Height: up to 4 ft (1 m)
Habit: shrub; compact, rounded
Site: shade, partial shade, especially in winter; moist, cool soil, pH 7 or lower; zone 5
Use: edging, hedge, foundation
Caution: protect from excessive wind

Recognition

Leaves: egg-shaped, wider above middle (obovate) to elongated, up to 1 in (25 mm) long; petioles glabrous
Flowers: clusters, mostly terminal; male flowers with rudimentary pistil nearly as long as calyx
Twigs: squarish, glabrous

Buxus sempervirens L. - common boxwood

Common boxwood, an import from the Mediteranean Region, graces the memorials and monuments of Washington, D.C., and many formal gardens. Although the habit or appearance of many of the cultivars of common boxwood is similar, especially if sheared in a similar fashion, hereditary differences among them enable this species to be successfully grown on a wide range of sites.

Planning

Height: up to 20 ft (6 m)
Habit: shrub, tree; rounded, multibranched, holding foliage entire length; May blooms fragrant
Site: sun, part shade, especially in winter; slightly above neutral to acidic, moist, cool soil; zone 5
Use: hedge, grouping, topiary

Recognition

Leaves: malodorous; elliptic to egg-shaped to oblong, up to 1.2 in (3 cm) long; petioles minutely hairy
Flowers: clusters axillary; male with rudimentary pistil half as long as sepals; stamens 4, longer than sepals; female with 6 sepals
Twigs: somewhat angled, minutely hairy; if bruised will emit an unpleasant odor

Pachysandra Michx. - pachysandra

Pachysandra is a genus of few species.

Recognition

Habit: subshrubs; evergreen or half-evergreen
Leaves: alternate; simple; toothed; 3 veins arising at base
Flowers: clusters (spikes) upright; flowers white; female at base of spike with 4 or more tepals, styles 2-3, spreading; male with 4 tepals; stamens 4, much longer than tepals
Fruits: dry, splitting, 2- to 3-horned (capsules) or fleshy (drupes)
Twigs: fleshy

Pachysandra procumbens Michx. - **Allegheny pachysandra**

Allegheny pachysandra, a native of the southeastern U.S., thrives in Maxwell Arboretum at UNL (Sutton 1998).

Planning

Height: up to 1 ft (30 cm)
Habit: subshrub; semi-evergreen to deciduous; procumbent, spreading by rhizomes; fragrant early spring blooms
Site: shade, partial shade; moist, neutral to acidic, organic, well-drained soil; zone 5
Use: groundcover

Recognition

Leaves: elliptic to wider above middle, prominently toothed above middle; glabrous above, slightly hairy beneath; petioles about 1 in (25 mm) long
Flowers: clusters (spikes) axillary, near base of stem; flowers white or pinkish
Fruits: capsules 3-horned

Pachysandra terminalis Sieb. & Zucc. - **Japanese pachysandra**

Japanese pachysandra forms a solid mat of vegetation on moist, acidic soil under trees.

Planning

Height: up to 1 ft (30 cm)
Habit: subshrub; evergreen; early spring bloom; upright, spreading by rhizomes
Site: shade; moist, high organic soils, pH 5.0-7.0; zone 4
Use: groundcover

Recongition

Leaves: whorled at top of stem; egg-shaped, sometimes wider above middle, coarsely toothed; surfaces glabrous, veination white; petioles about 0.8 in (2 cm) long
Flowers: clusters (spikes) terminal; flowers white
Fruits: berry-like, 2- to 3-horned, whitish, about 0.4 in (1 cm) long
Twigs: upright, greenish

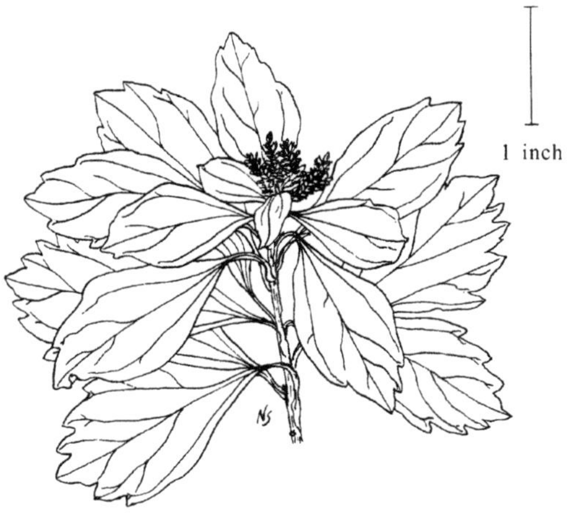

Pachysandra terminalis

Euphorbiaceae Juss. - **spurge family**

The spurge family of about 300 genera and 7,500 species is widely distributed in temperate and tropical regions (Cronquist 1981). A milky juice is found in many of the herbs, trees and shrubs of this varied family. Most of the woody plants are tropical.

Heavea braziliensis (Willd. ex Juss.) Mull., is the world's most important source of rubber. Cassava, *Manihot utilissima* Pohl, is the source of tapioca, and *Ricinus communis* L. provides castor oil. The spurge family also provides tung oil, vegetable tallow, dyes and lumber.

Euphorbia, the largest genus of the family with some 1,600 herbaceous species, is divided by some authorities into several genera. This divergence of opinion will not be addressed here. Common names of a few well-known plants: poinsettia, a symbol of the Christmas holiday season; leafy spurge, spotted spurge and a host of other spurges, some of which are troublesome weeds.

Securinega Juss. - securinega

About 10 species of *Securinega* grow in temperate and subtropical regions.

Securinega suffruticosa (Pall.) Rehd. - spoil-ax

Spoil-ax plants—native to Central China and northeastern Asia—growing in trials at the West Central Research and Extension Center at North Platte, Nebraska, and at the South Dakota Arboretum at Brookings are vigorous with attractive foliage, but are subject to winter injury (North Central Plant Introduction Station 1962).

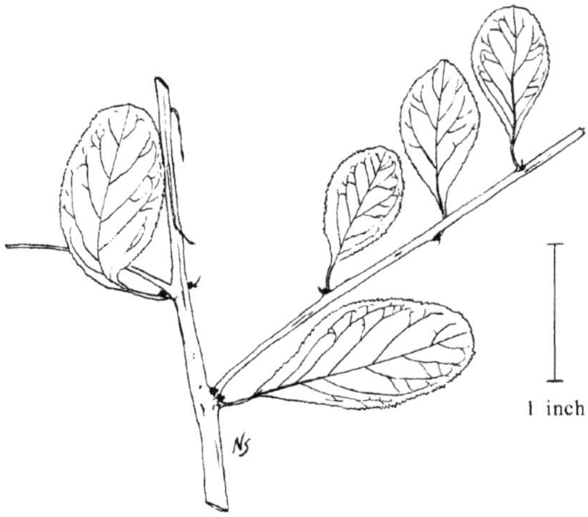

Securinega suffruticosa

Planning

Height: up to 10 ft (3 m)
Habit: shrub; deciduous; dense, upright; flowers not ornamental
Site: sun; pH and soil adaptable; drouth tolerant; zone 5
Use: border, windbreak

Recognition

Leaves: simple, alternate; egg-shaped to elliptic, up to 2.5 in (6 cm) long; tip pointed or rounded, base tapered, entire; bright green or yellowish green; petioles 0.1 in (3 mm) long
Flowers: male clusters of 5-10, female solitary, axillary; unisexual (dioecious or monoecious); flowers regular; tepals 5, greenish white; stamens 5; ovary 3-celled, styles 3
Fruits: dry, splitting (capsular schizocarps); subglobose, about 0.2 in (5 mm) across; greenish; persistent perianth at base
Twigs: slender, greenish, glabrous
Buds: egg-shaped; scales several

Order Rhamnales

The order Rhamnales consists of three families and about 1,700 species (Cronquist 1981).

Recognition

Habit: trees, shrubs, vines, seldom herbs; deciduous or evergreen
Leaves: alternate, sometimes opposite; simple, sometimes compound; stipules usually deciduous or wanting
Flowers: bisexual, sometimes unisexual, regular; petals distinct, sometimes united or wanting; stamens same number as the opposite petals or alternating with sepals
Fruits: fleshy (drupaceous) or dry, splitting (schizocarpic)

Rhamnaceae Juss. - buckthorn family

About 55 genera and nearly 900 species of the buckthorn family occur naturally throughout temperate and tropical regions. Seventy-five of the species are native to North America. The buckthorn family is a source of dyes and medicines and many species are valuable plants for landscapes and wildlife habitat.

Recognition

Habit: trees, shrubs, vines, rarely herbs; deciduous or evergreen
Leaves: simple, alternate or seldom opposite; usually with stipules, sometimes modified into spines
Flowers: solitary or clusters, terminal or axillary; usually bisexual; flowers small, greenish or whitish; sepals 4-5, petals 4-5 or wanting; stamens 4-5, distinct, alternate with sepals
Fruits: berry-like, nuts or capsules
Twigs: sometimes spiny

Ceanothus L. - ceanothus

Some 60 species of *Ceanothus* are natives of North America, mostly in California and parts of the dry, intermountain West. They are valuable food sources for wildlife and some are widely planted as ornamentals. Saponin, a glucoside poisonous to humans, is present in the flowers and fruits of many species but does not appear to harm browsing animals. Ceanothus flowers served as soap for Native Americans and pioneers. Most of the hybrids are best adapted to the dry, mild climate of southern California.

Recognition

Habit: shrubs, rarely small trees; deciduous or evergreen
Leaves: simple, alternate or opposite, usually toothed; 3 main veins; stipules small, deciduous
Flowers: clusters many-flowered; bisexual; flowers small; calyx urn-shaped, deeply 5-lobed, often incurved, colored; petals 5, hooded and clawed, spreading or recurved; stamens 5 opposite the petals
Fruits: capsules 3-lobed, hypanthium persistent after seed dispersal
Twigs: sometimes spiny

Ceanothus americanus L. - Jerseytea ceanothus

Jerseytea ceanothus, native to much of the eastern U.S. as well as southeastern Nebraska and eastern Kansas, is found on infertile, dry sites. It is adapted to these sites because of its massive root system and its capability to use atmospheric nitrogen. Because it is an attractive food for grazing and browsing animals, Jerseytea is most prevalent in ungrazed or lightly grazed prairies, open woods, and along roadsides. It supplied leaves for tea and roots for fuel at timber-scarce pioneer campsites (Gilmore 1977).

C. americanus var. *pitcherii* T. & G. is a native of the southeastern U.S. including eastern Texas and Kansas, east to Indiana (Stephens 1973). It is a parent of many ornamental hybrids. Variety *pitcherii* is easily confused with *C. ovatus*, a plant that grows in the same habitat, often side by side.

Rhamnaceae *Ceanothus*

Planning

Height: up to 3 ft (1 m)
Habit: shrub; deciduous; upright, multistemmed, dense; white blooms May-July on new growth
Site: sun, partial shade; pH and soil adaptable, tolerates drouth; zone 4
Use: naturalizing, border

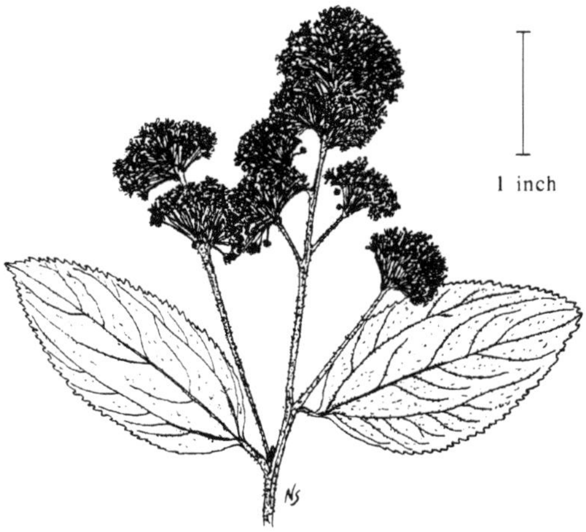

Ceanothus americanus

Recognition

Leaves: alternate; 1-3 in (3-8 cm) long, egg-shaped; sharp-pointed, base rounded, 3 main veins from base; margin minutely toothed (serrulate), gland-tipped, hairy (ciliate); dull green, variably hairy beneath; stipules small, linear, soon deciduous but sometimes persistent
Flowers: clusters (panicles) axillary, stalks leafless or sometimes subtended by 2 bracts; flowers white, odorless; calyx sharply incurved, white, smooth; petals pipe-shaped, tapered to slender thread-like base; stamens 5; ovary 3-lobed surrounded by 10-lobed disk; stigmas 3, linear
Fruits: capsules black, lobes crested; hypanthium cup-like; seeds dark red-brown, glossy, slightly keeled
Twigs: slender, densely hairy, gray-green; leaf scars half-round, somewhat raised, bundle scars 3; stipule scars narrow
Buds: sessile, egg-shaped; scales brown with long, pale hairs exserted from tip, inner scales hairy
Bark: greenish brown with splits showing brown inner bark
Roots: reddish

Ceanothus fendleri Gray - Fendler ceanothus

Fendler ceanothus is commonly found in the southwestern U.S. and also north into Colorado and the Black Hills of South Dakota. It usually grows on rocky soils in grasslands and open woodlands, especially among ponderosa pines. This valuable western browse plant should be useful in landscape plantings in the shortgrass prairie region.

Planning

Height: up to 3 ft (1 m)
Habit: shrub; tardily deciduous; thicket, spiny; summer blooms white
Site: sun; tolerates infertile, rocky soils and drouth; zone 4
Use: naturalizing, wildlife habitat

Recognition

Leaves: alternate; egg-shaped, up to 1 in (25 mm) long; tip pointed, base wedge-shaped, margin entire; dark green with few appressed hairs above, whitened with dense hairs below; stipules hairy, soon deciduous

Flowers: clusters terminal, similar to *C. americanus* except central disk is 6-lobed

Fruits: capsules dark brown, lobes slightly keeled; hypanthium saucer-shaped; seeds brown, glossy, obovoid

Twigs: rigid, gray-green, covered with gray hoary hairs (canescent) and coarse, granular, waxy blooms (pruinose); blister-like glands often scattered on surface; tips sharp (spinescent); leaf scars raised, circular

Buds: sessile, naked, consisting of densely hairy embryonic leaves, sometimes with a single outer, brown, scale-like structure

Ceanothus ovatus Desf. (*C. herbaceous* Raf.) inland ceanothus

The native range of inland ceanothus is very similar to that of Jerseytea ceanothus, but the inland is more commonly found in Kansas and Nebraska. The two species are easily confused unless studied in detail (see Jerseytea ceanothus). Dirr (1990) rates inland ceanothus with its showy white inflorescence superior to Jerseytea ceanothus for landscape plantings. It can be treated as a suffrutescent perennial, and can be pruned to the crown each autumn.

Variety *pubescens* (T. & G.) Shinners is distinguished by leaves with very hairy undersurfaces.

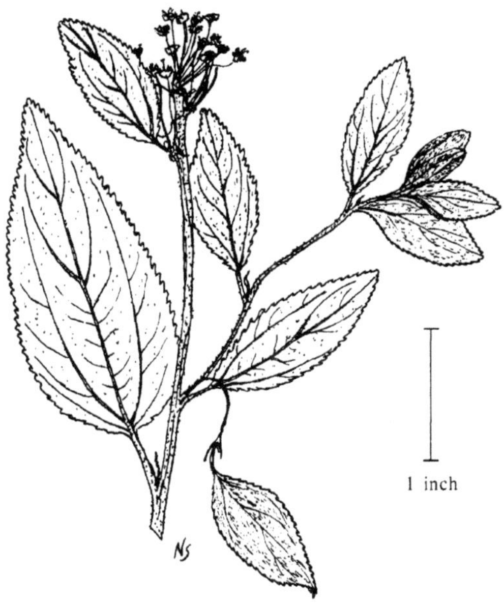

Ceanothus ovatus

Planning

Height: up to 3 ft (1 m)

Habit: shrub; deciduous; upright, dense; spring flowers white on new growth

Site: sun; neutral to alkaline soil, tolerates low fertility and drouth; zone 4

Use: border, naturalizing, wildlife habitat

Recognition

Leaves: alternate; about 2 in (5 cm) long, egg-shaped; tips variably pointed, base rounded, margin finely toothed with callous tips; surfaces dull, hairy or glabrous above, paler and variably hairy below; stipules hairy

Flowers: clusters terminal, stalks often bearing leaves, otherwise similar to *C. americanus*

Fruits: similar to *C. americanus* except capsule lobes without crests; fruiting clusters persist into the winter

Twigs: brittle, densely hairy, red-brown, otherwise very similar to *C. americanus*

Buds: narrowly egg-shaped, woolly, red-brown beneath the hairs

Ceanothus velutinus Dougl. - snowbrush

This native of the Black Hills and states to the west is the most widespread member of *Ceanothus*. Snowbush is the appropriate common name for this plant with white, fluffy flower masses. It is an excellent erosion control plant because of its dense, low-growing habit and nitrogen-fixing capability. Heat stimulates seed germination which allows snowbush to invade burned areas and become the dominant vegetation. The dense, compact growth of snowbrush inhibits reproduction of other plant species (U.S. Forest Service 1937).

Planning

Height: up to 5 ft (1.5 m)
Habit: shrub; evergreen; spreading, basal branches often procumbent; early summer blooms
Site: sun; tolerates dry, rocky soil; zone 4
Use: erosion control, wildlife habitat

Recognition

Leaves: alternate; oval, 2-3 in (5-8 cm) long; aromatic, thick, leathery, 3 main veins; tip and base rounded, margin finely toothed, gland tipped; upper surface hairy becoming glabrous, waxy and glossy above, minutely hairy below; stipules soon becoming brown but persistent
Flowers: clusters compact, grape-like; calyx lobes white, curved between petals over flower center; petals white, pipe-shaped, usually curved downward
Fruits: similar to *C. fendleri*
Twigs: rigid, olive green with very short, fine hairs (puberulent)
Buds: leaf axillary, naked, consisting of small, bract-like leaves; flower buds naked on hairy stalks (an embryonic thyrse loosely and partially covered by brown, hairy bracts)

Hovenia Thunb. - hovenia

Hovenia is comprised of a single species.

Hovenia dulcis Thunb. - Japanese raisin-tree

Japanese raisin-tree, a native of China, is a small, attractive tree suitable for yards or streets. A specimen in Maxwell Arboretum at UNL was removed after suffering severe dieback during several winters.

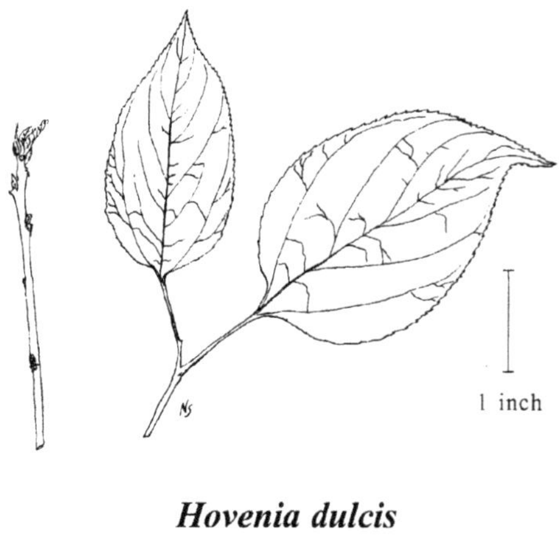

Hovenia dulcis

Planning

Height: up to 20 ft (6 m)
Habit: tree; deciduous; crown rounded; early summer blooms greenish white
Site: experimental; zone 6
Use: shade, street

Recognition

Leaves: deciduous, alternate; broad egg-shaped (ovate) or narrower, 3-nerved, 4-6 in (10-15 cm) long; long-pointed (acuminate), rounded base, margins coarsely toothed; surfaces shiny, sometimes hairy on veins below; petioles up to 2 in (5 cm)
Flowers: clusters (cymes) many-flowered; bisexual; sepals 5, petals 5, greenish white
Fruits: fleshy, drupe-like, about 0.3 in (8 mm) in diameter; seeds 3; fruit stalks red, fleshy, edible
Twigs: slender, zigzag, brown; lenticels small, round
Buds: terminal absent, lateral in leaf scar notch; scales 1-2, hairy
Bark: gray-brown, flat ridges, dark furrows

Rhamnus L. - buckthorn

About 100 species of buckthorn grow throughout temperate and tropical regions. Buckthorn fruits are valuable bird food and bark extracts are used for medicines.

Recognition

Habit: trees, shrubs; deciduous or evergreen
Leaves: alternate or opposite, margins entire to variously toothed
Flowers: clusters axillary; bisexual and unisexual on the same and different plants; sometimes without petals, greenish to whitish
Fruits: berry-like, fleshy, 2- to 4-seeded
Twigs: often spine-like (spinescent)
Buds: terminal sometimes absent, scaly or naked

Grouping Key for Rhamnus

(Rehder 1940)

Leaves opposite, subopposite, or sometimes alternate, deciduous or evergreen; flowers usually 4-parted, unisexual and bisexual on same plant; twigs spinescent, sometimes not; buds scaly, terminal wanting: **subgenus *Eurhamnus***

Leaves alternate, deciduous or evergreen; flowers 5-parted, bisexual; twigs unarmed; buds naked: **subgenus *Frangula***

Subgenus *Eurhamnus*

Rhamnus cathartica L. - European buckthorn

European buckthorn, a native of Europe and western and northern Asia, has made itself at home and unwelcome in much of the U.S. including the eastern prairie. This hardy plant thrives with benign neglect. Some 30 years ago, European buckthorn was the predominant shrub on the east campus of UNL. Birds were doing their bit to green the campus during a period of austerity for campus beautification.

European buckthorn is an attractive foliage plant that can serve as a screen or hedge and provide bountiful bird food. Medicinal products are made from its bark and fruits. Unfortunately, it serves as an alternate host for an oat rust disease and can become a nuisance weed.

Similar species are: *R. japonica* Maxim., *R. arguta* Maxim., *R. saxatilis* Jacq., and *R. dumetorum* Schneid.

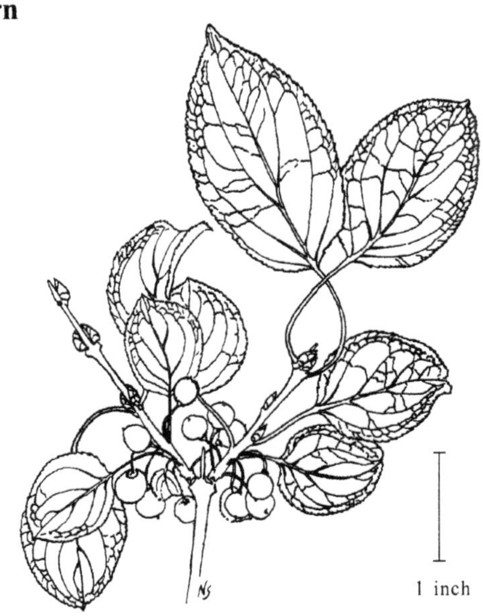

Rhamnus cathartica

Rhamnaceae *Rhamnus*

Planning

Height: up to 20 ft (6 m); National champion, 1972, 61 ft (18.5 m), Michigan
Habit: shrub, tree; deciduous; crown irregular to rounded; fruits persist into December
Site: sun, partial shade; pH and soil adaptable; zone 3
Use: border, naturalizing, wildlife habitat

Recognition

Leaves: opposite or subopposite; oval or widest above middle (obovate), 1-3 in (25-75 mm) long; abruptly pointed, base rounded to wedge-shaped (cuneate), margin with incurving glandular teeth; dark green above, surfaces glabrous; petioles slightly grooved above; stipules linear, often persistent
Flowers: clusters of 2 to 5; usually unisexual (dioecious); flowers yellow-green; calyx lobes 4; petals 4, curved around a stamen
Fruits: drupes berry-like, black, juicy; seeds 1-4
Twigs: glabrous, gray to yellow-brown, often ending as a spine
Buds: terminal a modified spine; lateral curved toward stem; scales overlapping, outer fringed with hairs
Bark: gray, smooth, becoming scaly; lenticels long, horizontal

Rhamnus davurica (*dahurica*) Pall. - dahurian buckthorn

Dahurian buckthorn is an import from northern China and Siberia and is very similar to European buckthorn in many characteristics including its ability to grow on a wide variety of sites. Naturalized plants have been identified in rural southeastern Nebraska (Sutton 1998). Dahurian buckthorn is adapted to the soils and climate of the Fort Collins area (Fairchild and Klett 1993).

Planning

Height: up to 25 ft (7.6 m)
Habit: shrub, tree; deciduous; crown open, spreading, with single or clustered stems; fruits persist into winter
Site: sun, partial shade; pH and soil adaptable; zone 3
Use: naturalizing, wildlife habitat, windbreak

Recognition

Leaves: usually opposite or subopposite; elliptic or oblong, 3-4 in (7-10 cm) long; tip pointed, base wedge-shapead, margin toothed with gland on incurved side; surfaces glabrous, yellow-green; petioles grooved above, stipules linear
Flowers and Fruits: very similar to those of *R. cathartica*
Twigs: glabrous, often with a gray waxy covering (pruinose), occasionally ending in a spine
Buds: lateral parallel to stem; scales often fringed
Bark: often silvery

Rhamnus davurica

Rhamnus lanceolata Pursh - lanceleaf buckthorn

Variety *glabrata* Gl. occupies open, wooded slopes and ravine banks in the open prairie as solitary plants and occasional thickets in the eastern portion of the Great Plains (Stephens 1973). The natural range of the species extends east into the Ohio River Valley and Tennessee.

Drab flowers and fruits dampen interest in this buckthorn's use for ornamental plantings. Although lanceleaf buckthorn provides valuable habitat for wildlife, it is seldom planted for that purpose.

Alternate branching and lack of thorns distinguishes this native from the naturalized buckthorns in the prairies.

Similar species with scaly buds and alternate branching include: *R. schneideri* Levl. & Vant., *R. imeretina* Kirchn., *R. fallax* Boiss., *R. alpina* L., *R. pumila* L., and *R. alnifolia* L'Her.

Planning

Height: up to 10 ft (3 m)

Habit: shrub; deciduous; dense with fine branches; early May blooms greenish yellow; early summer fruits

Site: sun; pH and soil adaptable, drouth tolerant; zone 4

Use: naturalizing, wildlife habitat

Recognition

Leaves: alternate; egg-shaped to elliptic, 2-3 in (50-75 mm) long; tip pointed, base broad wedge-shaped, margin finely toothed with incurved teeth; upper surface shiny, glabrous, hairy midrib below; petioles often reddish, glabrous, grooved above; stipules early deciduous

Flowers: solitary or clusters of 2 to 3, axillary; unisexual (dioecious); calyx 4-lobed; petals 4, greenish

Fruits: black, bloomy, a portion of calyx persistent; seeds 2

Twigs: glabrous, gray-brown; stipule scars prominent, leaf scars crescent-shaped, bundle scar 1

Buds: egg-shaped; scales overlapping, often hairy fringed (ciliate)

Bark: smooth, gray, often with light blotches; lenticels dark, horizontal

Subgenus *Frangula*

Rhamnus caroliniana Walt. - Carolina buckthorn

Carolina buckthorn—a small tree native to the southeastern U.S. as far west as southern Missouri and central Texas—grows on fertile bottomlands and limestone soils. Birds devour the Carolina's sweet, edible fruits.

Planning

Height: up to 30 ft (9 m); National champion, 1982, 43 ft (13 m), Tennessee

Habit: shrub, tree; deciduous; round to irregular; showy red to black fruits August through October

Site: sun, partial shade; pH and soil adaptable, prefers moist, fertile sites; zone 5

Use: naturalizing, wildlife habitat

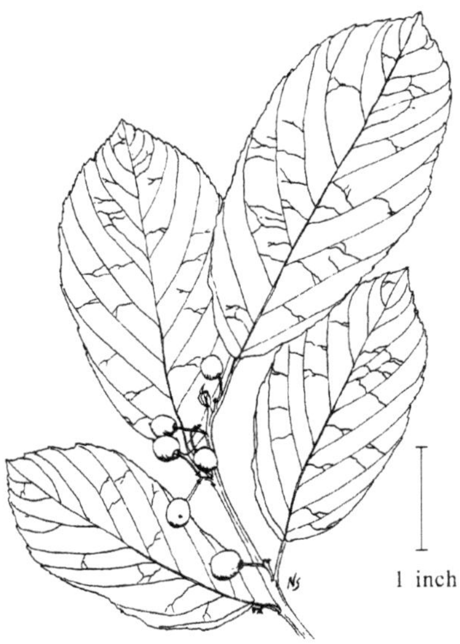

Rhamnus caroliniana

Rhamnaceae *Rhamnus*

Recognition

Leaves: alternate; elliptic, 2-5 in (5-13 cm) long; tip pointed, tapering or rounded to base; slightly toothed to entire; surfaces nearly glabrous, 8-10 vein pairs; petioles hairy, usually longer than flower stalks (peduncles)

Flowers: clustered (umbel-like cymes) at branchlet junctions, short-stalked; bisexual; calyx 5-lobed; petals 5, each folded around a stamen

Fruits: red ripening to black; seeds 2-4

Twigs: red-brown, hairy, becoming glabrous and gray; thornless

Buds: woolly; terminal longer than lateral

Bark: smooth, shallow-furrowed; gray with occasional dark blotches

Rhamnus frangula L. - glossy buckthorn

Glossy buckthorn, an import from Asia, Africa and Europe, has escaped from cultivation in much of the eastern U.S. 'Columnaris', often sold as "tall hedge," is well adapted to the soils and climate of PERC at Fort Collins with no disease or insect problems (Fairchild and Klett 1993). 'Asplenifolia', a narrow-leaved cultivar of fine texture, grows well in Fremont, Nebraska (Sutton 1998).

Planning

Height: up to 15 ft (4.5 m); National champion, 1991, 40 ft (12.2 m), Michigan

Habit: shrub, tree; deciduous; columnar to spreading; May blooms creamy green; summer fruits green, red and black all present at same time

Site: sun, partial shade; pH and soil adaptable; zone 4

Use: hedge, windbreak, wildlife habitat

Recognition

Leaves: alternate; oval or wider above middle, 1-3 in (25-75 mm) long; tip pointed, base rounded, margin entire; shiny dark green above, slightly hairy below

Flowers: solitary or fascicles of 2-10; bisexual; calyx 5-parted, petals 5, creamy green

Fruits: red to purple-black

Twigs: hairy, becoming glabrous; lenticels prominent, gray, rectangular

Buds: terminal present; naked, hairy

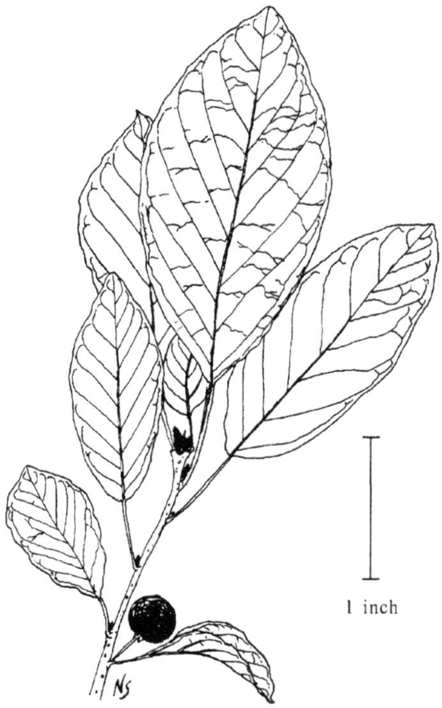

Rhamnus frangula

Zizyphus Mill. - zizyphus

About 40 species of *Zizyphus* grow in tropical and subtropical regions. One species is known to be adapted to a portion of the prairie region.

Zizyphus jujuba Mill. - jujube

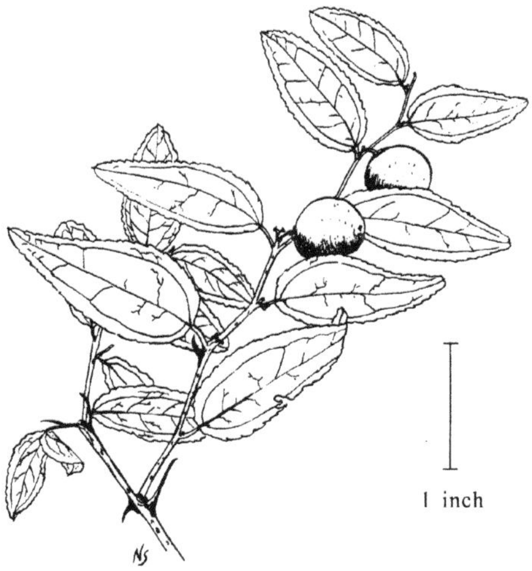

Zizyphus jujuba

Jujube—a native of Asia and subsequently naturalized in the southern U.S.—has been grown in China for its fruit for over 4,000 years, and later in southern Europe and northern Africa (Reich 1992). The jujube's fruits resemble dates in appearance, texture and flavor; sugar content is high.

Among the 400 jujube cultivars in China are 'Li', with fruits two inches (5 cm) in diameter, and 'Lang', with pear-shaped fruits of about the same size. 'Silverhill' produces relatively few suckers and thorns and bears fruits about one inch (2.5 cm) in diameter. Reich (1992) provides information on culture and utilization. Provenance plantations of this highly variable plant would be beneficial.

Jujube has been planted for wildlife habitat in the central and southern prairies. Plantings in Nebraska include a 1976 wildlife planting on the Laird Wolfe farm near Red Cloud and one at the Hans Burchardt Arboretum near Branched Oak Lake. A report by Wolfe (1989) states that the jujube at Red Cloud is sparse and ragged, but produces fruit that is apparently devoured by deer and other wildlife.

Planning

Height: up to 25 ft (7.6 m); National champion, 1989, 43 ft (13 m), Texas
Habit: shrub, tree; deciduous; sparse, suckering, thorny; extended blooming season
Site: sun; pH and soil adaptable, drouth tolerant; zone 5
Use: wildlife habitat, fruit production
Caution: disturbing the soil around the plants increases suckering

Recognition

Leaves: simple, alternate; 3-nerved from base, more or less egg-shaped, 1-2 in (25-50 mm) long; margin toothed (crenate-serrulate); surfaces glabrous; stipules spiny, 1 of the pair slender, about 1 inch (25 mm) long, the other decurved
Flowers: clusters of 2-6 axillary; bisexual; sepals 5; petals 5, yellow
Fruits: fleshy, drupaceous; red, becoming red-brown, shriveled
Twigs: glabrous, spiny
Buds: small; scales 2

Vitaceae Juss. - grape family

Most of the 11 genera and 700 species of Vitaceae are found in tropical and subtropical regions. Several species are important ornamental and food-producing plants in the temperate zone of the Northern Hemisphere. Most grapes are extremely susceptible to the herbicide 2,4-D. A minute concentration of this chemical can cause considerable injury, especially to fruits.

Recognition

Habit: vines (usually climbing by tendrils), rarely upright shrubs or small trees; deciduous or evergreen
Leaves: alternate; simple, less often compound; stipules usually deciduous
Flowers: clusters various, opposite a leaf or terminal, seldom axillary; bisexual or unisexual; perianth regular, usually 4- to 5-parted; calyx small, sometimes reduced to a collar; petals distinct, rarely adhering at base, early deciduous; stamens distinct, opposite petals; ovary superior, commonly somewhat surrounded by the disc
Fruits: berries; fleshy, sometimes dry

Ampelopsis Michx. - ampelopsis

About 20 species of *Ampelopsis*—grape-like vines native to North America and Asia—are cultivated primarily for their handsome fruits.

Recognition

Habit: vines climbing by forked tendrils; deciduous
Leaves: simple or compound
Flowers: clusters long-stalked, forking in pairs (dichotomous), opposite the leaves or terminal; bisexual; flowers small, greenish, usually 5-parted; petals free, calyx indistinct
Fruits: berries; seeds 1-4
Twigs: pith white
Bark: not shreddy; lenticels present

Ampelopsis aconitifolia Bge. - monkshood-vine

Monkshood-vine, a native of northern China, is a delicate, slender, fast-growing vine.

Planning

Height: up to 25 ft (7.5 m)
Habit: vine; fine textured; autumn fruits
Site: sun, partial shade; pH and soil adaptable; zone 5
Use: trellis, fence, rock pile

Recognition

Leaves: compound, long-stalked, 3- to 5-foliolate; leaflets sessile, much longer than wide, up to 2.7 in (7 cm) long; pinnately lobed, often nearly to midrib; margin toothed or entire; surfaces glabrous, long-hairy on veins beneath
Flowers: clusters glabrous, stems (peduncles) up to 2 in (5 cm) long
Fruits: berries about 0.2 in (6 mm) across, orange to blue
Twigs: glabrous

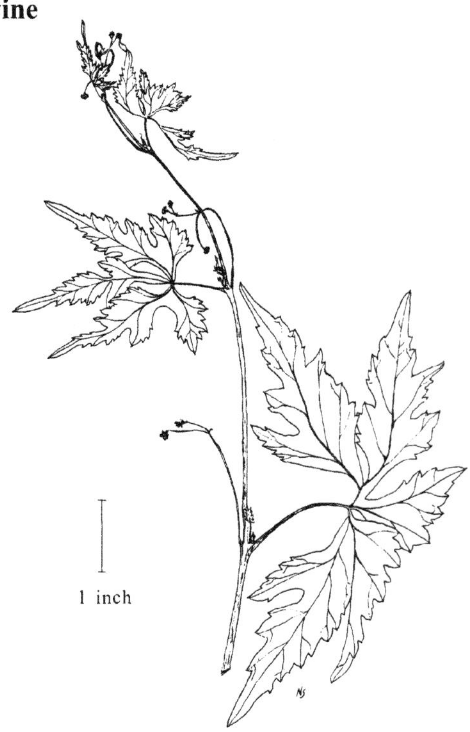

Ampelopsis aconitifolia

Ampelopsis brevipedunculata (Maxim.) Trautv. - **porcelain ampelopsis**

Porcelain ampelopsis, a native of Asia, provides rapid cover with few insect or disease problems and has become a weed in the eastern U.S. Best fruiting occurs when grown in full sunlight (Dirr 1990). Var. *maximowiczii* (Reg.) Rehd. is characterized by its deeply lobed leaves. The leaves of 'Elegans' are variegated.

A. humulifolia Bge., hops ampelopsis, is similar in appearance to poreclain ampelopsis, but the leaves of hops ampelopsis are thicker and firmer and whitish beneath.

Planning

Height: up to 25 ft (7.5 m)
Habit: vine; medium texture; autumn fruits colorful
Site: sun, partial shade; pH and soil adaptable; zone 5
Use: trellis, fence, rock pile

Recognition

Leaves: simple; broad egg-shaped, up to 5 in (13 cm) long; tip long-pointed, base heart-shaped; often variably 3-lobed, coarsely toothed; long-hairy beneath; petioles hairy, about as long as blade
Flowers: clusters hairy, long-stalked; flowers greenish
Fruits: berries about 0.3 in (8 mm) across, yellow to pale lilac and bright blue
Twigs: hairy when young, often angled

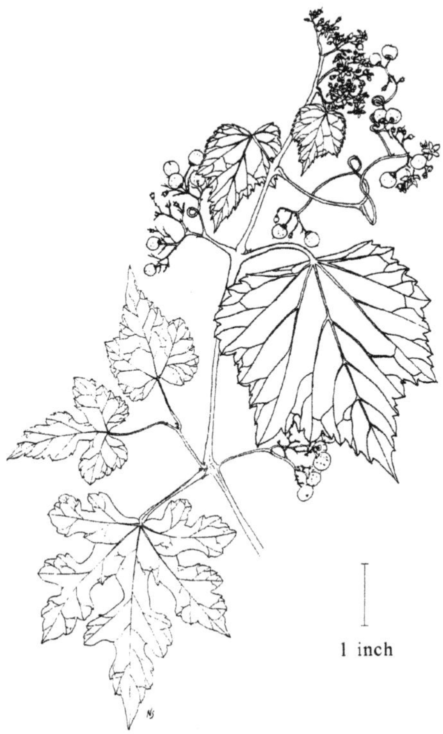

Ampelopsis brevipedunculata

Ampelopsis cordata Michx. - **heartleaf ampelopsis**

Heartleaf ampelopsis, also known as racoon-grape, is a native of the southeastern U.S. and grows naturally as far west as eastern Oklahoma, Kansas and southeastern Nebraska, where it thrives in fencerows, on rocky slopes, and along streambanks (Stephens 1973).

Planning

Height: up to 65 ft (20 m)
Habit: vine; early autumn fruits
Site: sun, partial shade; pH and soil adaptable; zone 5
Use: trellis, fence, rock pile

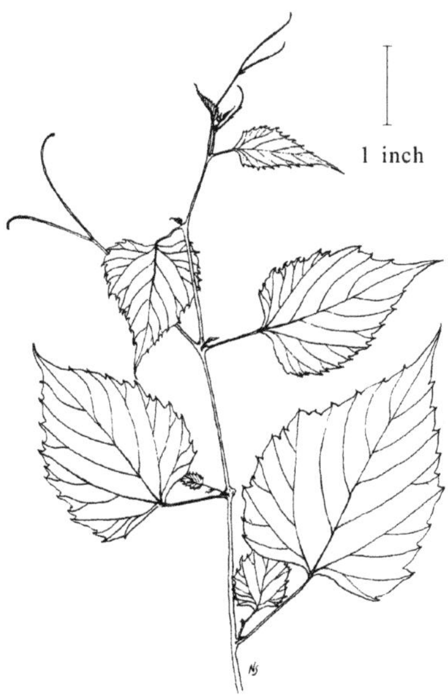

Ampelopsis cordata

Recognition

Leaves: simple; egg-shaped, up to 4 in (10 cm) long, coarsely sharp-toothed; tip pointed, base rounded or squared off; upper surface sparsely hairy becoming less so, sparsely hairy on veins beneath; petioles shorter than blades, often hairy

Flowers: clusters (cymes) green, glabrous, opposite a leaf; unisexual (dioecious); petals 5, green, early deciduous; male, stamens 5 opposite petals; female, stamens vestigial

Fruits: berries about 0.3 in (8 mm) across, varying on cluster from green to orange to rose-purple and turquoise blue; flesh milky, becoming dry

Twigs: glabrous, ridges extending along stem from leaf bases; leaf scars nearly circular

Buds: concealed beneath bark

Parthenocissus Planch. - virgin ivy

Ten species of *Parthenocissus* grow in Asia and North America. These vines, with few exceptions, attach themselves to smooth surfaces by tendrils with adhesive, disc-like tips. The North American species were named "ghost grapes" by the Ponca and Omaha Indian tribes (Gilmore 1977).

Recognition

Habit: vines; deciduous, rarely evergreen

Leaves: simple or compound, variously lobed

Flowers: usually bisexual; floral disc indistinct, petals usually 5, spreading; calyx very small

Fruits: berries dark blue to bluish black; seeds 1-4

Twigs: tendrils usually with adhesive tips; pith white

Buds: rounded; scales, 3 or 4 visible

Parthenocissus inserta (Kern.) K. Fritsch - thicket creeper

Thicket creeper—sometimes named woodbine—grows naturally throughout the prairie region in fencerows and wooded and uncultivated areas that are frequented by fruit-eating birds. Its natural range includes much of the U.S.

Thicket creeper cannot attach to walls because its tendrils lack the sucker discs characteristic of Virginia creeper and many other species.

Planning

Height: up to 30 ft (9 m)

Habit: vine; deciduous; late summer fruits on new growth; autumn foliage red-orange

Site: sun, partial shade; pH and soil adaptable; drouth tolerant; zone 3

Use: naturalizing, wildlife habitat, trellis, fence, rock pile

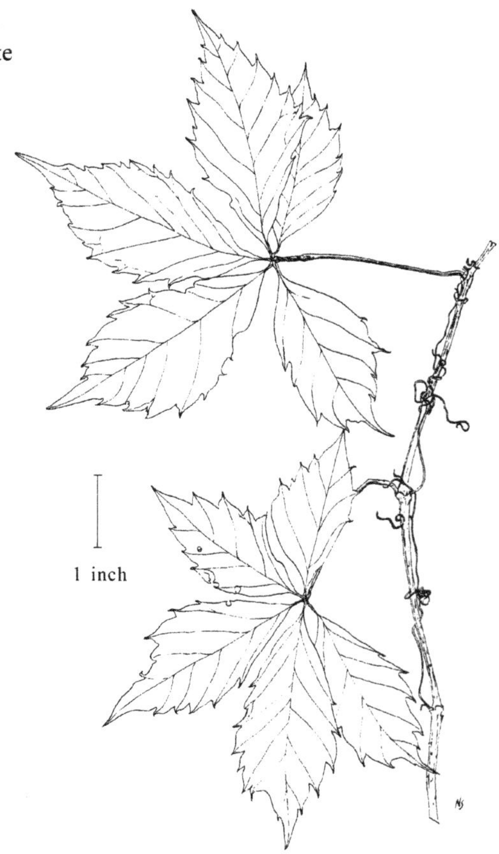

Parthenocissus inserta

Recognition

Leaves: palmately compound; leaflets 5 sometimes 7, egg-shaped, sometimes wider above the middle, coarsely toothed; tip pointed, base wedge-shaped; surfaces glabrous except hairs on veins, sometimes hairy beneath; petioles glabrous, bases enlarged; stipules early deciduous (caducous)
Flowers: clusters branch in pairs (dichotomous); bisexual; petals 5, yellow-green, reflexed, tips thick and incurved, pointed; calyx saucer-shaped with 5 broadly rounded lobes; stamens 5, erect
Fruits: berries nearly round, about 0.3 in (8 mm) across, deep purple with slight bloom; seeds 4
Twigs: glabrous, tendrils branched, seldom with sucker discs; lenticels prominent; leaf scars half-round, pith white
Buds: buried in scaly tissue, those at end of spurs egg-shaped, glabrous

Parthenocissus quinquefolia (L.) Planch. - Virginia creeper

Virginia creeper—a native of the eastern U.S. including the eastern prairies—grows in open woodlands and undisturbed areas. It tolerates extremes in soil, exposure and pollution (Dirr 1990), but Sutton (1998) has evidence that Virginia creeper is sensitive to environmental extremes. This tough, adaptable plant provides a low maintenance cover for many buildings. Tendrils tipped with adhesive discs provide the only support needed. The discs are so tenacious that they leave scars after their removal, especially on wood.

The underleaf surface of var. *hirsuta* (Donn) Fern. is hairy. The leaves of var. *engelmannii* (Graebn.) Rehd. are smaller than those of the species. Close examination of populations where Virginia creeper and thicket creeper overlap often reveals plants with characteristics of each species, an indication that natural hybridization occurs (Stephens 1973).

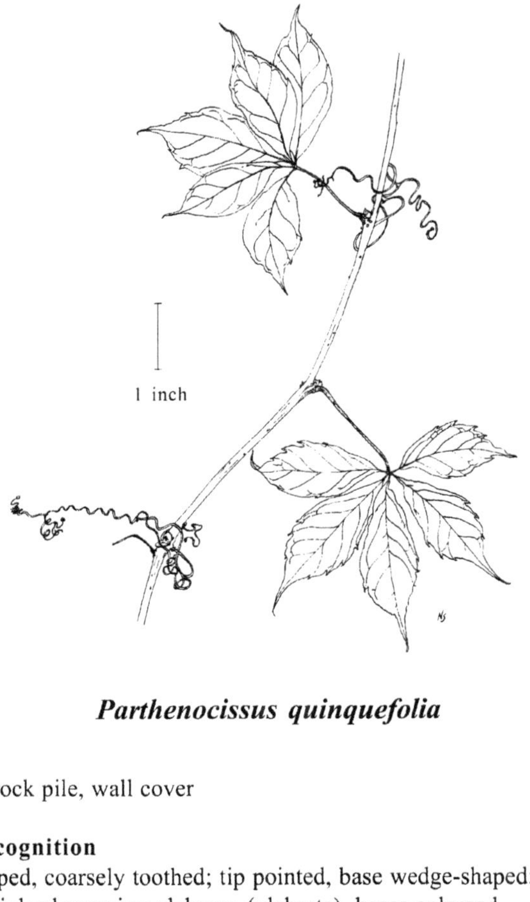

Parthenocissus quinquefolia

Planning

Height: up to 60 ft (18 m)
Habit: vine; deciduous; autumn foliage red-orange, fruit persists after leaf fall
Site: sun, shade; pH and soil adaptable; zone 4
Use: naturalizing, wildlife habitat, trellis, fence, rock pile, wall cover

Recognition

Leaves: palmately compound; leaflets 5, egg-shaped, coarsely toothed; tip pointed, base wedge-shaped; surfaces glabrous except on veins beneath; petioles becoming glabrous (glabrate), bases enlarged
Flowers: clusters with a well-defined axis; bisexual; petals 5, greenish to bronze, reflexed, tips incurved, pointed; calyx saucer-shaped, not lobed; stamens 5, erect
Fruits: berries nearly round, about 0.3 in (7 mm) across, dark purple with bloom; seeds 4
Twigs: finely hairy, tendrils branched, tipped with sucker discs; coarse adventitious roots without discs along old stems; leaf scars nearly circular; pith green to white
Buds: conical, red-brown, pointed

Parthenocissus tricuspidata (Sieb. & Zucc.) Planch. - **Boston ivy**

Boston ivy, a native of Japan and China, is nearly as adaptable to extreme sites as its American cousins. It provides an excellent, nearly trouble-free cover for masonry structures. Hoag (1965) states that some freeze-back can be expected in the northern plains, but new growth quickly fills the gaps each season.

Boston ivy cultivars include: 'Lowi', with deeply divided leaves and fine texture; 'Purpurea', with purplish leaves; 'Beverly Brooks', with large leaves and brilliant autumn foliage in some latitudes; and 'Robusta', with waxy, glossy green leaves that are often compound (trifoliate) (Dirr 1990).

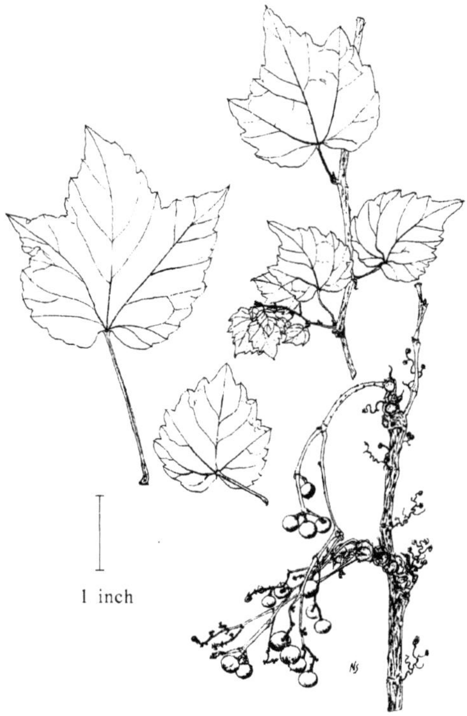

Parthenocissus tricuspidata

Planning

Height: up to 60 ft (18 m)
Habit: vine; deciduous; autumn foliage red to orange
Site: sun, partial shade; pH and soil adaptable; zone 5
Use: naturalizing, wildlife habitat, masonry wall cover
Caution: avoid south and west exposures in hot, dry climates

Recognition

Leaves: simple; broadly egg-shaped, usually 3-lobed and toothed, or sometimes compound (trifoliate) with stalked leaflets; glabrous above, hairy on veins beneath
Flowers: clusters (cymes) narrow, elongated; usually on short 2-leaved twigs
Fruits: berries bluish black, bloomy, globose, about 0.3 in (7 mm) across
Twigs: tendrils many-branched with sucker discs at tips

Vitis L. - **grape**

Most of the 60-odd grape species are natives of the temperate regions of the Northern Hemisphere. They provide valuable wildlife habitat, food for humans, and cover for trellises, arbors and loggia. Grapes were probably used for wines long before their "taming." Other important grape products are raisins, fresh fruits and juices. *V. vinifera* L., the wine grape, originated in the Caucasian region of eastern Europe and was among the first plants to be domesticated.

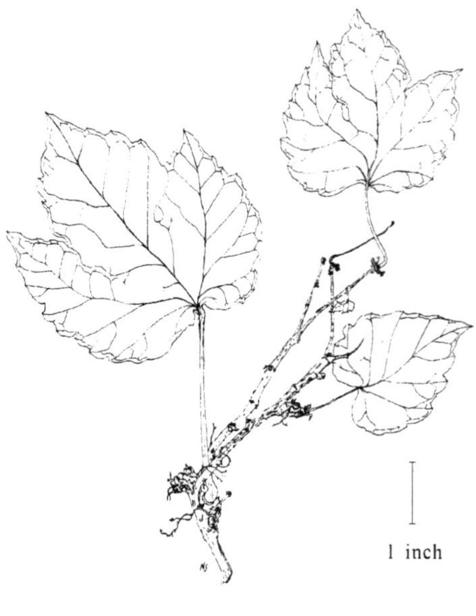

***Vitis* sp.**

V. labruska L., fox grape, a native of the eastern U.S., is the only indigenous grape grown commercially. The best known cultivar of fox grape is 'Concord', which originated on the farm of Ephraim Bull in Concord, Massachusetts (Hylander 1956). Many edible and hardy varieties are the result of selections from crosses between wine and fox grapes. The U.S. grape industry is centered in California and the Great Lakes region of Ohio and New York. However, small vineyards are appearing throughout the U.S., even in the prairies. Otherwise, grapes are primarily grown by the home gardener.

Extreme sensitivity of grapes to 2-4,D–an herbicide used to control broadleaf weeds in lawns and grain crops—discouraged fruit production in some areas, but its reduced use and more careful application in recent years has allowed successful home and commercial fruit production to become more widespread.

Recognition

Habit: vines; deciduous, rarely evergreen (never in the central and northern prairies)
Leaves: simple, rarely palmately compound; toothed, usually palmately lobed
Flowers: clusters opposite a leaf; unisexual (dioecious), also bisexual (polygamo-dioecious); perianth 5-parted; petals cohering at apex and falling in entirety at pollen shed (anthesis); sepals minute or obsolete
Fruits: berries pulpy; seeds 2-4
Twigs: with tendrils, usually forked; pith brown, usually interrupted at nodes by diaphragm
Bark: usually without lenticels on older branches, separating into long strips and fibers

Vitis acerifolia Raf. - bush grape

Bush grape grows in open or brushy areas in prairie ravines and on rocky bluffs in scattered locations in western Kansas, southeastern Colorado, Texas and New Mexico (Stephens 1973). This grape often forms dense thickets that control erosion and provide an excellent cover for wildlife.

Planning

Height: up to 20 ft (6 m)
Habit: vine; bushy, thicket forming; early August fruits
Site: sun; pH and soil adaptable; drouth tolerant; zone 5
Use: wildlife habitat, erosion control

Recognition

Leaves: simple; broadly egg-shaped to kidney-shaped, often tending to fold lengthwise, lobed or not lobed; sharp coarse toothed, valleys (sinuses) sharp, tips long-pointed, basal lobes rounded with broad sinus; cobwebby hairs above, short white hairs (tomentose) beneath; petioles up to 2 in (5 cm) long, cobwebby
Flowers: clusters (panicles) opposite a leaf on new growth; unisexual (dioecious); calyx saucer-shaped, low, flat, scarcely lobed; petals 5-6; stamens 5-6 (vestigial on female flowers)
Fruits: berries globose, about 0.3 in (7 mm) across, purple with bloom, strongly acrid, becoming sweet late, often after drying
Twigs: long-hairy becoming glabrous; tendrils scarce, short, simple or short-branched
Buds: egg-shaped, about 0.1 (3 mm) long; red-brown tipped with a white hair tuft
Bark: red-brown; not shredded first year, becoming long, thin, shreddy

Vitaceae *Vitis*

Vitis aestivalis Michx. - **summer grape**

Summer grape, a native of the eastern U.S., grows in eastern Kansas and extreme southeastern Nebraska where it usually is found on undisturbed dry upland sites such as fence rows, along bluffs and in open woods. 'Norton's Virginia' has been grown for table use.

Planning

Height: up to 30 ft (9 m)
Habit: vine; late August fruits
Site: sun, partial shade; pH and soil adaptable; drouth tolerant; zone 4
Use: trellis, arbor, jelly, wine, wildlife habitat

Recognition

Leaves: simple; egg-shaped to rounded, lobes 3-5, coarsely toothed, valleys (sinuses) rounded; tips short-pointed, basal lobes rounded with narrow sinus; upper surface veins sparsely hairy, light rusty cobwebby hairs beneath with straight hairs on veins; petioles up to 4.5 in (11.5 cm) long, often red, glabrous or sparsely hairy
Flowers: clusters dense, pyramidal, opposite a leaf on 2 successive nodes, third node without; unisexual (dioecious); petals 5, white, tips often reddish; stamens 5 (vestigial on female flowers)
Fruits: berries globose, about 0.5 in (12 mm) across; bloomy purple, sweet
Twigs: glabrous, purplish, finely ridged
Buds: egg-shaped, about 0.2 in (5 mm) long, often at right angles to twig; scales purplish, bloomy, hairy becoming glabrous

Vitis cinerea Engelm. - **winter grape**

Winter grape—a native of southeastern Nebraska, eastern Kansas, Oklahoma and Texas, east to Florida and Virginia—blooms in mid-June and ripens its fruits in late autumn. The small fruits are seldom picked for home use.

Planning

Height: up to 50 ft (15 m)
Habit: vine; mid-September fruits ripen in late autumn
Site: sun; pH and soil adaptable; zone 5
Use: wildlife habitat, jelly, wine

Recognition

Leaves: simple; broadly egg-shaped; no lobes or 2 small, coarsely toothed, somewhat rounded lobes, marginal hairs (ciliate); tips pointed, basal lobes broadly rounded with narrow sinus; upper surface with few cobwebby hairs, often wrinkled, white hairy beneath; petioles up to 4 in (10 cm) long, hairy
Flowers: clusters (panicles) opposite a leaf on new growth; unisexual (dioecious); petals 5-6, green; stamens 5-6 (vestigial on female flowers)
Fruits: berries globose, about 0.3 in (7 mm) across, deep purple with slight bloom; acrid until late autumn
Twigs: fine longitudinal lines (striate), patches of white matted hairs
Buds: conical, about 0.1 in (2 mm) long, red-brown with some matted hairs, especially at tip

Vitis riparia - riverbank grape

Riverbank grape, a native of the eastern U.S. as far west as Montana and New Mexico, grows on a variety of undisturbed sites where birds and animals have helpfully deposited its seeds. This grape's fruits are often picked for the kitchen, especially in the western reaches of its range where tame cultivars are difficult to grow.

Early settlers in the plains drove many miles to river canyons for the fruit. Riverbank grape provided their tables with fresh juice, jellies and, occasionally, a few bottles of wine. It has been cultivated on trellises. 'Beta', a cross with fox grape, is the most common of the cultivated grape varieties in the northern plains (Hoag 1965).

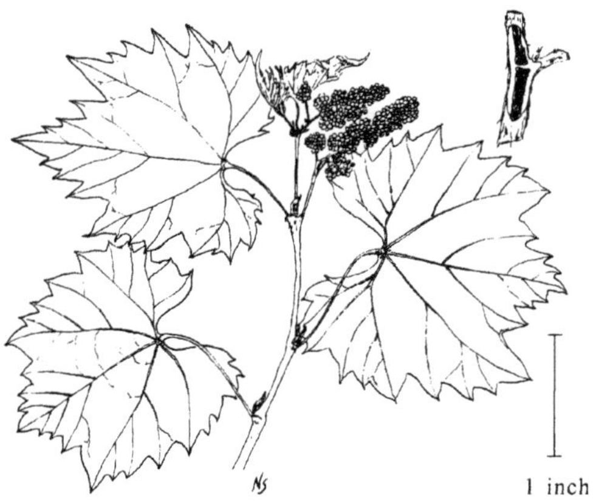

Vitis riparia

Planning

Height: up to 75 ft (23 m)
Habit: vine; high climber; July-August fruits
Site: sun; pH and soil adaptable; drouth tolerant; zone 3
Use: arbor, wildlife habitat, jelly, wine

Recognition

Leaves: simple; egg-shaped to nearly round, 2 lateral lobes; coarsely toothed, sides of teeth mostly concave, margins hairy (ciliate); tips long-pointed, basal lobes rounded with broad sinus; upper surface glabrous, hairy on veins beneath, tufts in vein axils; petioles up to 2.5 in (6 cm) long, becoming glabrous
Flowers: clusters (panicles) opposite a leaf on new growth; unisexual (dioecious); petals 5-6, yellow-green; stamens 5-6
Fruits: berries globose, about 0.4 in (1 cm) across; purple-blue with bloom, sweet, edible
Twigs: glabrous, minutely ridged, tendrils lacking on every third node; nodal diaphragm less than 0.05 in (1 mm) thick
Buds: egg-shaped, about 0.1 in (3 mm) long; outer scales keeled, red-brown, becoming glabrous, tan hair mat on inner scales often visible

Vitis vulpina L. - frost grape

Frost grape is a native of the eastern U.S., with the western boundary of its range extending from Wisconsin south to Texas, passing through southeastern Nebraska and much of Kansas. Frost grape can be distinguished from riverbank grape by its leaf character, a bark that is less shreddy than most species, its late ripening fruits with no bloom, and, if all else fails, its thick nodal diaphragm.

Planning

Height: up to 80 ft (25 m)
Habit: vine; September-October fruits
Site: sun, partial shade; pH and soil adaptable; zone 5
Use: arbor, wildlife habitat, jelly, wine

Recognition

Leaves: simple; nearly round to egg-shaped, shoulder-like lobes or none; coarsely toothed, teeth wider than long with convex sides; tips long-pointed, basal lobes rounded and often overlapping, sinus narrow; upper surface glabrous, short straight hairs on veins below; petioles up to 3 in (7.5 cm) long, glabrous

Flowers: clusters (panicles) loose, pendent, on new growth opposite a leaf on 2 successive nodes, the third without; unisexual (dioecious); petals 5, green; stamens 5 (vestigial on female flowers)

Fruits: berries globose, about 0.4 in (1 cm) across; black, slight bloom when immature becoming glossy; acrid until late autumn

Twigs: glabrous, gray-green becoming red-brown, finely ridged; nodal diaphragm about 0.1 in (2 mm) thick; tendrils about 8 in (20 cm) long, branched, lacking at every third node

Buds: egg-shaped, about 0.2 in (5 mm) long; outer scales keeled, dark brown, 1 large scale covering most of bud, inner scales covered with dense, brown, hairy mat

Bark: furrowed with very little shredding

Order Sapindales

Fifteen familes and about 5,400 species comprise this order (Cronquist 1981).

Recognition

Habit: woody plants, seldom herbs; deciduous or evergreen

Leaves: alternate, opposite or whorled; usually compound or deeply cleft or lobed, seldom simple and unlobed; stipules lacking on most, some minute and early deciduous

Flowers: bisexual or unisexual, regular or irregular; perianth commonly 4- to 5-parted; petals distinct, sometimes wanting; stamens distinct, sometimes filaments fused, commonly 8-10, sometimes 4-5 and sometimes more than twice as many as petals or sepals; ovary compound, usually superior

Fruits: various

Staphyleaceae Lindl. - bladdernut family

The bladdernuts, a family of 5 genera and about 50 species, are natives of the temperate Northern Hemisphere.

Recognition

Habit: trees, shrubs; deciduous

Leaves: opposite, seldom alternate; pinnately compound or trifoliate; leaflets generally toothed; stipules early deciduous or wanting

Flowers: axillary or terminal drooping clusters (panicles); bisexual or sometimes unisexual (dioecious); sepals distinct, overlapping (imbricate) or sometimes variably fused (connate); petals distinct, overlapping; stamens alternate with petals, ovary superior

Fruits: dry, inflated, usually splitting (capsules) or drupaceous

Staphylea L. - bladdernut

Eleven species have been recognized.

Staphylea trifolia L. - American bladdernut

Thickets of this white-flowered shrub are found in the woods and along streambanks from the northeastern U.S. and adjoining Canada, west to Minnesota and south through extreme eastern Nebraska, Kansas and Oklahoma, east to Georgia. American bladdernut is a candidate for understory plantings on the prairies where water is adequate (Richley 1988).

American bladdernut could be confused with waferash (*Ptelea trifoliata*) which occurs in the same general region, especially if the bladdernut's flowers or fruits are not visible. However, the terminal leaflet of the latter is sessile and its white, woolly buds are partially hidden.

Planning

Height: up to 10 ft (3 m)
Habit: shrub; upright, thicket-forming; April-May blooms on old wood
Site: partial shade; pH and soil adaptable, avoid dry sites; zone 4
Use: naturalizing, erosion control

Recognition

Leaves: opposite; compound (trifoliate); leaflets long-stemmed, egg-shaped, may be wider above middle, up to 3 in (8 cm) long; fine-toothed, tip long-pointed, base rounded; glabrous above, fine white hairs beneath; petioles hairy, up to 4 in (10 cm) long
Flowers: clusters (panicles) drooping; bisexual; petals 5, white, about 0.4 in (1 cm) long, narrowed at base, tips with jagged margin (erose); calyx bell-shaped, 5-lobed, pale green or white; stamens 5; ovary 3-lobed, styles 3, greenish, more or less joined
Fruits: capsules, up to 2 in (5 cm) long, inflated, papery; sections 3, each with free pointed tip, sparsely hairy; persistent into winter; seeds 1-4 per section
Twigs: flexible, glabrous, distinct offset at beginning of year's growth; leaf scars half-round
Buds: pair of laterals replaces true terminal; egg-shaped, obtuse, red-brown; scales with hairy margins (ciliate), 2 short scales at base

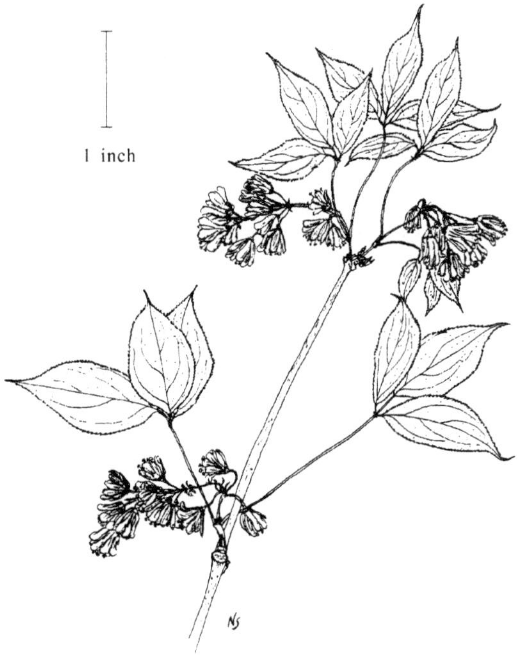

Staphylea trifolia

Sapindaceae Juss. - soapberry family

Most of the 140 genera and 1,500 species of the soapberry family are natives of tropical and subtropical regions.

Recognition

Habit: trees, shrubs, vines (woody or herbaceous); deciduous or evergreen
Leaves: mostly alternate; compound, seldom simple; usually without stipules
Flowers: clusters terminal or axillary, branched, seldom solitary, often slightly irregular; bisexual, unisexual or a mixture (polygamous); perianth 4- or 5-parted; sepals usually overlapping, distinct or sometimes fused below; petals distinct, overlapping, sometimes wanting; stamens 4-10, filaments often hairy
Fruits: diverse, fleshy or dry

Sapindaceae *Koelreuteria*

Koelreuteria Laxm. - goldenraintree

Only one of the six species of goldenraintree—all natives of Asia—is commonly planted in the prairie region.

Koelreuteria paniculata Laxm. - goldenraintree

Goldenraintrees growing in Nebraska are healthy and thriving on diverse sites extending from Nebraska City west to Gothenburg and north to Wayne (Sutton 1998). Dirr (1990) extolls goldenraintree as beautiful and adapted to extreme soil conditions without serious disease or insect problems. Although some authorities claim the wood is weak, Dirr (1990) reports that he as seen no injury to this tree after extreme ice storms that have damaged other trees. Goldenraintree suffered no more nor less harm than other trees in the October 1997 snowstorm in eastern Nebraska.

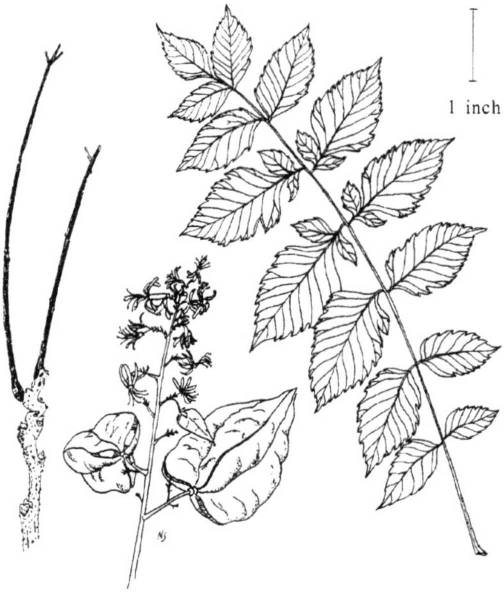

Koelreuteria paniculata

Planning

Height: up to 40 ft (12 m); Nebraska champion, 1980, 53 ft (16 m), Nebraska City

Habit: tree; deciduous; top wide, round; early summer blooms, autumn fruits

Site: sun, partial shade; pH and soil adaptable; tolerates drouth, air pollutants; zone 5

Use: specimen, shade, street tree

Recognition

Leaves: alternate, compound (pinnate or bipinnate); leaflets 7-17, variously egg-shaped, up to 3 in (8 cm) long; irregularly lobed or toothed; glabrous above, sometimes hairy on veins beneath

Flowers: clusters branched, up to 14 in (35 cm) long; bisexual; petals 4, yellow; calyx unequally 5-lobed; stamens usually 8

Fruits: capsules 3-parted, bladder-like, papery, up to 2 in (5 cm) long; seeds 3, pea-sized, black

Twigs: stout, glabrous, brown-red to gray; lenticels prominent; leaf scars shield-shaped; pith solid, white

Buds: terminal absent; laterals with 2 brown scales, glabrous, hairy beneath

Bark: brown, vertically furrowed

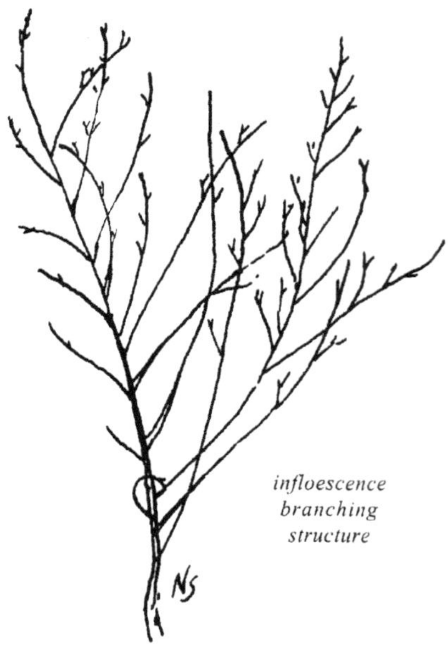

Koelreuteria paniculata

Sapindus L. - soapberry

Two of the widely distributed mostly tropical species of soapberry are natives of North America. Soapberry fruits contain poisonous saponin. When crushed they produce a soapy lather that substitutes for soap, hence the name soapberry. The hard black seeds have been drilled and used for beads.

Sapindus drummondii Hook. & Arn. - western soapberry

Western soapberry is an attractive, wind- and drouth-resistant tree that provides shade for difficult sites. Its heavy, hard wood is valued as fuel and has been used for making baskets and other small items. This tree of the arid southwest—found from south-central Kansas southwest to southern Arizona and northern Mexico—is truly a tree for the self-sufficient because of its many uses.

Planning

Height: up to 50 ft (15 m), National champion, 1984, 72 ft (22 m), Oklahoma
Habit: tree; deciduous; crown rounded; May-June blooms; autumn fruits persistent
Site: sun; pH and soil adaptable; drouth tolerant; zone 6
Use: shade, windbreak
Caution: fruits are poisonous

Recognition

Leaves: alternate; feather-like compound (pinnate), up to 18 in (45 cm) long; leaflets often alternate on central stalk (rachis), egg-shaped to much longer than wide, up to 2.5 in (6 cm) long; margin entire, tip long-pointed, base oblique; pale yellow-green, glabrous above, hairy on veins beneath
Flowers: clusters (panicles), branched, up to 9 in (23 cm) long; unisexual (dioecious); petals 5, white to yellowish, regular, round-tipped, narrowed to a claw with dense tuft of hairs at base, margin hairy fringed (ciliate); calyx lobes 5, distinct, cupped, glabrous, ciliate; stamens 7-8; ovary 3-lobed, style exserted before flower opens, stigma indistinctly 3-lobed
Fruits: drupaceous; nearly round, about 0.5 in (12 mm) across, golden, flesh thick, gummy; shrivels, darkens and persists through winter; seed usually 1, egg-shaped, about 0.3 in (9 mm) long, black, semi-glossy, hard
Twigs: brittle, short-hairy
Buds: often in pairs, one above the other (superposed), upper about 0.1 in (2 mm) across, lower smaller; scales brown
Bark: gray-brown, shallow fissured
Wood: hard, strong, heavy, yellow-brown

Xanthocerus Bge. - yellowhorn

The single yellowhorn species is a native of China.

Xanthocerus sorbifolium Bge. - yellowhorn

The discovery of a 10-foot specimen of yellowhorn in spectacular full bloom growing in a sunny Lincoln backyard created a flurry of excitement among plants people and a rush to get on the seed list. Only a handful of these northern China natives are known to be growing in Nebraska—the northernmost grows in Blair and another grows to the southwest in Indianola (Richley 1988).

Xanthocerus sorbifolium

Planning

Height: up to 25 ft (7.5 m)
Habit: tree, shrub; deciduous; upright, multi-stemmed; early May blooms showy
Site: sun; pH and soil adaptable; zone 4
Use: specimen, border

Recognition

Leaves: alternate; compound (pinnate), up to 1 foot (30 cm) long; leaflets narrow elliptic, usually opposite on the central stem (rachis), up to 2 in (5 cm) long; sharply toothed
Flowers: clusters (racemes) up to 10 in (25 cm) long; bisexual and unisexual on same plant (polygamous); sepals 5, overlapping (imbricate); petals 5, regular, egg-shaped, wider above middle (obovate), white, yellow blotch at base changes to red; disc thin, 5-lobed, each with yellow, upright, horn-like appendage half as long as stamen; stamens 8, half as long as petals
Fruits: capsules dry, splitting open, 3-valved; seeds dark brown, globose, pea-sized
Twigs: glabrous
Buds: egg-shaped; scales several

Hippocastanaceae DC. - horsechestnut family

The horsechestnut family is comprised of two genera, *Aesculus* and *Billia*. The latter is evergreen, found in an area from southern Mexico to northern South America (Cronquist 1981).

Aesculus L. - buckeye

About 13 species of trees and shrubs of *Aesculus* are widely distributed in temperate areas of North America, the Balkan Peninsula of Europe, and Asia.

The buckeyes are prized for their showy flowers and handsome foliage. In addition to large clusters of distinctive flowers, the palmately compound leaves are easily recognized. The unique, leathery capsules split open when ripe to expose one or more dark brown, shiny nuts (known as "conkers" by Europeans), each with a large pale spot reminiscent of a male deer's eye. The nuts somewhat resemble chestnut tree (*Castanea*) nuts.

The nuts of all species of *Aesculus* contain the poison aesculin which causes them to be shunned by most wildlife. However, squirrels eat them without ill effects. The leaves of some buckeye species are also toxic. Stephens (1973) states that roasting apparently destroys the poison because Native Americans are known to have eaten the roasted starchy nuts.

Recognition

Habit: trees, shrubs; deciduous
Leaves: opposite, palmately compound; leaflets 3-11, margins toothed (serrate); stipules wanting
Flowers: clusters (panicles or racemes) terminal, upright, many-flowered; bisexual, many functionally male (staminate); calyx bell- or tube-shaped, 5-lobed; petals 4 or 5, distinct, clawed, overlapping, unequal in size; stamens 6-8, distinct; ovary compound, superior (hypogynous)
Fruits: capsules smooth to spiny, splitting to shed 1-3 brown seeds with conspicuous light gray "eye" (hilum)
Twigs: stout, lenticels large; pith homogenous, white; leaf scars triangular to round, bundle scars conspicuous, usually 6-7
Buds: terminal present unless replaced with a fruit stalk scar; scales many, sometimes sticky, enlarging as leaves emerge and persisting until leaves are about half-grown

Aesculus glabra Willd. - Ohio buckeye

Ohio buckeye, a native of the Ohio River Valley, also occurs on moist lowlands to the west into southeastern Nebraska and south through eastern Kansas and Oklahoma to central Texas. It is sometimes called fetid buckeye because of the odor emitted from crushed plant parts.

Ohio buckeye, a tree for small yards, is a joy to watch in early spring as the large terminal buds swell and disgorge masses of bright purple-green leaves and spikes of creamy flowers. Fall color can be a handsome yellow-orange but is short-lived because the leaves soon drop after coloring. Although the wood is soft and light, it does not split easily and limb breakage is seldom a problem.

Dirr (1990) lists a large number of disease and insect problems associated with the Ohio buckeye, but healthy specimens on the UNL campus exhibit no evidence of serious pest injury. Ohio buckeye trees are well adapted to the soils and climate of PERC in Fort Collins with no insect or disease problems (Fairchild and Klett 1993). A twenty-foot specimen grows at 8,000 feet in southern Colorado's San Luis Valley (Sutton 1998).

A leaf rust caused by *Puccinia andropogonis* which uses bluestem grass species (*Andropogon* sp.) as an alternate host has been the most serious detriment to growth of young seedling Ohio buckeyes at Prairie Pines Arboretum near Lincoln where big bluestem (*A. gerardi*) abounds in the native prairies. This disease is less harmful after the trees grow to about 6 feet (2 m) tall.

Aesculus glabra

Planning

Height: up to 50 ft (15 m); Nebraska champion, 1985, 56 ft (17 m), Omaha; National champion, 1984, 144 ft (43.6 m), Kentucky

Habit: tree; crown rounded, dense; May blooms greenish yellow; autumn foliage yellow to orange-red

Site: sun, partial shade; pH and soil adaptable, prefers moist, slightly acidic soils; zone 3

Use: shade, naturalizing, specimen

Pests: leaf rust may cause early defoliation if *Andropogon* species are nearby

Recognition

Leaves: leaflets 5, rarely 7; elliptic, sometimes wider above the middle (obovate), sessile, 3-6 in (8-15 cm) long; apex long-pointed, base tapered, margin finely toothed and sparingly hairy (ciliate); surfaces glabrous, veins hairy beneath; petioles as long as leaves, enlarged at base

Flowers: clusters (panicles) of 50-60, terminal; bisexual; calyx bell-shaped, 5-lobed, greenish yellow, pink-tipped; petals 4, 2 upper spoon-shaped (spatulate), yellow with orange spots inside, often strongly recurved, outer side and margin covered with crinkly hairs; 2 lower petals egg-shaped, yellow with orange or red streak in center, crinkly and glandular hairs on the outside; stamens 7, exserted on long curved filaments, anthers yellow; ovary stalked, white, hairy, stigma red

Fruits: capsules nearly round, leathery, blunt-spiny; seeds 1-3, dark reddish brown, shiny with large pale scar

Twigs: rigid, glabrous, fetid odor when bruised

Buds: conical; scales strongly keeled, often toothed, sometimes waxy-bloomy

Bark: yellow-brown becoming dark brown, thick and furrowed

Wood: soft, white, tough, difficult to split

Aesculus hippocastanum L. - horsechestnut

Horsechestnut—a native of southwestern Asia—was imported to Europe in the 1600s and is now a common tree in France and England where trees over 300 years old and 125 feet (37 m) tall can be found. They often line formal alleés of French gardens.

The horsechestnut, though not a true chestnut, is reputed to be the "spreading chestnut tree" under which the village blacksmith labored in Longfellow's poem. The horsechestnut's opposite, palmately compound leaves, large sticky buds and showy flowers distinguish it from the true chestnut (*Castanea*) with its simple, alternate leaves, less showy flowers and dry buds.

Richley (1988) reports fewer disease problems on horsechestnut in southwest Nebraska than found on this species in more humid climates. Horsechesnut is being evaluated at PERC in Fort Collins, but the Japanese horse chestnut (*A. turbinata* Bl.) has failed (Fairchild and Klett 1993). Horsechestnut is hardy throughout Iowa, but it is subject to limb breakage in storms (Van Der Linden and Farrar 1984). Fall color is not a positive attribute of horsechestnuts growing on the campus of UNL.

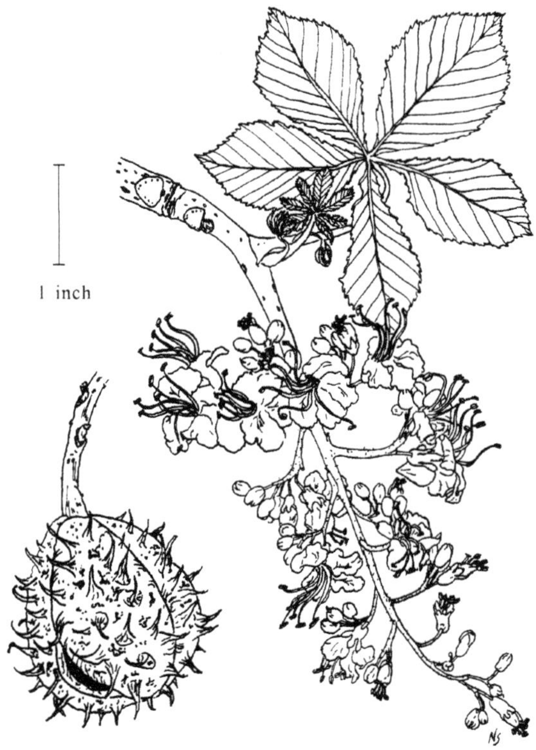

Aesculus hippocastanum

Several cultivars of horsechestnut have been selected with unusual leaf and form characteristics. For those who dislike litter, 'Baumannii' produces no fruit and double white, long-lasting flowers. Red horsechestnut, *A.* ×*carnea* Hayne, a cross between *A. pavia* and *A. hippocastanum*, is smaller in stature than horsechestnut and its flowers are red. Dirr (1990) states that red horsechestnut is less subject to disease than its namesake parent.

Planning

Height: up to 60 ft (18 m); Nebraska champion, 1987, 76 ft (23 m), West Point

Habit: tree; crown oval to rounded; May blooms creamy, autumn foliage sometimes yellowish, but leaves drop soon after frost

Site: sun, partial shade; pH and soil adaptable; zone 4

Use: shade for large areas, specimen

Recognition

Leaves: leaflets 7, sometimes 5; widest above middle, sessile, 4-8 in (10-20 cm) long, toothed; rusty-hairy near base

Flowers: clusters up to 12 in (30 cm) long; bisexual; petals 4-5, white with yellow-red blotch at base

Fruits: capsules nearly round, leathery, spiny, about 2 in (5 cm) in diameter; seeds 1-2, shiny, red-brown

Twigs: rigid, usually glabrous

Buds: red-brown, usually glistening sticky-resinous

Bark: smooth, becoming shallow-fissured

Aesculus octandra Marsh. (*A. flava* Ait.) - yellow buckeye

Although the native range of this larger "cousin" of Ohio buckeye is limited to the Appalachians of the Virginias, Georgia and Alabama and as far west as southern Illinois, yellow buckeye grows quite well in the eastern prairie on moist, fertile soils. It appears to be somewhat less susceptible to leaf diseases than Ohio buckeye.

Specimens of yellow and Ohio buckeye growing side by side on East Campus, UNL, provide an interesting comparison of the two species, especially in autumn. Leaves of the yellow buckeye remain green a few days later than those of Ohio buckeye and turn a much more brilliant yellow-red color that is effective much longer. Both trees bloom profusely, but yellow buckeye is much more fruitful. The yellow buckeye's gray bark remains smooth at over 8 inches (20 cm) diameter, while the Ohio buckeye's bark is distinctively striped on the small branches down to the main stem.

Aesculus octandra

Planning

Height: up to 50 ft (15 m); National champion, 1987, 145 ft (43.9 m), Tennessee
Habit: tree; crown rounded; May blooms colorful, autumn foliage yellow-red
Site: sun, partial shade; pH and soil adaptable; zone 4
Use: shade, specimen

Recognition

Leaves: leaflets 5, seldom 7; elliptic, tip long-pointed, base tapered, sessile; surfaces glabrous at maturity, margin toothed
Flowers: clusters up to 6 in (15 cm) long; bisexual; petals 4, yellow (red, pink or creamy), unequal, upper pair erect, lower pair larger; stamens usually 7, shorter than petals
Fruits: capsules nearly round, leathery, smooth; seeds usually 2, shiny, red-brown
Twigs: rigid, glabrous; no fetid odor
Buds: pointed; scales smooth, not strongly keeled
Bark: gray, smooth with fine scales developing on stems over 8 in (20 cm) in diameter

Aesculus parviflora Walt. - bottlebrush buckeye

Bottlebrush buckeye, native shrub of the southeastern U.S., is slow growing. Young plants on the UNL campus are thriving.

Planning

Height: up to 10 ft (3 m); National champion, 1989, 20 ft (6.1 m), North Carolina
Habit: shrub, tree; multistemmed, suckering; July blooms white, autumn foliage yellow
Site: sun, partial shade; pH and soil adaptable, prefers fertile, moist, slightly acidic site; zone 5
Use: grouping, border

Hippocastanaceae *Aesculus*

Recognition

Leaves: leaflets 5-7; elliptic to wider above middle, nearly sessile; glabrous above, hairy below; margin fine-toothed (crenate-serrulate)

Flowers: clusters cylindrical; bisexual and unisexual; petals 4, white; stamens usually 7, pinkish white, extending about 1 in (25 mm) beyond petals, anthers red

Fruits: capsules pear-shaped, smooth

Twigs: rigid; lenticels conspicuous, raised, light brown; leaf scars half-encircling bud

Buds: terminal present; scales usually 4, short-hairy, gray-brown

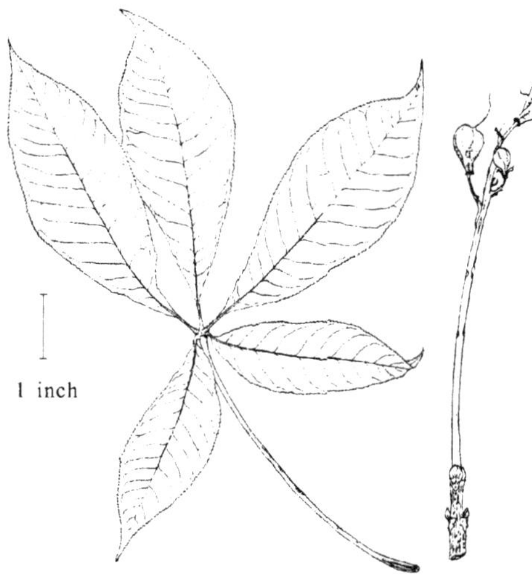

Aesculus parviflora

Aesculus pavia L. - red buckeye

Red buckeye, an understory tree or shrub that grows on fertile, moist soils in the southeastern U.S. as far north as Kentucky and southern Missouri, displays attractive red spring flowers. In May of 1990, magnificent red flowers graced a 15-foot (4.5 m) red buckeye growing in south Lincoln. Richley (1988) rates its performance good at Curtis in southwestern Nebraska.

Planning

Height: up to 15 ft (4.5 m); National champion, 1983, 64 ft (19.4 m), Michigan

Habit: shrub, tree; crown rounded; late May blooms bright red

Site: sun, shade; prefers moist, well-drained, slightly acidic soils; zone 5

Use: specimen, border

Recognition

Leaves: leaflets 5, short-stalked, 3-6 in (8-15 cm) long; usually widest above the middle, margin toothed; glabrous above, sometimes hairy beneath

Flowers: clusters (panicles) loose, up to 6 in (15 cm) long; calyx tubular, dark red; petals 4-5, very unlike, converging but not joined, glandular at margins; stamens 8, usually as long as petals

Fruits: capsules nearly round, smooth; seeds 1-2

Twigs: rigid, olive brown; lenticels raised, light brown; leaf scars triangular

Buds: brownish, glabrous

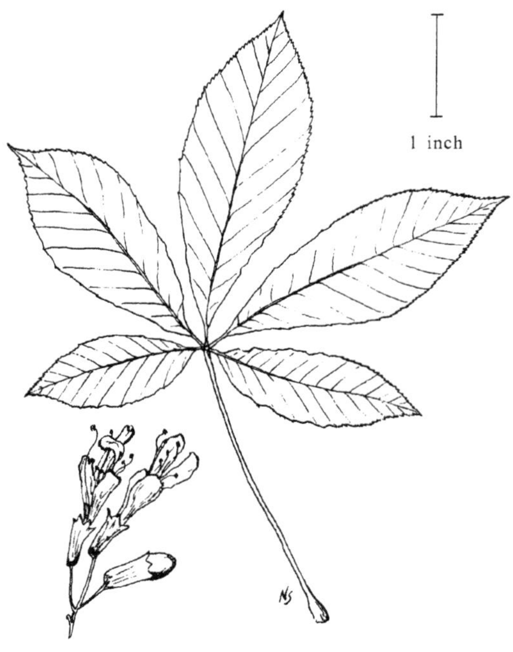

Aesculus pavia

Aceraceae Juss. - maple family

In addition to the well-known genus *Acer*, two species of *Dipteronia* Oliv., both natives of China, are also maple family members. The fruits of both genera are paired and bear conspicuous wings. The two nutlets of *Acer* are joined at the base, each with a wing, while the two joined nutlets of *Dipteronia* are completely surrounded by a wing. The pinnately compound leaves of the latter bear 7-15 leaflets.

Acer L. - maple

The maples provide a significant portion of the well-publicized brilliant autumn foliage that clothes a significant portion of the New England states, and sugar maple is the source of the tasty syrup that flavors flapjacks at an old-fashioned country breakfast. Many prairie dwellers with a longing for these amenities plant maple trees in attempt to bring a bit of Vermont to their homesteads. Unfortunately, the trees that provide brilliant fall colors on their native sites may be a disappointment at their new homes, and the prairie climate often fails to provide the proper temperature combinations in spring for sap to flow in quantity.

The majority of *Acer* species (124 species and 95 subspecies) are native of Asia and are small trees or shrubs (Gelderen et al. 1994). However, many of the 13 species native to North America are large trees that provide high quality wood products such as furniture and flooring. Leaf shapes and flower characteristics vary among species and the fruits are often a "key" to maple recognition.

Recognition

Habit: trees, occasionally shrubs; deciduous, rarely evergreen

Leaves: opposite; simple and palmately lobed or veined (undivided with parallel lateral veins in one species), or pinnately or palmately compound 3- to 7-foliate; mostly without stipules

Flowers: clusters (simple and compound racemes, umbels); bisexual or unisexual (most species polygamous or dioecious); flowers regular; calyx normally 5-parted, overlapping, distinct, seldom fused at base; petals 5, sometimes 4 or seldom none, overlapping, distinct; stamens most commonly 8, less often 4-12, filaments distinct, seldom poorly developed or wanting; ovary superior, styles distinct or fused at base

Fruits: dry, paired, non-splitting, winged, united at base, flattened, 1-seeded (double samara)

Twigs: circular in cross-section; lenticels variable, pith homogeneous; opposite leaf scars narrow, U- or V-shaped, often meet or are connected by a transverse ridge to nearly encircle the twig; bundle scars usually 3

Buds: terminal present unless replaced by fruits, lateral solitary or multiple (collateral), sometimes stalked; scales overlapping or not (valvate)

Maples of the World (Gelderen et al. 1994), a comprehensive treatment of genus *Acer*, is recommended for the serious scholar. Hybrids and cultivars are described. *Acer* is classified into 16 sections.

Grouping Key for *Acer*

(Gelderen et al. 1994)

NOTE: The species in parenthesis are mentioned in this book

Trees, shrubs; deciduous; leaves simple, 3-, sometimes 5- to 7-lobed or undivided, margins toothed (serrate); bud scales 2-, sometimes 3-paired; flower clusters terminal and axillary, 35-400 flowered; flowers 5-parted, perianths green-white, stamens mostly 8: **section Parviflora** (*A. spicatum*)

Trees, shrubs; deciduous or partly evergreen; leaves simple, 3- to 13-lobed or unlobed, margins toothed (serrate), sometimes entire; bud scales 4-paired; flower clusters terminal, sometimes axillary; flowers 5-parted, sepals red or green-red, petals mostly white, stamens 8: **section Palmata** (*A. circinatum*, *A. palmatum*, *A. japonicum*, *A. pseudosieboldianum*, *A. sieboldianum*)

Small trees, shrubs; leaves simple, 3-lobed, margins minutely toothed (serrulate); flower clusters upright; flowers, perianths red: **section Wardiana**

Shrubs, sometimes trees; deciduous; branches often white-striped, axillary buds stalked; leaves simple, undivided or 3- to 5-, sometimes 7-lobed, margins toothed (serrate); bud scales 2-paired, valvate, red or green-red; flower clusters terminal and axillary, 10-25 flowered; flowers 5-parted, perianth green-yellow or reddish, stamens 8: **section Macrantha** (*A. pensylvanicum*, *A. tegmentosum*)

Trees, shrubs; deciduous; leaves simple, undivided, 3- to 5-lobed, or partly trifoliate; bud scales 2- to 4-paired; flower clusters small, terminal, sometimes all axillary, 10-25 flowered; flowers 4- or 5-parted, yellowish: **section Glabra** (*A. glabrum*)

Trees, shrubs; deciduous; leaves compound, trifoliate or pinnate (5-7 foliate), margins entire or remotely toothed (dentate); bud scales 2- to 3-paired; flower clusters axillary from leafless buds, 15-50 flowered; flowers 4-parted, dioecious, stamens 4-6: **section Negundo** (*A. negundo* ssp. *negundo*, ssp. *californicum*, ssp. *interius*, ssp. *mexicanum*; *A. cissifolium*; *A. henryi*)

Small trees, shrubs; deciduous; leaves simple, undivided, margins toothed (doubly serrate); bud scales 9- to 13-paired, brown; flower clusters terminal and axillary, 10-20 flowered; flowers 4-parted, dioecious, perianth greenish yellow, stamens usually 6: **section Indivisa** (*A. carpinifolium*)

Trees, shrubs; deciduous; leaves simple, 3- to 5-lobed, margins coarsely toothed (crenate to dentate), sometimes entire; bud scales 5- to 13-paired, overlapping, gray-brown; flower clusters (corymbs) terminal and axillary; flowers 5-parted, perianths yellowish green, stamens usually 8, filaments exserted in male flowers: **section Acer** (*A. saccharum* ssp. *saccharum*, ssp. *grandidentatum*, ssp. *nigrum*; *A. pseudoplatanus*)

Key continued

Acer Key

continued from previous page

Trees; mostly evergreen; leaves simple, undivided or 3-lobed, usually white-waxy (glaucous) beneath, margins entire to toothed (lobulate-serrate); bud scales 4- to 8-paired, gray-brown; flower clusters (corymbs) terminal and axillary, 25- to 75-flowered; flowers 5-parted, perianths yellow-white, petals longer than sepals, stamens usually 8: **section Pentaphylla** (*A. buergerianum*)

Trees, shrubs; deciduous; bark sometimes exfoliating; leaves simple, trifoliate, margins nearly entire to coarsely toothed (serrate); bud scales 11- to 15-paired, dark gray-brown; flower clusters (racemes or corymbs) terminal and axillary, mostly 3-flowered up to 25-flowered; flowers 5- or 6-parted, perianths yellow-green, stamens 10-13: **section Trifoliata** (*A. griseum*, *A. maximowiczianum*, *A. triflorum*, *A. mandshuricum*)

Trees; deciduous; leaves simple, 3- to 5-lobed, margins lobulate or entire, sometimes toothed (serrate); bud scales 5- to 8- to 12-paired; flower clusters (racemes, corymbs) axillary, sometimes terminal, rachises elongated; flowers 5-parted, large, perianths green-yellow, stamens 8-12: **section Lithocarpa**

Trees, shrubs; deciduous; leaves simple, 3- to 5- to 7-lobed, rarely undivided, margins entire or remotely toothed (dentate to serrate), petioles with milky sap (lactiferous); bud scales 5- to 8- or 6- to 10-paired; flower clusters (corymbs) terminal and axillary; flowers 5-parted, stamens 5-8: **section Platanoides** (*A. campestre*, *A. miyabei*, *A. mono*, *A. platanoides*, *A. tenellum*, *A. truncatum*)

Shrubs, small trees; leaves simple, 3-lobed, margins coarsely tothed (serrate); bud scales 6- to 10-paired; flower clusters terminal or axillary; flowers, stamens mostly 5: **section Pubescentia**

Shrubs, small trees; deciduous; leaves simple, undivided or 3-lobed, margins toothed (serrate); bud scales 5- to 10-paired, gray-brown; flower clusters (corymbs) terminal and axillary; flowers 5-parted, perianths greenish white, stamens 8: **section Ginnala** (*A. tataricum* ssp. *ginnala*, ssp. *tataricum*)

Trees; deciduous; leaves simple, 3- to 5-lobed, white-waxy (glaucous) beneath, margins toothed (serrate); bud scales 4- to 7-paired, red; flower clusters axillary from leafless buds, mostly 5-flowered; flowers 5-parted, perianths red, sometimes green-red, petals sometimes wanting, sepals joined (connate), stamens mostly 5, anthesis before leaf development: **section Rubra** (*A. rubrum*, *A. saccharinum*)

Trees; evergreen, sometimes deciduous; leaves simple, undivided, margins entire; bud scales 7- to 11-paired; flower clusters axillary from leafless buds; flowers 5-parted, perianth yellowish, stamens 4-6 or 8-12: **section Hyptiocarpa**

Section Parviflora

Acer spicatum Lam. - mountain maple

Mountain maple is a shrubby understory tree native to northeastern Iowa, the Great Lakes region, the New England states, and southeastern Canada. It is seldom used in landscape plantings because of its exacting site requirements.

Planning

Height: up to 30 ft (9 m)
Habit: tree; deciduous; crown irregular, sparse; flowers after leaves; autumn foliage yellow to orange
Site: shade; soil moist, well-drained, acidic; zone 4
Use: naturalizing
Caution: experimental

Recognition

Leaves: simple, 3- to 5-lobed, 3-5 in (8-13 cm) long and wide; margin coarsely toothed, base heart-shaped (cordate); minutely hairy and wrinkled above, glabrous beneath; petioles 1.5-2.5 in (4-6 cm) long
Flowers: clusters narrow, spike-like, upright; flowers yellow-green, male and bisexual on the same plant
Fruits: wings slightly divergent, about 1 in (2.5 cm) long, turning red in autumn
Twigs: red becoming spotted with gray
Buds: scales 2, not overlapping (valvate)

Section Palmata

Acer palmatum ssp. *palmatum* Thunb. - Japanese maple

Japanese maple, a native of Japan, Korea and eastern China, can be grown on protected sites in the southern portion of the central prairie. Closely related species, also natives of the Orient, include *A. japonicum* Thunb., fullmoon maple, *A. pseudosieboldianum* Komar., purplebloom maple, and *A. sieboldianum* Miq., Siebold maple. Many cultivars differing in leaf shape and color, overall form, and habit of branching have been selected from these species.

Vine maple (*A. circinatum* Pursh), an inhabitant of the west coast of North America, is noted for its brilliant autumn colors, especially in the thickets of Oregon and Washington. It is ill-adapted to most prairie sites. Dirr (1990) could not establish it in his Illinois garden.

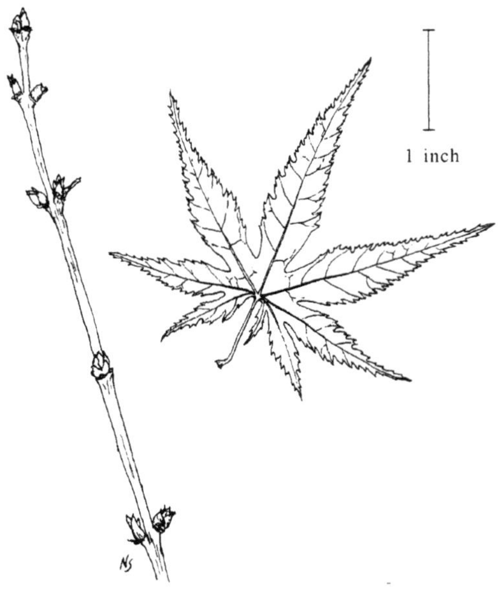

Acer palmatum* ssp. *palmatum

Planning

Height: up to 20 ft (6 m)
Habit: small tree, large shrub; deciduous; crown rounded, branches horizontal; May-June bloom purple; autumn foliage in yellows, bronze, purples and reds
Site: sun, shade; prefers moist, well-drained soils; zone 5
Use: specimen, border, grouping
Caution: experimental, protect from extreme winds; vulnerable to late spring frosts

Recognition

Leaves: deeply 5-9 lobed, long-pointed, toothed (doubly serrate); glabrous, bright green beneath; petioles glabrous, 0.5 to 2 in (12 to 50 mm) long
Flowers: clusters (corymbs) terminal on short leafy branchlets, glabrous; petals purplish
Fruits: wings widely divergent, incurved, turning red; nutlets glabrous, about 0.4 in (1 cm) long
Twigs: glabrous, green to red, often bloomy; leaf scar margin forms flaring collar
Buds: egg-shaped; outer scales 2, not overlapping (valvate), hidden by petiole base; frequently double terminal buds
Bark: gray

Section Macrantha

A. pensylvanicum L., striped maple, a native of the northeasten U. S. and the Appalachians, grows on sites similar to those where mountain maple is found. It is easily recognized by the conspicuous white vertical stripes on the green stems of young trees. It is not amenable to heat and drouth (Dirr 1990).

A. tegmentosum Maxim., a native to the banks of the Amur and Ussuri rivers in Russia, is very similar to striped maple.

Section Glabra

Acer glabrum Torr. - Rocky Mountain maple

Rocky Mountain maple is a native of the Rocky Mountains and the Black Hills and Pine Ridge of Sioux County, Nebraska. It grows along riverbanks and as undergrowth on moist, forested hillsides. Subspecies *douglasii* is found in the Cascade Range of Oregon and subspecies *siskiyouense* occurs in the Siskiyou Mountains of southern Oregon and northern California.

Seedlings from seed collected in Sioux County have been distributed to affiliate sites of the Nebraska Statewide Arboretum. This first extensive trial planting in Nebraska will determine its adaptability to a variety of sites.

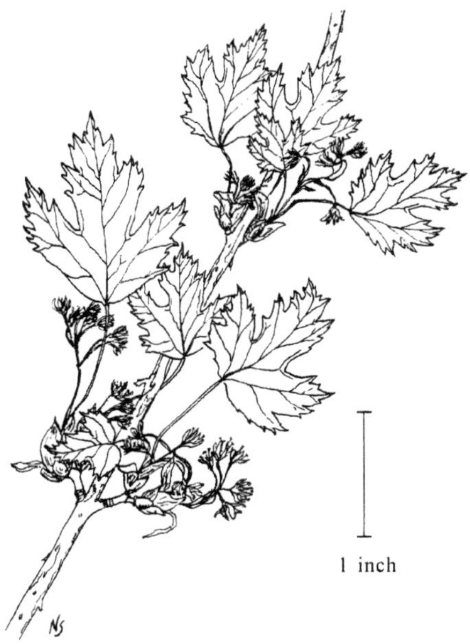

Acer glabrum

Planning

Height: up to 30 ft (9 m); National champion, 1989, 80 ft (24.4 m), Idaho
Habit: large shrub, small tree; deciduous; crown rounded, young stems purplish red; May flowers greenish yellow
Site: sun, part shade; pH and soil adaptable, prefers fertile, moist soils; zone 3
Use: patio, grouping, naturalizing

Recognition

Leaves: simple, 3-5 lobed, rarely compound (trifoliate), often wider than long, toothed (serrate or double serrate); surfaces glabrous; red glands on veins and at petiole junctions; petioles glabrous, grooved, often bright red, up to 6 in (15 cm) long

Flowers: clusters terminal, few-flowered, on short, leafy twigs; unisexual (normally dioecious); calyx tubular, 5-lobed, petals 5, greenish yellow

Fruits: divergent at less than 90 degrees, glabrous, often rose-colored in summer, about 0.8 in (2 cm) long

Twigs: glabrous, green, turning reddish brown; leaf scars crescent-shaped, encircle twig

Buds: sharp; scales 4 to 5 pairs (valvate), outer keeled, glabrous outside, hairy (densely villous) on inner surface, inner scales glossy, wine red

Bark: gray to red-brown, smooth, becoming slightly fissured showing light colored streaks of inner bark

Wood: hard, heavy, fine-grained

Section Negundo

Acer negundo ssp. *negundo* L. - boxelder

Boxelder is a maverick among the maples, a tree that has leaves resembling those of an ash. Ssp. *negundo* is native to the mixed forests of the eastern and middle regions of the U.S., westward to the Rocky Mountains and south to Guatemala (Gelderen et al. 1994). These authors also describe ssp. *californicum*, *interius* and *mexicanum*, which are all natives of the western U.S., thus making boxelder native over a range wider than any other North American maple. Within this species' natural range there is undoubtedly a wide variety of genotypes adapted to a great variety of sites, a situation that should make provenance research interesting and rewarding.

Despite weak limbs that are prone to breaking easily under pressure from wind and ice, boxelder was widely planted by the early white settlers of the prairies and plains. This hardy and adaptable tree provided shade, shelter from wind, and fuel, attributes that were much more important than beauty and grace in those trying times. Boxelder continues to be planted for shelter on the prairies of Canada under the name of Manitoba maple, and it is sometimes a component of shelter and landscape plantings on the western plains of the U.S. It is often considered to be a weed tree.

Many boxelder cultivars bear variegated leaves and are much in evidence in Europe according to Dirr (1990). Among the most common are: 'Auratum', with leaflets that are entirely yellow, sometimes lasting into autumn; 'Aureo-marginatum', with yellow-bordered leaflets; 'Elegans' and 'Variegatum', with leaflets edged with an irregular yellow margin; 'Flamingo', with brilliant pink new shoots that become green with a white border; and 'Sensation', with green leaves that become brilliant red in autumn. 'Sensation', introduced by J. Frank Schmidt and Son of Oregon, is a hardy tree suitable for street planting and has performed well in the Great Plains.

Boxelder is host to an insect of the same common name which, though harmless, bedevils tidy housewives with their presence in the home as they seek shelter from the cool of autumn. Boxelder bug, *Leptocoris trivittatus*, feeds on the sap of the leaves, fruits, and seeds of boxelder, apple, ash, cherry and plum. A "disease" named boxelder blight has plagued this tree species in the prairie region since the early 1950s, a date coinciding with the beginning of widespread use of the herbicide 2,4-D. It is evident that boxelder takes its place alongside grapevines as an indicator of very minute amounts of 2,4-D in the atmosphere.

A. henryi Pax and *A. cissifolium* (Sieb. & Zucc.) K. Koch, ivy-leaved maple, are small Asian trees also with trifoliate leaves that may be suitable for planting on prairie sites.

Planning

Height: up to 50 ft (15 m); Nebraska champion, 1984, 41 ft (12.4 m), North Platte; National champion, 1981, 110 ft (33.3 m), Michigan

Habit: tree; deciduous; crown rounded, often irregular with more than one trunk; autumn foliage yellow; hanging fruit clusters persist into winter

Site: sun, part shade; pH and soil adaptable, drouth tolerant; zone 2

Use: windbreak, shade, naturalizing

Recognition

Leaves: compound, trifoliate or pinnate; leaflets 3-7, ovate to oblong, irregularly toothed above middle to nearly entire, terminal leaflet often 3-lobed; glabrous above, hairy below becoming glabrous; petioles sometimes hairy, 2-3 in (5-8 cm) long, base enlarged

Flowers: unisexual (dioecious), petals absent; male in several-flowered clusters, calyx bell-shaped, 5 minute lobes, stamens 5, long exserted, anthers reddish or purplish; female in pendulous racemes, calyx lobes distinct, 5; stigmas 2, divergent, yellowish, 0.3 in (7 mm) long

Fruits: July-August, divergent at about 45 degrees

Twigs: green to reddish brown, glabrous with waxy whitish bloom, encircled by leaf scars, pungent when bruised

Buds: egg-shaped; 2 outer scales not overlapping (valvate), silky-hairy

Bark: fissured, ridges broad, rounded; scaly

Wood: light, weak

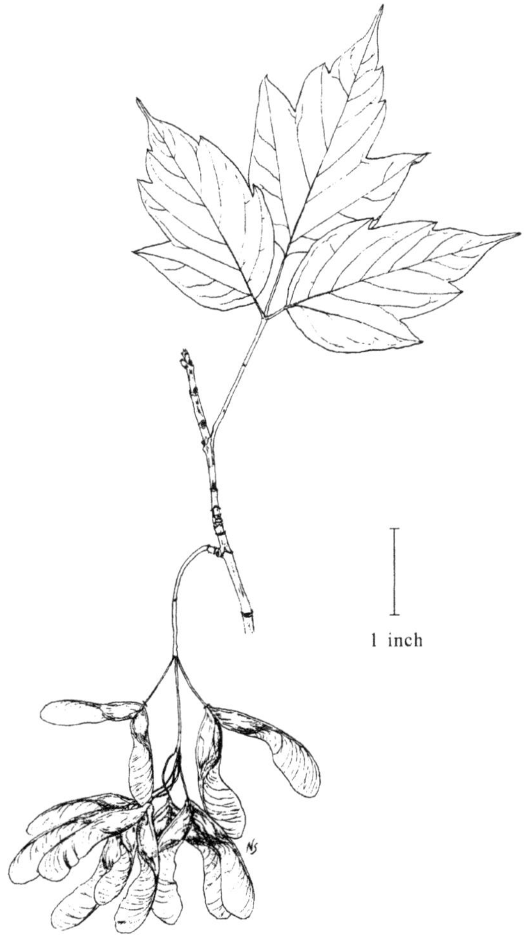

Acer negundo* ssp. *negundo

Section Indivisa

A. carpinifolium Sieb. & Zucc., hornbeam maple—a native of the high mountains of Japan—bears leaves similar to those of *Carpinus* species (hornbeams). This unique hardy maple is not readily available because viable seed is dificult to obtain (Gelderen et al. 1994).

Section Acer

Acer pseudoplatanus L. - planetree maple

Planetree maple, called sycamore maple in Europe, is a large native tree of Europe and western Asia that is seldom planted in North America except along the East Coast near the ocean. It tolerates salt-laden winds. Dirr (1990) states that planetree maple is adapted to soils ranging from high lime to acidic. High temperatures and desiccating winds are likely to be the most limiting factors that it will encounter in the central Great Plains, especially the western part. Large specimens grow on the Utah State University campus in Logan (Sutton 1998).

Dirr (1990) mentions these cultivars with unusual coloration of leaves and fruits: 'Atropurpureum' bearing leaves dark green above, purple beneath; 'Brilliantissimum', with leaves that are pink as they unfold, becoming pale cream and then yellow-green; 'Erythrocarpum' bearing red samara wings; and 'Leopoldii', with yellowish pink and purple-stained leaves. The form variegatum bears yellowish or yellowish white blotched or striped leaves.

Planning

Height: up to 50 ft (15 m)
Habit: tree; deciduous; crown rounded
Site: sun, part shade; pH and soil adaptable, subject to injury by stem cankers, tolerates air pollution; zone 5
Use: shade

Recognition

Leaves: simple, 5-lobed, margins coarsely and regularly toothed; surfaces glabrous except sometimes hairy on veins below
Flowers: clusters (panicles) terminal, pendulous, 25-150 flowered, up to 4 in (10 cm) long on short leafy branchlets; appearing with leaves; bisexual
Fruits: wings divergent at about 60 degrees, glabrous; nutlets egg-shaped, strongly veined; ripening in autumn
Twigs: glabrous, gray-brown, dull; leaf scars do not meet
Buds: egg-shaped; terminal large; 1 lateral above each leaf scar; scales overlapping, greenish, glabrous
Bark: gray to brown, flaking into squares; inner bark orange-brown

Acer saccharum ssp. saccharum Marsh. - sugar maple

The sugar maple is famous for the sugar and syrup produced from its sap as well as its brilliant autumn color. Floors, furniture and a variety of wood products in many homes are made from the heavy, hard, shock-resistant wood. The native range of sugar maple extends from eastern Kansas, Missouri and Iowa to the eastern seaboard, from Tennessee north to Novia Scotia. It is a major component of the eastern hardwood forest, but many scientists are concerned about the evidence of decline in the health of these maples due to acid rain.

The autumn coloration of sugar maple can be disappointing if the origin of planted trees is not compatible to the planting site. 'Green Mountain' in Maxwell Arboretum, Lincoln, was brilliant in mid-October, 1988, while on the same date the foliage of 'Columnare' growing nearby was an unassuming yellow-brown. Sugar maple trees of unknown origin growing in the vicinity of Lincoln display variably colored foliage from mid-September to mid-November. 'Bonfire' and 'Green Mountain' are adapted to the soils and climate of the Fort Collins area with only minor chlorosis and leaf scorch, and develop good fall color (Fairchild and Klett 1993).

A. saccharum ssp. *nigrum* (F. Michx.) Desmarais, black sugar maple, is recognized by some taxonomists as a distinct species. Black sugar maple leaves are leathery, they exhibit a hairy undersurface in early summer and many are only 3-lobed. Most notably, stipules are usually present at the petioles, a feature absent in all other subspecies of *A. saccharum* and in all of Aceeraceae. Hybrids complicate positive identification. Black sugar maple's native range is similar to that of sugar maple, though somewhat constricted on all sides. A small outlier has been reported in eastern South Dakota. Some authorities speculate that black sugar maple may be better adapted to drouth and heat than sugar maple is. Provenance plantings including both subspecies established at strategic locations in the prairie region would answer many questions about adaptability and genetic variability.

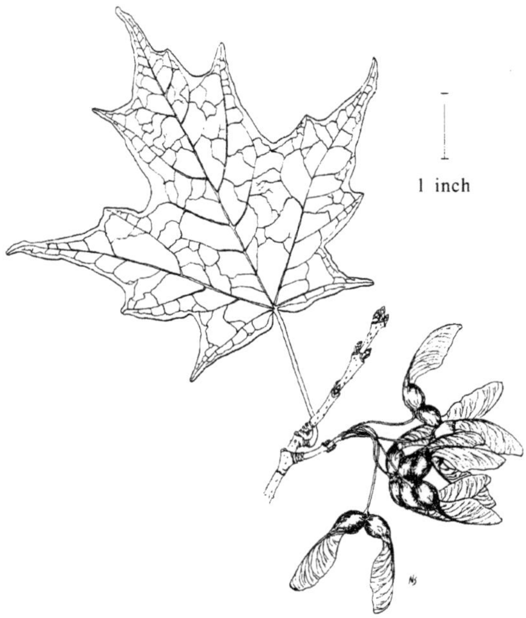

***Acer saccharum* ssp. saccharum**

A. saccharum ssp. *grandidentatum* (Torr. & Gray) Desmarais, bigtooth maple, is a small (sometimes shrubby) native tree of the Rocky Mountains, found primarily in the states of Utah, Arizona and New Mexico. It was introducted into cultivation by J. Frank Schmidt and Son of Oregon. It may be adapted to the central Great Plains on protected sites where moisture is adequate. A tree in Lincoln produced flowers and fruit at age 10 years (Sutton 1998).

The leaves of bigtooth maple are 3- to 5-lobed with large shallow spaces (sinuses) between, shiny glabrous above, waxy (glaucous) and sometimes hairy beneath. The U-shaped fruits are slightly smaller than those of subspecies *saccharum*, often rose-colored in summer and becoming green at maturity. Twigs are glabrous, bright red, and nearly encircled by narrow leaf scars. Buds are bright red and hairy. Autumn foliage is yellow to scarlet. Bigtooth maple produces excellent sap for sugaring (Barker et al. 1982).

Planning

Height: up to 65 ft (20 m); Nebraska champion, 1990, 64 ft (19.5 m), Stella; National champion, 1984, 93 ft (28.4 m), Connecticut
Habit: tree; deciduous; crown rounded to oval, dense; flowers yellow, appearing before or with the leaves; autumn foliage in various shades of yellow, orange and red
Site: sun, part shade; pH adaptable, prefers moist, fertile, well-drained soil; zone 4
Use: shade, specimen, street, wood, syrup
Caution: protect from excessive wind to prevent leaf tatter

Recognition

Leaves: usually 5-lobed, tips pointed, margins entire or variably toothed; surfaces glabrous, occasional hairy tufts in vein axils below

Flowers: clusters small, terminal (corymbose) on short leafy branchlets; bisexual and unisexual (monoecious or dioecious); petals none; calyx connate, 5-lobed, greenish to yellow; stamens exserted

Fruits: glabrous; U-shaped, wings nearly parallel

Twigs: glabrous, green, becoming reddish brown; leaf scars broadly U-shaped, narrow, extending halfway around the stem

Buds: solitary; cone-shaped, pointed; scales 4-8, overlapping, glabrous or nearly so; multiple flower buds absent

Bark: gray, deeply furrowed into long thin plates

Wood: tough, hard, heavy, sometimes with distinctive figure

Section Pentaphylla

Acer buergerianum Miq. - trident maple

Trident maple, a China import, has not been widely planted in the prairies. It is, however, an important species in China, Taiwan, Korea and Japan, where it is widely used as a garden and landscaping tree and as a bonsai subject (Gelderen et al. 1994). Dirr (1990) reports seeing beautiful specimens in Ohio and Massachusetts with little or no insect or disease problems. A tree in Lincoln (UNL city campus) seems to be well-adapted to that site. Autumn foliage can be striking, depending on the tree's locality.

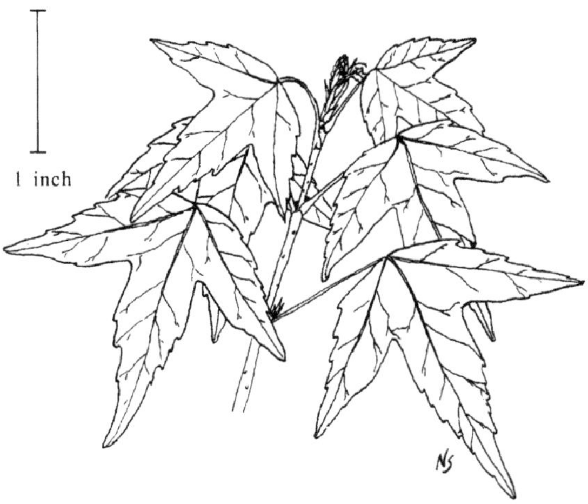

Acer buergerianum

Planning

Height: up to 30 ft (9 m), often much taller in its native habitat

Habit: tree, shrub; deciduous; crown oval to rounded or loose and open with several stems; flowers borne with the leaves; autumn foliage shades of yellow and red; often quite fruitful

Site: sun; tolerates slightly alkaline but prefers slightly acidic soils, tolerates drouth; zone 5

Use: shade, street

Pests: none serious

Recognition

Leaves: lobes 3, triangular, pointed forward, each with a central vein, all joined at the leaf base; entire or slightly toothed; base rounded to wedge-shaped; surfaces hairy, soon glabrous

Flowers: clusters (corymbs) terminal, upright, downy, on short leafy branchlets; flowers whitish to greenish yellow

Fruits: wings parallel or touching at tips; glabrous

Twigs: hairy, becoming glabrous second year

Buds: 4-angled, pyramid-shaped; scales sharply pointed, hairy-fringed

Bark: smooth, becoming platy, exfoliating

Section Trifoliata

Acer griseum (Franch.) Pax - paperbark maple

"One of the rarest and most beautiful of all small trees" is how Johnson (1973) describes paperbark maple, one of the trifoliate species with rich red-brown bark that peels away like the bark of some birches. The leaves and tiny fruits of this China native turn scarlet in autumn.

Dirr (1990) describes splendid specimens of *Acer griseum* growing on the campus at the University of Illinois and at the Arnold Arboretum where the trees are over 80 years old and the species "assumes a dignity unmatched by other trees." No botanical varieties or cultivars of paperbark maple have been named because of its remarkable uniformity. Planting stock is scarce because of low seed fertility.

A. triflorum Komar., three-flower maple, is another small tree that closely resembles *A. griseum*. The common name is a reference to its flowers that are borne in clusters of three. In addition to specimens at Arnold Arboretum, three-flower maple is also growing at the Minnesota Landscape Arboretum (Dirr 1990). It is adapted to a wide range of soil types. Unfortunately, much of the seed of *A. trifloruom* is empty (due to the lack of male flowers) and therefore scarce (Gelderen et al. 1994).

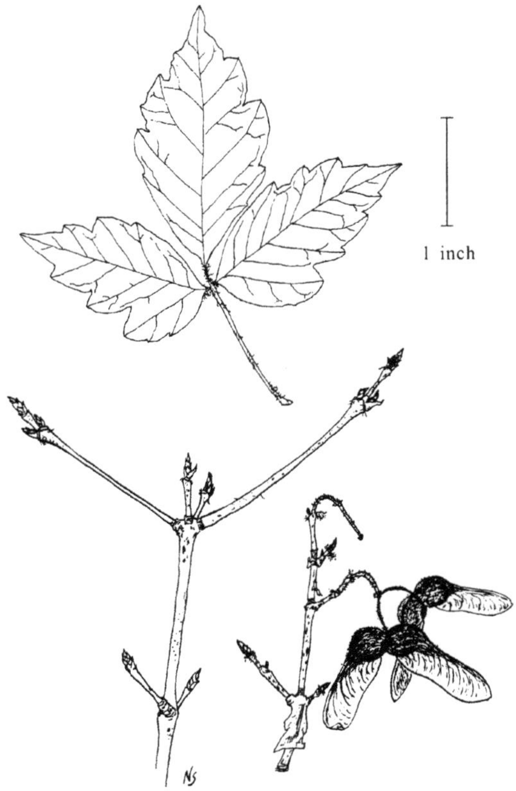

Acer griseum

Planning

Height: up to 30 ft (9 m)
Habit: tree; deciduous; crown oval to rounded;
autumn foliage bronze to shades of yellow and red; colorful peeling bark
Site: prefers partial shade; pH and soil adaptable, prefers moist, well-drained soils; zone 5
Use: specimen, patio
Pests: none serious

Recognition

Leaves: trifoliate; leaflets egg-shaped, about 2 in (5 cm) long, half as wide, middle short-stalked, laterals nearly sessile; margins coarsely toothed; hairy beneath, especially on veins, becoming glabrous; petioles hairy
Flowers: clusters (corymbs) or solitary, small, terminal, hairy, appearing after leaves; often monoecious
Fruits: wings diverge at nearly 90 degrees; nutlets woody, hairy (tomentose)
Twigs: hairy, becoming glabrous, exfoliating 2nd or 3rd year
Buds: small, dark; scales many, overlapping
Bark: cinnnamon-brown, exfoliating into thin papery flakes

Acer mandshuricum Maxim. - **Manchurian maple**

Manchurian maple, a native of Korea, eastern Siberia, southeastern Manchuria and adjoining provinces, exhibits a trait that most trees of northern latitudes do—early fall color at lower latitudes. Dirr (1990) describes specimens at Arnold Arboretum as "magnificent rose-red to red in early October." Manchurian maple is seldom available in the nursery trade because viable seed is scarce and vegetative propagation is difficult.

Planning

Height: up to 20 ft (6 m)

Habit: tree, shrub; deciduous; crown upright, rounded; flowers appear after the leaves; foliage often red in early autumn; bark smooth

Site: experimental; prefers slightly acidic soils; zone 5

Use: specimen, border

Recognition

Leaves: trifoliate; leaflets oblong, saw-toothed (obtuse-serrate); terminal leaflet about 4 in (10 cm) long, stalk (petiolule) about 0.7 in (2 cm) long; lateral leaflets shorter on stalks about 0.2 in (5 mm) long; surfaces glabrous except hairy on midrib beneath

Flowers: clusters of 3; monoecious; greenish yellow

Fruits: wings divergent at 90 degrees or more, glabrous; nutlets egg-shaped with net-like pattern (reticulate)

Twigs: glabrous, smooth, developing fissures 2nd year

Buds: brownish black, lateral appressed with slight spiral hook; scales overlapping, glabrous, angled

Bark: smooth (not peeling) gray

Acer maximowiczianum Miq. - **Nikko maple**

Nikko maple—incorrectly known by nursery operators as *A. nikoense* Maxim.—is a native of mountain forests in Japan, and is also widely distributed in central China (Gelderen et al. 1994). This maple is uncommon in the U.S. because plants are scarce due to infertile seeds.

Planning

Height: up to 50 ft (15 m)

Habit: tree; deciduous; crown round; leaves emerge bronze-colored, become green; yellow May flowers; autumn foliage in yellows, reds and purples

Site: experimental; prefers slightly acidic soils; zone 4

Use: specimen

Recognition

Leaves: trifoliate; leaflets egg-shaped to elliptic, 2-5 in (5-13 cm) long, middle short-stalked, lateral nearly sessile; tips pointed, margins entire or shallow-toothed, hairy (ciliate); long soft-hairy (villous) beneath, becoming nearly glabrous, veins hairy; petioles hairy

Flowers: fascicles terminal, usually 3-flowered, hairy, on short leafy branchlets; flowers yellow-green, dioecious (occasionally polygamodioecious)

Fruits: divergent at about 90 degrees, wings sometimes curving inward; nutlets thick, densely hairy, very woody

Twigs: somewhat purplish, long-hairy (villous), becoming glabrous 2nd year; lenticels prominent

Buds: pyramid-shaped; scales overlapping, densely hairy

Bark: gray to grayish brown, smooth, not peeling

Section Platanoides

Acer campestre L. - hedge maple

Hedge maple, a native of Europe (except Scandinavia), North Africa and western Asia grows on plains, along river banks and often occurs as undergrowth in hardwood forests. The numerous geographic forms were once described as separate species, subspecies, varieties, or forms, but they are not recognized by Gelderen et al. (1994). Among the many hedge maple cultivars are: 'Compactum', a dwarf multistemmed shrub; 'Fastigiatum', upright with corky branches and 5-lobed leaves; 'Postelense', with golden yellow leaves when young; 'Evelyn', a small- to medium-sized tree; and 'Schwerinii', which bears leaves that are purple when unfolding, becoming green (Dirr 1990).

Hedge maple can be pruned to form hedges, the common practice on the European continent, or lower branches can be removed to yield the conventional tree form. It is adapted to a wide range of soil conditions including compaction, dryness and alkalinity. Fairchild and Klett (1993) note that hedge maple is performing well at PERC in Fort Collins.

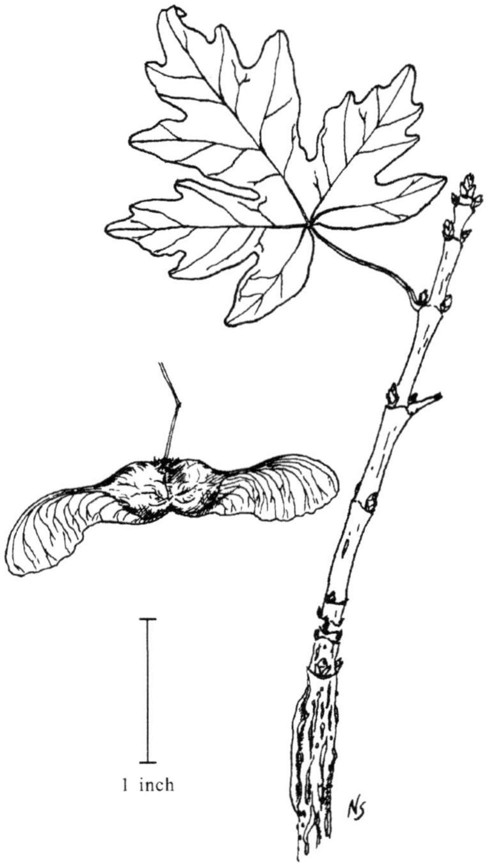

Acer campestre

Planning

Height: up to 30 ft (9 m)
Habit: tree, large shrub; deciduous; crown rounded, low-branched, dense; May flowers yellow-green; autumn foliage yellow
Site: sun, part shade; pH and soil adaptable, tolerates drouth and air pollution; zone 5
Use: shade, street, hedge, topiary

Recognition

Leaves: 3- to 5-lobed, palmate, wider than long; margin entire, more or less wavy; glabrous above, hairy below; petioles about as long as blade, sap milky
Flowers: clusters (corymbs) terminal, upright, hairy on short leafy branchlets; bisexual; yellow-green
Fruits: usually hairy, wings spreading horizontally; nutlets flat
Twigs: glabrous or hairy, becoming corky
Buds: dark; scales overlapping, tips often hairy and chaffy
Bark: brown-gray, corky, becoming scaly and fissured

Acer miyabei Maxim. - Miyabe maple

Miyabe maple, a native of Japan, is closely related to *A. campestre* which it resembles in some characteristics. But the Miyabe is usually easily distinguished from the latter by its larger, hairy leaves. It is rarely found in the nursery trade because it is difficult to propagate. Miyabe maple survives at the Minnesota Landscape Arboretum and several specimens are doing quite well at the Morton Arboretum (Dirr 1990).

Planning

Height: up to 30 ft (9 m)
Habit: tree; deciduous; crown becomes flat-topped; branches gray and hairy; autumn foliage yellow but short-lived
Site: sun; prefers slightly acidic soils; zone 5
Use: specimen, street

Recognition

Leaves: mostly 5-lobed with prominent indentations, tip pointed; surfaces hairy, becoming glabrous above, hairy tufts in vein axils below; petioles hairy, 4-7 in (10-18 cm) long, sap milky
Flowers: clusters (corymbs) terminal, many-flowered, on short, leafy branchlets, sometimes as short corymbs from leafless axillary buds; flowers yellow-green
Fruits: wings widely divergent, nutlet velvety hairy (velutinous)
Twigs: hairy, soon becoming glabrous, slightly corky with age
Buds: brown, lateral hidden by petioles; scales overlapping, hairy edged (ciliate)
Bark: gray, corky

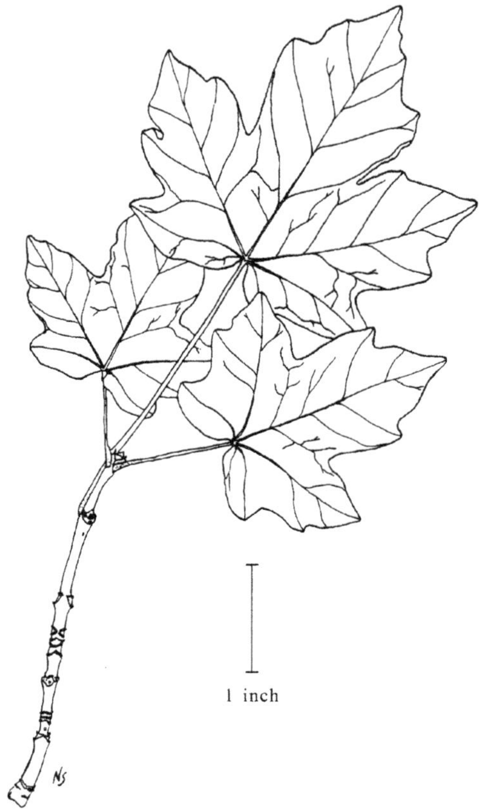

Acer miyabei

Acer platanoides L. - Norway maple

Norway maple, a native of northern and central Europe and the Baltic States, is quite variable in form and leaf character. Consequently, many Norway maple cultivars have been selected (Dirr 1990). 'Schwedleri', 'Deborah' and 'Fairview' exhibit foliage of various hues of red turning bronze to green in late summer. 'Crimson King' and 'Royal Red' maintain maroon foliage throughout the growing season.

Green-leaved 'Summershade' rates high in the Shade Tree Evaluation Trials at Wooster, Ohio, and is reputedly rapid-growing and heat resistant. Most of the cultivars mentioned above, as well as 'Columnar Broad', 'Columnare', and 'Superform', are adapted to the Fort Collins area, but 'Crimson Sentry' and 'Jade Glen' failed to establish at PERC (Fairchild and Klett 1993).

Planning

Height: up to 50 ft (15 m); Nebraska champion, 1989, 51 ft (15.5 m), Snyder; National champion, 1991, 137 ft (41.9 m), New York

Habit: tree; deciduous; crown broad, rounded (quite variable in form); yellow-cream to reddish flowers appear before the leaves; summer foliage green to purple (quite variable in leaf color); autumn foliage yellow

Site: sun, part shade; pH and soil adaptable, tolerates drouth, air pollution; zone 3

Use: shade

Caution: sensitive to 2,4-D; thin-barked young trees are sensitive to winter sun which can cause sunscald and frost cracks

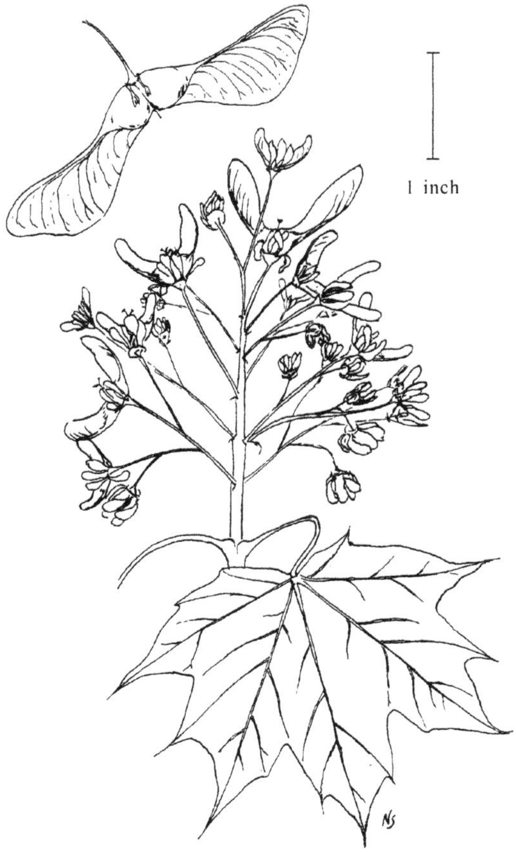

Acer platanoides

Recognition

Leaves: 5-lobed, sharp-pointed; margin toothed; base hastate or truncate; surfaces glabrous, hair tufts in vein axils below; petioles green, sap milky

Flowers: clusters (corymbs) many-flowered, erect, terminal on short leafy branchlets; flowers yellow-cream, sometimes reddish; sepals and petals distinct

Fruits: wings divergent at 180 degrees; ripening and falling in late summer; nutlets triangular, flattened, walls leathery with indistinct veins

Twigs: smooth, shiny, olive-brown, becoming gray-brown 2nd year; leaf scars encircle twig forming a sharp angle at meeting point

Buds: terminal egg-shaped; scales several, overlapping, shiny, glabrous, greenish maroon, hairy edges (ciliate); 2 smaller accessory buds beside terminal; lateral buds appressed to stem, scales 2-3

Bark: gray, shallow fissured, diamond-shape pattern, not peeling

Acer truncatum Bge. - purpleblow maple

Purpleblow maple, a native of northern China, Manchuria, Japan and Korea, is little known in the U.S. outside of a few arboreta. Dirr (1990) reports successful plantings at the Morton and Minnesota Landscape Arboretums. Hoag (1965) noted that exceptionally attractive specimens are growing at the Morden Experiment Station in Manitoba. A large tree in Colby, Kansas, has provided seeds for establishing additional plantings in that state (Sutton 1998). A healthy, attractive specimen grows on the UNL city campus.

The reddish purple emerging leaves (which give it the name purpleblow), the potential for colorful autumn foliage, and its heat and drouth tolerance are good reasons for considering purpleblow maple for yard and street plantings in the central and northern prairies. Among the few cultivars are 'Norwegian Sunset', an upright oval tree, and 'Pacific Sunset', an upright spreading tree, both hybrids between *A. truncatum* and *A. platanoides* (Dirr 1990).

A. mono Maxim., painted maple, is closely related and sometimes confused with purpleblow maple. One synonym accepted by some authorities is *A. truncatum* Bge. ssp. *mono* (Maxim) Murray. *A. tenellum* Pax is a rare shrubby maple closely allied to *A. mono*, differing mainly in size of leaves and bearing hairy petals (Gelderen et al. 1994).

Planning

Height: up to 20 ft (6 m)
Habit: tree, shrub; deciduous; crown round, branches densely packed and twisted; foliage emerges reddish purple; yellow-green flowers after the leaves; autumn foliage in shades of yellow, orange, and red
Site: sun; pH and soil adaptable; tolerant to heat and drouth; zone 4
Use: specimen, street

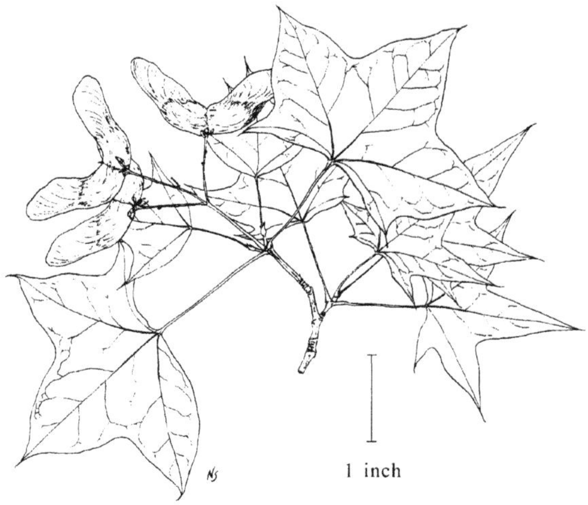

Acer truncatum

Recognition

Leaves: 5- to 7-lobed, tips pointed, margin entire, somewhat wavy; young leaves purplish, becoming olive-green with purple margins; surfaces glabrous, vein axils tufted beneath; petioles glabrous, sap milky
Flowers: clusters (corymbs) erect, terminal, on short leafy branchlets; flowers greenish yellow; petals and sepals distinct
Fruits: wings divergent at greater than 90 degrees; nutlet compressed, wings one and a half times longer
Twigs: glabrous, shiny red-brown, opposing leaf scars meet at a point
Buds: 4-sided, reddish brown; scales overlapping, glabrous
Bark: very rough and fissured, tinged with purple becoming gray-brown on larger stems

Section Ginnala

Section Ginnala includes one species with four subspecies, two of which are described here.

Acer tataricum ssp. *ginnala* (Maxim.) Wesmael - Amur maple

Amur maple, a native of northern China, Manchuria, North Korea and Japan, is a shrub commonly found in parks, gardens and windbreaks throughout the prairies. Hoag (1965) rates Amur as one of the hardiest and most valuable shrubby maples for the northern Great Plains. Many Amur maples at Prairie Pines Arboretum near Lincoln suffered severe dieback at about 20 years of age. However, fruit production is prolific and volunteer seedlings abound. A species specimen and cultivars 'Compacta' and 'Flame' are winter hardy in PERC at Fort Collins, but chlorosis is noticeable and especially damaging to the two cultivars (Fairchild and Klett 1993).

Amur maple is noted for its brilliant scarlet autumn coloration, but this can be disappointing. These maples at Prairie Pines Arboretum remain green into early November, turning pale yellow or brown as the weather dictates. Selection of cultivars adapted to the planting site is necessary if foliage coloration is desired. The cultivar 'Red Fruit' displays a brilliant red fruit that persists after leaf fall. 'Flame' is a release of the Plant Materials Center of the Natural Resources Conservation Service at Elsberry, Missouri. The fruit becomes bright red in summer, and autumn foliage is fiery red at that location.

A. tataricum L. ssp. *tataricum*, tatarian maple, is a close relative of Amur maple. This shrub or small tree usually bears unlobed leaves and its fruits are red during July and August. Tatarian maple grows in sunny, dry situations as forest undergrowth in eastern Europe as well as the Caucasus and Turkey. Sutton (1998) observed vigorous specimens (except for a slight tinge of chlorosis) growing at Colby, Kansas, and at Fort Collins, Colorado.

Acer tataricum* ssp. *ginnala

Planning

Height: up to 15 ft (4.5 m); Nebraska champion, 1989, 27 ft (8.2 m), Fremont
Habit: shrub, occasionally a small tree; deciduous; crown rounded; May flowers creamy; autumn foliage yellows to reds, fruits may be red
Site: sun, part shade; soil adaptable, prefers pH below 8.0; drouth tolerant; zone 2
Use: specimen, patio, container, grouping, naturalizing, wildlife habitat, windbreak

Recognition

Leaves: 3-lobed (some unlobed leaves always present), middle longer than 2 lateral; margin toothed (doubly serrate); surfaces glabrous, green, sometimes mottled with white or pink; petioles half as long as blade
Flowers: clusters (corymbs) terminal, on short leafy branchlets; bisexual; flowers yellowish white; fragrant
Fruits: glabrous; wings nearly parallel, often red becoming brown in September
Twigs: glabrous
Buds: inconspicuous
Bark: olive-gray

Section Rubra

Acer rubrum L. - red maple

Red maple, a native of southeastern Canada and the U.S. in the states east of Minnesota and Eastern Texas (excluding Iowa), is most at home in bottomland forests on fertile, deep soils, but it occasionally grows in upland forests on dry sites.

This maple is a good, moderately fast-growing yard and street tree where water is adequate but develops chlorosis on high pH soils. The young bark is thin and extremely suseptible to winter sunscald. A newly developed hybrid (*A. rubrum* × *A. saccharinum*), now called *A.* ×*freemanii*, is touted as having the hardiness and quick growth of silver maple and the foliage color of red maple (Sutton 1998). The Dutch cultivar 'Elegant' is one of these hybrids.

Though red maple is widely planted because of its potential for brilliant autumn color, it does not always deliver. Many cultivars have been selected (Dirr 1990, Gelderen et al. 1994). If fall color is an important objective for the landscape, trees of known performance selected from a nursery in the vicinity should be planted. Differences among cultivars is shown by the coloration in mid-October 1988 of three cultivars growing in Maxwell Arboretum at UNL. On that date a specimen of 'Scarlet Sentinel' displayed brilliant red foliage while 'Doric' and 'October Glory' remained green into November before coloring. A sharp freeze prior to inception of dormancy could spoil color development and late-clinging leaves can accumulate snow and cause branch breakage.

Planning

Height: up to 50 ft (15 m); National champion, 1984, 179 ft (54.3 m), Michigan
Habit: tree; deciduous; crown open, oval to irregular; April bloom red to yellowish before the leaves, impressive; immature fruits usually bright red; autumn foliage in shades of yellow to red, sometimes remaining green until freezing weather
Site: sun, part shade; prefers neutral to acidic moist soils; tolerates moderate air pollution; zone 3
Use: specimen, shade
Note: seeds will germinate immediately after ripening
Caution: sensitive to winter sun scald; intolerant of severe drouth

Acer rubrum

Recognition

Leaves: 3- to 5-lobed, often longer than wide; often red as they emerge; margin toothed (crenate-serrate); glabrous, dark green above, very pale beneath, sometimes hairy-veined; petioles often red, up to 4 in (10 cm) long
Flowers: clusters (fasciculate) from lateral leafless buds; unisexual, dioecious or sometimes monoecious; calyx 5-lobed, petals 5, yellow or bright red, very small; ovary glabrous
Fruits: glabrous, immature nutlet usually bright red, ripening late spring; wings divergent at about 60 degrees
Twigs: green, becoming red, shiny, not odoriferous when crushed
Buds: terminal egg-shaped, blunt; lateral smaller, solitary or multiple (collateral), slightly stalked; flower buds globose, in clusters; scales overlapping, red
Bark: gray, smooth, becoming shallow fissured in long narrow scaly plates
Wood: moderately heavy and hard

Acer saccharinum L. - silver maple

Silver maple is a tree commonly found along major streams in eastern Nebraska, Kansas and Oklahoma. Its native range extends eastward into all the states of the U.S. bordering the Atlantic Ocean and into southern Quebec and Ontario of Canada. This fastest growing maple tolerates a wide range of soil conditions where water is adequate.

Silver maples have been widely planted for quick shade throughout much of the prairies with reasonable success except on soils of high pH. Brittle branches are easily broken by wind and ice. 'Blair', a Nebraska selection with a strong branching pattern, is reputed to be somewhat resistant to breakage. 'Silver queen' is a nearly fruitless cultivar of good form. Leaves of 'Laciniatum' and 'Weiri' are deeply dissected.

Silver maples will sprout from adventitious buds near the trunk base during their early life to form multiple stems, necessitating periodic pruning for good form. Prolific fruit production provides food for birds and squirrels and plenty of seedlings if moisture conditions are good in late spring. Silver maple can be tapped for syrup production. Wood is used for furniture, crates, pallets and fuel. Its fast growth and sprouting capability make silver maple a candidate for energy plantations, but not for homesites or streets.

Acer saccharinum

Planning

Height: up to 60 ft (18 m); Nebraska champion, 1988, 84 ft (25.6 m), Oakland; National champion, 1972, 125 ft (37.9 m), Michigan

Habit: tree; deciduous; crown open and spreading; March flowers ahead of leaves, greenish to rose-colored; autumn foliage greenish, yellow, or sometimes tinged red

Site: sun, part shade; pH (except very high) and soil adaptable, prefers moist soils; zone 3

Use: shade, lumber, windbreak

Note: branches easily broken; seeds will germinate immediately after ripening

Recognition

Leaves: deeply 5-lobed, sometimes divided to the base, pointed, central lobe 3-lobulate; margin irregularly toothed; green, glabrous above, hairy, white to silvery (glaucous) below, becoming glabrous; petioles up to 5 in (13 cm) long, thin

Flowers: clusters small from lateral leafless buds; bisexual and unisexual (monoecious or dioecious); greenish to rose-colored; calyx 5-lobed, hidden by bud scales, petals none; ovary hairy

Fruits: hairy when young; wings divergent at 90 degrees, curving in (falcate), up to 2 in (5 cm) long; maturing late spring, germinating immediately

Twigs: reddish, glabrous, lustrous (very similar to *A. rubrum*), odor rank when bruised

Buds: leaf solitary, terminal or lateral, egg-shaped, pointed; flower cluster near twig and spur ends or at base of current growth, globose; scales red, hairy-fringed (ciliate), slightly keeled

Bark: silvery gray, breaking into long thin scaly plates, loose at ends, shaggy

Wood: brittle, moderately heavy and hard

Anacardiaceae Lindl. - **sumac family**

The sumac family—600 species in 60 to 80 genera—is largely found in tropical and subtropical regions. Although some species can cause dermatitis on susceptible people, many are a source of important products to world commerce and add beauty to landscapes as well. Some familiar species of the sumac family include the cashew *(Anacardium occidentale* L.), a native of Haiti; the pistachio (*Pistacia vera* L.), a native of the eastern Mediterranean region; and the mango (*Mangifera indica* L.) from Asia, which is widely cultivated in warm regions for its delicious fruit. Several sumac species are a source of resins, lacquers and varnishes.

Recognition

Habit: trees, shrubs, vines; deciduous, evergreen

Leaves: alternate, rarely opposite or whorled; compound, seldom simple; stipules wanting or rarely present and inconspicuous

Flowers: clusters (panicles); bisexual or unisexual, regular; perianth 3- to 5-parted, sepals usually fused below, petals distinct, sometimes wanting; stamens 5-10, filaments distinct, rarely fused at base; ovary compound, styles distinct or united

Fruits: fleshy, 1-seeded (drupaceous), resinous, waxy or oily

Cotinus Adans. - **smoketree**

One of two species of *Cotinus* is a native of the eastern U.S. and the other is a native of Europe and Asia.

Recognition

Habit: shrubs, trees; deciduous

Leaves: simple, alternate; margins entire

Flowers: clusters (panicles) large, loose, terminal; bisexual and unisexual on same plant (polygamous) or dioecious; stalks (pedicels) of sterile flowers long, plumose: petals 5, oblong, twice as long as sepals; stamens shorter than petals; ovary superior, styles 3, lateral

Fruits: dry drupe obliquely egg-shaped

Cotinus coggygria Scop. - **smoketree**

The smoky floral display of the exotic smoketree is its main claim to fame. Purple-leaved varieties have the potential for colorful autumn foliage which adds interest and contrast to landscapes. Cultivars vary in their ability to withstand the rigors of the northern prairie.

'Royal Purple' is subject to dieback in the north (North Central Plant Introduction Station 1963), but Hoag (1965) states that it has merit as a suffrutescent shrub for color accent in herbaceous borders in the northern plains. Purple-leaved 'Nordine' is hardier than 'Royal Purple'; 'Velvet Cloak' retains purple leaves through much of the summer and often displays a spectacular reddish purple color in the fall; and 'Daydream' has green foliage and bears dense, fluffy flower clusters.

Cotinus coggygria

Planning

Height: up to 15 ft (4.5 m)
Habit: shrub; upright, open, often wider than tall; early summer blooms, colorful autumn foliage
Site: sun, part shade; pH and soil adaptable; drouth tolerant; zone 5
Use: border, grouping

Recognition

Leaves: mostly oval to wider above middle (obovate), up to 3 in (8 cm) long; tip rounded or with shallow notch (emarginate), base abruptly contracted; surfaces glabrous, bluish green; veins distinct, parallel; petioles up to 1.6 in (4 cm) long
Flowers: clusters up to 8 in (20 cm) long; sterile pedicels numerous with long purplish or greenish hairs; polygamous or dioecious
Fruits: about 0.2 in (4 mm) across
Twigs: brown or purplish, bloomy, glabrous, strong odor when crushed; bloomy purple color around leaf scars; lenticels numerous
Buds: less than 0.2 (4 mm) long, sessile, pointed; scales several, overlapping, red-brown

Cotinus obovatus Raf. (*C. americanus* Nutt.) - American smoketree

The American smoketree is a rare native tree of the rocky limestone soils of Tennessee, Alabama, the Ozark Mountains and eastern Oklahoma, and is more plentiful as a shrub on the Edwards Plateau in south-central Texas. Today's scarcity of the American smoketree is due in part to its use as a source of yellow dye during the Civil War.

Although American smoketree flowers are not as spectacular as those of the exotic smoketree, its autumn foliage may produce the most intense color of all American shrubby trees (Dirr 1990). Autumn leaves of 'Red Leaf' are especially brilliant (Snyder 1980). Plantings north of Omaha (Rassmussen 1989) and east of Lincoln (Sutton 1998) have displayed excellent fall color. Provenance research and test plots are needed to determine the potential value of American smoketree for prairie landscapes.

Planning

Height: up to 30 ft (9 m); National champion, 1989, 39 ft (11.8 m), Indiana
Habit: tree, shrub; upright; spring blooms, brilliant autumn foliage
Site: sun; pH and soil adaptable; drouth tolerant; zone 5
Use: border, grouping

Recognition

Leaves: egg-shaped, mostly wider above middle (obovate), up to 4.7 in (12 cm) long; tip rounded, base rounded to wedge-shaped; margin slightly wavy; light purple and hairy when young, becoming glabrous and dark olive-green above, hairy beneath; petioles up to 2 in (5 cm) long
Flowers: clusters up to 6 in (15 cm) long; sterile pedicels bearing inconspicuous pale purplish or brownish hairs; usually dioecious, flowers few
Fruits: about 0.1 in (2-3 mm) across
Twigs: purple turning green, reddish brown with age, glabrous; aromatic when bruised; sap milky, gummy; lenticels corky, white
Buds: terminal about 0.2 in (5 mm) long, lateral smaller, often clustered near twig end; scales hairy, red-brown, 4 visible
Bark: thin, gray-brown with oblong scales

Rhus L. - sumac

About 120 nearly worldwide species represent genus *Rhus*. Fifteen of these are natives of North America.

Recognition

Habit: shrubs, trees; deciduous, evergreen

Leaves: simple, 3-foliate or pinnately compound, alternate

Flowers: clusters terminal; dioecious or bisexual and unisexual on the same plant (polygamous); petals 5, overlapping (imbricate), longer than imbricate sepals; stamens 5, inserted below a brownish disk; ovary superior, styles 3, usually separate

Fruits: berry-like drupes, rounded to egg-shaped; often reddish and densely hairy

Buds: tiny, roundish, appearing to be without scales, dense-hairy

Rhus aromatica Ait.- fragrant sumac

Fragrant sumac is a quite variable species that is a native of much of the eastern U.S. as well as the Great Plains. Attempts to identify and describe these variants has established a number of varieties, a subspecies (infraspecific taxa), and another species (*R. trilobata* Nutt.). Rehder (1940) describes the added species as having smaller leaflets with rounded teeth and an undersurface which is slightly hairy when young, becoming hairless (glabrate). However, Weber (1990) considers this added species to be a subspecies [*R. aromatica* ssp. *trilobata* (Nutt.) W. Weber].

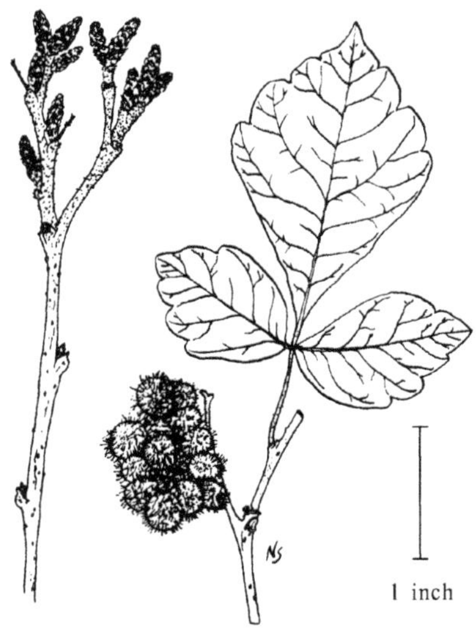

Rhus aromatica

Stephens (1973) concludes that additional research on the physical characteristics of fragrant sumac and the relationship of flowering to leaf emergence is necessary to properly identify and name the entities. We have elected to follow the recommendation of Stephens and recognize only one highly variable species. Fairchild and Klett (1993) have included plants identified as *R. aromatica* and *R. trilobata* in the trials at PERC in Fort Collins, where both species are adapted.

An upright plant of eastern Kansas origin—identified as *R. aromatica* var. *serotina* (Greene) Rehd.—has been selected and named 'Konza' by the Natural Resources Conservation Service Plant Materials Center at Manhattan, Kansas. It is a superior plant with good growth rate and form, resistance to diseases and insects, and potential for excellent wildlife habitat. 'Konza' is adapted to Nebraska and the Great Plains to the south.

A selection named 'Big Horn' by the Plant Materials Center at Los Lunas, New Mexico—identified as *R. trilobata* Nutt.—originated in the Big Horn Basin of Wyoming. Test plots show that 'Big Horn' is well adapted to the dry plains east of the Rocky Mountains including South Dakota. 'Green Globe', a rounded, dense shrub; 'Gro-low', a low, wide-spreading shrub with glossy foliage; and 'Amber Autumn', a prostrate cultivar collected near Littlefield, Texas, are also selections of fragrant sumac (Dirr 1990).

Planning

Height: up to 10 ft (3 m)

Habit: shrub; deciduous; dense, spreading, sprawling to upright; colorful autumn foliage; fruits persist through winter

Site: sun, part shade; pH and soil adaptable; drouth tolerant; zone 3

Use: wildlife habitat, erosion control, windbreak, naturalizing, grouping

Recognition

Leaves: compound (trifoliate); leaflets sessile, egg-shaped to broadly so; lobed to coarsely toothed, tips pointed or rounded; surfaces variably hairy to glabrous; petioles about 0.5 in (12 mm) long

Flowers: clusters dense, terminal, developing from catkins of previous season; polygamo-dioecious; yellowish

Fruits: clustered; each fruit globose, each about 0.2 in (5 mm) across, red or orange-red, not juicy, variably hairy; seeds 1

Twigs: glabrous or hairy, often red glandular when young; bearing winter catkins up to 0.3 in (8 mm) long; leaf scars elongated, raised, U-shaped; bundle scars many

Buds: covered by old leaf base; scales hidden by hairs

Bark: dark brown, smooth, soon becoming cracked and slightly fissured; lenticels transverse, prominent

Rhus copallina L. - flameleaf sumac

Flameleaf sumac, a native of most of the eastern U.S. as well as southeastern Nebraska and an area extending south into eastern Texas, grows along forest edges, fence rows, gullies and waste places on dry soils. Variety *latifolia* is a native of the western part of its range and variety *copallina* grows in the eastern U.S. (Stephens 1973).

Winter dieback has been severe on flameleaf sumac specimens at Prairie Pines Arboretum near Lincoln. Evaluation of comprehensive provenance plantations could provide useful information about flameleaf sumac's adaptability to prairie sites.

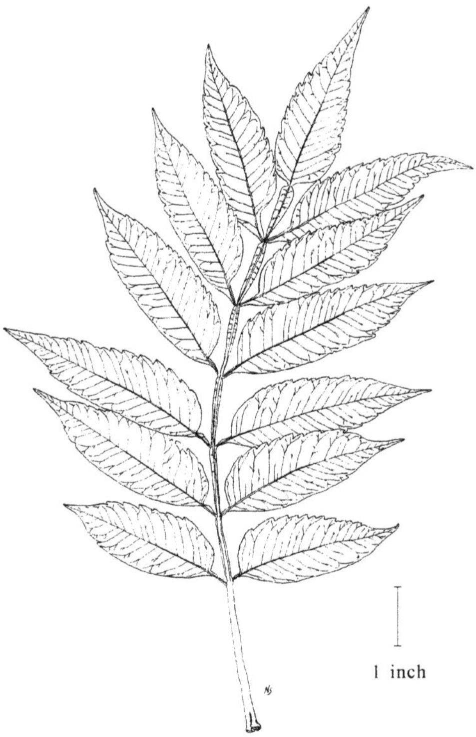

Rhus copallina

Planning

Height: up to 20 ft (6 m); National champion, 1991, 33 ft (10.1 m), Michigan

Habit: shrub, tree; deciduous; dense, becoming open, irregular; suckering; colorful autumn foliage and fruits

Site: sun; pH and soil adaptable; zone 5

Use: grouping, naturalizing, wildlife habitat, erosion control

Recognition

Leaves: pinnately compound, stalk (rachis) winged; leaflets 9-23, elliptical to egg-shaped up to 3 in (8 cm) long; margin entire or barely toothed near apex, hairy (ciliate); tip pointed, base tapered to obtuse; upper surface shiny, dark green, hairy beneath; petioles up to 3 in (8 cm) long; stipules wanting

Flowers: clusters (panicles) terminal, pyramidal, up to 8 in (20 cm) long on new growth; stalks hairy; unisexual (dioecious); stamens rudimentary on female

Fruits: clustered, drooping, conical; each fruit dark red, globular, about 0.2 in (4 mm) across, hairy

Twigs: curly haired; roughened by raised lenticels; leaf scars U-shaped, reaching to near top of bud, vascular bundles indistinct

Buds: terminal absent; lateral tiny, egg-shaped, pale-hairy

Bark: gray, shallow fissured; lenticels prominent, horizontally elongated

Anacardiaceae *Rhus*

Rhus glabra L. - smooth sumac

Smooth sumac, a pioneer that invades abandoned croplands and marches along rail corridors and streambanks, can also be found in prairies and along woodland edges on a wide variety of soils throughout much of North America. It is especially abundant east of the Rockies.

Smooth sumac is blessed with a vigorous, rhizomatous root system that forms circular, nearly grass-free islands that can be havens for animals during prairie fires. Plains Indians prepared an astringent broth from the fruit of smooth sumac to allay internal and external hemorrhaging; gathered the red autumn leaves for smoking; and harvested the roots for making yellow dye (Gilmore 1977). The leather tanning industry has utilized tannin extracted from smooth sumac leaves and bark.

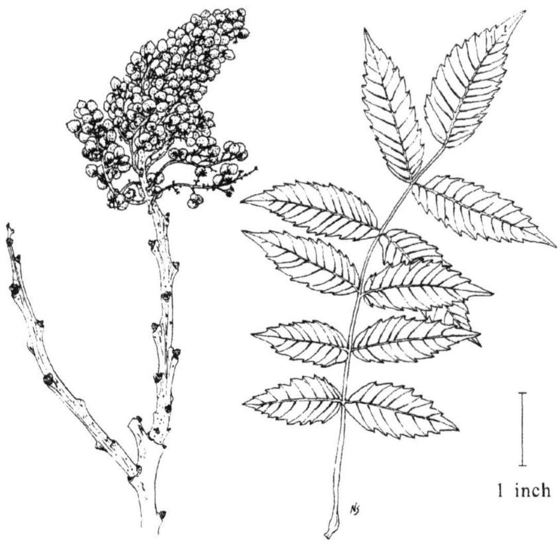

Rhus glabra

Soil fertility and water determine the ultimate height of native smooth sumac colonies. They seldom exceed 10 feet (3 m) in the western prairies. Smooth sumac's latitude of origin determines date of dormancy and leaf coloration and also affects its ultimate height. A healthy planting of smooth sumac more than 15 feet (5 m) tall screens a parking lot on the East Campus of UNL. Its autumn beauty appears in mid-October, an indication that the origin of its plants is a latitude considerably south of the campus, whereas native stands in the Lincoln area provide mid-September color. Variety *laciniata* is named for its deeply cut leaflets and *cis-montana* is a smaller variety of smooth sumac.

Planning

Hieght: up to 15 ft (4.5 m); National champion, 1970, 28 ft (8.5 m), Idaho
Habit: shrub, tree; deciduous; upright, colonial; autumn foliage colorful; fruit clusters persist over winter
Site: sun; pH and soil adaptable; drouth tolerant; zone 3
Use: naturalizing, wildlife habitat, screen, erosion control, grouping

Recognition

Leaves: pinnately compound, stalk (rachis) glabrous or with a narrow line of hairs on upper surface; leaflets 15-23, up to 3.5 in (9 cm) long, elliptic; coarsely toothed, tip pointed, base obliquely rounded; upper surface shiny, hairy at base and along mid-vein, whitened and glabrous beneath; petioles up to 3.5 in (9 cm) long, often hairy and red above
Flowers: clusters (panicles) pyramidal, terminal on new growth; male loose, up to 10 in (25 cm) long, female dense, up to 6 in (15 cm) long; dioecious; calyx lobes 5, distinct; petals 5, egg-shaped, yellow-green, line of hairs through center on inside
Fruits: clusters erect, dense, pyramidal; each fruit (drupe) globose, dark red, coarsely hairy, about 0.2 in (4 mm) across, not juicy
Twigs: glabrous, brownish, sometimes bloomy; sap milky, sticky; lenticels gray, leaf scars narrow encircling bud, vascular bundles indistinct
Buds: terminal absent, lateral conical, developing inside petiole base; hairy, gray or buff
Bark: gray-brown, roughened with raised horizontal lenticels, shallow fissured with age

Rhus typhina L. - staghorn sumac

The range of this native of the northeastern U.S. extends south to Tennessee and west to eastern Minnesota. It grows along woodland edges and disturbed areas such as roadsides on a variety of soils. Staghorn sumac resembles smooth sumac except for its velvety hairy, antler-like branches—its "trademark." Two female cultivars, 'Laciniata' and 'Dissecta' bear deeply divided leaflets. 'Laciniata' growing in PERC at Fort Collins suffered dieback during severe winters, and chlorosis varied from minor to severe (Fairchild and Klett 1993).

Planning

Height: up to 20 ft (6 m); National champion, 1985, 61 ft (18.5 m), Alabama

Habit: shrub, tree; deciduous; open, upright, becoming gaunt and scraggly; rhizomatous, forming colonies; colorful autumn foliage; red fruits persist through winter

Site: sun, part shade; pH and soil adaptable; drouth tolerant; zone 3

Use: naturalizing, grouping, erosion control, wildlife habitat

Recognition

Leaves: pinnately compound, leaflets 13-27, up to 5 in (13 cm) long; lance-oblong, curved, tip pointed, base rounded, toothed; hairy beneath becoming glabrous; petioles about 2 in (5 cm) long

Flowers: clusters (panicles) greenish, terminal, pyramidal; female dense, male loose; polygamous or dioecious

Fruits: clusters dense, pyramidal; each fruit globose, hairy, not juicy, crimson

Twigs: densely velvety hairy hiding lenticels; sap milky, resinous; leaf scars not elevated, c-shaped, nearly encircling buds

Buds: terminal absent; lateral hairy

Bark: thin, smooth, developing small scales with age

Rhus typhina

Toxicodendron Mill. - poison sumac

Approximately 20 species of poison sumac are natives of North America and Asia. Rehder (1940), Porter (1967) and others place some of these species in the genus *Rhus*. They consider *Toxicodendron* to be a genus of poisonous plants in the spurge family (Euphorbiaceae) that are native to South Africa.

Recognition

Habit: trees, shrubs, vines
Leaves: compound (trifoliate or pinnate)
Flowers: clusters axillary, drooping, loose; style single, unbranched
Fruits: white or grayish, glabrous or slightly fuzzy, shiny; wax-producing; stones ridged
Twigs: sap poisonous

Toxicodendron radicans (L.) Kuntze **ssp.** ***negundo*** (Greene) Gillis - **poisonivy**

Poisonivy (also named *Rhus radicans* L.) is often inadvertantly discovered but never forgotten once an outbreak of a persistent, itching skin rash is known. Poisonivy often bears the brunt of eradication efforts and seldom, if ever, is purposely planted and cultivated. However, it is an important food source for wildlife. The fruits are eaten by flickers and other woodpeckers, catbirds, chicadees, quail, pheasants and turkey. They "plant" the seed and this insures a continuing food supply. Deer and rabbits browse the twigs. Intensive browsing by goats and sheep is sometimes used to eradicate poison ivy.

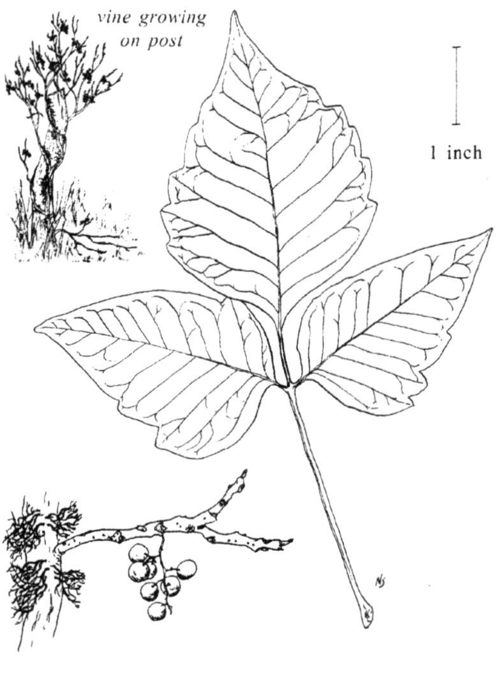

Toxicodendron radicans

Urushiol, a yellow, slightly volatile oil, is the toxic chemical found in the sap of poisonivy. It is also a component of the Japanese lacquer-tree, *T. vernicifluum* (Stokes) Mill., a source of the famous black varnish of the Orient. In addition to avoiding the smoke of burning poisonivy, some precautions to lessen the danger of "ivy" poisoning are: (1) Avoid shrubs or vines with clusters of white, cream-colored or ivory fruit; (2) avoid shrubs or vines with three leaflets ("Leaflets three, let it be"); and (3) avoid poison sumac, *T. vernix* (L.) Kuntze, a shrub or small tree found in the lowlands east of the Mississippi River, which bears alternate compound leaves of seven or more leaflets with entire margins, flowers in axillary clusters, and smooth fruits (U.S. Forest Service 1937).

Poisonivy and closely related species are found throughout most of North America in woodlands and along fence rows and roadsides on a wide variety of soils. Western poisonivy, *T. rydbergii* (Small) Greene, occurs in the western prairies, much to the dismay of hay makers. It is a low shrub with stems seldom-branched. Pacific poisonoak, *T. diversilobum* (Torr. & Gray) Greene, is the plant to avoid in the west coast states of the U.S.

Planning

Height: up to 65 ft (20 m)
Habit: shrub, vine; deciduous; climbing by aerial roots, colonial; autumn foliage colorful
Site: sun, shade; pH and soil adaptable; drouth tolerant; zone 3
Use: wildlife habitat
Caution: toxic sap can cause skin rash

Recognition

Leaves: compound, 3-foliate; leaflets egg-shaped, 2-5 in (5-13 cm) long; tip pointed, base obtuse, margin small-lobed or coarsely toothed; glabrous above, hairy below; petioles hairy
Flowers: clusters (panicles) axillary on new growth: unisexual (dioecious); petals yellow-green, recurved; calyx lobes 5, united at base, green; stamens 5, erect; pistil 1 with single style, stigma 3-lobed
Fruits: clusters grape-like; each fruit globose, 0.2 in (5 mm) across, glabrous, creamy white, apex slightly indented, style persistent
Twigs: woolly hairy, lenticels raised; aerial roots red, becoming brown
Buds: terminal 0.2 in (5 mm) long; laterals shorter, flattened, variously pointed, tawny hairy
Bark: becomes flaky with large size

Simaroubaceae DC. - quassia family

The quassia family is a native of the tropics and subtropics of the Northern Hemisphere with a few species extending their range into the temperate regions. Six genera are natives of North America.

Recognition

Habit: trees, shrubs
Leaves: alternate, rarely opposite, pinnately compound, seldom simple; stipules none
Flowers: clusters (panicles or spikes) axillary or terminal; unisexual, sometimes bisexual; regular, perianth 3- to 5-parted, sepals fused near base, sometimes distinct, petals distinct; stamens usually 10, filaments distinct; ovary superior (hypogynous)
Fruits: usually dry (capsules or samaras)

Ailanthus Desf. - ailanthus

Ten ailanthus species are recognized—all are natives of Asia and northern Australia.

Ailanthus altissima (Mill.) Swingle - tree-of-heaven

The tree-of-heaven today is rarely planted intentionally in North America. An 1875 law bans it from the District of Columbia because the "stinking ash" is a "nuisance injurious to health." This law was a reaction to the discovery that the tree was not desirable for planting along Washington's broad avenues. Tree-of-heaven—a native of China where it is considered to be a medicinal plant—grows in the streets of Beijing and is cultivated in Australia, New Zealand, southern Africa and Morocco.

The tree-of-heaven is adaptable to sites where many other plants fail. It withstands the grime and soot of the inner city and even survives the rigors of the western Nebraska climate. Beautiful specimens of tree-of-heaven grow in large cities such as Omaha and Lincoln but often suffer freezing injury on tender twigs where soil fertility and moisture are optimal for tender growth late in the growing season. If soils are dry in late summer or temperature decline is gradual from early autumn into winter, allowing the twigs and buds to mature and harden before freezing temperatures arrive, these trees will withstand sub-zero winter temperatures without injury.

Planning

Height: up to 50 ft (15 m); Nebraska champion, 1984, · 35 ft (10.7 m), Omaha; National champion, 1972, 64 ft (19.5 m), New York (Long Island)
Habit: tree; deciduous; upright branches coarse, crown becoming broad; root sprouts prevalent; fruits sometimes red in autumn, persistent and brown through winter
Site: sun, part shade; pH and soil adaptable; tolerates pollution; pest resistant; zone 5
Caution: subject to cold injury where site favors lush fall growth

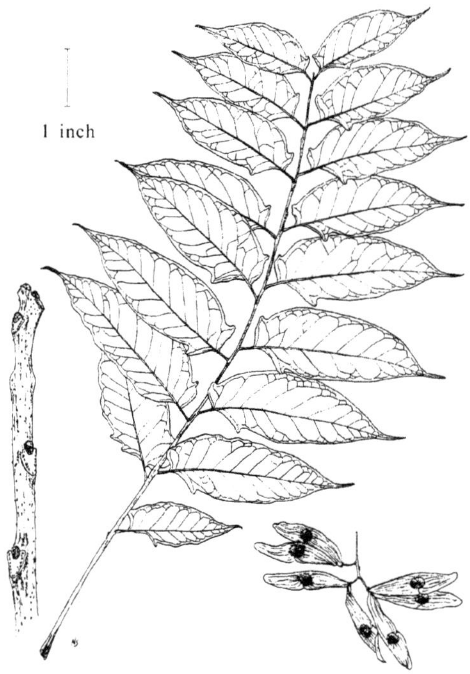

Ailanthus altissima

Recognition

Leaves: alternate, pinnately compound, up to 3 ft (90 cm) long; leaflets paired, much longer than wide, up to 6 in (15 cm) long, one-third as wide; margins hairy (ciliate), entire except coarse glandular teeth near base; upper surface nearly hairless (glabrate), hairy beneath; petioles swollen at base, velvety hairy
Flowers: clusters (panicles) terminal on new growth; unisexual (dioecious), sometimes bisexual; male unpleasantly odorous; calyx broadly bell-shaped, shallow 5-lobed; petals 5, greenish, hairy; stamens 10, ovaries 5, styles 5, twisted and fused
Fruits: winged (samara), oblong, twisted, yellow-brown to reddish; seed off-center
Twigs: coarse, rigid, velvety downy, rank odor when bruised; lenticels prominent, raised, reddish; leaf scars heart-shaped, bundle scars many in a row around the edge
Buds: terminal absent; lateral partly concealed in leaf scar notch, dome-shaped, brown, downy
Bark: gray, smooth becoming shallow-fissured
Wood: soft, weak

Rutaceae Juss. - **citrus family**

The citrus family—about 150 genera and 1500 species of trees, shrubs and a few herbs—is widely distributed throughout tropical and subtropical regions with a few members native to the temperate zone. North America is home to five genera: one in Mexico and four in the U.S.

Transluscent glands filled with aromatic oils present in the leaves of all family members are the source of flavorings, incense and perfumes. Oranges, limes, lemons and grapefruits are well-known citrus fruits.

Recognition

Habit: trees, shrubs, herbs; evergreen, deciduous
Leaves: alternate, rarely opposite; usually compound; often bearing translucent glands; stipules none
Flowers: clustered, seldom solitary; bisexual, seldom unisexual; regular, usually 4- to 5-parted; sepals distinct or fused at base; petals distinct, alternate with sepals, rarely wanting; stamens various, ovary usually superior
Fruits: dry splitting (capsules), berry or berry-like, or dry, flattened and winged

Evodia Forst. - evodia

Only one of about 50 species of Evodia—natives of eastern and southeastern Asia, Australia and Polynesia—is occasionally planted in the central prairie.

Evodia daniellii (Benn.) Hemsl. - Korean evodia

This native of northern China and Korea is a small mid-summer flowering tree seldom planted in the U.S. A specimen grows in the City Park Arboretum of Blair, Nebraska (Steyer 1990); Dirr (1990) reports specimens growing at Morton Arboretum in Lisle, Illinois. Dirr recommends Korean evodia as an excellent tree for small landscapes. It is a tree that should be included in more trials throughout the central prairie.

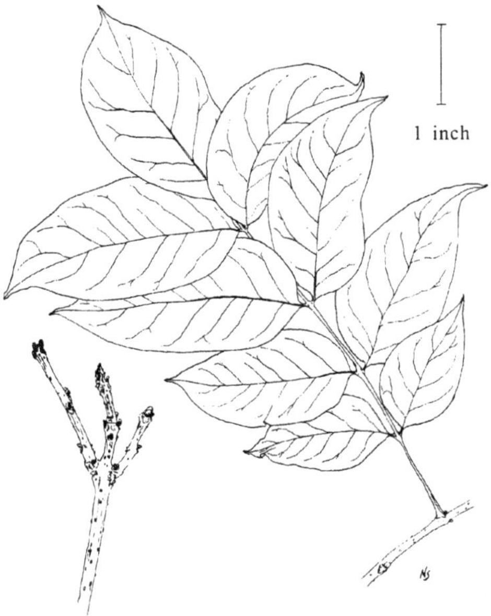

Evodia daniellii

Planning

Height: up to 25 ft (7.5 m)
Habit: tree; deciduous; crown round, wide; white summer blooms on new growth
Site: sun; pH adaptable, prefers moist, fertile soil; zone 5
Use: shade, bee pasture

Recognition

Leaves: opposite; pinnately compound; leaflets 7-11, short-stalked or sessile, up to 5 in ((13 cm) long; egg-shaped; margin with small, rounded teeth (crenulate); tip long-pointed, base rounded; shiny dark green above, lighter below, hairy on midrib and vein axils; petioles about 2 in (5 cm) long
Flowers: clusters (corymbs) up to 6 in (15 cm) across; unisexual; petals 4, white; sepals overlapping; stamens 4-5
Fruits: dry pod with a short, usually hooked beak, splitting from the top, about 0.4 in (8 mm) long, red to black; seeds 1-2, ellipsoid, shiny brownish black, about 0.1 in (3 mm) long
Twigs: hairy at first, becoming glabrous, red-brown with large pale lenticels; pith continuous; leaf scars broadly crescent-shaped, bundle traces 3
Buds: solitary, egg-shaped; scales, 1 pair, indistinct
Bark: smooth, gray, lenticels raised

Phellodendron Rupr. - phellodendron

The eight phellodoendron species are natives of eastern Asia. Dirr (1990) examined several phellodendron species growing in Arnold Arboretum and could not detect enough difference in leaf, bud and stem character to differentiate among them.

Phellodendron amurense Rupr. - Amur corktree

Male and female 25-year-old specimens of Amur corktree growing on the East Campus of UNL exhibit pleasing dark green summer foliage but do not add significant color to the autumn landscape. Nevertheless, they have been pest free and no other problems have emerged except for occasional twig dieback.

Beautiful specimens of Amur corktree are growing in the Morton and Arnold Arboreta, but in Wichita, Kansas, leaf scorch often defoliates Amur corktrees during August, and they averaged less than 5 inches (13 cm) height growth per year over a 10-year test period (Dirr 1990). Three Amur corktrees growing at PERC in Fort Collins are adapted to the climate and soil (Fairchild and Klett 1993).

P. sachalinense (Fr. Schmidt) Sarg., Sakhalin corktree, grows in the Minnesota Landscape Arboretum. Dirr (1990) reports that it is very fast growing and hardier than the Amur corktree. Leaflet margins of Sakhalin corktree are not hairy, hairs are sparse on the flower clusters, and the trunk bark is dark brown and slightly fissured (Rehder 1940).

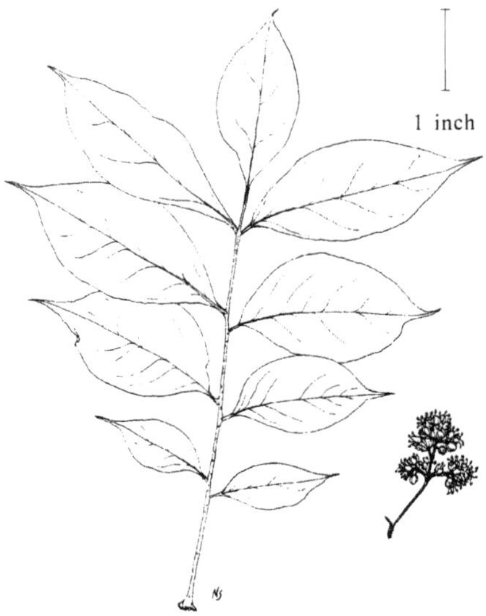

Phellodendron amurense

Planning

Height: up to 30 ft (9 m)
Habit: tree; deciduous; crown broadly rounded; May flowers inconspicuous; autumn foliage yellow, of short duration; fruits persist into winter
Site: sun, part shade; pH and soil adaptable, tolerates drouth, air pollution; zone 4
Use: shade

Recognition

Leaves: opposite; pinnately compound, leaflets 5-13, narrow-oblong to egg-shaped, about 3 in (9 cm) long, half as wide, margin smooth or sometimes hairy; tip long-pointed, base narrowed or rounded; shiny dark green above, glabrous below except along midrib, glandular dotted
Flowers: clusters (panicles) terminal; unisexual (dioecious); yellow-green; 5- to 8-parted, petals much longer than sepals; male with 5-6 stamens, longer than petals, pistil rudimentary; female with 5-6 sterile stamens (staminodes), stigma 5-lobed
Fruits: drupe, black, subglobose, about 0.4 in (1 cm) across, odiferous when bruised
Twigs: glabrous, brown; lenticels pale, inner bark yellow-green; leaf scars raised, horseshoe-shaped, enclosing bud
Buds: hidden by petiole base, solitary, compressed from sides; silky reddish hairs mask 2 scales
Bark: light gray, deep-fissured, corky

Ptelea L. - hoptree

The hoptree genus—seven to ten species—are natives of North America and Mexico.

Ptelea trifoliata L. - waferash

Waferash—sometimes called hoptree because its bitter fruits have sometimes been substituted for hops in brewing beer—is a native of the eastern U.S. west to Minnesota and extending southwest to Colorado and Texas.

'Aurea', which bears rich yellow young leaves that fade to lime green in August, would be an effective shrub border (Dirr 1990). 'Glauca' leaves are blue-green.

Hoptree (*P. polyadenia* Greene)—a native of the southwestern U. S. from Oklahoma to Arizona—is not widely available in nurseries. It differs from waferash by exhibiting more hairs on leaves and twigs. The hoptree and waferash are adapted to the soils and climate at PERC in Fort Collins (Fairchild and Klett 1993).

Planning

Height: up to 10 ft (3 m); National champion, 1991, 34 ft (10.4 m), Connecticut
Habit: shrub, tree; deciduous; crown rounded, spreading; June blooms greenish white, fragrant on new growth; autumn foliage yellow; fruit persists into winter
Site: shade, sun; pH and soil adaptable, prefers fertile, moist sites; zone 3
Use: naturalizing, border

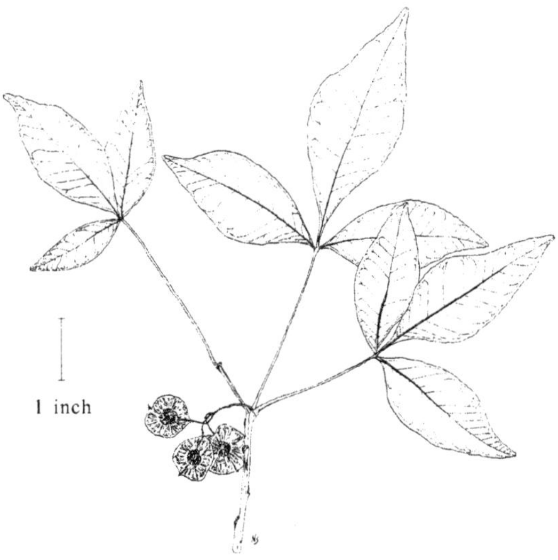

Ptelea trifoliata

Recognition

Leaves: compound (trifoliate); leaflets nearly stalkless (subsessile), 2-5 in (5-13 cm) long; oblong egg-shaped, apex pointed, lateral leaflets oblique at base; margins entire or obscure-toothed; glabrous, shiny dark green above, translucent glands visable under hand lens; petioles 2-4 in (5-10 cm) long
Flowers: clusters (cymes) terminal on short lateral twigs; unisexual (dioecious); about 0.4 in (1 cm) across; petals 4, pale yellow, finely hairy; calyx lobes 4, joined at base; stamens 4 opposite calyx lobes (vestigial on female); ovary winged laterally (vestigial on male)
Fruits: samaras flat, winged all around, 0.6-1.0 in (15-25 mm) across, tiny tip (mucronate)
Twigs: red-brown, sparingly hairy becoming glabrous; odoriferous when bruised; leaf scars U-shaped, bundle scars 3
Buds: globose, inconspicuous, partly hidden by leaf scar, silvery-silky
Bark: dark gray, smooth; lenticels prominent, elongated horizontal
Wood: heavy, hard, yellow-brown

Zanthoxylum L. - pricklyash

The genus *Zanthoxylum* includes over 200 species, mostly found in the tropics, but some grow in the temperate zones of North America and Asia. Six species reach tree size in the U.S., five of which are native only in the southern states. Spicy, bitter tasting oils that are present in the fruits, leaves and bark of all species, are used in medicines.

Zanthoxylum americanum Mill. - common pricklyash

The thicket-forming common pricklyash—a native of the eastern portion of the Great Plains from North Dakota through Oklahoma and much of the eastern U.S.—plagues workers and hikers in open woods with its sharp-spined branches. Pricklyash could be confused with blacklocust except for its red, woolly buds; clusters of round, aromatic fruits in winter; and its inconspicuous greenish yellow flowers of early spring.

Although seldom cultivated for landscape purposes, the bitter oil xanthoxylin, most concentrated in the bark and roots of the pricklyash, is used in medical preparations. The plant provided early settlers with a "cure" for toothaches and rheumatism.

Three species of Asian origin that may be adapted to part of the central prairie are: *Z. piperitum* (L.) DC., the compact and dense Japanese pricklyash; *Z. simulans* Hance, flatspine pricklyash, a spreading shrub or small tree that bears reddish fruit; and *Z. schinifolium* Sieb. & Zucc., peppertree, with red fruit and solitary spines (Dirr 1990). Flowers of these Asian species are borne in terminal clusters that appear after the leaves emerge.

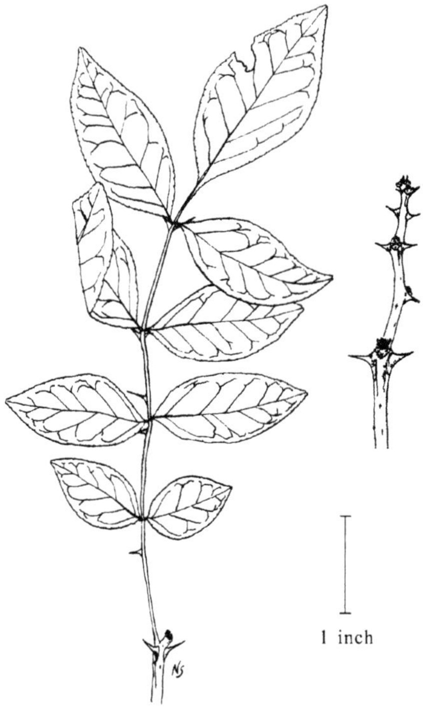

Zanthoxylum americanum

Planning

Height: up to 20 ft (6 m); National champion, 1985, 28 ft (8.5 m), Michigan
Habit: shrub, tree; deciduous; thicket-forming; flowers appear before the leaves
Site: shade, sun; pH and soil adaptable; zone 3
Use: barrier, naturalizing

Recognition

Leaves: alternate; compound, aromatic; leaflets 5-11, opposite, nearly sessile, egg-shaped, tip pointed, base rounded, 1-2 in (25-50 mm) long; margins slightly toothed (crenate) or entire; dark green, glandular dotted, slightly wrinkled above, hairy beneath; petioles rusty hairy
Flowers: clusters axillary on short spurs; unisexual (dioecious); calyx none, petals 5-6, green, red-tipped; stamens 5-6, alternate with petals; pistils 2-5, ovaries distinct, hairy, styles fused (connate)
Fruits: dry; rounded 2-valved follicles in dense clusters; each fruit 0.15 in (4 mm) in diameter, pitted, spice-scented; seeds shiny black, oily, pitted, aromatic
Twigs: glabrous, sometimes hairy, lemon-like odor when bruised; paired spines at each node, broad-based, recurved, 0.5 in (12 mm) long; lenticels inconspicuous; leaf scars half rounded or triangular, bundle scars 3
Buds: rounded, often one above another (superposed); scales indistinct, covered with red, woolly hairs
Bark: smooth, becoming slightly fissured with age

Order Apiales

Two families and about 3,700 species are included in the order Apiales (Cronquist 1981).

Recognition

Habit: woody, herbaceous; deciduous, evergreen
Leaves: usually alternate, sometimes compound or simple; petioles commonly broad and sheathing
Flowers: clusters (umbels), sometimes heads or otherwise; usually bisexual; calyx teeth very small or sometimes wanting, petals usually distinct, seldom wanting; stamens alternate with petals, seldom more numerous than petals; ovary usually inferior (epigynous), carpels 2-5, styles usually as many as carpels
Fruits: dry, splitting (schizocarps) or drupes or berries

Araliaceae - ginseng family

The ginsengs—a family of tropical and temperate regions—is divided into about 70 genera and 700 species. It is known for the ginseng root (*Panax ginseng* C.A. Mey.) which is used for medicinal purposes, and the rice paper plant (*Tetrapanax*) of the Orient—grown as an ornamental in the U.S.—from which thin paper is produced.

Recognition

Habit: trees, shrubs, vines, rarely herbs; deciduous, evergreen
Leaves: alternate, rarely otherwise; compound, less often simple; petioles often sheathing at base
Flowers: clusters (umbels), seldom otherwise; bisexual, sometimes unisexual; mostly regular, perianth attached to upper part of ovary (epigynous); calyx small-toothed at summit of ovary, or wanting, petals usually 5 and distinct; stamens usually alternate with petals; ovary inferior, rarely otherwise
Fruits: drupe or berry, rarely otherwise
Twigs: often prickly, sometimes climbing

Acanthopanax Miq. - **acanthopanax**

About 20 species of acanthopanax are natives of eastern Asia and the Himalayas. In addition to those described below, *A. henryi* (Oliv.) Harms may be adapted to parts of the central prairie (Dirr 1990). It is an upright multistemmed shrub that forms an impenetrable thicket.

Recognition

Habit: trees, shrubs; deciduous, rarely evergreen
Leaves: alternate; palmately compound
Flowers: solitary or clustered, terminal; bisexual or sometimes with unisexual flowers on same plant (polygamous); calyx teeth minute, petals 5, rarely 4; stamens as many as petals; styles 2-5, united or free
Fruits: drupes, black; seeds 2-5
Twigs: prickly, rarely unarmed
Buds: egg-shaped; scales several

Acanthopanax sessiliflorus (Rupr. & Maxim.) Seemen - **hardy aralia**

Hardy aralia—a native of northern China, Manchuria and Korea—is hardy in the northern plains but has little else to recommend it except as a barrier or as wildlife habitat (Hoag 1965). Trial plantings of hardy aralia were rated good in the central Great Plains except on alkaline soils (North Central Plant Introduction Station 1962).

Planning

Height: up to 10 ft (3 m)
Habit: shrub; deciduous; upright, coarse; spreads by suckers
Site: sun, shade; soil adaptable; tolerates pollution; zone 3
Use: wildlife, border
Caution: avoid alkaline soils

Recognition

Leaves: leaflets 3, sometimes 5, very short-stalked, oblong, often wider above the middle (oblong-ovate), up to 6 in (15 cm) long; tip long-pointed, base wedge-shaped, irregularly toothed (serrate); surfaces nearly glabrous or sometimes rough to touch (scabrid); petioles about 1.5 in (4 cm) long
Flowers: heads globose, dense, nearly sessile; bisexual; dull purplish, styles 2
Fruits: in a globose head; ellipsoid, about 0.5 in (12 mm) long, black
Twigs: unarmed or few prickles, short

Acanthopanax sieboldianus Mak. - **Siebolds acanthopanax**

Siebolds acanthopanax, a native of Japan, seldom fruits because typically only female plants are available for cultivation. Dirr (1990) recommends Siebolds acanthopanax for city planting because it adapts to unfavorable sites. Richley (1988) reports a tendency for it to die back during winter in southwestern Nebraska. Hoag (1965) recommends this shrub for planting only in sheltered sites in the northern plains.

Planning

Height: up to 10 ft (3 m)
Habit: shrub; deciduous; erect with arching stems, suckering; June blooms; fruits few
Site: sun, shade; pH and soil adaptable; drouth tolerant; zone 5
Use: barrier, screen, wildlife habitat
Note: tolerates pruning

Recognition

Leaves: leaflets 5 to 7, subsessile, wider above the middle (obovate to oblong-obovate), up to 1.5 in (4 cm) long; tip pointed, base wedge-shaped, toothed (crenate-serrate); bright green, glabrous; petioles up to 3 in (8 cm) long
Flowers: clusters (umbels) spherical, on a long slender stalk from short lateral spurs; unisexual (usually dioecious); greenish white; styles 5, fused nearly to apex
Fruits: nearly round, about 0.3 in (7 mm) across, 2- to 5-seeded
Twigs: light brown, warty; lenticels prominent; 1-2 prickles beneath each leaf scar or unarmed; leaf scars U-shaped, somewhat raised
Buds: solitary, ovoid; scales 3, exposed

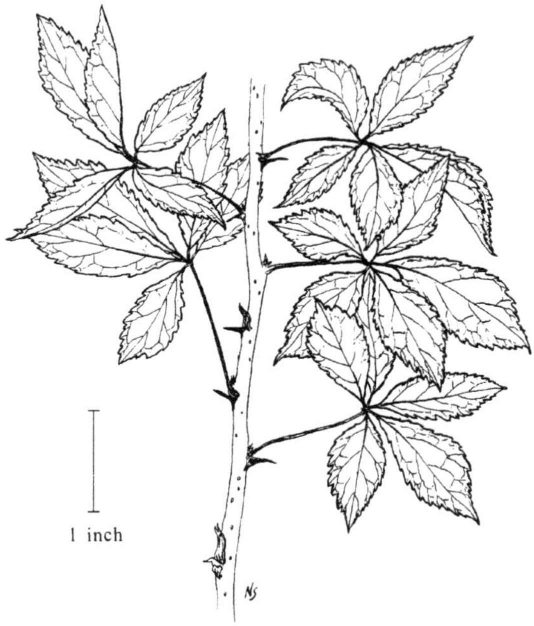

Acanthopanax sessiliflorus

Aralia L. - aralia

Thirty species of aralia occur in North America and Asia.

Recognition

Habit: trees, shrubs, herbs; deciduous
Leaves: compound (pinnate to thrice pinnate)
Flowers: clusters (umbels) flat-topped, terminal; bisexual, sometimes unisexual; sepals minute, petals overlapping; stamens 5
Fruits: berry-like drupes
Twigs: stout, often armed with stout prickles

Aralia elata (Miq.) Seemen - Japanese aralia

Japanese aralia, native of Japan, Korea and Manchuria, is reduced to a large shrub in the northern plains due to dieback (Hoag 1965). It is seldom planted in the prairies. Cultivars with variegated leaves include 'Aureo-variegata' with an irregular golden-yellow border, and 'Variegata' with a creamy white border (Dirr 1990).

Planning

Height: up to 50 ft (15 m)
Habit: shrub, tree; branches coarse; terminal inflorescences large
Site: sun, shade; soil adaptable; zone 4
Use: novelty; border

Recognition

Leaves: bipinnate, sometimes over 3 ft (1 m) long; leaflets egg-shaped to much longer than wide, subsessile, up to 4 in (10 cm) long; long-pointed, toothed; dark green above, waxy bloom beneath (glaucescent); veins straight ending in teeth
Flowers: cluster broad, main axis short, several spreading subclusters over 1 ft (30 cm) long
Twigs: prickly (usually)

Aralia spinosa - devils walkingstick

Devils walkingstick—a native of the eastern U.S. from Pennsylvania south to Florida and southwest to southeastern Missouri and eastern Texas—grows along streams as an understory tree on moist, fertile soils, but adapts to other sites to earn the reputation as a plant that "thrives with neglect" (Dirr 1990).

The large leaves and lacy flowers of devils walkingstick form a handsome cover at the top of the plant throughout much of the summer. The juicy black fruits are devoured by a variety of wildlife. Fruits, bark and roots have been used in home remedies as stimulants (Elias 1980).

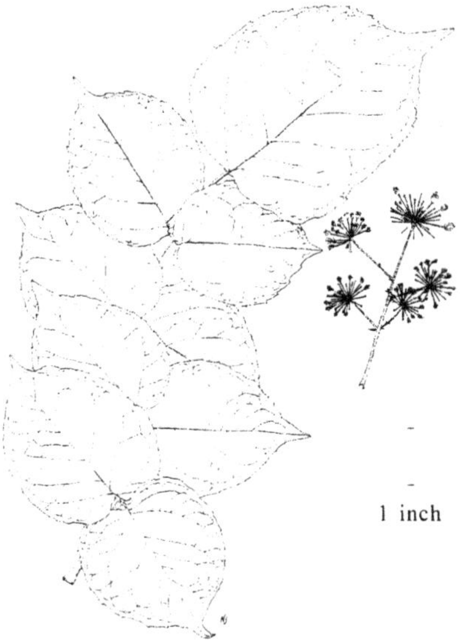

Aralia spinosa

Planning

Height: up to 20 ft (6 m)
Habit: shrub, tree; branches coarse; develops basal shoots; lacy summer bloom long-lasting
Site: sun, part shade; pH and soil adaptable; zone 5
Use: border, wildlife habitat

Recognition

Leaves: bi- to tri-pinnately compound, up to 32 in (80 cm) long, leaf stem (rachis) prickly; leaflets egg-shaped, about 3 in (8 cm) long, distinctly stalked; long-pointed, toothed; waxy bloom beneath, veins curving before reaching margin; petioles up to 10 in (25 cm) long
Flowers: clusters at branch ends, up to 18 in (45 cm) across; whitish
Fruits: about 0.2 in (6 mm) across
Twigs: stout, glabrous; prickles stout; lenticels conspicuous, elongated; leaf scars u-shaped, almost encircling twig, bundle scars many, in a row; pith white, round, very large
Buds: terminal conical, several-scaled; lateral triangular, flattened, appressed, solitary
Bark: thin, fissured, spiny; inner bark bright yellow

Hedera L. - ivy

The five species of ivy are natives of Europe, North Africa and Asia.

Hedera helix - English ivy

English ivy has been cultivated since ancient times as an ornamental for interior and exterior landscapes. This ivy will grow in full sun, but it is likely to suffer injury in winter from exposure to sun and wind. Richley (1988) reports that English ivy grows well on the east and north faces of buildings in southwest Nebraska. Hardy cultivars include 'Baltica', 'Bulgaria', 'Thorndale' and 'Hebron'. The many more English ivy cultivars and varieties that have been described are beyond the scope of this book.

English ivy fruits are supposedly poisonous but seldom seen because they only develop on old plants, usually high atop trees or buildings.

Planning

Height: up to 6 in (15 cm) as groundcover, climbs to 100 ft (30 m)
Habit: vine; evergreen; clings to walls by aerial rootlets
Site: sun, shade; pH adaptable; zone 5
Use: masonry wall cover, groundcover

Recognition

Leaves: alternate; simple, 3- to 5-lobed on juvenile shoots; egg-shaped to rhombic on flowering branches; lustrous dark green above, veins often whitish; margin entire, tip pointed, base tapered
Flowers: clusters terminal; bisexual; greenish white; calyx 5-toothed, petals 5; stamens 5
Fruits: drupes, berry-like; about 0.2 in (6 mm) across, black; seeds 3-5
Twigs: greenish becoming brownish; develop aerial rootlets
Buds: egg-shaped; scales several

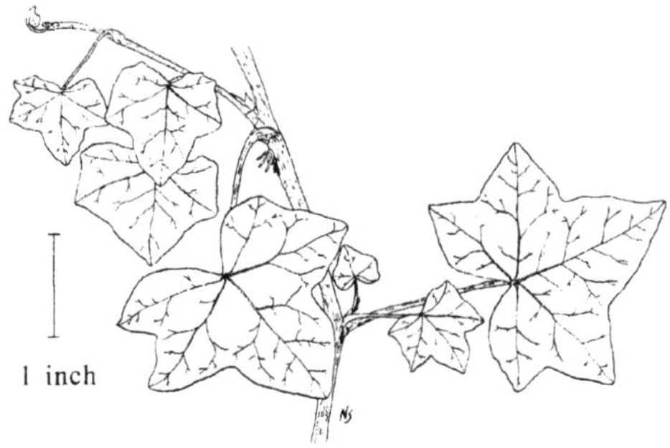

Hedera helix

Kalopanax Miq. - **kalopanax**

One species of kalopanax is native to eastern Asia.

Kalopanax pictus (Thunb.) Nakai - **castor-aralia**

Castor-aralia is a tree not commonly found in prairie landscape plantings. Dirr (1990) reports numerous volunteer castor-aralia seedlings growing at Arnold Arboretum, and it is hardy at the University of Wisconsin where it survived the extremely cold winter of 1976-77.

Planning

Height: up to 50 ft (15 m)
Habit: tree; deciduous; crown oval, coarse; summer blooms attract bees; autumn fruits relished by birds
Site: sun; pH and soil adaptable, prefers moist, fertile soil; zone 5
Use: specimen, shade, wildlife habitat
Note: long-lived, pest-free

Recognition

Leaves: simple, palmately 5-7 shallow-lobed, about 10 in (25 cm) wide; margin minutely saw-toothed (serrulate); lustrous, dark green and glabrous above, hairs on vein axils beneath; petioles often longer than blade
Flowers: clusters (umbels) grouped to form a large compound terminal cluster up to 12 in (30 cm) across; bisexual; white; 5-parted; stamens longer than petals
Fruits: drupes, subglobose; black, about 0.2 in (5 mm) across, crowned by slender style; seeds 2
Twigs: coarse, stout, glabrous; prickles numerous, broad-based, stout
Buds: glabrous; scales 2-3
Bark: blackish, deeply ridged and furrowed

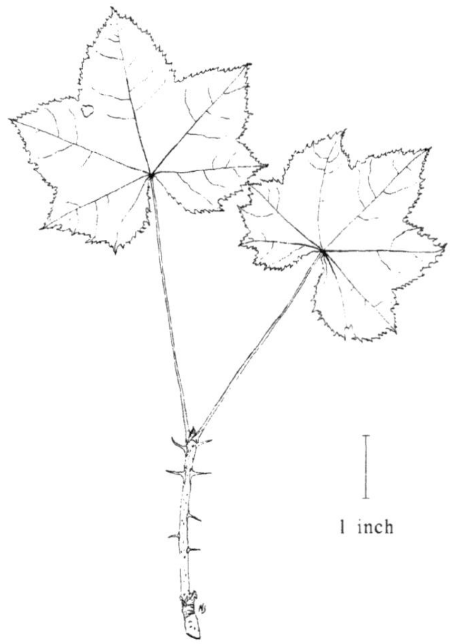

Kalopanax pictus

Subclass Asteridae

Subclass Asteridae consists of 11 orders and nearly 60,000 species. About one-third of these species belong to the family Asteraceae, the largest family in the dicotyledons (Cronquist 1981).

Order Gentianales

Six families and about 5,500 species are placed in the order Genitanalee (Cronquist 1981).

Recognition

Habit: trees, shrubs, herbs; deciduous, evergreen
Leaves: opposite, sometimes whorled; simple; margins usually entire
Flowers: bisexual, sometimes unisexual; regular, commonly showy; petals united at base (sympetalous), 5-parted, sometimes 4-parted; stamens alternate with corolla lobes, filaments attached to corolla tube; ovary superior (hypogynous), compound, style 1
Fruits: usually dry (capsules or follicles); seeds numerous

Apocynaceae Juss. - dogbane family

Only one species of the dogbanes—a family of about 200 genera and 2,000 species—is useful in the prairies.

Vinca L. - vinca

Vinca minor L. - periwinkle

Periwinkle—a native of Europe and western Asia, though not truly a woody plant—provides a uniform, hardy, all-year groundcover with persistent stems and leaves that remain green all winter, especially if protected by shade and/or snow during the dormant season. Periwinkle enlarges its area by forming roots at the nodes of trailing stems. Rooting is encouraged by covering the nodes with soil. An interplanting of bulb-forming perennials such as daffodils adds color and interest to a periwinkle "carpet" in early spring.

A few of the many periwinkle cultivars are: 'Alba', bearing white flowers; 'Aureo-variegata', bearing blue flowers and yellow-blotched leaves; 'Argenteo-variegata', bearing blue flowers and leaves with creamy white margins; 'Bowles Variety', bearing large azure blue flowers; and 'Multiplex' and 'Rosea Plena', both bearing double flowers (Dirr 1990). *V. major*, a larger-leafed and taller species than *V. minor*, prefers moist, well-drained soils and shade (Dirr 1990).

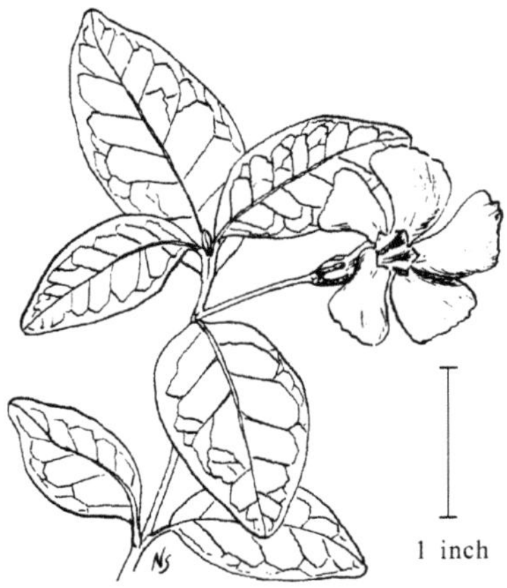

Vinca minor

Planning

Height: up to 6 in (15 cm)
Habit: vine; evergreen; prostrate, mat-forming; April-May blooms
Site: shade, part sun; pH 5.5-7.5; soil adaptable; zone 4
Use: groundcover, window box

Recognition

Leaves: opposite, variably elliptic, tip pointed to obtuse, base tapered; lustrous dark green above, surfaces hairless; petioles about 0.1 in (3 mm) long, exuding milky juice when broken during active growth

Flowers: solitary on upright shoots; calyx lobes 5, narrow, about 0.1 in (3 mm) long; corolla a slender tube with lobes extending at right angles (salverform), about 0.8 in (2 cm) across, lavender blue; stamens attached to middle of tube

Fruits: dry, splitting along 1 suture (follicle), cylindric, about 0.3 in (8 mm) long

Twigs: flexible, green

Buds: minute, egg-shaped, green

Order Solanales

Solanales consists of eight families and about 5,000 species (Cronquist 1981).

Recognition

Habit: herbs, shrubs, vines, seldom trees

Leaves: alternate, sometimes opposite; simple or compound; margins entire or toothed or lobed; stipules wanting

Flowers: bisexual; mostly regular, petals united at the base (sympetalous), 5-parted, seldom otherwise; stamens generally alternate with corolla lobes, attached to the corolla tube; ovary compound, superior, rarely half-inferior

Fruits: most commonly dry and splitting or berries; seeds several to many

Solanaceae Juss. - potato family

The potato family, sometimes known as the nightshade family, is found in a region centered in South America but it also grows worldwide in temperate and warm regions. Included among the 2,000 or more family species are valuable food-producing plants (potato, eggplant, tomato, paprika, cayenne, red pepper) and those that are best known for their narcotic and poisonous compounds (tobacco, belladonna, jimson weed, deadly nightshade). Petunias are popular cultivated potato family ornamentals. Woody species are a small minority of plants in the potato family.

Recognition

Habit: herbs, shrubs, vines, trees; deciduous, evergreen

Leaves: mostly alternate; simple and entire or variously cleft, sometimes compound; stipules wanting

Flowers: solitary or clusters; mostly bisexual and regular; usually 5-parted, stamens alternate with corolla lobes, rarely 2 or 4 fertile, remainder infertile (staminodes); ovary compound, superior, style terminal, stigma mostly 2-lobed

Fruits: dry or a berry

Twigs: sometimes thorny

Lycium L. - box-thorn

About 100 box-thorn species—found worldwide—are mostly thorny but are valued for their showy, scarlet berries.

Lycium halimifolium Mill. - matrimonyvine

Matrimonyvine, a native of southeastern Europe and western Asia that persists around abandoned farmsteads throughout much of the prairie region, has been replaced in many landscapes by more decorative plants. Its flowers are short-lived, fruits are sparse, stems are often thorny, and it has been known to poison livestock.

North Central Plant Introduction Station (1969) records show that matrimonyvine grows fast and is hardy in the northern Great Plains; Hoag (1965) reports that it persists on dry, exposed sites.

Chinese wolfberry, *L. chinense* Mill., bears larger leaves than matrimonyvine and flowers with a corolla tube shorter than the lobes; it is usually thornless. Hoag (1965) reports that Chinese wolfberry dies back during winter in the northern plains.

Lycium halimifolium

Planning

Height: up to 10 ft (3 m)

Habit: shrub; deciduous; spreading, vine-like, somewhat spiny, suckering; summer blooms; autumn red berries (sparse)

Site: shade, partial shade; pH and soil adaptable; drouth tolerant; zone 3

Use: barrier, naturalizing, wildlife habitat

Recognition

Leaves: alternate; simple, often clustered on short spurs; elliptic, sometimes wider above the middle, up to 2.5 in (6 cm) long; margin entire, tip rounded to pointed, base wedge-shaped extending downward onto the petiole (decurrent); surfaces glabrous; petioles flat-topped

Flowers: axillary, 1-3 on each axil; bisexual; corolla tubular, 5-lobed; tube longer than lobes, narrowed below middle, lilac purple fading to pink, dark purple lines in tube throat; calyx lobes 3-5; stamens 5, yellow anthers exserted

Fruits: berries, ellipsoid to wider above middle, about 0.4 in (1 cm) long, bright red

Twigs: flexible, glabrous, occasionally spiny; leaf scars narrow, upper edge straight, bundle scar 1

Buds: concealed in a cluster of leaf bases and scars

Bark: gray-brown, smooth, becoming flaky

Solanum L. - nightshade

Over a thousand species of herbs, vines and a few trees are members of *Solanum*. They include the white potato, the eggplant and the Jerusalem cherry.

Solanum dulcamara L. - bitter nightshade

Bitter nightshade, a native of Europe, is naturalized in the eastern U.S. The plant's ornamental scarlet fruits are poisonous.

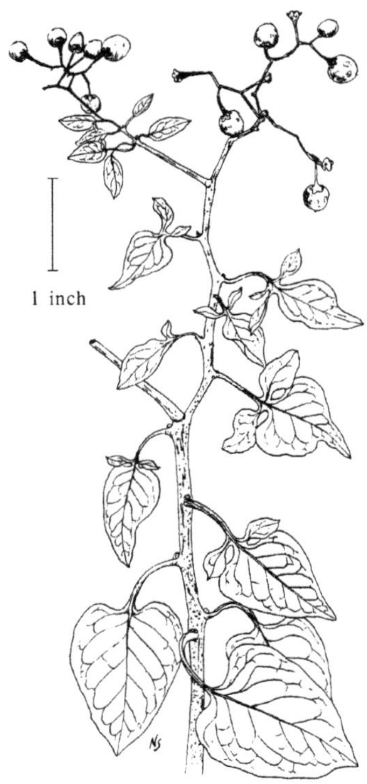

Solanum dulcamara

Planning

Height: up to 8 ft (2.5 m)

Habit: vine-like halfshrub; climber, dies back in winter to woody base; summer blooms, autumn fruits red

Site: sun, partial shade; soil adaptable; zone 5

Use: screen

Recognition

Leaves: alternate; simple or sometimes lobed on either side at base or compound with 3 leaflets; egg-shaped; apex taper-pointed, base heart-shaped, margin entire; sparingly hairy surfaces; petioles up to 1 in (25 mm) long

Flowers: clusters (cymes) long-stalked, drooping; bisexual; corolla tube short, lobes reflexed, violet with 2 green spots near base; calyx toothed, persistent; stamens 5

Fruit: berries, egg-shaped, about 0.4 in (1 cm) long, red

Twigs: glabrous or hairy

Order Lamiales

Four families and about 7,800 species are included in Lamiales (Cronquist 1981).

Recognition

Habit: herbs, shrubs, vines, trees

Leaves: opposite, sometimes alternate or whorled; simple, occasionally compound; margin entire or toothed; without stipules

Flowers: clusters; bisexual, rarely unisexual; petals fused at the base (sympetalous), irregular or regular, 5-parted; stamens attached to tube; alternate with lobes, but often only 2 or 4; ovary compound

Fruits: nutlet clusters, less often drupes or capsules

Verbenaceae J. St.-Hil. - vervain family

Members of the vervain family—about 100 genera and 2,600 species—are mostly of tropical origin. Among them is teak (*Tectona grandis* F.L.), a tree valued for its heavy, durable wood. Most of the prairie vervains are herbaceous.

Recognition

Habit: herbs, shrubs, trees; deciduous, evergreen

Leaves: opposite, rarely alternate or whorled; simple or compound

Flowers: clusters (usually cymes or panicles); bisexual, seldom unisexual; corolla oblique or 2-lipped, rarely regular, 4 or 5 overlapping lobes; calyx usually 4- or 5-toothed, sometimes irregular; stamens 4, often in pairs of different lengths, rarely 5 or 2, sometimes staminodes present; ovary compound, sometimes shallowly 4-lobed, style terminal, slender

Fruits: drupes or berries, often separating into drupelets

Callicarpa L. - beautyberry

Only a few of the 40 beautyberry species are found outside of the tropics and subtropics. Most of the hardier species are natives of China and Japan. Since their blooms are borne on the current season's growth, beautyberry plants can be treated as herbaceous perennials. Most of the previous season's growth can be removed each spring without reducing fruiting.

C. americana L., is a native of the southeastern U.S., Mexico and the West Indies, but is not adapted to the central and northern prairies. *C. bodinieri* Levl., a native of China, failed to overwinter at PERC in Fort Collins (Fairchild and Klett 1993).

Recognition

Habit: shrubs, trees; deciduous, evergreen

Leaves: opposite, simple, usually toothed (serrate)

Flowers: clusters (cymes) axillary; calyx bell-shaped, 4-lobed; corolla tube short, 4-lobed, regular; stamens 4 of equal length, exserted

Fruits: berry-like drupes, nearly round; stones 2-4

Twigs: often with star-shaped or scurfy hairs

Buds: naked, in pairs, one above the other (superposed)

Callicarpa dichotoma (Lour.) K. Koch - **purple beautyberry**

Purple beautyberry is a native of China and Korea. A young specimen growing in Maxwell Arboretum at UNL is thriving in a very shady location, but fruit production is light, probably due to the dense shade. Although the twigs may be injured by severe cold, purple beautyberry will respond to severe pruning and provide a colorful autumn display if moisture is adequate. Var. *albifructus* bears white fruit. Dirr (1990) considers this graceful plant to be a more desirable ornamental shrub than *C. japonica*.

Callicarpa dichotoma

Planning

Height: up to 6 ft (2 m)
Habit: shrub; deciduous; rounded, arching; summer blooms on new growth; autumn fruits showy
Site: sun, partial shade; pH and soil adaptable; zone 5
Use: border, mass
Note: generally free of troublesome pests

Recognition

Leaves: borne in one plane along stem; elliptic, up to 3 in (8 cm) long; tip long-pointed, base tapered, coarsely toothed except towards base and at tip; glabrous above, glandular and sparingly hairy beneath; petioles about 0.1 (3 mm) long
Flowers: pink; stalks longer than petioles
Fruits: lilac purple or white, about 0.1 in (3 mm) across
Twigs: scurfy-hairy
Buds: brown, hairy

Callicarpa japonica Thunb. - **Japanese beautyberry**

Although Japanese beautyberry is considered to be more resistant to cold injury than other species of *Callicarpa*, this advantage is of little importance because it blooms on new wood. It is not hardy in the Fort Collins area (Fairchild and Klett 1993). 'Leucocarpa' is white-fruited.

Planning

Height: up to 6 ft (2 m)
Habit: shrub; deciduous; rounded with arching branches; July blooms inconspicuous; October fruits showy
Site: sun, partial shade; pH and soil adaptable; zone 5
Use: border, grouping
Note: generally free of troublesome pests

Recognition

Leaves: not in one plane along stem; elliptic to lance-shaped, up to 5 in (13 cm) long; tip long-pointed, base tapered, fine toothed from near the base (serrulate); glabrous above, glandular beneath; petioles about 0.2 in (5 mm) long
Flowers: pink or white, stalks (peduncles) longer than petioles
Fruits: violet, about 0.2 in (4 mm) across
Twigs: woolly-hairy, becoming glabrous
Buds: often stalked; round or tapering to tip and base (fusiform oblong)

Caryopteris Bge. - caryopteris

Six species of caryopteris are natives of eastern Asia.

Recognition

Habit: shrubs, subshrubs, herbs; deciduous
Leaves: opposite; simple, 1-3 in (2-8 cm) long; margin toothed or entire; petioles short
Flowers: clusters (cymes) axillary; calyx bell-shaped, deeply 5-lobed; corolla blue, tube short, 5-lobed, one lobe larger and fringed; stamens 4, pairs of unequal length (didynamous), exserted
Fruits: dry, separating into 4 small-winged nutlets

Caryopteris incana (Thunb.) Miq. - common bluebeard

The bluebeard's blue flowers blended with gray-green foliage, all pleasantly scented, are welcome additions to a shrub border or perennial flower garden. Common bluebeard is more showy than the other caryopteris species, but is less hardy (Dirr 1940). However, where cold injury is a problem, it can be treated as a herbaceous perennial because flowering is on new growth. Fairchild and Klett (1993) state that it is performing well at PERC in Fort Collins.

The hybrid bluebeard, *C. ×cladonensis* Simmonds, is a cross between the hardy *C. mongholica* Bge. and the less hardy but showier, *C. incana*. Cultivars include the original clone, 'Arthur Simmonds'; 'Azure', common in Nebraska; 'Blue Mist', with powder blue flowers; and 'Dark Knight', bearing deep blue-purple blooms (Dirr 1990). The latter cultivar is performing well in trials at PERC in Fort Collins (Fairchild and Klett 1993).

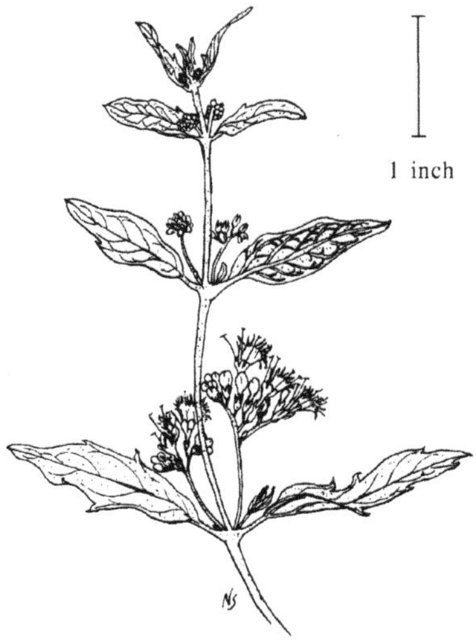

Caryopteris ×cladonensis

Planning

Height: up to 2 ft (60 cm)
Habit: shrub; spreading; summer and early fall blooms, fragrant
Site: sun; pH and soil adaptable; zone 5
Use: border, grouping

Recognition

Leaves: much longer than wide, wider below middle, tapering to tip, up to 2 in (5 cm) long; broad, wedge-shaped to rounded at base, margin coarsely toothed; dull green and hairy above, silvery hairy beneath; petioles 0.2-0.7 in (5-20 mm) long; aromatic
Flowers: clusters many-flowered, slender-stalked in the axils of upper leaves; corolla hairy outside
Twigs: 4-sided, greenish red, hairy; nodes flattened

Lamiaceae Lindl. - mint family

Most of the 200 genera and 3,200 species of the mint family are aromatic herbs growing naturally throughout much of the world in warm, temperate regions. Those valued for their medicinal and culinary uses include catnip (*Nepeta cataria* L.), horehound (*Marrubium*), mint (*Mentha*) and thyme (*Thymus*). Many of the few shrubs in the mint family die back to a woody base (suffrutescent), especially in the prairies, and are classified as subshrubs or semi-woody. Flowers are borne on new stems. Two subshrubs most likely to be of interest in the prairies are described here.

Recognition

Habit: herbs, occasionally shrubs and trees; deciduous, evergreen
Leaves: opposite or whorled, usually simple
Flowers: usually bisexual; calyx usually persistent on fruit, 5-lobed from nearly regular to 2-lipped; corolla irregular, usually 2-lipped with 5 overlapping lobes; stamens 4, 2 long and 2 short (didynamous) or only 2
Fruits: 4 nutlets
Twigs: stems usually 4-angled

Elsholtzia Willd. - elsholtzia

Most of the 20 elsholtzia species are aromatic herb or subshrub natives of Asia.

Elsholtzia stauntonii Benth. - Staunton elsholtzia

Staunton elshotzia is valued for its flowers late in the growing season. It is most likely to succeed in gardens in the southern part of the central prairie where it can be cultivated as a perennial herb.

Planning

Height: up to 5 ft (1.5 m)
Habit: subshrub; deciduous; upright, wide; late summer and autumn blooms
Site: sun; pH and soil adaptable; zone 5
Use: border, grouping, cut flower

Recognition

Leaves: opposite; simple, much longer than wide (lanceolate), up to 5 in (13 cm) long; coarse-toothed except at ends; dark green and glabrous above, pale and densely glandular beneath, aromatic
Flowers: clusters terminal, spike-like, one-sided, up to 8 in (20 cm) long; calyx 5-toothed; corolla lilac purple, 2-lipped, upper with a slight notch, lower 3-lobed; stamens 4, exserted
Fruits: nutlets egg-shaped, glabrous
Twigs: circular in cross-section, hairy

Lavandula L. - lavender

Most of the 20 lavender species are natives of the Mediterranean region.

Lavandula angustifolia Moench - common lavender

Common lavender, a subshrub in southerly climates, becomes an herbaceous perennial in most of the central prairie. This aromatic plant is a source of oil of lavender and is a desirable component of the herb garden. It requires protection from winter winds and excessive water will kill the plant (Dirr 1990).

Planning

Height: up to 1.5 ft (45 cm)
Habit: subshrub; evergreen; upright, wide; summer blooms
Site: sun; well-drained, neutral to slightly alkaline soil; drouth tolerant; zone 5
Use: herb garden, border, grouping

Lavandula angustifolia

Buddlejaceae *Buddleja*

Recognition

Leaves: opposite, up to 1.5 in (4 cm) long; much longer than wide (linear to lanceolate); margin rolled (revolute), entire; apex obtuse; woolly white (lanate), especially beneath, aromatic
Flowers: spikes up to 2.5 in (6 cm) long on stalks up to 6 in (15 cm) long; calyx tubular, 5-toothed; corolla lavender, also white and shades of pink and blue, tubular with short 2-lipped limb; stamens 4
Fruits: nutlets smooth
Twigs: hairy

Order Scrophulariales

Cronquist (1981) includes 12 families and more than 11,000 species in Order Scrophulariales.

Recognition

Habit: herbs, shrubs, trees; deciduous, evergreen
Leaves: opposite, alternate, or whorled; simple, sometimes compound; most without stipules
Flowers: mostly bisexual; irregular, corolla united (sympetalous), often 2-lipped (bilabiate), less often regular and 5-lobed, or wanting; stamens attached to corolla tube, variable in number, sometimes staminodial; ovary compound, superior or sometimes partly or wholly inferior
Fruits: commonly dry and splitting (capsules) or sometimes drupaceous

Buddlejaceae Wilhelm - butterflybush family

Ten genera and 150 species—mostly shrubs and trees of tropical and subtropical origin—are placed in the butterflybush family (Cronquist 1981).

Recognition

Habit: shrubs, trees, sometimes herbs; deciduous or evergreen
Leaves: usually opposite; simple; margin toothed or lobed, sometimes entire; stipules seldom well developed
Flowers: clusters various; usually bisexual, sometimes unisexual (dioecious); regular, perianth usually 4-parted; corolla united (sympetalous); stamens usually as many as corolla lobes
Fruits: capsules, rarely berries or drupes

Buddleja L. - butterflybush

About 100 species of butterflybush grow in tropical and temperate regions of Asia, the Americas and South Africa.

Recognition

Habit: trees, shrubs, rarely herbs; deciduous, evergreen
Leaves: opposite, rarely alternate; margin entire or toothed (serrate); stipules usually reduced to a stipular line; petioles short
Flowers: clusters various; bisexual; perianth 4-parted; calyx bell-shaped; corolla tubular or bell-shaped; stamens 4
Fruits: capsules 2-valved, splits along or in partitions; calyx and usually the corolla are persistent; seeds very small and numerous
Buds: superposed; pointed; scales, usually 2 visible

Buddleja alternifolia Maxim. - **alternate-leaf butterflybush**

Although the alternate-leafed species is the hardiest of the butterflybushes, it may suffer severe winter injury to the tops causing bloom to be seriously reduced or eliminated because flowers are formed on previous year's wood.

Buddleja alternifolia

Planning

Height: up to 20 ft (6 m)
Habit: shrub, tree; deciduous; rounded, shoots arching, lax; May-June purple lilac blooms on old wood
Site: sun; pH and soil adaptable; zone 5
Use: border, specimen

Recognition

Leaves: alternate; much longer than wide, up to 4 in (10 cm) long; margin entire, somewhat wavy, apex pointed to somewhat obtuse, base tapered; dull dark green above, grayish white scurfy hairy beneath; petioles less than 0.1 in (1 mm) long
Flowers: clusters short-stalked, axillary clusters, up to 1 in (25 mm) long; calyx grayish hairy with wide, egg-shaped lobes; corolla bright lilac purple, fragrant, tube 0.3 in (8 mm) long
Fruits: capsules about 0.2 in (4 mm) long
Twigs: scurfy, soon becoming glabrous, yellow-brown, 4-angled
Buds: gray-green, scurfy; scales minutely leaf-like, outer much longer curling over bud tip

Buddleja davidii Franch. - **orange-eye butterflybush**

Despite its tendency to die back during severe winters, orange-eye butterflybush is often planted to provide fragrant, colorful, late summer blooms that attract butterflies and bees. On fertile soils orange-eye butterflybush will reach a height of 8 feet (2.5 m) and bloom profusely a second and third time after being cut back to ground level.

Numerous cultivars sport flowers ranging from white to purples and reds. They include: 'Black Knight', bearing very dark purple flowers; 'Charming', with pink flowers; and 'Royal Red' and 'White Bouquet'.

Buddleja davidii

Planning

Height: up to 15 ft (4.5 m)
Habit: shrub; deciduous; multistemmed, arching; fragrant blossoms summer to frost on new wood
Site: sun; pH and soil adaptable; zone 5
Use: border, grouping, cut flower

Recognition

Leaves: opposite; much longer than wide, up to 10 in (25 cm) long; tip sharp long-pointed, base tapered, margin fine-toothed; gray-green to dark green above, white-woolly beneath; petioles very short

Flowers: clusters (panicles) terminal, nodding or upright, up to 10 in (25 cm) long; calyx hairy with egg-shaped lobes; corolla usually lilac (also white to purple) with orange at mouth, tube straight, about 0.4 in (1 cm) long, glabrous outside, lobes uniform; stamens inserted about the middle of the tube

Fruits: about 0.3 in (8 mm) long

Twigs: slightly angled, hairy becoming glabrous

Buds: hairy, grayish green, foliaceous

Oleaceae Hoffmanns. & Link - **olive family**

As many as 30 genera and 600 species have been recognized in the olive family, which grows in temperate and tropical regions, especially in southeastern Asia and Australia. The edible olive, *Olea europeae* L., the valuable timber trees of the genus *Fraxinus*, and several shrubby genera that control erosion, protect and feed wildlife, and that beautify landscapes are all members of this diverse family.

Recognition

Habit: trees, shrubs; deciduous, rarely evergreen

Leaves: opposite, rarely alternate; simple or compound; without stipules

Flowers: bisexual or unisexual (sometimes dioecious); regular, flower parts commonly in 4s; calyx rarely wanting; corolla typically united at the base (sympetalous) or frequently absent; stamens commonly 2; ovary compound, superior, style 1, terminal, or wanting, stigma 2-lobed

Fruits: various—capsules, berries or berry-like, nuts or winged samaras

Abeliophyllum Nakai - **abelialeaf**

The single abelialeaf species is a native of Korea.

Abeliophyllum distichum Nakai - **Korean abelialeaf**

The Korean abelialeaf's white flowers, sometimes tinged with pink, appear as early as late March. This small plant received good marks as an ornamental and hardy shrub in a 10-year trial at South Dakota State University in Brookings (Klett 1975). A healthy young abelialeaf grows in partial shade in Maxwell Arboretum at UNL. According to Dirr (1990), Korean abelialeaf is adapted to a wide range of soil conditions in either full sun or light shade.

Planning

Height: up to 5 ft (1.5 m)

Habit: shrub; deciduous; branches arching; fragrant, white blossoms before leaves

Site: sun, light shade; soil adaptable; zone 5

Use: hedge, grouping

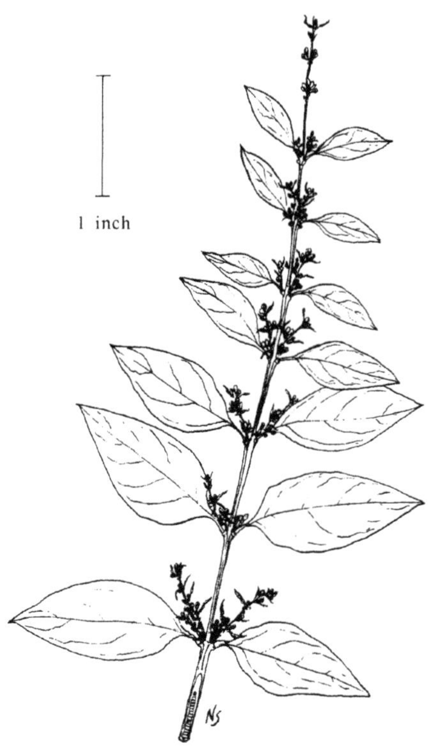

Abeliophyllum distichum

Recognition

Leaves: opposite; simple, spreading in 2 ranks, egg-shaped; margin entire; surfaces variably long hairy, dark green above; petioles about 0.2 in (5 mm) long

Flowers: clusters (racemes) short; bisexual; corolla white, measuring 0.6 in (15 mm) across, petals united at base, tube as long as lobes; stamens 2

Fruits: capsules 2-celled, winged all around, 1 in (25 mm) across

Twigs: glabrous, green, 4-sided with prominent ridges, becoming shreddy 2nd year; pith chambered

Buds: leaf buds narrow, pointed, 0.1 (2 mm) long, glabrous, brown; flower buds egg-shaped, wider above middle, about 0.1 (2 mm) long, prominent through winter

Chionanthus L. - fringetree

The two species of the fringetree are widely planted as ornamentals.

Recognition

Habit: trees, shrubs; deciduous

Leaves: opposite; simple; margin entire; stipules wanting

Flowers: clusters (panicles) large, loose; unisexual (dioecious) and bisexual (polygamo-dioecious); calyx 4-parted; corolla white, 4-parted, divided nearly to base; stamens 2; pistil 1

Fruits: drupes, dark blue; seeds 1

Chionanthus retusus Lindl. & Paxt. - Chinese fringetree

Chinese fringetree can be grown as a small tree or a multistemmed shrub. The flowers, fruits, foliage and bark of this native of China, Korea and Japan are all highly ornamental. It is most likely to succeed in plantings in the southern portion of the central prairie.

Planning

Height: 15-30 ft (5-9 m)

Habit: shrub, tree; crown rounded, spreading; spring blossoms and autumn fruits highly ornamental; bark handsome

Site: sun; soil adaptable; zone 6

Use: specimen, backyard wildlife habitat

Recognition

Leaves: elliptical to egg-shaped, 1.5-4.0 in (4-10 cm) long; tip pointed to rounded, sometimes with shallow notch, base wedge-shaped to rounded; hairy on veins beneath, becoming glabrous

Flowers: clusters (panicles) broad, 2.5-4.0 in (6-10 cm) long; calyx lobes much longer than wide (lanceolate)

Fruits: drupes ellipsoid, about 0.5 in (12 mm) long, dark blue

Twigs: hairy when young, becoming glabrous

Bark: polished light brown, sometimes exfoliating

Chionanthus virginicus L. - white fringetree

White fringetree—a native of the southeastern U.S.—is hardy in Maine and Minnesota and tolerates air pollution (Dirr 1990). It grows on sandy to fertile loamy soils from streambanks to ridge tops among hardwood forests of red oak, yellow-poplar, American basswood, hickories, dogwoods, and hawthorns.

A young 12-foot (3.6 m) white fringetree on city campus at UNL bloomed profusely, but its life was short for reasons unknown. Richley (1988) reported that white fringetree grows in southwestern Nebraska on protected sites. Blooms of male trees are more effective than those of female trees because of their longer petals. The fruits of white fringetree are relished by birds.

Planning

Hieght: 15-25 ft (5-8 m)
Habit: shrub, tree; crown spreading; spring blossoms
Site: sun, partial shade; prefers soils with pH slightly acid to normal; zone 4
Use: specimen, border, backyard wildlife habitat

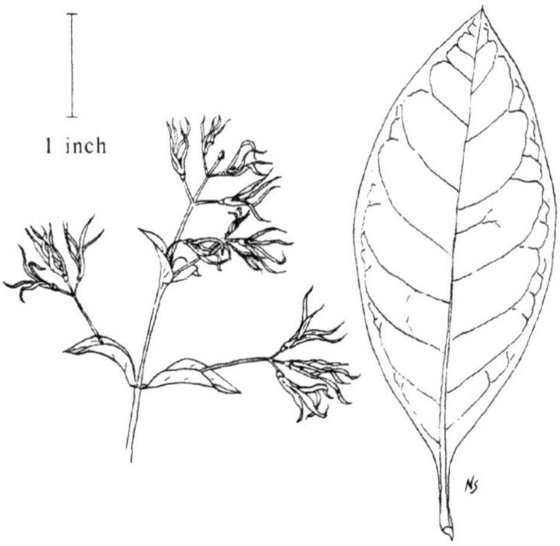

Chionanthus virginicus

Recognition

Leaves: narrow, elliptic to oblong, 3-8 in (8-20 cm) long; tip pointed, base wedge-shaped, margin somewhat wavy; green above, paler below and hairy, especially on veins, becoming glabrous
Flowers: clusters (panicles) loose, open, 4-8 in (10-20 cm) long; calyx lobes triangular; male flowers bear sterile pistils
Fruits: drupes blue with whitish bloom, about 0.7 in (18 mm) long
Twigs: sometimes hairy when young, slightly 4-angled; lenticels conspicuous, warty, elongated; leaf scars half-round, elevated, bundle scar 1, U-shaped
Buds: terminal egg-shaped, brown, pointed; scales 6-8, keeled; lateral often superposed

Forestiera Poir. - **forestiera**

Four or five forestiera species of the approximately 20 species native to the Western Hemisphere grow in the southern U.S. All are shrubs, but sometimes become small trees on good sites. Forestiera fruits provide food for birds and small mammals, but the plants are considered to be of little value as ornamentals because the apetalous flowers are not showy.

Forestiera neo-mexicana Gray - **New Mexico forestiera**

New Mexico forestiera, a native of the southwestern U.S. including southwestern Colorado, has been planted experimentally in the central Great Plains as a component of windbreak and wildlife plantings. A specimen grows in Maxwell Arboretum at UNL. Specimens growing in a shaded location at PERC in Fort Collins maintain a dense, attractive growth habit, are winter hardy, and have been insect and disease free (Fairchild and Klett 1993). Sutton (1990) has observed this forestiera doing well as a sheared hedge in full sun.

F. acuminata (Michx.) Poir., swamp privet, is a shrub that sometimes reaches tree size. It is a native of swampy sites in the southeastern U.S. west to southeastern Kansas. The yellow stamens of flowers that have no perianth provide subdued color on the plants as they emerge before the leaves. The elliptic leaves are about 3.5 inches (9 cm) long and the ovoid to cylindric purple fruits are about one inch (25 mm) long.

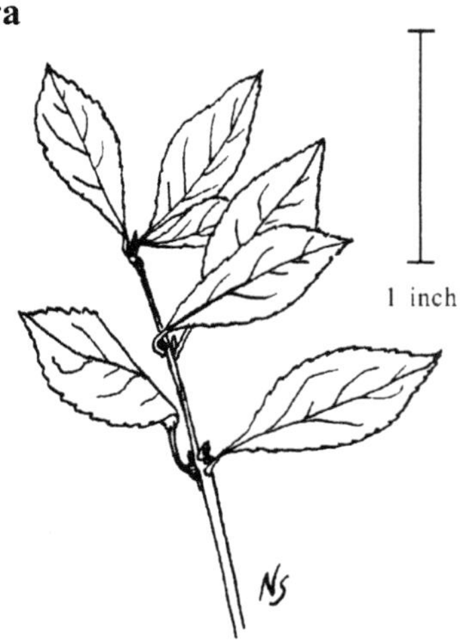

Forestiera neo-mexicana

Planning

Height: up to 10 ft (3 m)
Habit: shrub; deciduous; upright; yellow stamens of male flowers provide color before the leaves appear; fruits ripen late summer
Site: sun, partial shade; pH and soil adaptable; drouth tolerant; zone 5
Use: border, hedge, windbreak, wildlife habitat, naturalizing

Recognition

Leaves: opposite or in axillary clusters; broadest above the middle (oblong-obovate); margin entire or minutely toothed; grayish green, not hairy (glabrous); petioles short
Flowers: clusters lateral; unisexual, occasionally bisexual (polygamo-dioecious); petals and sepals wanting; stamens 2-4, yellow, often sterile on pistillate flowers
Fruits: drupes short ellipsoid, 0.2 in (5 mm) long, blue-black with whitish bloom
Twigs: gray with prominent leafing spurs
Bark: golden tan

Forsythia Vahl - forsythia

One of the first shrubs in the spring to display bright yellow flowers is very likely to be one of the six or seven forsythia species native to southeastern Europe and eastern Asia. However, these harbingers of spring often can fail to appear. For example, an unknown forsythia cultivar that seldom fails to bloom in the city of Lincoln has bloomed erratically over a 20-year period at rural Prairie Pines Arboretum.

Freezing temperatures may occur in rural areas but not in a nearby city in fall and spring. The above-mentioned failure to bloom cannot be explained by extreme low winter temperatures because some of the better flower displays of the forsythia growing in the rural site occurred after very severe winters. Gradually declining fall temperatures preceded these cold winters. This indicates that the injury to the flower buds occurs in the fall before the buds are hardened off by temperatures not nearly as low as those often attained in dead of winter.

'Vermont Sun'—a selection of *F. mandschurica*—though not widely planted in the U.S., is one of the hardiest (Dirr 1990). 'Vermont Sun' was released in 1978 after testing at the University of Vermont where flower buds proved to be hardy at -25 degrees F (-31 degrees C).

The large number of forsythia hybrids and cultivars is beyond the scope of this book. Up-to-date information about the forsythias is best obtained from current periodicals.

Recognition

Habit: shrubs; deciduous
Leaves: opposite; simple, occasionally 3-parted or 3-foliate; margins entire or toothed, surfaces rarely hairy
Flowers: axillary before the leaves; bisexual; calyx deeply 4-lobed, persistent; corolla yellow with 4 oblong lobes longer than the bell-shaped (campanulate) tube; stamens 2, style 1, stigma 2-celled
Fruits: capsules dry, splitting, 2-celled; seeds winged, many
Twigs: rarely hairy, pith hollow or chambered (lamellate), sometimes solid at the nodes
Buds: conical, light yellow-brown; accessory buds often above true lateral buds (superposed) and often laterally branched; scales often loosely arranged

Oleaceae *Forsythia*

Forsythia ×intermedia Zab. - border forsythia

The many cultivars of border forsythia—*F. suspensa* × *F. viridissima*—are the forsythias most often planted. These cultivars—selected from seedlings, propagated from sport branches, or resulting from colchicine treatment—are quite variable in form and flower character.

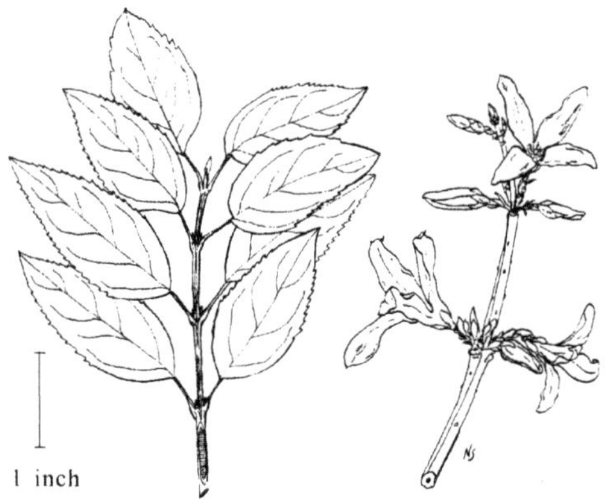

Forsythia ×intermedia

Cultivars include 'Spectabilis', a vigorous, profusely flowering shrub that is the standard by which all others are judged, and 'Arnold Giant', a colchicine-induced tetraploid of a seedling of 'Spectabilis' that has been crossed with 'Spectabilis' to produce 'Beatrix Farrand' and 'Karl Sax' (Dirr 1990). Several other cultivars including 'Arnold Dwarf' have originated from combinations of the aforementioned plants or other species.

'Arnold Dwarf', 'Spectabilis' and 'Beatrix Farand' growing at PERC in Fort Collins suffer winter dieback and flower unreliably but are well adapted to alkaline soils, and insects and diseases are not problems (Fairchild and Klett 1993).

Planning

Height: up to 10 ft (3 m)
Habit: shrub; upright arching; early spring flowers
Site: sun, partial shade; pH and soil adaptable; zone 5
Use: border, grouping, specimen
Note: floral display is best in full sun

Recognition

Leaves: often 3-parted on vigorous shoots, egg-shaped to oblong; tip pointed, base wedge-shaped to rounded; margin usually toothed above the middle
Flowers: clusters; corolla various shades of yellow; calyx lobes oblong, sometimes as long as corolla tube
Twigs: often slightly 4-angled; pith chambered between nodes, solid at nodes
Buds: several grouped at a node

Forsythia ovata Nakai - early forsythia

Though not as spectacular as other forsythias, early forsythia from Korea is the earliest to bloom. It has been used in breeding programs to combine its small size and cold hardiness with the more profuse flowering of other species.

Cultivars include 'Ottawa', vigorous with heavy floral production, and 'French's Florence', with greater flower bud hardiness (Dirr 1990). 'Nakoi' and 'Sunrise' suffered severe injury to twigs and buds during cold winters at Fort Collins, but produced a nice display during mild winters with no pest problems (Fairchild and Klett 1993).

Planning

Height: up to 6 ft (2 m)
Habit: shrub; stiff, spreading; very early spring blooms
Site: sun; pH and soil adaptable; zone 5
Use: borders, specimen, grouping

Recognition

Leaves: broadly egg-shaped; margin toothed or entire, tip abruptly pointed, base variable; petioles about 0.4 in (1 cm) long
Flowers: usually solitary, bright yellow with tinge of green; calyx lobes broad egg-shaped, half the length of corolla tube
Twigs: yellowish, becoming gray; pith chambered

Forsythia suspensa (Thunb.) Vahl - weeping forsythia

Three varieties of weeping forsythia provide a wide range of form and flower characteristics. Var. *sieboldii* Zab. has graceful spreading branches. Varieties *fortunei* (Lindl.) Rehd. and *atrocaulis* Rehd. exhibit more erect habits. Several other varieties have been described. Weeping forsythis is one of the parents of border forsythia (Rehder 1940).

Planning

Height: up to 10 ft (3 m)
Habit: shrub; variably pendulous, spreading; April blossoms less profuse than border forsythia's blossoms
Site: sun; soil and pH adaptable; zone 6
Use: border, grouping, draped over walls

Recognition

Leaves: often 3-foliate or 3-parted, egg-shaped to oblong; tip pointed, base broadly wedge-shaped to rounded, margin toothed (serrate); petioles 0.5 in (12 mm) long
Flowers: single, often in 3s; golden yellow; calyx lobes oblong, about as long as corolla tube
Twigs: slightly 4-angled, spreading to pendulous; pith solid at nodes, hollow in between

Forsythia viridissima Lindl. - greenstem forsythia

Greenstem forsythia is one of the parents of border forsythia. The dwarf cultivar 'Bronxensis' bears yellow-green flowers and reaches a height of about 2 feet (60 cm), which makes it well suited for growing in containers (Dirr 1990).

Planning

Height: up to 10 ft (3 m)
Habit: shrub; stiff, upright; blooms later than weeping forsythia
Site: sun; pH and soil adaptable; zone 6
Use: border, grouping

Recognition

Leaves: seldom 3-parted, oblong, usually broadest about the middle; tip pointed, base wedge-shaped to rounded, margin toothed above middle or sometimes entire; petioles about 0.4 in (1 cm)
Flowers: single, sometimes in 3s; bright yellow with greenish tinge; calyx lobes oval, about half length of corolla tube
Twigs: green; pith chambered full length on young twigs, becoming hollow

Fraxinus L. - ash

Sixty-five to 70 ash species, mostly trees, are natives of the Northern Hemisphere. In addition to their value for shade and ornament, several of the larger species produce heavy, hard, straight-grained wood that is well suited for specialty uses such as tool handles, baseball bats, tennis racket frames, boat building and oars, toys, and cooperage (barrels). Limited supplies of ash trees capable of producing high quality wood and today's advanced plastics technology have conspired to minimize many of these uses. Ash trees are also prized for firewood because the wood is easily split and has a high heat value.

Oleaceae *Fraxinus*

In addition to the species we describe, several other native and introduced species may be adapted to the southern portion of the central prairie. *F. profunda* (Bush) Bush, pumpkin ash, which is a native from Illinois and Ohio throughout the southeastern U.S., is a tree that grows in swamps and along river bottoms. *F. cuspidata* Torr., fragrant ash, a tree native to the southwestern U.S., bears fragrant petaled flowers.

Among the European and Asian species, F. ornus L., flowering ash, a tree with showy fragrant flowers, provides a substance used in the preparation of laxatives. *F. angustifolia* Vahl, narrowleaf ash, and a closely relates species, *F. oxycarpa* Willd., may be adapted to the southern portion of the central prairie. 'Raywood', a selection of the latter species, and *F. holotricha* Koehne, 'Moraine', both failed to overwinter at PERC in Fort Collins (Fairchild and Klett 1993).

Recognition

Habit: trees, rarely shrubs; usually deciduous
Leaves: opposite, odd-pinnately compound (first leaves of seedlings are simple)
Flowers: bisexual, more commonly unisexual (dioecious or polygamous); calyx 4-lobed or wanting, petals usually wanting; stamens 2-4; ovary compound, 2-celled
Fruits: nutlet with terminal wing (samara); seed 1, elongated
Twigs: sometimes hairy; pith homogeneous, white; leaf scars conspicuous, half-round to concave, bundle scars numerous
Buds: terminal rounded, rough-downy; lateral smaller, solitary or one above another (superposed)
Bark: variable
Wood: heavy, hard

Fraxinus americana L. - white ash

White ash is native to the states of the U.S. east of southeast Nebraska south through eastern Kansas into Texas. Although this ash grows most commonly on fertile, well-drained soils supplied with plentiful water, it occupies such a wide range of site conditions that great genetic variability within the species is inevitable. Observations of the early growth of a white ash provenance plantation established near Hastings, Nebraska, in cooperation with the North Central Forest Experiment Station confirm this statement. Data collected from this plantation and other similar plantations will provide a measure of this variability. In the meantime, vigorous white ash specimens, including those with outstanding fall coloration, can be selected from this plantation for propagation and release to the public for planting in the immediate region.

At present our choices of white ash for landscape planting is limited to mostly male cultivars marketed primarily for their outstanding autumn color. Unfortunately, this characteristic changes with site, resulting in disappointments, not only in leaf color but also in growth rate and longevity. For instance, 'Champaign County' white ash failed to survive winter temperatures at Fort Collins, whereas 'Autumn Purple' is well adapted to that area and displayed good fall color (Fairchild and Klett 1993). 'Autumn Purple' also adds a handsome reddish purple color to mid-October landscapes in Lincoln.

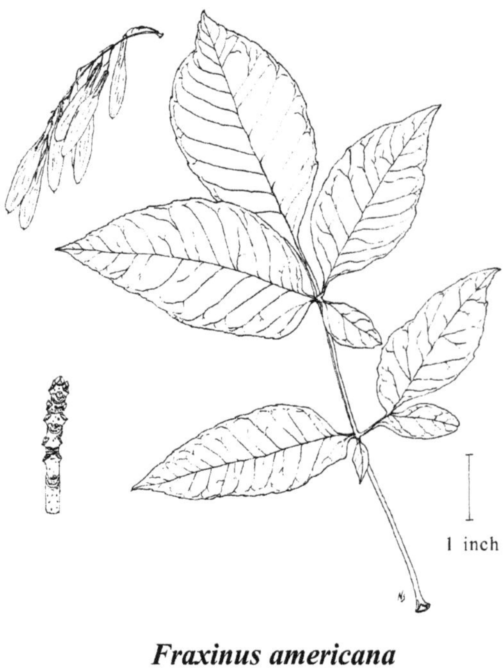

Fraxinus americana

Planning

Height: up to 70 ft (21 m); Nebraska champion, 1987, 71 ft (21.5 m), Falls City; National champion, 1983, 95 ft (29 m), New York
Habit: tree; crown spreading; autumn foliage may be colorful
Site: sun, partial shade when young; pH and soil adaptable; zone 4
Use: specimen, shade, windbreak, lumber

Recognition

Leaves: leaflets usually 5-9, oval or egg-shaped to oblong; tip short-pointed, base wedge-shaped to rounded, margin entire to round-toothed; dark green, sometimes varying shades of reddish purple in autumn, whitish beneath and sometimes sparsely hairy, veins hairy to sparsely hairy or glabrous; leaf stalks glabrous, slightly grooved
Flowers: unisexual (dioecious), appearing before leaves; petals wanting; male in dense clusters, calyx slightly lobed, 2-3 stamens; female in elongate clusters, calyx deeply lobed
Fruits: clusters dense; each samara flattened, 1-2 in (25-50 mm) long; wing terminal or extending along no more than one-third of fruit body, widest near a pointed or notched tip
Twigs: circular in cross-section, green becoming grayish; leaf scars half-round, usually somewhat narrowed and concave above; bundle scars small, numerous, often indistinct
Buds: broadly egg-shaped, often nearly black, scurfy
Bark: deep narrow fissures, firm narrow ridges
Wood: tough, strong, heavy, nearly white

Fraxinus anomala Torr. - singleleaf ash

Singleleaf ash, a native of western Colorado, Utah, Arizona and southeastern California, grows along streams and gulches in a region of scant precipitation; it is often shrubby. Test plantings in the southwestern part of the central prairie are recommended to determine its potential as a landscape plant for that area. Budded singleleaf ashes on green ash root stock are small trees at the UNDARC arboretum near Ithica, Nebraska (Sutton 1998)

Planning

Height: up to 25 ft (8 m); National champion, 1973, 24 ft (7.3 m), Colorado (Mesa County)
Habit: tree, shrub; crown round-topped, contorted
Site: sun; neutral to alkaline soils; drouth tolerant; zone 6
Use: specimen, windbreak

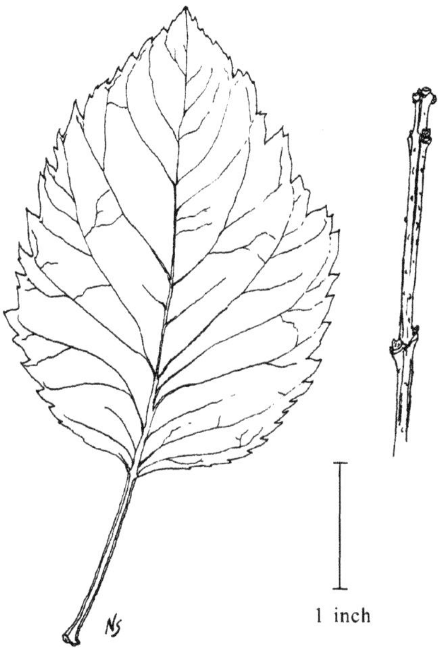

Fraxinus anomala

Recognition

Leaves: simple (rarely compound with 2-5 leaflets), almost circular to broadly egg-shaped; tip rounded to short-pointed, base rounded to broadly wedge-shaped, margin entire or with rounded or pointed teeth; dark to yellow-green above, paler and becoming glabrous beneath; petioles 0.5-1.5 in (12-38 mm) long, rusty hairy when young
Flowers: clusters short, appearing after the leaves; unisexual, sometimes bisexual; petals wanting; calyx cup-shaped, minutely 4-toothed
Fruits: samaras flattened, about 0.5 in (12 mm) long; wing surrounds seed cavity, broadest near rounded or slightly notched tip
Twigs: 4-sided to slightly winged, nearly glabrous; leaf scars crescent-shaped to nearly round
Buds: egg-shaped, orange-hairy
Bark: shallow furrowed with narrow scaly ridges
Wood: heavy, hard, close-grained

Fraxinus excelsior L. - European ash

European ash, the tallest hardwood tree of Europe and Asia Minor, has been cultivated for centuries in its native lands. It is not widely planted in America where native species of similar form and color are generally better adapted. The wide natural range of this ash accounts for the varied reports on hardiness and growth rate in the U.S.

A 'Kimberly' European ash growing at PERC in Fort Collins suffers dieback of new growth in cold winters (Fairchild and Klett 1993). 'Hessei' is one of the highest rated shade trees in the Ohio Shade Tree Evaluation Tests and it performs well in Wichita, Kansas, but is very susceptible to stem borers (Dirr 1990). 'Hessei' trees surrounded by concrete and asphalt maintain good vigor in a Lincoln parking lot (Bagley 1996) and on UNL east campus (Sutton 1998).

The leaves of 'Hessei' are simple and toothed. They keep their dark green color too late in fall to exhibit consistent autumn color in the middle of the central prairie. Other cultivars of European ash exhibit various traits of growth, leaf color and hardiness.

Fraxinus excelsior

Planning

Height: up to 60 ft (18 m); Nebraska champion ('Hessei'), 1989, 24 ft (7.3 m), Lincoln
Habit: tree; crown rounded, spreading; autumn foliage yellow
Site: sun; prefers moist, loamy, calcareous soils; zone 5
Use: shade, windbreak
Caution: very susceptible to stem borers

Recognition

Leaves: compound ('Hessei' simple); leaflets usually 11-13, nearly stalkless except terminal; long egg-shaped, wider below middle, tip short-pointed, base rounded, margin toothed; dark green above, paler beneath, variously hairy on veins; leaf stalks glabrous, grooved above
Flowers: unisexual and bisexual on same plant (polygamous); petals and sepals wanting; anthers shorter than filaments
Fruits: samara narrow oblong; wing 1.0-1.5 in (25-40 mm) long
Twigs: slightly flattened at nodes, glabrous; leaf scar slightly concave
Buds: egg-shaped, blunt, black, hairy

Fraxinus mandshurica Rupr. - Manchurian ash

This ash of northeast Asia is not widely cultivated. A young Manchurian ash in the Horning State Farm Arboretum is a vigorous, well-formed tree. Another specimen of this ash at PERC in Fort Collins is well adapted to the soils and climate of that site (Fairchild and Klett 1993). 'Mancana' is a Canadian release (Morgenson 1988).

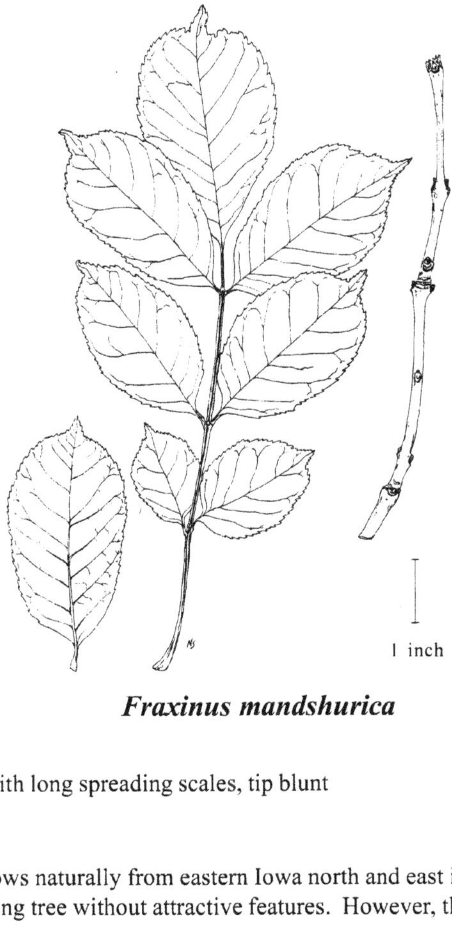

Fraxinus mandshurica

Planning

Height: up to 40 ft (12 m)
Habit: tree; crown dense, upright, oval; autumn foliage golden yellow
Site: sun; pH and soil adaptable; zone 4
Use: specimen, shade, street

Recognition

Leaves: leaflets 5-7, egg-shaped to oblong, wider above the middle; tip short-pointed, base wedge-shaped to rounded, margin toothed; dark green, paler beneath, white, rigid hairs on larger veins; leaf stalks glabrous, conspicuously grooved
Flowers: dioecious, appearing before the leaves; petals and sepals wanting
Fruits: samara wings 1.0-1.4 in (25-35 mm) long
Twigs: somewhat 4-sided, glabrous; leaf scars concave
Buds: egg-shaped, nearly black; terminal enclosed with long spreading scales, tip blunt

Fraxinus nigra Marsh. - black ash

Black ash, a tree of swamps and moist habitats, grows naturally from eastern Iowa north and east into Canada and Newfoundland. This ash is a slow-growing tree without attractive features. However, the burls frequently found on black ash trunks are reponsible for a handsome figured wood know to cabinet makers as "curly ash."

Planning

Height: up to 50 ft (15 m)
Habit: tree; crown narrow
Site: sun; prefers wet soils; zone 3
Use: naturalizing

Recognition

Leaves: leaflets 7-11, nearly stalkless (sessile), oblong to much longer than wide; tip long-pointed, tapering to rounded base, margin toothed; dark green above, paler beneath, without hairs both sides except tufts at leaflet base; leaf stalks stout, flattened or narrowly grooved on top
Flowers: unisexual and bisexual (polygamo-dioecious), appearing with or before leaves; male in compact clusters, female in open clusters; petals and sepals wanting
Fruits: samara wings 1.0-1.5 in (25-40 mm) long, flattened, broadest above middle extending along seed cavity to base, about 0.4 in (1 cm) wide, apex usually notched
Twigs: round, hairy, becoming glabrous; leaf scars half round with straight upper margin to nearly circular
Buds: egg-shaped, dark brown to black, scurfy
Bark: gray, shallow-furrowed, ridges scaly
Wood: coarse-grained, soft, easily split

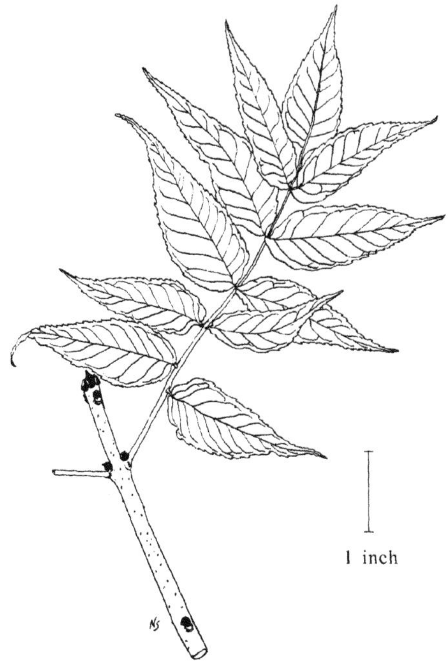

Fraxinus nigra

Fraxinus pennsylvanica Marsh. - green ash

Green ash, a native of the eastern U.S. and southern Canada including a large part of the Great Plains, is considered by the Omaha and Pawnee tribes to be a sacred tree with mystical powers. Native Americans fashioned pipe stems and arrow shafts from the wood (Gilmore 1977). Green ash has provided lumber and excellent firewood from the date of European settlement of the prairies to the present.

Green ash was planted with excellent success throughout the prairies in tree claims established under the Timber Culture Act of 1873, and for shade and protection of farm and urban dwellings as well. It received high ratings in a survey of trees surviving in the drouth-ravaged Dust Bowl (Pool 1939). Green ash was widely planted in shelterbelts throughout the Prairie States Forestry Project of 1935 to 1942. Read (1958) reported outstanding survival of this species in those plantings. He stated that it was one of the best trees from the standpoint of adaptation to tough sites and effectiveness as a wind barrier.

Fraxinus pennsylvanica

Green ash is a slow-to-moderately-fast-growing species adapted to a wide range of site conditions. Seed origin is very important to its performance. For instance, green ash grown from seed collected from natural stands in the Great Plains region grow faster on sites north of their origin than trees that are native to near the planting site. Conversely, they grow more slowly than trees native to a planting site when planted south of their native stands (Ying and Bagley 1976b). However, if trees are planted so far north of their native stand that they remain actively growing late into the fall, they can be injured by sharp freezes or branches can be broken by a heavy snowfall before the leaves drop.

In a provenance study that included seed origins from throughout the entire natural range of green ash, the tallest trees of most plantations in that study at age 6 years originated in southern Ontario and the central prairie region of the U.S. (Steiner et al. 1988). Winter injury was prevalent among sources from the south of that region in many of those plantations.

Provenance research shows the importance of knowledge of the origin of a tree to be planted. The seed origin of most cultivars is not known. For instance, 'Marshall Seedless', which was selected by a Nebraska nursery, may not be native to Nebraska.

Test plantings of 'Cardan' indicate that it is an excellent cultivar for the northern Great Plains (Morgenson 1988). 'Emerald', one of the most unusual green ash cultivars, is seedless with larger than normal dark green leaves and an orange-tinted bark. This cultivar at PERC in Fort Collins developed an unattractive, irregular crown while 'Kindred', 'Marshall Seedless' and 'Summit' were more attractive (Fairchild and Klett 1993). The authors reported that all green ash specimens at PERC are well adpated to alkaline soils and are winter hardy.

Planning

Height: up to 70 ft (21 m); Nebraska champion, 1986, 84 ft (25.5 m), Ponca; National champion, 1981, 131 ft (40 m), Michigan

Habit: tree; crown spreading; golden autumn foliage is short-lived

Site: sun, will grow in partial shade when young; tolerates alkalinity, drouth and short periods of flooding; zone 3

Use: shade, street, windbreak, lumber

Recognition

Leaves: leaflets usually 5-9, egg-shaped to much longer than wide; tip long-pointed, base wedge-shaped to rounded, margin toothed; dark green to yellow-green above, paler and glabrous or variably hairy beneath; leaf stalks glabrous, grooved above

Flowers: unisexual (dioecious), male in dense clusters (panicles), female in open clusters; corolla wanting; calyx 2- to 4-lobed

Fruits: clusters (pendant panicles); samaras 1-2 in (25-50 mm) long; wings terminal, extending at least halfway along fruit body, often abruptly widened above seed, about 0.2 in (5 mm) wide

Twigs: round, green and glabrous to hairy when young, becoming glabrous and gray; leaf scars half round, rarely slightly concave above

Buds: egg-shaped, brownish, more narrow than leaf scars, often with small bud at base; scales, 2 pairs visible

Bark: dark gray, fissures and ridges narrow

Wood: heavy, tough, strong, slightly yellowish

Fraxinus quadrangulata Michx. - blue ash

Blue ash, a native of the central states from Michigan south to Georgia and Alabama and extending westward into eastern Oklahoma, southeastern Kansas, and southeastern Iowa, is not widely planted and little is known of its adaptability to various sites. Pioneers prepared a blue dye by macerating blue ash bark in water (Harlow and Harrar 1968). A large, old blue ash grows on the East Campus of UNL.

Planning

Height: up to 60 ft (18 m); Nebraska champion, 1981, 56 ft (17 m), UNL, Lincoln; National champion, 1970, 90 ft (27.4 m), Kentucky
Habit: tree; crown narrow, rounded; autumn foliage yellow
Site: sun; pH and soil adaptable; zone 5
Use: shade

Recognition

Leaves: leaflets usually 7 (sometimes 5-11), short-stalked to sessile, much longer than wide; tip long-pointed, base usually unequal, wedge-shaped to rounded, margin toothed; dark green becoming yellow-green above, paler and glabrous to hairy beneath, midrib often hairy; leaf stalks slender, slightly 2-ridged
Flowers: clusters (panicles) short, many-flowered, before the leaves; usually bisexual; calyx minute, petals wanting; stamens 2
Fruits: samaras 1-2 in (25-50 mm) long; wing extending full length of seed cavity, often widest near notched or rounded tip, about 0.3 in (8 mm) wide
Twigs: 4-angled with corky ridges, glabrous to short hairy; leaf scars crescent-shaped
Buds: gray to red-brown, hairy
Bark: gray, shallow furrows with flat-topped ridges, becoming scaly

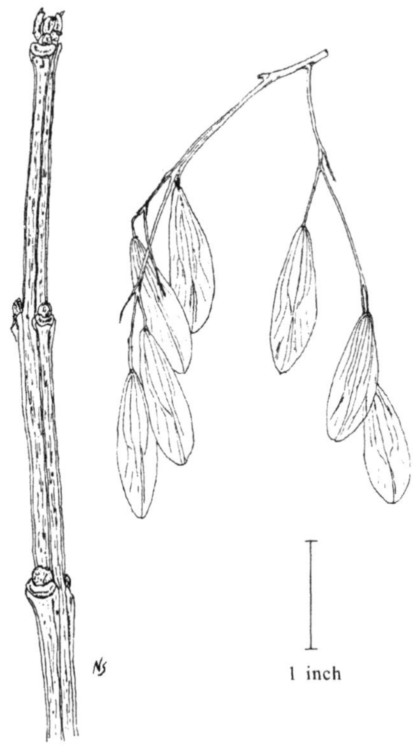

Fraxinus quadrangulata

Fraxinus zanthoxyloides (Don) DC. - Afghan ash

A specimen of this native of the Himalayas and Afghanistan is growing north of Omaha with no apparent problems (Rassmussen 1989).

Planning

Height: up to 15 ft (4.5 m)
Habit: shrub, tree
Site: experimental

Recognition

Leaves: leaflets 5-9, subsessile; oval to oblong, tip variable, margin toothed (crenate-serrate); pilose on midrib beneath; leaf stalk (rachis) narrowly winged, slightly hairy
Flowers: clusters (panicles) dense; bisexual; calyx persistent on fruits; petals wanting
Fruits: samaras narrow oblong, about 1.5 in (4 cm) long, wing rounded or notched

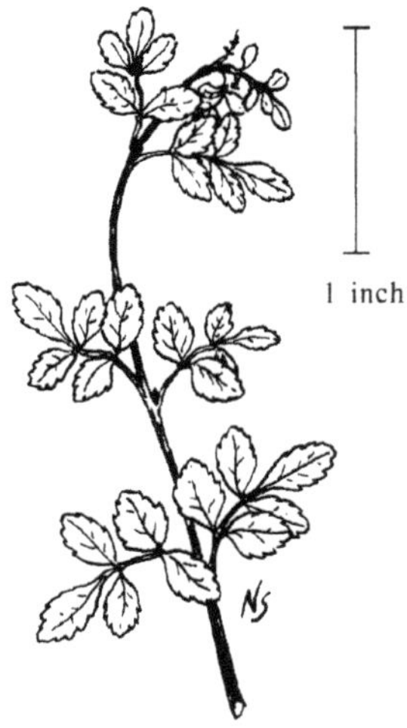

Fraxinus zanthoxyloides

Ligustrum L. - privet

About 50 species of *Ligustrum* are native to eastern Asia, Malaysia and Australia. These fast-growing shrubs and small trees, cultivated for their handsome foliage and white flowers, adapt to shearing and are often used as formal hedges.

Recognition

Habit: shrubs, small trees; deciduous or evergreen
Leaves: opposite; simple; margins entire
Flowers: clusters (panicles) terminal, rarely from lateral leafless buds; bisexual; white; calyx bell-shaped (campanulate), 4-toothed; corolla tube with 4 spreading lobes; stamens 2; ovary compound, style cylindric, not exceeding stamens
Fruits: berry-like drupes, usually black or blue-black; seeds 1-4

Ligustrum amurense Carr. - Amur privet

Selections of Amur privet from northern China are popular in the central prairie. It is adapted to a wide range of site conditions and low temperatures. Satisfactory performance can be expected from Amur privet in the southeastern part of the central Great Plains (North Central Plant Introduction Station 1956).

Planning

Height: up to 15 ft (4.5 m)
Habit: shrub; deciduous; upright, dense; May-June blossoms unpleasantly fragrant
Site: partial shade, sun; pH and soil adaptable; pollution and drouth tolerant; zone 4
Use: hedge, border

Recognition

Leaves: elliptic to oblong; tip pointed to rounded, base rounded to broadly wedge-shaped, margin fringed with minute hairs (ciliolate); surfaces glabrous except on midrib below; petioles 0.1 in (3 mm) long, usually hairy
Flowers: clusters (panicles) hairy, upright, branching; corolla creamy white, tube 2-3 times as long as lobes; calyx smooth or slightly hairy at base; anthers slightly exserted, not reaching middle of corolla lobes
Fruits: black, slightly bloomy, about 0.3 in (8 mm) long; persistent
Twigs: purplish, hairy, becoming gray

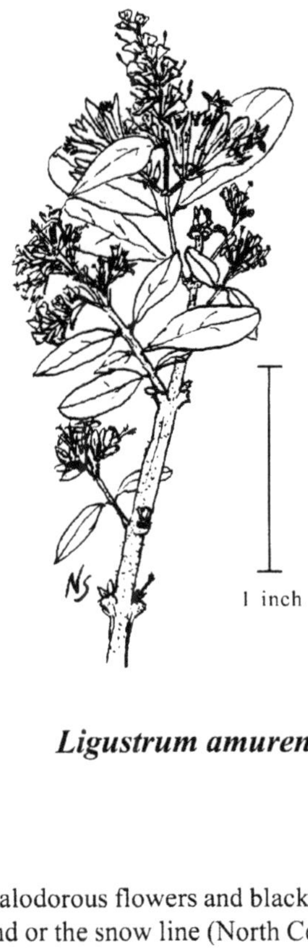

Ligustrum amurense

Ligustrum obtusifolium Sieb. & Zucc. - border privet

Border privet, a large multistemmed Japanese shrub, bears showy, malodorous flowers and black fruits that persist into winter. Plants in North Dakota trials froze to the ground or the snow line (North Central Plant Introduction Station 1965).

Variety *regelianum* (Koehne) Rehd. is a low-growing horizontally spreading shrub with branches and leaves regularly spaced in a flat plane, and 'Constitution' is a compact, dense form of border privet (Dirr 1990). Ibolium privet (*L. ovalifolium* Hassk. × *L. obtusifolium*) is a handsome plant bearing glossy dark green foliage.

Planning

Height: up to 12 ft (3.5 m)
Habit: shrub; deciduous; many-stemmed, branches spreading; June blossoms unpleasantly fragrant; autumn foliage russet to purplish
Site: sun, partial shade; pH adaptable; zone 4
Use: border, hedge, mass

Recognition

Leaves: oblong to widest above middle (obovate); tip pointed to rounded, base rounded to broadly wedge-shaped; dark green above, hairy beneath or only on midrib; petioles about 0.1 in (3mm) long, hairy
Flowers: clusters hairy, nodding; calyx hairy; corolla white, about 0.4 in (1 cm) long; anthers nearly as long as corolla lobes
Fruits: black, slightly bloomy, about 0.2 in (6 mm) long; persistent
Twigs: green, slight purplish tinge, hairy, becoming gray

Ligustrum vulgare L. - European privet

European privet, brought from Europe for hedging, is one of the most commonly planted shrubs in the eastern U.S. where it has escaped and became naturalized. Anthracnose twig blight can be a serious problem. However, test plantings show European privet can be useful in the southeastern part of the central Great Plains (North Central Plant Introduction Station 1957, 1977).

'Lodense', a dwarf cultivar, is useful for parterres that are small, interwoven and formal hedges, and 'Pyramidale' is a pyramidal form used for hedges. 'Cheyenne', the most hardy of the cultivars, is useful as far north and west as Cheyenne, Wyoming (North Central Plant Introduction Station 1956); however, it is not hardy at the Minnesota Landscape Arboretum (Dirr 1990).

L. ×*vicaryi* (*L. ovalifolium* Hassk. 'Aureum' × *L. vulgare*), golden privet, is vase-shaped and bears golden yellow leaves in sunny locations. Golden privet on the East Campus of UNL died back to the ground during one of Nebraska's unusual winters. This hybrid is most likely to succeed south of eastern Nebraska. 'Storzinger Strain' is reputed to be a hardy selection of golden privet (Dirr 1990).

Planning

Height: up to 15 ft (4.5 m)
Habit: shrub; deciduous, half-evergreen; much-branched, wide spreading; June blossoms unpleasantly fragrant
Site: sun, partial shade; pH and soil adaptable; zone 5
Use: border, hedge

Recognition

Leaves: oblong egg-shaped, apex variable; dark green, 4-5 pairs of indistinct veins; petioles 0.1-0.4 in (3-10 mm) long
Flowers: clusters (panicles) dense; corolla white, tube shorter than lobes; anthers exceeding the tube, shorter than lobes
Fruits: black, lustrous, about 0.3 in (8 mm) long; persistent
Twigs: green, minutely hairy, becoming glabrous

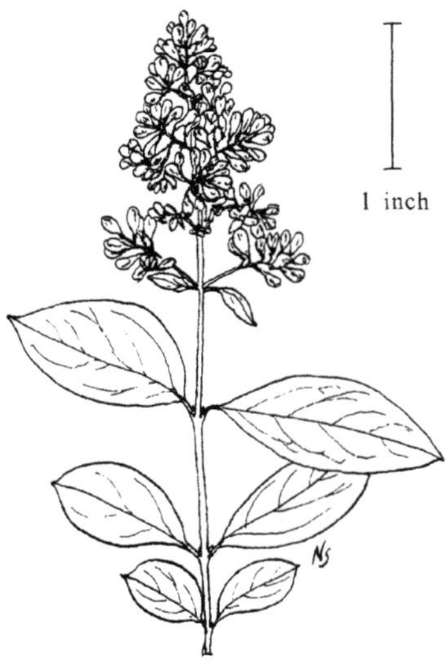

Ligustrum vulgare

Syringa L. - lilac

About 30 species of lilacs—shrubs and small trees with showy, fragrant flowers—belong to the genus *Syringa* a native of southeastern Europe and Asia. Hoag (1965) states that these widely grown lilac species are "perhaps the most valuable flowering shrubs for the northern plains when hardiness, adaptability to high lime soils, exposure, and floriferous nature are considered." Their beauty and lifespan is increased by annual removal of old canes.

Countless cultivars fill the pages of garden catalogs. We will mention only a few of the some 3,000 named cultivars. The book **Lilacs, the Genus Syringa** is a source of extensive information about lilac cultivars (Fiala 1988). Hoag (1965) suggests that of all of the hardy and useful species of lilac, the most valuable are common lilac (*S. vulgaris*), Chinese lilac (*S.* ×*chinensis*), Preston hybrids (*S.* ×*prestoniae*), and Meyer lilac (*S. meyeri*).

One of the largest collections of rare and beautiful lilacs in the western prairies can be viewed at Meadowlark Hill, the home of Max and Darlene Peterson, 14 miles southeast of Ogallala, Nebraska. Visitors are welcome to this Perkins County wheat farm whose first lilacs were planted in 1969. Another large lilac collection can be seen in the Maxwell Arboretum at UNL.

Recognition

Habit: shrubs, trees; deciduous
Leaves: opposite; simple, entire or sometimes lobed or seldom pinnately compound
Flowers: clusters (panicles) terminal or lateral; bisexual; calyx bell-shaped, 4-toothed, small, persistent; corolla a slender tube abruptly expanded to 4 spreading valvate lobes (salverform); stamens 2; ovary 2-celled, style with 2-lobed stigma
Fruits: capsules oblong, leathery, splitting on back; seeds 2 in each cell, winged
Buds: terminal often wanting, egg-shaped; scales several

Syringa ×chinensis Willd. - Chinese lilac

Chinese lilac, a graceful, dense, non-suckering hybrid, is delicate and profuse in flower. Authorities differ on parentage. It is the result of crossing *S. vulgaris* with *S. oblata* or S. ×*persica* or *S. laciniata* Mill. It has been sometimes incorrectly labeled Persian lilac (see *S.* ×*persica*). 'Alba' bears nearly white light pink flowers, and 'Saugeana' has lilac-red flowers (Dirr 1990).

Chinese lilac is winter hardy, free from insects, and consistently produces abundant and showy flowers at PERC in Fort Collins (Fairchild and Klett 1993).

Planning

Height: up to 12 ft (3.5 m)
Habit: shrub; branches slender, arching; May blossoms fragrant
Site: sun; pH and soil adaptable, drouth tolerant; zone 4
Use: specimen, hedge, grouping, windbreak
Pests: mildew resistant, usually insect free

Recognition

Leaves: simple; egg-shaped to much longer than wide, wider below the middle; apex long-pointed, base wedge-shaped or rounded, margin entire; petioles about 0.4 in (1 cm) long
Flowers: clusters loose; corolla purple, lilac or white, tube about 0.3 in (8 mm) long, lobes 4, variable

Syringa meyeri Schneid. - Meyer lilac

The handsome Meyer lilac, introduced from northern China, will commence blooming when it reaches about one foot (30 cm) tall and requires little maintenance. 'Palibin', a compact form of the species, is sometimes incorrectly referred to as S. *palibiniana* by the nursery trade (Dirr 1990).

Planning

Height: up to 6 ft (2 m)
Habit: shrub; uniform, dense, mounded; May blossoms bountiful, fragrant
Site: sun; pH and soil adaptable; drouth tolerant; zone 3
Use: border, grouping
Note: low maintenance, mildew resistant

Recognition

Leaves: simple; elliptic, sometimes wider above middle, up to 1.7 in (4 cm) long; tip acute to rounded, base broad wedge-shaped, margin wavy; dark green, hairy on veins beneath near the base, 2-3 pairs of veins from base nearly reach apex; petioles about 0.3 in (8 mm) long
Flowers: clusters dense, up to 4 in (10 cm) long, in pairs at branch ends; calyx violet, not hairy; corolla violet, tube about 0.6 in (15 mm) long, lobes pointed, spreading, about 0.1 in (3 mm) long; anthers inserted much below mouth of tube
Fruits: warty
Twigs: slightly 4-sided, glabrous or minutely hairy
Buds: terminal wanting

Syringa microphylla Diels - littleleaf lilac

Littleleaf lilac, an introduction from northern China, has many of the attributes of Meyer lilac. 'Superba' bears pink-red flowers.

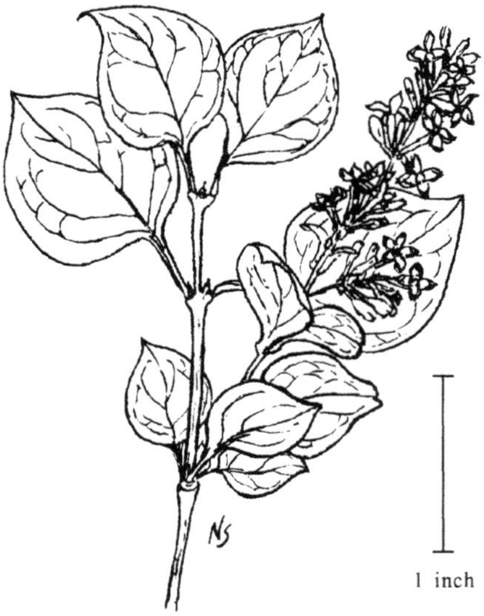

Syringa microphylla

Planning

Height: up to 6 ft (2 m)
Habit: shrub; dense, broad spreading; late May blossoms fragrant
Site: sun; pH and soils adaptable; heat tolerant; zone 5
Use: border, hedge, grouping

Recognition

Leaves: simple; elliptic to egg-shaped, about 1.2 in (3 cm) long; tip abruptly pointed or rounded, base broadly wedge-shaped to rounded; slightly hairy to glabrous above, densely hairy beneath, becoming hairy only on veins or nearly glabrous; petioles about 0.3 in (8 mm) long
Flowers: clusters hairy, about 2 in (5 cm) long; calyx hairy; corolla rosy lilac, tube about 0.4 in (1 cm) long, lobes pointed; anthers inserted much below mouth of tube
Fruits: warty, often curved
Twigs: hairy at first
Buds: terminal wanting

Syringa oblata Lindl. - early lilac

Early lilac, a native of Korea, is similar to the common lilac but blooms 1-2 weeks earlier and is one of the few lilacs that provides reddish fall color. Var. *dilata* (Nikai) Rehd. is more commonly grown than the species.

Planning

Height: up to 12 ft (4 m)
Habit: shrub, small tree; coarse; late April blossoms; autumn foliage colorful
Site: sun; soil and pH adaptable; zone 4
Use: border, wildlife habitat, windbreak

Recognition

Leaves: simple; orbicular-ovate to kidney-shaped, often broader than long; tip abruptly pointed, base heart-shaped; dark green, sometimes bluish; petioles about 0.8 in (2 cm) long
Flowers: clusters broad, about 3.5 in (9 cm) long; calyx slightly glandular; corolla pale to purple lilac, tube about 0.4 in (1 cm) long, lobes spreading, rounded; anthers reaching slightly above middle of tube
Fruits: smooth, compressed
Twigs: glabrous
Buds: terminal wanting

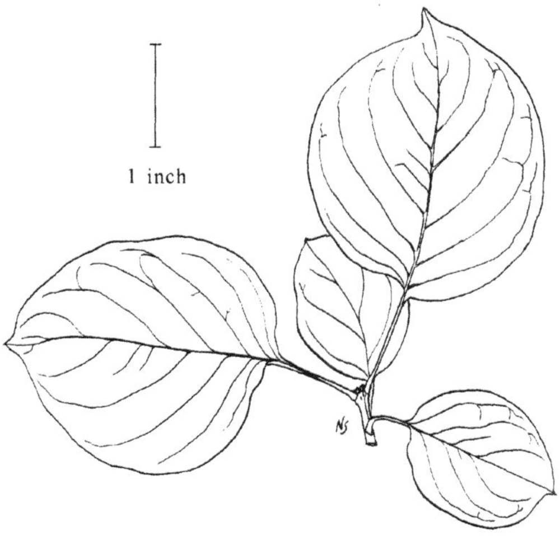

Syringa oblata

Syringa patula (Palib.) Nakai - **Manchurian lilac**

The soft pubescence of the leaves and twigs of Manchurian lilac gives this native of northern China and Korea a satiny appearance. Manchurian lilac is frequently confused with Meyer and littleleaf lilac, and is sometimes sold as *S. palibiniana* (Dirr 1990). 'Miss Kim' is low growing with 3 inch (8 cm) clusters of icy blue flowers; it produces a good and reliable flower display with no cultural problems at PERC in Fort Collins (Fairchild and Klett 1993).

Planning

Height: up to 9 ft (3 m)
Habit: shrub; upright; May-June blossoms fragrant; autumn foliage may be colorful
Site: sun; pH and soil adaptable; zone 4
Use: border, grouping

Recognition

Leaves: simple; elliptic to egg-shaped, up to 5 in (12 cm) long; tip long-pointed, base broadly wedge-shaped or rounded; slightly hairy to glabrous above, densely hairy beneath or only on midrib and veins; petioles about 0.3 in (8 mm) long
Flowers: clusters hairy, 4-6 in (10-15 cm) long; calyx usually hairy; corolla lavender purple, tube about 0.3 in (9 mm) long with short, often hooded lobes; anthers violet, not reaching tube mouth
Fruit: warty, pointed
Twigs: slightly hairy to glabrous, often glandular
Buds: terminal wanting

Syringa patula

Syringa pekinensis Rupr. - Pekin lilac

Pekin lilac is very similar to Japanese tree lilac (*S. reticulata*) but its growth is slower and on many of the trees the shiny bark exfoliates. Pekin lilac at PERC in Fort Collins formed attractive clumps of dense foliage and a good, consistent floral display (Fairchild and Klett 1993). It deserves more trials throughout the prairies.

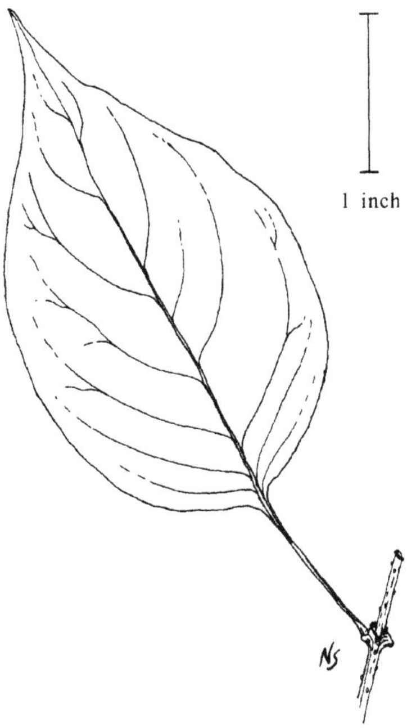

Syringa pekinensis

Planning

Height: up to 20 ft (6 m)
Habit: shrub, tree; multistemmed, upright, rounded; June blossoms
Site: sun; pH and soil adaptable; zone 4
Use: specimen, screen, massing, border

Recognition

Leaves: simple; egg-shaped to much longer than wide, up to 4 in (10 cm) long; tip long-pointed, base broadly wedge-shaped or rounded; dark green above, veins not more than slightly elevated; petioles about 0.8 in (2 cm) long
Flowers: clusters hairless, 3-6 in (8-15 cm) long; corolla creamy white, tube nearly as long as calyx, each flower about 0.2 in (6 mm) across; stamens about as long as lobes (limb), anthers exserted
Fruits: warty or smooth, pointed
Bark: reddish brown, exfoliating on some trees; lenticels prominent, horizontal

Syringa ×persica L. - Persian lilac

Persian lilac, considered to be a hybrid because of its sterility, is often confused with Chinese lilac, another hybrid. According to Hoag (1965) these two hybrids have a common parent, *S. laciniata* Mill.; the other parent of Persian lilac is *S. afghanica* Schneid. Several flowering types have been selected from the population of this graceful and hardy shrub, including the white-flowered 'Alba'.

Planning

Height: up to 9 ft (3 m)
Habit: shrub; upright, arching; May blossoms fragrant
Site: sun; pH and soil adaptable; zone 4
Use: border

Recognition

Leaves: simple; variable, sometimes slightly lobed; margin entire, about 2 in (5 cm) long, apex pointed, base wedge-shaped; dark green; petioles about 0.3 (8 cm) long
Flowers: clusters loose, 2-3 in (5-8 cm) long; corolla white, red-purple to lavender, tube about 0.4 in (1 cm) long

Syringa ×prestoniae McKelvey - **Preston lilacs**

These late-blooming hybrids, the product of breeding by a Canadian horticulturist, are crosses between *S. reflexa* Schneid. and *S. villosa*. The many Preston selections are similar to *S. villosa* except for their superior flowers with a wide variety of colors ranging from light pink on 'Coral', bluish on 'Desdemona', violet on 'Jessica', lilac on 'Patience', and magenta on 'Redwine' (Dirr 1990). 'James MacFarlane' at PERC in Fort Collins is vigorous despite minor chlorosis, and it produces pretty but not abundant flower clusters (Fairchild and Klett 1993).

Planning

Height: up to 12 ft (4 m)
Habit: shrub; erect, stout branches, few suckers; May-June flowers
Site: sun; pH and soil adaptable; zone 3
Use: border, mass, hedge, screen, specimen

Recognition

Leaves: simple; elliptic to oblong, 3-6 in (8-15 cm) long; glabrous above, usually hairy beneath
Flowers: purple-lilac to pink

Syringa reticulata (Bl.) Hara - **Japanese tree lilac**

Japanese tree lilac is named *S. amurensis* (Rupr.) Rupr. var. *japonica* (Maxim.) Fr. & Sav. according to many authors. It is described by Hoag (1965) as an outstanding species among the late-blooming lilacs for planting in the northern plains, and Dirr (1990) lauds it as the most trouble-free lilac. Although its fragrance is not as pleasant as that of other lilacs, the large white flower clusters provide an excellent show in June.

A few large Japanese tree lilacs grow throughout Nebraska, and it is well adapted to the Fort Collins area (Fairchild and Klett 1993). A large specimen grows in Douglas, Wyoming (Sutton 1990).

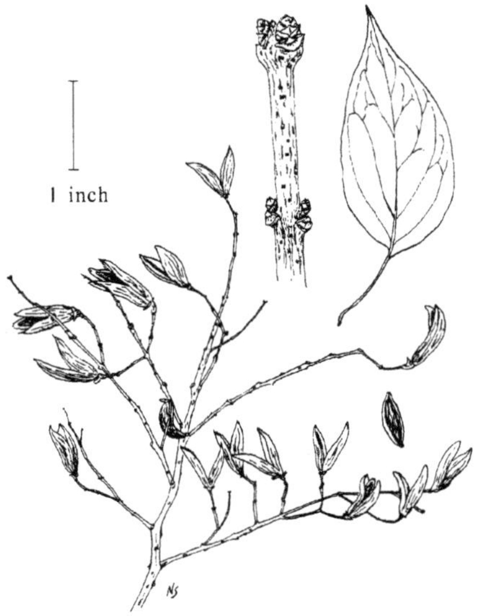

Syringa reticulata

Planning

Height: up to 30 ft (9 m)
Habit: shrub, tree; multistemmed or single stem, stiff spreading branches; June blossoms
Site: sun; pH and soil adaptable; zone 4
Use: specimen, street tree, grouping, shade

Recognition

Leaves: simple; egg-shaped or broadly so, up to 5 in (13 cm) long; margin entire, tip pointed, base rounded or nearly heart-shaped; dark green above, slightly hairy or glabrous beneath, veins elevated (reticulate); petioles 0.5-1.0 in (12-25 mm) long
Flowers: clusters terminal, 6-12 in (15-30 cm) long; calyx about as long as corolla tube; petals white; stamens about twice as long as petal lobes, anthers exserted
Fruits: warty, up to 0.8 in (2 cm) long
Twigs: stout, shiny, glabrous, brown; lenticels prominent
Buds: terminal frequently wanting; subglobose; scales, 4 pairs
Bark: reddish brown, exfoliating; lenticels prominent, horizontal

Syringa villosa Vahl - late lilac

The late lilac from China is one of the least attractive lilac species (Hoag 1965). It has largely been replaced by its hybrids (see *S. ×prestoniae*).

Planning

Height: up to 10 ft (3 m)
Habit: shrub; erect, stout branches, dense, wide-spreading; late May blossoms
Site: sun; pH and soil adaptable; zone 3
Use: border

Recognition

Leaves: simple; broadly elliptic to oblong, up to 7 in (17 cm) long; tip pointed, base narrow wedge-shaped; dull dark green above, slight waxy bloom beneath, usually hairy near midrib, veins impressed, somewhat rugose; petioles 0.4-0.8 in (1-2 cm) long
Flowers: clusters pyramidal, short-stalked, dense, usually hairy; corolla white to rosy lilac, tube cylindric, about 0.5 in (12 mm) long, lobes rounded, spreading; anthers not reaching tube mouth
Fruits: obtuse, about 0.6 in (15 mm) long, smooth or nearly so

Syringa vulgaris L. - common lilac

Common lilac, an introduction from southern Europe, marks the locations of many pioneer settlements throughout the prairies. Despite attacks by borers and scale, many of these hardy shrubs have survived 100 years or more because of their ability to reproduce by suckers and seedlings. Though widely planted today in windbreaks and wildlife habitat plantings, common lilac for the garden has deferred to the hundreds of cultivars that have been been developed by plant breeders, beginning with Victor Lemoine of Nancy, France, originator of the "French Hybrids."

We have space to mention only a few of the many single- and double-flowered cultivars that are available today. White flowers are represented by 'Mont Blanc' and 'Edith Cavell', violet by 'de Miribel' and 'Marechal Lannes', blue by 'President Lincoln' and 'Ami Schott', lilac by 'Christophe Columb' and 'Victor Lemoine', pink by 'Lucie Baltet' and 'Montaigne', magenta by 'Congo' and 'Paul Thirion', and purple by 'Night' and 'Paul Hariot'.

Although 'President Lincoln' is adapted to the soils and climate of Fort Collins, its flower display is not reliable or showy (Fairchild and Klett 1993).

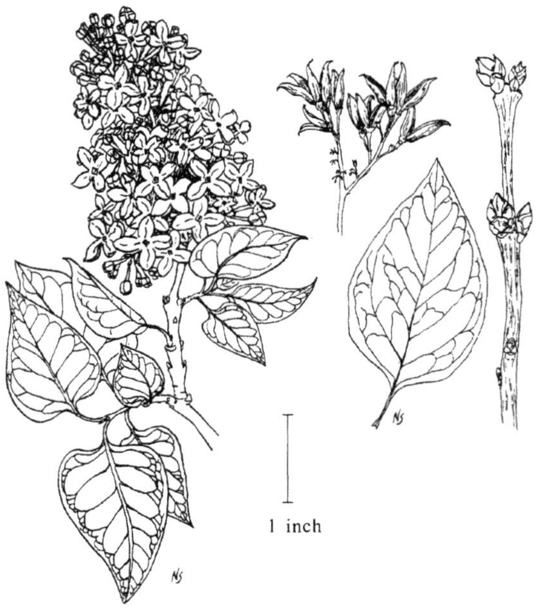

Syringa vulgaris

Planning

Height: up to 15 ft (4.5 m)
Habit: shrub; upright, leggy, suckering; May blossoms fragrant
Site: sun; pH and soil adaptable; drouth tolerant; zone 4
Use: border, grouping, windbreak, wildlife habitat

Recognition

Leaves: simple; egg-shaped or broadly so, up to 5 in (13 cm) long; tip pointed (acuminate), base nearly heart-shaped to broadly wedge-shaped; dark green, glabrous; petioles 0.7-1.2 in (15-30 mm) long
Flowers: clusters long, usually in pairs from terminal buds; corolla lilac, tube slender, about 0.4 in (1 cm) long, lobes rounded, spreading; anthers barely below the tube mouth
Fruits: about 0.5 in (12 mm) long, pointed, smooth
Twigs: stout, angled, glabrous, brown
Buds: terminal wanting

Bignoniaceae Juss. - bignonia family

The bignonia family of over 100 genera and more than 700 species, many of which are woody vines, is mostly tropical. Many are planted for their spectacular flowers and a few yield useful lumber.

Recognition

Habit: trees, shrubs, vines, rarely herbs; deciduous, evergreen
Leaves: usually opposite; compound, or occasionally simple; without stipules
Flowers: solitary or clusters; bisexual; flower parts attached near base of ovary (superior or hypogynous); calyx 5-lobed or sometimes 2-lipped; petals united (sympetalous), bell- or funnel-shaped, nearly regular to strongly 2-lipped; stamens usually 4 in 2 pairs with sterile 5th, sometimes 5 or 2, inserted on corolla tube; ovary compound, stigma 2-lipped
Fruits: capsules, often elongated; seldom fleshy and not splitting (indehiscent); seeds mostly flat, usually winged in capsules

Campsis Lour. - trumpet-vine

The two species of trumpet-vine—one a native of North America and the other of eastern Asia—are deciduous shrubs capable of climbing with the aid of aerial rootlets. *C. grandiflora* (Thunb.) Loisel. is not adapted to the prairies, but C. ×*tagliabuana* (Vis.) Rehd., a cross between the two species from which many cultivars have been selected, might be grown successfully in the southeastern part of the central prairie.

Campsis radicans (L.) Seem. - trumpet-vine

Trumpet-vine, a native of the eastern U.S. including eastern Kansas, Oklahoma and Texas, is a very vigorous, easily propagated vine with large trumpet-shaped flowers bearing sumptuous nectaries. It can be grown as a small tree with the assistance of a stake supporting it until the trunk becomes strong enough to stand alone. Trumpet-vine is often promoted as a source of food for hummingbirds. Unfortunately, these birds may not be present during the peak of bloom.

Trumpet-vine in leaf is an attractive cover for trellises, arbors and masonry walls, but it is a tangle of dead-looking twigs during the winter. Aerial roots can damage wood siding, fill eave troughs and lift shingles. A large part of the upper branches may suffer winter injury, but don't despair, new shoots will appear. Trumpet-vine is a plant for large landscape spaces where naturalizing is the theme.

'Crimson Trumpet' and 'Praecox' bear red flowers, and var. *flava* (Bosse) Rehd. is a particularly handsome form with yellow to orange-yellow flowers (Dirr 1990).

Planning

Height: up to 30 ft (9 m)

Habit: shrub (clinging vine); deciduous; suckering; July-August blossoms on new wood

Site: sun, partial shade; pH 6.5-7.5; soil adaptable; zone 5

Use: screen on trellis, wall, or other structure, wildlife habitat, erosion control

Recognition

Leaves: opposite; compound, odd-pinnate; leaflets 7-11, egg-shaped; coarsely toothed, tip pointed, base rounded to wedge-shaped; upper surface dark green, glabrous, hairy on veins beneath; petioles grooved above

Flowers: clusters (cymes or panicles) terminal; calyx tubular-bell-shaped, 5 triangular lobes much shorter than the tube; corolla funnel-shaped, orange outside, enlarged above the calyx with spreading, 5-lobed oblique scarlet limb; stamens 4 in pairs of unequal length (didynamous), not exserted; ovary 2-celled

Fruits: capsules dry, splitting, beaked; seeds numerous, compressed, 2-winged

Twigs: pale yellow, glabrous; form adventitious roots

Bark: yellow, becoming shreddy

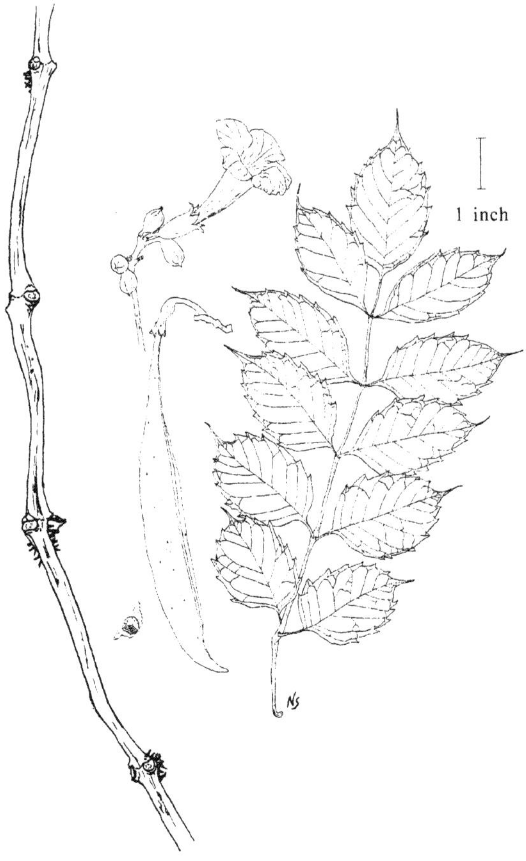

Campsis radicans

Catalpa Scop. - catalpa

Two of the 12 catalpa species that grow naturally in North America, the West Indies, and eastern Asia are natives of the U.S. The West Indian species are evergreen. Catalpas are easily recognized by their showy bell-shaped flowers, broad simple leaves, and the long cylinder-shaped capsules that persist into winter.

Recognition

Habit: trees; deciduous, evergreen (West Indies species)

Leaves: opposite, sometimes whorled; simple; sometimes coarsely lobed, margins entire; 3- to 5-veined at base, a purple gland often at vein axils beneath

Flowers: clusters (panicles or racemes) terminal; calyx 2-lipped or irregular; corolla bell-shaped, 2 small upper and 3 large lower lobes; stamens, 2 fertile, included; style longer than stamens

Fruits: capsules long, cylinder-shaped, separating into 2 valves; seeds numerous, oblong, a tuft of long white hairs at each end

Buds: terminal wanting; scales several

Catalpa ovata Don - **Chinese catalpa**

This native catalpa of China naturalizes in the northeastern U.S. where it appears to be better adapted to cold than our native species (Viertel 1970). It is not well known in the prairies. A small Chinese catalpa tree in the UNL Maxwell Arboretum blooms and sets fruits. Additional trial plantings will be necessary to determine its adaptability to the prairie region.

Several named clones have been selected from a group of hybrids (*C. bignonioides* × *C. ovata*) including 'Purpurea' with black-purple emerging shoots that gradually change to dark green (Dirr 1990). *C.* ×*japonica*, another of these hybrids, and *C. bungei* C. A. Mey., a small, bushy tree of northern China, are being evaluated at PERC in Fort Collins (Fairchild and Klett 1993).

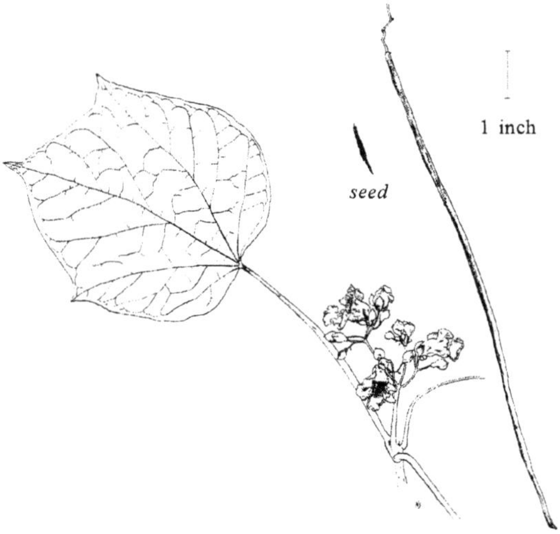

Catalpa ovata

Planning

Height: up to 30 ft (9 m)
Habit: tree; deciduous; crown oval; June-July blossoms
Site: sun; experimental; zone 5
Use: specimen, shade

Recognition

Leaves: whorls of 3 or sometimes opposite; broadly egg-shaped, up to 10 in (25 cm) long; often with 3-5 shallow lobes, tip abruptly taper-pointed; dull green above, lighter beneath and variably hairy on the veins; petioles up to 5 in (13 cm) long
Flowers: corolla yellowish white, up to 0.8 in (2 cm) long with purple dots and 2 narrow splotches of bright yellow-orange inside
Fruits: capsules up to 12 in (30 cm) long, 0.3 in (8 mm) in diameter
Twigs: stout, glabrous, rarely a few hairs
Buds: globose, brown

Catalpa speciosa Warder - **northern catalpa**

The original, very limited range of northern catalpa centers on the "bootheel" of Missouri and extends along the Mississippi and Ohio river floodplains, from southwestern Indiana to northeastern Arkansas. The early European settlers planted catalpa for fence posts, railroad ties and lumber. Nebraska Governor Robert W. Furnas, owner of a Brownville nursery, promoted northern catalpa and osage orange for post plantations.

Remnants of early catalpa plantations dot our landscape in groves and fencerows, and wildings can be found on floodplains. An occasional catalpa will dominate the yard of an older home in town or country. Although seldom planted today, this fast-growing tree can be useful in large areas such as parks and farmsteads. Leaves fall with the first frost exposing a rather coarse framework described as ugly or picturesque, depending upon the viewer's taste.

Bignoniaceae *Catalpa*

Southern catalpa, *C. bignonioides* Walt., a native of four southeastern states, has been widely planted and has become naturalized in the eastern U.S. Although considered to be a smaller tree than its northern relative, the National champion southern catalpa (Illinois 1991) is a sizeable tree at 10.3 ft (3.15 m) in diameter and 52 ft (15.9 m) tall. The number of southern catalpas growing in the prairies is unknown since we seldom distinguish between these two catalpa species because of their slight morphological differences.

Southern catalpa is considered to be better than northern catalpa for ornamental plantings because of its smaller potential size and larger flower clusters. Its leaf tips are shorter, individual flowers are slightly smaller, and pods are smaller in diameter and length than those of northern catalpa. Cultivars selected from southern catalpa include 'Aurea', with leaves of rich yellow, and 'Nana', a dwarf, bushy form (Dirr 1990).

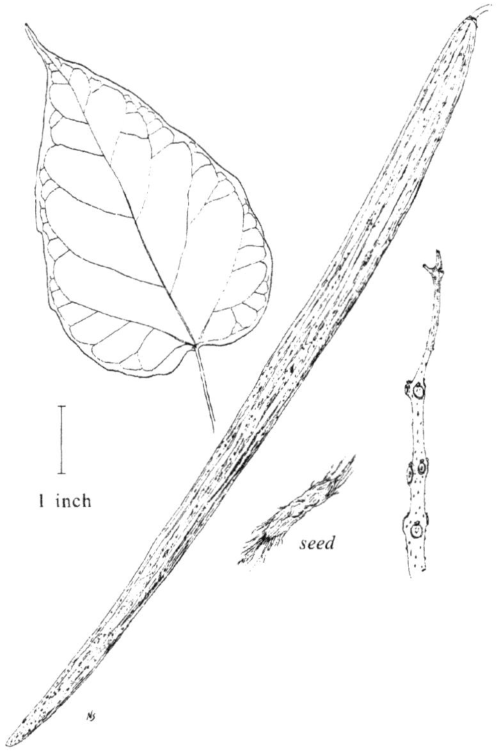

Catalpa speciosa

Planning

Height: up to 75 ft (23 m); Nebraska champion, 1991, 60 ft (18.3 m), Lincoln; National champion, 1990, 107 ft (32.7 m), Michigan

Habit: tree; deciduous; crown oval to irregular; June blossoms; fruits persist into winter

Site: sun, partial shade; pH and soil adaptable; zone 5

Use: specimen, shade, posts, lumber

Note: the seasoned wood is only somewhat durable when in contact with moist soil

Recognition

Leaves: whorls of 3 or sometimes opposite; broadly egg-shaped, up to 12 in (30 cm) long; margin entire, occasionally 1-3 irregular teeth, apex long taper-pointed; yellow-green, glabrous above, soft hairy beneath; petioles about 4 in (10 cm) long, sparingly hairy

Flowers: calyx lobes 2, egg-shaped with a short point (apiculate), purplish; corolla tube about 2 in (5 cm) long, lobes wavy, white with yellow and purple spots on inner surface; stamens, 2 fertile, 3 rudimentary

Fruits: capsules up to 24 in (60 cm) long, 0.8 in (2 cm) in diameter; seeds flattened, wings rounded

Twigs: stout; leaf scars crater-like, center depressed; bundle scars many, conspicuous, in a closed ring

Buds: solitary, globose; scales red-brown, lighter on margin, fringed with hairs (ciliate), keeled

Wood: light brown, soft, durable

Chilopsis Don - desert willow

The single species of *Chilopsis* is a native of the southwestern U.S., ranging from southern Texas to southeastern California and adjacent Mexico.

Chilopsis linearis (Cav.) Sweet - desert-willow

Desert-willow, an inhabitant of desert washes and grasslands, grows on gravelly and rocky soils in association with the joshua tree (*Yucca brevifolia* Engelm.), the mesquite (*Prosopis* L. sp.), and associated desert vegetation.

Desert-willow was planted in the shrub row of many shelterbelts of the Prairie States Forestry Project of the 1930s from northern Texas to south central Kansas. Read (1958) states that it became too tall and open to be an effective shrub in these plantings. At age 16 years approximately half of the desert willows in the shelterbelt project were living and averaged 10 feet (3 m) in height on sandy sites. Survival and growth rates were less for those planted on soils of medium to heavy texture.

Desert-willow has been grown successfully in Denver by Professor Allen R. Taylor of Colorado University (Sutton 1990). Trial plantations of desert willow provenances may enable us to discover plants adapted to the arid sites of the southwestern part of the central prairie.

Planning

Height: up to 30 ft (9 m); National champion, 1976, 56 ft (17 m), Arizona
Habit: shrub, tree; deciduous; crown narrow, upright; June-August blossoms
Site: sun; prefers sandy soils; drouth tolerant; zone 6
Use: specimen, border

Recognition

Leaves: alternate; simple; linear to narrowly lance-shaped, up to 6 in (15 cm) long; apex long-pointed, base tapered, margin entire; glabrous; petioles very short or sessile
Flowers: clusters (racemes) terminal, unbranched, many-flowered, hairy, up to 2.5 in (6 cm) long; calyx cup-shaped, 2-lipped; corolla funnel-shaped, slightly 2-lipped, upper 2-lobed, lower 3-lobed, pink, white or purple, yellow or purple lines on inner surface; stamens 4, in 2 pairs
Fruits: capsules narrow, cylindrical up to 12 in (30 cm) long; seeds narrow, flattened, winged with long hairs at ends
Twigs: slender, glabrous to hairy
Buds: terminal absent, lateral tiny; scales overlapping, hairy
Bark: dark brown, splitting into broad, scaly ridges

Paulownia Sieb & Zucc. - paulownia

About a dozen species of paulownia are natives of Asia. They closely resemble the catalpas and are placed in the Bignoniaceae family (Cronquist 1981; Little 1953). However, some authorities, including Rehder (1940), place them in the closely related Scrophulariaceae family. Only one species is commonly grown in North America.

Paulownia tomentosa (Thunb.) Sieb. & Zucc. - royal paulownia

Royal paulownia, a native of China, has been cultivated in Japan and Europe for hundreds of years. The Japanese value its lightweight wood for musical instruments, furniture and many specialty items. Royal paulownia has escaped from cultivation and grows as wildings in the southeastern U.S. These trees are capable of producing high quality logs that command high prices in Japan. Unfortunately, it is difficult to produce the desired wood quality in plantations.

Royal paulownia is often advertised by mail order nurseries as the wonder tree that produces instant shade. Unscrupulous dealers fail to mention its intolerance to temperatures below 0 degrees F (-18 degrees C). Flower buds are especially vulnerable to freezing temperatures.

Several royal paulownia trees were planted in Lincoln in the 1970s. The appearance of fragrant, purple flowers in May after a milder-than-normal winter caused a flurry of queries about the identity of these trees. Over 15 years have passed without a repetition of flowering. A dearth of blooms and many dead twigs on these paulownias caused many to be removed.

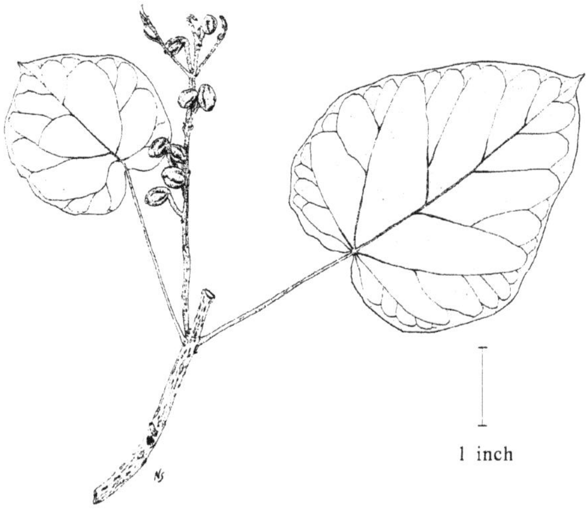

Paulownia tomentosa

Planning

Planning: up to 50 ft (15 m); National champion, 1989, 64 ft (19.4 m), Indiana
Habit: tree; deciduous; crown rounded, dense; spring bloom seldom except in the south, occurs before the leaves
Site: sun, partial shade; pH and soil adaptable; zone 6
Use: shade, specimen, lumber
Caution: flower buds vulnerable to freezing temperatures

Recognition

Leaves: opposite; simple; broadly egg-shaped, sometimes with 3 shallow lobes, up to 12 in (30 cm) long; margin entire, tip pointed, base heart-shaped; sparsely hairy above, densely so beneath; petioles up to 8 in (20 cm) long
Flowers: clusters (panicles) terminal, upright, pyramidal; calyx 5-lobed; corolla tube bell-shaped, purple with yellow stripes inside, about 2 in (5 cm) long, spreading 5-lobed limb; stamens 4
Fruits: capsules egg-shaped, beaked, about 2 in (5 cm) long, densely glandular hairy; seeds numerous, winged
Twigs: stout, brittle, densely soft-hairy, becoming glabrous; lenticels prominent; pith chambered or hollow between nodes; leaf scars nearly round, often notched at top
Buds: terminal absent; lateral leaf buds buried in bark, visible scales blunt, hairy; flower buds formed in the summer prior to blooming; brown, hairy, spherical, very noticeable during winter
Bark: thin, flaky

Order Rubiales

One family is included in Rubiales (Cronquist 1981).

Recognition

Habit: woody, herbaceous
Leaves: commonly opposite; simple; margins entire
Flowers: clusters (mostly cymose); usually bisexual; calyx mostly 4- to 5-lobed, sometimes wanting; petals mostly regular, united (sympetalous); stamens alternate with the lobes, ovary inferior (epigynous)
Fruits: diverse

Rubiaceae Juss. - madder family

The madder family—comprised of about 450 genera and 6,500 species—provides us with coffee, quinine, dyes and flowers (including gardenias). Most species are tropical, but some grow in temperate and cold climates. Although most North American species are herbs, woody plants are present in seven genera.

Recognition

Habit: trees, shrubs, herbs

Leaves: opposite or whorled; stipules present though sometimes reduced to a stipular line connecting opposite leaves

Flowers: clusters or dense heads, rarely solitary; bisexual, rarely unisexual; stamens 4-5, attached within corolla tube or at the mouth; ovary compound, style slender, terminal

Fruits: diverse

Cephlanthus L. - buttonbush

Seven *Cephlanthus* species are natives of Asia and Africa and one is a native of North America.

Cephalanthus occidentalis L. - buttonbush

Buttonbush, a native of lowland areas of the eastern U.S. including southeastern Nebraska, also grows in Texas, Mexico, and west into southern California. Although it grows naturally on wet sites, buttonbush can also be found on well-drained sites. It is useful for protecting shorelines and provides valuable food and cover for wildlife, especially waterfowl. A long blooming period provides a valuable nectar source for bees.

Planning

Height: up to 20 ft (6 m); National champion, 1991, 34 ft (10.4 m), Michigan

Habit: shrub, rarely a tree; deciduous; upright with rounded spreading crown; midsummer blossoms fragrant; summer foliage glossy, fall foliage yellow; fruits persist into winter

Site: sun, partial shade; moist to wet soils but withstands dry periods; pH adaptable; zone 4

Use: naturalizing, wildlife habitat, erosion control, bee pasture

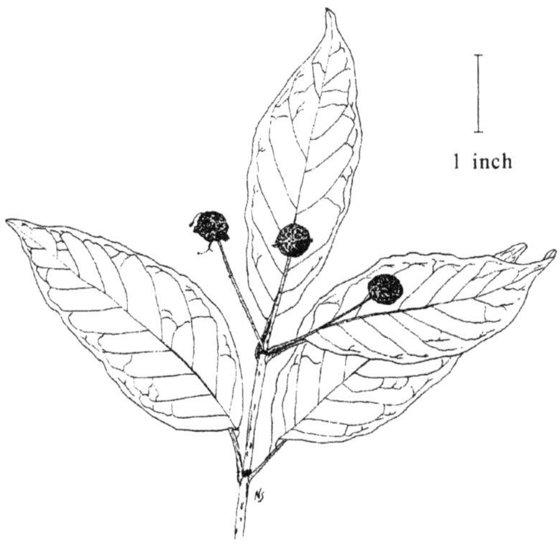

Cephalanthus occidentalis

Recognition

Leaves: opposite or in whorls of 3; egg-shaped to lance-shaped, about 4 in (10 cm) long; margin entire, tip pointed, base variable; lustrous dark green above, variably hairy beneath, mostly on veins; petioles glabrous, about 0.3 in (8 mm) long; stipules triangular, about 0.1 in (3 mm) long, becoming brown

Flowers: clusters globular, many-flowered, about 1 in (25 mm) across including exserted styles, stalks green, glabrous; calyx conical, larger at apex, 4-lobed, green; corolla funnel-shaped, 4-lobed, white; stamens 4, styles white, about 0.6 in (15 mm) long, exserted

Fruits: nutlets conical in globose heads up to 1 in (25 mm) across, persistent into winter; seeds 2, yellowish, capped with white aril

Twigs: circular in cross-section, red-brown, glossy, lenticels conspicuous; leaf scars circular to shield-shaped, bundle scars 3 to several in a crescent line; pith yellow, continuous

Buds: terminal absent; lateral remote from leaf scar, partially embedded, solitary or in pairs one above the other (superposed)

Bark: brown, becoming scaly ridged

Mitchella L. - partridgeberry

The single species of partridgeberry is a native of southeastern Canada and the eastern U.S. as far west as Minnesota and eastern Texas.

Mitchella repens L. - partridgeberry

The delicate partridgeberry grows under a forest canopy along with mosses and lichens. When transplanted it must be moved with as much soil as possible, kept moist, and abundantly supplied with acid leafmold or peat (Dirr 1990). Partridgeberry can be an attractive addition to rock gardens in the northern prairie.

Planning

Height: up to 4 in (10 cm)
Habit: subshrub; evergreen; creeping, ground-hugging; blossoms in late spring, early summer, fragrant; fruits red, persistent
Site: shade; moist, acidic soil; zone 4
Use: groundcover, rock garden

Recognition

Leaves: opposite, broadly egg-shaped to orbicular, 0.3 to 0.8 in (8-20 mm) long; tip rounded, base rounded; lustrous and glabrous above, often variegated with whitish lines; petioles about as long as blade
Flowers: pairs short-stemmed with ovaries united; calyx 4-toothed; corolla funnel-shaped with 4 spreading, recurved lobes, white or pinkish, sometimes with a purple tinge; stamens inserted on throat; stigma 4-lobed
Fruits: drupes berry-like, united pairs, about 0.3 in (7 mm) across, red, crowned by 2 persistent calyxes
Twigs: slender, usually glabrous; develop adventitious roots

Order Dipsacales

Dipsacales is composed of four familes and about 1,000 species (Cronquist 1981).

Recognition

Habit: herbs, shrubs, seldom small trees; deciduous, evergreen
Leaves: opposite, sometimes whorled, usually simple; stipules mostly wanting
Flowers: clusters (cymes or heads); mostly bisexual; corolla united (sympetalous), regular or irregular, usually 5-lobed; stamens alternate with lobes or fewer, attached high in corolla tube; ovary compound, inferior (epigynous), style terminal
Fruits: diverse

Caprifoliaceae Juss. - honeysuckle family

Members of the honeysuckle family—15 genera and 400 species—are found in tropical and temperate areas throughout the world. The greatest number of species grow in eastern North America and eastern Asia. Honeysuckle species are grown for their attractive flowers and fruits. They provide food and cover for wildlife, and a few supply food for man.

Recognition

Habit: shrubs, small trees, rarely herbs; deciduous, evergreen
Leaves: opposite, simple (compound in *Sambucus*), normally without stipules
Flowers: clusters (cymes) branched, rounded to flat-topped; bisexual; calyx 4- or 5-lobed; corolla tubular to wheel-shaped (rotate), lobes overlapping or valvate, regular or irregular, sometimes 2-lipped (bilabiate); stamens 4-5, alternate with the lobes, attached to the tube; ovary compound, inferior (epigynous)
Fruits: berries, berry-like drupes, achenes or capsules

Abelia R. Br. - abelia

About 25 abelia species are natives of Eastern Asia and Mexico. Only one, the hybrid described below, is adapted to the more southerly part of the central prairie. The abelias are characterized by their handsome tubular to bell-shaped flowers borne throughout much of the growing season, and their dry fruits (achenes).

Abelia ×grandiflora (Andre) Rehd. - glossy abelia

Glossy abelia (*A. chinensis* R. Br. × *A. uniflora* R. Br.) may be killed back to the ground or snow line when temperatures fall to -5 degrees F (-20 degrees C) according to Dirr (1990). 'Prostrata' and 'Sherwood' are dense and compact forms.

Planning

Height: up to 6 ft (2 m)
Habit: shrub; semi-evergreen; dense, multistemmed, arching branches; profuse summer blossoms on new growth
Site: sun, partial shade; acidic, moist soil; zone 6
Use: grouping, border, erosion control

Abelia ×grandiflora

Caprifoliaceae *Diervilla*

Recognition

Leaves: opposite; simple; egg-shaped up to 1.4 in (35 mm) long; tip pointed, tapering to rounded base, margins toothed; shiny dark green above, bearded near base of midrib beneath

Flowers: clusters (panicles) leafy; sepals 2-5, often partly united, purplish, persistent; corolla funnel-shaped, about 0.8 in (2 cm) long and wide, 5-lobed, pinkish white

Fruits: dry, one-seeded (achenes)

Twigs: hairy, reddish, bark shredding with age; leaf scars connected by stipular scars

Buds: egg-shaped; scales loose, 2 pairs

Diervilla Adans. - bush-honeysuckle

The three species of bush-honeysuckle, natives of southeastern and south central Canada and the eastern U.S., are hardy, thicket-forming plants that tolerate a wide range of soil conditions. They are valuable for controlling erosion on steep slopes.

Recognition

Habit: shrubs; deciduous

Leaves: opposite; simple; toothed (serrate); stipules wanting

Flowers: clusters 3- to many-flowered; calyx lobes 5, awl-shaped; corolla tube funnel-shaped, 2-lipped; stamens 5, styles exserted

Fruits: capsules dry, thin-walled, beaked, imperfectly splitting, crowned by shriveled calyx; seeds many, tiny

Buds: sessile, oblong; scales, several pairs, pointed

Diervilla lonicera Mill. - dwarf bush-honeysuckle

Colonies of dwarf bush-honeysuckle grow on rocky sites in Canada and the U.S. It is less vigorous than *D. sessilifolia* in the Minnesota Landscape Arboretum (Dirr 1990), and it is also being evaluated at PERC in Fort Collins (Fairchild and Klett 1993). This hardy, infrequently planted shrub deserves trials on other prairie sites.

Planning

Height: up to 4 ft (122 cm)

Habit: shrub; spreading by stolons; summer blooms inconspicuous

Site: shade, sun; prefers soils slightly acidic; drouth tolerant; zone 3

Use: naturalizing, erosion control

Recognition

Leaves: egg-shaped to oblong, up to 4 in (10 cm) long; margin with minute hairy fringe (ciliolate), tip long-pointed; petioles about 0.2 in (5 mm) long

Flowers: usually in 3s; corolla yellow, limb nearly as long as tube

Fruits: about 0.3 in (8 mm) long

Twigs: nearly round in cross-section, glabrous or nearly so

Diervilla sessilifolia Buckl. - southern bush-honeysuckle

Southern bush-honeysuckle is growing successfully at the UNL Agriculture Research Development Center near Ithaca. *D.* ×*splendens* (Carr.) Kirchn. (*D. lonicera* × *D. sessilifolia*) resembles this species but is reported to be hardier (Bailey Hortorium 1976). *D. rivularis* Gatt. resembles southern bush-honeysuckle except that the former's leaves are hairy on both surfaces and the limbs of its flowers are about as long as its tubes.

Planning

Height: up to 5 ft (150 cm)
Habit: shrub; spreading by stolons; summer blooms
Site: shade, sun; pH and soil adaptable; drouth tolerant; zone 5
Use: steep banks, erosion control, wildlife habitat

Recognition

Leaves: about 4 times as long as wide (ovate-lanceolate), up to 6 in (15 cm) long; tip long-pointed; glabrous except on midrib above; petioles short (sub-sessile)
Flowers: clusters of 3 to 7; corolla sulfur yellow, limb shorter than tube
Fruits: up to 0.5 in (12 mm) long
Twigs: 4-angled, glabrous or nearly so
Buds: oblong, appressed to stem, sometimes one above another (superposed)

Diervilla sessilifolia

Kolkwitzia Graebn. - beautybush

The single beautybush species is a native of China.

Kolkwitzia amabilis Graebn. - beautybush

Beautybush is a leggy, arching shrub that is handsome in bloom and trouble free except for the need of an occasional pruning. It is a reliable bloomer that is suitable for a large garden. Richley (1988) states that beautybush is a good background shrub for southwestern Nebraska.

Planning

Height: up to 12 ft (4 m)
Habit: shrub; upright, arching; May-June blooms on old wood
Site: sun, partial shade; soil and pH adaptable; drouth tolerant; zone 5
Use: specimen, border

Kolkwitzia amabilis

Recognition

Leaves: opposite; simple; egg-shaped, up to 3 in (8 cm) long; margin shallow-toothed to nearly entire with hairy fringe (ciliate), tip long-pointed, base rounded; slighly hairy above, shaggy-hairy on veins below; petioles about 0.1 in (2 mm) long, shaggy-hairy
Flowers: paired clusters (corymbs), hairy; sepals 5, narrow, shaggy-hairy, spreading, persistent on fruit; corolla bell-shaped, 5-lobed, about 0.6 in (15 mm) long, pink, yellow in throat; stamens 4 about as long as tube
Fruits: capsules dry, splitting, bristly, long persistent
Twigs: hairy, becoming glabrous, encircled by leaf scars
Buds: solitary, egg-shaped; scales, 3-4 pairs
Bark: brownish, becoming shreddy

Lonicera L. - honeysuckle

About 180 species of honeysuckle are natives throughout the northern hemisphere. They are typically tall, leggy shrubs bearing attractive flowers and fruits that are suitable for landscapes, windbreaks and wildlife habitat plantings. The goal of much of the selection and breeding of the honeysuckles is the development of cultivars of compact habit with pest resistance.

Many varieties and hybrids of honeysuckle have been introduced to the marketplace. Some lack adequate documentation as to parentage which causes considerable name confusion. One example is *Lonicera* ×'Freedom' introduced by the University of Minnesota. We mention only a few of the species, hybrids, and cultivars that can be grown in the prairies.

Honeysuckles are easily propagated and established without serious pests that will affect their performance. However, one of the most troublesome insects is the leaf-folding aphid, *Hyadaphis tataricae*, which attacks tender shoots and causes witches-broom, but breeding programs have developed resistant cultivars. Although honeysuckles adapt to a wide range of soil conditions, most species require full sunlight for good health. When they become overgrown or ragged, the best treatment is to prune them back to ground level to facilitate the development of new shoots.

Lonicera ×'Freedom', a hybrid of undocumented parentage, is a large shrub that bears white flowers followed by red fruits that are readily eaten by birds (Fairchild and Klett 1993). These authors state that some winter dieback occurs, but the plant does not show chlorosis, and blooming and fruit set are reliable.

Recognition

Habit: shrubs, rarely tree-like; deciduous, rarely half-evergreen or evergreen

Leaves: opposite; simple; distinct or fused (connate); margins entire, rarely lobed; short-petioled or sometimes sessile

Flowers: axillary, peduncled pairs or sessile whorls; calyx 5-toothed; corolla slender or short, 2-lipped or nearly equally 5-lobed; stamens 5

Fruits: berry-like; seeds few to many

Twigs: encircled or nearly encircled by leaf scars; pith solid and white or wanting

Grouping Key for *Lonicera*

Flowers and Twigs

Shrubs; flowers in pairs, leaves distinct; twig pith solid: ***alpigena, canadensis, coerulea, fragrantissima, involucrata, hispida, maximowiczii, syringantha***

Shrubs; flowers in pairs, leaves distinct; twig pith wanting (hollow): ***×bella, korolkowii, maackii, morrowii, tatarica, xylosteum***

Vines; flowers in pairs, leaves distinct; twig pith wanting (hollow): ***henryi, japonica***

Vines; flowers in whorls at branch ends, first leaf pair below flowers usually joined (connate); twig pith wanting (hollow): ***sempervirens, dioica***

Flower Stalks

Flower stalks (peduncles) longer than petioles: ***alpigena, ×bella, dioica, fragrantissima, involucrata, korolkowii, morrowii, sempervirens, tatarica, xylosteum***

Flower stalks about same length as petioles or shorter: ***maackii***

Lonicera alpigena L. - Alps honeysuckle

Alps honeysuckle, a native of the mountains of central and southern Europe, is most noted for its drooping, scarlet fruits. 'Nana', a slow-growing, dense shrub with glossy dark green foliage, growing at PERC in Fort Collins, is not reliably fruitful, but it is adapted to the climate and soils of the site (Fairchild and Klett 1993).

Planning

Height: up to 9 ft (3 m)
Habit: shrub; deciduous; erect, much-branched; spring bloom not showy; summer fruits ornamental
Site: sun; soil and pH adaptable but prefers moist, well-drained soil; zone 4
Use: specimen, border, foundation

Recognition

Leaves: distinct; elliptic to oblong, about 3 in (8 cm) long; apex pointed, base rounded to wedge-shaped; dark green, lustrous; petioles about about 0.5 in (12 mm) long
Flowers: pairs on upright, slender stalks about 1 in (25 mm) long; bracts about twice as long as ovaries; corolla 2-lipped, greenish yellow, tinged brown-red, tube swollen on one sie at base (gibbous), much shorter than limb; stamens about as long as limb, style slightly longer, anthers red
Fruits: cherry-like, red, up to 0.5 in (12 mm) long, often united at base
Twigs: slightly hairy or not hairy, glandular; pith white, solid

Lonicera coerulea L. - bearberry honeysuckle

Two varieties of bearberry honeysuckle, a variable Eurasian native, perform well in the northern part of our region. Var. *dependens* is a hardy shrub with a rounded form that is suitable for planting in North Dakota (Herman 1985); Hoag (1965) states that var. *edulis* is an attractive shrub that performs well in the northern plains.

A related species, *L. canadensis* Marsh. (fly honeysuckle), is a low-growing shrub bearing red fruits that is a native of southeastern Canada and the northeastern U.S. Its leaves are hairy-fringed (ciliate) and its awl-shaped bracts are shorter than the calyx.

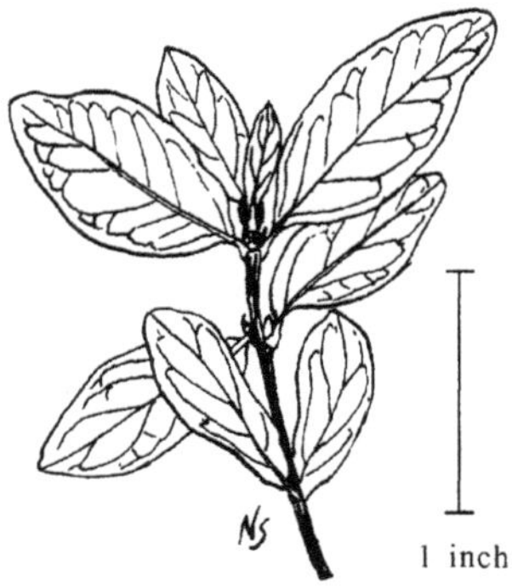

Lonicera coerulea

Planning

Height: up to 5 ft (150 cm)
Habit: shrub; deciduous; dense, spreading; May blossoms; summer fruits
Site: sun; pH and soil adaptable; zone 2
Use: border, foundation, wildlife habitat

Recognition

Leaves: distinct; egg-shaped, oval or oblong, up to 3 in (8 cm) long; tip pointed to rounded, base rounded; bright green, hairy to nearly glabrous on veins below
Flowers: pairs on short nodding stalks; bracts awl-shaped, exceeding the short, hair-fringed calyx; corolla yellowish white, tubular funnel-shaped, tube swollen on 1 side (gibbous), usually longer than the oblong lobes; stamens extending beyond lobes; style longer than stamens, glabrous
Fruits: pairs sub-globose to ellipsoid due to a fleshy cupula tightly enclosing ovaries, dark blue, bloomy
Twigs: glabrous or hairy, yellowish or red-brown; pith solid
Buds: spreading; scales, 2 outer
Bark: flaky

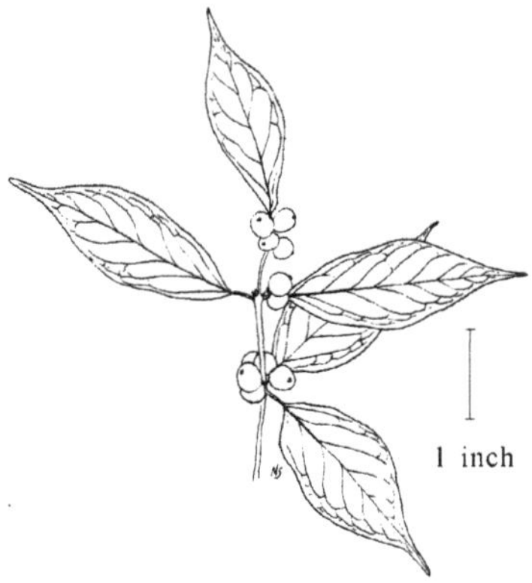

Lonicera canadensis

Caprifoliaceae *Lonicera*

Lonicera dioica L. - limber honeysuckle

Limber honeysuckle, a sprawling vine-like shrub, is native to much of the prairies, growing on wooded bluffs and rock ledges. Limber honeysuckle also grows in the eastern U.S. and across southern Canada. Var. *glaucescens* (Rydb.) Butters, with its hairy corolla and leaves hairy beneath, is common in Nebraska and northward (Stephens 1973) although Sutton (1998) has never seen it in the wild.

Although the flowers are not striking, limber honeysuckle foliage is handsome and the fruits are relished by birds. It is one of the few vine-like honeysuckles that is hardy throughout our region, but it is not widely available in the nursery trade. Significant genetic diversity is possible among plants of this species due to its widespread natural range. Desirable cultivars might be discovered through provenance research.

Limber honeysuckle is sometimes confused with a very similar species, *L. prolifera* (Kirchn.) Rehd., which grows in the same areas. Its connate leaves are bloomy above and nearly circular or with rounded ends (Stephens 1973).

Planning

Height: up to 10 ft (3 m)
Habit: vine-like shrub; deciduous; sprawling; May-June blossoms, summer fruits
Site: sun, partial shade; pH and soil adaptable; zone 2
Use: grouping, naturalizing, wildlife habitat

Recognition

Leaves: distinct, except those directly below the flowers are fused (connate); apex rounded, tipped with abrupt point (mucronate), base tapered to rounded; yellow-green above, white bloomy and sometimes hairy beneath; petioles sessile or semi-clasping, less than 0.1 in (2 mm) long (shorter than peduncles)
Flowers: clusters of 1-3 whorls terminal, each consisting of 2 clusters of 3 flowers subtended by a triangular bract less than 0.1 in (1.5 mm) long; calyx lobes 5, triangular, acute, green, less than 0.1 in (1 mm) long; corolla tubular, swollen at base (gibbous), 2-lipped, the upper 4-lobed, orange to purple, sometimes hairy inside and out; stamens 5, inserted at mouth, as long as lobes; style longer than stamens
Fruits: clustered, red
Twigs: slightly twining, flexible, glabrous; pith usually hollow, sometimes white and broken; leaf scars triangular, joining around the twig
Buds: pointed, often spreading
Bark: shreddy

Lonicera fragrantissima Lindl. & Paxt. - winter honeysuckle

Winter honeysuckle, unlike most honeysuckles, bears its fragrant flowers on the old wood. Birds devour the fruits of this native of eastern China soon after ripening. Prune, if necessary, after the fruits are gone.

L. pileata Oliv., a native of China, is another half-evergreen, low shrub that may be adapted to the prairies. It bears lustrous dark green foliage, white flowers and violet purple fruits.

Planning

Height: up to 10 ft (3 m)
Habit: shrub; half-evergreen; wide-spreading, irregular; fragrant March-April blossoms precede leaves; early June fruits; November leaf drop
Site: sun, partial shade; pH and soil adaptable; zone 5
Use: naturalizing, border, hedge, wildlife habitat

Recognition

Leaves: distinct; elliptic to broadly egg-shaped, up to 3 in (8 mm) long; apex rounded to pointed, margin fringed with bristly hairs (setose-ciliate); blue-green to dark green above, midrib bristly beneath; petioles 0.1 in (2 mm) long (shorter than peduncles)
Flowers: pairs axillary; corolla tubular, swollen at base, creamy white, glabrous, 2-lipped; ovaries more or less fused together (connate); quite fragrant
Fruits: red
Twigs: glabrous, bloomy when young; pith solid, white
Buds: terminal wanting; lateral with 2 outer scales

Lonicera hispida Pall. - hispid honeysuckle

Hispid honeysuckle—a native of Turkestan and western China—may prove to be a valuable shrub for planting in the western prairie because of its ability to survive on difficult sites. A specimen in the Hans Burchardt Arboretum near Branched Oak Lake in Lancaster County, Nebraska, has thrived for several years with minimal care (J. Morgenson 1992). Further test plantings should be rewarding.

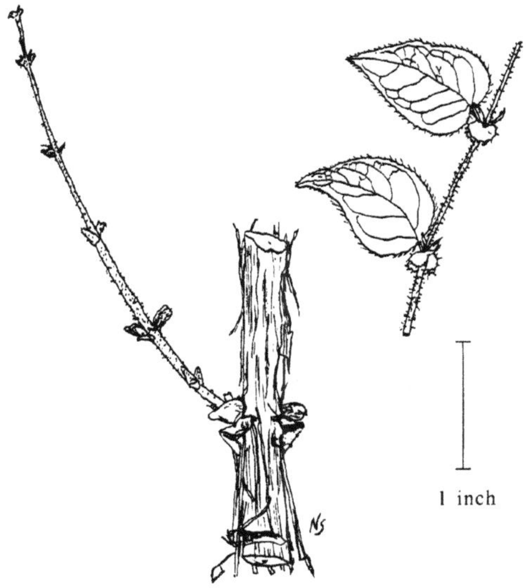

Lonicera hispida

Planning

Height: up to 5 ft (1.5 m)
Habit: shrub; deciduous; upright
Site: sun; pH and soil adaptable; drouth tolerant; zone 4
Use: border, wildlife habitat

Recognition

Leaves: distinct; elliptic to egg-shaped; apex pointed, base rounded, margin fringed with stiff hairs (setose-ciliate); dark green and sparsely hairy above, slight waxy bloom beneath, usually stiff hairs on veins (hispid)
Flowers: nodding, stalks hairy; bracts egg-shaped, up to 1 in (25 mm) long, enveloping the ovaries, bractlets wanting; corolla white or yellowish white, tubular to funnel-shaped, up to 1.2 in (3 cm) long, pouchy at the base, enlarged above the middle, shaggy hairy (pilose), limb much shorter than the tube; stamens nearly as long as limb; ovary oblong, style exserted
Fruits: red, oblong, up to 0.6 in (15 mm) long
Twigs: sparse stiff hairs (hispid); pith solid, white
Buds: terminal wanting; 2 outer scales pointed
Bark: shreddy

Lonicera involucrata (Richards.) Banks ex Spreng. - bearberry honeysuckle

Bearberry honeysuckle, a native of the north country—Canada and Alaska—and the Rocky Mountains, grows in moist, open woodlands. A small, healthy specimen of this species is growing in Lincoln under partial shade with the soil heavily mulched (Sutton 1998).

Var. *flavescens* (Dipp.) Rehd. is a native of Oregon, Wyoming and Utah. Its corolla is swollen on the side instead of at the base. Another less showy native of the west, *L. utahensis* S. Wats., is a small, fine-textured shrub that bears red fruits. Both of these plants deserve further trials, especially in the northern prairie.

Planning

Height: up to 3 ft (1 m)
Habit: shrub; deciduous; spreading; June blossoms; showy fruits
Site: sun, partial shade; pH 6.0-7.0; moist, well-drained soil; zone 3
Use: naturalizing, wildlife habitat

Recognition

Leaves: distinct; elliptic to oblong, sometimes wider above middle, up to 4.5 in (11 cm) long; apex long-pointed, base tapered; bright green, usually glabrous; petioles up to 0.4 in (1 cm) long (shorter than peduncles)
Flowers: pairs on upright stalks about 1 in (25 mm) long; bracts egg-shaped, pointed, glandular-hairy, reaching to middle of corolla; bractlets exceeding the ovaries; corolla tubular, swollen at base (saccate), lobes 5, egg-shaped, upright, yellow, slight red tinge, glandular-hairy outside; stamens as long as limb
Fruits: shiny, purple-black, conspicuous, surrounded by reflexed bractlets and bracts
Twigs: glabrous; pith white, solid

Lonicera involucrata

Lonicera japonica Thunb. - Japanese honeysuckle

Japanese honeysuckle, a native of Asia, has escaped from cultivation in the U.S. to become an undesirable weed, especially in the south. Stephens (1973) reports that wild plants grow in central Nebraska and eastern Kansas. However, if Japanese honeysuckle is planted where it can be controlled, it is an attractive, fast-growing twining shrub that provides excellent wildlife habitat and erosion control.

Several cultivars provide interesting foliage and flowers (Dirr 1990). 'Halliana' is a vigorous vine with pure white flowers that become yellow. Interesting foliage is provided by 'Aureo-reticulata' with its yellow-netted leaves, and var. *purpurea* is a vigorous vine with purple-tinted, dark green foliage.

L. henryi Hemsl. is a twining, half-evergreen shrub from western China that may be successfully grown on protected sites in the central prairie. It is distinguished from Japanese honeysuckle by its twigs (densely covered with stiff hairs) and its awl-shaped bracts on the flower clusters.

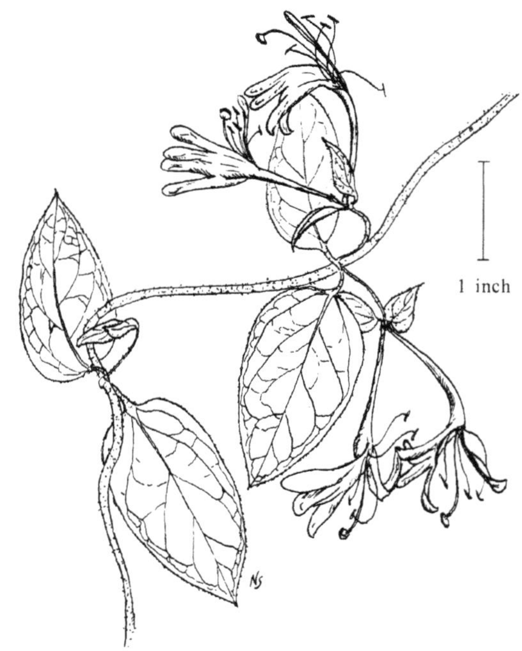

Lonicera japonica

Planning

Height: up to 30 ft (9 m)
Habit: shrub (vine); deciduous to evergreen; twining, dense, aggressive; June-September blossoms fragrant
Site: sun; pH and soil adaptable; zone 5
Use: groundcover, erosion control, wildlife habitat

Recognition

Leaves: distinct; egg-shaped, up to 3 in (8 cm) long; apex pointed, base tapering or ending abruptly (truncate); dark green and variably hairy above becoming nearly glabrous, variably hairy beneath, especially on lower midrib; petioles hairy, about 0.3 in (8 mm) long, usually shorter than flower stalks (peduncles)
Flowers: pairs axillary; bracts 2, leaf-like, near peduncle summit; bractlets 2, 0.05 in (1 mm) long; calyx 5-lobed, ciliate; corolla white, cream or pinkish, becoming yellow before falling, tube hairy on both sides, about as long as the 2 lips; one lip narrow and oblong, the other 4-lobed; style and stamens longer than limb
Fruits: black, subtended by persistent bracts and bractlets
Twigs: twining, green, flexible, hairy; pith white, continuous or hollow
Buds: 0.05 in (1 mm) long, red-brown, tips spreading
Bark: yellow-brown, long papery shreds

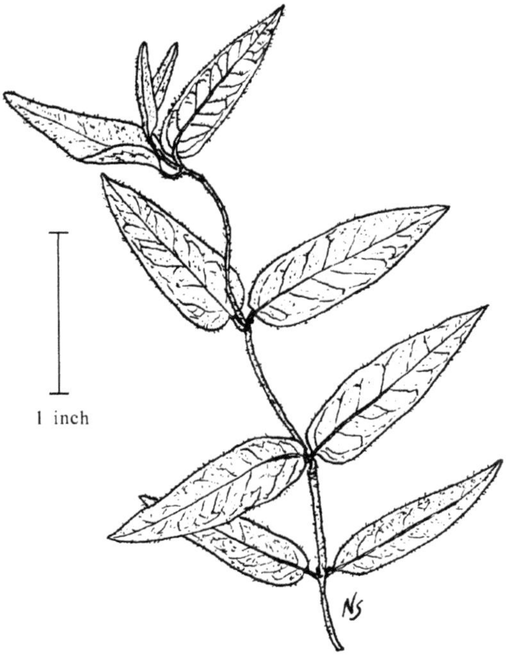

Lonicera henryi

Lonicera korolkowii Stapf - blueleaf honeysuckle

A gray-blue shrub bearing abundant flowers and fruit describes blueleaf honeysuckle, a native of central Asia. Richley (1988) reports it to be relatively free of aphids and well adapted to southwestern Nebraska. Blueleaf honeysuckle is fruitful and adapted to the soils and climate of the Fort Collins area (Fairchild and Klett 1993). A young specimen of this species growing at Prairie Pines Arboretum near Lincoln is exceptionally healthy.

Var. *zabelii* (Rehd.) Rehd. bears rose-pink flowers, is hardy to -35 degrees F (-37 degrees C), and is little affected by chlorosis (North Central Plant Introduction Station 1970). 'Aurora' bears profuse purplish pink flowers; 'Floribunda' produces abundant flowers; and 'Freedom', of uncertain parentage, introduced by the University of Minnesota, bears white flowers and red fruits and is resistant to the Russian aphid (Dirr 1990). Fairchild and Klett (1993) state that 'Freedom' is adapted to the soils and climate of the Fort Collins area.

Hoag (1965) states that a plant advertised as Zabel's honeysuckle does not always bear bluish foliage and flowers of the same form as those of *L. korolkowii*, and therefore it is apparently not var. *zabelli*. The cultivar 'Zabelii' is thought to be a selection of *L. tatarica* (Dirr 1990).

Caprifoliaceae *Lonicera*

Planning

Height: up to 10 ft (3 m)
Habit: shrub; deciduous; spreading, arching; blue-gray foliage; late May blossoms, summer fruits
Site: sun, partial shade; pH and soil adaptable; drouth tolerant; zone 4
Use: border, screen, wildlife habitat, windbreak, naturalizing

Recognition

Leaves: distinct; egg-shaped to elliptic, up to 1.5 in (4 cm) long; apex round to short-pointed, base tapered to rounded; surfaces more or less hairy, bluish green; petioles hairy, up to 0.2 in (6 mm) long (shorter than peduncles)
Flowers: pairs on hairy stalks up to 1 in (25 mm) long; bracts about as long as ovaries; bractlets about one-third as long as ovaries, fused at the base (connate); corolla rose-colored, rarely white, tubular, slightly swollen at base (gibbous), 2-lobed, lateral lobes of upper lip divided to the middle; stamens and style shorter than limb; ovaries distinct
Fruits: bright red
Twigs: hairy; pith white to brown, soon disappearing

Lonicera korolkowii

Lonicera maackii Maxim. - Amur honeysuckle

Amur honeysuckle, a native of Manchuria and Korea, is a pestiferous weed in some localities (a good indicator of its adaptability). However, it is not thorny and is easily eradicated, especially while small. Perhaps it is a more desirable occupant of waste areas than giant ragweed and marijuana plants.

Although wind tears its leaves, causing them to be unsightly, Amur honeysuckle is adapted to the northern plains (Hoag 1965). Dirr (1990) states that Amur honeysuckle shows amazing shade tolerance, but shaded plants were short-lived in Prairie Pines Arboretum (Bagley 1996). 'Cling Red', an introduction by the NRCS Elsberry Plant Materials Center, matures fruits in mid-September that persist into winter until taken by birds. This cultivar is adapted to most of the central Great Plains and much of the eastern U. S.

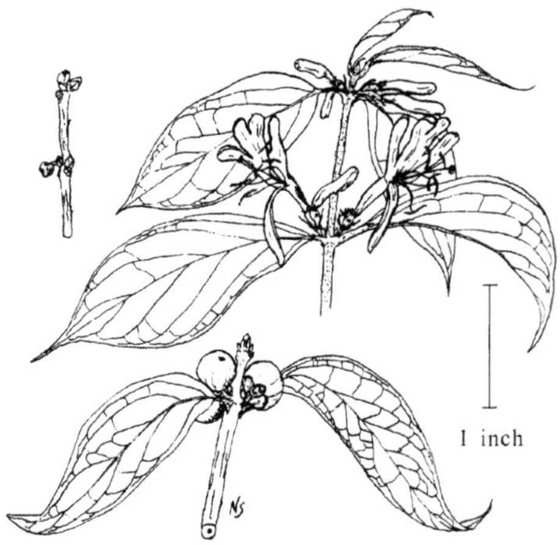

Lonicera maackii

Planning

Height: up to 15 ft (4.5 m)
Habit: shrub; deciduous; spreading, leggy; late May blossoms; autumn fruits
Site: sun, partial shade; pH and soil adaptable; zone 2
Use: windbreak, screen, wildlife habitat
Caution: plants may appear in places where not wanted

Recognition

Leaves: distinct; egg-shaped to lance-shaped, up to 3 in (8 cm) long; apex long-pointed, base tapering to rounded; dark green and usually glabrous above, hairy beneath, especially on veins; petioles hairy, about as long as fruiting stalks

Flowers: pairs, peduncles hairy; bracts linear, longer than ovaries; bractlets fused into pairs, up to as long as ovaries; calyx divided to the middle; corolla white, turning yellow, tubular, thin, not gibbous, 2-lipped, limb longer than tube and divided to about the middle; stamens and style shorter than limb

Fruits: dark red

Twigs: hairy, becoming glabrous second year; pith wanting

Lonicera maximowiczii Reg. var. *sachalinensis* Fr. Schmidt - Sachalin honeysuckle

Sachalin honeysuckle, a northeast Asia native, is adapted to the northern Great Plains (Hoag 1965, Herman 1985). A young plant in Prairie Pines Arboretum is healthy but much slower growing than a nearby specimen of *L. korolkowii* of the same age (Bagley 1996). Sachalin honeysuckle is a handsome plant with foliage superior to many of the other honeysuckle species (Dirr 1990).

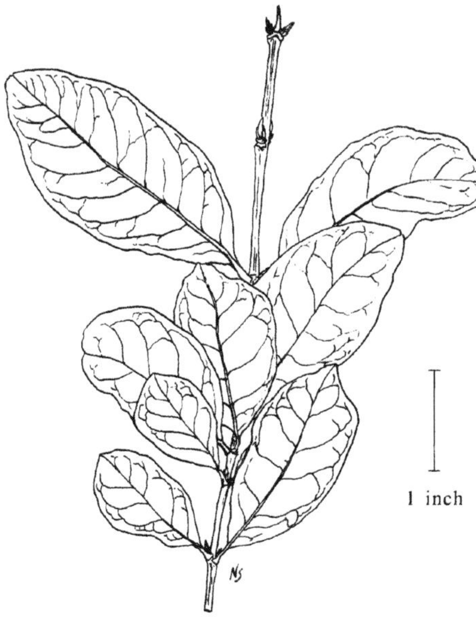

Lonicera maximowiczii* var. *sachalinensis

Planning

Height: up to 6 ft (2 m)

Habit: shrub; deciduous; rounded, dense; May blooms

Site: sun; pH and soil adaptable; zone 3

Use: border, screen, wildlife habitat

Recognition

Leaves: distinct; elliptic, up to 2.7 in (7 cm) long; apex obtuse, base tapered to rounded; reddish becoming bright green, whitish bloom on underside; petioles bear a rudimentary wing that forms a groove above, about 0.2 in (5 mm) long, shorter than peduncles

Flowers: pairs; bracts awl-shaped, about 0.3 times as long as ovary; bractlets rounded, shorter than bracts, hairy fringed; corolla purple to red, tube with swollen base (gibbous), 2-lipped, limb longer than tube; stamens slightly longer than limb, style as long as limb

Fruits: red

Twigs: glabrous, shiny brown, encircled by leaf scars; pith solid, white

Buds: divergent, narrow, pointed; scales glabrous, overlapping, 4-5 pairs

Lonicera morrowii A. Gray - Morrow honeysuckle

Morrow honeysuckle is one of the better honeysuckles for planting in the northern plains (Hoag 1965). Though seldom seen in cultivation, Dirr (1990) states it is "better than most honeysuckles as the foliage hugs the ground." Morrow honeysuckle, a native of Japan, is very similar to *L. tatarica* L., differing mainly in its form and flower color.

Belle honeysuckle (*L.* ×*bella* Zab.), a cross between Morrow and Tatarian honeysuckles, is a useful plant, but it is not outstanding enough to supplant either parent (Hoag 1965). Dirr (1990) states that Belle honeysuckle does not rival *L. tatarica* for abundance and showiness of fruit. Several cultivars have been selected that differ mainly in flower color. The red fruits ripen in July and persist into autumn.

Caprifoliaceae *Lonicera*

Planning

Height: up to 10 ft (3 m)
Habit: shrub; deciduous; broad, dense; May-June blossoms, summer fruits
Site: sun; pH and soil adaptable; drouth tolerant; zone 3
Use: border, screen, wildlife habitat, windbreak

Lonicera morrowii

Recognition

Leaves: distinct; elliptic to oblong, up to 2 in (5 cm) long; apex tipped with minute point, base rounded; somewhat hairy above when young, hairy beneath; petioles about 0.1 in (2 mm) long, shorter than peduncles
Flowers: pairs; bracts hairy, usually extending beyond calyx; bractlets hairy about as long as ovary; corolla white, turning yellow, tube slender with swollen base (gibbous), hairy outside, 2-lipped, upper lip divided to base; stamens shorter than limb
Fruits: dark red, rarely yellow
Twigs: soft-hairy; pith brown, dissolving

Lonicera sempervirens L. - trumpet honeysuckle

Trumpet honeysuckle, presumably a native of the eastern U.S. west to Nebraska and Texas, has escaped cultivation and become naturalized in so many areas that its exact original natural range is unknown. Hoag (1965) states that its blooms are not reliable in the northern plains. Several cultivars bearing flowers of different colors are available.

A hybrid, 'Dropmore Scarlet Trumpet', released by Dr. F. L. Skinner at Dropmore, Manitoba, produces sterile, rosy orange flowers throughout the summer and is sufficiently hardy to be highly recommended for the northern plains (Hoag 1965). It is a cross between trumpet honeysuckle and *L. hirsuta* Eaton.

'Goldflame' and 'Summer King' are selections of everblooming honeysuckle (*L.* ×*heckrottii* Rehd.). Trumpet honeysuckle is probably one of the parents of this handsome vine of dubious origin. Everblooming honeysuckle comes by that name because it blooms throughout the summer. Its carmine flower buds change to pink and expose yellow inside the corolla as they open.

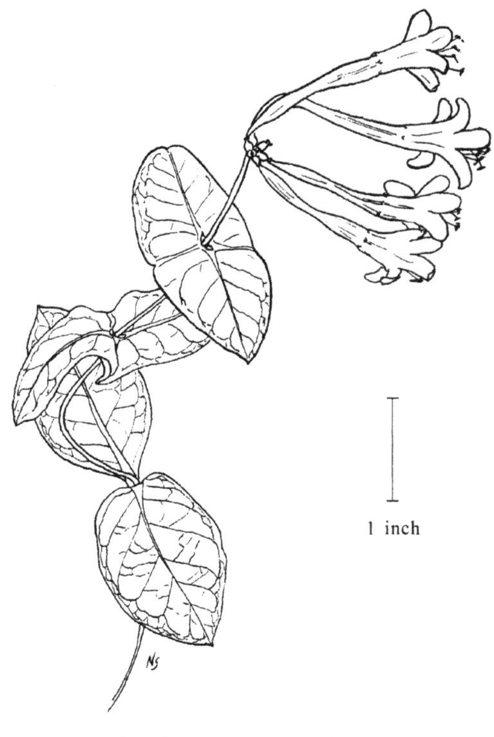

Lonicera sempervirens

Planning

Height: up to 15 ft (5 m)
Habit: shrub; deciduous, evergreen; vine-like, twining, dense; late May-June blossoms, late summer-fall fruits
Site: partial shade; pH and soil adaptable; zone 4
Use: groundcover, trellis, fences, walls, wildlife habitat

Recognition

Leaves: distinct, except 1-2 pairs fused directly below the flowers (connate); oval to broadly elliptic; apex pointed, base tapered; yellow-green above, whitish bloom beneath (glaucous), thick and leathery; petioles about 0.3 in (8 mm) long, flattened above, slightly clasping
Flowers: clusters 3-6 on new growth, terminal; hypanthium with 5 short, rounded lobes; corolla red, orange, or purple tinged, yellow inside, tube sometimes swollen on one side, often abruptly widened above, lobes 5, nearly equal with rounded tips; stamens 5, yellow filaments attached to tube on lower half and slightly exserted; style exserted
Fruits: bright red with minute, black tip, glabrous; seeds 1
Twigs: yellow-brown, becoming shredded second year; pith white, dissolving
Buds: narrow egg-shaped; scales pointed, keeled and spreading at tips
Bark: shredding into long fibrous strips

Lonicera syringantha Maxim. - lilac honeysuckle

Lilac honeysuckle, an introduction from northwestern China, has fared well in low-maintenance trials at UNARDC near Ithaca and at the Hans Burchardt Arboretum near Branched Oak Lake northwest of Lincoln.

Planning

Height: up to 10 ft (3 m)
Habit: shrub; deciduous; upright; flowers fragrant
Site: sun; pH and soil adaptable; zone 5
Use: border, screen, wildlife habitat

Recognition

Leaves: distinct; often in whorls of 3, elliptic to oblong up to 1 in (25 mm) long; apex variable; surfaces glabrous
Flowers: pairs axillary; bracts linear oblong, longer than calyx; calyx teeth narrow, pointed, often longer than ovary; corolla bell-shaped about 0.6 in (15 mm) long, glabrous, pinkish white to rosy lilac, limb nearly regular, 5-lobed, lobes 0.3 as long as tube; ovaries distinct
Fruits: red-orange, glossy
Twigs: slender, glabrous; pith white, solid

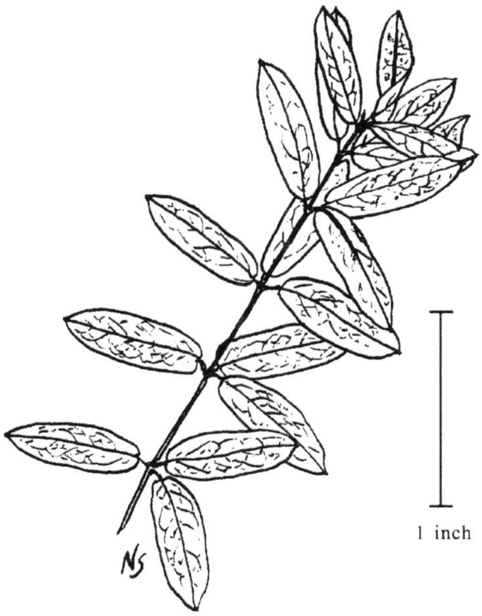

Lonicera syringantha

Caprifoliaceae *Lonicera*

Lonicera tatarica L. - Tatarian honeysuckle

Tatarian honeysuckle was planted in many of the shelterbelt shrub rows in the northern portion of the Prairie States Forestry Project. Read (1958) noted that the effectiveness of this Asian shrub as a windbreak diminished with age due to its tendency to become leggy. However, Tatarian honeysuckle continues to be a component of many windbreak and wildlife habitat plantings in the northern prairie.

The highly variable Tatarian honeysuckle with flowers ranging from white to rose and fruits varying from golden orange to deep red has yielded a bonanza of cultivars. Among them are the white-flowered 'Parvifolia'; the red-flowered 'Arnold Red'; the orange-fruited 'Morden Orange'; and the dwarf 'Nana' (Dirr 1990).

Hoag (1965) reports that 'Cascade' of Canadian origin is outstanding for its profuse production of deep red fruits on young bushes. Pink-flowered 'Sibirica' is adapted to the Fort Collins area where it consistently produces flowers and fruit (Fairchild and Klett 1993).

Many plantings of Tatarian honeysuckle have been rather short-lived for a variety of reasons. Selection of cultivars with emphasis on vigor, adaptation to specific sites, and insect and disease resistance should be rewarding.

Lonicera tatarica

Planning

Height: up to 10 ft (3 m)
Habit: shrub; deciduous: upright, multistemmed, becoming leggy; May blossoms, summer fruits
Site: sun; pH and soil adaptable; drouth tolerant; zone 3
Use: border, screen, windbreak, wildlife habitat, naturalizing

Recognition

Leaves: distinct; egg-shaped to oblong up to 2 in (5 cm) long; apex rounded to pointed, base tapered to rounded, margin sparsely hairy-fringed; upper surface yellow-green, glabrous, sparse hairs beneath; petioles about 0.2 in (5 mm) long, broadly grooved above, shorter than peduncles
Flowers: pairs axillary; hypanthium 5-lobed, glabrous; bracts 2, spreading, linear, about as long as calyx; bractlets distinct, roundish about 0.4 times as long as ovary; corolla pink to white, tube gradually widening upward (funnelform), hairy inside, swollen at base (gibbous), 2-lipped, upper lobe deeply divided into 3-4 spreading segments, longer than tube; stamens 5, exserted, filaments adhering to tube on lower half, hairy toward the base
Fruits: single or in pairs with bases fused, orange-red; seeds 3-6
Twigs: glabrous; pith brown, disappearing
Buds: broadly egg-shaped, often one above the other (superposed), glabrous; lower scales valvate
Bark: gray-brown with flat, thin scales

Lonicera xylosteum L. - European fly honeysuckle

European fly honeysuckle, though not spectular in flower, is successfully planted in the northern plains (Hoag 1965). 'Emerald Mound' or 'Nana' is a low-growing, mounded honeysuckle with bluish green foliage that Dirr (1990) extolled as the most handsome of honeysuckles after viewing a mass planting at the Minnesota Landscape Arboretum. 'Miniglobe' from the Morden Research Station resembles 'Emerald Mound' but is hardier and more compact, according to Dirr.

L. ×*xylosteoides* Tausch (*L. tatarica* × *L. xylosteum*) 'Miniglobe' is a slow-growing small shrub with a dense rounded form adapted to the soils and climate of the Fort Collins area (Fairchild and Klett 1993). 'Claveyi', sometimes marketed as 'Clavey's Dwarf', is adapted to the northern plains (Hoag 1965). The true identity of this release by Clavey Nurseries is somewhat clouded. Hoag does not identify the origin of 'Clavey's Dwarf', but Dirr (1990) infers that it is a selection of *L. xylosteum*. Hortus III (Bailey Hortorium 1976) lists it as a cultivar of *L.* ×*xylosteoides*. Another release of Clavey Nurseries, 'Hedge King', is a narrow, upright plant suitable for hedges (Dirr 1990) but is difficult to establish, according to Herman (1985).

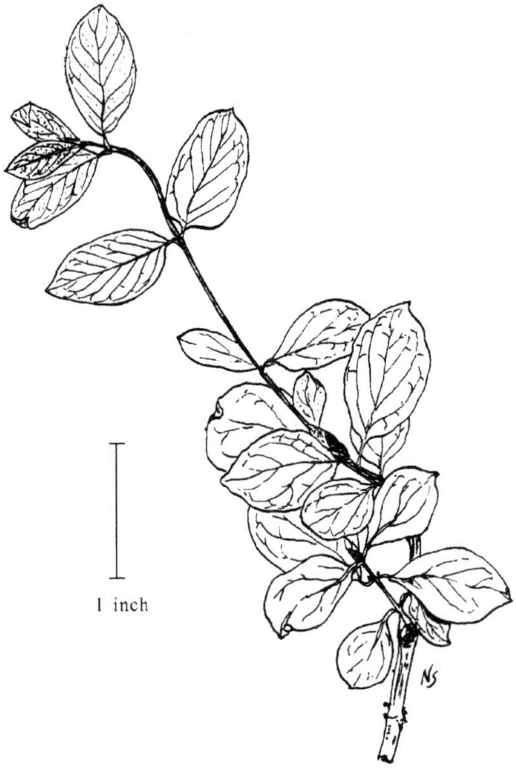

Lonicera xylosteum

Planning

Height: up to 10 ft (3 m)
Habit: shrub; deciduous; wide arching; May blossoms, summer fruits
Site: sun; pH and soil adaptable; zone 3
Use: grouping, border, specimen, foundation, wildlife habitat

Recognition

Leaves: distinct; egg-shaped, sometimes wider above middle, up to 2.5 in (6 cm) long; apex pointed, base tapered to rounded; dark to gray-green and somewhat hairy above, hairy beneath; petioles up to 0.3 in (8 mm) long, shorter than peduncles

Flowers: pairs axillary on hairy peduncles; bracts awl-shaped, shorter than ovary; bractlets broad, about half as long as ovary, hairy and glandular; corolla white to yellow-white, often with reddish tinge, becoming yellow, tube hairy outside, swollen at base (gibbous), 2-lipped; stamens nearly as long as limb, filaments hairy toward base, style shorter than stamens

Fruits: dark red

Twigs: hairy, sometimes glabrous; pith brown, disappearing

Buds: elongated, pointed; scales hairy, margins long-hairy (ciliate)

Sambucus L. - elder

Seven of the 30 species of elder, sometimes called elderberry, are natives of North America. The others, mostly temperate plants, are found in Europe amd Asia. Although short-lived, the elders reproduce readily from roots and seed and are fast growing in fertile, moist soil. They provide valuable food for wildlife but are seldom used in landscape plantings because of their coarse, rather unkempt appearance, even with constant attention. Elder fruits and flowers are edible and have been used for food and drink, especially by Native Americans and pioneers. Stems have provided the makings for musical instruments and toys.

Recognition

Habit: shrubs, small trees, rarely herbs; deciduous

Leaves: opposite; compound (odd-pinnate); with or without stipules; leaflet margins toothed (serrate); with or without basal appendages (stipels)

Flowers: clusters (corymbs or panicles) compound, many-flowered, terminal; perianth usually 5-parted; calyx limb minute; corolla wheel-shaped (rotate), lobes egg-shaped to oblong; stamens with short filaments; ovary 3- to 5-celled, style short, 3- to 5-lobed

Fruits: berry-like drupes; seeds 3-5

Twigs: stout; pith large

Buds: scales, several pairs

Sambucus coerulea Raf. - blue elder

Blue elder, a native of western North America, grows from southern British Columbia and Montana to Mexico and southwestern Texas where it is found along streams, fence rows and rocky outcrops. Its adaptation to a wide latitudinal range of varied site conditions undoubtedly marks a quite variable species. Blue elder (var. *neo-mexicana*) attains tree stature more commonly in the southern part of its range.

The natural range of blackbead elder, *S. melanocarpa* Gray, is very similar to that of blue elder. It ranges in form and size from a compact shrub to a small tree in Oregon where the national champion is 42 ft (12.7 m) tall. The pith of two-year-old twigs becomes brown.

Planning

Height: up 40 ft (12 m); National champion, 1979, 40 ft (12 m), California

Habit: shrub, tree; upright; summer blossoms, autumn fruits

Site: sun, partial shade; pH and soil adaptable; tolerates dry soil; zone 5

Use: naturalizing, wildlife habitat, fruit

Recognition

Leaves: leaflets 5-7, sometimes 9, oblong to lance-shaped, broadest at middle, up to 6 in (15 cm) long; coarsely toothed, lower sometimes 3-parted; bright green, hairy becoming glabrous

Flowers: clusters flat-topped, up to 6 in (15 cm) across; corolla yellowish white

Fruits: about 0.3 in (8 mm) across, blue-black whitened by waxy covering

Twigs: stout, green, bloomy, sometimes hairy when young; pith thick, white

Buds: tapering to a point; scales 4-6, chestnut brown

Sambucus canadensis L. - American elder

American elder—a native of eastern North America from Nova Scotia to Florida, west to eastern Nebraska and Texas—occupies fertile, moist soils along streams, woodland margins and fence rows. Although the plants are short-lived and seldom reach tree stature in the prairies, they are perpetuated by new shoots from the roots.

Cultivars include 'Acutiloba', with deeply divided leaflets; 'Maxima', with large leaves and flower clusters; and 'Rubra', with red fruits (Dirr 1990).

Planning

Height: up to 7 ft (2 m), National champion, 1987, 16 ft (4.9 m), Virginia
Habit: shrub, tree; upright, suckering; late spring bloom
Site: sun, partial shade; pH and soil adaptable; tolerates dry soil; zone 3
Use: naturalizing, wildlife habitat, edible fruit

Sambucus canadensis

Recognition

Leaves: leaflets 5-7, elliptic to egg-shaped, up to 4.7 in (12 cm) long; sharply toothed, tip long-pointed, base slightly rounded, uneven, lower leaflets occasionally 3-parted; dark green, shiny, glabrous above, slightly hairy beneath, especially on midrib; petioles, leaflet stalks (petiolules), and rachis grooved above; stiples linear at base of leaflet stalks
Flowers: clusters flat-topped, up to 7 in (18 cm) across; calyx lobes 5, white, triangular; corolla 5-lobed, creamy white; stamens 5, attached at corolla base between petals; style obsolete, stigmas 3, white
Fruits: about 0.2 in (6 mm) across, deep purple, smooth, glossy
Twigs: stout, glabrous; pith white, three-fourths of stem; leaf scars shield-shaped, bundle scars 5
Buds: egg-shaped, glabrous
Bark: yellow-brown, not fissured

Sambucus racemosa L. ssp. *pubens* (Michx.) Koehne - scarlet elder

Scarlet elder has been described as a North American species (*S. pubens* Michx.) but is now considered a subspecies of European red elder. It grows in Montana and South Dakota, southeastward to Tennessee, north to New Foundland, and west to Alberta, Canada. Var. *microbotrys* (Rydb.) Kearney & Peebles—a native of Colorado and westward—is very similar to scarlet elder except that its leaves are glabrous and its buds are brown. The leaflets of 'Dissecta' are deeply divided (Dirr 1990).

Planning

Height: up to 25 ft (7.5 m); National champion, 1989, 27 ft (8.2 m), Michigan
Habit: shrub; upright, non-suckering; late spring blossoms, late summer fruits
Site: sun, partial shade; zone 3
Use: wildlife habitat, naturalizing

Recognition

Leaves: leaflets 5-7, lance-shaped, oblong or elliptic, up to 4 in (10 cm) long; sharply toothed, tip long-pointed; yellow-green with sparse hairs above, hairy below
Flowers: clusters pyramidal, about 1.5 in (4 cm) across and about as high; calyx lobes triangular, reflexed at anthesis, greenish white; corolla yellowish white, lobes oblong, reflexed; stamens inserted on corolla base, alternate with petals; ovary green, style obsolete, stigmas 3, whitish
Fruit: about 0.2 in (6 mm) across, bright red, semi-glossy, old flower parts persistent; seeds 2-4
Twigs: short-hairy becoming glabrous; pith yellow-brown to brown
Buds: egg-shaped or wider above middle, short-stalked, nearly glabrous; lower scales brownish purple, deciduous, inner scales purple with white hairy margins (ciliate)

Symphoricarpos Dunham. - snowberry

The snowberries, with the exception of one Asian species, inhabit the temperate regions of North America south into Mexico. Most of the twelve to fifteen species are natives of the West. Many of the snowberries are so similar that it is difficult to differentiate betwen them.

Snowberries are grown for their showy, persistent fruits that provide winter food for birds. These hardy, trouble-free shrubs tolerate drouth and thrive on a wide range of soils but suffer from foliar diseases under high humidity. Snowberries are important forage plants for livestock, deer and elk in the Intermountain Region of the West (U. S. Forest Service 1937).

Recognition

Habit: shrubs, upright to prostrate; deciduous
Leaves: opposite; simple; without stipules
Flowers: clusters axillary or terminal; calyx 4- to 5-toothed; corolla bell-shaped to tubular, 4- to 5-lobed, white or pink; stamens 4-5, inserted on tube, included or slightly longer than tube
Fruits: berry-like drupes, persistent; nutlets 2
Twigs: often hollow
Bark: becoming shreddy

Symphoricarpos albus var. *albus* (L.) Blake - common snowberry

Common snowberry, the most widely distributed species of the genus, is a native shrub of the northern Great Plains. It also grows southeast to North Carolina, north to Nova Scotia, and northwest to Oregon and Alaska. Saponin, a poisonous drug, is found in the leaves of common snowberry, but not in sufficient quantity to be injurious to browsing animals (U. S. Forest Service 1937).

Variety *laevigatus* (Fern.) Blake, a shrub larger than the species, bears fruits up to 0.6 in (15 mm) long (Stephens 1973). It becomes unsightly from leaf diseases in areas of high humidity.

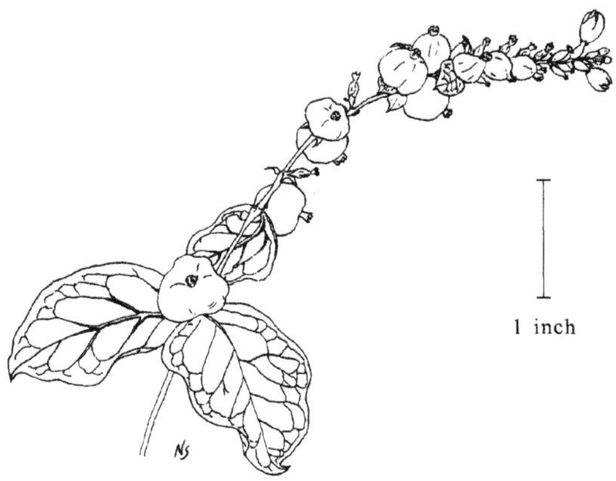

Symphoricarpos albus* var. *albus

Planning

Height: up to 2 ft (60 cm)
Habit: shrub; upright, not strongly rhizomatous; inconspicuous June blossoms; August fruits persist into winter
Site: sun, shade; pH and soil adaptable; zone 3
Use: naturalizing, erosion control, wildlife habitat

Recognition

Leaves: broad-ovate to elliptic, up to 2 in (5 cm) long; margin entire, sometimes hairy (ciliate), occasionally toothed or lobed, tip pointed; scattered hairs above becoming glabrous, hairy beneath; petioles about 0.1 in (3 mm) long, hairy, flattened above, margins joining opposing petioles
Flowers: clusters (racemes) of 2-5, terminal or in axils of upper leaves; 2 bracts below ovary, margins hairy; calyx lobes 5, triangular; corolla bell-shaped, pink or white, long-hairy inside, lobes 5, erect or slightly spreading; stamens 5, not extending beyond the corolla
Fruits: egg-shaped, up to 0.4 in (1 cm) long, white, smooth; calyx lobes persistent
Twigs: yellow-brown; fine, curled hairs on surface; pith brown, often disappearing
Buds: egg-shaped, about 0.1 in (2 mm) long, pointed; outer scales slightly keeled, curly-hairy

Symphoricarpos occidentalis Hook. - **western snowberry**

Western snowberry grows on dry sites in prairies and open woods from British Columbia to Ontario and southwest to New Mexico. When not in fruit it is easily confused with the red-fruited *S. orbiculatus* (described below).

Planning

Height: up to 3 ft (1 m)
Habit: shrub; strongly rhizomatous; inconspicuous summer blossoms; autumn fruits
Site: sun, partial shade; soil and pH adaptable; drouth tolerant; zone 2
Use: erosion control, wildlife habitat, naturalizing

Recognition

Leaves: broad-ovate to elliptic, up to 2.7 in (7 cm) long; apex pointed or obtuse, margin entire or with coarse, rounded teeth, variably hairy (ciliate); surfaces somewhat hairy becoming glabrous; petioles about 0.1 in (3 mm) long, variably hairy, flattened above
Flowers: clusters (spikes or racemes) of 10-20, terminal or axillary; flowers sessile or nearly so; 2 bracts at base of each ovary; calyx lobes 5, deltoid, green with hairy margins; corolla bell-shaped, white or pink, glabrous outside, long-hairy inside, lobes 5, egg-shaped; stamens 5, inserted at top of corolla tube, extending beyond the tube; style 0.2 in (6 mm) long, white, with long white hairs at the middle
Fruits: globose, about 0.3 in (8 mm) across, white, occasional brownish dots; calyx lobes persistent
Twigs: hairy, becoming glabrous; pith brown, often disappearing; leaf scars half-round, bundles not distinct
Buds: egg-shaped, about 0.1 in (2 mm) long; scales slightly hairy, brown, tips often spreading
Bark: peeling into rigid flakes

Symphoricarpos orbiculatus Moench - **coralberry**

Coralberry, a native of the eastern U.S. including most states east of Minnesota, southeastern Nebraska and eastern Texas, is often dubbed buckbrush, but coralberry is preferred because of its fruit color. Its ability to form large colonies aids its ability to become a troublesome pasture weed even though tender young shoots are broused by domestic livestock, deer and elk. The persistent coralberry fruits provide winter food for birds, especially bob-white quail. Plains Indians made a coralberry leaf potion for treating inflamed eyes (Gilmore 1977).

'Hancock', a selection of *S.* ×*chenaultii* Rehd. (*S. microphyllus* Kunth × *S. orbiculatus*), is a low-growing shrub with abundant pinkish flowers and pink fruits that tend to be white on their shady side. 'Mother of Pearl' and 'Magic Berry', cultivars of *S.* ×*doorenbosii* (*S. albus* var. *laevigatus* × *S.* ×*chenaultii*), bear white-pink cheeked and rose-pink fruits, respectively (Dirr 1990).

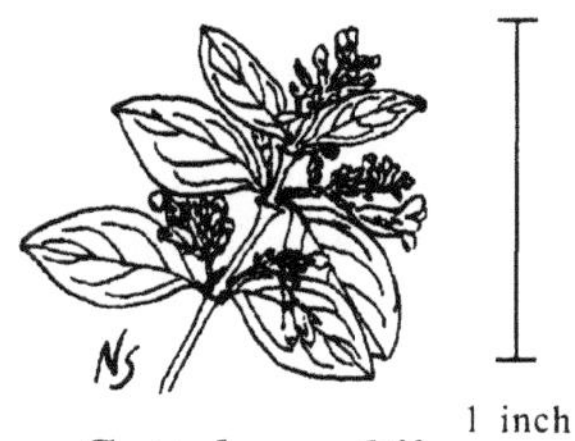

S.* ×*chenaultii

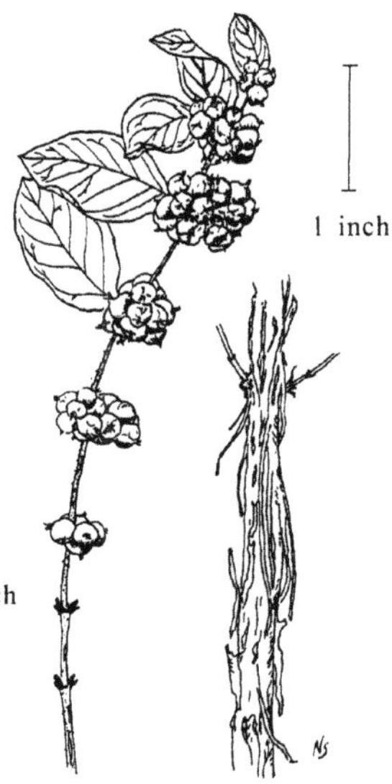

Symphoricarpos orbiculatus

Planning

Height: up to 3 ft (1 m)
Habit: shrub; arching, vigorous suckering; early July blossoms inconspicuous; autumn fruits persist
Site: sun, shade; pH and soil adaptable; drouth tolerant; zone 3
Use: wildlife habitat, erosion control, groundcover

Recognition

Leaves: egg-shaped to oval, up to 2 in (5 cm) long; apex rounded or obtuse, margin entire or rarely coarse-toothed, with hairy fringe (ciliate); upper surface dull yellow-green, lower hairy; petioles hairy

Flowers: clusters (spikes) short in the leaf axils, peduncle sharply downturned; flowers sessile; bracts, 1 large amd 2 small at base of each ovary; calyx lobes short, pointed, green; corolla narrow bell-shaped, greenish white or purplish, glabrous outside, long hairy inside, lobes 5, obtuse; stamens 5, inserted on upper part of tube alternating with lobes, filaments hairy at base, yellow anthers within the tube; style 0.1 in (2 mm) long, long white hairs at middle

Fruits: clusters tight; fruit sessile, globose, about 0.2 in (5 mm) across, red, smooth; calyx lobes persistent

Twigs: white-hairy, becoming glabrous; pith brownish; leaf scars usually roughened and indistinct, bundle scar 1

Buds: egg-shaped, gray-brown, pointed

Bark: brown, peeling into papery strips; inner bark red-brown

Symphoricarpos oreophilus Gray - mountain snowberry

Mountain snowberry, a resident of dry sites in the Rocky Mountain foothills, is cultivated in gardens of the Intermountain Region (Sutton and Johnson 1974). *S. utahensis* Rydb., another snowberry, is a native of Colorado and Utah that might be successful in our western plantations (Sutton 1998).

Planning

Height: up to 5 ft (1.5 m)

Habit: shrub; slender spreading branches; inconspicuous summer blossoms

Site: sun, partial shade; pH and soil adaptable; drouth tolerant; zone 3

Use: erosion control, wildlife habitat, naturalizing

Recognition

Leaves: elliptic, up to 1.5 in (4 cm) long; tip usually pointed, margin entire or toothed; upper surface fine-hairy, lower whitish-hairy to glabrous

Flowers: clusters (spikes) few-flowered or axillary pairs; corolla white to pinkish, tubular, about 0.4 in (1 cm) long, tube 4-5 times as long as lobes; anthers reaching to middle of corolla lobes

Fruits: egg-shaped, about 0.3 in (8 mm) long, white; seeds 2, pointed at one end

Twigs: finely hairy to glabrous

Viburnum L. - viburnum

The viburnums inhabit wooded sites in temperate and subtropical regions throughout the Northern Hemisphere. Many of the species that are native to Europe and Asia have been cultivated in gardens for centuries and were introduced into North American landscapes during colonial times. Many viburnums are all-season plants with beautiful flowers in spring and attractive foliage throughout much of the year. Some species display eye-catching fruits from late summer into winter that are valuable wildlife food. Fruitfulness for many viburnums is dependent upon cross-pollination between different clones of a species.

The primary goal of much of the prior breeding and selection of viburnum cultivars has been the development of plants with desirable ornamental qualities; pest resistance and site adaptation have often been secondary considerations. We have virtually ignored viburnums when selecting plants for inclusion in windbreaks and wildlife habitat plantings for much of the prairie. The great genetic diversity within this genus presents a great opportunity for breeding and selection to identify plants suitable for many uses on a wide range of site conditions.

Fifteen of the 100 to 150 viburnum species are natives of North America. Many of these grow naturally in the prairie region. Among the species not described herein are shade-tolerant *V. acerifolium* L., mapleleaf viburnum, and *V. alnifolium* Marsh., hobblebush, both of which are best adapted to natural landscapes on moist sites. They may be worthy of trial on cool prairie sites.

Recognition

Habit: shrubs, small trees; evergreen, deciduous
Leaves: opposite; simple; entire, toothed or lobed; with or without stipules
Flowers: clusters (cymes) large, spreading, round to flat-topped; calyx minutely 5-toothed; corolla wheel- to bell-shaped or tubular, 5-lobed, white or pinkish; stamens 5; ovary 1-celled, sometimes sterile
Fruits: drupes; seeds 1
Buds: naked (foliar) or scaly

Grouping Key for *Viburnum* (Bagley 1996)

Leaves

Leaves with star-shaped hairs (stellate tomentose); twigs stellate tomentose; buds naked (foliaceous): ***carlesii, lantana, burejaeticum, rhytidophyllum, macrocephalum, ×rhytidophylloides***

Leaves with 3 or more veins arising from one point (palmate), usually lobed; buds with scales: ***acerifolium, opulus, trilobum, sargentii***

Leaf veins usually rising along a central midrib (penninerved):

Buds with no scales (naked, foliaceous): ***alnifolium, ×burkwoodii, burejaeticum, carlesii, ×juddii, lantana, macrocephalum, ×rhytidophylloides, rhytidophyllum***

Buds with 2 or more scales: ***cassinoides, dentatum, dilatatum, edule, nudum, plicatum, prunifolium, rafinesquianum, rufidulum, setigerum, sieboldii***

Leaf veins netted, joining without reaching teeth (anastomosing): ***burejaeticum, cassinoides, lentago, macrocephalum, nudum, prunifolium***

Leaf veins nearly straight ending in teeth: ***carlesii, dilatatum, lantana, plicatum, setigerum***

Flowers

Sterile marginal flowers: ***alnifolium, macrocephalum, opulus, plicatum, sargentii, trilobum***

All flowers bisexual: ***acerifolium, ×burkwoodii, burejaeticum, carlesii, cassinoides, dentatum, dilatatum, edule, ×juddii, lantana, lentago, prunifolium, rafinesquianum, rufidulum, ×rhytidophylloides, rhytidophyllum, setigerum, sieboldii***

Flowers fragrant: ***×burkwoodii, carlesii, rhytidophyllum***

Flowers malodorous: ***cassinoides, lantana, sargentii, sieboldii***

Grouping Key for *Viburnum* (Rehder 1940)

Leaves not lobed, veins rising along a central midrib (penninerved), stipules wanting; flower clusters panicle-like with opposite branches; fruits oblong-obovate, blue-black, stones slightly compressed with 1 groove: **section Thyrsosma (*V. sieboldii*)**

Leaves toothed (denticulate) or entire, not lobed, penninerved, star-shaped hairs (stellate-tomentose); flower clusters flat-topped; fruits black or red to black, stones much flattened with 5 grooves; twigs stellate-tomentose, buds naked (foliaceous): **section Lantana (*V. carlesii, V. lantana, V. rhytidophyllum*)**

Leaves toothed (dentate), penninerved straight veins, stellate-tomentose; flower clusters with sterile marginal flowers; fruits red to blue-black, stones compressed with 1 groove; twigs stellate-tomentose when young; buds with 2 scales: **section Pseudopulus (*V. plicatum*)**

Leaves toothed (serrulate) or entire, penninerved veins curving not reaching the teeth (anastomosing); flower clusters without sterile marginal flowers; fruits blue-black, stones convex on back with 3 grooves; buds with 1 pair of scales: **section Lentago (*V. cassinoides, V. lentago, V. prunifolium, V. rufidulum*)**

Leaves toothed (dentate), penninerved straight veins ending in teeth, sometimes 3-nerved and lobed; flower clusters with wheel-shaped corolla (rotate); fruits red to scarlet, stones variably grooved; buds with 2 pairs of outer scales: **section Odontotinus (*V. dentatum, V. dilatatum, V. rafinesquianum, V. setigerum*)**

Leaves lobed, rarely not, 3- to 5-nerved at base, with stipules; flower clusters with sterile marginal flowers; fruits red, stones flattened, not or scarcely grooved; buds with 2 united (connate) outer scales: **section Opulus (*V. edule, V. opulus, V. sargentii, V. trilobum*)**

Note: The species in parentheses are described in detail.

Section Thyrsosma

Viburnum sieboldii Miq. - Siebold viburnum

Siebold viburnum, an import from Japan, provides a succession of colors throughout the growing season—creamy white flowers, lustrous dark green foliage, and red fruits on rose red stalks—which makes it a favorite shrub or small tree to Wyman (1969) and Dirr (1990).

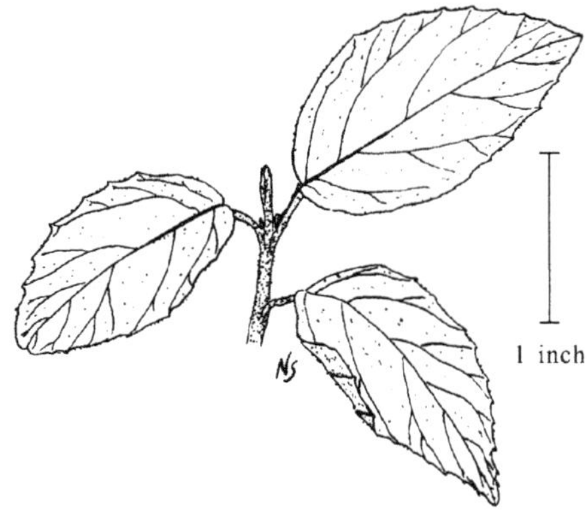

Viburnum sieboldii

Planning

Height: up to 20 ft (6 m)
Habit: shrub, tree; deciduous; upright, stiff, spreading; late May blossoms malodorous; late summer fruits red; autumn fruit stalks rose red
Site: sun, light shade; pH adaptable, avoid drouth stress; zone 5
Use: specimen, border, wildlife habitat

Recognition

Leaves: fetid when crushed; elliptic, sometimes wider above middle, up to 5 in (13 cm) long; tip slightly pointed to rounded, base tapered, coarsely toothed (crenate-serrate) with 7-10 veins ending in teeth; dark green, glabrous above, sparsely star-shaped hairy beneath, chiefly on veins; petioles up to 0.4 in (1 cm) long, sparsely hairy, grooved above, broad, almost winged

Flowers: clusters (paniculate cymes) pyramidal to subglobose, up to 4 in (10 cm) long and broad, stalks rose red; bisexual; corolla creamy white

Fruits: oblong to egg-shaped, about 0.4 in (1 cm) long

Twigs: grayish brown, sparsely star-shaped hairy (stellate); leaf scars connect around stem

Buds: angled; scales not overlapping (valvate), slightly hairy; leaf buds smaller than flower buds

Section Lantana

Viburnum carlesii Hemsl. - Koreanspice viburnum

The semi-snowball Koreanspice viburnum, noted for its attractive fragrant flowers, has provided many cultivars and has been one of the parents of several valuable hybrids. 'Compactum' is one of the best of the smaller clones, and 'Aurora', 'Charis' amd 'Diana' are known for their strongly fragrant flowers (Dirr 1990).

A semi-evergreen viburnum, *V. ×burkwoodii* Burkwood, is tolerant of a wide range of climatic conditions and is rated as one of the best of the fragrant viburnums (Dirr 1990). Its parents are *V. carlesii* and an evergreen shrub, *V. utile* Hemsl. A Burkwood viburnum on the East Campus of UNL survived for over 35 years. However, this hybrid has failed to survive at PERC in Fort Collins (Fairchild and Klett 1993). Its cultivars include 'Anne Russell', 'Fulbrook', and 'Mohawk', backcrosses of Burkwood viburnum with *V. carlesii*. *V. ×juddii* Rehd. (*V. carlesii × bitchiuense* Mak.) is thriving in partial shade at Prairie Pines Arboretum near Lincoln.

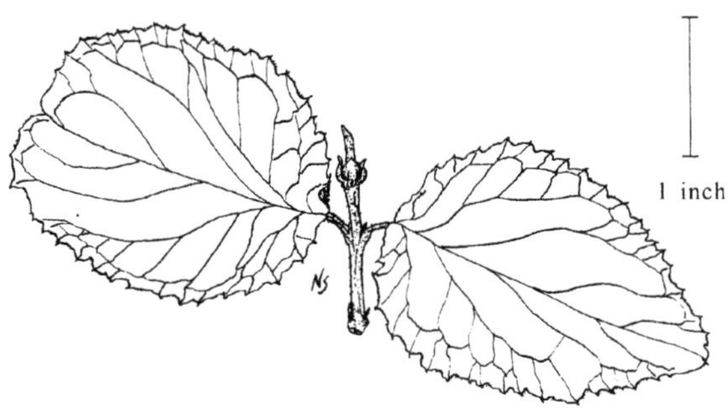

Viburnum carlesii

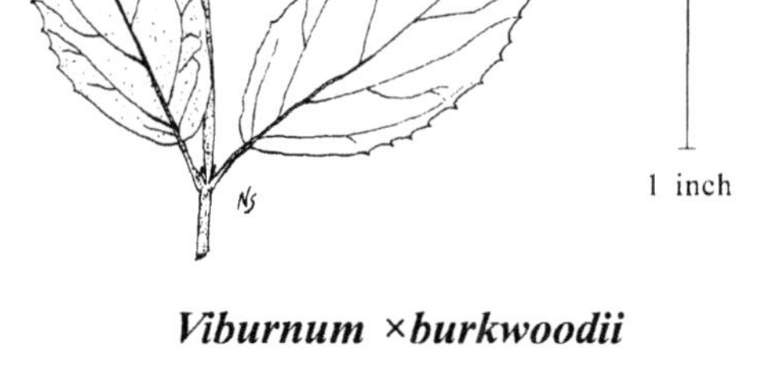

Viburnum ×burkwoodii

Planning

Height: up to 7 ft (2 m)

Habit: shrub; deciduous; upright, spreading; fragrant early May blossoms appearing with the leaves; autumn foliage reddish but not dependable

Site: sun, partial shade; prefers slightly acidic, fertile soils; zone 4

Use: border, foundation, grouping, specimen

Recognition

Leaves: egg-shaped to elliptic, up to 4 in (10 cm) long; margins irregularly toothed, veins ending in teeth; tip pointed, base usually rounded; star-like hairs above, more dense beneath; petioles about 0.3 in (8 mm) long

Flowers: clusters dense, hemispherical; up to 3 in (8 cm) across; bisexual; corolla cylindric with lobes at right angles (salver-shaped), reddish in bud, opening white; stamens inserted above the middle of corolla

Fruits: egg-shaped, flattened, about 0.4 in (1 cm) across, changing from red to black

Twigs: star-shaped hairy (stellate) when young

Buds: naked, hairy; flower buds up to 0.5 in (12 mm) wide

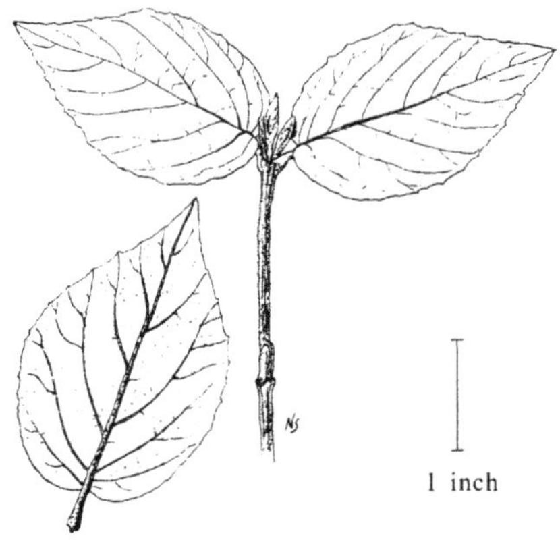

Viburnum ×juddii

Viburnum lantana L. - wayfaring tree

The wayfaring tree, a native of Europe and Asia, is as tough as its wrinkled, leathery leaves portray. It flourishes on neglect, thrives on sunny sites, and withstands long drouth periods. Judicious pruning can convert this shrub into a small tree. The wayfaring tree and its cultivars, 'Mohican' and 'Rugosum', are attractive, fruitful and well-adapted to the soils and climate of Fort Collins (Fairchild and Klett 1993).

'Mohican', a seedling from Poland, was selected for its compact form, thick, dark green leaves and orange-red fruits that are effective for about a month before turning black (Dirr 1990). The leaves of 'Rugosum' are darker green and thicker textured than the species type (Fairchild and Klett 1993).

Manchurian viburnum, *V. burejaeticum* Reg. & Herd., is similar to the wayfaring tree in appearance and growth characteristics. It has been grown successfully on the plains of Montana and the Canadian prairies. Both of these species deserve more trials in windbreak and wildlife habitat plantings throughout the prairie, especially in the west.

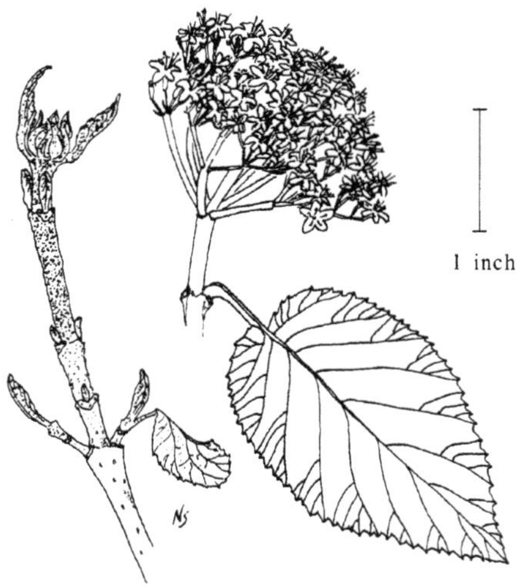

Viburnum lantana

Planning

Height: up to 15 ft (4.5 m)

Habit: shrub; deciduous; upright, almost tree-like; May blossoms; autumn foliage color inconsistent; fruits persist into winter

Site: sun; pH and soil adaptable; drouth tolerant; zone 4

Use: border, screen, grouping, specimen, wildlife habitat, windbreak

Recognition

Leaves: egg-shaped, up to 5 in (13 cm) long; uniformly toothed (denticulate), strongly netted vein pattern ending in teeth, tip pointed to obtuse, base rounded to heart-shaped; surface wrinkled, star-shaped hairs sparingly spaced above, dense beneath; petioles about 0.5 in (12 mm) long, grooved above, hairy
Flowers: clusters flat-topped; bisexual, odorless; calyx glabrous or slightly hairy; corolla white, wheel-shaped (rotate) with lobes longer than tube; anthers yellow, visible
Fruits: nearly round, about 0.3 in (8 mm) long, becoming red, finally black
Twigs: scurfy, gray-brown
Buds: naked, white-hairy, narrow, oblong; flower buds larger, up to 0.5 in (12 mm) across

Viburnum rhytidophyllum Hemsl. - leatherleaf viburnum

Although its large, creamy white flower clusters and red and black fruits are effective, leatherleaf viburnum is often planted primarily because of its handsome foliage. The leaves of this import from China remain green all or almost all of the year, depending upon the minimum temperatures and amount of protection afforded. Several specimens are growing on the campus of UNL.

The hybrid, *V.* ×*rhytidophylloides* (*V. lantana* × *V. rhytidophyllum*), bears similar foliage but withstands colder climates. The leaves of this shrub—dubbed lantanaphyllum viburnum—are broader and less wrinkled than those of leatherleaf viburnum and are usually deciduous. Its flowers and fruits are similar to those of the parents. Good fruiting of lantanaphyllum viburnum is dependent upon the presence of its parents as pollinators.

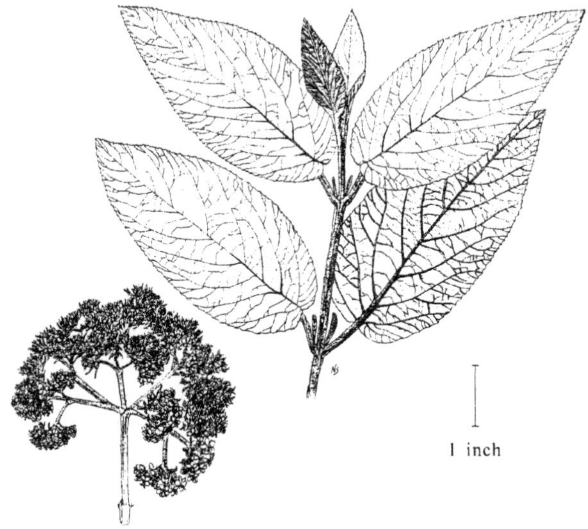

Viburnum* ×*rhytidophylloides

'Alleghany', a selection of lantanaphyllum viburnum, grows successfully at the University of Minnesota Landscape Arboretum. It is also adapted to the soils and climate of the Fort Collins area, where its showy fruits ripen unevenly so that red and black are present in the same cluster and persist into winter (Fairchild and Klett 1993). 'Willowwood' is another cultivar that performs well in the Midwest (Dirr 1990).

Planning

Height: up to 15 ft (4.5 m)
Habit: shrub; semi-evergreen; multistemmed, spreading; mid-May blossoms slightly fragrant; autumn foliage green; fruits red, becoming black
Site: shade to partial shade; pH adaptable; zone 5
Use: specimen, grouping, border
Caution: protect from winter sun and wind

Recognition

Leaves: much longer than wide, broadest below the middle (ovate-lanceolate), up to 7 in (18 cm) long; tip pointed or rounded, base rounded to heart-shaped (subcordate), entire or obscurely toothed; dark green, glabrous and wrinkled above, prominent net of veins beneath (reticulate), densely woolly; petioles up to 1.2 in (3 cm) long

Flowers: clusters flat-topped, stellate tomentose, formed in autumn and remaining naked through winter before blooming; bisexual; flowers yellowish white

Fruits: ellipsoidal, about 0.3 in (8 mm) long, red, ripening black

Twigs: gray to brown, star-shaped hairy (stellate-tomentose), becoming glabrous

Buds: naked; flower buds in semi-rounded clusters up to 2 in (5 cm) across, evident through winter; leaf buds rusty, densely woolly

Section Pseudopulus

Viburnum plicatum Miq. var. *tomentosum* Rehd. - doublefile viburnum

Doublefile viburnum, a native of Japan and China, is described by Dirr (1990) as "the most elegant of flowering shrubs." The horizontally tiered branching, handsome in winter, is accentuated in spring by flat-topped clusters of snow white flowers. Unfortunately, most forms are marginally hardy in the central prairies. Dieback occurs on the doublefile viburnum 'Mariesii' at UNL.

Older publications such as Rehder (1941) and Hoag (1965) use the synonym *V. tomentosum* for doublefile viburnum, and the variety that bears sterile flowers only in globose clusters as var. *sterile* K. Koch. They describe var. *Mariesii* Veitch as bearing marginal sterile flowers larger than those of the species. According to Dirr (1990) the species *V. plicatum*, Japanese snowball viburnum, bears only sterile flowers and should not be considered a species since it cannot produce fruit. He considers 'Mariesii' to be a cultivar.

Selections of var. *tomentosum* include: 'Mariesii', with ray flowers up to 1.75 in (4.5 cm) in diameter on long peduncles extending above the foliage; 'Lanarth', with ray flowers up to 2 in (5 cm) across; 'Pink Beauty' and 'Roseum' with pink petals; and 'Shasta', a 1979 introduction from the U.S. National Arboretum with wide-spreading horizontally accentuated branches and immense inflorescences but which failed to overwinter at PERC in Fort Collins (Fairchild and Klett 1993).

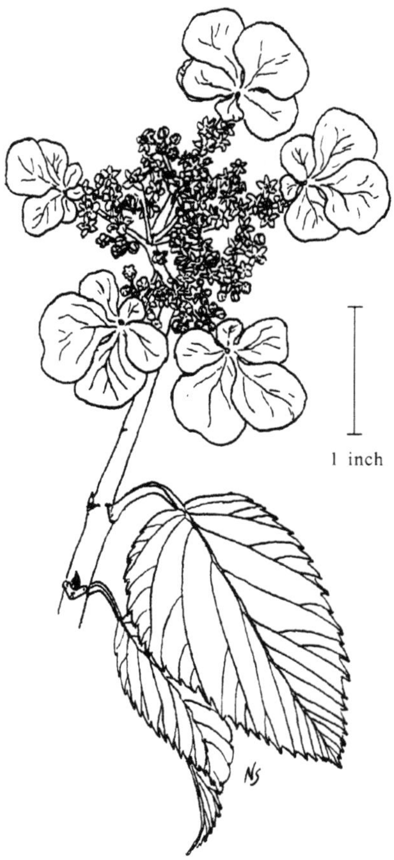

Viburnum plicatum* var. *tomentosum

Planning

Height: up to 10 ft (3 m)

Habit: shrub; deciduous; branches horizontally tiered, broad; May blossoms; autumn foliage red-purple

Site: partial to full shade; pH 5.5-7.0; requires moist well-drained soils; zone 6

Use: grouping, screen, border

Recognition

Leaves: egg-shaped to elliptic, sometimes wider above middle, up to 4 in (10 cm) long; tips pointed, base rounded to broad wedge-shaped, toothed, veins nearly straight ending in teeth; nearly glabrous above, star-shaped hairs beneath (stellate); petioles about 0.8 in (2 cm) long

Flowers: clusters terminal, flat-topped, above foliage, peduncles with star-shaped (stellate) hairs; flowers white, odorless, outer ring sterile and showy, up to 2 in (5 cm) across, inner fertile and inconspicuous

Fruits: ellipsoid, about 0.3 in (8 mm) long, red, becoming black

Twigs: gray or brownish, star-shaped hairs, becoming glabrous

Buds: naked, hairy; flower buds with paired scales, not overlapping (valvate), hairy

Section Lentago

Viburnum cassinoides L. - witherod viburnum

Witherod viburnum, a native of eastern North America among pioneer plant species on rocky, woodland sites, is spectacular in fruit. Smooth witherod, *V. nudum* L., a native of more southerly sites in the eastern U.S., is very similar in appearance to *V. cassinoides* except for its shiny leaves with entire margins. Both species should be tested more widely in the central prairie. Specimens of each are growing on the UNL campus.

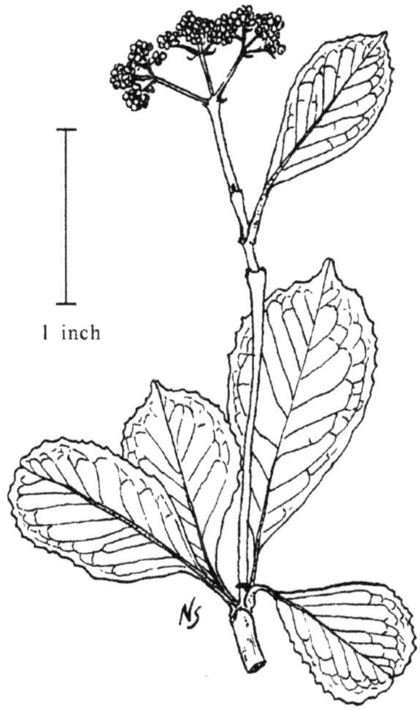

Viburnum cassinoides

Planning

Height: up to 10 ft (3 m)

Habit: shrub; deciduous; compact, arching; June blossoms malodorous; fruit colors progressive from green, pink, red, blue to black; autumn foliage colorful reds

Site: sun, partial shade; pH 5.5-6.5; zone 4

Use: border, grouping, naturalizing, wildlife habitat, edible fruit

Recognition

Leaves: egg-shaped to elliptic, up to 3.5 in (9 cm) long; tips pointed, obscurely toothed; dull green above, nearly glabrous, scurfy beneath; petioles up to 0.8 in (2 cm) long

Flowers: clusters flat-topped; bisexual; flowers white

Fruits: blue-black, nearly round, about 0.3 in (8 mm) across

Twigs: gray-brown, scurfy, becoming glabrous

Buds: scales meet at edges (valvate); flower buds terminal with bulbous base

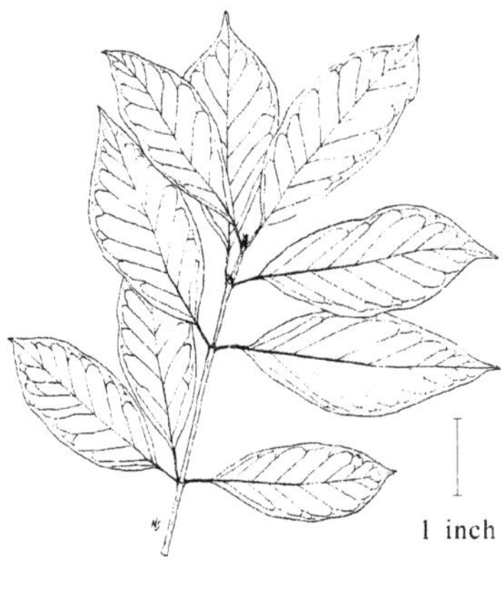

Viburnum nudum

Viburnum lentago L. - nannyberry

Nannyberry is a native of North America from New England to Manitoba, south to eastern Nebraska, and east to West Virginia. It grows on a wide variety of sites including open woodlands, streambanks and hillsides where soils range from sands to clays and dry to moist. Trial plantings have been successful in eastern Nebraska and to the west of Kearney, Nebraska, with careful site modification (North Central Plant Introduction Station 1959). An attractive specimen is well adapted to the soils and climate of PERC in Fort Collins (Fairchild and Klett 1993). Nannyberry is a good candidate for windbreak and wildlife habitat plantings in much of our region.

Glossy leaves with abruptly long-pointed tips and winged petioles combined with the long-pointed flask-shaped flower buds provide distinctive characteristics for easy identification of nannybery. It might be confused with some specimens of the variable black haw (*V. rufidulum*) if the rusty red hairs of the latter happen to be more sparse than normal.

Viburnum lentago

Planning

Height: up to 25 ft (8 m), National champion, 1976, 40 ft (12.2 m), Michigan
Habit: shrub, small tree; deciduous: arching, suckering; May blooms; variably colored autumn foliage
Site: sun, shade; pH and soil adaptable; drouth tolerant; zone 3
Use: border, naturalizing, specimen, wildlife habitat, windbreak
Note: keep sucker growth in check to develop a small tree

Recognition

Leaves: egg-shaped to narrowly so, up to 4 in (10 cm) long; finely toothed (serrate), veins curving and closing before reaching margin, tip abruptly long-pointed, base tapering to rounded; shiny above, glabrous except sometimes scurfy on veins beneath, thin and leathery; petiole grooved above, winged with wavy margin, up to 0.6 in (15 mm) long
Flowers: clusters stalkless, flat-topped, up to 5 in (13 cm) across; bisexual; corolla white, lobes divided nearly to base; calyx glabrous, greenish white or pinkish; stamens visible, anthers yellow
Fruits: ellipsoid, about about 0.5 in (12 mm) long, bluish black, bloomy, sweet, juicy
Twigs: glabrous, bloomy, brownish; leaf scars narrow
Buds: flower—terminal, long narrow with bulge near base, tip long tapered, often curved; leaf—narrow, long-pointed, slightly curved, one of opposing pair often larger than the other; scales 2, not overlapping (valvate)

Viburnum prunifolium L. - blackhaw viburnum

Blackhaw viburnum—a native of extreme eastern Kansas, east to the Atlantic coast—grows on a variety of soils along woodland margins and streambanks. It is similar in general appearance to nannyberry (*V. lentago*), though the branches are stiffer and tend to assume horizontal layers.

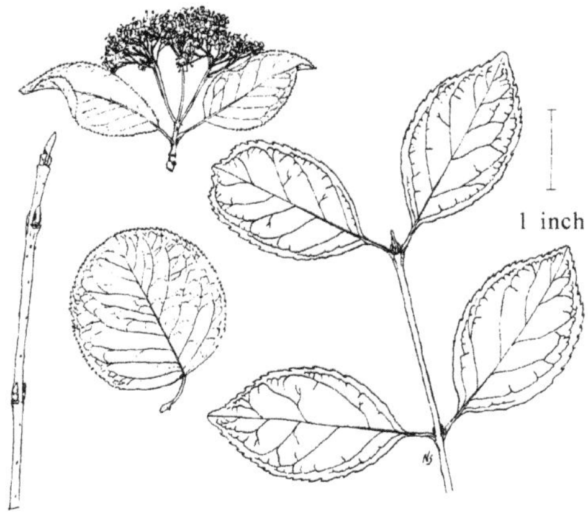

Viburnum prunifolium

Planning

Height: up to 15 ft (4.5 m)
Habit: shrub, tree; deciduous; multistemmed or a round-headed tree; May blossoms; autumn foliage red-bronze
Site: sun, shade; pH and soil adaptable; drouth tolerant; zone 5
Use: specimen, grouping, border, wildlife habitat, windbreak

Recognition

Leaves: egg-shaped to elliptic, up to 3 in (8 cm) long; tip pointed to obtuse, base tapering to rounded, margins finely toothed, many teeth incurved, veins curved and closing (anastomosing); surfaces glabrous; petioles glabrous, reddish, grooved above, not winged or narrowly so, about 0.5 in (12 mm) long
Flowers: clusters rounded; bisexual; flowers white, odorless; anthers yellow
Fruits: ellipsoid about 0.5 in (12 mm) long, pinkish, becoming blue-black and bloomy, mealy, sweet and edible; seeds oval, flat; stalks red, glabrous
Twigs: glabrous, gray-brown; lenticels orange
Buds: glabrous or scurfy, pinkish gray; scales paired, not overlapping, narrow; flower buds broadly flask-shaped, short-pointed

Viburnum rufidulum Raf. - rusty blackhaw

Rusty blackhaw is similar to blackhaw viburnum (*V. prunifolium*) except for the dense rusty hairs on many of its plant parts. It is native to the southeastern U.S. as far west as eastern Kansas. At the northern limits of its range rusty blackhaw is a shrub, but it can become a small tree in the south. Northern provenances of this species might be adapted to southerly sites in the central prairie. A specimen grows at the Folsom Children's zoo in Lincoln (Sutton 1998).

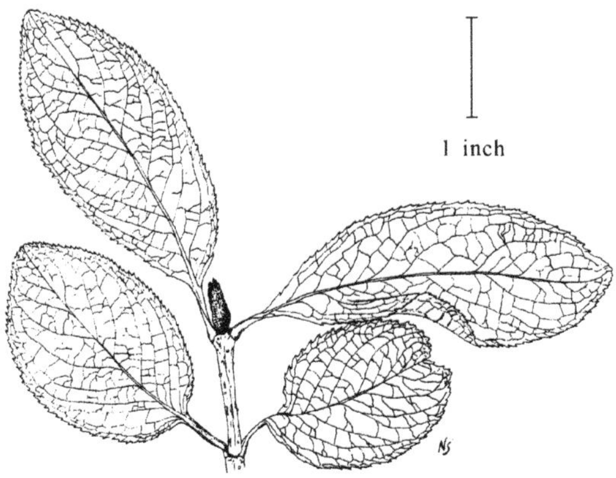

Viburnum rufidulum

Planning

Height: up to 30 ft (9 m)
Habit: shrub, tree; deciduous; open, irregular; May blossoms; autumn fruits reddish, becoming black, sweet and edible
Site: sun, partial shade; pH 5.5-7.0; drouth tolerant; zone 5
Use: border, grouping, naturalizing, wildlife habitat, windbreak

Recognition

Leaves: egg-shaped, oval to broad elliptic, often wider above the middle, up to 3 in (8 cm) long; tip pointed to rounded, base variable, margin finely toothed; shiny dark green above, rusty hairy beneath, veins curved and closing (anastomosing); petioles rusty hairy, stipules wanting

Flowers: clusters flat-topped or rounded; bisexual; flowers odorless; hypanthium cylindric, green, calyx lobes 5, distinct; corolla wheel-shaped (rotate), white, petals 5; stamens 5 attached to corolla between lobes, anthers yellow

Fruits: ellipsoid, flattened, about 0.6 in (15 mm) long, red becoming dark blue to black, whitened with bloom (glaucous); stone oval, flat with prominent tip

Twigs: light brown and red-brown, hairy, becoming gray-brown and glabrous; leaf scars crescent-shaped, raised, bundle scars 3

Buds: flower—swollen abruptly just below middle; 1 pair of scales, not overlapping (valvate), red-brown, densely woolly (tomentose); leaf buds—smaller, not swollen

Section Odontotinus

Viburnum dentatum L. - arrowwood viburnum

The natural range of arrowwood viburnum skirts just the eastern edge of the Great Plains from Canada to eastern Texas and includes most of the eastern U.S. and southeastern Canada. Within this area it consorts with several other species with similar characteristics, such as *V. molle* Michx., *V. bracteatum* Rehd., *V. pubescens* Pursh, *V. rafinesquianum* Schult. and *V. scabrellum* Chapm. Since none of these have spectacular fruits or outstanding flowers, they have been largely overlooked for landscape plantings.

Arrowwood viburnum and related species are collectively adapted to a wide variety of sites and provide excellent habitat for wildlife. The usually bountiful fruit crops are devoured by birds soon after ripening. This group of viburnums with blue-black fruits deserves more attention from plant scientists. Provenance testing should be very rewarding.

Planning

Height: up to 15 ft (4.5 m)

Habit: shrub; deciduous; dense, arching; May blossoms odorless; autumn foliage variable

Site: sun, partial shade; soil adaptable, becomes chlorotic at high ph; drouth tolerant; zone 3

Use: border, screen, windbreak, wildlife habitat

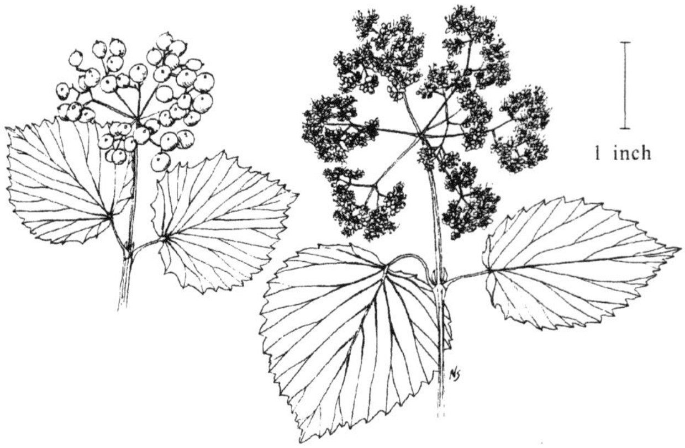

Viburnum dentatum

Recognition

Leaves: roundish to egg-shaped, mostly 2-3 in (5-8 cm) long; coarsely toothed (dentate), tip pointed, base heart-shaped to rounded; surfaces glabrous; petioles about 0.5 in (12 mm) long with scattered hairs

Flowers: clusters flat-topped, up to 3 in (8 cm) across, stalks glabrous, slender; bisexual; petals white; stamens longer than corolla, anthers yellow

Fruits: nearly round, about 0.2 in (5 mm) across, blue-black

Twigs: green, changing to brown and grayish, glabrous, angled; leaf scars narrow crescent-shaped

Buds: somewhat appressed; scales overlapping, glabrous, green to brownish

Viburnum dilatatum Thunb. - **linden viburnum**

This uncommon Asian viburnum deserves more attention throughout our region. The potential for selecting linden viburnum plants suitable for the prairies is shown by the variation displayed in its cultivars. The dwarf 'Catskill' and the round-mounded 'Erie' and 'Iroquois' are all highly fruitful; 'Mt. Airy' bears large, bright red fruits and lustrous dark green foliage; and 'Xanthocarpum' bears yellow-orange fruits (Dirr 1990).

Viburnum dilatatum

Planning

Height: up to 10 ft (3 m)

Habit: shrub; deciduous; upright, open and leggy to dense and compact; May blossoms; autumn fruits bright red, persistent into winter; autumn foliage colorful

Site: sun, partial shade; pH and soil adaptable; zone 5

Use: specimen, border, grouping, wildlife habitat

Recognition

Leaves: roundish to egg-shaped, sometimes wider above middle, up to 4.7 in (12 cm) long; coarsely toothed, veins ending in teeth, tip pointed, base heart-shaped to rounded; surfaces hairy; petioles about 0.4 in (1 cm) long

Flowers: clusters hairy, flat-topped, up to 4.7 in (12 cm) across; bisexual; corolla white, hairy; stamens longer than corolla

Fruits: broadly egg-shaped, about 0.3 (8 mm) long, bright red

Twigs: densely stiff-hairy, becoming glabrous; lenticels prominent, orange

Buds: scales overlapping, slightly hairy, brownish with a tinge of red

Viburnum rafinesquianum Schult. - **downy arrowwood**

Downy arrowwood, a native of much of eastern North America, grows on a variety of soils on ridges and hillsides in woodlands and open areas. Downy arrowwood is thriving at PERC in Fort Collins (Fairchild and Klett 1993). Test plantings are recommended throughout the central and northern prairies.

Planning

Height: up to 6 ft (2 m)

Habit: shrub; deciduous; open, spreading, dense in sun; May-June blossoms; autumn foliage reddish yellow

Site: sun, shade; pH and soil adaptable; prefers moist soil but drouth tolerant; zone 3

Use: border, hedge, grouping, wildlife habitat, specimen

Recognition

Leaves: egg-shaped to elliptic, up to 2.7 in (7 cm) long; tip pointed, base variable, margin coarsely toothed and hairy (ciliate); upper surface sometimes sparsely hairy, lower variably long-hairy, veins branch directly to each tooth; petioles hairy, sharply grooved above, up to 0.2 in (6 mm) long
Flowers: clusters terminal, rounded; bisexual; odorless; hypanthium green, cylindric, calyx lobes 5, triangular; corolla wheel-shaped (rotate), petals 5, white; stamens 5, erect or spreading
Fruits: ellipsoid to nearly round, about 0.4 in (1 cm) long, dark purple to black, glossy; sepals and style persistent; stone oval, flattened, dark red-brown
Twigs: bloomy, brown to gray-brown; leaf scars crescent-shaped, bundle traces 3
Buds: terminal egg-shaped flanked by 2 small lateral buds; scales paired, margins hairy (ciliate)

Viburnum setigerum Hance - **tea viburnum**

Tea viburnum, a China native from which leaf tea has been brewed, is the most handsome in fruit of all viburnums (Dirr 1990). 'Auranticum' bears orange-colored fruit.

Planning

Height: up to 12 ft (3.5 m)
Habit: shrub; deciduous; multistemmed, leggy; late May blossoms; autumn foliage sometimes colorful; fruits bright red
Site: sun, light shade; soil and pH adaptable; zone 6
Use: border, windbreak, wildlife habitat

Recognition

Leaves: egg-shaped to oblong, up to 6 in (15 cm) long; tip long-pointed, base rounded, widely spaced teeth (remotely denticulate), veins prominent ending in teeth; surfaces glabrous except silky hairs on veins beneath; petioles glabrous or sparsely hairy, up to 0.8 in (2 cm) long
Flowers: clusters flat-topped, up to 2 in (5 cm) across; bisexual; odorless; calyx purple, glabrous; corolla white; stamens half as long as petals
Fruits: egg-shaped, about 0.4 in (1 cm) long, red
Twigs: glabrous, smooth, grayish
Buds: scales overlapping, 3-5 visible, green with red tips, glabrous

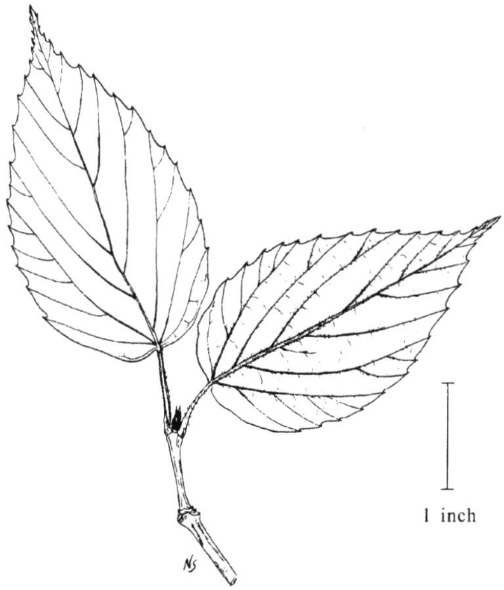

Viburnum setigerum

Section Opulus

Viburnum edule (Michx.) Raf. - **mooseberry**

Mooseberry, a native of the north woods of North America, ranges across the continent from New Foundland to Alaska, dips down to Oregon, and across to Pennsylvania including the Black Hills and the Colorado Rockies. Its preferred habitat is moist, shady, wooded hillsides. Mooseberry may be useful for planting in the northern prairies on shady sites.

Planning

Height: up to 5 ft (1.5 m)

Habit: shrub; deciduous; open, erect to decumbent; early June blossoms; autumn foliage red; sparsely fruitful

Site: full shade or partial shade; pH 6.0-7.0; moist; zone 2

Use: naturalizing, wildlife habitat

Recognition

Leaves: 3-lobed and egg-shaped or undivided and elliptic, up to 4 in (10 cm) long; tip pointed, base variable, coarsely toothed with marginal hairs (ciliate); glabrous above, hairy beneath, glandular along veins, often a large gland on blade near petiole; petioles about 1 in (25 mm) long, grooved above, sparsely hairy

Flowers: clusters loose on short, lateral branches with 2 leaves; bisexual; hypanthium cylindric to bell-shaped, green, lobes very short; corolla white, bell-shaped with spreading lobes; stamens 5, inserted midway on corolla tube, style stout, 0.05 in (1 mm) long

Fruits: ellipsoid, about 0.4 in (1 cm) long, bright, glossy red, translucent, soft, acrid, juicy

Twigs: yellow-brown, becoming grayish, glabrous, 3 ridges extending down from leaf scars; pith white, large; leaf scars deeply U-shaped, bundle scars 3

Buds: purplish red; scales in pairs, completely joined (connate); terminal—ellipsoid, keeled on 2 sides, about 0.3 in (8 mm) long; lateral—wider above middle (obovoid), apex rounded, keeled, stalked

Viburnum opulus L. - European cranberrybush

European cranberrybush (*V. opulus* L.) has been cultivated for centuries and is more widely planted in the U.S. than American cranberrybush. Hoag (1965) states that the European shrubs produce highly unpalatable berries and their foliage does not turn red in autumn in the northern Great Plains. Also, Hoag notes that the snowball aphid attacks European cranberrybush while avoiding the American shrubs.

Specimens purported to be European cranberrybush growing at Prairie Pines Arboretum near Lincoln show some fall coloration and have not been plagued with aphids. However, a large and varied bird population there agrees with the observation that the fruits are not palatable because fruits on heavily laden branches hang untouched until they shrivel and drop in late winter.

Cultivars of *V. opulus* include: 'Roseum', the Guilder rose, with a showy inflorescence of all sterile flowers; 'Compactum', an extremely dense, short plant bearing excellent flowers and fruit; and 'Nanum', a dwarf form that supposedly never produces flowers.

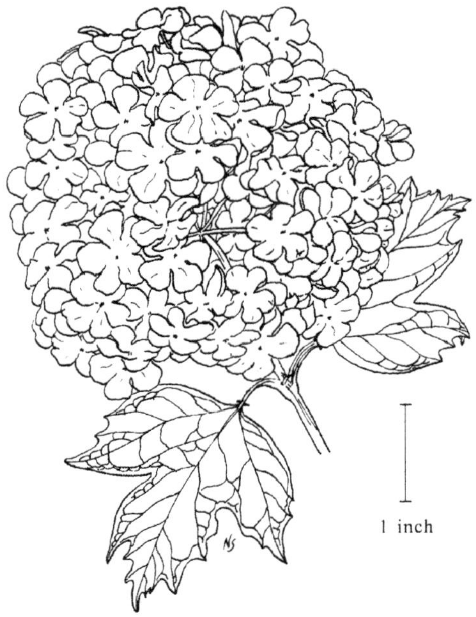

***Viburnum opulus* 'Roseum'**

V. trilobum and *V. opulus*—two species with similar broad natural latitudinal and longitudinal distributions on their respective continents—so closely resemble each other that scientists differ in their descriptions and classifications of the two plants. Some taxonomists consider the American shrubs to be *V. opulus* L. var. *americanum* Ait. while others prefer *V. opulus* ssp. *trilobum* Clausen as the proper clasification.

Rehder (1941) uses grooves and glands on the petioles as key differences between plants of the two species. He states that petiole grooves are shallow and glands are small and stalked on *V. trilobum*, with an added note that the middle lobe of the leaf is sometimes elongated and entire. By contrast, the petiole grooves of *V. opulus* are narrow with a few large disc glands and the lobes are more toothed. Hoag (1965) makes a similar statement about the glands and adds that the leaf lobes of *V. trilobum* are sharp and long-pointed (acuminate) while leaf lobes of *V. opulus* are obtuse to short-acuminate. Dirr (1990) makes a statement that though the buds of the two species are similar, those of *V. trilobum* are sometimes sticky and that the twigs have a waxy appearance. He also states that the twigs of *V. opulus* are ribbed.

Since variability among the plants of each of these two populations makes positive identification difficult, we propose they be treated as a single species. If so, then it follows that the American species should be considered *V. opulus* L. var. *americanum* as proposed by Stephens (1973). Observations at PERC reinforce this premise because the two shrubs and their several cultivars are all well adapted to the soils and climate of the Fort Collins area with slight differences in pest resistance and chlorotic tendencies noted among the cultivars (Fairchild and Klett 1993).

Range-wide provenance research incorporating populations of the American and European shrubs would allow us to identify pest-resistant plants from which we could select those with desirable form, flower and fruit characteristics. Research plantations established on varied sites throughout the prairies would permit selection of cultivars best adapted to each site. A system of certification for correct identity would minimize labeling errors.

Planning

Height: up to 13 ft (4 m)

Habit: shrub; deciduous; upright, round-topped; late May blossoms odorless; autumn foliage red-purple; fruits bright red persisting into winter

Site: sun, part shade; pH and soil adaptable, avoid drouth stress; zone 3

Use: screen, windbreak, wildlife habitat, specimen, grouping, naturalizing

Recognition

Leaves: 3-lobed, 3- to 5-nerved at base, more or less round in outline, up to 4 in (10 cm) long; lobe tips obtuse to short-pointed, base variable, coarsely toothed or entire, mostly glabrous above, variably hairy below, often with hairs on veins; petioles about 0.8 in (2 cm) long with a few large disc-like glands, narrow grooved above; linear stipules at base of petiole

Flowers: clusters terminal, flat-topped; outer ring of flowers sterile with expanded petals, inner fertile and inconspicuous; hypanthium of bisexual flowers bell-shaped, calyx lobes barely discernable; corolla white, tube bell-shaped, petals recurved; stamens 5, alternate with petals, long exserted, anthers yellow

Fruits: clusters pendant, each globose about 0.4 in (1 cm) long, bright red, glossy

Twigs: green becoming gray-brown, ribbed, glabrous; lenticels not noticeable; leaf scars U-shaped, bundle scars 3

Buds: stalked, egg-shaped; scales not overlapping (valvate), glabrous, brown, outer pair totally fused (connate)

Viburnum sargentii Koehne - Sargent viburnum

Sargent viburnum, an import from northeast Asia that is a close relative and look-alike of *V. opulus*, is purported to be more vigorous and resistant to aphids than the latter (Dirr 1990). Cultivars include: 'Flavum', with yellow anthers and fruits; 'Onondaga', with velvety maroon young foliage; and 'Susquehanna', an upright, fruitful plant with dark green foliage. 'Flavum' is adapted to the soils and climate of PERC at Fort Collins (Fairchild and Klett 1993).

Planning

Height: up to 15 ft (4.5 m)
Habit: shrub; deciduous; multistemmed, upright, coarse; May blossoms malodorous; autumn foliage yellow to red; edible yellow to red fruits persist through winter
Site: sun, partial shade; pH and soil adaptable; prefers moist sites; zone 4
Use: border, screen, grouping, wildlife habitat, windbreak, specimen

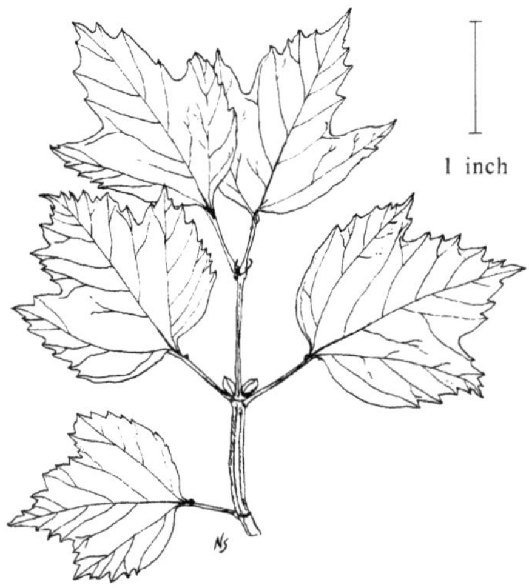

Viburnum sargentii

Recognition

Leaves: 3-lobed, broadly ovate in outline; upper leaves often with middle lobe entire and elongated, sometimes long and narrow without lobes; lobe tips pointed, base rounded or squared-off (truncate), often coarsely toothed; thick textured, glabrous above, slightly hairy beneath; petioles up to 1.8 in (45 mm) long with prominent disc-like glands
Flowers: clusters flat-topped; ray flowers sterile; corolla white; anthers purple, rarely yellow
Fruits: nearly round, about 0.3 in (8 mm) long; bright red, sometimes yellow
Twigs: yellow-brown, stout, appear waxy, sometimes hairy; lenticels prominent
Buds: scales 2, fused at edges (connate), glabrous, shiny, shades of green and red-brown

Viburnum trilobum Marsh. (*V. opulus* L. var. *americanum*) - American cranberrybush

American cranberrybush is a native of southern Canada from British Columbia to Newfoundland and the northern U.S. from Pennsylvania to Washington including the Black Hills of South Dakota. It is not the traditional cranberry bush that provides the red berry so essential to our Thanksgiving dinners. However, the tart fruits, which resemble those of *Vaccinium macrocarpum* Ait. in size and color, have been used occasionally for making jams. Cultivars of American cranberrybush include the compact forms—'Compactum' and 'Alfredo'—and those that have been selected for their large and edible fruits—'Andrews', 'Hahs' and 'Wentworth'.

Planning and Recognition

See *V. opulus*

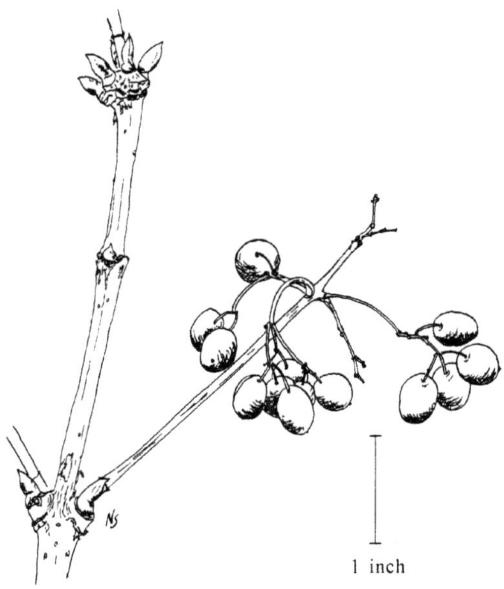

Viburnum trilobum

Weigela Thunb. - weigela

Weigela species, natives of east Asia, were once classified as belonging to the genus *Diervilla*. Many cultivars have been selected from twelve species and their hybrids.

Recognition

Habit: shrubs; deciduous

Leaves: opposite; simple; toothed; short-petioled or subsessile, without stipules

Flowers: single to several in axillary clusters (cymes), lateral on last year's twigs except occasional summer-fall flowers on current growth; bisexual; calyx lobes 2-5, distinct or united below; corolla white, pink, purple or crimson, tubular, funnel- or bell-shaped, tube much longer than 5 broad lobes; stamens 5, much shorter than corolla, style sometimes exserted

Fruits: dry, splitting (capsules); seeds small, angular, often winged

Buds: narrow, scales pointed

Weigela florida (Sieb. & Zucc.) A. DC. - cardinalbush

Flower color has been the principal criteria for selecting cultivars of cardinalbush, the most widely planted of the weigelas. Removal of dead branches is a ritual of spring at the mid-latitudes of the central prairie. Red-flowered 'Vanicek', sometimes called 'Newport Red', is one of the hardiest cultivars (Dirr 1990; Spangler and Ripperda 1977). 'Dropmore Pink' can be grown in the northeastern prairie with protection from wind (Hoag 1965).

Dirr (1990) rates 'Mount Blanc' as the best white-flowered cultivar and 'Red Prince' as a hardy red-flowered introduction from Iowa State University. Other cultivars bear flowers of pink, rose-carmine, magenta rose and other shades of red as well as variegated foliage.

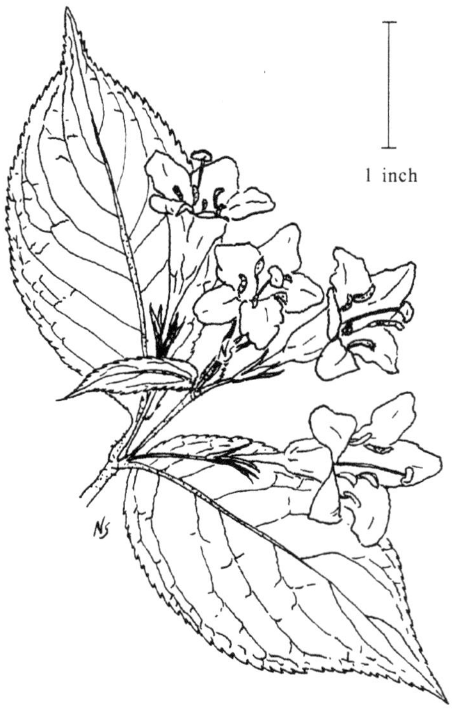

Weigela florida

Planning

Height: up to 10 ft (3 m)

Habit: shrub; spreading, rounded, branches arching; May-June blossoms slightly fragrant, sporadic blooms later in season on new growth

Site: sun; pH and soil adaptable; zone 5

Use: border, grouping, specimen

Recognition

Leaves: elliptic to egg-shaped, up to 4 in (10 cm) long, sometimes wider above middle; tip long-pointed, base tapered to rounded, margin with incurving teeth; glabrous above except on midrib, hairy on veins beneath; petioles subsessile, clasping

Flowers: single or in axillary clusters, each about 1.2 in (3 cm) long; calyx sparsely hairy to nearly glabrous, united below middle, lobes linear about 0.4 in (8 mm) long; corolla abruptly narrowed below middle, lobes rounded, spreading, rosy pink, paler within

Fruits: capsules glabrous; seeds not winged

Twigs: row of hairs each side from node to node, greenish brown, scurfy; pith white to light brown

Buds: divergent, narrow, greenish brown, sparsely hairy

Weigela praecox (Lemoine) Bailey - **praecox weigela**

Field tests for praecox weigela by the North Central Plant Introduction Station (1975) discovered that the cultivar 'Le Printemps' was hardy to -29 degrees F (-34 degrees C) but flower production was reduced. It was also injured by dry weather.

Planning

Height: up to 6 ft (2 m)
Habit: shrub; spreading, rounded; summer blossoms
Site: sun; pH and soil adaptable; zone 5
Use: border, grouping

Recognition

Leaves: elliptic to egg-shaped, sometimes wider above middle; tip long-pointed, margin toothed; hairy above, soft hairy below; petioles subsessile
Flowers: nodding, 3-5 on short lateral twigs; calyx long-hairy (pilose), lobes linear, united below middle; corolla abruptly narrowed below middle, rosy or purplish pink, yellow in throat, hairy outside
Fruits: glabrous; seeds not winged
Twigs: glabrous or long-hairy (pilose)

Order Asterales

Asteraceae is the only family in asterales.

Asteraceae Dumort. - aster family

The diverse aster family, one of the largest in the plant kingdom, includes over 20,000 species in about 1,100 genera grouped into 13 tribes (Cronquist 1981). Most species are herbaceous and inhabit temperate and subtropical regions. The aster family provides us with lettuce, artichokes, sunflower seeds, safflower oil, and a host of blooming ornamentals such as asters, daisies and chrysanthemums. A number of plants that we classify as troublesome weeds, such as the cocklebur and some of the thistles, are also members of this family. Some of the few woody species of the aster family are natives of the semi-arid western plains. Hence, they are useful components of landscapes on xeric sites.

Baccharis salicina Torr. & Gray, willow baccharis, is a low, dense native shrub that grows on sandy, often saline soils of floodplains in the southwestern part of the central prairie. Although it is not a highly decorative shrub, it might be useful for natural landscapes on sandy, saline soils. The dense upright branches of willow baccharis provide good nesting sites for redwing blackbirds where water is nearby (Stephens 1973). Its simple gray-green leaves, toothed or entire, are alternate on scurfy, ridged twigs. The unisexual dioecious flowers are borne in many-flowered, small, dense heads arranged on terminal panicles.

Hymenoxys scaposa (DC.) Parker, bitterweed, a low-branching subshrub native of the western central prairie, is adapted to high, dry limestone soils (Stephens 1973). The narrow, linear, simple leaves are clustered alternately near the twig end. The lower portion of the branches and main stem are barren except for dried leaf bases of previous years. Summer's yellow flowers are borne on solitary heads.

Hymenoxys scaposa

Recognition

Habit: herbs, subshrubs, seldom trees; deciduous, evergreen

Leaves: alternate, opposite, or rarely whorled; simple, variously divided, sometimes compound; entire or toothed; without stipules

Flowers: heads surrounded by a whorl of bracts (involucre) appearing to be a single flower but actually a number of small flowers; bisexual or unisexual; perianth 4- or 5-parted; calyx late-developing and reduced to scales, awns, hairs or wanting; corolla tubular and regular or with an elongated limb on one side (ligulate), rarely 2-lipped (bilabiate); stamens 4-5, inserted on corolla tube, filaments usually free but anthers united in a ring around the style; ovary inferior (epigynous), style 1, stigmas usually 2

Fruits: dry, non-splitting, 1-seeded (achenes), usually crowned by persistent calyx (pappus) that is sometimes deciduous or wanting

Twigs: often with milky juice

Artemisia L. - sagebrush

Artemisia—an inhabitant of northern temperate regions of the hemisphere—is a genus of about 200 species composed of aromatic herbs, subshrubs or shrubs. It is not to be confused with the *Salvia* species used in cookery. Several sagebrush species are moderately nutritious wildlife and livestock browse. The abundance of pollen produced by them can cause severe cases of hayfever for those who are allergic. Native Americans used sagebrush species for their fragrance and medicinal value, such as compresses, poultices, toilet paper, deodorants, nasal decongestants and mosquito repellant smoke.

Although *Artemisia* flowers are inconspicuous, the gray-green foliage has been effectively used by designers to imbue a regional flavor in southwestern landscapes. Many sagebrush species are adapted to semiarid alkaline sites. They may become rank, disease-prone weeds on fertile soils in humid climates.

Recognition

Habit: herbs, subshrubs, shrubs; deciduous

Leaves: alternate, simple; entire, toothed or lobed, aromatic

Flowers: heads inconspicuous, in terminal clusters (leafy spikes or panicles); involucre composed of overlapping bracts, receptacle not chaffy; without ray flowers; all flowers or only the marginal ones fertile

Fruits: without pappus arising from summit

Artemisia abrotanum L. - southernwood

Southernwood, a native of Eurasia and long cultivated for its aromatic odor, has become naturalized in eastern North America. The species and the dwarf variety *nanum* are hardy throughout the central prairie and tolerate slight alkalinity without injury (North Central Plant Introduction Station 1960, 1961).

Planning

Height: up to 6 ft (2 m)

Habit: subshrub; multi-branched; autumn blossoms yellow

Site: sun; pH 6.5-8.5; dry, well-drained soil; zone 4

Use: naturalizing, ornamental foliage, wildlife habitat

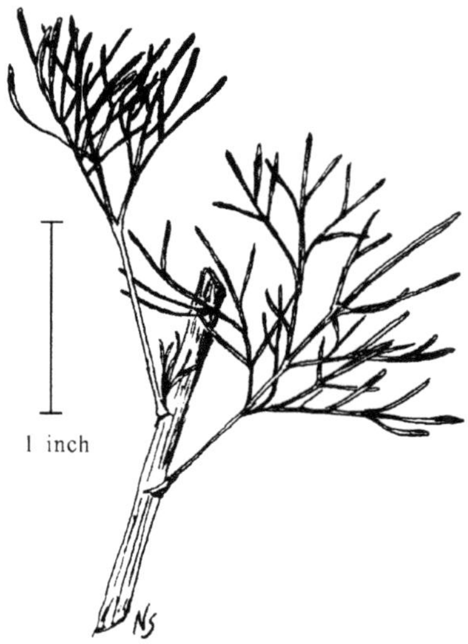

Artemisia abrotanum

Recognition

Leaves: up to 2.5 in (6 cm) long, lobed with long, rounded segments about 0.4 in (1 cm) long; glabrous or slightly hairy, green, fragrant

Flowers: heads several-flowered, about 0.2 in (5 mm) across on terminal spikes; bisexual except marginal female, all fertile; yellow; encircling bracts (involucre) hairy, nearly hemispherical with outer bracts long and narrow, tips pointed, and inner bracts wider above the middle

Twigs: green, dying back in winter

Artemisia cana Pursh - silver sagebrush

Silver sagebrush is a native of southwestern Canada and much of the western U.S. It can be found in the western Dakotas and northwest Nebraska, usually in valleys on sandy soils that tend to be occasionally moist (Stephens 1973).

Planning

Height: up to 3 ft (1 m)

Habit: shrub; densely branched, often at ground level; September blossoms yellow, very small

Site: sun; pH 6.5-8.0; zone 3

Use: naturalizing, ornamental foliage, wildlife browse

Recognition

Leaves: linear, up to 3 in (8 cm) long, about 0.2 in (5 mm) wide; tapering to both ends, widening at base, entire; surfaces covered with dense hairs mostly attached at middle with ends free (malpighiaceous); aromatic; sessile

Flowers: clusters (panicles) of heads terminal, each subtended by involucre of 8-10 bracts, further subtended by leaf-like bract; bisexual; calyx wanting; corolla yellow, tubular, lobes 5, acute, erect to recurved; stamens 5, alternate with corolla lobes

Fruits: achenes angular with 5-6 heavy ribs

Twigs: white, densely hairy (malpighiaceous)

Buds: naked with hairy embryonic leaves showing

Bark: thin, shredding into long fibrous strips

Wood: hard, fine-grained; sapwood yellow with dark brown rays

Artemisia filifolia Torr. - sand sagebrush

Sand sagebrush inhabits sandy soils of the southwestern U.S., extending north into eastern Colorado and Wyoming and western Kansas, Nebraska and South Dakota. Since it is disliked by animals and seldom browsed unless no other food is available, sand sagebrush survives and increases on overgrazed pastures, thereby aiding to control erosion. It provides protection for small mammals and birds.

Many dollars have been spent in attempts to eradicate sand sagebrush from rangeland. On the other hand, urban dwellers see the ornamental qualities of this fine-textured shrub, especially if a low water requirement plant is desired. Sand sagebrush planted at the Hans Burchardt Arboretum near Lincoln, far from its native habitat, continues to grow (Sutton 1998).

Artemisia filifolia

Planning

Height: up to 3 ft (1 m)

Habit: shrub; upright, dense, spreading from ground level; August blossoms inconspicuous; many leaves cling through winter

Site: sun; pH 6.5-8.0; well-drained soil; drouth tolerant; zone 4

Use: naturalizing, ornamental foliage, wildlife habitat

Recognition

Leaves: thread-like (filiform), about 1.5 in (4 cm) long, entire or with thread-like lobes; gray-green, appressed hairy; sessile

Flowers: clusters (panicles) terminal, leafy, up to 8 in (20 cm) long, with many densely flowered branches, overlaping cupped involucre bracts densely woolly; calyx wanting; 2-4 outer flowers female, with thin tubular corolla; 3-5 inner flowers male, with funnel-form corolla, yellow with 5 lobes, stamens 5

Fruits: achenes faintly 4-ribbed, glabrous; floral parts often persist

Twigs: gray-hairy (canescent), becoming less so; leaf bases persistent

Buds: nearly round, gray-hairy; scales strongly cupped

Bark: furrowed, shreddy

Artemisia frigida Willd. - **fringed sagewort**

Fringed sagewort occupies an enormous area in the Northern Hemisphere. The specific epithet, *frigida*, refers to the frigid regions of eastern Siberia where it was first discovered and named. Fringed sagewort grows in abundance on the high plains along the eastern slopes of the Rockies out into western Nebraska and north into Canada. It invades overgrazed ranges and moves into abandoned cropland. However, it is fair to good forage for livestock, deer and elk.

Fringed sagewort yields a fragrant essential oil containing a high percentage of borneol camphor and cineol with antiseptic value. It is a possible source of several commercial products of high potential value. Boiled plant parts of the sagewort were used by the Indians and early western explorers to obtain a potion used to treat colds, and as a diuretic and mild cathartic. Fringed sagewort is a possible alternative crop for agriculture (U. S. Forest Service 1937).

Planning

Height: up to 2 ft (50 cm)

Habit: subshrub; spreading at base, then upright, multiplies from rootstocks; leaves silvery; flowers scented (camphor-like), inconspicuous

Site: sun, moderate shade; pH and soil adaptable; drouth tolerant; zone 2

Use: naturalizing, wildlife habitat, erosion control, chemical production

Recognition

Leaves: clustered near base and scattered above, up to 0.2 in (5 mm) long, divided into linear rounded segments, aromatic; surfaces silvery-hairy; lower leaves petioled, upper sessile

Flowers: clusters (panicles) open, composed of globe-shaped densely white-hairy heads surrounded by several-rowed series of bracts (involucre), outer leaf-like, inner with papery margins; outer flowers female, slender tubular, yellow; inner (disc) flowers bisexual

Twigs: annual erect, slender, leafy, densely gray-hairy, fragrant; lower stems perennial, woody, much-branched, rooting in contact with soil

Artemisia tridentata Nutt. - **big sagebrush**

The specific epithet of big sagebrush, *tridentata*, refers to the three teeth at the apex of its leaf. It grows from extreme western Nebraska and North and South Dakota west to the eastern part of the west coastal states and British Columbia on semi-arid lands. Big sagebrush is the most abundant shrub in western North America and is increasing due to continued overgrazing.

This deep-rooted plant grows best on light, pervious, alkali-free soils that have a good water supply. Tall, dense stands of big sagebrush were an indicator to western settlers of soils suitable for cultivation. Although the pungent leaves are bitter, they are equal to alfalfa meal in protein and are higher in carbohydrates and fat. Big sagebrush is valuable forage for sheep and it serves as browse and habitat for wildlife (U. S. Forest Service 1937).

Big sagebrush provided an invaluable source of fuel to Native Americans and early explorers. It has been used as a tea substitute, tonic, eye-wash and antiseptic for wounds, but is much despised by persons allergic to the pollen that swirls in yellow clouds from it on a windy September day.

Artemisia tridentata

Planning

Height: up to 15 ft (4.5 m)
Habit: shrub; leaves dry and cling over winter; upright stems branched at ground level; September blossoms yellow, inconspicuous
Site: sun; prefers well-drained alkali-free soils; drouth tolerant; zone 4
Use: naturalizing, wildlife habitat

Recognition

Leaves: alternate; solitary or in groups; wedge-shaped, wider near apex, about 0.3 in (7 mm) wide and 8 in (20 cm) long, 3 rounded teeth at tip; surfaces covered with densely appressed silvery hairs (malpighiaceous), glandular; sessile
Flower: clusters (panicles) terminal, open, up to 6 in. (15cm) long; flowers bisexual, 5-8 in heads about 0.1 in (3 mm) across, surrounded by 8-18 bracts (involucre); calyx wanting, corolla tubular, yellow; stamens 5
Fruits: achenes glabrous or hairy, cylindric-top-shaped, 4- or 5-angled or ribbed, about 0.1 in (3 mm) long, glandular
Twigs: green becoming brown second year, densely hairy; leaf scars concealed by old leaf bases
Buds: egg-shaped, densely hairy
Bark: gray-brown, shreddy

Chrysothamnus Nutt. - rabbitbrush

About 70 species of rabbitbrush frequent alkaline plains and streambanks in semiarid regions of western North America, from southwestern Canada to Mexico and extending east to western Nebraska. They are most abundant in the Great Basin of Utah and Nevada. Most have little or no forage value and are indicators of range deterioration (U. S. Forest Service 1937).

Indians discovered rubber in some rabbitbrush species by chewing the stems, thereby creating a crude chewing gum. The latex-rich milky sap was seriously considered as an alternate rubber source during World War II. The showy yellow autumn flowers of some rabbitbrush species have prompted local use as ornamentals. *C. speciosus*, tall rabbitbrush, is especially handsome in bloom. It offers possibilities for cultivation (U. S. Forest Service 1937).

Recognition

Habit: small shrubs; deciduous

Leaves: alternate, rarely opposite, linear, sometimes larger above the middle; entire or minutely finely sharp-toothed; sessile or nearly so

Flowers: clusters terminal; flowers small in small heads of 3-20, bisexual; surrounded by several rows of dry, papery or leathery overlapping bracts (involucre); ray florets absent; corolla yellow, tubular

Fruits: hairy crowned by a ring of soft white to grayish white hairs (pappus)

Twigs: greenish, or with dense woolly hairs

Chrysothamnus nauseosus (Pall.) Brit. - rubber rabbitbrush

Rubber rabbitbrush contains a high grade rubber—chrysil—which vulcanizes easily. The disagreeable flavor of the foliage gives it the specific epithet *nauseosus* (U. S. Forest Service 1937). Greenplume *Chrysothamnus nauseosus*, ssp. *graveolens* (Nutt.) Piper, is the most common form found in the north central plains (Stephens 1973). Its natural range extends from British Columbia and Alberta south to Texas and eastern California.

Rubber rabbitbrush and its subspecies represent an untapped resource for naturalized landscaping in the western prairies because of its drouth tolerance and adaptation to a wide range of soil conditions including alkaline. Sutton (1998) has seen striking autumn landscapes of natural stands of rabbitbrush formed by gray stems accented with yellow flowers, backed by red-orange chokecherry foliage.

Chrysothamnus nauseosus

Planning

Height: up to 6 ft (2 m)

Habit: shrub; dense, ascending, narrow; foliage gray-green to whitish, late summer yellow blossoms

Site: sun; drouth tolerant; pH 6.5-8.5, soil adaptable; zone 3

Use: naturalizing, erosion control, wildlife habitat

Recognition

Leaves: alternate, linear, up to 3 in (8 cm) long, somewhat trough-shaped; surfaces covered with a fine felt of hairs; base broad, semi-clasping; sessile

Flowers: heads narrow, cone-shaped, 4- to 6-flowered, clustered at end of stems; bisexual; corolla yellow, tubular, sharply 5-toothed, lobes erect or slightly spreading; pappus slightly feathery attached to top of ovary, white, turning buff, nearly as long as corolla tube

Fruits: achenes sparsely hairy, cylindric, 4-6 small ridges; crowned by a ring of straight, soft, white hairs (pappus)

Twigs: covered with dense, gray-green to whitish felted hairs

Buds: egg-shaped, often concealed by old leaf bases

Bark: gray-brown, fissured, slighty shreddy

Chrysothamnus parryi (A. Gray) Greene **ssp.** ***howardi*** (Parry) H. Hall & Clem. **- Howard rabbitbrush**

The natural range of Howard rabbitbrush is limited to the western prairies. This low shrub appears gray from a distance because of the dense, long, matted hairs covering the foliage and twigs. Latex is not present in the leaves and stems.

Planning

Height: up to 1.5 ft (45 cm)

Habit: shrub; low, spreading; late summer yellow blossoms

Site: sun; pH and soil adaptable; drouth tolerant; zone 4

Use: naturalizing, erosion control, wildlife habitat

Recognition

Leaves: alternate; linear, up to 2 in (5 cm) long, somewhat v-shaped in cross-section; surfaces gray-green, dense mat of long hairs on both sides; narrows just above broad base, sessile

Flowers: heads of 4-6 florets in a terminal leafy cluster, persistent over winter; corolla yellow, cylindric, lobes 5, spreading, triangular, sharp-pointed; pappus of white, barbed hairs

Fruits: achenes densely hairy, tapering from a pointed base (obconic); crowned by a spreading tawny pappus

Twigs: covered with a whitish or greenish hairy mat

Buds: broadly egg-shaped, concealed beneath leaf bases; scales green, succulent

Bark: brown, loose and scaly

Chrysothamnus pulchellus (A. Gray) Greene **- rabbitbush**

Rabbitbush, a shrub of the southwestern U. S., is found naturally in southwestern Kansas and southeastern Colorado on sandy soils. Foliage provides food for rabbits and mice, and kangaroo rats eat the achenes (Stephens 1973).

Planning

Height: up to 2 ft (60 cm)

Habit: shrub; rounded, open, lower branches reclining except at tips (decumbent); summer blossoms white, inconspicuous; foliage greenish

Site: sun; tolerates drouth and sandy soils; zone 5

Use: naturalizing, wildlife habitat

Recognition

Leaves: alternate; linear, up to 1.5 in (4 cm) long, midrib prominent; tip short, abrupt, margins rough-hairy (scabrous-ciliate) or slightly barbed; surfaces green, granular; sessile, base broad

Flowers: heads of 5 florets in flat-topped clusters (cymes); involucre bracts in 5 rows, 5-6 per row, white with green tip; 5 flowers in each involucre, bisexual; corolla tubular, white, lobes 5, sharp-pointed, erect; pappus white, slightly barbed, longer than corolla

Fruits: achenes minutely hairy or glabrous, 4-angled; pappus straight, dense, buffy, spreading

Twigs: yellowish or green, becoming gray-brown, glabrous

Buds: hidden under persistent leaf bases

Chrysothamnus visidiflorus (Nutt.) Hook. - twistleaf rabbitbrush

Twistleaf rabbitbrush, a native of dry prairies and hillsides in the foothills of southwestern Canada and western U.S., has also been found in a small area of Sioux County, Nebraska (Stephens 1973). It provides fair to good forage for sheep and cattle (U. S. Forest Service 1937).

Chrysothamnus visidiflorus

Planning

Height: up to 3 ft (1 m)

Habit: shrub; rounded; late summer yellow blossoms

Site: sun; pH 6.5-8.5, soil adaptable; tolerates dry sites; zone 4

Use: naturalizing, erosion control

Recognition

Leaves: alternate; linear, up to 1.5 in (4 cm) long, usually twisted, 1 main vein and 2 indistinct from base; tip pointed, margin with small forward-pointing barbs; yellow-green, glabrous, sometimes minutely hairy below; base broad, sessile

Flowers: heads in terminal flat-topped clusters (cymes); flowers bisexual, 5 per involucre of 5 rows of bracts, keeled, yellow; corolla yellow, tubular, lobes 5, oblong, sharp-pointed, spreading; pappus of stiff, barbed, white hairs, equaling corolla

Fruits: achenes appressed hairy, wedge-shaped, 5-ribbed, somewhat flattened; crowned with yellow-brown pappus

Twigs: pale green or whitish, scattered hairs or glabrous

Buds: conical, brown, somewhat resinous; 2 outer scales short, spreading at tip

Bark: light brown, shreddy

Gutierrezia Lag. - **broomweed**

This genus was named after Piedro Gutierrez of the Madrid Botanical Garden, Madrid, Spain. About 20 species, all confined to the New World, have been described.

Gutierrezia sarothrae (Pursh) Brit. & Rusby - **broom snakeweed**

Broom snakeweed is native throughout the western half of the prairies, including the plains of Washington, California and Mexico. It is an agressive pioneer that invades areas where the climax vegetation has been depleted, except on saline or alkaline soils where it is not well adapted (U. S. Forest Service 1937).

Broom snakeweed is poor forage and will cause sickness in animals if eaten in large quantities. However, it, like many other weedy plants, can play an important role on land depleted of more useful vegetation by protecting the soil from wind and water erosion. Quail and kangaroo rats relish the broom snakeweed's achenes (Stephens 1973).

Planning

Height: up to 2 ft (.6 m)
Habit: subshrub; deciduous; dense, branching close to ground; late summer yellow blossoms
Site: sun; soil adaptable except not on highly saline or alkaline sites; drouth toleant; zone 3
Use: naturalizing, erosion control, wildlife habitat

Recognition

Leaves: alternate; simple; linear to thread-like, up to 1.5 in (4 cm) long; midrib prominent, margin inrolled; surfaces glutinous, dotted (punctate), with scattered, short, barb-like projections; broad base sessile and clasping
Flowers: heads few-flowered forming terminal clusters (cymes); heads surrounded by 5 rows of overlapping bracts, 4 per row; bracts linear, sharp, margins pale (scarious) midrib green; ray flowers female, corolla yellow, tubular, usually expanding into a ray about 0.1 in (2 mm) long; disc flowers bisexual, tubular, yellow; pappus chaff-like thin scales, short
Fruits: achenes cylindric, finely hairy, tipped with short, papery, chaffy scales (pappus)
Twigs: erect, green, becoming brown, unbranched except at the flowering top, slightly sticky, dying to near the base but remaining throughout the winter
Buds: naked in a cluster of tiny leaves
Bark: brown, split into long thin plates, becoming shreddy

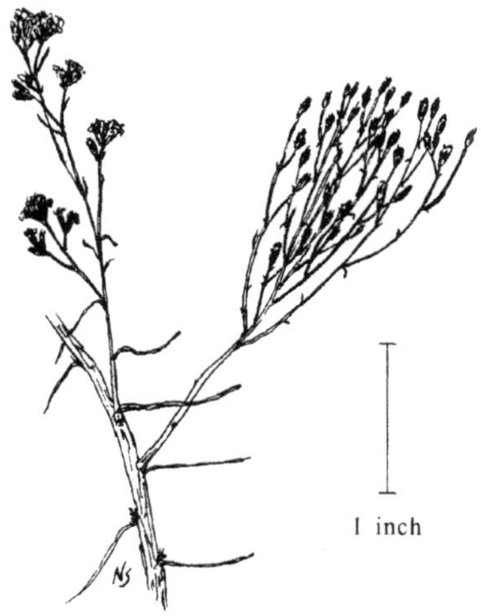

Gutierrezia sarothrae

Class Liliopsida - monocots

Liliopsida is divided into 5 subclasses, 19 orders, 65 families and about 50,000 species (Cronquist 1981). Most monocots are herbaceous. The few that are tree-like include bamboos, palms, yuccas, and dracaenas.

Recognition

Habit: herbaceous, woody
Leaves: usually parallel-veined, often sheathing at base, commonly oblong or linear in shape; petioles seldom developed
Flowers: bisexual (perfect) or unisexual (imperfect), parts usually in 3's or multiples of 3; cotyledon 1
Stems: without central pith or annular layers

Subclass - Liliidae

Cronquist (1981) divides Liliidae into two orders, 19 families and about 25,000 species. More than 80 percent of the species belong to two families: Liliaceae and Orchidaceae.

Recognition

Leaves: usually alternate, simple, entire or toothed
Flowers: usually bisexual, often showy
Fruits: most commonly capsules

Order Liliales

Liliales consists of 15 families and about 8000 species (Cronquist 1981).

Agavaceae Endl. - century-plant family

Agavaceae is divided into 18 genera and nearly 600 species (Cronquist 1981). These species, separated from Liliaceae on the basis of growth habit, are included in Liliaceae Adans. by some authorities (Rehder 1940; Elias 1980; Stephens 1973). Members of the century-plant family are natives of warm, arid regions.

The leaf fibers of the century plant (*Agave*) of tropical America are used in ropemaking. They are also the source of alcoholic drinks—pulque and mescal (Hylander 1956).

Recognition

Habit: shrubs, trees, herbs
Leaves: perennial, crowded in dense rosettes at stem ends or at the ground level on a short main stem, usually leathery or firm-succulent, often spine-pointed; margins often prickly, veins commonly obscure
Flowers: clusters (racemes, panicles or heads), massive, often dense, terminal or axillary; bisexual or sometimes unisexual (dioecious or polygamo-dioecious); perianth in 2 cycles of 3, petal-like, often thick and fleshy, distinct or joined below; stamens 6, filaments distinct; ovary superior (hypogynous) or inferior (epigynous), stigmas 3
Fruits: capsules that split into a cavity or berries; seeds flattened

Yucca L. - yucca

About 40 species of yucca are natives of the warm, dry parts of North and Central America and the West Indies. Many hybrids are cultivated, especially in Europe.

Recognition

Habit: shrubs, trees; evergreen
Leaves: long, sword-shaped, crowded into dense, terminal clusters
Flowers: clusters terminal, large, erect; bisexual; tepals white, sometimes tinged with purple, distinct or nearly so; stamens 6, pistil 6-sided
Fruits: non-splitting capsules, cylindrical, rounded or 6-sided, leathery or fleshy angled; seeds numerous

Yucca glauca Nutt. - soapweed

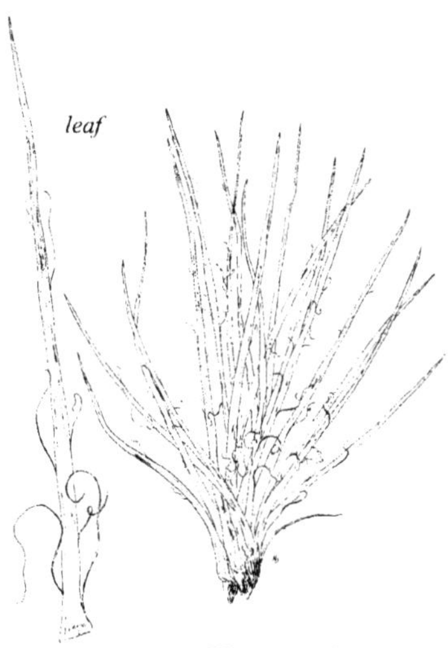

Yucca glauca
(not to scale)

Soapweed grows naturally throughout much of the western Great Plains. Native Americans used the soapweed's mascerated roots for soap and its leaves supplied sharp tips and thread-like fibers that served as needles and thread (Gilmore 1977).

Large soapweed populations occur in overgrazed pastures on sandy soils where it is well-equipped to become established and thrive. However, heavy winter grazing by mature cows can reduce the soapweed population (Stubbendieck et al. 1989). Cattle avoid the tough, fibrous leaves but eat the tender stems. Cows turned into a pasture populated with soapweed during the bloom period will run from plant to plant, vying with each other for a feast of the tasty flowers.

Yucca glauca

Planning

Height: up to 3 ft (1 m)
Habit: short-stemmed shrub, single or branched at base; leaves persisting for several years; spring blossoms white
Site: sun; prefers sandy soil or well-drained loams, pH adaptable; drouth tolerant (xerophytic); zone 4
Use: specimen, grouping, border, naturalizing

Recognition

Leaves: alternate; simple; 1-2 ft (30-60 cm) long, about 0.4 in (1 cm) wide; bases completely cover stem, persisting several years; stiff, concave, sharp-pointed, margin white, entire, from which white threads strip and curl, rolled inward at tip; base broadened, white; gray-green, glabrous, parallel veins inconspicuous
Flowers: clusters (racemes) narrow, up to 3 ft (1 m) tall; wide-spaced leaf-like bracts along the stalk; flowers bisexual, drooping; perianth cup-shaped, sepals 3, greenish white, petals 3, white, otherwise similar; stamens 5-6, filaments white, club-shaped, hairy, anthers yellow; ovary cylindric, green, glabrous, 3-celled, style with 3 swollen divisions, green
Fruits: capsules cylindric, erect, hard; 3 carpels splitting at base, each divided into 2 sections with short tips; seeds many, black, semi-glossy, somewhat triangular
Trunk: woody, short, stout, up to 1 ft (30 cm) high and 4 in (10 cm) thick

Yucca smalliana Fern. - Adam's needle

Adam's needle is a native of the southeastern U. S. that has escaped and become naturalized in eastern Kansas and Nebraska (Great Plains Flora Association 1986).

Y. filamentosa L., often confused with *Y. smalliana* and known by the same common name, bears leaves up to 4 in (10 cm) wide that are abruptly narrowed and infolded at the apex; the central axil of the panicle is normally glabrous and bears larger flowers than the latter (Dirr 1940). Dirr lists cultivars 'Ivory Tower', with summer blooming ivory-white flowers and 'Starburst', with long narrow green leaves striped with creamy yellow and tinged with pink in cool weather.

Planning

Height: up to 6 ft (2 m)

Habit: short-stemmed shrub; leaves basal, radiating upward from a short woody base (caudex); evergreen; June blossoms white

Site: sun; soil and pH adaptable; drouth tolerant (xerophytic); zone 4

Use: specimen, grouping, border

Recognition

Leaves: alternate; simple; 1-2 ft (30-60 cm) long, about 1.2 in (3 cm) wide; stiff, apex long-tapered; margin entire with long, white, peeling, usually curling, threads; margins at tip rolled in and united, forming stiff spiny point; surface gray-green, rough (scabrous); base white, sessile

Yucca filamentosa

Flowers: clusters (panicles) branched, 3-6 ft (1-2 m) tall, panicle axis and branches minutely hairy (puberulent); bracts about 1 in (25 mm) long at base of each branch, pointed; perianth broadly bell-shaped, sepals and petals distinct, cupped, glabrous, about 2 in (5 cm) long; stamens 6, filaments white, flattened, covered with small nipples (papillose), anthers yellow; ovary 3-celled, green, glabrous, styles 3, fused, stigma 3-lobed

Fruits: cylindric, constricted at middle, each section ridged along back, 3 carpels splitting at apex; seeds in 6 columns, black, flat, somewhat triangular

Trunk: woody, short, stout, concealed by leaves

Smilacaceae Vent. **- catbrier family**

The catbrier family contains about 12 genera that are natives of the North Temperate Zone and tropical and subtropical regions of the Southern Hemisphere (Cronquist 1981). Cronquist separates the catbriers from traditional inclusion in the Liliaceae because of their different characteristics in habit, leaf, and vascular systems.

Recognition

Habit: climbing herbaceous or woody vines, less often erect herbs or branching shrubs; deciduous, evergreen

Leaves: usually alternate; simple; blade expanded with 3-7 somewhat parallel, curved-convergent main veins connected by a network of smaller veins; petioles with tendrils commonly arising from junctions with short stipular flanges or open sheaths

Flowers: commonly clusters, axillary; unisexual (dioecious) or bisexual; perianth parts (tepals) 6, in 2 series, more or less petal-like; stamens usually 6, ovary usually superior

Fruits: berries

Smilax L. - greenbrier

About 350 species of greenbrier have been recognized, most of which are climbers. *Smilax bona-nox* L. is a native of Mexico and the southern U.S. as far north as southeastern Kansas. It is distinguished by flat, green prickles, many fine branchlets, and an impenetrable mass of branches formed from sprouts of underground stems (Stephens 1973).

A statement by Rydberg (1932) that *S. glauca* Walt. was found in Nebraska was a mistake attributed to his discovery of a variant of *S. hispida* (Great Plains Flora Association 1986). The species of *Smilax* exhibit considerable variation in the appearance of leaves and prickles on the stems. *S. ecirrata* (Engelm.) S. Wats. and *S. herbacea* L. also are natives of portions of the Great Plains (Great Plains Flora Association 1986).

Smilax hispida Muhl. - bristly greenbrier

Bristly greenbrier climbs to great heights through the canopy of trees in woodlands of the prairie region. Its fruits provide food for birds and they in turn insure a plentiful supply of greenbrier vines by spreading seeds under convenient trees.

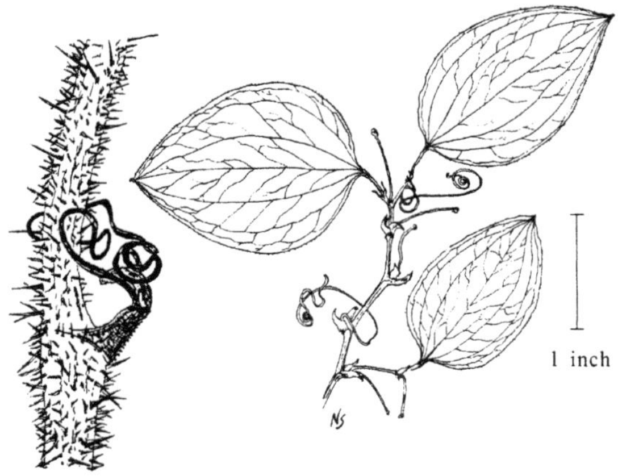

Smilax hispida

Planning

Height: up to 30 ft (9 m)
Habit: vine climbing by tendrils; deciduous; sparsely branching, not stoloniferous; spring blossoms greenish on new growth
Site: shade; soil and pH adaptable; zone 4
Use: wildlife habitat, naturalizing

Recognition

Leaves: alternate; simple; egg-shaped, variably pointed, base heart-shaped to wedge-shaped, 5- to 9-nerved, margin slightly roughened; surfaces glabrous, upper dark green, lustrous, lower pale; petioles grooved above, often bent and twisted at summit; stipules adnate, may be prolonged into a tendril
Flowers: clusters (umbels) axillary; unisexual (dioecious); perianth (tepals) 6-parted, greenish to bronze; male flowers—tepals recurved, filaments flattened, green, pistil vestigial; female flowers—tepals spreading, ovary green, egg-shaped, style very small, stigmas stout, recurved
Fruits: black, globose, smooth, about 0.3 in (8 mm) across; seeds 1-2, dark red, smooth, glossy
Twigs: green with minute white dots, few or no prickles on outer branches
Buds: green, conical, hidden by old leaf base
Trunk: green, smooth, hard; prickles dense, decreasing toward tip, nearly black, slender, terete; no true bark, wood and cambium

Glossary

Botanical terms are defined in this glossary that are seldom or never used in our text, but are included to aid those readers of articles that may not define such words therein. **Webster's New World Dictionary of the American Language**, College Edition, published by The World Publishing Company also defines many botanical terms.

A-. A prefix denoting a lack of, as in "apetalous"—lacking petals.

Achene (akene). A dry, one-seeded, indehiscent fruit.

Acicular. Needle-like.

Acuminate. Gradually tapering to a long point; attenuated.

Acute. Tapering to a point to form an angle less than 90 degrees.

Adherent. Describing dissimilar organs touching, but not fused.

Adnate. Unlike parts are fused together.

Adventitious. Arising from an unusual place, as when roots or shoots develop along a stem at places other than at nodes.

Aggregate fruit. A cluster formed by the fusion of separate pistils of the same flower; characteristic of *Rubus* species. See also **Multiple fruit**

Allopatric. Occuring in different areas.

Ament. See **Catkin**

Anastomosing. Interlacing, as when veins of a leaf join to enclose the space between.

Androecium. Stamens, as a group.

Anemophily. Refers to flowers pollinated by wind.

Anther. Pollen-bearing portion of the stamen.

Anthesis. The period when pollen is released from the anther; the time of blooming.

Apetalous. Lacking petals.

Apiculate. Abruptly pointed.

Apomixus (apogamy). The development of a plant without the union of sexual organs or cells.

Appressed. Flattened against, as a bud against a stem.

Arcuate. Describes leaf veins curving to parallel the margin; characteristic of *Cornus* species.

Areole. A spot on a stem or leaf often bearing a tuft of spines; characteristic of *Opuntia* species.

Aril. A fleshy appendage, often spongy or gelatinous, that grows from the point of seed attachment and sometimes envelops the seed; characteristic of *Taxus* species.

Articulate. Having nodes, joints or points of separation.

Asepalous. Without sepals.

Auriculate. Bearing ear-like appendages (auricles); characteristic of the base of petals and the leaves of *Quercus robur*.

Awn. Bristle-like appendage.

Axil. The angle at the point of attachment, as between a leaf and a stem.

Axillary. Referring to within the axis, as flowers that are borne in leaf axils.

Baccate. Berry-like; pulpy, fleshy.

Berry. A simple (developed from a single pistil), fleshy, usually non-splitting fruit with one or more seeds; characteristic of the fruits of *Vitis* species.

Bifid. Two-cleft.

Bilabiate. Two-lipped.

Bisexual. A flower that bears both stamens and pistils.

Bloom. A waxy or powdery coating, usually whitish and easily removed.

Boss. A raised usually pointed projection.

Bract. A modified leaf subtending a flower or a flower cluster.

Bud. An appendage in the embryo stage that gives rise to leaves, flowers and/or new shoots; the stage at which the growing points of a plant are in dormancy.

Bud scale. A modified leaf or other structure protecting the embryonic tissues of a bud.

Bullate. A blistered or puckered surface found between leaf veins.

Bundle scar. The broken ends of woody vascular strands, as seen on a leaf scar.

Caducous. Falling early or prematurely, as stipules at the base of petioles do.

Caliper. A term used in the nursery trade to denote the diameter of a woody stem above the root swell; used to determine the size of seedling trees and shrubs.

Calyx. The outer floral envelope composed of sepals. See also **Perianth**

Glossary

Campanulate. Bell-shaped.

Canescent. Covered with gray, hoary hairs.

Capitate. A dense, head-like cluster.

Capsule. A simple, dry fruit with two or more carpels splitting along two or more lines of a suture; characteristic of *Catalpa* species.

Carpel. A simple pistil or one unit of a compound pistil; also the cone scale of a conifer fruit.

Catkin. An elongate, unisexual flower cluster lacking petals (ament); descriptive of Salicaceae species.

Cespitose (caespitose). Forming a low mat or tuft.

Ciliate. Describes a margin fringed with hairs.

Clavate. Club-shaped.

Clone. A group of plants, obtained by vegetative means such as grafting, budding or rooting, that are all descended from a single individual.

Collateral. Describes accessory, subordinate buds arranged on either side of lateral buds.

Complete. Describes a flower having sepals, petals, stamens and one or more pistils.

Compound. Having several parts; describes a leaf with two or more leaflets or a pistil with two or more carpels.

Cone. A flower cluster or fruit with overlapping scales; the common name for the fruits of conifers.

Coniferous. Cone-bearing.

Connate. Referring to like structures that are fused together, as anthers of Caesalpiniaceae and Fabaceae.

Cordate. Heart-shaped with a basal notch.

Coriaceous. Leathery texture.

Corolla. The inner floral envelope (petals). See also **Perianth**

Corymb. An indeterminate flower cluster consisting of a central stem bearing flowers with stems of different lengths to form a flat or somewhat rounded cluster. The outer flowers open first; characteristic of *Viburnum* species.

Cotyledons. The primary leaves of the embryo contained within the seed sometimes emerging at germination as the first seedling leaves.

Crenate. Describes a margin that has rounded to blunt teeth.

Crenulate. Finely crenate.

Cruciform. Cross-shaped.

Cultivar. A single entity within a species resulting from selection or breeding under cultivation. Example: *Picea pungens* 'Koster' or *P. pungens* cv. Koster. The name may include up to three words; no Latin terms after January 1, 1959, can be used. See also **Variety**

Cuneate. Wedge-shaped base.

Cuspidate. Tipped with an abrupt, sharp, often rigid point.

Cyme. A determinate flower cluster consisting of a central stem bearing flowers on stems of different lengths to form a flat to rounded cluster. The central flowers open first; characteristic of *Sambucus* species.

DBH. Diameter breast-high—4.5 ft (1.4 m) above the ground.

Deciduous. Not persistent; dropping as a regular occurence, as leaves do in autumn.

Decumbent plant. Reclines on the ground with ascending tips.

Decompound. More than once compounded.

Decurrent. Extending down the stem; describes a leaf blade extending down the twig, as wings below the point of attachment.

Decussate. Alternating in pairs at right angles to each other; characteristic of the leaves of *Thuja* species.

Dehiscent. Splits open, as the opening of an anther or fruit is split by valves or slits.

Deliquescent. A many-branched axis; characteristic of *Ulmus americana*.

Deltoid. Triangular with a nearly straight base, with sides often slightly curved toward the apex; descriptive of *Populus deltoides* leaves.

Dentate. A margin with sharp teeth pointing outward, perpendicular to the margin.

Denticulate. Finely dentate.

Determinate. The terminal (or central) flower of a cluster opens first, thereby arresting prolongation.

Diadelphous. Describes stamens that are formed into two groups by union of their filaments; characteristic of many species of Fabaceae.

Dichogamy. Female-male flowering sequence.

Dichotomous. Branching by pairs.

Dicotyledonous. Bearing two cotyledons.

Didynamous. Bearing four stamens in pairs of unequal length.

Dimorphic. Occurring in two forms.

Glossary

Dioecious. Male and female flowers are produced on separate plants.

Disk flowers. Tubular flowers in the central part of the head; characteristic of the flowers of Asteraceae.

Distal. Toward the apex; opposite of proximal.

Diurnal. Recurring every day; blossoms open only during the day.

Dorsal. The back, outer, or lower face of an organ; the side away from the axis.

Double flower. More than the usual number of petals.

Drupaceous. Drupe-like fruit; not necessarily a true drupe.

Drupe. Simple, one-seeded, fleshy fruit with a bony inner wall enclosing the seed; a stone fruit, descriptive of *Prunus* fruits.

Duodichogamy. Male-female flowering sequence.

Echinate. Armed with prickles.

Eglandular. Lacking glands.

Ellipsoid. Football-shaped; shaped like a three-dimensional ellipse.

Elliptical. Resembling an ellipse; longer than wide, with rounded ends.

Emarginate. A shallow notch in the apex, as in a leaf or petal.

Entire. A margin wholly without teeth or lobes.

Epi-. A prefix meaning upon.

Epigynous flower. The ovary is inferior; the other parts are inserted upon or near the summit of the ovary.

Epiphyte. An independent plant growing upon another plant without a connection to the ground.

Erose. Irregularly toothed or eroded; jagged.

Espalier. A plant trained to grow in one plane, often against a wall.

Estipulate. Lacking stipules.

Excavated pith. Hollow between the stem nodes.

Excurrent. A prolonged axis forming an undivided stem; descriptive of the branching habit of *Pinus* species.

Exserted. Extending beyond, as stamens may extend beyond the corolla.

Extrorse. Facing outward from the center.

Falcate. Sickle- or scythe-shaped; curved.

False terminal bud. See **Pseudo-terminal bud**

Fascicle. A cluster; used to describe *Pinus* leaves (needles).

Fastigiate. Erect and closely spaced branches; columnar.

Filament. The stalk of a stamen; supports the anther.

Fluted. Alternating ridges and depressions, as a trunk that has rounded lengthwise ridges.

Foliaceous. Leaf-like in appearance and texture, usually green in color.

Follicle. A dry fruit with one carpel (simple ovary) splitting along one side.

Forma. A subdivision of a species that may occur naturally, seldom breeds true, and does not develop a natural population. Members often differs in a single character only, such as flower color.

Fruit. A ripened ovary with adnate parts, characteristic of the Angiosperms; loosely used to denote the seed-bearing structures (cones) of Gymnosperms.

Fruticose. Shrubby.

Funnelform. The corolla tube of the flower gradually widens, resembling a funnel.

Fusiform. Spindle-shaped; tapering to the ends from a thickened middle.

Gamopetalous. Petals united, at least at their base.

Genus. A group of species possessing well-marked common traits, especially a similarity of flowers and fruits.

Gibbous. A swelling or protuberance, as on the side of the corolla tube of some *Lonicera* species.

Glabrate. Nearly or becoming devoid of hairs at maturity.

Glabrous. Smooth; lacking hairs.

Glaucous. Covered with a grayish or bluish waxy or powdery substance, as the "bloom" on the fruits (grapes) of *Vitis* species.

Globose. Spherical, or nearly so.

Gymnospermous. Bearing naked seeds.

Gynoecium. The female element of a flower; a collective term for the pistil or pistils.

Halophyte. A plant usually associated with saline soils.

Hardy. Generally used to describe a plant capable of enduring winter stress.

Hastate. Having the shape of an arrowhead with divergent lobes at the base.

Head. A dense spherical or flat-topped cluster of stemless or very short-stemmed flowers on a short or broadened axis.

Hilum. The point of attachment of a seed, marked by a scar.

Hip. The enlarged hypanthium that is the fruit of *Rosa* species.

Hirsute. Covered with stiff, bristly hairs.

Hoary. Densely grayish white hairy.

Holotype. The specimen used or designated in the naming of a plant.

Homogeneous. Of uniform texture, as the pith of stems.

Hybrid. A plant that results from crossing different species or subspecies but not from crossing minor variants of a species; designated by an "×" preceding the specific epithet (the second name of the binomial name).

Hypanthium. A saucer-, cup-, rod-shaped, or sometimes tubular expansion of the axis that produces floral organs (sepals, petals and stamens) from its upper margin. Examples: it may be cup-shaped and surrounding but free from the gynoecium, as in *Prunus* species; it may be cup-like and expanding to become the hip of *Rosa* species; or it may surround the ovary and become adnate to it to become the fruit, as in *Malus*. See also **Receptacle**

Hypogynous flower. The ovary is superior; above the point of attachment of other flower parts.

Imbricate. Overlapping, similar to the way roof shingles are arranged.

Imperfect flower. Organs of one sex are abortive or wanting.

Incomplete flower. Lacking one or more parts.

Indehiscent fruit. Not opening by valves or slits.

Indeterminate. The flowers of a cluster open progressively from the base upwards, or from the outer to the inner, as in a corymb; growth continues as the flowers open.

Indigenous. Native and original to an area.

Inflorescence. A flower cluster.

Internode. The portion of stem between two adjacent nodes.

Involucre. A cluster of bracts subtending a flower or cluster of flowers, such as those subtending the heads of many Asteraceae species.

Irregular. Having dissimilar parts, such as flowers with petals of different shapes; not symmetrical; zygomorphic.

Keel. A central ridge on the outer (dorsal) surface; the united petals of a flower common to many species of Fabaceae.

Key. A winged, indehiscent fruit; characteristic of *Acer* species. See **Samara**

Labiate corolla. Lipped.

Laciniate. Cut into lobes separated by deep, narrow, irregular incisions.

Lactiferous. Having milky sap.

Lamina. Expanded portion of a leaf, sepal or petal; blade.

Lanate. Woolly.

Lanceolate. Much longer than wide, broadest below the middle and tapering to the apex.

Lateral bud. Borne above the point of leaf attachment; not terminal on the twig.

Leaflet. A single division of a compound leaf.

Legume. A dry fruit from a simple ovary usually splitting at maturity along two sutures; characteristic of Fabaceae species.

Lenticel. A corky spot on a twig or branchlet that provides gaseous exchange between the inner tissues and the atmosphere.

Liana. Tropical jungle woody vine.

Limb. The expanded part of a united (gamopetalous) petal or sepal.

Linear. Long and narrow, margins nearly parallel.

Lingulate. Strap-shaped.

Lobe. A protuding segment of a leaf or other organ.

Locule. A cell or compartment within an ovary.

Lunate. Crescent-shaped, as the moon at the quarter phase.

Lustrous. Slight metallic gloss.

Malpighiaceous hairs. Appressed, straight and attached near the middle; characteristic on parts of some Cornaceae species.

Marcescent. Withering, but persisting in place.

Membranous. Thin, somewhat flexible and more or less translucent.

Mesophyte. A plant having medium water requirements.

Midrib. The central vein of a leaf or other part.

Monadelphous stamens. United into one group by union of filaments.

Monocotyledonous. Bearing one cotyledon.

Monoecious. Male and female unisexual flowers borne on the same plant.

Monotype. Only one species in a genus.

Mucro. A short abrupt tip, as on a leaf or cone scale.

Mucronate. Terminated with a mucro; bristle-tipped.

Multiple fruit. Formed by the fusion of pistils of several flowers inserted on a common receptacle; characteristic of fruits of *Morus* species. See also **Aggregrate fruit**

Naked bud. Lacking scales.

Node. A joint on a stem; a point of origin of a bud.

Nut. A dry, nonsplitting, 1-celled and 1-seeded fruit with a bony, woody, leathery or papery wall, partially or wholly encased in an involucre or husk.

Oblanceolate. Broadest above the middle; inversely lanceolate.

Obovate. Inversely ovate; broadest above the middle.

Obovoid. Appearing as an inverted egg.

Odd-pinnate. Referring to a compound leaf with a terminal leaflet.

Odoriferous. Aromatic but not pleasantly so.

Orbicular. Circular.

Oval. Broadly elliptical; the width greater than one-half the length.

Ovary. The basal portion of a pistil containing one or more ovules.

Ovate. Egg-shaped in outline; broadest below the middle.

Ovoid. Appearing as an upright egg.

Ovule. The structure in the ovary that becomes a seed after fertilization.

Palmate. Digitate; parts diverge from a common base; characteristic of leaflets of *Aesculus*; also describes leaf veins.

Panicle. An elongated, indeterminate flower cluster whose axis bears branching flower stalks; a branching raceme.

Papilionaceous. Describing flowers of Fabaceae having a standard (banner), two wings and two keel petals comprising the corolla.

Pappus. The modified calyx (hairs, bristles, awns, etc.) of Asteraceae species borne on the ovary and usually persisting on the summit of the achene.

Parthenocarpic. Describing a seedless fruit developed without fertilization.

Pectinate. Comblike.

Pedicel. A stalk of a single flower or fruit in a cluster.

Peduncle. A primary stalk supporting either a cluster of flowers or fruits; the stalk of a single flower or fruit.

Peltate. Shield-shaped; the supporting stalk is attached near the middle rather than the edge of a *Taxodium* species cone scale.

PERC. The W. D. Holley Plant Environmental Research Center at Colorado State University, Fort Collins.

Perfect flower. Functioning male and female sex organs (stamens and pistils) are present.

Perianth. The floral envelope of a flower; a collective term referring to the calyx and corolla as a unit, or either if only one is present.

Pericarp. The wall of the ripened ovary.

Perigynous flower. The ovary is "half-inferior"; the calyx, corolla and stamens are borne on the edge of a cup-shaped hypanthium that is not fused to the ovary.

Petal. A unit of the corolla.

Petaloid. Resembling a petal, as a sepal of the color and texture of a petal.

Petiole. Leaf stalk.

Petiolule. Leaflet stalk.

Phyllotaxy. The arrangement of plant parts (leaves, flowers, etc.) on an axis.

Pilose. Shaggy with soft hairs.

Pinnate. Like a feather.

Pinnately compound leaf. Segments arranged along each side of a common axis or rachis; feather-like.

Pistil. The female reproductive part (gynoecium); one to many in a flower.

Pistillate. Refers to a flower with no functional stamens.

Pith. The central tissue of a twig; usually softer and of different color than the wood (xylem).

Pod. A general term for a dry, splitting fruit.

Pollen. The microspores produced by the anther of a stamen.

Polygamo-dioecious. Describes plants bearing bisexual flowers as well as bearing male or female flowers on separate plants.

Polygamo-monoecious. Describes plants bearing both bisexual and some staminate flowers.

Polygamous. Describes plants bearing both unisexual and bisexual flowers.

Polypetalous flower. The corolla consists of separate petals.

Polysepalous flower. The calyx consists of separate sepals.

Pome. A simple fruit, with flesh derived largely from the adnate hypanthium; tasty on some species of the subfamily Maloideae.

Procumbent plant. Lies flat on the ground; the stem does not root at nodes or tip.

Protogynous flower. The stigma is receptive of pollen before the anthers on the same flower mature and release pollen.

Protrandrous flower. The anthers mature before the stigma is receptive.

Proximal. Toward the base.

Pruinose. Describes a waxy bloom that is granular or dust-like.

Pseudo-terminal bud. A false terminal bud; the uppermost lateral bud of a twig with a leaf scar on one side and a branch scar opposite.

Puberulent. Minute hairs that require a lens to detect.

Pubescent. Bearing soft, short hairs.

Punctate. Bearing pits or dots that may be translucent.

Pyriform. Pear-shaped.

Raceme. An indeterminate flower cluster on an elongated axis with flowers on simple pedicels.

Rachilla. A rachis branch; a secondary axis.

Rachis. The primary axis of an elongated flower cluster or of a pinnately compound leaf.

Ray flower. Located on the margin of a head, often bears a strap-shaped corolla; characteristic of Asteraceae species.

Receptacle. The axis to which the various flower parts are attached. It may be enlarged to accomodate a head, as in Asteraceae species. See **Hypanthium**

Regular. Describes flowers that have uniform petals and sepals; actinomorphic or symmetrical.

Reniform. Kidney-shaped.

Repand. Having a wavy margin.

Reticulate. Net-like.

Rhizome. A modified underground stem, usually horizontal; differs from a root by the presence of nodes.

Rosette. A basal leaf cluster on a very short stem.

Rostrate. Beaked.

Rotate corolla. Wheel-shaped with an inconspicuous tube and a widely spreading limb flaring at right angles to the flower axis.

Rugose. Wrinkled.

Runner. Horizontal above-ground stem that may root at the nodes or apex.

Saccate. Bag-shaped.

Sagitate. Shaped like an arrowhead.

Salverform corolla or calyx. Having a slender tube and a widely spreading limb at right angles to the tube. Compare with "Rotate"

Samara. Simple, dry, nonsplitting fruit, usually 1-seeded and 1-winged; characteristic of *Fraxinus*, *Ptelea* and *Acer* species.

Scabrous. Rough to the touch due to the presence of stiff hairs or other projections.

Scarious. Thin and dry; not green.

Schizocarp. A fruit that splits apart, having a seed in each half; characteristic of *Acer* species. See **Samara**

Schlerophyllous leaf. Leathery or tough.

Scurfy. Covered with minute scales.

Seed. A mature ovule; an embryo with a protective coat.

Sepal. One of a set of the outer floral leaves; a part of the calyx. See also **Perianth**

Sericeus. Silky with soft hairs.

Glossary

Serotinous. Describes cones that remain closed long after maturity; characteristic of fruits of *Pinus banksiana*.

Serrate. Having sharp teeth pointing toward the tip; saw-toothed.

Serrulate. Minutely serrate.

Sessile. Lacking a stalk.

Seta. A bristle.

Silique. A simple, dry, splitting fruit with two carpels that separate, leaving a thin, persistent partition on the plant; characteristic of fruits of Brassicaceae species.

Simple. Not compound; unbranched; e.g., a fruit derived from a single flower and a single pistil, a leaf having a blade in one piece, or a pistil with a single carpel.

Sinus. The space between two lobes.

Spatulate. Oblong, sometimes broader toward the upper end, with a rounded apex.

Species. The basic unit for classification of living organisms. A natural group composed of similar individuals which produce similar offspring from generation to generation.

Spike. An elongated cluster of stalkless or subsessile flowers; unbranched.

Sporophyll. A reproductive organ that produces spores; the stamen and carpel of a flowering plant.

Stamen. The male pollen-producing organ of a flower consisting of the filament and anther; a microsporophyll. See also **Androecium**

Staminate. Referring to the male unisexual flower or plant.

Staminode (staminodium). A sterile stamen.

Standard. The uppermost large petal (banner) of flowers of Fabaceae species.

Stellate. Star-shaped; describes hairs on leaves and other plant parts.

Sterigmata. Woody peglike outgrowths from a twig to which leaves are attached; characteristic of *Picea* species.

Stigma. The generally enlarged and terminal part of the pistil on which pollen adheres and germinates.

Stipil. A stipule-like appendage at the base of a leaflet.

Stipule. A leafy appendage attached to the twig at the base of a leaf stalk, usually found in pairs, one on each side. They are sometimes deciduous before the leaves fall. See also **Caducous**

Stipule scar. A mark on the twig at the point of stipule attachment; often very small and difficult to detect.

Stolon. See **Runner**

Stoloniferous. Bearing stolons (runners).

Stoma (stomate). A specialized orifice in the leaf epidermis that permits gaseous interchange.

Striate. Covered with fine longitudinal lines or ridges.

Strigose. Covered with straight appressed hairs.

Strobilus (strobile). A seed-bearing cone, as of Pinaceae.

Style. The stalk-like part of a pistil connecting the ovary and stigma.

Subopposite. Leaf pairs not found at exactly the same level on the twig; characteristic of *Rhamnus cathartica* leaves.

Subspecies. A morphologically distinct population within a species that is used by some authors to designate geographical variation. A subspecies may have evolved differently from growing in different habitats or areas, but will normally be able to interbreed if the opportunity arises. See also **Variety**

Subulate. Awl-shaped.

Sucker. A shoot arising from the roots or the stem at the ground surface.

Suffrutescent. A plant with a woody base, but otherwise herbaceous; subshrub.

Superposed. Inserted one over the other; descriptive of buds common on *Juglans* species.

Symmetrical flower. Having the same number of parts in the calyx and corolla.

Sympatric. Occuring in one area.

Sympatry. Exhibited by two or more species that share the same site without interbreeding.

Sympetalous. See **Gamopetalous**

Syncarpous pistil. Carpels united.

Taxon. A unit of classification (form, variety, subspecies, species, genus, etc).

Tendril. A stem or leaf modified to form a slender, twining structure.

Tepal. A segment of an undifferentiated perianth; the calyx and corolla can not be distinguished; characteristic of *Magnolia* flowers.

Terete. Cylindrical or circular in cross-section.

Ternate. Found in threes.

Glossary

Tetradynamous flower. Having six stamens, four longer than the other two; characteristic of some Brassicaceae species.

Tomentose. Covered with soft, curly, matted hairs; woolly.

Torus. See **Receptacle**

Truncate. The base or apex appears to be abruptly cut off, descriptive of a leaf or petal.

Tubular. Tube-like; descriptive of a calyx or corolla.

Turbinate. Top-shaped; an inverted cone.

Twig. A woody plant shoot of the current season.

Umbel. A flat-topped or rounded flower cluster with stalks (pedicels) of nearly equal length attached to the top of the main stalk (peduncle); characteristically blooming from the outside toward the center, indeterminate.

Umbo. A conical projection arising above the surface; descriptive of a protuberance on pine cone scales.

Unarmed. Lacking prickles or spines.

Uncinate. Hooked at the end; descriptive of some spines.

UNARDC. The University of Nebraska Agricultural Research and Development Center at Ithaca.

Undulate. Wavy.

Unisexual. Of one sex only.

Urceolate corolla. Urn-shaped; united petals with a bulbous tube, a narrowed neck, and a very small limb.

Utricle. One-seeded fruit with a thin wall, often opening by a lid.

Valvate. Meeting at the edges, not overlapping; often descriptive of bud scales such as those of *Viburnum lentago*.

Variegated. Striped or mottled; descriptive of leaves marked with a color other than green.

Variety. Individuals of a species found in the wild that are distinct in form or function and generally breed true. See also **Subspecies**; **Cultivar**

Vascular bundle scar. Spots on a leaf scar indicating the conducting tissues.

Vasiform. Elongated funnel-shaped.

Vein. A bundle of visible transporting tissue in a leaf.

Velutinous. Covered with a heavy coating of straight, dense hairs.

Venation. Vein pattern.

Ventral. Inner face of an organ; the opposite of dorsal.

Ventricose. A one-sided swelling; descriptive of some corolla tubes.

Vernal. Pertaining to, or appearing in the spring.

Verrucose. A wart-like surface.

Versatile. Attached to the middle, free to swing; descriptive of some anthers.

Verticil. A whorl.

Verticillate. Arranged in whorls.

Villous (villose). Covered with long, fine hairs, not matted.

Virgate. Wand-like; describes a slender, straight leafy stem bearing only short or no branches.

Viscid. Sticky; foreign particles will adhere.

Whorl. A group of three or more parts at a node; characteristic of *Catalpa* leaves.

Wing. A flat, thin appendage to a fruit; also, a lateral petal of a papilionaceous corolla, characteristic of flowers of Fabaceae species.

Woolly. Having long and matted hairs.

Xerophyte. A plant adapted to very dry sites.

Zygomorphic flower. Not radially symmetrical; having parts of the same kind of different sizes or shapes so as to be bilaterally symmetrical instead.

Cited and Selected References

Allard, R. W. 1960. Principles of plant breeding. John Wiley & Sons, Inc., New York. 485 p.

Allen, S. W., and G. W. Sharpe. 1960. An introduction to American forestry. 3rd Ed. McGraw-Hill Book Co., New York. 466 p.

American Forestry Association. 1992. National register of big trees. American Forests special section. 47 p.

Angle, K. J. 1985. Variation in drought resistance of four provenances of bur oak (*Quercus macrocarpa* Michx.). Master's Thesis, University of Nebraska, Lincoln. 140 p.

Bagley, W. T. 1973. Hybrid poplar clones compared. Jour. Forestry 71 (1):26-7.

Bagley, W. T. 1979. Hackberry—a hardy plant often overlooked. American Nurseryman 150(3):15, 88-89.

Bagley, W. T. 1988. Agroforestry and windbreaks, p. 583-598. In J. R. Brandle, D. L. Hintz and J. W. Sturrock [ed.], Windbreak Technology. Elsevior, Amsterdam—Oxford—New York—Tokyo.

Bagley, W. T. 1998. Personal observations of the author.

Bailey Hortorium. 1976. Hortus III. Macmillan Co., New York. 1116 p.

Barker, P. A., D. C. Freeman, and K. T. Harper. 1982. Variation in the breeding system of *Acer grandidentatum*. Forest Sci. 28(3):563-72.

Bates, C. G. 1944. The windbreak as a farm asset. U. S. Dept. of Agr. Farmers' Bull. No. 1405. 22 p.

Beales, P. 1992. Roses. H. Holt and Co., Inc., New York. 472 p.

Bey, C. 1973. Growth of black walnut trees in eight midwestern states—a provenance test. U. S. Forest Service, Res. Paper N.C. 91. 7 p.

Bowers, C. G. 1960. Rhododendrons and Azaleas. Macmillan Co., New York. 525 p.

Brandle, J. R., B. B. Johnston, and T. Akeson. 1992. Field windbreaks: are they economical? Jour. of Prod. Agr. 5(3):393-398.

Braun, E. L. 1969. The woody plants of Ohio. Hafner Publishing Co., New York. 362 p.

Callaway, D. 1994. World of magnolias. Timber Press, Inc., Portland, Ore. 260 p.

Cathey, H. M. 1990. USDA Plant Hardiness Zone Map. U. S. Agricultural Research Service, U. S. National Arboretum, Misc. Pub. No. 1475.

Chadwick, L. C. and R. A. Keen. 1976. A study of the genus *Taxus*. Res. Bull. 1086, Ohio Agr. Res. and Development Center, Wooster. 56 p.

Clark, R. C. 1989. Chinkapin oak: satisfaction for the homeowner, skuldeggary for the botanist. The Morton Arboretum Quarterly 25(1):11-16.

Cook, D. I. and D. F. Van Haverbeke. 1971. Trees and shrubs for noise abatement. Neb. Agr. Exp. Sta. Bull. 246. 75 p.

Cronquist, A. 1981. An integrated system of classification of flowering plants. Columbia Univ. Press, New York. 1262 p.

Dicke, S. G. and W. T. Bagley. 1980. Variability of *Quercus macrocarpa* Michx. in an eastern Nebraska provenance study. Silvae Genetica 29(5-6):171-76.

Dickman, D. I. and K. W. Stuart. 1983. The culture of poplars of eastern North America. Michigan State Univ., East Lansing. 168 p.

Dirr, M. A. 1990. Manual of woody landscape plants. 4th ed. Stipes Publishing Co., Champaign, Ill. 1,007 p.

Droze, W. H. 1977. Trees, prairies and people: a history of tree plantings in the plains states. Texas Woman's Univ. Press, Denton. 313 p.

Duncan, W. H. and M. B. Duncan. 1988. Trees of the southeastern United States. Univ. of Georgia Press, Athens. 322 p.

Elias, T. S. 1980. The complete trees of North America—field guide and natural history. Book Division, Times Mirror Magazines, Inc., New York (Van Nostrand Reinhold Co., New York). 948 p.

Emerson, R. A. 1906. The relation of early maturity to hardiness in trees. Nebr. Agr. Exp. Sta. Report 19:101-10.

Epping, J. E., and E. R. Hasselkuss. 1987. Spotlight on shrub roses. American Nurseryman 170 (2):27-39.

Fairchild, D. W. and J. E. Klett. 1993. Woody landscape plants for the high plains. Colorado State Univ. Tech. Bull. LTB93-1. 109 p.

Farnsworth, D. H., G. E. Gatherum, J. J. Jokela, H. B. Kriebel, D. T. Lester, C. Merritt, S. S. Pauley, R. A. Read, R. L. Saidak, and J. W. Wright. 1972. Geographic variation in Japanese larch in North Central United States plantations. Silvae Genetica 2(3-4):65-148.

Farrar, J. 1990. Field guide to wildflowers of Nebraska and the Great Plains. Nebraskaland Magazine, Nebr. Game and Parks Commission, Lincoln. 215 p.

Fiala, Fr. J. L. 1988. Lilacs, the genus *Syringa*. Timber Press, Inc., Portland, Ore. 266 p.

Finke, Rich and Luann. 1966. Finke Gardens and Nursery Notes. Lincoln. 4 p.

Flemmer, W., III. 1985. A nurseryman picks 40 of the finest. Horticulture 63(7):14-18.

Flint, H. S. 1983. Landscape plants for eastern North America: Exclusive of Florida and the immediate Gulf coast. John Wiley and Sons, New York. 677 p.

Fowells, H. A. 1965. Silvics of forest trees of the United States. U. S. Forest Service Agr. Handbook 271. 762 p.

Freeman, C. 1992. Personal communication with author.

Cited and Selected References

Froiland, S. G. 1962. The genus *Salix* in the Black Hills of South Dakota. U. S. Forest Service Tech. Bull. 1269. 75 p.

Galle, F. C. 1987. Azaleas. Timber Press, Inc., Portland, Ore. 519 p.

Garman, E. H. 1957. The occurrence of spruce in the interior of British Columbia. Dept. of Lands and For., British Columbia For. Serv. Tech. Pub. T-49. 31 p.

Gelderen, D. M. van and J. R. P. van Hoey Smith. 1989. Conifers, 2nd Ed. Timber Press, Inc., Portland, Ore. 356 p.

Gelderen, D. M. van, P. C. de Jong, and H. J. Oterdoom. 1994. Maples of the World. Timber Press, Inc., Portland, Ore. 458 p.

Gilmore, M. R. 1977. Uses of plants by the Indians of the Missouri River Region. Univ. of Nebr. Press, Lincoln. 109 p.

Gleason, H. A. and A. Cronquist. 1963. Manual of vascular plants of northeastern United States and adjacent Canada. D. Van Nostrand Co., Inc., Princeton, N. J. 810 p.

Gleason, H. A. and A. Cronquist. 1964. The natural geography of plants. Columbia University Press, New York and London. 420 p.

Graves, W. R. 1994. Urban soil temperatures and their potential impact on tree growth. Jour. of Arboriculture 20(1):24-27.

Great Plains Flora Association. 1977. Atlas of the flora of the Great Plains. Iowa State Univ. Press, Ames. 600 p.

Great Plains Flora Association. 1986. Flora of the Great Plains. The Univ. Press of Kansas, Lawrence, 1392 p.

Gustafson, W. A., Jr. 1990. Nut tree cultivars for Nebraska. Coop. Ext. Service, Univ. of Nebr., E. C. 90-1762. 8 p.

Hall, R. B., G. A. Miller, T. L. Robison and O. U. Onokpise. 1983. Developing *Alnus* for intensive culture. U. S. Forest Service Tech. Report NC-91:35-45.

Hansen, N. E. 1901. Ornamentals for South Dakota. Bull. 72, S. Dak. Agr. Exp. Station. 206 p.

Hansen, N. E. 1903. Elements of prairie horticulture. Part VI, Bull. 81, S. Dak. Agr. Exp. Station. p. 67-110.

Hansen, N. E. 1904. The western sandcherry. Bull. 87, S. Dak. Agr. Exp. Station. 64 p.

Harlow, W. M. 1946. Fruit key to northeastern trees. Dover Publications, Inc., New York. 50 p.

Harlow, W. M. 1954. Twig key to deciduous woody plants of Eastern North America. 4th ed. Dover Publications, Inc., New York. 56 p.

Harlow, W. M., and E. S. Harrar. 1968. Textbook of dendrology. 5th ed. McGraw-Hill Book Co., New York. 512 p.

Hasselkuss, E. R. 1993. Personal communication with author.

Hensley, D. L., S. C. Weist, C. E. Long, and F. D. Gibbons. Evaluation of ten landscape trees for the midwest. Jour. of Environmental Hort. 9(3):149-55.

Hepting, G. H. 1971. Diseases of forest and shade trees of the United States. U. S. Forest Service, Agr. Handbook No. 386. 658 p.

Herman, D. E. 1985. Shrubs for northern plains landscapes. American Nurseryman 161(12):62-69.

Herman, D. E. 1989. Evaluation of *Lonicera* taxa for honeysuckle aphid susceptibility, winter hardiness, and plant use. Proc. Intermountain Nursery Assoc., U. S. Forest Service, Rocky Mt. For. and Range Exp. Sta. Gen. Tech. Report 184:119-26.

Herman, D. E. 1991. What's New (J. S. Higginbotham). American Nurseryman 173(4):44-45.

Hightshoe, G. L. 1987. Native trees, shrubs and vines for urban and rural America. Van Nostrand Rheinhold, New York. 819 p.

Hoag, D. G. 1965. Trees and shrubs for the northern plains. North Dakota Institute for Regional Studies, Fargo (Lund Press, Inc., Minneapolis, Minn.). 376 p.

Hylander, C. J. 1956. The world of plant life. The Macmillan Co., New York. 653 p.

Jaynes, R. A., ed. 1979. Nut tree culture in North America. Northern Nut Growers Association, Inc., Hamden, Conn. 466 p.

Johnson, Hugh. 1973. The international book of trees. Mitchell Beazley Publishers Ltd., London. 288 p.

Johnson, W. T. and H. H. Lyon. 1991. Insects that feed on trees and shrubs. Cornell Univ. Press, Ithaca, N. Y. 556 p.

Kaittinen, H., O. Muona, K. Karkkainen, and Z. Borzan. 1991. Serbian spruce, a narrow endemic, contains much genetic variation. Can. Jour. For. Res. 21:363-67.

Kelly, G. W. 1970. A guide to the woody plants of Colorado. Pruett Publishing, Boulder. 172 p.

Kincaid, W. H., W. B. Kurtz, and H. E. Garrett. 1982. A silvicultural-economic model for black walnut. In U. S. Forest Service, Gen. Tech. Report NC 174:122-27.

Klett, J. E. 1975. Deciduous shrubs for the Great Plains. American Horticulturist 56(3):34-36.

Klett, J. E. 1977. Superior groundcover junipers for the Great Plains. American Horticulturist 58(3):26-28.

Kossuth, S. V. and G. H. Fechner. 1973. Incompatibility between *Picea pungens* Engelm. and *Picea engelmannii* Parry. For. Sci. 19:50-60.

Krussmann, G. 1985. Manual of cultivated conifers. Timber Press, Portland, Ore. 361 p.

Kuhns, M. 1991. Nebraska's champion tree register. Nebr. Coop. Ext. Serv., Univ. of Nebr. EC 91-1759. 11 p.

Cited and Selected References

Kuhns, M. 1992. Tree identification manual. Nebr. Coop. Ext. Serv., Univ. of Nebr. 4-H 332. 67 p.

Leach, D. G. 1961. Rhododendrons of the world. Charles Scribner's Sons, New York. 544 p.

Little, E. L., Jr. 1953. Check list of native and naturalized trees of the United States. U. S. Forest Service, Agr. Handbook No. 41. 472 p.

Little, E. L., Jr. 1971. Atlas of the United States trees, Vol. 1: Conifers and important hardwoods. U. S. Forest Service, Misc. Pub. No. 1146. 9 p., 200 maps.

Little, E. L., Jr. 1980a. The Audubon Society field guide to North American trees, Eastern Region. Knopf, New York. 714 p.

Little, E. L., Jr., 1980b. The Audubon Society field guide to North American trees, Western Region. Knopf, New York. 639 p.

Long, G. 1965. Response of *Quercus macrocarpa* provenances to four daylengths. Master's Thesis, Univ. of Nebr., Lincoln. 47 p.

Loucks, W. L. 1984. Kansas man-made forests—an overview. Kansas Forest Res. Conf. Paper. 4 p.

MacDonald, G. B. 1954. Fifty years of professional forestry training at I. S. C.: evolution of the Ames foresters. The Ames Forester 41:4-18.

Miller, D. R., N. J. Rosenberg, and W. T. Bagley. 1974. Microclimate modification with shelterbelts. Jour. of Soil and Water Conservation 29(1):41-44.

Miller, D. R., N. J. Rosenberg, and W. T. Bagley. 1975. Wind reduction and aerodynamic drag by a highly permeable tree shelterbelt. Agricultural Meteorology 14:321-33.

Miller, V. J. 1954. Woody ornamentals and their use. Extension Service, Univ. of Nebr. E. C. 54-1200. 33 p.

Miller, V. J. 1988. Personal communication with author.

Mitchell, A. F. 1974. A field guide of the trees of Britain and Northern Europe. Collins, London. 415 p.

Mitchell, A. F. 1987. The trees of North America. Facts on File Publications, New York and London. 208 p.

Morgenson, Greg. 1988. Recent releases of trees and shrubs for conservation and ornamental use in the upper Midwest. Abstracts, Trees for Nebraska Conference, Lincoln.

Morgenson, Greg. 1992. Personal communication with author.

Morgenson, Jon. 1992. Personal communication with author.

Nebraska Statewide Arboretum. 1982. Common and scientific names of Nebraska plants: native and introduced. Univ. of Nebr. Publication 101. 85 p.

Newman, S. E. 1981. Seed germination of *Ostrya virginiana*. Master's Thesis, Univ. of Nebr. 152 p.

North Central Plant Introduction Station. 1954-1984. Unpublished reports. Ames, Iowa.

Pair, J. C. 1988. Woody ornamental evaluations. Kansas State Univ., Wichita Hort. Res. Center Progress Report 548.

Peattie, D. C. 1950. A natural history of trees of Eastern and Central North America. Houghton Mifflin Co., Boston. 606 p.

Pellet, H. 1994. Selecting trees for cold climates. Jour. of Arboriculture 20(1):28-32.

Petersen, N.F. 1912. Flora of Nebraska. Woodruff Printing Co., Lincoln, Nebr. 220 p.

Peterson, G. W. 1976. Disease of Russian-olive caused by *Botryodiplodia theobromae*. Plant Disease Reporter 60:490-94.

Peterson, G. W. 1981. Pine and juniper diseases in the Great Plains. U. S. Forest Service, Rocky Mt. For. and Rang Exp. Sta., Gen.Tech. Report RM-86. 47 p.

Peterson, Max. 1987. Woody plants on a Perkins County farmstead. Personal communication with author.

Polunin, Oleg. 1976. Trees and bushes of Europe. Oxford Univ. Press. 208 p.

Pool, R. J. 1939. Some reactions of the vegetation in the towns and cities of Nebraska to the great drought.Bull. Torrey Bot. Club 66:457-64.

Pool, R. J. 1951. Handbook of Nebraska trees. Rev. Bull. 32, Conservation and Survey Div., Univ. of Nebr. 179 p.

Porter, C. L. 1967. Taxonomy of flowering plants. W. H. Freeman Co., San Francisco. 472 p.

Preston, R. J., Jr. 1968. Rocky Mountain trees. 3rd. rev. ed. Dover Publications, Inc., New York. 285 p.

Preston, R. J., Jr. and V. G. Wright. 1976. Identification of southeastern trees in winter. N. Carolina Agr. Ext. Serv. Ag. 42. 113 p.

Rassmussen, Edward. 1989. Personal communication with author.

Raulston, J. C. and K. E. Tripp. 1994. Exploring the complexities of plant hardiness. Arnoldia 54(3):22-31.

Read, R. A. 1958. The Great Plains shelterbelt in 1954. Agr. Exp. Sta., Univ. of Nebr. Sta. Bull. 441. 125 p.

Read, R. A. 1970. A new tree for eastern Nebraska. Nebr. Agr. Exp. Sta. Quarterly 17(3):10-11.

Read, R. A. 1971. Scots pine in eastern Nebraska: a provenance study. U. S. Forest Service, Rocky Mt. For. and Range Exp. Sta. Res. Paper RM-78. 13 p.

Read, R. A. 1976. Austrian (European black) pine in eastern Nebraska: a provenance study. U. S. Forest Service, Rocky Mt. For. and Range Exp. Sta. Res. Paper 180. 8 p.

Read, R. A. 1983. Ten-year performance of ponderosa pine provenances in the Great Plains of North America. U. S. Forest Service, Rocky Mt. For. and Range Exp. Sta. Res. Paper RM-250. 17 p.

Read, R. A. and J. A. Sprackling. 1976. Douglas-fir in eastern Nebraska: a provenance study. U. S. Forest Service, Rocky Mt. For. and Range. Exp. Sta. Res. Paper RM-178. 10 p.

Cited and Selected References

Rehder, Alfred. 1940. Manual of cultivated trees and shrubs. 2nd ed. Macmillan Co., New York. 996 p.

Reich, L. 1992. Jujube, the Chinese date. American Horticulturist 71(2):25-27, 40.

Richley, Tim. 1988. Evaluation of woody plants in southwestern Nebraska. Personal communicaton with author.

Rocky Mountain Forest and Range Experiment Station and University of Nebraska Cooperative Extension Service. 1986. Common insect pests of trees in the Great Plains. Univ. of Nebr. Coop. Ext. Serv. EC 86-1548 (Great Plains Agr. Council Publ. No. 119). 44 p.

Rushmore, F. M. 1961. Silvical characteristics of beech, *Fagus grandifolia*. U. S. Forest Service, Northeast For. Exp. Sta. Paper 161. 26 p.

Rydberg, P. A. 1932. Flora of the prairies and plains of central North America. New York Botanical Garden, New York. 969 p.

Santamour, F. S. and F. G. Meyer. 1977. Clarifying monarch birch's origins and characteristics. American Nurseryman 145(12):7, 88-94.

Sargent, C. S. 1933. Manual of the trees of North America. 2nd ed. Houghton Mifflin Co., Boston and New York. 910 p.

Schlarbaum, S. E. and W. T. Bagley. 1981. Interspecific genetic variation of *Quercus rubra*, Northern red oak. Silvae Genetica 30(2-3):50-56.

Schmidt, Gabor. 1994. Personal communication with author.

Scott, Nancy. 1998. Personal communication with author.

Settergren, Carl and R. E. McDermott. 1968. Trees of Missouri. Univ. of Missouri, Agr. Exp. Sta. B767. 123 p.

Shigo, A. L. 1994. Tree anatomy. Shigo and Trees Associates, Durham, N. H. 104 p.

Sinclair, W. A., H. H. Lyon, and W. T. Johnson. 1987. Diseases of trees and shrubs. Cornell Univ. Press, Ithaca, N. Y. 574 p.

Smith, J. R. 1953. Tree crops: a permanent agriculture. Devin-Adair Co., New York. 408 p.

Snyder, L. C. 1980. Trees and shrubs for northern gardens. Univ. of Minn. Press, Minneapolis. 411 p.

Spangler, R. L. and J. Ripperda. 1977. Landscape plants for the central and northeastern United States. Burgess Publishing Co., Minneapolis. 506 p.

Sprackling, J. A. and R. A. Read. 1975a. Jack pine provenance study in eastern Nebraska. U. S. Forest Service, Rocky Mt. For. and Range Exp. Sta. Res. Paper RM-143. 8 p.

Sprackling, J. A. and R. A. Read. 1975b. Red pine provenance study in eastern Nebraska. U. S. Forest Service, Rocky Mt. For. and Range Exp. Sta. Res. Paper RM-144. 7 p.

Sprackling, J. A. and R. A. Read. 1976. Eastern white pine in eastern Nebraska: A provenance study of southern Appalachian origins. U. S. Forest Service, Rocky Mt. For. and Range Exp. Sta. Res. Paper RM-179. 8 p.

Steiner, K. C., M. W. Williams, D. H. Hayes, R. B. Hall, R. T. Eckert, W. T. Bagley, W. A. Lemmien, D. F. Karnosky, K. K. Carter, and F. C. Cech. 1988. Juvenile performance in a range-wide provenance test of *Fraxinus pennsylvanica* Marsh. Silvae Genetica 37(3-4):104-11.

Stephens, H. A. 1973. Woody plants of the North Central Plains. Univ. Press of Kansas, Lawrence. 530 p.

Steyer, R. 1990. Personal communication with author.

Stoeckeler, J. H. 1962. Shelterbelt influences on Great Plains field environment and crops. U. S. Forest Service Prod. Res. Report No. 62. 26 p.

Straight, Rich. 1993. Personal communication with author.

Strickler, J. 1971. Forests on the Prairie. Coop. Ext. Serv., Kansas State Univ. C-300, revised. 16 p.

Stubbendieck, J., S. L. Hatch, and K. J. Hirrsh. 1986. North American Range Plants. 3rd. ed. Univ. of Nebr. Press, Lincoln. 465 p.

Stubbendieck, J., J. T. Nichols, and C. H. Butterfield. 1989. Nebraska range and pasture forbs and shrubs. Univ. of Nebr. Coop. Ext. E. C. 89-118. 152 p.

Sutton, R. K. and C. W. Johnson. 1974. Landscape plants from Utah's mountains. Utah State Univ. Coop. Ext. Circ. EC-358. 136 p.

Sutton, R. K. 1987. A comparison of stem density and basal area versus groundcover for two types of closed canopy coniferous woodlands. Unpublished manuscript.

Sutton, R. K. 1998. Personal observations of author.

Thirgood, J. V. 1971. The historical significance of oak. U. S. Forest Service, Northeastern For. Exp. Sta. Oak Symposium Proc. 16 p.

U. S. Forest Service. 1937. Range plant handbook. U. S. Govt. Printing Office, Washington. Not paged.

U. S. Forest Service. 1975. Forest Nursery Diseases in the United States. Agr. Handbook No. 470. 125 p.

Van Der Linden, P. J. and D. R. Farrar. 1984. Forest and shade trees of Iowa. Iowa State Univ. Press., Ames. 133 p.

Van Haverbeke, D. F. 1968. A population analysis of *Juniperus* in the Missouri River Basin. Univ. of Nebr. Studies, New Series 38. 82 p.

Van Haverbeke, D. F. 1983. Seventeen-year performance of *Pinus flexilis* and *P. strobiformis* progenies in eastern Nebraska. Silvae Genetica 32(3-4):71-76.

Van Haverbeke, D. F. 1984. Genetic variation in blue spruce: a test of populations in Nebraska. U. S. Forest Service, Rocky Mt. For. and Range Exp. Sta. Res. Paper RM-253. 6 p.

Cited and Selected References

Van Haverbeke, D. F. and R. M. King. 1990. Genetic variation in Great Plains *Juniperus*. U. S. Forest Service, Rocky Mt. For. and Range Exp. Sta. Res. Paper RM-292. 8 p.

Viertel, A. T. 1970. Trees, shrubs and vines: a pictoral guide to the ornamental woody plants of the northern United States. Syracuse Univ. Press, Syracuse, N. Y. Not paged.

Ware, G. H. 1991. *Ulmus davidiana*. Plant profiles. The Public Garden 6(1):41.

Weber, W. A. 1990. Colorado flora: eastern slope. Univ. Press of Colo., Niwot. 306 p.

Welch, H. J. 1991. The conifer manual: vol. 1. Kluwer Academic Publishers, Dordrecht / Boston / London. 436 p.

Widrlechner, M. P., E. R. Hasselkaus, D. E. Herman, J. K. Iles, J. C. Pair, E. T. Paparozzi, R. E. Schatzki and D. K. Wildung. 1992. Performance of landscape plants from Yugoslavia in the North Central United States. Jour. of Environmental Hort. 10(4):192-98.

Wolfe, L. G. 1989. Evaluation of *Ziphyphus jujuba* in a wildlife planting at Red Cloud, Nebraska. Personal communication with author.

Woodruff, N. P. 1954. Shelterbelt and surface barrier effects on wind velocities, evaporation, househeating, snowdrifting. Kansas Agr. Exp. Sta. Tech. Bull. 77. 27 p.

Woodruff, N. P., R. A. Read and W. S. Chepil. 1959. Influence of a field windbreak on summer wind movement and air temperature. Kansas Agr. Exp. Sta. Tech. Bull. 100. 24 p.

Woodruff, N. P., D. W. Fryear and L. Lyles. 1963. Reducing wind velocity with field shelterbelts. Kansas Agr. Exp. Sta. Tech. Bull. 131. 26 p.

Wright, J. W. 1955. Species crossability in spruce in relation to distribution and taxonomy. Forest Sci. 1:319-49.

Wyman, Donald. 1965. Trees for American gardens. Macmillan Co., New York. 502 p.

Wyman, Donald. 1969. Shrubs and vines for American gardens. Macmillan Co., New York. 613 p.

Ying, C. C. and W. T. Bagley. 1976a. Genetic variation of eastern cottonwood in an eastern Nebraska provenance study. Silvae Genetica 25(2):67-73.

Ying, C. C. and W. T. Bagley. 1976b. Performance of green ash provenances of the Great Plains Region. 10th Central States Tree Improvement Conference Proc.

Ying, C. C. and W. T. Bagley. 1977. Variation in rooting capability of *Populus deltoides*. Silva Genetica 26(5-6):204-7.

Index

B

H

S

www.ingramcontent.com/pod-product-compliance
Ingram Content Group UK Ltd.
Pitfield, Milton Keynes, MK11 3LW, UK
UKHW051450290926
15471UKWH00008B/1131

9 781930 665507